UNITED STATES
DEPARTMENT OF AGRICULTURE

NATIC   **DATE DUE**   :RVICE

D1523423

GAYLORD   PRINTED IN U.S.A.

rary
Avenue
1

GOVERNMENT PRINTING OFFICE
WASHINGTON: 2006

For sale by the Superintendent of Documents, U.S. Government Printing Office
Internet: bookstore.gpo.gov   Phone: toll free (866) 512-1800;   DC area (202) 512-1800
Fax: (202) 512-2250 Mail: Stop IDCC, Washington, DC 20402-0001

ISBN 0-16-076178-6

Agricultural Statistics 2006

Agricultural Statistics, 2006 was prepared under the direction of FORESTINE CHAPMAN, Agricultural Statistics Board, National Agricultural Statistics Service. ROSE M. PETRONE was responsible for coordination and technical editorial work.

The USDA and NASS invite you to explore their information on the Internet. The USDA Home Page address is **http://www.usda.gov/** and the NASS Home Page address is: **http://www.usda.gov/nass/.**

For information on NASS products you may call the **Agricultural Statistics Hotline, 1–800–727–9540 or send e-mail to nass@nass.usda.gov.**

The cooperation of the many contributors to this publication is gratefully acknowledged. Source notes below each table credit the various Government agencies which collaborated in furnishing information.

CONTENTS

Introduction

Agricultural Statistics is published each year to meet the diverse need for a reliable reference book on agricultural production, supplies, consumption, facilities, costs, and returns. Its tables of annual data cover a wide variety of facts in forms suited to most common use.

Inquiries concerning more current or more detailed data, past and prospective revisions, or the statistical methodology used should be addressed directly to the agency credited with preparing the table. Most of the data were prepared or compiled in the U.S. Department of Agriculture.

The historical series in this volume have been generally limited to data beginning with 1995 or later.

Foreign agricultural trade statistics include Government as well as non-Government shipments of merchandise from the United States and Territories to foreign countries. They do not include U.S. shipments to the U.S. Armed Forces abroad for their own use or shipments between the States and U.S. Territories. The world summaries of production and trade of major farm products are prepared by the U.S. Department of Agriculture from reports of the U.S. Department of Commerce, official statistics of foreign governments, other foreign source materials, reports of U.S. Agricultural Attachés and Foreign Service Officers, and the result of office research.

Statistics presented in many of the tables represent actual counts of the items covered. Most of the statistics relating to foreign trade and to Government programs, such as numbers and amounts of loans made to farmers, and amounts of loans made by the Commodity Credit Corporation, etc., are data of this type. A large number of other tables, however, contain data that are estimates made by the Department of Agriculture.

The estimates for crops, livestock, and poultry made by the U.S. Department of Agriculture are prepared mainly to give timely current State and national totals and averages. They are based on data obtained by sample surveys of farmers and of people who do business with farmers. The survey data are supplemented by information from the Censuses of Agriculture taken every five years and check data from various sources. Being estimates, they are subject to revision as more data become available from commerical or Government sources. Unless otherwise indicated, the totals for the United States shown in the various tables on area, production, numbers, price, value, supplies, and disposition are based on official Department estimates. They exclude States for which no official estimates are compiled.

DEFINITIONS

"Value of production" as applied to crops in the various tables, is derived by multiplying production by the estimated season average price received by farmers for that portion of the commodity actually sold. In the case of fruits and vegetables, quantities not harvested because of low prices or other economic factors are not included in value of production. The word "Value" is used in the inventory tables on livestock and poultry to mean value of the number of head on the inventory date. It is derived by multiplying the number of head by an estimated value per head as of the date.

The word "Year" (alone) in a column heading means calendar year unless otherwise indicated. "Ton" when used in this book without qualifications means a short ton of 2,000 pounds.

WEIGHTS, MEASURES, AND CONVERSION FACTORS

The following table on weights, measures, and conversion factors covers the most important agricultural products, or the products for which such information is most frequently asked of the U.S. Department of Agriculture. It does not cover all farm products nor all containers for any one product.

The information has been assembled from State schedules of legal weights, various sources within the U.S. Department of Agriculture, and other Government agencies. For most products, particularly fruits and vegetables, there is a considerable variation in weight per unit of volume due to differences in variety or size of commodity, condition and tightness of pack, degree to which the container is heaped, etc. Effort has been made to select the most representative and fairest average for each product. For those commodities which develop considerable shrinkage, the point of origin weight or weight at harvest has been used.

The approximate or average weights as given in this table do not necessarily have official standing as a basis for packing or as grounds for settling disputes. Not all of them are recognized as legal weight. The table was prepared chiefly for use of workers in the U.S. Department of Agriculture who have need of conversion factors in statistical computations.

WEIGHTS, MEASURES, AND CONVERSION FACTORS
(See explanatory text just preceding this table)

WEIGHTS AND MEASURES

Commodity	Unit[1]	Approximate net weight U.S.	Approximate net weight Metric	Commodity	Unit[1]	Approximate net weight U.S.	Approximate net weight Metric
		Pounds	Kilograms			Pounds	Kilograms
Alfalfa seed	Bushel	60	27.2				
Apples	....do	48	21.8	Celery	Crate[8]	60	27.2
Do	Loose pack	38–42	17.2–19.1	Cherries	Lug (Campbell)[9]	16	7.3
Do	Tray pack	40–45	18.1–20.4	Do	Lug	20	9.1
Do	Cell pack	37–41	16.8–18.6	Clover seed	Bushel	60	27.2
Apricots	Lug (brentwood)[2]	24	10.9	Coffee	Bag	132.3	60
Western	4–basket crate[3]	26	11.8	Corn:			
Artichokes:				Ear, husked	Bushel	[10]70	31.8
Globe	Ctn, by count and loose pack	20–25	9.1–11.3	Shelled	....do	56	25.4
				Meal	....do	50	22.7
Jerusalem	Bushel	50	22.7	Oil	Gallon	[7]7.7	3.5
Asparagus	Crate (NJ)	30	13.6	Syrup	....do	11.72	5.3
Avocados	Lug[4]	12–15	5.4–6.8	Sweet	Wirebound crate	50	22.7
Bananas	Fiber folding box[5]	40	18.1	Do	Ctn, packed 5 oz. ears	50	22.7
Barley	Bushel	48	21.8	Do	WDB crate, 4½–5 oz. (from FL & NJ)	42	19.1
Beans:							
Lima, dry	....do	56	25.4	Cotton	Bale, gross	[11]500	227
Other, dry	....do	60	27.2	Do	Bale, net	[11]480	218
	Sack	100	45.4	Cottonseed	Bushel	[12]32	14.5
Lima unshelled	Bushel	28–32	12.7–14.5	Cottonseed oil	Gallon	[7]7.7	3.5
Snap	....do	28–32	12.7–14.5	Cowpeas	Bushel	60	27.2
Beets:				Cranberries	Barrel	100	45.4
Topped	Sack	25	11.3	Do	¼–bbl. box[13]	25	11.3
Bunched	½ crate 2 dz-bchs	36–40	16.3–18.1	Cream, 40–percent butterfat	Gallon	8.38	3.80
Berries frozen pack:				Cucumbers	Bushel	48	21.8
Without sugar	50–gal. barrel	380	172	Dewberries	24–qt. crate	36	16.3
3 + 1 pack	....do	425	193	Eggplant	Bushel	33	15.0
2 + 1 pack	....do	450	204	Eggs, average size	Case, 30 dozen	47.0	21.3
Blackberries	12, ½-pint basket	6	2.7	Escarole	Bushel	25	11.3
Bluegrass seed	Bushel	14–30	6.4–13.6	Figs, fresh	Box single layer[14]	6	2.7
Broccoli	Wirebound crate	20–25	9.1–11.3	Flaxseed	Bushel	56	25.4
Broomcorn (6 bales per ton)	Bale	333	151	Flour, various	Bag	100	45.4
Broomcorn seed	Bushel	44–50	20.0–22.7	Do	Ctn or Crate, Bulk	30	13.6
Brussels sprouts	Ctn, loose pack	25	11.3	Garlic	Ctn of 12 tubes or 12 film bag pkgs 12 cloves each	10	4.5
Buckwheat	Bushel	48	21.8				
Butter	Block	55,68	25,30.9				
Cabbage	Open mesh bag	50	22.7				
Do	Flat crate (1¾ bu)	50–60	22.7–27.2	Grapefruit:			
Do	Ctn, place pack	53	24.0	Florida and Texas	½–box mesh bag	40	18.1
Cantaloups	Crate[6]	40	18.1	Florida	1⅗ bu. box	85	38.6
Carrots	Film plastic Bags, mesh sacks & cartons holding 48 1 lb. film bags	55	24.9	Texas	1⅘ bu. box	80	36.3
				California and Arizona	Box[15]	[16]67	30.4
Without tops	Burlap sack	74–80	33.6–36.3	Grapes:			
Castor beans	Bushel	41	18.6	Eastern	12–qt. basket	20	9.1
Castor oil	Gallon	[7]8	3.6	Western	Lug	28	12.7
Cauliflower	W.G.A. crate	50–60	22.7–27.2	Do	4–basket crate[17]	20	9.1
Do	Fiberboard box wrapper leaves removed film-wrapped, 2 layers	23–35	10.4–15.9	Hempseed	Bushel	44	20.0
				Hickory nuts	....do	50	22.7
				Honey	Gallon	11.84	5.4
				Honeydew melons	⅔ Ctn	28–32	12.7–14.5
				Hops	Bale, gross	200	90.7

See footnotes on page ix.

WEIGHTS AND MEASURES—Continued

Commodity	Unit [1]	Approximate net weight		Commodity	Unit [1]	Approximate net weight	
		U.S.	Metric			U.S.	Metric
		Pounds	*Kilograms*			*Pounds*	*Kilograms*
Horseradish				Do	Std box, 4/5 bu	45–48	20.4–21.8
roots	Bushel	35	15.9	Do	Ctn, Tight-fill		
Do	Sack	50	22.7		pack	36–37	16.3–16.7
Hungarian millet				Peas:			
seed	Bushel	48–50	21.8–22.7	Green,			
Kale	Ctn or crate	25	11.3	unshelled	Bushel	28–30	12.7–13.6
Kapok seed	.....do	35–40	15.9–18.1	Dry	.....do	60	27.2
Lard	Tierce	375	170	Peppers, green	.....do	25–30	11.3–13.6
Lemons:				Do	1½ bu carton	28	12.7
California and				Perilla seed	Bushel	37–40	16.8–18.1
Arizona	Box [18]	76	34.5	Pineapples	Carton	40	18.1
Do	Carton	38	17.2	Plums and			
Lentils	Bushel	60	27.2	prunes:	Ctn & lugs	28	12.7
Lettuce, iceberg	Iceberg, carton			Do	½-bu. basket	30	13.6
	packed 24	43–52	19.5–23.6	Popcorn:			
Lettuce, hot-				On ear	Bushel	[10]70	31.8
house	24-qt. basket	10	4.5	Shelled	.....do	56	25.4
Limes (Florida)	Box	88	39.9	Poppy seed	.....do	46	20.9
Linseed oil	Gallon	[7]7.7	3.5	Potatoes	Bushel	60	27.2
Malt	Bushel	34	15.4	Do	Barrel	165	74.8
Maple syrup	Gallon	11.02	5.0	Do	Box	50	22.7
Meadow fescue				Do	.....do	100	45.4
seed	Bushel	24	10.9	Quinces	Bushel	48	21.8
Milk	Gallon	8.6	3.9	Rapeseed	.....do	50–60	22.7–27.2
Millet	Bushel	48–60	21.8–27.2	Raspberries	½-pint baskets	6	2.7
Molasses:				Redtop seed	Bushel	50–60	22.7–27.2
edible	Gallon	11.74	5.3	Refiners' syrup	Gallon	11.45	5.2
inedible	.....do	11.74	5.3	Rice:			
Mustard seed	Bushel	58–60	26.3–27.2	Rough	Bushel	45	20.4
Oats	.....do	32	14.5	Do	Bag	100	45.4
Olives	Lug	25–30	11.3–13.6	Do	Barrel	162	73.5
Olive oil	Gallon	[7]7.6	3.4	Milled	Pocket or bag	100	45.4
Onions, dry	Sack	50	22.7	Rosin	Drum, net	520	236
Onions, green				Rutabagas	Bushel	56	25.4
bunched	Ctn, 24-dz bchs	10–16	4.5–7.3	Rye	.....do	56	25.4
Oranges:				Sesame seed	.....do	46	20.9
Florida	Box	90	40.8	Shallots	Crate (4–7 doz.		
Texas	Box	85	38.5		bunches)	20–35	9.1–15.9
California and				Sorgo:			
Arizona	Box [15]	75	34.0	Seed	Bushel	50	22.7
Do	Carton	38	17.2	Syrup	Gallon	11.55	5.2
Orchardgrass				Sorghum			
seed	Bushel	14	6.4	grain [19]	Bushel	56	25.4
Palm oil	Gallon	[7]7.7	3.5	Soybeans	.....do	60	27.2
Parsnips	Bushel	50	22.7	Soybean oil	Gallon	[7]7.7	3.5
Peaches	.....do	48	21.8	Spelt	Bushel	40	18.1
Do	2 layer ctn or			Spinach	.....do	18–20	8.2–9.1
	lug	22	10.0	Strawberries	24-qt. crate	36	16.3
Do	¾-Bu, Ctn/crate	38	17.2	Do	12-pt. crate	9–11	4.1–5.0
Peanut oil	Gallon	[7]7.7	3.5	Sudangrass			
Peanuts,				seed	Bushel	40	18.1
unshelled:				Sugarcane:			
Virginia type	Bushel	17	7.7	Syrup			
Runners,				(sulfured or			
South-east-				un-sulfured)	Gallon	11.45	5.2
ern	.....do	21	9.5	Sunflower seed	Bushel	24–32	10.9–14.5
Spanish:				Sweetpotatoes	.....do	[20]55	24.9
South-				Do	Crate	50	22.7
eastern	.....do	25	11.3	Tangerines:			
South-				Florida	Box	95	43.1
western	.....do	25	11.3	Arizona	Box	75	34.0
Pears:				California	Box	75	34.0
California	Bushel	48	21.8				
Other	.....do	50	22.7				

See footnotes on page ix.

WEIGHTS AND MEASURES—Continued

Commodity	Unit [1]	Approximate net weight		Commodity	Unit [1]	Approximate net weight	
		U.S.	Metric			U.S.	Metric
		Pounds	*Kilograms*			*Pounds*	*Kilograms*
Timothy seed	Bushel	45	20.4	Turnips:			
Tobacco:				Without tops ..	Mesh sack	50	22.7
Maryland	Hogshead	775	352	Bunched	Crate [6]	70–80	31.8–36.3
Flue-cured	.....do	950	431	Turpentine	Gallon	7.23	3.3
Burley	.....do	975	442	Velvetbeans			
Dark air-cured	.....do	1,150	522	(hulled)	Bushel	60	27.2
Virginia fire-				Vetch seed	.....do	60	27.2
cured	.....do	1,350	612	Walnuts	Sacks	50	22.7
Kentucky and				Water 60° F	Gallon	8.33	3.8
Tennessee				Watermelons	Melons of aver-		
fire-cured	.....do	1,500	680		age or me-		
Cigar-leaf	Case	250–365	113–166		dium size	25	11.3
Do	Bale	150–175	68.0–79.4	Wheat	Bushel	60	27.2
Tomatoes	Crate	60	27.2	Various com-			
Do	Lug box	32	14.5	modities	Short ton	2,000	907
Do	2-layer flat	21	9.5	Do	Long ton	2,240	1,016
Tomatoes, hot-				Do	Metric ton	2,204.6	1,000
house	12-qt. basket	20	9.1				
Tung oil	Gallon	[7]7.8	3.5				

See footnotes on page ix.

To Convert From Avoirdupois Pounds

To	Multiply by
Kilograms	0.45359237
Metric tons	0.00045359237

Conversion Factors

1 Metric ton=2,204.622 pounds
1 Kilogram=2.2046 pounds
1 Acre=0.4047 hectares
1 Hectare=2.47 acres
1 Square mile=640 acres=259 hectares
1 Gallon=3.7853 liters

CONVERSION FACTORS

Commodity	Unit	Approximate equivalent
Apples	1 pound dried	7 pounds fresh; beginning 1943, 8 pounds fresh
Do	1 pound chops	5 pounds fresh
Do	1 case canned[21]	1.4 bushels fresh
Applesauce	......do[21]	1.2 bushels fresh
Apricots	1 pound dried	6 pounds fresh
Barley flour	100 pounds	4.59 bushels barley
Beans, lima	1 pound shelled	2 pounds unshelled
Beans, snap or wax	1 case canned[22]	0.008 ton fresh
Buckwheat flour	100 pounds	3.47 bushels buckwheat
Calves	1 pound live weight	0.611 pound dressed weight (1999 average)
Cattle	......do	0.607 pound dressed weight (1999 average)
Cane syrup	1 gallon	5 pounds sugar
Cherries, tart	1 case canned[21]	0.023 ton fresh
Chickens	1 pound live weight	0.72 pound ready-to-cook weight
Corn, shelled	1 bushel (56 lbs.)	2 bushels (70 pounds) of husked ear corn
Corn, sweet	1 case canned[22]	0.030 ton fresh
Cornmeal:		
Degermed	100 pounds	3.16 bushels corn, beginning 1946
Nondegermed	......do	2 bushels corn, beginning 1946
Cotton	1 pound ginned	3.26 pounds seed cotton, including trash[23]
Cottonseed meal	1 pound	2.10 pounds cottonseed
Cottonseed oil	......do	5.88 pounds cottonseed
Dairy products:		
Butter	......do	21.1 pounds milk
Cheese	......do	10 pounds milk
Condensed milk, whole	......do	2.3 pounds milk
Dry cream	......do	19 pounds milk
Dry milk, whole	......do	7.6 pounds milk
Evaporated milk, whole	......do	2.14 pounds milk
Malted milk	......do	2.6 pounds milk
Nonfat dry milk	......do	11 pounds liquid skim milk
Ice cream[24]	1 gallon	15 pounds milk
Ice cream[24] (eliminating fat from butter and concentrated milk).	......do	12 pounds milk
Eggs	1 case	47 pounds
Eggs, shell	......do	41.2 pounds frozen or liquid whole eggs
Do	......do	10.3 pounds dried whole eggs
Figs	1 pound dried	3 pounds fresh in California; 4 pounds fresh elsewhere
Flaxseed	1 bushel	About 2½ gallons oil
Grapefruit, Florida	1 case canned juice[22]	0.64 box fresh fruit
Hogs	1 pound live weight	0.737 pound dressed weight, excluding lard (1999 average)
Linseed meal	1 pound	1.51 pounds flaxseed
Linseed oil	......do	2.77 pounds flaxseed
Malt	1 bushel (34 lbs.)	1 bushel barley (48 lbs.)
Maple syrup	1 gallon	8 pounds maple sugar
Nuts:		
Almonds, imported	1 pound shelled	3½ pounds unshelled
Almonds, California	......do	2.22 pounds unshelled through 1949; 2 pounds thereafter
Brazil	......do	2 pounds unshelled
Cashews	......do	4.55 pounds unshelled
Chestnuts	......do	1.19 pounds unshelled
Filberts	......do	2.22 pounds unshelled through 1949; 2.5 pounds thereafter
Pecans:		
Seedling	......do	2.78 pounds unshelled
Improved	......do	2.50 pounds unshelled
Pignolias	......do	1.3 pounds unshelled
Pistachios	......do	2 pounds unshelled
Walnuts:		
Black	......do	5.88 pounds unshelled
Persian (English)	......do	2.67 pounds unshelled
Oatmeal	100 pounds	7.6 bushels oats, beginning 1943
Oranges, Florida	1 case canned juice[22]	0.53 box fresh
Peaches, California, freestone	1 pound dried	5⅓ pounds fresh through 1918; 6 pounds fresh for 1919–28; and 6½ pounds fresh from 1929 to date
Peaches, California, clingstone	......do	7½ pounds fresh
Peaches, clingstone	1 case canned[21]	1 bushel fresh
Do	......do	0.0230 ton fresh
Peanuts	1 pound shelled	1½ pounds unshelled
Pears	1 pound dried	6½ pounds fresh
Pears, Bartlett	1 case canned[22]	1.1 bushels fresh
Do	......do	0.026 ton fresh

See footnotes on page ix.

CONVERSION FACTORS—Continued

Commodity	Unit	Approximate equivalent
Peas, green	1 pound shelled	2½ pounds unshelled
Do	1 case canned[22]	0.009 ton fresh (shelled)
Prunes	1 pound dried	2.7 pounds fresh in California; 3 to 4 pounds fresh elsewhere
Raisins	1 pound	4.3 pounds fresh grapes
Rice, milled (excluding brewers)	100 pounds	152 pounds rough or unhulled rice
Rye flour	......do	2.23 bushels rye, beginning 1947
Sheep and lambs	1 pound live weight	0.504 pound dressed weight (1999 average)
Soybean meal	1 pound	1.27 pounds soybeans
Soybean oil	......do	5.49 pounds soybeans
Sugar	1 ton raw	0.9346 ton refined
Tobacco	1 pound farm-sales weight ..	Various weights of stemmed and unstemmed, according to aging and the type of tobacco. (See circular 435, U.S. Dept. of Agr.)
Tomatoes	1 case canned[22]	0.018 ton fresh
Turkeys	1 pound live weight	0.80 pound ready-to-cook weight
Wheat flour	100 pounds	2.30 bushels wheat[25]
Wool, domestic apparel shorn	1 pound greasy	0.48 pounds scoured
Wool, domestic apparel pulled	......do	0.73 pound scoured

[1] Standard bushel used in the United States contains 2,150.42 cubic inches; the gallon, 231 cubic inches; the cranberry barrel, 5,826 cubic inches; and the standard fruit and vegetable barrel, 7,056 cubic inches. Such large-sized products as apples and potatoes sometimes are sold on the basis of a heaped bushel, which would exceed somewhat the 2,150.42 cubic inches of a bushel basket level full. This also applies to such products as sweetpotatoes, peaches, green beans, green peas, spinach, etc.

[2] Approximate inside dimensions, 4⅝ by 12½ by 16⅛ inches.

[3] Approximate inside dimensions, 4½ by 16 by 16⅛ inches.

[4] Approximate dimensions, 4½ by 13½ by 16⅛ inches.

[5] Approximate inside dimensions, 13 by 12 by 32 inches.

[6] Approximate inside dimensions, 13 by 18 by 21⅝ inches.

[7] This is the weight commonly used in trade practices, the actual weight varying according to temperature conditions.

[8] Approximate inside dimensions, 9¾ by 16 by 20 inches.

[9] Approximate inside dimensions, 4⅛ by 11½ by 14 inches.

[10] The standard weight of 70 pounds is usually recognized as being about 2 measured bushels of corn, husked, on the ear, because it required 70 pounds to yield 1 bushel, or 56 pounds, of shelled corn.

[11] For statistical purposes the bale of cotton is 500 pounds or 480 pounds net weight. Prior to Aug. 1, 1946, the net weight was estimated at 478 pounds. Actual bale weights vary considerably, and the customary average weights of bales of foreign cotton differ from that of the American square bale.

[12] This is the average weight of cottonseed, although the legal weight in some States varies from this figure of 32 pounds.

[13] Approximate inside dimensions, 9¼ by 10½ by 15 inches.

[14] Approximate inside dimensions, 1¾ by 11 by 16⅛ inches.

[15] Approximate inside dimensions, 11½ by 11½ by 24 inches.

[16] Beginning with the 1993-94 season, net weights for California Desert Valley and Arizona grapefruit were increased from 64 to 67 pounds, equal to the California other area net weight, making a 67 pound net weight apply to all of California.

[17] Approximate inside dimensions, 4¾ by 16 by 16⅛ inches.

[18] Approximate inside dimensions, 9⅞ by 13 by 25 inches.6 by 16 by 16⅛ inches.

[19] Includes both sorghum grain (kafir, milo, hegari, etc.) and sweet sorghum varieties.

[20] This average of 55 pounds indicates the usual weight of sweetpotatoes when harvested. Much weight is lost in curing or drying and the net weight when sold in terminal markets may be below 55 pounds.

[21] Case of 24 No. 2½ cans.

[22] Case of 24 No. 303 cans.

[23] Varies widely by method of harvesting.

[24] The milk equivalent of ice cream per gallon is 15 pounds. Reports from plants indicate about 81 percent of the butterfat in ice cream is from milk and cream, the remainder being from butter and concentrated milk. Thus the milk equivalent of the milk and cream in a gallon of ice cream is about 12 pounds.

[25] This is equivalent to 4.51 bushels of wheat per barrel (196 pounds) of flour and has been used in conversions, beginning July 1, 1957. Because of changes in milling processes, the following factors per barrel of flour have been used for earlier periods: 1790–1879, 5 bushels; 1880–1908, 4.75 bushels, 1909–17, 4.7 bushels; 1918 and 1919, 4.5 bushels; 1920, 4.6 bushels; 1921–44, 4.7 bushels; July 1944–Feb. 1946, 4.57 bushels; March 1946–Oct. 1946, average was about 4.31 bushels; and Nov. 1946–June 1957, 4.57 bushels.

CHAPTER I
STATISTICS OF GRAIN AND FEED

This chapter contains tables for wheat, rye, rice, corn, oats, barley, sorghum grain, and feedstuffs. Estimates are given of area, production, disposition, supply and disappearance, prices, value of production, stocks, foreign production and trade, price-support operations, animal units fed, and feed consumed by livestock and poultry.

Table 1-1.—Total grain: Supply and disappearance, United States, 1996–2005 [1]

Year [2]	Supply				Disappearance			Ending stocks
	Beginning stocks	Production	Imports	Total	Domestic use	Exports	Total disappear-ance	
	Million metric tons	Million metric tons	Million metric tons	Million metric tons	Million metric tons	Million metric tons	Million metric tons	Million metric tons
1996	25.8	335.5	5.9	367.2	244.5	82.4	326.9	40.3
1997	40.3	336.3	5.9	382.5	245.9	77.5	323.4	59.1
1998	59.1	349.2	6.4	414.6	248.0	88.4	336.5	78.1
1999	78.1	334.8	5.8	418.7	252.8	89.9	342.8	76.0
2000	76.0	342.4	5.7	424.0	256.9	89.3	346.2	77.8
2001	77.8	324.5	6.1	408.4	254.8	85.6	340.5	68.0
2002	68.0	297.0	5.3	370.3	250.3	74.5	324.8	45.5
2003	45.5	348.0	4.8	398.3	263.7	90.0	353.6	44.7
2004 [3]	44.7	388.6	4.7	438.1	253.1	85.2	338.3	99.7
2005 [4]	75.2	365.9	4.7	445.8	254.1	84.7	338.8	107.0

[1] Aggregate data on corn, sorghum, barley, oats, wheat, rye, and rice. [2] The marketing year for corn and sorghum begins September 1; for oats, barley, wheat, and rye, June 1; and for rice, August 1. [3] Preliminary. [4] Projected as of January 12, 2006; World Agricultural Supply and Demand Estimates. Totals may not add due to independent rounding.

ERS, Market and Trade Economics Division, (202) 694–5296.

Table 1-2.—Wheat: Area, yield, production, and value, United States, 1996–2005

Year	Area		Yield per harvested acre	Production	Marketing year average price per bushel received by farmers [2]	Value of production [2]
	Planted [1]	Harvested				
	1,000 acres	1,000 acres	Bushels	1,000 bushels	Dollars	1,000 dollars
1996	75,105	62,819	36.3	2,277,388	4.30	9,782,238
1997	70,412	62,840	39.5	2,481,466	3.38	8,286,741
1998	65,821	59,002	43.2	2,547,321	2.65	6,780,623
1999	62,664	53,773	42.7	2,295,560	2.48	5,586,675
2000	62,549	53,063	42.0	2,228,160	2.62	5,771,786
2001	59,432	48,473	40.2	1,947,453	2.78	5,412,834
2002	60,318	45,824	35.0	1,605,878	3.56	5,637,416
2003	62,141	53,063	44.2	2,344,760	3.40	7,929,039
2004	59,674	49,999	43.2	2,158,245	3.40	7,283,324
2005	57,229	50,119	42.0	2,104,690	3.40	7,140,357

[1] Includes area seeded in preceding fall for winter wheat. [2] Includes allowance for loans outstanding and purchases by the Government valued at the average loan and purchase rate, by States, where applicable.

NASS, Crops Branch, (202) 720–2127.

GRAIN AND FEED

Table 1-3.—Wheat, by type: Area, yield, production, and value, United States, 1996–2005

Year	Area		Yield per harvested acre	Production	Marketing year average price per bushel received by farmers [2]	Value of production [2]
	Planted [1]	Harvested				
			Winter wheat			
	1,000 acres	*1,000 acres*	*Bushels*	*1,000 bushels*	*Dollars*	*1,000 dollars*
1996	51,445	39,574	37.1	1,469,618	4.33	6,396,217
1997	47,985	41,340	44.6	1,845,528	3.23	5,948,655
1998	46,449	40,126	46.9	1,880,733	2.52	4,740,361
1999	43,281	35,436	47.8	1,693,130	2.29	3,863,641
2000	43,313	35,002	44.6	1,561,723	2.51	3,883,640
2001	40,943	31,165	43.4	1,353,119	2.72	3,661,591
2002	41,766	29,742	38.2	1,137,001	3.41	3,810,235
2003	45,384	36,753	46.7	1,716,721	3.27	5,597,974
2004	43,350	34,462	43.5	1,499,434	3.32	4,948,510
2005	40,433	33,794	44.4	1,499,129	3.30	4,924,953
			Durum wheat			
	1,000 acres	*1,000 acres*	*Bushels*	*1,000 bushels*	*Dollars*	*1,000 dollars*
1996	3,630	3,556	32.6	116,090	4.67	541,993
1997	3,310	3,177	27.6	87,783	4.92	422,497
1998	3,805	3,728	37.0	138,119	3.15	452,860
1999	4,035	3,569	27.8	99,322	2.73	284,677
2000	3,937	3,572	30.7	109,805	2.66	301,356
2001	2,910	2,789	30.0	83,556	3.08	269,391
2002	2,913	2,709	29.5	79,960	4.05	329,936
2003	2,915	2,869	33.7	96,637	3.97	396,905
2004	2,561	2,363	38.0	89,893	3.85	347,336
2005	2,760	2,716	37.2	101,105	3.55	362,010
			Other spring wheat [3]			
	1,000 acres	*1,000 acres*	*Bushels*	*1,000 bushels*	*Dollars*	*1,000 dollars*
1996	20,030	19,689	35.1	691,680	4.20	2,844,028
1997	19,117	18,323	29.9	548,155	3.53	1,915,589
1998	15,567	15,148	34.9	528,469	3.00	1,587,402
1999	15,348	14,768	34.1	503,108	2.88	1,438,357
2000	15,299	14,489	38.4	556,632	2.85	1,586,790
2001	15,579	14,519	35.2	510,778	2.90	1,481,852
2002	15,639	13,373	29.1	388,917	3.82	1,497,245
2003	13,842	13,441	39.5	531,402	3.62	1,934,160
2004	13,763	13,174	43.2	568,918	3.51	1,987,478
2005	14,036	13,609	37.1	504,456	3.65	1,853,394

[1] Seeded in preceding fall for winter wheat.   [2] Obtained by weighting State prices by quantity sold.   [3] Includes small quantities of Durum wheat grown in other States.

NASS, Crops Branch, (202) 720–2127.

Table 1-4.—Wheat: Stocks on and off farms, United States, 1996–2005

Year beginning September	All wheat							
	On farms				Off farms [1]			
	Sept. 1	Dec. 1	Mar. 1	Jun. 1	Sept. 1	Dec. 1	Mar. 1	Jun. 1
	1,000 bushels	1,000 bushels	1,000 bushels	1,000 bushels	1,000 bushels	1,000 bushels	1,000 bushels	1,000 bushels
1996	824,500	584,150	320,750	154,560	899,696	634,660	501,069	289,047
1997	794,350	604,000	399,920	224,210	1,281,998	1,015,242	766,644	498,268
1998	885,720	680,200	471,220	277,710	1,499,595	1,215,481	979,191	668,208
1999	888,060	647,400	424,680	226,780	1,556,983	1,236,344	991,841	722,968
2000	808,390	623,420	384,750	197,270	1,544,280	1,182,705	953,648	678,912
2001	696,850	517,890	338,500	216,830	1,458,964	1,105,565	871,268	560,282
2002	578,200	384,800	236,300	132,110	1,170,787	935,069	670,333	359,306
2003	687,320	491,925	257,890	131,880	1,351,652	1,028,359	762,727	414,559
2004	790,600	531,020	304,710	161,275	1,147,807	899,306	679,681	378,825
2005	721,360	513,000	NA	NA	1,201,931	916,531	NA	NA

Year beginning September	Durum wheat [2]							
	On farms				Off farms [1]			
	Sept. 1	Dec. 1	Mar. 1	Jun. 1	Sept. 1	Dec. 1	Mar. 1	Jun. 1
	1,000 bushels	1,000 bushels	1,000 bushels	1,000 bushels	1,000 bushels	1,000 bushels	1,000 bushels	1,000 bushels
1996	79,700	66,100	33,100	17,800	22,410	19,541	21,855	12,938
1997	51,000	37,000	22,000	13,380	36,712	30,280	20,473	12,448
1998	88,000	75,300	58,200	37,500	37,908	33,300	30,372	17,302
1999	96,900	74,500	51,700	30,300	39,830	35,449	29,617	19,532
2000	85,700	72,000	44,200	29,100	37,573	32,306	28,616	16,073
2001	63,300	49,600	30,200	20,600	33,779	26,997	21,690	12,390
2002	66,000	50,800	31,700	15,100	26,854	25,917	25,149	13,008
2003	58,000	41,400	24,800	13,600	29,241	25,569	19,447	12,712
2004	65,600	51,800	35,200	24,100	25,508	26,805	20,496	13,494
2005	70,200	57,700	NA	NA	31,135	23,684	NA	NA

[1] Includes stocks at mills, elevators, warehouses, terminals, and processors.   [2] Included in all wheat.   NA-not available.
NASS, Crops Branch, (202) 720–2127.

Table 1-5.—Wheat: Supply and disappearance, by class, United States, 2001–2005 [1]

Item	Year beginning June				
	2001	2002	2003	2004	2005
	Million bushels	Million bushels	Million bushels	Million bushels	Million bushels
All wheat:					
Stocks, June 1	876	777	491	546	540
Production	1,947	1,606	2,345	2,158	2,098
Supply [2]	2,931	2,460	2,899	2,775	2,718
Exports [3]	962	850	1,158	1,063	1,000
Domestic disappearance	1,192	1,119	1,194	1,172	1,188
Stocks, May 31	777	491	546	540	530
Hard red winter:					
Stocks, June 1	411	363	188	227	193
Production	766	620	1,071	856	925
Supply [2]	1,178	984	1,260	1,084	1,119
Exports [3]	349	308	510	388	435
Domestic disappearance	465	488	522	503	509
Stocks, May 31	363	188	227	193	175
Soft red winter:					
Stocks, June 1	135	78	55	64	88
Production	397	321	380	380	309
Supply [2]	535	412	457	466	418
Exports [3]	200	105	138	122	80
Domestic disappearance	258	253	256	256	248
Stocks, May 31	78	55	64	88	90
Hard red spring:					
Stocks, June 1	210	230	145	157	159
Production	475	351	500	525	467
Supply [2]	746	605	654	690	639
Exports [3]	217	258	272	314	275
Domestic disappearance	299	202	225	217	245
Stocks, May 31	230	145	157	159	119
Durum:					
Stocks, June 1	45	33	28	26	38
Production	84	80	97	90	100
Supply [2]	163	143	145	145	167
Exports [3]	49	33	46	31	35
Domestic disappearance	81	82	73	76	74
Stocks, May 31	33	28	26	38	58
White:					
Stocks, June 1	75	73	75	72	63
Production	226	233	297	306	298
Supply [2]	309	317	383	390	376
Exports [3]	147	147	192	207	175
Domestic disappearance	89	94	119	120	112
Stocks, May 31	73	75	72	63	89

[1] Data except production are approximations.　[2] Total supply includes imports.　[3] Imports and exports include flour and products in wheat equivalent.

ERS, Market and Trade Economics Division, (202) 694–5285.

Table 1-6.—Wheat: Area, yield, and production, by States, 2003–2005

State	Area planted [1]			Area harvested			Yield per harvested acre			Production		
	2003	2004	2005	2003	2004	2005	2003	2004	2005	2003	2004	2005
	1,000 acres	1,000 acres	1,000 acres	1,000 acres	1,000 acres	1,000 acres	Bushels	Bushels	Bushels	1,000 bushels	1,000 bushels	1,000 bushels
AL	150	120	100	75	60	45	42.0	48.0	50.0	3,150	2,880	2,250
AZ	119	105	85	119	103	81	100.1	96.7	99.5	11,912	9,963	8,060
AR	700	670	220	570	620	160	50.0	53.0	52.0	28,500	32,860	8,320
CA	870	680	570	525	420	369	69.5	86.2	76.3	36,510	36,200	28,155
CO	2,630	2,315	2,570	2,229	1,714	2,219	35.1	27.4	24.4	78,160	46,880	54,035
DE	50	50	52	47	47	51	41.0	58.0	70.0	1,927	2,726	3,570
FL	20	18	18	12	15	8	41.0	45.0	45.0	492	675	360
GA	380	330	280	230	190	140	46.0	45.0	52.0	10,580	8,550	7,280
ID	1,190	1,250	1,260	1,130	1,190	1,200	74.9	85.5	83.8	84,660	101,710	100,590
IL	850	920	630	810	900	600	65.0	59.0	61.0	52,650	53,100	36,600
IN	460	450	360	430	440	340	69.0	62.0	72.0	29,670	27,280	24,480
IA	25	28	20	21	24	15	61.0	55.0	50.0	1,281	1,320	750
KS	10,500	10,000	10,000	10,000	8,500	9,500	48.0	37.0	40.0	480,000	314,500	380,000
KY	500	530	390	350	380	300	62.0	54.0	68.0	21,700	20,520	20,400
LA	155	180	110	140	165	100	41.0	50.0	48.0	5,740	8,250	4,800
MD	165	160	155	145	145	140	37.0	59.0	66.0	5,365	8,555	9,240
MI	680	660	600	660	640	590	68.0	64.0	66.0	44,880	40,960	38,940
MN	1,877	1,728	1,820	1,825	1,636	1,745	57.8	54.8	41.0	105,482	89,605	71,470
MS	150	160	70	125	135	65	49.0	53.0	50.0	6,125	7,155	3,250
MO	960	1,050	590	870	930	540	61.0	52.0	54.0	53,070	48,360	29,160
MT	5,440	5,470	5,340	5,200	5,025	5,235	27.4	34.5	36.8	142,330	173,165	192,480
NE	1,900	1,850	1,850	1,820	1,650	1,760	46.0	37.0	39.0	83,720	61,050	68,640
NV	12	14	14	7	9	8	78.4	106.7	100.6	549	960	805
NJ	31	28	28	26	24	23	42.0	47.0	53.0	1,092	1,128	1,219
NM	500	490	450	140	300	270	30.0	26.0	36.0	4,200	7,800	9,720
NY	130	105	100	120	100	95	53.0	53.0	54.0	6,360	5,300	5,130
NC	530	600	560	410	460	435	36.0	50.0	57.0	14,760	23,000	24,795
ND	8,630	8,195	9,090	8,500	7,775	8,835	37.3	39.4	34.4	317,090	306,650	303,765
OH	1,060	920	860	1,000	890	830	68.0	62.0	71.0	68,000	55,180	58,930
OK	6,700	6,200	5,700	4,600	4,700	4,000	39.0	35.0	32.0	179,400	164,500	128,000
OR	1,115	1,000	955	1,080	955	895	49.6	58.6	59.8	53,540	55,980	53,560
PA	175	140	150	165	135	145	43.0	49.0	54.0	7,095	6,615	7,830
SC	200	190	170	185	180	165	39.0	44.0	52.0	7,215	7,920	8,580
SD	3,078	3,270	3,315	2,797	2,798	3,193	42.3	46.0	41.8	118,391	128,610	133,420
TN	430	400	240	270	280	150	50.0	49.0	56.0	13,500	13,720	8,400
TX	6,600	6,300	5,500	3,450	3,500	3,000	28.0	31.0	32.0	96,600	108,500	96,000
UT	177	143	163	137	132	148	41.4	44.4	48.0	5,677	5,856	7,099
VA	210	210	180	160	180	160	46.0	55.0	63.0	7,360	9,900	10,080
WA	2,400	2,330	2,280	2,345	2,275	2,225	59.4	63.1	62.6	139,345	143,500	139,300
WV	12	8	7	7	5	5	41.0	52.0	60.0	287	260	300
WI	212	247	208	180	231	182	68.3	55.6	56.4	12,300	12,852	10,262
WY	168	160	169	151	141	152	27.1	26.6	30.7	4,095	3,750	4,665
US	62,141	59,674	57,229	53,063	49,999	50,119	44.2	43.2	42.0	2,344,760	2,158,245	2,104,690

[1] Includes area planted preceding fall.

NASS, Crops Branch, (202) 720–2127.

Table 1-7.—Wheat: Supply and disappearance, United States, 1996–2005

Year beginning June	Supply				Disappearance						Ending stocks May 31
	Beginning stocks	Production	Imports [1]	Total	Domestic use				Exports [1]	Total disappearance	
					Food	Seed	Feed [2]	Total			
	Million bushels	Million bushels	Million bushels	Million bushels	Million bushels	Million bushels	Million bushels	Million bushels	Million bushels	Million bushels	Million bushels
1996	376	2,277	92	2,746	891	102	308	1,301	1,002	2,302	444
1997	444	2,481	95	3,020	914	92	251	1,257	1,040	2,298	722
1998	722	2,547	103	3,373	909	81	391	1,381	1,046	2,427	946
1999	946	2,296	95	3,336	929	92	279	1,300	1,086	2,386	950
2000	950	2,228	90	3,268	950	79	300	1,330	1,062	2,392	876
2001	876	1,947	108	2,931	926	83	182	1,192	962	2,154	777
2002	777	1,606	77	2,460	919	84	116	1,119	850	1,969	491
2003	491	2,345	63	2,899	912	80	203	1,194	1,158	2,353	546
2004	546	2,158	71	2,775	907	79	187	1,172	1,063	2,235	540
2005 [3]	540	2,098	80	2,718	910	78	200	1,188	1,000	2,188	530

[1] Imports and exports include flour and other products expressed in wheat equivalent. [2] Approximates feed and residual use and includes negligible quantities used for distilled spirits. [3] Preliminary. Totals may not add due to independent rounding.

ERS, Market and Trade Economics Division, (202) 694–5296.

GRAIN AND FEED

Table 1-8.—Wheat, by type: Area, yield, and production, by States, 2003–2005

State	Area planted [1]			Area harvested			Yield per harvested acre			Production		
	2003	2004	2005	2003	2004	2005	2003	2004	2005	2003	2004	2005
	1,000 acres	*1,000 acres*	*1,000 acres*	*1,000 acres*	*1,000 acres*	*1,000 acres*	*Bushels*	*Bushels*	*Bushels*	*1,000 bushels*	*1,000 bushels*	*1,000 bushels*
Winter wheat												
AL	150	120	100	75	60	45	42.0	48.0	50.0	3,150	2,880	2,250
AZ	4	5	5	4	4	2	103.0	90.0	80.0	412	360	160
AR	700	670	220	570	620	160	50.0	53.0	52.0	28,500	32,860	8,320
CA	740	560	495	410	320	300	61.0	85.0	72.0	25,010	27,200	21,600
CO	2,600	2,300	2,550	2,200	1,700	2,200	35.0	27.0	24.0	77,000	45,900	52,800
DE	50	50	52	47	47	51	41.0	58.0	70.0	1,927	2,726	3,570
FL	20	18	18	12	15	8	41.0	45.0	45.0	492	675	360
GA	380	330	280	230	190	140	46.0	45.0	52.0	10,580	8,550	7,280
ID	760	750	770	720	700	730	80.0	90.0	91.0	57,600	63,000	66,430
IL	850	920	630	810	900	600	65.0	59.0	61.0	52,650	53,100	36,600
IN	460	450	360	430	440	340	69.0	62.0	72.0	29,670	27,280	24,480
IA	25	28	20	21	24	15	61.0	55.0	50.0	1,281	1,320	750
KS	10,500	10,000	10,000	10,000	8,500	9,500	48.0	37.0	40.0	480,000	314,500	380,000
KY	500	530	390	350	380	300	62.0	54.0	68.0	21,700	20,520	20,400
LA	155	180	110	140	165	100	41.0	50.0	48.0	5,740	8,250	4,800
MD	165	160	155	145	145	140	37.0	59.0	66.0	5,365	8,555	9,240
MI	680	660	600	660	640	590	68.0	64.0	66.0	44,880	40,960	38,940
MN	25	27	20	23	25	15	42.0	40.0	36.0	966	1,000	540
MS	150	160	70	125	135	65	49.0	53.0	50.0	6,125	7,155	3,250
MO	960	1,050	590	870	930	540	61.0	52.0	54.0	53,070	48,360	29,160
MT	1,900	1,900	2,150	1,820	1,630	2,100	37.0	41.0	45.0	67,340	66,830	94,500
NE	1,900	1,850	1,850	1,820	1,650	1,760	46.0	37.0	39.0	83,720	61,050	68,640
NV	7	6	8	3	3	5	83.0	110.0	110.0	249	330	550
NJ	31	28	28	26	24	23	42.0	47.0	53.0	1,092	1,128	1,219
NM	500	490	450	140	300	270	30.0	26.0	36.0	4,200	7,800	9,720
NY	130	105	100	120	100	95	53.0	53.0	54.0	6,360	5,300	5,130
NC	530	600	560	410	460	435	36.0	50.0	57.0	14,760	23,000	24,795
ND	130	245	310	120	225	285	49.0	44.0	39.0	5,880	9,900	11,115
OH	1,060	920	860	1,000	890	830	68.0	62.0	71.0	68,000	55,180	58,930
OK	6,700	6,200	5,700	4,600	4,700	4,000	39.0	35.0	32.0	179,400	164,500	128,000
OR	970	820	830	940	780	780	51.0	61.0	61.0	47,940	47,580	47,580
PA	175	140	150	165	135	145	43.0	49.0	54.0	7,095	6,615	7,830
SC	200	190	170	185	180	165	39.0	44.0	52.0	7,215	7,920	8,580
SD	1,650	1,650	1,550	1,430	1,250	1,490	43.0	45.0	44.0	61,490	56,250	65,560
TN	430	400	240	270	280	150	50.0	49.0	56.0	13,500	13,720	8,400
TX	6,600	6,300	5,500	3,450	3,500	3,000	28.0	31.0	32.0	96,600	108,500	96,000
UT	160	130	145	125	120	135	41.0	43.0	47.0	5,125	5,160	6,345
VA	210	210	180	160	180	160	46.0	55.0	63.0	7,360	9,900	10,080
WA	1,850	1,800	1,850	1,800	1,750	1,800	65.0	67.0	67.0	117,000	117,250	120,600
WV	12	8	7	7	5	5	41.0	52.0	60.0	287	260	300
WI	205	240	200	175	225	175	69.0	56.0	57.0	12,075	12,600	9,975
WY	160	150	160	145	135	145	27.0	26.0	30.0	3,915	3,510	4,350
US	45,384	43,350	40,433	36,753	34,462	33,794	46.7	43.5	44.4	1,716,721	1,499,434	1,499,129
Durum wheat												
AZ	115	100	80	115	99	79	100.0	97.0	100.0	11,500	9,603	7,900
CA	130	120	75	115	100	69	100.0	90.0	95.0	11,500	9,000	6,555
ID [2]	....	....	20	....	....	20	....	....	88.0	....	....	1,760
MN [3]	2	1	....	2	1	....	58.0	55.0	....	116	55	....
MT	640	570	590	630	545	585	23.0	33.0	28.0	14,490	17,985	16,380
ND	2,000	1,750	1,980	1,980	1,600	1,950	29.5	33.0	35.0	58,410	52,800	68,250
SD	28	20	15	27	18	13	23.0	25.0	20.0	621	450	260
US	2,915	2,561	2,760	2,869	2,363	2,716	33.7	38.0	37.2	96,637	89,893	101,105
Other spring wheat												
CO	30	15	20	29	14	19	40.0	70.0	65.0	1,160	980	1,235
ID	430	500	470	410	490	450	66.0	79.0	72.0	27,060	38,710	32,400
MN	1,850	1,700	1,800	1,800	1,610	1,730	58.0	55.0	41.0	104,400	88,550	70,930
MT	2,900	3,000	2,600	2,750	2,850	2,550	22.0	31.0	32.0	60,500	88,350	81,600
NV	5	8	6	4	6	3	75.0	105.0	85.0	300	630	255
ND	6,500	6,200	6,800	6,400	5,950	6,600	39.5	41.0	34.0	252,800	243,950	224,400
OR	145	180	125	140	175	115	40.0	48.0	52.0	5,600	8,400	5,980
SD	1,400	1,600	1,750	1,340	1,530	1,690	42.0	47.0	40.0	56,280	71,910	67,600
UT	17	13	18	12	12	13	46.0	58.0	58.0	552	696	754
WA	550	530	430	545	525	425	41.0	50.0	44.0	22,345	26,250	18,700
WI	7	7	8	5	6	7	45.0	42.0	41.0	225	252	287
WY	8	10	9	6	6	7	30.0	40.0	45.0	180	240	315
US	13,842	13,763	14,036	13,441	13,174	13,609	39.5	43.2	37.1	531,402	568,918	504,456

[1] Includes area planted preceding fall.   [2] Estimates began in 2005.   [3] Estimates discontinued in 2005.

NASS, Crops Branch, (202) 720–2127.

Table 1-9.—Wheat: Support operations, United States, 1996–2005

| Marketing year beginning June 1 | Income support payment rates per bushel [1] | Program price levels per bushel | | Put under loan [4] | | Acquired by CCC under loan program [5] | Owned by CCC at end of marketing year [6] |
		Loan [2]	Target [3]	Quantity	Percentage of production		
	Dollars	Dollars	Dollars	Million bushels	Percent	Million bushels	Million bushels
1996/1997 ...	0.87	2.58	NA	194	8.5	0	93
1997/1998 ...	0.63	2.58	NA	264	10.6	2	94
1998/1999 ...	0.99	2.58	NA	363	14.3	30	128
1999/2000 ...	1.27	2.58	NA	154	6.7	13	104
2000/2001 ...	1.23	2.58	NA	181	8.1	27	97
2001/2002 ...	1.01	2.58	NA	197	10.1	17	99
2002/2003 ...	0.52/0.00	2.80	3.86	120	7.5	2	66
2003/2004 ...	0.52/0.00	2.80	3.86	186	7.9	3	61
2004/2005 ...	0.52/0.00	2.75	3.92	178	8.2	6	55
2005/2006 ...	0.52/0.00	2.75	3.92	....	....	....	....

[1] Payment rates for the 1995/96 and prior crop years were calculated according to the deficiency payment/production adjustment program provisions. Payment rates for the 1996/97 through 2001/2002 crops were calculated according to the Production Flexibility Contract (PFC) program provisions of the Federal Agriculture Improvement and Reform Act of 1996 (1996 Act) and include supplemental PFC payment rates for 1998 through 2001. Payment rates for the 2002/2003 and subsequent crops are calculated according to the Direct and Counter-cyclical program provisions, following enactment of the Farm Security and Rural Investment Act of 2002 (2002 Act). Payment rates are rounded to the nearest cent. Beginning with 2002/2003, the first entry is the direct payment rate and the second entry is the counter-cyclical payment rate.   [2] The national average loan rate was also known as the price support rate prior to enactment of the 1996 Act.   [3] Between the 1996/97 and 2001/2002 marketing years, target prices were no longer applicable; however, target prices were reestablished under the 2002 Act.   [4] Represents loans made, purchases, and purchase agreements entered into. Purchases and purchase agreements are no longer authorized for the 1996 and subsequent crops following enactment of the 1996 Act. Percentage of production is on a grain basis. Excludes quantity on which loan deficiency payments were made.   [5] Acquisition of all loans forfeited during the marketing year. For 2004/05, as of October 25, 2005.   [6] Includes 147 million bushels in Food Security Reserve for 1993/94, 141 million in 1994/95, 118 million in 1995/96, 93 million in 1996/97 through 2001/02, 66 million in 2002/03, 59 million in 2003/04 and 52 million in 2004/05. (The Food Security Reserve became the Food Security Commodity Trust in July of 1999 and the Bill Emerson Humanitarian Trust in July of 2002.)   NA-not applicable.
FSA, Food Grains, (202) 720–5653.

Table 1-10.—Wheat: Marketing year average price and value, by States, crop of 2003, 2004, and 2005

| State | Marketing year average price per bushel | | | Value of production | | |
	2003	2004	2005 [1]	2003	2004	2005 [1]
	Dollars	Dollars	Dollars	1,000 dollars	1,000 dollars	1,000 dollars
AL	3.20	3.55	3.00	10,080	10,224	6,750
AZ	4.64	4.25	4.20	55,082	42,073	33,756
AR	3.08	3.49	3.30	87,780	114,681	27,456
CA	3.54	3.80	3.55	137,399	135,618	99,005
CO	3.32	3.25	3.35	260,106	152,399	180,894
DE	3.10	3.05	2.95	5,974	8,314	10,532
FL	3.00	3.45	3.00	1,476	2,329	1,080
GA	3.05	3.45	2.95	32,269	29,498	21,476
ID	3.49	3.61	3.40	294,269	365,222	337,876
IL	3.20	3.19	3.20	168,480	169,389	117,120
IN	3.21	3.24	3.15	95,241	88,387	77,112
IA	2.85	3.05	3.10	3,651	4,026	2,325
KS	3.15	3.25	3.30	1,512,000	1,022,125	1,254,000
KY	3.17	2.96	3.20	68,789	60,739	65,280
LA	3.30	3.40	3.20	18,942	28,050	15,360
MD	3.15	3.10	3.00	16,900	26,521	27,720
MI	3.25	3.01	3.15	145,860	123,290	122,661
MN	3.66	3.32	3.60	386,130	298,045	257,076
MS	3.34	3.37	3.30	20,458	24,112	10,725
MO	3.09	3.24	3.40	163,986	156,686	99,144
MT	3.73	3.61	3.60	527,394	623,324	696,537
NE	3.22	3.23	3.25	269,578	197,192	223,080
NV	3.45	3.60	3.30	1,914	3,533	2,638
NJ	3.10	3.30	3.15	3,385	3,722	3,840
NM	3.30	3.15	3.10	13,860	24,570	30,132
NY	2.43	2.80	3.05	15,455	14,840	15,647
NC	2.85	3.10	3.05	42,066	71,300	75,625
ND	3.63	3.40	3.55	1,149,746	1,042,884	1,074,981
OH	3.20	3.16	3.20	217,600	174,369	188,576
OK	3.31	3.32	3.35	593,814	546,140	428,800
OR	3.70	3.69	3.45	197,580	206,539	180,934
PA	3.31	3.40	3.40	23,484	22,491	26,622
SC	3.00	3.20	2.70	21,645	25,344	23,166
SD	3.46	3.37	3.60	408,188	432,114	480,861
TN	3.17	3.48	3.35	42,795	47,746	28,140
TX	3.06	3.34	3.30	295,596	362,390	316,800
UT	4.00	3.84	3.65	22,756	22,427	25,707
VA	2.98	2.95	2.90	21,933	29,205	29,232
WA	3.75	3.68	3.45	521,163	524,493	476,005
WV	3.13	3.04	3.05	898	790	915
WI	3.20	2.65	2.85	39,439	34,171	29,376
WY	3.40	3.20	3.30	13,878	12,012	15,395
US	3.40	3.40	3.40	7,929,039	7,283,324	7,140,357

[1] Preliminary.
NASS, Crops Branch, (202) 720–2127.

GRAIN AND FEED

Table 1-11.—Wheat: Area, yield, and production in specified countries, 2002/2003–2004/2005 [1]

Continent and country	Area [2]			Yield per hectare			Production		
	2002/ 2003	2003/ 2004	2004/ 2005 [3]	2002/ 2003	2003/ 2004	2004/ 2005 [3]	2002/ 2003	2003/ 2004	2004/ 2005 [3]
	1,000 hectares	1,000 hectares	1,000 hectares	Metric tons	Metric tons	Metric tons	1,000 metric tons	1,000 metric tons	1,000 metric tons
North America:									
Canada	8,836	10,467	9,862	1.83	2.25	2.62	16,198	23,552	25,860
Mexico	630	600	510	5.13	4.50	4.55	3,230	2,700	2,320
UnitedStates	18,544	21,474	20,234	2.36	2.97	2.90	43,705	63,814	58,738
Total	28,010	32,541	30,606	2.25	2.77	2.84	63,133	90,066	86,918
South America:									
Argentina	5,900	5,700	6,100	2.08	2.46	2.62	12,300	14,000	16,000
Bolivia	114	110	114	0.99	0.91	1.03	113	100	117
Brazil	2,043	2,464	2,756	1.43	2.37	2.12	2,925	5,851	5,845
Chile	416	420	420	4.32	4.58	4.43	1,797	1,922	1,860
Colombia	15	15	15	2.13	2.13	2.20	32	32	33
Ecuador	20	19	18	0.65	0.63	0.67	13	12	12
Paraguay	160	350	300	2.06	2.00	1.83	330	700	550
Peru	139	140	128	1.35	1.35	1.18	187	189	151
Uruguay	137	118	180	1.50	2.76	2.94	206	326	530
Total	8,944	9,336	10,031	2.00	2.48	2.50	17,903	23,132	25,098
Guatemala	1	1	1	1.00	1.00	1.00	1	1	1
Europe:									
Austria	284	267	284	5.05	4.37	5.97	1,434	1,166	1,695
Belgium-Luxembourg	212	210	235	8.16	8.24	8.51	1,729	1,730	2,000
Cyprus	6	6	5	2.17	2.17	1.60	13	13	8
Czech Republic	849	648	863	4.55	4.07	5.84	3,867	2,638	5,042
Denmark	580	630	675	6.99	7.46	7.09	4,056	4,701	4,787
Estonia	66	70	76	2.24	2.07	2.43	148	145	185
Finland	170	180	225	3.35	3.77	3.48	570	679	782
France	5,231	4,917	5,240	7.44	6.23	7.58	38,934	30,641	39,712
Germany	3,015	2,964	3,112	6.90	6.50	8.17	20,818	19,260	25,427
Greece	600	775	825	2.97	1.65	1.64	1,783	1,280	1,350
Hungary	1,100	1,112	1,150	3.55	2.61	5.04	3,910	2,900	5,800
Ireland	103	95	102	8.42	8.36	9.43	867	794	962
Italy	2,415	2,266	2,353	3.13	2.81	3.73	7,547	6,362	8,783
Latvia	152	140	170	3.42	3.34	2.90	520	468	493
Lithuania	335	337	355	3.64	3.57	4.03	1,218	1,204	1,430
Malta	1	1	2	5.00	5.00	5.00	5	5	10
Netherlands	135	135	135	7.83	9.11	9.26	1,057	1,230	1,250
Poland	2,414	2,308	2,311	3.85	3.40	4.28	9,304	7,858	9,892
Portugal	230	174	189	1.80	0.93	1.33	413	161	251
Slovakia	403	303	362	3.86	3.07	4.88	1,555	930	1,765
Slovenia	36	36	32	4.86	3.42	4.59	175	123	147
Spain	2,406	2,221	2,152	2.84	2.71	3.30	6,822	6,019	7,108
Sweden	340	411	400	6.21	5.55	6.02	2,112	2,283	2,409
United Kingdom	1,996	1,837	1,990	8.00	7.78	7.78	15,973	14,288	15,473
Total	23,079	22,043	23,243	5.41	4.85	5.88	124,830	106,878	136,761
Other Europe:									
Albania	100	100	100	2.90	3.00	3.00	290	300	300
Bosnia-Hercegovina	110	72	110	2.70	2.71	2.27	297	195	250
Bulgaria	1,150	750	950	3.00	2.27	3.79	3,450	1,700	3,600
Croatia	219	195	200	4.31	4.31	4.20	943	840	840
Macedonia (Skopje)	120	100	120	2.08	2.20	2.33	250	220	280
Norway	65	65	65	4.31	4.31	4.31	280	280	280
Romania	2,190	1,500	1,800	1.96	1.33	3.61	4,300	2,000	6,500
Serbia and Montenegro	694	600	636	3.23	2.27	4.34	2,240	1,360	2,758
Switzerland	100	100	100	6.10	6.00	6.00	610	600	600
Total E. Europe	4,748	3,482	4,081	2.67	2.15	3.78	12,660	7,495	15,408

See footnotes at end of table.

Table 1-11.—Wheat: Area, yield, and production in specified countries, 2002/2003–2004/2005 [1]—Continued

Continent and country	Area [2]			Yield per hectare			Production		
	2002/ 2003	2003/ 2004	2004/ 2005 [3]	2002/ 2003	2003/ 2004	2004/ 2005 [3]	2002/ 2003	2003/ 2004	2004/ 2005 [3]
	1,000 hec- tares	*1,000 hec- tares*	*1,000 hec- tares*	*Metric tons*	*Metric tons*	*Metric tons*	*1,000 metric tons*	*1,000 metric tons*	*1,000 metric tons*
Fmr. Soviet Union:.									
Armenia	105	115	130	2.71	1.87	2.62	285	215	340
Azerbaijan	650	600	610	2.60	2.58	2.58	1,690	1,550	1,575
Belarus	378	350	550	2.69	2.00	2.60	1,017	700	1,430
Georgia	130	110	100	1.54	2.05	1.85	200	225	185
Kazakhstan	11,500	11,300	11,800	1.10	0.97	0.84	12,600	11,000	9,950
Kyrgyzstan	500	435	415	2.61	2.53	2.41	1,306	1,100	1,000
Moldova	420	150	310	2.86	1.07	2.74	1,200	160	850
Russian Fed.	25,700	22,150	24,200	1.97	1.54	1.87	50,550	34,100	45,300
Tajikistan	325	290	300	1.68	2.28	2.17	545	660	650
Turkmenistan ...	700	850	800	2.86	2.59	3.06	2,000	2,200	2,450
Ukraine	6,750	2,450	5,900	3.05	1.47	2.97	20,556	3,600	17,500
Uzbekistan	1,200	1,450	1,400	4.17	3.72	3.71	5,000	5,400	5,200
Total	48,358	40,250	46,515	2.00	1.51	1.86	96,949	60,910	86,430
Middle East:									
Iran	6,200	6,500	6,600	2.00	2.08	2.12	12,400	13,500	14,000
Iraq	1,800	1,800	1,800	1.00	1.11	1.22	1,800	2,000	2,200
Israel	70	70	70	2.56	2.67	1.83	179	187	128
Jordan	50	13	55	1.40	1.15	0.91	70	15	50
Lebanon	20	20	20	3.00	3.00	3.00	60	60	60
Saudi Arabia	446	446	320	4.48	4.71	5.00	2,000	2,100	1,600
Syria	1,600	1,700	1,700	2.81	2.76	2.53	4,500	4,700	4,300
Turkey	8,550	8,600	8,600	1.96	1.95	2.09	16,800	16,800	18,000
Yemen	89	89	90	1.48	1.39	1.39	132	124	125
Total	18,825	19,238	19,255	2.02	2.05	2.10	37,941	39,486	40,463
Africa:									
Algeria	2,165	2,760	1,998	0.69	1.08	1.30	1,502	2,970	2,602
Angola	5	5	5	0.80	0.80	0.80	4	4	4
Chad	2	2	4	2.00	1.50	1.25	4	3	5
Congo, Dem.Rep.	10	10	10	1.80	1.80	1.10	18	18	11
Egypt	1,008	1,029	1,092	6.25	6.26	6.07	6,300	6,443	6,630
Eritrea	25	20	24	0.12	0.20	0.21	3	4	5
Ethiopia	1,700	1,700	1,700	1.12	1.18	1.35	1,900	2,000	2,300
Kenya	140	111	120	2.14	1.77	1.64	300	196	197
Lesotho	29	23	23	0.52	0.57	0.87	15	13	20
Libya	250	250	250	0.50	0.50	0.50	125	125	125
Morocco	2,626	2,989	3,064	1.28	1.72	1.81	3,357	5,147	5,540
Mozambique	1	3	3	1.00	1.00	1.00	1	3	3
Nigeria	35	35	35	1.43	1.57	1.71	50	55	60
South Africa, Rep.	941	748	830	2.47	2.06	2.02	2,320	1,540	1,680
Sudan	100	110	170	2.50	3.64	2.75	250	400	467
Tanzania, United Rep. ..	55	65	70	1.36	1.15	1.07	75	75	75
Tunisia	755	900	974	0.56	1.78	1.77	420	1,600	1,722
Zambia	12	25	33	6.25	5.40	2.58	75	135	85
Zimbabwe	38	30	35	3.95	3.00	4.00	150	90	140
Total	9,897	10,815	10,440	1.70	1.93	2.08	16,869	20,821	21,671

See footnotes at end of table.

Table 1-11.—Wheat: Area, yield, and production in specified countries, 2002/2003–2004/2005 [1]—Continued

Continent and country	Area [2]			Yield per hectare			Production		
	2002/ 2003	2003/ 2004	2004/ 2005 [3]	2002/ 2003	2003/ 2004	2004/ 2005 [3]	2002/ 2003	2003/ 2004	2004/ 2005 [3]
	1,000 hec- tares	1,000 hec- tares	1,000 hec- tares	Metric tons	Metric tons	Metric tons	1,000 metric tons	1,000 metric tons	1,000 metric tons
Asia:									
Afghanistan	1,742	2,300	2,200	1.54	1.90	1.82	2,686	4,360	4,000
Bangladesh	707	567	600	2.14	2.21	1.83	1,510	1,253	1,100
Bhutan	13	13	13	1.54	1.54	1.54	20	20	20
Burma	79	83	106	1.22	1.29	1.32	96	107	140
China, Peop. Rep.	23,910	22,000	21,626	3.78	3.93	4.25	90,290	86,490	91,950
India	25,900	24,860	26,620	2.77	2.62	2.71	71,810	65,100	72,060
Japan	207	212	212	4.00	4.03	4.06	828	855	860
Korea, Dem. Rep.	95	95	100	2.05	2.26	2.20	195	215	220
Korea, Rep.	2	2	4	3.00	5.00	3.25	6	10	13
Mongolia	218	220	300	0.68	0.82	0.50	149	180	150
Nepal	640	640	640	1.64	1.64	1.64	1,050	1,050	1,050
Pakistan	8,057	8,094	8,200	2.26	2.37	2.32	18,226	19,192	19,000
Total	61,570	59,086	60,621	3.04	3.03	3.14	186,866	178,832	190,563
Oceania:									
Australia	11,070	13,024	12,200	0.92	2.01	1.76	10,132	26,231	21,500
New Zealand	56	56	56	6.34	6.07	6.07	355	340	340
Total	11,126	13,080	12,256	0.94	2.03	1.78	10,487	26,571	21,840
World Total ...	214,049	209,870	217,050	2.65	2.64	2.88	566,938	554,190	625,150

[1] Years shown refer to years of harvest. Harvests of Northern Hemisphere countries are combined with those of the Southern Hemisphere which immediately follow; thus the crop harvested in the Northern Hemisphere in 1994 is combined with estimates for the Southern Hemisphere Harvests, which begin late in 1994 and end early in 1995.   [2] Harvested area as far as possible.   [3] Preliminary.

FAS, Production Estimates and Crop Assessment Division, (202) 720–0888. Prepared or estimated on the basis of official statistics of foreign governments, other foreign source materials, reports of U.S. Agricultural Counselors, Attachés and Foreign Service Officers, results of office research, and related information.

Table 1-12.—Wheat and flour: United States imports,1995–2004

Year beginning June	Wheat grain [1]	Flour (wheat equivalent)	Other products (wheat equivalent) [2]	Total wheat, flour, and other products
	1,000 bushels	1,000 bushels	1,000 bushels	1,000 bushels
1995	47,753	6,687	13,493	67,933
1996	71,727	6,386	14,220	92,333
1997	73,245	6,055	15,623	94,923
1998	79,766	7,423	15,815	103,004
1999	72,408	7,116	14,986	94,511
2000	66,313	8,863	14,649	89,825
2001	82,615	9,907	15,029	107,551
2002	49,741	11,946	15,687	77,374
2003	37,156	11,363	14,508	63,026
2004	44,525	11,525	14,927	70,072

[1] Starting January 1989, Census ceased reporting wheat suitable for milling and unfit for human consumption.   [2] Includes macaroni, semolina, and similar products. Beginning in 1988/89 total wheat grain is reported under the suitable for milling column.

ERS, Market and Trade Economics Division, (202) 694–5285.

Table 1-13.—Wheat, flour, and products: [1] International trade, 2002/2003–2004/2005 [2]

Country	2002/2003	2003/2004	2004/2005 [3]
	1,000 metric tons	*1,000 metric tons*	*1,000 metric tons*
Principal exporters:			
Argentina	6,276	7,346	13,502
Australia	10,946	15,096	15,826
Canada	9,393	15,526	15,142
India	5,350	5,425	1,500
Kazakhstan	6,238	4,110	2,700
Russia	12,621	3,114	7,951
Syria	800	1,200	700
Turkey	839	854	2,217
Ukraine	6,569	66	4,351
EU-25	19,940	10,931	14,400
Other Europe	1,657	166	1,295
Others	6,442	8,390	4,860
Subtotal	87,071	72,224	84,444
United States	22,834	32,295	28,464
Total	109,905	104,519	112,908
Principal importers:			
Algeria	6,079	3,933	5,398
Bangladesh	1,335	1,945	2,000
Bolivia	356	271	376
Brazil	6,631	5,559	5,309
Chile	421	442	321
China	418	3,749	6,747
Colombia	1,166	1,246	1,248
Cuba	819	727	836
Ecuador	347	514	416
Egypt	6,327	7,295	8,150
Ethiopia	611	782	431
India	19	8	20
Indonesia	3,984	4,535	4,661
Iran	1,561	246	200
Iraq	1,579	1,925	3,010
Israel	1,691	951	1,549
Japan	5,579	5,751	5,744
Jordan	1,147	595	718
Kenya	656	419	474
Korea, North	400	400	400
Korea, South	4,052	3,434	3,591
Libya	1,421	1,057	1,508
Malaysia	1,195	1,329	1,425
Mexico	3,161	3,644	3,717
Morocco	2,720	2,414	2,300
Nigeria	2,304	2,383	3,014
Pakistan	181	47	1,416
Peru	1,157	1,488	1,449
Philippines	3,230	2,975	2,500
Russia	1,045	1,026	1,197
South Africa	1,024	911	1,407
Sri Lanka	1,019	886	1,150
Sudan	860	995	1,500
Taiwan	1,003	1,216	1,150
Thailand	895	1,253	1,100
Tunisia	2,167	781	1,075
Turkey	1,217	1,056	500
UAE	1,010	1,135	1,200
Uzbekistan	254	229	200
Venezuela	961	1,538	1,504
Vietnam	875	830	1,225
Yemen	1,772	1,635	1,900
EU-25	13,921	5,912	7,200
Other Europe	1,921	4,214	1,775
United States	1,958	1,760	1,946
Subtotal	92,449	85,441	94,957
Other Countries	15,232	16,415	15,066
Unaccounted	2,224	2,663	2,885
Total	109,905	104,519	112,908

[1] Flour and products reported in terms of grain equivalent.   [2] Year beginning July 1.   [3] Preliminary.

FAS, Grain and Feed Division, (202) 720–6219. www.fas.usda.gov/grain/default.html. Prepared or estimated on the basis of official statistics from foreign governments, other foreign source materials, reports of U.S. Agricultural Counselors, Attachés and Foreign Service Officers, results of office research, and related information.

GRAIN AND FEED

Table 1-14.—Wheat and flour: [1] United States exports by country of destination, 2002/2003 and 2004/2005

Country of destination	Year [2]		
	2002/2003	2003/2004	2004/2005
	1,000 metric tons	*1,000 metric tons*	*1,000 metric tons*
Wheat:			
Japan	3,081	3,185	2,856
Mexico	2,431	2,902	2,812
Nigeria	1,672	2,164	2,634
China	88	1,466	1,786
Egypt	856	3,980	1,765
Philippines	1,489	1,213	1,751
Korea, South	1,216	1,465	1,309
Taiwan	842	1,049	970
Venezuela	539	795	810
Colombia	736	733	788
Yemen, South	521	521	647
Peru	400	984	626
Italy	556	880	549
Israel	404	648	464
Pakistan	160	14	419
Thailand	402	450	414
Cuba	77	327	399
Ethiopia	297	539	365
Dominican Republic	246	313	330
Guatemala	344	207	289
Iraq	57	246	281
South Africa	48	475	270
Spain	151	343	251
Algeria	179	481	239
Costa Rica	205	196	234
Other Countries	4,845	5,592	3,758
Total	22,104	31,588	27,946
Wheat flour:			
Canada	27	50	60
Mexico	34	48	41
Bolivia	27	19	32
Tajikistan	35	34	28
Haiti	9	29	20
Chad	10	2	10
Kenya	4	7	7
Iraq	17	2	6
Moldova	0	17	6
West Bank	0	0	6
Bahamas	2	3	3
Ethiopia	0	72	3
Israel	0	21	3
Netherlands Antilles	0	4	3
Honduras	0	0	2
Panama	0	2	2
Azerbaijan	1	10	1
Colombia	0	1	1
Dominican Republic	2	2	1
Other Countries	182	69	4
Total	364	400	258

[1] Flour reported in terms of grain equivalent.   [2] Year beginning Jul 1.

FAS, Grain and Feed Division, (202) 720–6219. www.fas.usda.gov/grain/default.html. Compiled from reports of the U.S. Department of Commerce.

Table 1-15.—Rye: Area, yield, production, disposition, and value, United States, 1996–2005

Year	Area		Yield per harvested acre	Production	Marketing year average price per bushel received by farmers [2]	Value of production [2]
	Planted [1]	Harvested				
	1,000 acres	*1,000 acres*	*Bushels*	*1,000 bushels*	*Dollars*	*1,000 dollars*
1996	1,457	345	25.9	8,936	3.70	33,118
1997	1,400	316	25.7	8,132	3.75	30,120
1998	1,566	418	29.1	12,161	2.50	30,404
1999	1,582	383	28.8	11,038	2.27	25,084
2000	1,329	296	28.3	8,386	2.60	21,830
2001	1,328	250	27.6	6,896	2.86	19,752
2002	1,355	263	24.7	6,488	3.32	21,549
2003	1,348	319	27.1	8,634	2.93	25,336
2004	1,380	300	27.5	8,255	3.22	26,551
2005	1,433	279	27.0	7,537	3.32	25,053

[1] Area planted in preceding fall.   [2] Preliminary.
NASS, Crops Branch, (202) 720–2127.

Table 1-16.—Rye: Supply and disappearance, United States, 1996–2005

Year begin-ning June	Supply				Disappearance							Ending stocks May 31
	Begin-ning stocks	Produc-tion	Imports	Total	Domestic use					Exports	Total dis-appear-ance	
					Food	Seed	Industry	Feed [1]	Total			
	1,000 bushels	*1,000 bushels*	*1,000 bushels*	*1,000 bushels*	*1,000 bushels*	*1,000 bushels*	*1,000 bushels*	*1,000 bushels*	*1,000 bushels*	*1,000 bushels*	*1,000 bushels*	*1,000 bushels*
1996 ..	898	8,936	4,327	14,161	3,459	3,000	2,000	4,916	13,375	32	13,407	754
1997 ..	754	8,132	5,562	14,448	3,298	2,000	3,000	5,306	13,604	80	13,684	764
1998 ..	764	12,161	3,322	16,247	3,639	3,000	3,000	4,392	14,031	33	14,064	2,183
1999 ..	2,449	11,038	3,424	16,911	3,300	3,000	3,000	5,736	15,036	286	15,322	1,589
2000 ..	1,589	8,386	3,230	13,205	3,300	3,000	3,000	2,325	11,625	390	12,015	1,190
2001 ..	1,190	6,896	4,945	13,031	3,300	3,000	3,000	2,970	12,270	193	12,463	568
2002 ..	568	6,488	6,140	13,196	3,300	3,000	3,000	3,329	12,629	122	12,751	445
2003 ..	445	8,634	3,286	12,365	3,300	3,000	3,000	2,425	11,725	60	11,785	584
2004 ..	584	8,255	5,626	14,465	3,300	3,000	3,000	4,237	13,537	145	13,682	783
2005 [2]	783	7,537	4,500	12,820	3,300	3,000	3,000	2,900	12,200	100	12,300	520

[1] Residual, approximates total feed use.   [2] Preliminary. Totals may not add due to independent rounding.
ERS, Market and Trade Economics Division, (202) 694–5302.

Table 1-17.—Rye: Area, yield, and production, by States, 2003–2005

State	Area planted [1]			Area harvested			Yield per harvested acre			Production		
	2003	2004	2005	2003	2004	2005	2003	2004	2005	2003	2004	2005
	1,000 acres	*1,000 acres*	*1,000 acres*	*1,000 acres*	*1,000 acres*	*1,000 acres*	*Bush-e ls*	*Bush-els*	*Bush-els*	*1,000 bush-els*	*1,000 bush-els*	*1,000 bush-els*
GA	270	250	270	50	25	30	16.0	24.0	27.0	800	600	810
ND [2]	18	25	....	15	20	....	50.0	39.0	....	750	780	....
OK	260	300	310	70	90	70	22.0	18.0	20.0	1,540	1,620	1,400
SD [2]	20	20	....	14	11	....	48.0	59.0	....	672	649	....
Oth Sts [3]	780	785	853	170	154	179	28.7	29.9	29.8	4,872	4,606	5,327
US	1,348	1,380	1,433	319	300	279	27.1	27.5	27.0	8,634	8,255	7,537

[1] Includes area planted preceding fall. [2] Beginning in 2005, ND and SD are no longer published individually. [3] For 2003 and 2004, Other States include IL, KS, MI, MN, NE, NY, NC, PA, SC, TX, and WI. For 2005, Other States include IL, KS, MI, MN, NE, NY, NC, ND, PA, SC, SD, TX, and WI.
NASS, Crops Branch, (202) 720–2127.

Table 1-18.—Rye: Marketing year average price and value, by States, crop of 2003, 2004, and 2005

State	Marketing year average price per bushel			Value of production		
	2003	2004	2005 [1]	2003	2004	2005 [1]
	Dollars	*Dollars*	*Dollars*	*1,000 dollars*	*1,000 dollars*	*1,000 dollars*
GA	4.00	4.00	4.00	3,200	2,400	3,240
ND [2]	2.12	2.00	....	1,590	1,560	....
OK	3.90	4.20	4.05	6,006	6,804	5,670
SD [2]	2.25	2.90	....	1,512	1,882	....
Oth Sts [3]	2.67	3.02	3.05	13,028	13,905	16,143
US	2.93	3.22	3.32	25,336	26,551	25,053

[1] Preliminary. [2] Beginning in 2005, ND and SD are no longer published individually. [3] For 2003 and 2004, Other States include IL, KS, MI, MN, NE, NY, NC, PA, SC, TX, and WI. For 2005, Other States include IL, KS, MI, MN, NE, NY, NC, ND, PA, SC, SD, TX, and WI.
NASS, Crops Branch, (202) 720-2127.

Table 1-19.—Rye: Area, yield, and production in specified countries, 2002/2003–2004/2005 [1]

Continent and country	Area [2]			Yield per hectare			Production		
	2002/ 2003	2003/ 2004	2004/ 2005 [3]	2002/ 2003	2003/ 2004	2004/ 2005 [3]	2002/ 2003	2003/ 2004	2004/ 2005 [3]
	1,000 hec-tares	*1,000 hec-tares*	*1,000 hec-tares*	*Metric tons*	*Metric tons*	*Metric tons*	*1,000 metric tons*	*1,000 metric tons*	*1,000 metric tons*
North America:									
Canada	77	147	157	1.74	2.22	2.57	134	327	404
United States ...	106	129	129	1.56	1.70	1.70	165	219	219
Total	183	276	286	1.63	1.98	2.18	299	546	623
South America:									
Argentina	57	39	61	1.40	0.95	1.46	80	37	89
Brazil	3	3	3	1.00	1.00	1.33	3	3	4
Chile	1	1	1	2.00	2.00	2.00	2	2	2
Total	61	43	65	1.39	0.98	1.46	85	42	95

See footnotes at end of table.

Table 1-19.—Rye: Area, yield, and production in specified countries,
2002/2003–2004/2005 [1]—Continued

Continent and country	Area [2]			Yield per hectare			Production		
	2002/ 2003	2003/ 2004	2004/ 2005 [3]	2002/ 2003	2003/ 2004	2004/ 2005 [3]	2002/ 2003	2003/ 2004	2004/ 2005 [3]
	1,000 hec-tares	*1,000 hec-tares*	*1,000 hec-tares*	*Metric tons*	*Metric tons*	*Metric tons*	*1,000 metric tons*	*1,000 metric tons*	*1,000 metric tons*
European Union:									
Austria	47	40	46	3.64	3.33	4.63	171	133	213
Belgium-Luxem-bourg	1	1	3	8.00	5.00	3.33	8	5	10
Czech Republic	35	42	59	3.40	3.79	5.32	119	159	314
Denmark	46	35	32	5.00	4.83	4.81	230	169	154
Estonia	18	13	9	2.28	1.69	1.89	41	22	17
Finland	30	30	32	2.43	2.43	1.97	73	73	63
France	29	27	33	4.79	4.07	5.00	139	110	165
Germany	728	531	625	5.04	4.29	6.13	3,666	2,277	3,830
Greece	15	15	13	2.27	1.73	1.31	34	26	17
Hungary	50	50	50	2.00	1.34	2.70	100	67	135
Italy	3	3	3	3.33	2.33	2.67	10	7	8
Latvia	41	40	48	2.44	2.20	1.90	100	88	91
Lithuania	75	60	56	2.27	2.45	2.52	170	147	141
Netherlands	4	5	4	4.25	4.00	5.00	17	20	20
Poland	1,560	1,479	1,550	2.46	2.14	2.76	3,831	3,172	4,281
Portugal	36	30	29	0.94	0.90	1.00	34	27	29
Slovakia	40	25	32	2.43	2.48	4.06	97	62	130
Slovenia	1	1	1	3.00	3.00	4.00	3	3	4
Spain	102	108	91	1.71	1.64	1.82	174	177	166
Sweden	24	24	25	5.33	4.92	5.36	128	118	134
United Kingdom	10	10	10	4.50	4.50	4.50	45	45	45
Total	2,895	2,569	2,751	3.17	2.69	3.62	9,190	6,907	9,967
Other Europe:									
Albania	10	10	10	1.00	1.00	1.00	10	10	10
Bosnia-Hercegovina	5	5	5	2.60	2.00	2.40	13	10	12
Bulgaria	10	10	10	1.50	1.50	1.50	15	15	15
Croatia	3	3	2	2.33	2.00	3.00	7	6	6
Macedonia (Skopje)	7	7	7	1.57	1.57	1.57	11	11	11
Norway	3	3	3	3.67	3.67	3.67	11	11	11
Romania	20	20	20	2.50	2.50	2.50	50	50	50
Serbia and Montenegro ..	6	6	6	1.67	1.67	1.67	10	10	10
Switzerland	3	3	3	6.67	6.67	6.67	20	20	20
Total	67	67	66	2.19	2.13	2.20	147	143	145
Fmr. Soviet Union:.									
Belarus	709	700	750	2.26	1.71	1.87	1,600	1,200	1,400
Kazakhstan	70	70	70	0.71	0.71	0.71	50	50	50
Russian Fed.	3,750	2,350	2,000	1.91	1.79	1.43	7,150	4,200	2,850
Ukraine	750	400	725	2.01	1.56	2.21	1,511	625	1,600
Total	5,279	3,520	3,545	1.95	1.73	1.66	10,311	6,075	5,900
Asia:									
Turkey	147	147	150	1.73	1.63	1.60	255	240	240
Africa:									
South Africa	20	20	20	0.15	0.15	0.15	3	3	3
Oceania:									
Australia	35	35	35	0.57	0.57	0.57	20	20	20
World Total ...	8,686	6,680	6,910	2.34	2.09	2.46	20,302	13,980	16,980

[1] Years shown refer to years of harvest. Harvests of Northern Hemisphere countries are combined with those of the Southern Hemisphere which immediately follow: thus the crop harvested in the Northern Hemisphere in 1994 is combined with estimates for the Southern Hemisphere harvests, which begin late in 1994 and end early in 1995.   [2] Harvested area as far as possible.   [3] Preliminary.

FAS, Production Estimates and Crop Assessment Division, (202) 720–0888. Prepared or estimated on the basis of official statistics of foreign Governments, other foreign source materials, reports of Agricultural Counselors, Attachés, and Foreign Service Officers, results of office research, and related information.

GRAIN AND FEED

Table 1-20.—Rice, rough: Area, yield, production, and value, United States, 1996–2005 [1]

Year	Area planted	Area harvested	Yield per acre	Production	Marketing year average price per cwt. received by farmers	Value of production
	1,000 acres	*1,000 acres*	*Pounds*	*1,000 cwt.*	*Dollars*	*1,000 dollars*
1996	2,824.0	2,804.0	6,120	171,599	9.96	1,690,270
1997	3,125.0	3,103.0	5,897	182,992	9.70	1,756,136
1998	3,285.0	3,257.0	5,663	184,443	8.89	1,654,157
1999	3,531.0	3,512.0	5,866	206,027	5.93	1,231,207
2000	3,060.0	3,039.0	6,281	190,872	5.61	1,049,961
2001	3,334.0	3,314.0	6,496	215,270	4.25	925,055
2002	3,240.0	3,207.0	6,578	210,960	4.49	979,628
2003	3,022.0	2,997.0	6,670	199,897	8.08	1,628,948
2004	3,347.0	3,325.0	6,988	232,362	7.33	1,701,822
2005	3,384.0	3,364.0	6,636	223,235	7.80	1,789,225

[1] Sweet rice yield and production included in 2003 as short grain but not in previous years.
NASS, Crops Branch, (202) 720–2127.

Table 1-21.—Rice, rough: Stocks on and off farms, United States, 1997–2006

Year beginning previous December	On farms			Off farms [1]		
	Dec. 1	Mar. 1	Aug. 1	Dec. 1	Mar. 1	Aug. 1
	1,000 cwt.	*1,000 cwt.*	*1,000 cwt.*	*1,000 cwt.*	*1,000 cwt.*	*1,000 cwt.*
1997	32,719	16,003	428	86,350	64,141	21,365
1998	33,470	21,205	1,136	90,873	66,846	19,855
1999	35,584	22,290	1,560	85,394	57,960	15,066
2000	50,185	27,212	1,141	89,191	63,025	20,829
2001	38,085	18,715	921	95,842	67,305	21,097
2002 [2]	52,680	31,725	5,180	101,881	81,783	26,629
2003 [2]	53,220	27,505	1,225	103,850	75,073	18,846
2004 [2]	43,165	18,325	571	92,154	69,515	18,944
2005 [2]	57,545	37,590	2,815	109,151	81,193	28,822
2006 [2]	58,630	NA	NA	101,518	NA	NA

[1] Stocks at mills and in attached warehouses, in warehouses not attached to mills, and in ports or in transit.   [2] Preliminary.   NA-not available.
NASS, Crops Branch, (202) 720–2127.

Table 1-22.—Rice, by length of grain: Area, yield, and production, United States, 1996–2005

Year	Area harvested			Yield per acre			Production		
	Long grain	Medium grain	Short grain	Long grain	Medium grain	Short grain	Long grain	Medium grain	Short grain
	1,000 acres	*1,000 acres*	*1,000 acres*	*Pounds*	*Pounds*	*Pounds*	*1,000 cwt.*	*1,000 cwt.*	*1,000 cwt.*
1996	1,967.0	822.0	15.0	5,777	6,922	7,127	113,629	56,901	1,069
1997	2,309.0	776.0	18.0	5,391	7,357	7,867	124,485	57,091	1,416
1998	2,568.0	656.0	33.0	5,426	6,616	5,185	139,328	43,404	1,711
1999	2,718.0	742.0	52.0	5,587	6,811	6,969	151,863	50,540	3,624
2000	2,189.0	814.0	36.0	5,882	7,311	7,228	128,756	59,514	2,602
2001	2,697.0	591.0	26.0	6,213	7,801	6,192	167,555	46,105	1,610
2002	2,512.0	668.0	27.0	6,260	7,815	5,615	157,243	52,201	1,516
2003	2,310.0	644.0	43.0	6,451	7,481	6,293	149,011	48,180	2,706
2004	2,571.0	705.0	49.0	6,630	8,325	6,588	170,445	58,689	3,228
2005	2,734.0	575.0	55.0	6,493	7,375	6,000	177,527	42,408	3,300

NASS, Crops Branch, (202) 720–2127.

Table 1-23.—Rice, rough, by length of grain: Stocks in all positions, United States, 1997–2006

Year begin- ning previous Decem- ber	Long grain				Medium grain				Short grain			
	Dec. 1	Mar. 1	Aug. 1	Oct. 1 [1]	Dec. 1	Mar. 1	Aug. 1	Oct. 1 [1]	Dec. 1	Mar. 1	Aug. 1	Oct. 1 [1]
	1,000 cwt.	1,000 cwt.	1,000 cwt.	1,000 cwt.	1,000 cwt.	1,000 cwt.	1,000 cwt.	1,000 cwt.	1,000 cwt.	1,000 cwt.	1,000 cwt.	1,000 cwt.
1997	68,687	47,871	10,839	([3])	49,015	31,665	10,723	4,334	1,367	608	231	([3])
1998	78,329	53,329	9,863	([3])	44,477	33,596	10,640	4,368	1,537	1,126	488	([3])
1999	84,346	57,636	10,947	261	34,774	21,329	5,037	861	1,858	1,285	642	290
2000	96,383	62,755	12,511	160	39,601	25,381	8,299	3,287	3,392	2,101	1,160	680
2001	82,718	51,428	8,305	116	48,438	32,504	12,841	5,066	2,771	2,088	872	732
2002	109,953	83,723	22,743	434	42,525	28,515	8,477	2,691	2,083	1,270	589	363
2003	113,897	75,733	11,673	59	40,918	25,529	7,760	2,688	2,255	1,316	638	407
2004	93,881	59,671	8,035	169	38,736	26,562	10,887	4,261	2,702	1,607	593	370
2005	112,799	79,994	19,026	172	51,005	36,761	11,791	4,413	2,892	2028	820	470
2006 [2] ..	124,485	NA	NA	NA	32,802	NA	NA	NA	2,861	NA	NA	NA

[1] California only. [2] Preliminary. [3] Not published to avoid disclosing individual reports. NA-not available.
NASS, Crops Branch, (202) 720–2127.

Table 1-24.—Rough and milled rice (rough equivalent): Supply and disappearance, United States, 1996–2005 [1]

Year begin- ning August	Supply				Disappearance					Ending stocks July 31
	Begin- ning stocks	Produc- tion	Imports [2]	Total	Food, in- dustrial, & resid- ual [3]	Seed	Total	Exports [2]	Total dis- appear- ance	
	Million cwt.	Million cwt.	Million cwt.	Million cwt.	Million cwt.	Million cwt.	Million cwt.	Million cwt.	Million cwt.	Million cwt.
1996 ...	25.0	171.6	10.5	207.2	97.7	3.9	101.6	78.3	179.9	27.2
1997 ...	27.2	183.0	9.3	219.5	99.9	4.1	103.9	87.7	191.6	27.9
1998 ...	27.9	184.4	10.6	223.0	109.7	4.4	114.0	86.8	200.9	22.1
1999 ...	22.1	206.0	10.1	238.2	118.1	3.8	121.9	88.8	210.7	27.5
2000 ...	27.5	190.9	10.9	229.2	113.4	4.1	117.5	83.2	200.7	28.5
2001 ...	28.5	215.3	13.2	256.9	119.3	4.0	123.3	94.7	218.0	39.0
2002 ...	39.0	211.0	14.8	264.8	109.7	3.7	113.4	124.6	238.0	26.8
2003 ...	26.8	199.9	15.0	241.7	110.8	4.1	115.0	103.1	218.0	23.7
2004 ...	23.7	232.4	13.2	269.3	117.0	4.2	121.2	110.4	231.6	37.7
2005 [4]	37.7	223.2	13.5	274.4	123.0	3.9	126.9	121.0	247.9	26.5

All data updated as of February 2006. Totals may not add due to independent rounding.
[1] Consolidated supply and disappearance of rough and milled rice. Milled rice data converted to a rough basis using annu- ally derived extraction rates as factors. [2] Trade data from Bureau of the Census. [3] The residual includes unaccounted losses in drying, processing, and handling. [4] Preliminary.
ERS, Market and Trade Economics Division, (202) 694–5292.

Table 1-25.—Rice, by length of grain: Area, yield, and production, by States, 2003–2005

State	Area harvested			Yield per acre			Production		
	2003	2004	2005[1]	2003	2004	2005[1]	2003	2004	2005[1]
	Long grain								
	1,000 acres	*1,000 acres*	*1,000 acres*	*Pounds*	*Pounds*	*Pounds*	*1,000 cwt.*	*1,000 cwt.*	*1,000 cwt.*
AR	1,290.0	1,400.0	1,533.0	6,600	6,980	6,650	85,140	97,720	101,945
CA	7.0	7.0	9.0	6,900	7,300	7,100	483	511	639
LA	430.0	520.0	515.0	5,870	5,400	5,900	25,241	28,080	30,385
MS	234.0	234.0	263.0	6,800	6,900	6,400	15,912	16,146	16,832
MO	170.0	194.0	213.0	6,130	6,800	6,600	10,421	13,192	14,058
TX	179.0	216.0	201.0	6,600	6,850	6,800	11,814	14,796	13,668
US	2,310.0	2,571.0	2,734.0	6,451	6,630	6,493	149,011	170,445	177,527
	Medium grain								
	1,000 acres	*1,000 acres*	*1,000 acres*	*Pounds*	*Pounds*	*Pounds*	*1,000 cwt.*	*1,000 cwt.*	*1,000 cwt.*
AR	164.0	154.0	101.0	6,700	7,000	6,720	10,988	10,780	6,787
CA	458.0	535.0	463.0	7,840	8,800	7,550	35,907	47,080	34,957
LA	20.0	13.0	10.0	5,780	5,000	5,980	1,156	650	598
MO	1.0	1.0	1.0	6,300	6,900	6,600	63	69	66
TX	1.0	2.0	0.0	6,600	5,500	0	66	110	0
US	644.0	705.0	575.0	7,481	8,325	7,375	48,180	58,689	42,408
	Short grain								
	1,000 acres	*1,000 acres*	*1,000 acres*	*Pounds*	*Pounds*	*Pounds*	*1,000 cwt.*	*1,000 cwt.*	*1,000 cwt.*
AR	1.0	1.0	1.0	6,000	6,000	6,000	60	60	60
CA	42.0	48.0	54.0	6,300	6,600	6,000	2,646	3,168	3,240
US	43.0	49.0	55.0	6,293	6,588	6,000	2,706	3,228	3,300

[1] Preliminary.
NASS, Crops Branch, (202) 720–2127.

Table 1-26.—Rice: Area, yield, and production, by States, 2003–2005 [1]

State	Area planted			Area harvested			Yield per harvested acre			Production		
	2003	2004	2005[2]	2003	2004	2005[2]	2003	2004	2005[2]	2003	2004	2005[2]
	1,000 acres	*1,000 acres*	*1,000 acres*	*1,000 acres*	*1,000 acres*	*1,000 acres*	*Pounds*	*Pounds*	*Pounds*	*1,000 cwt.*	*1,000 cwt.*	*1,000 cwt.*
AR	1,466.0	1,561.0	1,643.0	1,455.0	1,555.0	1,635.0	6,610	6,980	6,650	96,188	108,560	108,792
CA	509.0	595.0	528.0	507.0	590.0	526.0	7,700	8,600	7,380	39,036	50,759	38,836
LA	455.0	538.0	530.0	450.0	533.0	525.0	5,870	5,390	5,900	26,397	28,730	30,983
MS	235.0	235.0	265.0	234.0	234.0	263.0	6,800	6,900	6,400	15,912	16,146	16,832
MO	176.0	196.0	216.0	171.0	195.0	214.0	6,130	6,800	6,600	10,484	13,261	14,124
TX	181.0	222.0	202.0	180.0	218.0	201.0	6,600	6,840	6,800	11,880	14,906	13,668
US	3,022.0	3,347.0	3,384.0	2,997.0	3,325.0	3,364.0	6,670	6,988	6,636	199,897	232,362	223,235

[1] Sweet rice acreage included with short grain.   [2] Preliminary.
NASS, Crops Branch, (202) 720–2127.

Table 1-27.—Rice: Marketing year average price and value, by States, crop of 2003–2005

State	Marketing year average price per cwt.			Value of production		
	2003	2004	2005 [1]	2003	2004	2005 [1]
	Dollars	*Dollars*	*Dollars*	*1,000 dollars*	*1,000 dollars*	*1,000 dollars*
AR	7.70	7.13	7.45	740,648	774,033	810,500
CA	10.40	7.34	10.50	405,974	372,571	407,778
LA	7.68	7.77	7.40	202,729	223,232	229,274
MS	7.34	7.48	7.55	116,794	120,772	127,082
MO	7.20	6.98	7.50	75,485	92,562	105,930
TX	7.35	7.96	7.95	87,318	118,652	108,661
US	8.08	7.33	7.80	1,628,948	1,701,822	1,789,225

[1] Preliminary.
NASS, Crops Branch, (202) 720–2127.

Table 1-28.—Rice, milled, by length of grain: Stocks in all positions, United States, 1997–2006

Year beginning previous Dec.	Whole kernels (head rice)											
	Long grain				Medium grain				Short grain			
	Dec. 1	Mar. 1	Aug. 1	Oct. 1 [1]	Dec. 1	Mar. 1	Aug. 1	Oct. 1 [1]	Dec. 1	Mar. 1	Aug. 1	Oct. 1 [1]
1997	3,358	2,622	2,312	10	1,448	1,079	788	584	56	85	34	66
1998	3,443	3,871	3,228	37	1,248	1,080	728	313	46	35	100	25
1999	2,980	2,361	2,159	21	1,613	1,114	657	301	30	62	131	49
2000	2,732	2,506	2,160	14	770	758	644	540	63	46	30	66
2001	3,624	2,470	2,287	26	1,348	1,164	1,207	342	67	84	87	57
2002	3,796	3,222	2,788	23	986	622	1,032	388	62	110	72	53
2003	4,390	3,656	2,739	9	1,674	1,351	543	277	58	59	60	30
2004	3,338	2,862	1,622	8	2,000	2,194	547	322	114	122	77	31
2005	3,089	2,796	2,629	*	917	1,925	804	363	31	69	56	*
2006 [2] ..	3,305	NA	NA	NA	1,247	NA	NA	NA	75	NA	NA	NA

Year beginning previous Dec.	Broken kernels [3]											
	Second heads				Screenings				Brewers			
	Dec. 1	Mar. 1	Aug. 1	Oct. 1 [1]	Dec. 1	Mar. 1	Aug. 1	Oct. 1 [1]	Dec. 1	Mar. 1	Aug. 1	Oct. 1 [1]
1997	553	623	467	97	49	20	13	....	187	182	205	19
1998	642	711	524	273	27	27	33	....	276	155	182	14
1999	662	612	588	30	29	92	102	....	102	103	140	12
2000 [2] ..	608	937	652	238	64	85	61	....	348	238	285	46
2001 [2] ..	1,006	1,035	667	403	66	3	72	....	251	228	117	31
2002 [2] ..	825	648	696	246	123	139	133	....	115	72	209	16
2003 [2] ..	1,026	1,190	1,066	587	91	146	62	....	242	225	104	12
2004 [2] ..	968	1,199	515	167	71	22	77	....	125	114	113	13
2005 [2] ..	460	512	619	*	21	40	28	....	123	89	152	*
2006 [2] ..	795	NA	NA	NA	198	NA	NA	....	320	NA	NA	NA

[1] California only. [2] Preliminary. [3] Screenings included in second heads in California. * Not published to avoid disclosing individual operations. NA-not available.
NASS, Crops Branch, (202) 720–2127.

Table 1-29.—Rice, rough: Support operations, United States, 1996–2005

Marketing year beginning August 1	Income support payment rates per cwt [1]	Program price levels per cwt		Put under loan [4]		Acquired by CCC under loan program [5]	Owned by CCC at end of marketing year
		Loan [2]	Target [3]	Quantity	Percentage of production		
	Dollars	Dollars	Dollars	Million cwt	Percent	Million cwt	Million cwt
1996/1997	2.77	6.50	NA	68.9	40.1	0.0	0.0
1997/1998	2.71	6.50	NA	67.6	36.9	0.0	0.0
1998/1999	4.37	6.50	NA	80.3	43.5	0.0	0.0
1999/2000	5.64	6.50	NA	110.8	53.6	0.1	0.0
2000/2001	5.42	6.50	NA	97.4	51.0	0.0	0.0
2001/2002	4.49	6.50	NA	128.0	59.5	0.0	0.0
2002/2003	2.35/0.00	6.50	10.50	132.8	62.5	0.0	0.0
2003/2004	2.35/---	6.50	10.50	91.2	45.6	0.0	0.0
2004/2005	2.35	6.50	10.50	147.3	63.8	0.6	0.0
2005/2006	2.35	6.50	10.50	....	....	....	....

[1] Payment rates for the 1995/96 & prior crop years were calculated according to the deficiency payment/production adjustment program. Payment rates for the 1996/97 through 2001/2002 crops were calculated according to the Production Flexibility Contract (PFC) program provisions of the Federal Agriculture Improvement and Reform Act of 1996 (1996 Act) and include supplemental PFC payment rates for 1998 through 2001. Payment rates for the 2002/03 and subsequent crops are calculated according to the Direct and Counter-cyclical program provisions following enactment of the Farm Security and Rural Investment Act of 2002 (2002 Act). Payment rates are rounded to the nearest cent. Beginning with 2002/03, the first entry is the direct payment rate and the second entry is the counter-cyclical payment rate. [2] The national average loan rate was also known as the price support rate prior to enactment of the 1996 Act. [3] Between the 1996/97 and 2001/02 marketing years, target prices were no longer applicable; however, target prices were reestablished under the 2002 Act. [4] Represents loans made, purchases, and purchase agreements entered into. Purchase and purchase agreements are no longer authorized for the 1996 and subsequent following enactment of the 1996 Act. Percentage of production is on a grain basis. Excludes quantity on which loan deficiency payments were made. [5] Acquisition of all loans forfeited during the marketing year. For 2004/05, as of November 1, 2005. NA-not applicable.

FSA, Food Grains Analysis Group, (202) 720-3134.

Table 1-30.—Rice: United States exports (milled basis), by country of destination, 2002–2004 [1]

Country of destination	Year		
	2002	2003	2004
	1,000 metric tons	1,000 metric tons	1,000 metric tons
Mexico	530	574	515
Haiti	273	339	219
Canada	163	171	215
Nicaragua	106	101	106
Honduras	103	88	100
Costa Rica	87	95	134
El Salvador	83	74	48
Guatemala	50	44	55
Total Western Hemisphere [2]	1,614	2,135	1,768
European Union:			
United Kingdom	100	116	92
Germany	60	53	49
Netherlands	56	56	34
France	38	22	22
Spain	33	23	16
Italy	28	0	0
Belgium-Luxembourg	24	23	11
Total Europe [2]	452	342	248
Japan	308	336	380
Turkey	154	131	46
Saudi Arabia	92	102	94
Total Asia, Middle East, and Oceania [2]	865	987	801
Ghana	96	110	101
South Africa	73	62	0
Cote D'ivoire	31	60	49
Liberia	13	14	6
Niger	9	0	2
Total Africa [2]	313	314	220
World total [2]	3,265	3,798	3,062

[1] Year beginning Jan 1. [2] Includes countries not shown.

FAS, Grain and Feed Division, (202) 720–6219. www.fas.usda.gov/grain/default.html.

Table 1-31.—Rice, milled: Area, yield, and production in specified countries, 2002/2003–2004/2005 [1]

Continent and country	Area [2]			Yield per hectare			Production		
	2002/ 2003	2003/ 2004	2004/ 2005 [3]	2002/ 2003	2003/ 2004	2004/ 2005 [3]	2002/ 2003	2003/ 2004	2004/ 2005 [3]
	1,000 hec- tares	*1,000 hec- tares*	*1,000 hec- tares*	*Metric tons*	*Metric tons*	*Metric tons*	*1,000 metric tons*	*1,000 metric tons*	*1,000 metric tons*
North America:									
Mexico	57	58	66	2.33	3.45	3.03	133	200	200
United States	1,298	1,213	1,346	5.04	5.29	5.50	6,536	6,420	7,407
Total	1,355	1,271	1,412	4.92	5.21	5.39	6,669	6,620	7,607
South America:									
Argentina	133	174	165	3.51	4.26	4.14	467	742	683
Bolivia	141	152	145	1.95	1.41	1.79	275	215	260
Brazil	3,186	3,732	3,905	2.21	2.33	2.30	7,050	8,709	8,976
Chile	28	25	20	3.21	2.96	3.20	90	74	64
Colombia	450	480	495	2.89	2.85	2.79	1,300	1,367	1,380
Ecuador	244	230	215	1.84	1.83	2.00	450	420	430
Guyana	116	130	130	2.48	2.51	2.51	288	326	326
Paraguay	28	31	28	2.64	2.71	2.64	74	84	74
Peru	326	280	310	4.90	4.88	4.52	1,597	1,367	1,400
Suriname	40	52	51	2.48	2.35	2.41	99	122	123
Uruguay	153	186	180	4.14	4.75	4.72	634	884	850
Venezuela	90	105	125	3.22	3.38	3.40	290	355	425
Total	4,935	5,577	5,769	2.56	2.63	2.60	12,614	14,665	14,991
Central America:									
Costa Rica	49	50	57	2.51	2.44	2.28	123	122	130
El Salvador	6	5	3	4.00	3.80	4.00	24	19	12
Guatemala	15	15	15	2.00	2.00	2.00	30	30	30
Honduras	4	4	4	1.50	2.50	1.75	6	10	7
Nicaragua	111	113	111	1.72	1.57	1.59	191	177	176
Panama	75	75	75	2.65	2.67	1.88	199	200	141
Total	260	262	265	2.20	2.13	1.87	573	558	496
Carribean:									
Cuba	198	205	185	2.27	2.22	2.15	450	455	397
Dominican Rep.	111	97	82	3.13	3.25	3.41	347	315	280
Haiti	40	40	40	1.63	1.63	1.63	65	65	65
Trinidad	1	1	1	3.00	3.00	3.00	3	3	3
Total	350	343	308	2.47	2.44	2.42	865	838	745
European Union:									
France	18	20	21	3.50	3.30	2.76	63	66	58
Greece	22	22	25	5.27	5.27	5.20	116	116	130
Hungary	5	5	3	2.00	2.00	2.67	10	10	8
Italy	219	220	230	3.97	3.78	4.06	870	831	934
Portugal	25	25	26	4.08	4.08	4.00	102	102	104
Spain	112	117	121	5.09	5.11	5.21	570	598	630
Total	401	409	426	4.32	4.21	4.38	1,731	1,723	1,864
Other Europe:									
Bulgaria	2	2	2	1.50	1.50	1.50	3	3	3
Macedonia (Skopje)	2	3	3	2.50	2.67	3.33	5	8	10
Romania	6	6	6	1.17	1.17	1.17	7	7	7
Total	10	11	11	1.50	1.64	1.82	15	18	20
Fmr. Soviet Union:									
Kazakhstan	68	85	81	1.90	2.09	2.21	129	178	179
Kyrgyzstan	5	5	5	2.40	2.40	2.40	12	12	12
Russian Fed.	129	155	130	2.43	1.89	2.35	314	293	306
Tajikistan	18	18	18	1.83	2.11	1.94	33	38	35
Turkmenistan	45	60	60	1.16	1.17	1.17	52	70	70
Ukraine	19	22	21	2.58	2.50	2.38	49	55	50
Uzbekistan	54	105	61	1.87	1.90	1.93	101	200	118
Total	338	450	376	2.04	1.88	2.05	690	846	770
Middle East:									
Iran	610	560	630	3.11	3.89	3.49	1,900	2,180	2,200
Iraq	12	28	50	1.42	1.18	1.34	17	33	67
Turkey	70	70	80	3.34	3.86	4.06	234	270	325
Total	692	658	760	3.11	3.77	3.41	2,151	2,483	2,592

See footnotes at end of table.

Table 1-31.—Rice, milled: Area, yield, and production in specified countries, 2002/2003–2004/2005 [1]—Continued

Continent and country	Area [2]			Yield per hectare			Production		
	2002/ 2003	2003/ 2004	2004/ 2005 [3]	2002/ 2003	2003/ 2004	2004/ 2005 [3]	2002/ 2003	2003/ 2004	2004/ 2005 [3]
	1,000 hec- tares	1,000 hec- tares	1,000 hec- tares	Metric tons	Metric tons	Metric tons	1,000 metric tons	1,000 metric tons	1,000 metric tons
Africa:									
Algeria	1	1	1	1.00	1.00	1.00	1	1	1
Angola	10	8	12	1.20	1.25	1.08	12	10	13
Benin	20	20	20	1.30	1.75	2.50	26	35	50
Burkina	45	45	45	1.31	1.42	1.38	59	64	62
Cameroon	20	20	20	2.55	2.10	2.45	51	42	49
Chad	105	95	80	0.88	0.91	0.78	92	86	62
Congo, Dem. Rep.	415	415	400	0.46	0.46	0.47	189	189	189
Cote d'Ivoire	460	340	500	0.85	0.82	0.99	390	280	495
Egypt	588	630	635	6.30	6.40	6.50	3,705	4,030	4,128
Gambia, The	12	16	16	1.00	1.19	1.56	12	19	25
Ghana	123	125	125	1.10	1.20	1.32	135	150	165
Guinea	522	525	525	1.00	1.04	1.05	520	548	553
Guinea-Bissau	65	55	65	0.88	0.78	0.91	57	43	59
Kenya	10	17	17	2.40	1.94	1.76	24	33	30
Liberia	120	120	120	0.55	0.55	0.80	66	66	96
Madagascar	1,200	1,220	1,200	1.39	1.50	1.62	1,664	1,824	1,939
Malawi	55	55	55	1.15	1.05	0.80	63	58	44
Mauritania	18	18	18	2.83	3.00	3.39	51	54	61
Morocco	8	8	8	5.00	5.00	5.00	40	40	40
Mozambique	180	180	180	0.73	0.66	0.66	132	119	119
Niger	25	25	25	2.16	1.64	2.04	54	41	51
Nigeria	1,660	1,800	1,850	1.33	1.22	1.24	2,200	2,200	2,300
Senegal	76	87	95	1.47	1.72	1.73	112	150	164
Sierra Leone	200	200	200	0.69	0.75	1.41	138	150	282
Somalia	1	1	1	1.00	1.00	1.00	1	1	1
Sudan	9	6	6	1.33	2.17	0.83	12	13	5
Swaziland	2	2	2	1.50	1.50	1.50	3	3	3
Tanzania, United Rep.	450	570	585	0.93	0.83	0.73	419	475	425
Togo	45	45	45	0.91	0.89	1.13	41	40	51
Zambia	15	15	15	0.47	0.47	0.47	7	7	7
Total	6,460	6,664	6,866	1.59	1.62	1.67	10,276	10,771	11,469
Asia:									
Afghanistan	135	145	145	1.93	2.01	2.07	260	291	300
Bangladesh	10,777	10,902	11,000	2.34	2.40	2.35	25,187	26,152	25,900
Brunei	1	1	1	1.00	1.00	1.00	1	1	1
Burma	6,200	6,300	6,000	1.74	1.70	1.60	10,788	10,730	9,570
Cambodia	2,000	2,240	2,150	1.20	1.32	1.21	2,400	2,960	2,600
China Peoples Republic	28,200	26,508	28,379	4.33	4.24	4.42	122,180	112,462	125,363
India	40,400	42,400	42,300	1.78	2.08	2.02	71,820	88,280	85,310
Indonesia	11,500	11,900	11,650	2.91	2.94	2.94	33,411	35,024	34,250
Japan	1,688	1,665	1,701	4.79	4.26	4.67	8,089	7,091	7,944
Korea, Democratic Pe	585	585	585	2.48	2.50	2.63	1,450	1,460	1,540
Korea, Republic of	1,053	1,016	1,001	4.68	4.38	5.00	4,927	4,451	5,000
Laos	780	750	770	1.85	1.90	1.97	1,446	1,425	1,518
Malaysia	667	672	675	2.13	2.19	2.10	1,418	1,470	1,420
Mali	370	385	370	1.25	1.61	1.52	462	620	561
Nepal	1,545	1,545	1,559	1.78	1.92	1.84	2,752	2,968	2,864
Pakistan	2,201	2,460	2,500	2.03	1.97	1.97	4,479	4,848	4,920
Philippines	4,100	4,094	4,100	2.06	2.25	2.30	8,450	9,200	9,445
Sri Lanka	806	810	789	2.55	2.35	2.50	2,058	1,900	1,974
Taiwan	307	272	237	4.14	4.28	4.27	1,271	1,164	1,011
Thailand	10,158	10,315	9,815	1.69	1.75	1.73	17,198	18,011	17,000
Vietnam	7,463	7,468	7,420	2.88	2.96	3.05	21,527	22,082	22,627
Total	130,936	132,433	133,147	2.61	2.66	2.71	341,574	352,590	361,118
Oceania:									
Australia	38	65	50	7.37	5.88	4.36	280	382	218
World total	145,775	148,143	149,390	2.59	2.64	2.69	377,809	391,494	401,890

[1] Crop year beginning Aug. 1. Crops harvested in the Northern Hemisphere during the latter part of the year are combined with those harvested in Asia principally from November to May, and in the Southern Hemisphere harvested during the first part of the following year.   [2] Harvested area as far as possible.   [3] Preliminary.

FAS, Production Estimates and Crop Assessment Division, (202) 720–0888. Prepared or estimated on the basis of official statistics of foreign governments, other foreign source materials, reports of U.S. Agricultural Counselors, Attachés and Foreign Service Officers, results of office research, and related information.

Table 1-32.—Rice, milled equivalent: [1] International trade, 2003–2005 [2]

Country	2003	2004	2005 [3]
	1,000 metric tons	*1,000 metric tons*	*1,000 metric tons*
Exports:			
Argentina	170	249	350
Australia	141	131	125
Burma	388	130	150
China	2,583	880	750
Egypt	579	826	1,100
India	4,421	3,172	4,500
Pakistan	1,958	1,986	2,350
Thailand	7,552	10,137	7,250
Uruguay	675	804	650
Vietnam	3,795	4,295	5,000
EU	220	187	175
Others	1,234	1,229	1,516
Subtotal	23,716	24,026	23,916
United States	3,834	3,090	3,800
Total	27,550	27,116	27,716
Imports:			
Bangladesh	1,112	801	800
Brazil	1,063	762	500
Canada	242	285	250
China	258	1,122	500
Colombia	75	58	50
Costa Rica	95	136	140
Cote d'Ivoire	750	750	800
Cuba	371	639	850
El Salvador	74	51	75
Ghana	357	425	400
Guinea	350	350	350
Haiti	345	269	300
Honduras	88	102	100
Indonesia	2,750	650	900
Iran	900	950	950
Iraq	672	889	1,000
Jamaica & Dep	51	35	50
Japan	654	706	650
Korea, North	633	465	600
Korea, South	179	188	120
Malaysia	500	700	650
Mexico	582	521	550
Nigeria	1,448	1,369	1,600
Nicaragua	102	109	110
Peru	32	88	115
Philippines	1,300	1,100	1,900
Russia	385	350	350
Saudi Arabia	1,150	1,500	1,250
Senegal	750	850	1,100
Singapore	375	346	375
South Africa	725	818	750
Sri Lanka	29	215	100
Syria	190	200	250
Taiwan	135	158	125
Turkey	340	153	250
Uzbekistan	25	25	25
UAE	80	80	80
Yemen	250	275	250
EU	950	1,079	1,050
Other Europe	209	230	220
United States	458	477	410
Subtotal	21,034	20,276	20,895
Other Countries	4,615	5,105	4,812
Unaccounted	1,901	1,735	2,009
World Total	27,550	27,116	27,716

[1] Includes milled, semi-milled, broken, and rough rice in terms of milled equivalent. [2] Year beginning Jan 1. [3] Preliminary.

FAS, Grain and Feed Division, (202) 720–6219. www.fas.usda.gov/grain/default.html. Prepared or estimated on the basis of official statistics frpm foreign governments, other foreign source materials, reports of U.S. Agricultural Counselors, Attachés and Foreign Service Officers, results of office research, and related information.

Table 1-33.—Food grains: Average price, selected markets and grades, 1998–2005 [1]

Calendar year [2]	Kansas City			Minneapolis (rail)			Portland Wheat No. 1 Soft White	St. Louis Wheat, No. 2 Soft Red Winter (truck)
	Wheat, No. 1 Hard Winter, Ordinary Protein (rail)	Wheat, No. 1 Hard Winter, 13% protein (rail)	Wheat, No. 2 Soft Red Winter (rail)	Wheat, No. 1 Hard Amber Durum (milling) (rail)	Wheat, No. 1 Dark Northern Spring (rail), 14% protein	Rye, No. 2, 20 day delivery (truck)		
	Dollars per bushel	Dollars per bushel	Dollars per bushel	Dollars per bushel	Dollars per bushel	Dollars per bushel	Dollars per bushel	Dollars per bushel
1998	3.70	3.96	3.38	5.97	4.31	....	3.81	3.43
1999	3.08	3.47	2.68	4.05	3.83	....	2.98	2.41
2000	2.87	3.41	2.53	4.23	3.61	....	3.02	2.39
2001	3.34	3.44	2.79	4.98	....	2.69	3.43	2.61
2002	3.94	3.99	3.50	4.97	4.15	3.57	3.89	3.28
2003	3.86	3.97	3.60	5.30	4.26	3.09	3.69	3.47
2004	4.14	4.28	3.97	5.32	4.63	3.49	4.07	3.66
2005	4.10	4.17	3.92	....	4.89	3.63	3.72	3.06

Calendar year [2]	Chicago Wheat, No. 2 Soft Red Winter (rail)	Denver Wheat, No. 1 Hard Winter (rail)	S.W. Louisiana Milled Rice		Arkansas Milled Rice		Texas Milled Rice
			Medium	Long	Medium	Long	Long
	Dollars per bushel	Dollars per bushel	Dollars per cwt.	Dollars per cwt.	Dollars per cwt.	Dollars per cwt.	Dollars per cwt.
1998	3.29	3.33	18.70	18.27	18.34	18.46	18.94
1999	2.46	2.79	18.45	15.31	19.96	15.41	17.02
2000	2.19	2.29	13.10	12.22	13.56	12.24	14.82
2001	2.04	2.89	....	....	....	....	....
2002	2.30	3.53	....	....	....	....	....
2003	3.40	3.35	....	20.82	....	21.51	22.91
2004	3.36	3.53	19.36	16.47	19.22	17.22	18.65
2005	3.01	3.37	16.55	14.22	15.94	14.32	16.53

[1] Simple average of daily prices. [2] For wheat and rye, crop year begins in June. For rice, crop year begins in August.
AMS, Livestock and Grain Market News branch, (202) 720–6231.

Table 1-34.—Corn: Area, yield, production, and value, United States, 1996–2005

Year	Area planted, all purposes	Corn for grain					Corn for silage		
		Area harvested	Yield per harvested acre	Production	Marketing year average price per bushel	Value of production	Area harvested	Yield per harvested acre	Production
	1,000 acres	1,000 acres	Bushels	1,000 bushels	Dollars	1,000 dollars	1,000 acres	Tons	1,000 tons
1996 ...	79,229	72,644	127.1	9,232,557	2.71	25,149,013	5,607	15.4	86,581
1997 ...	79,537	72,671	126.7	9,206,832	2.43	22,351,507	6,054	16.1	97,192
1998 ...	80,165	72,589	134.4	9,758,685	1.94	18,922,084	5,913	16.1	95,479
1999 ...	77,386	70,487	133.8	9,430,612	1.82	17,103,991	6,037	15.8	95,633
2000 ...	79,551	72,440	136.9	9,915,051	1.85	18,499,002	6,082	16.8	102,156
2001 ...	75,702	68,768	138.2	9,502,580	1.97	18,878,819	6,142	16.6	101,992
2002 ...	78,894	69,330	129.3	8,966,787	2.32	20,882,448	7,122	14.4	102,293
2003 ...	78,603	70,944	142.2	10,089,222	2.42	24,476,803	6,583	16.3	107,378
2004 ...	80,929	73,631	160.4	11,807,086	2.06	24,381,294	6,101	17.6	107,293
2005 [1]	81,759	75,107	147.9	11,112,072	1.90	21,040,707	5,920	18.0	106,311

[1] Preliminary.
NASS, Crops Branch, (202) 720–2127.

Table 1-35.—Corn: Stocks on and off farms, United States, 1997–2006

Year beginning previous December	On farms				Off farms [1]			
	Dec. 1	Mar. 1	Jun. 1	Sep. 1 [2]	Dec. 1	Mar. 1	Jun. 1	Sep. 1 [2]
	1,000 bushels	1,000 bushels	1,000 bushels	1,000 bushels	1,000 bushels	1,000 bushels	1,000 bushels	1,000 bushels
1997	4,800,000	2,870,000	1,501,000	475,000	2,102,974	1,624,128	995,550	408,161
1998	4,822,000	2,975,000	1,830,000	640,000	2,424,756	1,964,898	1,209,757	667,803
1999	5,320,000	3,570,000	2,257,000	797,000	2,731,846	2,128,428	1,359,225	989,977
2000	5,195,000	3,300,000	2,029,800	793,000	2,844,443	2,301,895	1,556,138	924,549
2001	5,550,000	3,600,000	2,230,800	753,150	2,979,634	2,442,999	1,693,158	1,145,958
2002	5,275,000	3,355,000	2,020,600	586,800	2,989,715	2,440,263	1,576,290	1,009,626
2003	4,800,000	2,940,000	1,620,200	484,900	2,837,971	2,191,873	1,364,718	601,773
2004	5,286,000	3,030,000	1,540,000	438,000	2,667,775	2,241,459	1,430,140	520,091
2005	6,144,000	4,137,000	2,462,300	820,500	3,308,488	2,619,334	1,858,513	1,293,472
2006 [3]	6,325,000	NA	NA	NA	3,487,835	NA	NA	NA

[1] Includes stocks at mills, elevators, warehouses, terminals, and processors. [2] Old crop only. [3] Preliminary. NA-not available.
NASS, Crops Branch, (202) 720–2127.

Table 1-36.—Corn: Area, yield, and production, by States, 2003-2005

State	Area planted for all purposes			Corn for grain								
				Area harvested			Yield per harvested acre			Production		
	2003	2004	2005[1]	2003	2004	2005[1]	2003	2004	2005[1]	2003	2004	2005[1]
	1,000 acres	*1,000 acres*	*1,000 acres*	*1,000 acres*	*1,000 acres*	*1,000 acres*	*Bush- els*	*Bush- els*	*Bush- els*	*1,000 bushels*	*1,000 bushels*	*1,000 bushels*
AL ...	220	220	220	190	195	200	122.0	123.0	119.0	23,180	23,985	23,800
AZ ...	47	53	50	22	27	22	190.0	180.0	195.0	4,180	4,860	4,290
AR ...	365	320	240	350	305	230	140.0	140.0	131.0	49,000	42,700	30,130
CA ...	530	540	540	140	150	110	160.0	175.0	172.0	22,400	26,250	18,920
CO ..	1,080	1,200	1,100	890	1,040	950	135.0	135.0	148.0	120,150	140,400	140,600
CT ...	30	30	28	(2)	(2)	(2)	(2)	(2)	(2)	(2)	(2)	(2)
DE ...	170	160	160	162	153	154	123.0	152.0	143.0	19,926	23,256	22,022
FL ...	75	70	65	39	32	28	82.0	90.0	94.0	3,198	2,880	2,632
GA ..	340	335	270	290	280	230	129.0	130.0	129.0	37,410	36,400	29,670
ID	190	230	235	50	75	60	140.0	170.0	170.0	7,000	12,750	10,200
IL	11,200	11,750	12,100	11,050	11,600	11,950	164.0	180.0	143.0	1,812,200	2,088,000	1,708,850
IN	5,600	5,700	5,900	5,390	5,530	5,770	146.0	168.0	154.0	786,940	929,040	888,580
IA	12,300	12,700	12,800	11,900	12,400	12,500	157.0	181.0	173.0	1,868,300	2,244,400	2,162,500
KS ...	2,900	3,100	3,650	2,500	2,880	3,450	120.0	150.0	135.0	300,000	432,000	465,750
KY ...	1,170	1,210	1,250	1,080	1,140	1,180	137.0	152.0	132.0	147,960	173,280	155,760
LA ...	520	420	340	500	410	330	134.0	135.0	136.0	67,000	55,350	44,880
ME ...	28	28	26	(2)	(2)	(2)	(2)	(2)	(2)	(2)	(2)	(2)
MD ..	480	490	470	410	425	400	123.0	153.0	135.0	50,430	65,025	54,000
MA ...	20	20	20	(2)	(2)	(2)	(2)	(2)	(2)	(2)	(2)	(2)
MI	2,250	2,200	2,250	2,030	1,920	2,020	128.0	134.0	143.0	259,840	257,280	288,860
MN ..	7,200	7,500	7,300	6,650	7,050	6,850	146.0	159.0	174.0	970,900	1,120,950	1,191,900
MS ..	550	460	380	530	440	365	135.0	136.0	129.0	71,550	59,840	47,085
MO ..	2,900	2,950	3,100	2,800	2,880	2,970	108.0	162.0	111.0	302,400	466,560	329,670
MT ...	68	70	65	17	15	17	140.0	143.0	148.0	2,380	2,145	2,516
NE ...	8,100	8,250	8,500	7,700	7,950	8,250	146.0	166.0	154.0	1,124,200	1,319,700	1,270,500
NV ...	4	4	5	(2)	(2)	(2)	(2)	(2)	(2)	(2)	(2)	(2)
NH ...	15	15	15	(2)	(2)	(2)	(2)	(2)	(2)	(2)	(2)	(2)
NJ ...	80	86	80	61	72	62	113.0	143.0	122.0	6,893	10,296	7,564
NM ...	130	125	140	48	58	55	180.0	180.0	175.0	8,640	10,440	9,625
NY ...	1,000	980	990	440	500	460	121.0	122.0	124.0	53,240	61,000	57,040
NC ...	740	820	750	680	740	700	106.0	117.0	120.0	72,080	86,580	84,000
ND ...	1,450	1,800	1,410	1,170	1,150	1,200	112.0	105.0	129.0	131,040	120,750	154,800
OH ...	3,300	3,350	3,450	3,070	3,110	3,250	156.0	158.0	143.0	478,920	491,380	464,750
OK ..	230	250	290	190	200	250	125.0	150.0	115.0	23,750	30,000	28,750
OR ..	51	58	53	30	28	25	170.0	170.0	160.0	5,100	4,760	4,000
PA ...	1,450	1,400	1,350	890	980	960	115.0	140.0	122.0	102,350	137,200	117,120
RI	2	2	2	(2)	(2)	(2)	(2)	(2)	(2)	(2)	(2)	(2)
SC ...	240	315	300	215	295	285	105.0	100.0	116.0	22,575	29,500	33,060
SD ...	4,400	4,650	4,450	3,850	4,150	3,950	111.0	130.0	119.0	427,350	539,500	470,050
TN ...	710	680	650	620	615	595	131.0	140.0	130.0	81,220	86,100	77,350
TX ...	1,830	1,830	2,050	1,650	1,680	1,850	118.0	139.0	114.0	194,700	233,520	210,900
UT ...	55	55	55	13	12	12	155.0	155.0	163.0	2,015	1,860	1,956
VT ...	100	95	95	(2)	(2)	(2)	(2)	(2)	(2)	(2)	(2)	(2)
VA ...	470	500	490	330	360	360	115.0	145.0	118.0	37,950	52,200	42,480
WA ...	130	170	150	70	105	80	195.0	200.0	205.0	13,650	21,000	16,400
WV ..	48	48	45	27	29	28	115.0	131.0	109.0	3,105	3,799	3,052
WI ...	3,750	3,600	3,800	2,850	2,600	2,900	129.0	136.0	148.0	367,650	353,600	429,200
WY ...	85	90	80	50	50	49	129.0	131.0	140.0	6,450	6,550	6,860
US ...	78,603	80,929	81,759	70,944	73,631	75,107	142.2	160.4	147.9	10,089,222	11,807,086	11,112,072

[1] Preliminary.   [2] Not estimated.
NASS, Crops Branch, (202) 720–2127.

Table 1-37.—Corn: Supply and disappearance, United States, 1996-2005

Year beginning Sep- tember 1	Supply				Disappearance					Ending stocks Aug. 31		
	Begin- ning stocks	Produc- tion	Imports	Total	Domestic use			Exports	Total dis- appear- ance	Privately held[1]	Govern- ment	Total
					Feed and re- sidual	Food, seed, and in- dustrial	Total					
	Million bushels	*Million bushels*	*Million bushels*	*Million bushels*	*Million bushels*	*Million bushels*	*Million bushels*	*Million bushels*	*Million bushels*	*Million bushels*	*Million bushels*	*Million bushels*
1996	426	9,233	13	9,672	5,277	1,714	6,991	1,797	8,789	881	2	883
1997	883	9,207	9	10,099	5,482	1,805	7,287	1,504	8,791	1,304	4	1,308
1998	1,308	9,759	19	11,085	5,471	1,846	7,318	1,984	9,298	1,775	12	1,787
1999	1,787	9,431	15	11,232	5,664	1,913	7,578	1,937	9,515	1,704	14	1,718
2000	1,718	9,915	7	11,639	5,842	1,957	7,799	1,941	9,740	1,891	8	1,899
2001	1,899	9,503	10	11,412	5,864	2,046	7,911	1,905	9,815	1,590	6	1,596
2002	1,596	8,967	14	10,578	5,563	2,340	7,903	1,588	9,491	1,083	4	1,087
2003	1,087	10,089	14	11,190	5,795	2,537	8,332	1,900	10,232	958	0	958
2004[2]	958	11,807	11	12,776	6,162	2,686	8,848	1,814	10,662	2,113	1	2,114
2005[3]	2,114	11,112	10	13,236	6,000	2,960	8,960	1,850	10,810	2,425	1	2,426

[1] Includes quantity under loan and farmer–owned reserve.   [2] Preliminary.   [3] Projected as of January 12, 2006, World Agricultural Supply and Demand Estimates. Totals may not add due to independent rounding.
ERS, Market and Trade Economics Division, (202) 694–5296.

GRAIN AND FEED

Table 1-38.—Corn: Utilization for silage, by States, 2003–2005

State	Silage								
	Area harvested			Yield per acre			Production		
	2003	2004	2005[1]	2003	2004	2005[1]	2003	2004	2005[1]
	1,000 acres	1,000 acres	1,000 acres	Tons	Tons	Tons	1,000 tons	1,000 tons	1,000 tons
AL	20	10	15	12.0	17.0	16.0	240	170	240
AZ	24	25	27	28.0	27.0	27.0	672	675	729
AR	8	5	5	15.0	17.0	12.0	120	85	60
CA	385	385	425	26.0	26.0	26.0	10,010	10,010	11,050
CO	90	110	110	21.0	22.5	23.0	1,890	2,475	2,530
CT	28	27	26	17.5	21.5	20.0	490	581	520
DE	5	6	5	16.0	17.0	19.0	80	102	95
FL	28	33	28	19.0	17.0	19.0	532	561	532
GA	45	45	35	17.0	16.0	19.0	765	720	665
ID	135	150	170	26.0	26.5	26.5	3,510	3,975	4,505
IL	110	110	115	15.0	20.0	15.0	1,650	2,200	1,725
IN	150	140	100	19.0	20.5	20.0	2,850	2,870	2,000
IA	330	230	230	20.0	19.5	18.5	6,600	4,485	4,255
KS	280	170	150	11.0	15.0	16.0	3,080	2,550	2,400
KY	80	65	65	18.0	17.5	15.0	1,440	1,138	975
LA	10	5	5	16.0	12.0	18.0	160	60	90
ME	25	25	24	18.0	19.5	18.5	450	488	444
MD	65	60	65	16.0	20.0	17.0	1,040	1,200	1,105
MA	17	17	17	19.0	22.0	21.5	323	374	366
MI	210	265	220	16.0	18.0	17.5	3,360	4,770	3,850
MN	475	400	400	14.0	16.0	16.0	6,650	6,400	6,400
MS	10	15	10	15.0	14.0	16.0	150	210	160
MO	80	50	110	10.5	14.5	13.0	840	725	1,430
MT	49	51	46	24.0	22.0	24.0	1,176	1,122	1,104
NE	300	230	200	13.0	16.5	15.5	3,900	3,795	3,100
NV	4	4	5	23.0	22.0	23.0	92	88	115
NH	14	14	14	19.5	21.0	20.5	273	294	287
NJ	18	13	17	15.0	20.0	16.0	270	260	272
NM	80	66	84	23.0	25.0	24.0	1,840	1,650	2,016
NY	550	470	520	17.5	17.0	17.0	9,625	7,990	8,840
NC	55	75	45	16.0	19.0	17.0	880	1,425	765
ND	220	215	170	6.8	8.7	11.0	1,496	1,871	1,870
OH	170	190	160	19.0	17.0	17.0	3,230	3,230	2,720
OK	24	30	27	18.0	19.0	18.0	432	570	486
OR	20	30	28	22.0	25.0	26.0	440	750	728
PA	550	400	380	14.5	18.0	18.0	7,975	7,200	6,840
RI	2	2	2	18.0	20.0	20.0	36	40	40
SC	7	12	12	15.0	16.0	15.0	105	192	180
SD	470	450	420	8.5	11.0	11.0	3,995	4,950	4,620
TN	60	55	50	17.0	19.0	19.0	1,020	1,045	950
TX	120	110	130	18.0	23.0	20.0	2,160	2,530	2,600
UT	41	42	42	21.0	22.0	22.0	861	924	924
VT	91	90	90	18.5	19.5	20.5	1,684	1,755	1,845
VA	135	135	125	17.5	20.0	17.0	2,363	2,700	2,125
WA	60	65	70	25.0	26.0	27.0	1,500	1,690	1,890
WV	19	18	16	15.5	17.0	15.5	295	306	248
WI	880	950	880	16.0	14.0	17.0	14,080	13,300	14,960
WY	34	36	30	22.0	22.0	22.0	748	792	660
US	6,583	6,101	5,920	16.3	17.6	18.0	107,378	107,293	106,311

[1] Preliminary.
NASS, Crops Branch, (202) 720–2127.

Table 1-39.—Corn for grain: Marketing year average price and value, by States, crop of 2003, 2004, and 2005

State	Marketing year average price per bushel			Value of production		
	2003	2004	2005[1]	2003	2004	2005[1]
	Dollars	Dollars	Dollars	1,000 dollars	1,000 dollars	1,000 dollars
AL	2.36	2.48	2.35	54,705	59,483	55,930
AZ	3.28	3.03	2.90	13,710	14,726	12,441
AR	2.37	2.39	2.10	116,130	102,053	63,273
CA	2.90	2.65	2.75	64,960	69,563	52,030
CO	2.49	2.23	2.25	299,174	313,092	316,350
DE	2.87	2.19	2.05	57,188	50,931	45,145
FL	2.55	2.30	2.00	8,155	6,624	5,264
GA	2.45	2.20	2.15	91,655	80,080	63,791
ID	2.94	2.82	2.65	20,580	35,955	27,030
IL	2.42	2.14	1.95	4,385,524	4,468,320	3,332,258
IN	2.53	1.99	1.80	1,990,958	1,848,790	1,599,444
IA	2.37	1.99	1.85	4,427,871	4,466,356	4,000,625
KS	2.51	2.12	2.10	753,000	915,840	978,075
KY	2.53	2.24	2.05	374,339	388,147	319,308
LA	2.40	2.45	2.25	160,800	135,608	100,980
MD	2.83	2.17	2.00	142,717	141,104	108,000
MI	2.37	1.97	1.70	615,821	506,842	491,062
MN	2.35	1.94	1.75	2,281,615	2,174,643	2,085,825
MS	2.28	2.43	2.15	163,134	145,411	101,233
MO	2.46	2.03	1.90	743,904	947,117	626,373
MT	2.65	2.42	2.40	6,307	5,191	6,038
NE	2.39	2.02	1.85	2,686,838	2,665,794	2,350,425
NJ	2.81	2.20	2.00	19,369	22,651	15,128
NM	2.96	2.40	2.50	25,574	25,056	24,063
NY	2.82	2.37	2.05	150,137	144,570	116,932
NC	2.68	2.44	2.25	193,174	211,255	189,000
ND	2.37	1.88	1.80	310,565	227,010	278,640
OH	2.45	2.04	1.80	1,173,354	1,002,415	836,550
OK	2.60	2.53	2.40	61,750	75,900	69,000
OR	3.08	2.75	2.60	15,708	13,090	10,400
PA	2.96	2.25	2.20	302,956	308,700	257,664
SC	2.70	2.30	2.10	60,953	67,850	69,426
SD	2.28	1.82	1.70	974,358	981,890	799,085
TN	2.37	2.17	1.95	192,491	186,837	150,833
TX	2.59	2.60	2.50	504,273	607,152	527,250
UT	2.99	2.56	2.35	6,025	4,762	4,597
VA	2.57	2.17	2.10	97,532	113,274	89,208
WA	3.00	2.97	2.75	40,950	62,370	45,100
WV	2.72	2.20	2.00	8,446	8,358	6,104
WI	2.35	2.15	1.85	863,978	760,240	794,020
WY	2.50	2.48	2.45	16,125	16,244	16,807
US	2.42	2.06	1.90	24,476,803	24,381,294	21,040,707

[1] Preliminary.
NASS, Crops Branch, (202) 720–2127.

Table 1-40.—Corn: Area, yield, and production in specified countries, 2002/2003–2004/2005 [1]

Continent and country	Area[2]			Yield per hectare			Production		
	2002/ 2003	2003/ 2004	2004/ 2005[3]	2002/ 2003	2003/ 2004	2004/ 2005[3]	2002/ 2003	2003/ 2004	2004/ 2005[3]
	1,000 hec- tares	1,000 hec- tares	1,000 hec- tares	Metric tons	Metric tons	Metric tons	1,000 metric tons	1,000 metric tons	1,000 metric tons
North America:									
Canada	1,283	1,230	1,067	7.01	7.80	8.25	8,999	9,600	8,802
Mexico	7,030	7,690	7,755	2.74	2.83	2.92	19,280	21,800	22,630
United States	28,057	28,710	29,798	8.12	8.93	10.07	227,767	256,278	299,917
Total	36,370	37,630	38,620	7.04	7.64	8.58	256,046	287,678	331,349
Central America:									
Costa Rica	7	8	8	1.71	1.88	1.75	12	15	14
El Salvador	296	249	228	1.90	2.55	2.67	563	635	608
Guatemala	860	860	860	1.22	1.23	1.24	1,050	1,060	1,070
Honduras	370	335	212	1.36	1.57	1.73	505	525	366
Nicaragua	450	474	390	1.11	1.24	1.14	500	589	444
Panama	15	16	16	3.07	3.00	3.00	46	48	48
Total	1,998	1,942	1,714	1.34	1.48	1.49	2,676	2,872	2,550

See footnotes at end of table.

Table 1-40.—Corn: Area, yield, and production in specified countries, 2002/2003–2004/2005 [1]—Continued

Continent and country	Area [2]			Yield per hectare			Production		
	2002/ 2003	2003/ 2004	2004/ 2005 [3]	2002/ 2003	2003/ 2004	2004/ 2005 [3]	2002/ 2003	2003/ 2004	2004/ 2005 [3]
	1,000 hec- tares	*1,000 hec- tares*	*1,000 hec- tares*	*Metric tons*	*Metric tons*	*Metric tons*	*1,000 metric tons*	*1,000 metric tons*	*1,000 metric tons*
South America:									
Argentina	2,450	2,300	2,700	6.33	6.52	7.22	15,500	15,000	19,500
Bolivia	280	310	280	2.32	1.81	2.39	650	560	670
Brazil	12,956	12,440	11,470	3.43	3.38	3.05	44,500	42,000	35,000
Chile	110	119	128	10.82	11.10	11.02	1,190	1,321	1,410
Colombia	553	565	568	2.07	2.31	2.30	1,145	1,305	1,309
Ecuador	170	110	113	1.65	2.27	2.65	280	250	300
Guyana	5	5	5	1.00	1.00	1.00	5	5	5
Paraguay	525	425	450	2.86	2.00	2.67	1,500	850	1,200
Peru	483	525	580	2.68	2.76	2.76	1,293	1,448	1,600
Uruguay	39	45	64	4.59	4.96	3.13	179	223	200
Venezuela	400	390	400	2.88	3.00	3.00	1,150	1,170	1,200
Total	17,971	17,234	16,758	3.75	3.72	3.72	67,392	64,132	62,394
Carribean:									
Cuba	75	65	40	0.93	0.92	0.88	70	60	35
Dominican Rep.	30	32	28	1.27	1.25	1.25	38	40	35
Haiti	350	350	350	0.86	0.86	0.86	300	300	300
Jamaica & Dep.	4	4	4	0.50	0.50	0.50	2	2	2
Trinidad	3	3	3	1.67	1.67	1.67	5	5	5
Total	462	454	425	0.90	0.90	0.89	415	407	377
European Union:									
Austria	171	172	178	9.69	8.38	9.21	1,657	1,442	1,640
Belgium-Luxembourg	47	53	50	10.81	9.70	13.00	508	514	650
Czech Republic	71	85	90	8.68	5.60	6.13	616	476	552
France	1,831	1,654	1,822	8.98	7.24	8.99	16,440	11,980	16,378
Germany	399	463	462	9.37	7.24	9.09	3,738	3,354	4,200
Greece	140	155	140	9.64	10.32	9.79	1,350	1,600	1,370
Hungary	1,150	1,100	1,200	5.22	4.18	6.83	6,000	4,600	8,200
Italy	1,112	1,163	1,199	9.49	6.96	9.16	10,554	8,100	10,983
Netherlands	24	25	25	9.46	10.00	10.00	227	250	250
Poland	319	356	412	6.15	5.29	5.69	1,962	1,884	2,344
Portugal	139	138	132	5.72	5.77	6.02	795	796	795
Slovakia	138	150	147	5.07	2.01	5.86	700	301	862
Slovenia	45	45	46	7.78	4.98	7.78	350	224	358
Spain	463	476	480	9.64	8.94	9.93	4,463	4,255	4,766
Total	6,049	6,035	6,383	8.16	6.59	8.36	49,360	39,776	53,348
Other Europe:									
Albania	60	60	60	3.33	3.33	3.33	200	200	200
Bosnia-Hercegovina	228	230	225	4.00	2.83	4.00	912	650	900
Bulgaria	280	400	350	3.57	2.50	4.29	1,000	1,000	1,500
Croatia	389	380	400	6.17	5.50	5.50	2,400	2,090	2,200
Macedonia (Skopje)	45	45	45	3.33	3.33	3.33	150	150	150
Romania	2,666	2,700	3,000	2.74	2.60	4.00	7,300	7,020	12,000
Serbia and Montenegro	1,196	1,200	1,202	4.67	3.17	5.22	5,585	3,800	6,274
Switzerland	25	25	25	8.80	8.80	8.80	220	220	220
Total	4,889	5,040	5,307	3.63	3.00	4.42	17,767	15,130	23,444
Fmr. Soviet Union:									
Azerbaijan	30	30	30	3.33	3.33	5.00	100	100	150
Georgia	200	200	200	2.00	2.25	2.00	400	450	400
Kazakhstan	135	100	100	3.22	3.00	3.00	435	300	300
Kyrgyzstan	70	75	75	6.11	5.87	6.00	428	440	450
Moldova	450	380	500	2.67	2.63	3.00	1,200	1,000	1,500
Russian Fed.	550	700	900	2.82	3.00	3.83	1,550	2,100	3,450
Tajikistan	13	25	25	4.23	3.80	3.20	55	95	80
Turkmenistan	25	25	25	2.00	2.00	2.00	50	50	50
Ukraine	1,189	2,000	2,300	3.52	3.43	3.83	4,180	6,850	8,800
Uzbekistan	35	35	35	4.29	4.14	3.43	150	145	120
Total	2,697	3,570	4,190	3.17	3.23	3.65	8,548	11,530	15,300
Africa:									
Algeria	1	1	1	1.00	1.00	1.00	1	1	1
Angola	815	985	1,070	0.52	0.56	0.54	425	550	575
Benin	600	600	800	1.25	1.33	1.50	750	800	1,200
Botswana	30	8	20	0.33	0.13	0.25	10	1	5
Burkina	250	260	260	2.10	2.85	2.12	525	740	550
Burundi	115	115	115	1.00	1.09	1.30	115	125	150
Cameroon	350	350	350	2.23	2.36	2.43	780	825	850
Cape Verde	30	15	15	0.17	1.00	0.27	5	15	4
Central African Republic	90	80	90	0.89	1.00	1.22	80	80	110
Chad	130	120	130	0.65	1.08	0.69	85	130	90
Congo (Brazzaville)	30	30	30	0.67	0.67	0.67	20	20	20
Congo, Democratic Rep.	1,350	1,350	1,350	0.78	0.81	0.85	1,050	1,100	1,150
Cote d'Ivorie	600	600	600	0.98	0.88	0.92	590	525	550
Egypt	700	672	680	8.57	8.54	8.50	6,000	5,740	5,780

See footnotes at end of table.

Table 1-40.—Corn: Area, yield, and production in specified countries, 2002/2003–2004/2005 [1] —Continued

Continent and country	Area [2]			Yield per hectare			Production		
	2002/ 2003	2003/ 2004	2004/ 2005 [3]	2002/ 2003	2003/ 2004	2004/ 2005 [3]	2002/ 2003	2003/ 2004	2004/ 2005 [3]
	1,000 hec- tares	*1,000 hec- tares*	*1,000 hec- tares*	*Metric tons*	*Metric tons*	*Metric tons*	*1,000 metric tons*	*1,000 metric tons*	*1,000 metric tons*
Africa—Continued									
Eritrea	11	13	10	0.36	0.38	0.30	4	5	3
Ethiopia	1,780	2,000	1,865	1.17	1.50	1.82	2,085	3,000	3,400
Gambia, The	18	10	10	1.17	3.00	3.50	21	30	35
Ghana	700	700	720	1.40	1.26	1.39	980	880	1,000
Guinea	90	90	90	1.00	1.11	1.11	90	100	100
Guinea-Bissau	15	15	15	1.67	1.80	2.33	25	27	35
Kenya	1,500	1,500	1,500	1.62	1.67	1.33	2,430	2,500	2,000
Lesotho	90	150	105	0.39	0.50	0.57	35	75	60
Madagascar	190	190	190	0.92	0.79	0.89	175	150	170
Malawi	1,490	1,570	1,535	1.33	1.26	1.11	1,980	1,985	1,705
Mauritania	7	10	10	0.86	1.00	0.50	6	10	5
Morocco	250	250	250	0.80	0.80	0.80	200	200	200
Mozambique	1,360	1,300	1,270	0.92	1.11	1.13	1,250	1,440	1,435
Nigeria	3,500	3,700	3,700	1.49	1.49	1.76	5,200	5,500	6,500
Rwanda	80	75	75	1.44	1.07	1.20	115	80	90
Senegal	110	175	180	0.73	1.14	1.67	80	200	300
Sierra Leone	10	10	10	1.00	1.00	1.00	10	10	10
Somalia	260	270	220	0.77	0.65	0.57	200	175	125
South Africa, Republic	3,650	3,300	3,500	2.65	2.94	3.43	9,675	9,700	12,000
Swaziland	68	68	55	1.00	1.01	1.22	68	69	67
Tanzania, United Rep.	1,600	2,700	2,950	1.69	0.86	1.09	2,700	2,320	3,230
Togo	380	380	380	1.32	1.28	1.47	500	485	560
Tunisia	1	1	1	1.00	1.00	1.00	1	1	1
Uganda	600	600	600	1.25	1.29	1.29	750	775	775
Zambia	575	670	630	1.04	1.55	1.93	600	1,040	1,213
Zimbabwe	1,355	1,365	1,200	0.59	0.66	0.46	800	900	550
Total	24,781	26,298	26,582	1.63	1.61	1.75	40,416	42,309	46,604
Asia:									
Afghanistan	100	105	100	3.00	2.95	2.50	300	310	250
Bhutan	45	45	45	1.56	1.56	1.56	70	70	70
Burma	310	300	308	2.13	2.37	2.63	660	710	810
Cambodia	72	90	90	2.07	2.11	2.00	149	190	180
China, People's Republic of	24,634	24,068	25,446	4.92	4.81	5.12	121,300	115,830	130,300
India	6,300	7,420	7,000	1.76	1.98	1.94	11,100	14,720	13,600
Indonesia	3,050	3,200	3,300	2.00	1.98	1.97	6,100	6,350	6,500
Japan	1	1	1	1.00	1.00	1.00	1	1	1
Korea, Democratic People's Rep	495	495	495	3.33	3.48	3.49	1,650	1,725	1,730
Korea, Rep. of	17	17	18	4.29	4.12	4.33	73	70	78
Malaysia	23	24	24	3.04	3.00	3.00	70	72	72
Mali	280	445	200	1.30	1.15	2.25	365	510	450
Nepal	880	880	880	1.70	1.70	1.70	1,500	1,500	1,500
Pakistan	875	875	900	1.46	1.46	1.44	1,275	1,275	1,300
Philippines	2,375	2,485	2,400	1.87	1.97	2.10	4,430	4,900	5,050
Taiwan	10	10	10	5.00	5.30	5.00	50	53	50
Thailand	1,134	1,110	1,130	3.75	3.69	3.54	4,250	4,100	4,000
Vietnam	810	900	920	2.86	3.11	3.04	2,313	2,800	2,800
Total	41,411	42,470	43,267	3.76	3.65	3.90	155,656	155,186	168,741
Middle East:									
Iran	120	120	130	5.42	5.42	5.38	650	650	700
Iraq	50	50	50	1.20	1.20	1.20	60	60	60
Jordan	1	1	1	1.00	1.00	1.00	1	1	1
Lebanon	2	2	2	1.00	1.00	1.00	2	2	2
Saudi Arabia	3	3	3	1.33	1.33	1.33	4	4	4
Syria	50	50	50	2.50	2.50	2.50	125	125	125
Turkey	550	625	700	3.82	4.48	4.29	2,100	2,800	3,000
Yemen	32	32	35	1.28	1.25	1.14	41	40	40
Total	808	883	971	3.69	4.17	4.05	2,983	3,682	3,932
Oceania:									
Australia	60	79	80	5.17	4.96	5.00	310	392	400
New Zealand	15	15	15	9.67	9.67	9.67	145	145	145
Total	75	94	95	6.07	5.71	5.74	455	537	545
World total	137,511	141,650	144,315	4.38	4.40	4.91	601,714	623,339	708,584

[1] Years shown refer to years of harvest. Harvests of Northern Hemisphere countries are combined with those of the Southern Hemisphere which immediately follow; thus the crop harvested in the Northern Hemisphere in 1994 is combined with estimates for the Southern Hemisphere harvest, which begins late in 1994 and ends early in 1995. [2] Harvested area as far as possible. [3] Preliminary.

FAS, Production Estimates and Crop Assessment Division, (202) 720–0888. Prepared or estimated on the basis of official statistics of foreign governments, other foreign source materials, reports of U.S. Agricultural Counselors, Attachés and Foreign Service Officers, results of office research, and related information.

Table 1-41.—Corn: International trade, 2003–2005 [1]

Country	2002/2003	2003/2004	2004/2005 [2]
	1,000 metric tons	*1,000 Metric tons*	*1,000 metric tons*
Exports:			
Argentina	12,349	10,439	14,200
Brazil	3,181	5,818	1,431
Canada	306	367	200
China	15,244	7,553	7,589
Hungary	516	548	500
Romania	144	93	650
South Africa	1,141	797	1,517
Thailand	137	726	396
Ukraine	811	1,238	2,300
EU	1,995	459	200
Others	1,296	2,281	2,006
Subtotal	37,120	30,319	30,989
United States	40,924	48,808	45,200
Total	78,044	79,127	76,189
Imports:			
Algeria	1,643	1,765	2,000
Brazil	521	677	500
Canada	3,846	2,033	2,400
Chile	933	1,043	1,000
China	29	2	2
Colombia	2,112	1,999	2,100
Costa Rica	514	583	600
Cuba	279	469	550
Dominican Republic	906	824	1,100
Ecuador	304	457	450
Egypt	4,848	3,743	5,300
El Salvador	394	476	500
Guatemala	513	513	650
Indonesia	1,633	1,436	500
Iran	2,157	1,857	2,600
Israel	776	1,377	1,300
Japan	16,863	16,781	16,485
Jordan	406	371	350
Korea, North	144	89	275
Korea, South	8,786	8,783	8,638
Malaysia	2,408	2,401	2,400
Mexico	5,269	5,739	6,000
Morocco	1,054	1,183	1,400
Peru	917	1,041	1,200
Philippines	68	52	100
Russia	99	496	200
Saudi Arabia	1,424	1,621	1,200
South Africa	617	495	131
Syria	919	941	1,650
Taiwan	4,681	4,951	4,500
Tunisia	734	784	650
Turkey	1,475	1,023	150
Venezuela	675	687	100
Vietnam	344	36	300
Zimbabwe	625	233	1,000
EU-25	4,327	5,752	2,500
United States	374	341	250
Subtotal	73,617	73,054	71,031
Other Countries	3,552	3,851	4,157
Unaccounted	875	2,222	1,001
World Total	78,044	79,127	76,189

[1] Year beginning Oct 1.   [2] Preliminary.

FAS, Grain and Feed Division, (202) 720–6219. www.fas.usda.gov/grain/default.html. Prepared or estimated on the basis of official statistics from foreign governments, other foreign source materials, reports of U.S. Agricultural Counselors, Attachés and Foreign Service Officers, results of office research, and related information.

Table 1-42.—Corn: Support operations, United States, 1996-2005

Marketing year beginning September 1	Income support payment rates per bushel [1]	Program price levels per bushel		Put under loan [4]		Acquired by CCC under loan program [5]	Owned by CCC at end of marketing year
		Loan [2]	Target [3]	Quantity	Percentage of production		
	Dollars	Dollars	Dollars	Million bushels	Percent	Million bushels	Million bushels
1996/1997	0.25	1.89	NA	970	10.5	0	2
1997/1998	0.49	1.89	NA	1,141	12.4	2	4
1998/1999	0.56	1.89	NA	1,775	18.2	24	12
1999/2000	0.73	1.89	NA	1,421	15.1	23	14
2000/2001	0.70	1.89	NA	1,394	14.1	31	8
2001/2002	0.58	1.89	NA	1,395	14.7	24	6
2002/2003	0.28/0.00	1.98	2.60	1,367	15.2	1	4
2003/2004	0.28/0.00	1.98	2.60	1,327	13.4	0	0
2004/2005	0.29/---	1.95	2.63	1,366	11.6	6	0.2
2005/2006	0.28/---	1.95	2.63				

[1] Payment rates for the 1995/96 and prior crop years were calculated according to the deficiency payment/production adjustment program provisions. Payment rates for the 1996/97 through 2001/02 crops were calculated according to the Production Flexibility Contract (PFC) program provisions of the Federal Agriculture Improvement and Reform Act of 1996 (1996 Act) and include supplemental PFC payment rates for 1998 through 2001. Payment rates for the 2002/03 and subsequent crops are calculated according to the Direct and Counter-cyclical program provisions, following enactment of the Farm Security and Rural Investment Act of 2002 (2002 Act). Payment rates are rounded to the nearest cent. Beginning with 2002/03, the first entry is the direct payment rate and the second entry is the counter-cylical payment rate. [2] The national average loan rate was also known as the price support rate prior to enactment of the 1996 Act. [3] Between the 1996/97 and 2001/02 marketing years, target prices were no longer applicable; however, target prices were reestablished under the 2002 Act. [4] Represents loans made, purchases, and purchase agreements entered into. Purchases and purchase agreements are no longer authorized for the 1996 and subsequent crops following enactment of the 1996 Act. Percentage of production is on a grain basis. Excludes quantity on which loan deficiency payments were made. [5] Acquisition of all loans forfeited during the marketing year. For 2004/05, as of October 25, 2005. NA-not applicable.
FSA, Feed Grains & Oilseeds Analysis Group, (202) 720-8838.

Table 1-43.—Corn: United States exports, specified by country of destination, 2002/2003-2004/2005 [1]

Country of destination	Year		
	2002/2003	2003/2004	2004/2005
	1,000 metric tons	1,000 metric tons	1,000 metric tons
Japan	14,387	14,973	15,015
Mexico	5,269	5,739	5,920
Taiwan	4,139	4,758	4,449
Egypt	2,904	3,120	3,738
Canada	3,836	2,033	2,236
Korea, South	274	3,946	2,211
Colombia	1,607	1,804	1,967
Syria	538	840	1,288
Dominican Republic	905	823	1,007
Algeria	1,009	1,158	962
Morocco	105	762	781
Guatemala	466	513	633
Costa Rica	514	581	567
El Salvador	388	456	510
Cuba	279	443	444
Israel	313	1,154	393
Ecuador	176	388	322
Honduras	203	228	306
Panama	272	293	278
Jordan	46	207	264
Peru	42	157	251
Lebanon	88	254	238
Tunisia	144	623	234
Jamaica	218	241	224
Saudi Arabia	222	402	126
Other	2,822	3,406	975
Total	40,924	48,809	45,223

[1] Year beginning Oct. 1. Compiled from U.S. Census data.
FAS, Grain and Feed Division, (202) 720-6219.www.fas.usda.gov/grain/default.html.

Table 1-44.—Oats: Area, yield, production, and value, United States, 1996–2005

Year	Area		Yield per harvested acre	Production	Marketing year average price per bushel received by farmers [2]	Value of production
	Planted [1]	Harvested				
	1,000 acres	1,000 acres	Bushels	1,000 bushels	Dollars	1,000 dollars
1996	4,638	2,655	57.7	153,245	1.96	313,910
1997	5,068	2,813	59.5	167,246	1.60	273,284
1998	4,891	2,752	60.2	165,768	1.10	199,475
1999	4,668	2,445	59.6	145,628	1.12	174,307
2000	4,473	2,325	64.2	149,165	1.10	175,432
2001	4,401	1,911	61.5	117,602	1.59	197,181
2002	4,995	2,058	56.4	116,002	1.81	212,078
2003	4,597	2,220	65.0	144,383	1.48	224,910
2004	4,085	1,787	64.7	115,695	1.48	178,327
2005 [2]	4,246	1,823	63.0	114,878	1.58	187,275

[1] Relates to the total area of oats sown for all purposes, including oats sown in the preceding fall.   [2] Preliminary.
NASS, Crops Branch, (202) 720–2127.

Table 1-45.—Oats: Stocks on and off farms, United States, 1996–2005

Year beginning September	On farms				Off farms [1]			
	Sep. 1	Dec. 1	Mar. 1	Jun. 1	Sep. 1	Dec. 1	Mar. 1	Jun. 1
1996	93,400	80,650	56,200	33,100	38,459	45,218	39,330	33,576
1997	105,950	83,200	58,800	34,500	48,972	61,051	52,418	39,498
1998	110,300	81,500	61,700	40,700	51,502	61,835	50,850	40,678
1999	97,300	79,800	53,300	36,000	51,151	53,872	48,500	40,031
2000	101,200	86,900	55,800	32,050	49,177	57,237	54,128	40,677
2001	74,800	58,100	40,200	28,650	41,592	56,117	53,158	34,552
2002	70,500	52,500	35,000	20,600	41,212	51,284	47,879	29,233
2003	82,100	64,400	45,600	27,500	49,637	54,900	49,414	37,348
2004	74,300	60,400	43,500	25,350	41,458	44,513	38,946	32,592
2005 [2]	71,700	60,100	NA	NA	41,803	35,635	NA	NA

[1] Inlcudes stocks at mills, elevators, warehouses, terminals, and processors.   [2] Preliminary.   NA-not available.
NASS, Crops Branch, (202) 720–2127.

Table 1-46.—Oats: Supply and disappearance, United States, 1996–2005

Year beginning June 1	Supply				Disappearance					Ending stocks May 31		
	Beginning stocks	Production	Imports	Total	Domestic use			Exports	Total disappearance	Privately held [1]	Government	Total
					Feed and residual	Food, seed and industrial	Total					
	Million bushels	Million bushels	Million bushels	Million bushels	Million bushels	Million bushels	Million bushels	Million bushels	Million bushels	Million bushels	Million bushels	Million bushels
1996 ..	66	153	97	317	172	76	248	3	250	67	0	67
1997 ..	67	167	98	332	185	72	256	2	258	74	0	74
1998 ..	74	166	108	348	195	69	264	2	266	81	0	81
1999 ..	81	146	99	326	179	68	248	2	250	76	0	76
2000 ..	76	149	106	331	189	68	257	2	258	73	0	73
2001 ..	73	118	96	286	148	72	220	3	223	63	0	63
2002 ..	63	116	95	274	150	72	222	3	224	50	0	50
2003 ..	50	144	90	284	144	73	217	2	219	65	0	65
2004 [2]	65	116	88	268	134	74	208	2	210	58	0	58
2005 [3]	58	115	75	248	120	74	194	3	197	51	0	51

[1] Includes quantity under loan and farmer-owned reserve.   [2] Preliminary.   [3] Projected as of January 12, 2006, World Agricultural Supply and Demand Estimates. Totals may not add due to independent rounding.
ERS, Market and Trade Economics Division, (202) 694–5296.

Table 1-47.—Oats: Support operations, United States, 1996–2005

Marketing Year beginning June 1	Income support payment rates per bushel [1]	Program price levels per bushel		Put under loan [4]		Acquired by CCC under loan program [5]	Owned by CCC at end of marketing year
		Loan [2]	Target [3]	Quantity	Percentage of production		
	Dollars	Dollars	Dollars	Million bushels	Percent	Million bushels	Million bushels
1996/1997	0.03	1.03	NA	1.5	1.0	0.0	0.0
1997/1998	0.03	1.11	NA	2.4	1.4	0.1	0.0
1998/1999	0.05	1.11	NA	4.6	2.8	0.8	0.0
1999/2000	0.06	1.13	NA	1.7	1.2	0.1	0.0
2000/2001	0.06	1.16	NA	1.7	1.1	0.1	0.0
2001/2002	0.05	1.21	NA	1.7	1.5	0.0	0.0
2002/2003	0.02/0.00	1.35	1.40	2.0	1.7	0.0	0.0
2003/2004	0.02/0.00	1.35	1.40	5.2	3.6	0.3	0.0
2004/2005	0.02/0.00	1.33	1.44	3.3	2.8	0.1	0.0
2005/2006	0.02/0.00	1.33	1.44	....	....	....	....

[1] Payment rates for the 1995/96 and prior crop years were calculated according to the deficiency payment/production adjustment program provisions. Payment rates for the 1996/97 through 2001/02 crops were calculated according to the Production Flexibility Contract (PFC) program provisions of the Federal Agriculture Improvement and Reform Act of 1996 (1996 Act) and include supplemental PFC payment rates for 1998 through 2001. Payment rates for the 2002/03 and subsequent crops are calculated according to the Direct and Counter-cyclical program provisions, following enactment of the Farm Security and Rural Investment Act of 2002 (2002 Act). Payment rates are rounded to the nearest cent. Beginning with 2002/2003, the first entry is the direct payment rate and the second entry is the counter-cyclical payment rate. [2] The national average loan rate was also known as the price support rate prior to enactment of the 1996 Act. [3] Between the 1996/97 and 2001/02 marketing years, target prices were no longer applicable; however, target prices were reestablished under the 2002 Act. [4] Represents loans made, purchases, and purchase agreements entered into. Purchases and purchase agreements are no longer authorized for the 1996 and subsequent crops following enactment of the 1996 Act. Percentage of production is on a grain basis. Excludes quantity on which loan deficiency payments were made. [5] Acquisition of all loans forfeited during the marketing year. For 2004/05, as of October 25, 2005. NA-not applicable.

FSA, Feed Grains & Oilseeds Analysis Group, (202) 720–8838.

Table 1-48.—Oats: Area, yield, and production, by States, 2003–2005

State	Area planted [1]			Area harvested			Yield per harvested acre			Production		
	2003	2004	2005	2003	2004	2005	2003	2004	2005	2003	2004	2005
	1,000 acres	1,000 acres	1,000 acres	1,000 acres	1,000 acres	1,000 acres	Bushels	Bushels	Bushels	1,000 bushels	1,000 bushels	1,000 bushels
AL [2]	....	....	50	....	....	20	....	....	55.0	....	....	1,100
CA	260	240	270	35	25	20	80.0	85.0	75.0	2,800	2,125	1,500
CO	100	75	75	15	20	15	65.0	55.0	75.0	975	1,100	1,125
GA	100	90	75	30	25	20	56.0	50.0	60.0	1,680	1,250	1,200
ID	120	90	90	25	20	20	65.0	72.0	64.0	1,625	1,440	1,280
IL	60	55	60	50	35	40	89.0	70.0	79.0	4,450	2,450	3,160
IN	25	25	20	15	12	9	70.0	75.0	69.0	1,050	900	621
IA	220	220	210	130	140	125	83.0	72.0	79.0	10,790	10,080	9,875
KS	140	120	100	70	40	40	65.0	43.0	59.0	4,550	1,720	2,360
ME	27	34	32	26	32	28	78.0	80.0	70.0	2,028	2,560	1,960
MI	90	80	90	75	65	75	70.0	68.0	61.0	5,250	4,420	4,575
MN	350	310	310	265	190	205	71.0	70.0	62.0	18,815	13,300	12,710
MO	30	26	35	18	13	20	67.0	50.0	65.0	1,206	650	1,300
MT	120	105	90	45	40	35	44.0	60.0	53.0	1,980	2,400	1,855
NE	220	140	150	90	50	60	73.0	68.0	73.0	6,570	3,400	4,380
NY	85	65	95	70	50	75	63.0	65.0	54.0	4,410	3,250	4,050
NC	55	55	50	22	25	23	59.0	70.0	73.0	1,298	1,750	1,679
ND	620	490	490	360	220	240	59.0	64.0	59.0	21,240	14,080	14,160
OH	80	65	80	60	50	60	66.0	63.0	60.0	3,960	3,150	3,600
OK	70	50	45	25	15	10	36.0	37.0	41.0	900	555	410
OR	60	50	40	20	20	18	75.0	97.0	78.0	1,500	1,940	1,404
PA	140	130	140	110	110	110	59.0	55.0	55.0	6,490	6,050	6,050
SC	40	40	35	20	20	20	56.0	55.0	59.0	1,120	1,100	1,180
SD	420	380	380	230	170	180	68.0	82.0	72.0	15,640	13,940	12,960
TX	625	680	690	140	160	110	45.0	40.0	43.0	6,300	6,400	4,730
UT	65	60	50	6	8	7	82.0	78.0	73.0	492	624	511
VA [2]	....	....	14	....	....	3	....	....	61.0	....	....	183
WA	35	20	25	15	7	8	50.0	88.0	75.0	750	616	600
WI	380	340	400	230	210	215	67.0	65.0	64.0	15,410	13,650	13,760
WY	60	50	55	23	15	12	48.0	53.0	50.0	1,104	795	600
US	4,597	4,085	4,246	2,220	1,787	1,823	65.0	64.7	63.0	144,383	115,695	114,878

[1] Relates to the total area of oats sown for all purposes, including oats sown in the preceding fall. [2] Estimates began in 2005.

NASS, Crops Branch, (202) 720–2127.

Table 1-49.—Oats: Marketing year average price and value of production, by States, crop of 2003, 2004, and 2005

State	Marketing year average price per bushel			Value of production		
	2003	2004	2005[1]	2003	2004	2005[1]
	Dollars	Dollars	Dollars	1,000 dollars	1,000 dollars	1,000 dollars
AL[2]			1.60			1,760
CA	2.00	2.00	1.30	5,600	4,250	1,950
CO	2.06	2.02	2.20	2,009	2,222	2,475
GA	1.70	1.70	1.60	2,856	2,125	1,920
ID	1.50	1.25	1.30	2,438	1,800	1,664
IL	1.66	1.57	1.80	7,387	3,847	5,688
IN	1.90	1.80	1.80	1,995	1,620	1,118
IA	1.54	1.49	1.70	16,617	15,019	16,788
KS	1.45	1.25	1.30	6,598	2,150	3,068
ME	1.10	1.20	1.10	2,231	3,072	2,156
MI	1.65	1.72	1.90	8,663	7,602	8,693
MN	1.39	1.31	1.45	26,153	17,423	18,430
MO	1.65	1.65	1.80	1,990	1,073	2,340
MT	1.70	1.70	1.75	3,366	4,080	3,246
NE	1.53	1.51	1.60	10,052	5,134	7,008
NY	1.81	1.80	1.60	7,982	5,850	6,480
NC	1.90	2.00	2.05	2,466	3,500	3,442
ND	1.33	1.18	1.40	28,249	16,614	19,824
OH	1.78	1.78	1.80	7,049	5,607	6,480
OK	1.80	1.70	1.80	1,620	944	738
OR	1.98	1.85	2.20	2,970	3,589	3,089
PA	1.81	1.89	2.05	11,747	11,435	12,403
SC	1.75	2.05	1.75	1,960	2,255	2,065
SD	1.43	1.49	1.50	22,365	20,771	19,440
TX	2.20	1.91	2.35	13,860	12,224	11,116
UT	2.30	1.95	1.80	1,132	1,217	920
VA[2]			2.10			384
WA	1.63	1.50	1.65	1,223	924	990
WI	1.45	1.52	1.50	22,345	20,748	20,640
WY	1.80	1.55	1.60	1,987	1,232	960
US	1.48	1.48	1.58	224,910	178,327	187,275

[1] Preliminary. [2] Estimates began in 2005.
NASS, Crops Branch, (202) 720–2127.

Table 1-50.—Oats: Area, yield, and production in specified countries, 2002/2003–2004/2005 [1]

Continent and country	Area[2]			Yield per hectare			Production		
	2002/ 2003	2003/ 2004	2004/ 2005[3]	2002/ 2003	2003/ 2004	2004/ 2005[3]	2002/ 2003	2003/ 2004	2004/ 2005[3]
	1,000 hec- tares	1,000 hec- tares	1,000 hec- tares	Metric tons	Metric tons	Metric tons	1,000 metric tons	1,000 metric tons	1,000 metric tons
North America:									
Canada	1,379	1,575	1,315	2.11	2.34	2.80	2,911	3,691	3,683
Mexico	40	40	60	1.63	2.38	1.25	65	95	75
United States	833	898	725	2.02	2.33	2.32	1,684	2,096	1,683
Total	2,252	2,513	2,100	2.07	2.34	2.59	4,660	5,882	5,441
South America:									
Argentina	300	231	277	1.67	1.51	1.83	500	348	508
Brazil	267	300	299	1.46	1.38	1.37	390	413	411
Chile	105	125	125	4.00	3.40	3.40	420	425	425
Colombia	3	3	3	1.33	1.33	1.33	4	4	4
Ecuador	1	1	1	1.00	1.00	1.00	1	1	1
Uruguay	12	22	19	1.33	1.36	1.37	16	30	26
Total	688	682	724	1.93	1.79	1.90	1,331	1,221	1,375
European Union:									
Austria	32	34	30	3.66	3.71	4.63	117	126	139
Belgium-Luxembourg	8	10	10	5.50	4.50	4.50	44	45	45
Czech Republic	61	77	59	2.75	3.04	3.85	168	234	227
Denmark	55	50	61	5.02	5.20	5.05	276	260	308
Estonia	38	40	39	1.63	1.55	1.92	62	62	75
Finland	451	424	326	3.38	3.05	3.07	1,525	1,294	1,002
France	153	139	128	5.05	4.09	4.72	773	569	604
Germany	233	262	227	4.36	4.59	5.22	1,016	1,202	1,186
Greece	45	45	45	1.38	1.31	2.00	62	59	90
Hungary	60	60	71	2.30	1.70	3.04	138	102	216

See footnotes at end of table.

Table 1-50.—Oats: Area, yield, and production in specified countries, 2002/2003–2004/2005 [1]—Continued

Continent and country	Area [2]			Yield per hectare			Production		
	2002/ 2003	2003/ 2004	2004/ 2005 [3]	2002/ 2003	2003/ 2004	2004/ 2005 [3]	2002/ 2003	2003/ 2004	2004/ 2005 [3]
	1,000 hec- tares	1,000 hec- tares	1,000 hec- tares	Metric tons	Metric tons	Metric tons	1,000 metric tons	1,000 metric tons	1,000 metric tons
Europe, cont.:									
Ireland	19	20	21	7.00	7.75	7.29	133	155	153
Italy	151	148	147	2.18	2.07	2.29	329	306	336
Latvia	47	40	50	1.70	1.95	2.24	80	78	112
Lithuania	55	48	53	1.78	2.40	2.23	98	115	118
Netherlands	2	5	5	6.50	3.00	5.00	13	15	25
Poland	605	527	520	2.46	2.24	2.75	1,486	1,182	1,430
Portugal	57	54	57	1.07	0.69	0.88	61	37	50
Slovakia	21	30	22	2.05	1.97	2.55	43	59	56
Slovenia	2	2	2	3.00	3.00	3.00	6	6	6
Spain	455	496	477	2.01	1.78	2.14	916	881	1,019
Sweden	296	275	231	3.99	3.96	4.02	1,181	1,089	929
United Kingdom	126	122	108	5.98	6.14	5.83	753	749	630
Total	2,972	2,908	2,689	3.12	2.97	3.26	9,280	8,625	8,756
Other Europe:									
Albania	20	20	20	0.75	0.75	0.75	15	15	15
Bosnia-Hercegovina	26	26	26	2.12	2.12	2.12	55	55	55
Bulgaria	35	35	35	1.43	1.29	1.43	50	45	50
Croatia	24	20	20	2.08	1.70	2.00	50	34	40
Macedonia (Skopje)	3	3	3	1.33	1.33	1.33	4	4	4
Norway	90	90	90	4.44	4.44	4.44	400	400	400
Romania	225	225	225	1.56	1.56	1.56	350	350	350
Serbia and Montenegro	65	65	65	1.92	1.69	1.85	125	110	120
Switzerland	7	7	7	5.00	5.00	5.00	35	35	35
Total	495	491	491	2.19	2.13	2.18	1,084	1,048	1,069
Fmr. Soviet Union:.									
Belarus	262	250	250	2.19	2.00	3.08	575	500	770
Georgia	7	7	7	0.71	0.71	0.71	5	5	5
Kazakhstan	100	100	140	1.00	1.00	1.00	100	100	140
Kyrgyzstan	3	3	3	2.67	1.67	1.67	8	5	5
Russian Fed.	4,400	3,700	3,550	1.30	1.41	1.39	5,700	5,200	4,950
Ukraine	500	550	510	1.89	1.68	1.96	943	925	1,000
Total	5,272	4,610	4,460	1.39	1.46	1.54	7,331	6,735	6,870
Turkey	154	154	155	1.88	1.85	1.87	290	285	290
Africa:.									
Algeria	50	70	70	0.70	1.07	1.14	35	75	80
Morocco	40	40	40	0.75	0.75	0.75	30	30	30
South Africa, Rep. of	700	700	700	0.06	0.06	0.06	45	45	45
Total	790	810	810	0.14	0.19	0.19	110	150	155
Asia:									
China, People's Republic	500	500	500	1.20	1.20	1.20	600	600	600
Japan	1	1	1	2.00	2.00	2.00	2	2	2
Total	501	501	501	1.20	1.20	1.20	602	602	602
Oceania:									
Australia	911	1,076	900	1.05	1.83	1.22	957	1,965	1,100
New Zealand	20	20	20	3.75	3.75	3.75	75	75	75
Total	931	1,096	920	1.11	1.86	1.28	1,032	2,040	1,175
World total	14,118	13,770	12,850	1.83	1.93	2.00	25,845	26,590	25,730

[1] Years shown refer to years of harvest. Harvests of Northern Hemisphere countries are combined with those of the Southern Hemisphere which immediately follow; thus the crop harvested in the Northern Hemisphere in 1994 is combined with estimates for the Southern Hemisphere harvest, which begins late in 1994 and ends early in 1995.　[2] Harvested area as far as possible.　[3] Preliminary.

FAS Production Estimates and Crop Assessment Division, (202) 720–0888. Prepared or estimated on the basis of official statistics of foreign governments, other foreign source materials, reports of U.S. Agricultural Counselors, Attachés, and Foreign Service Officers, results of office research, and related information.

GRAIN AND FEED

Table 1-51.—Barley: Area, yield, production, and value, United States, 1996–2005

Year	Area		Yield per harvested acre	Production	Marketing year average price per bushel received by farmers	Value of production
	Planted [1]	Harvested				
	1,000 acres	*1,000 acres*	*Bushels*	*1,000 bushels*	*Dollars*	*1,000 dollars*
1996	7,094	6,707	58.5	392,433	2.74	1,080,940
1997	6,706	6,198	58.1	359,878	2.38	861,620
1998	6,325	5,854	60.1	351,569	1.98	685,734
1999	4,983	4,573	59.5	271,996	2.13	578,425
2000	5,801	5,200	61.1	317,804	2.11	647,966
2001	4,951	4,273	58.1	248,329	2.22	535,110
2002	5,008	4,123	55.0	226,906	2.72	605,635
2003	5,348	4,727	58.9	278,283	2.83	755,140
2004	4,527	4,021	69.6	279,743	2.48	698,184
2005 [2]	3,875	3,269	64.8	211,896	2.45	505,962

[1] Barley sown for all purposes, including barley sown in the preceding fall. [2] Preliminary.
NASS, Crops Branch, (202) 720–2127.

Table 1-52.—Barley: Stocks on and off farms, United States, 1996–2005

Year beginning September	On farms				Off farms [1]			
	Sep. 1	Dec. 1	Mar. 1	June 1	Sep. 1	Dec. 1	Mar. 1	June 1
	1,000 bushels	*1,000 bushels*	*1,000 bushels*	*1,000 bushels*	*1,000 bushels*	*1,000 bushels*	*1,000 bushels*	*1,000 bushels*
1996	191,700	135,700	82,060	43,715	122,078	110,522	90,840	65,735
1997	195,500	134,500	81,900	43,700	131,347	109,561	98,299	75,533
1998	193,500	149,000	86,900	52,000	132,674	121,535	113,808	89,653
1999	165,100	108,000	59,300	25,400	130,145	119,721	110,917	85,924
2000	151,700	111,500	58,600	28,850	142,341	117,369	103,544	77,409
2001	134,800	92,400	46,000	23,210	110,564	102,587	95,748	68,919
2002	131,300	83,400	36,730	14,860	92,419	86,601	86,710	54,480
2003	141,900	97,200	51,700	28,320	99,730	100,679	101,186	91,988
2004	175,300	130,700	79,680	41,100	114,777	115,276	111,001	87,317
2005 [2]	137,400	103,650	NA	NA	117,511	104,335	NA	NA

[1] Includes stocks at mills, elevators, warehouses, terminals, and processors. [2] Preliminary. NA-not available.
NASS, Crops Branch, (202) 720–2127.

Table 1-53.—Barley: Supply and disappearance, United States, 1996–2005

Year beginning June 1	Supply				Disappearance					Ending stocks May 31		
	Beginning stocks	Production	Imports	Total	Domestic use			Exports	Total disappearance	Privately held [1]	Government	Total
					Feed and residual	Food, seed, and industrial	Total					
	Million bushels	*Million bushels*	*Million bushels*	*Million bushels*	*Million bushels*	*Million bushels*	*Million bushels*	*Million bushels*	*Million bushels*	*Million bushels*	*Million bushels*	*Million bushels*
1996	100	392	37	529	217	172	389	31	419	109	0	109
1997	109	360	40	510	144	172	316	74	390	119	0	119
1998	119	352	30	501	160	170	330	29	359	142	0	142
1999	142	272	28	441	130	172	302	28	330	111	0	111
2000	111	318	29	458	136	159	294	58	352	106	0	106
2001	106	248	24	379	104	156	260	26	286	92	0	92
2002	92	227	18	337	84	154	238	30	268	69	0	69
2003	69	278	21	368	72	157	229	19	248	120	0	120
2004 [2] ...	120	280	12	412	116	145	261	23	284	128	0	128
2005 [3] ...	128	212	10	350	70	140	210	30	240	110	0	110

[1] Includes quantity under loan and farmer–owned reserve. [2] Preliminary. [3] Projected as of January 12, 2006, World Agricultural Supply and Demand Estimates. Totals may not add due to independent rounding.
ERS, Market and Trade Economics Division, (202) 694–5296.

Table 1-54.—Barley: Area, yield, and production, by States, 2003-2005

State	Area planted[1]			Area harvested			Yield per harvested acre			Production		
	2003	2004	2005[2]	2003	2004	2005[2]	2003	2004	2005[2]	2003	2004	2005[2]
	1,000 acres	1,000 acres	1,000 acres	1,000 acres	1,000 acres	1,000 acres	Bushels	Bushels	Bushels	1,000 bushels	1,000 bushels	1,000 bushels
AZ	32	40	34	30	38	30	118.0	110.0	100.0	3,540	4,180	3,000
CA	100	110	100	58	75	60	64.0	60.0	63.0	3,712	4,500	3,780
CO	85	80	60	82	77	59	109.0	118.0	130.0	8,938	9,086	7,670
DE	25	29	29	21	26	27	59.0	80.0	81.0	1,239	2,080	2,187
ID	750	680	630	720	650	600	66.0	92.0	87.0	47,520	59,800	52,200
KS	9	15	19	8	12	14	57.0	28.0	42.0	456	336	588
KY	9	9	10	8	8	9	75.0	77.0	83.0	600	616	747
ME	28	23	23	27	22	22	65.0	60.0	60.0	1,755	1,320	1,320
MD	43	42	46	36	39	41	57.0	73.0	86.0	2,052	2,847	3,526
MI	15	14	15	14	12	11	56.0	51.0	47.0	784	612	517
MN	190	130	125	170	115	90	75.0	68.0	43.0	12,750	7,820	3,870
MT	1,150	1,000	900	850	830	700	40.0	59.0	56.0	34,000	48,970	39,200
NE[3]	6	6	....	4	3	....	50.0	54.0	....	200	162	....
NV	5	4	4	3	2	2	80.0	105.0	85.0	240	210	170
NJ	4	3	3	3	2	2	45.0	63.0	71.0	135	126	142
NY	15	14	17	13	10	15	50.0	53.0	49.0	650	530	735
NC	20	23	24	14	15	19	56.0	64.0	78.0	784	960	1,482
ND	2,050	1,600	1,200	1,980	1,480	1,060	60.0	62.0	54.0	118,800	91,760	57,240
OH	7	5	6	6	4	5	58.0	50.0	60.0	348	200	300
OR	70	75	65	60	66	45	64.0	73.0	45.0	3,840	4,818	2,025
PA	75	65	55	65	55	47	61.0	62.0	72.0	3,965	3,410	3,384
SD	75	70	65	55	50	47	53.0	63.0	49.0	2,915	3,150	2,303
UT	45	50	40	35	40	24	80.0	86.0	80.0	2,800	3,440	1,920
VA	75	55	60	45	40	45	62.0	74.0	87.0	2,790	2,960	3,915
WA	320	250	215	310	245	205	47.0	70.0	61.0	14,570	17,150	12,505
WI	55	45	55	35	30	30	55.0	55.0	53.0	1,925	1,650	1,590
WY	90	90	75	75	75	60	93.0	94.0	93.0	6,975	7,050	5,580
US	5,348	4,527	3,875	4,727	4,021	3,269	58.9	69.6	64.8	278,283	279,743	211,896

[1] Includes area planted in the preceding fall.   [2] Preliminary.   [3] Estimates discontinued in 2005.
NASS, Crops Branch, (202) 720-2127.

Table 1-55.—Barley: Marketing year average price and value, by States, crop of 2003, 2004, and 2005

State	Marketing year average price per bushel			Value of production		
	2003	2004	2005[1]	2003	2004	2005[1]
	Dollars	Dollars	Dollars	1,000 dollars	1,000 dollars	1,000 dollars
AZ	2.84	2.80	2.75	10,054	11,704	8,250
CA	2.77	2.65	2.55	10,282	11,925	9,639
CO	3.05	2.82	2.85	27,261	25,623	21,860
DE	1.60	1.78	1.35	1,982	3,702	2,952
ID	3.15	3.02	2.90	149,688	180,596	151,380
KS	2.15	1.87	1.95	980	628	1,147
KY	1.80	2.02	1.95	1,080	1,244	1,457
ME	1.30	1.58	1.55	2,282	2,086	2,046
MD	1.80	1.92	1.50	3,694	5,466	5,289
MI	1.70	1.80	1.80	1,333	1,102	931
MN	2.52	2.22	1.95	32,130	17,360	7,547
MT	2.93	2.85	2.85	99,620	139,565	111,720
NE[2]	1.90	1.80	....	380	292	....
NV	3.30	3.30	3.05	792	693	519
NJ	1.95	2.10	2.00	263	265	284
NY	2.09	2.05	1.90	1,359	1,087	1,397
NC	1.90	2.10	1.75	1,490	2,016	2,594
ND	2.59	2.12	1.80	307,692	194,531	103,032
OH	1.80	1.90	1.70	626	380	510
OR	2.47	1.94	2.10	9,485	9,347	4,253
PA	2.13	2.20	1.80	8,445	7,502	6,091
SD	2.37	2.02	2.00	6,909	6,363	4,606
UT	2.30	2.21	2.10	6,440	7,602	4,032
VA	1.55	1.73	1.60	4,325	5,121	6,264
WA	2.66	2.02	2.15	38,756	34,643	26,886
WI	1.90	2.00	1.80	3,658	3,300	2,862
WY	3.46	3.41	3.30	24,134	24,041	18,414
US	2.83	2.48	2.45	755,140	698,184	505,962

[1] Preliminary.   [2] Estimates discontinued in 2005.
NASS, Crops Branch, (202) 720-2127.

Table 1-56.—Barley: Area, yield, and production in specified countries, 2002/2003–2004/2005 [1]

Continent and country	Area [2]			Yield per hectare			Production		
	2002/ 2003	2003/ 2004	2004/ 2005 [3]	2002/ 2003	2003/ 2004	2004/ 2005 [3]	2002/ 2003	2003/ 2004	2004/ 2005 [3]
	1,000 hec- tares	1,000 hec- tares	1,000 hec- tares	Metric tons	Metric tons	Metric tons	1,000 metric tons	1,000 metric tons	1,000 metric tons
North America:									
Canada	3,348	4,446	4,050	2.24	2.77	3.26	7,489	12,328	13,186
Mexico	320	350	350	2.41	2.57	2.57	770	900	900
United States	1,669	1,913	1,627	2.96	3.17	3.74	4,940	6,059	6,080
Total	5,337	6,709	6,027	2.47	2.87	3.35	13,199	19,287	20,166
South America:									
Argentina	248	330	270	2.19	3.03	3.28	543	1,000	886
Bolivia	90	90	90	0.78	0.78	0.78	70	70	70
Brazil	114	137	137	2.67	2.64	2.68	304	362	367
Chile	18	17	15	2.78	3.53	4.00	50	60	60
Colombia	10	10	10	2.00	2.00	2.00	20	20	20
Ecuador	35	35	35	1.00	1.00	1.00	35	35	35
Peru	154	152	150	1.30	1.11	1.33	200	168	200
Uruguay	103	118	137	1.72	2.75	2.92	177	324	400
Total	772	889	844	1.81	2.29	2.41	1,399	2,039	2,038
European Union:									
Austria	201	212	191	4.28	4.15	5.26	861	880	1,004
Belgium-Luxem- bourg	57	50	50	7.00	6.00	7.00	399	300	350
Cyprus	45	50	50	2.84	2.58	1.44	128	129	72
Czech Republic	488	550	469	3.67	3.76	4.97	1,793	2,069	2,331
Denmark	828	714	707	4.98	5.29	5.14	4,120	3,776	3,634
Estonia	132	125	128	1.89	2.02	2.26	249	253	289
Finland	522	540	550	3.33	0.61	0.00	1,739	330	
France	1,643	1,749	1,630	6.69	5.69	6.77	10,988	9,956	11,032
Germany	1,970	2,075	1,979	5.55	5.11	6.57	10,928	10,596	12,993
Greece	105	155	155	1.91	2.00	2.00	201	310	310
Hungary	372	380	350	2.96	2.17	4.00	1,100	825	1,400
Ireland	176	186	180	5.47	6.44	7.28	963	1,197	1,310
Italy	343	310	307	3.47	3.29	3.80	1,190	1,021	1,167
Latvia	137	100	136	1.91	2.46	2.04	262	246	278
Lithuania	365	308	293	2.39	2.92	2.93	871	900	859
Malta	1	1	1	2.00	2.00	3.00	2	2	3
Netherlands	57	60	50	5.53	6.17	6.20	315	370	310
Poland	1,051	1,016	1,014	3.21	2.79	3.52	3,369	2,832	3,571
Portugal	11	11	13	1.82	1.18	1.38	20	13	18
Slovakia	220	270	240	3.16	2.98	3.82	695	804	916
Slovenia	13	11	15	3.69	3.64	4.00	48	40	60
Spain	3,102	3,111	3,170	2.70	2.79	3.35	8,362	8,694	10,609
Sweden	408	363	370	4.36	4.26	4.58	1,778	1,546	1,693
United Kingdom	1,101	1,078	1,010	5.57	5.91	5.76	6,128	6,370	5,815
Total	13,348	13,425	13,058	4.23	3.98	4.60	56,509	53,459	60,024
Other Europe:									
Albania	10	10	10	3.00	3.00	3.00	30	30	30
Bosnia- Hercegovina	23	20	20	2.39	2.25	2.50	55	45	50
Bulgaria	350	270	280	2.86	1.67	3.39	1,000	450	950
Croatia	48	45	50	3.56	2.33	3.60	171	105	180
Macedonia (Skopje)	50	50	50	2.40	2.40	2.40	120	120	120
Norway	175	175	175	3.71	3.71	3.71	650	650	650
Romania	296	317	400	2.23	1.70	3.50	660	540	1,400
Serbia and Monte- negro	120	120	110	2.33	1.83	4.27	280	220	470
Switzerland	45	45	45	6.67	6.67	6.67	300	300	300
Total E. Europe	1,117	1,052	1,140	2.92	2.34	3.64	3,266	2,460	4,150

See footnotes at end of table.

Table 1-56.—Barley: Area, yield, and production in specified countries, 2002/2003–2004/2005 [1]—Continued

Continent and country	Area [2]			Yield per hectare			Production		
	2002/ 2003	2003/ 2004	2004/ 2005 [3]	2002/ 2003	2003/ 2004	2004/ 2005 [3]	2002/ 2003	2003/ 2004	2004/ 2005 [3]
	1,000 hec- tares	*1,000 hec- tares*	*1,000 hec- tares*	*Metric tons*	*Metric tons*	*Metric tons*	*1,000 metric tons*	*1,000 metric tons*	*1,000 metric tons*
Fmr. Soviet Union:									
Armenia	35	55	65	3.14	1.27	1.62	110	70	105
Azerbaijan	132	130	160	2.20	2.23	2.19	290	290	350
Belarus	675	800	660	2.49	2.25	3.03	1,681	1,800	2,000
Georgia	40	35	35	1.38	1.43	1.71	55	50	60
Kazakhstan	1,750	1,800	1,700	1.26	1.17	0.88	2,200	2,100	1,500
Kyrgyzstan	65	90	100	2.55	2.22	2.35	166	200	235
Moldova	100	60	120	2.50	0.83	2.50	250	50	300
Russian Fed.	10,250	10,500	10,000	1.82	1.71	1.72	18,700	18,000	17,200
Tajikistan	30	50	50	1.20	1.02	1.80	36	51	90
Turkmenistan	20	20	20	1.00	1.00	1.00	20	20	20
Ukraine	4,153	4,600	4,500	2.50	1.49	2.47	10,364	6,850	11,100
Uzbekistan	75	110	75	2.67	1.36	2.67	200	150	200
Total	17,325	18,250	17,485	1.97	1.62	1.90	34,072	29,631	33,160
Middle East:									
Iran	1,400	1,400	1,400	1.43	1.43	1.43	2,000	2,000	2,000
Iraq	1,300	1,300	1,300	0.77	0.88	0.96	1,000	1,150	1,250
Israel	10	1	1	0.80	2.00	2.00	8	2	2
Jordan	66	6	50	1.03	1.17	0.60	68	7	30
Lebanon	11	13	13	1.55	1.54	1.54	17	20	20
Saudi Arabia	17	0	0	5.88	0.00	0.00	100	0	0
Syria	1,400	1,300	1,000	0.66	0.85	0.90	920	1,100	900
Turkey	3,550	3,450	3,500	2.03	2.00	2.11	7,200	6,900	7,400
Yemen	40	40	40	1.00	0.93	1.00	40	37	40
Total	7,794	7,510	7,304	1.46	1.49	1.59	11,353	11,216	11,642
Africa:									
Algeria	401	783	916	1.04	1.56	1.43	416	1,220	1,314
Egypt	55	55	50	3.09	3.09	3.20	170	170	160
Eritrea	40	45	52	0.25	0.20	0.21	10	9	11
Ethiopia	1,285	1,200	1,130	0.93	1.18	1.39	1,200	1,410	1,575
Kenya	70	70	70	1.07	1.07	1.07	75	75	75
Libya	280	280	280	0.30	0.30	0.30	85	85	85
Morocco	2,002	2,267	2,324	0.83	1.16	1.19	1,669	2,620	2,760
South Africa, Rep. of	72	85	83	2.54	2.82	2.23	183	240	185
Tunisia	150	550	560	0.60	1.27	1.10	90	700	617
Zimbabwe	3	5	5	5.67	6.00	5.00	17	30	25
Total	4,358	5,340	5,470	0.90	1.23	1.24	3,915	6,559	6,807
Asia:									
Afghanistan	236	275	275	1.46	1.49	1.45	345	410	400
Bangladesh	7	7	7	0.71	0.71	0.71	5	5	5
Bhutan	5	5	5	1.00	1.00	1.00	5	5	5
China, Peoples Rep. of	914	950	900	3.63	3.58	3.61	3,322	3,400	3,250
India	750	750	755	2.00	1.88	1.93	1,500	1,410	1,460
Japan	64	64	60	3.39	3.11	3.27	217	199	196
Korea, Rep. of	79	61	70	3.80	3.61	3.71	300	220	260
Nepal	40	40	40	1.00	1.00	1.00	40	40	40
Pakistan	160	160	160	1.03	1.03	1.66	165	165	265
Total	2,255	2,312	2,272	2.62	2.53	2.59	5,899	5,854	5,881
Oceania:									
Australia	3,864	4,404	3,800	1.00	2.34	1.84	3,865	10,287	7,000
New Zealand	80	80	80	5.00	5.00	5.00	400	400	400
Total	3,944	4,484	3,880	1.08	2.38	1.91	4,265	10,687	7,400
World total	56,249	59,971	57,480	2.38	2.38	2.66	133,877	142,559	152,992

[1] Years shown refer to year of harvest. Harvests of Northern Hemisphere countries are combined with those of the Southern Hemisphere which immediately follow; thus the crop harvested in the Northern Hemisphere in 1994 is combined with estimates of the Southern Hemisphere harvests, which begin late in 1994 and end early in 1995.   [2] Harvested area as far as possible.   [3] Preliminary.

FAS, Production Estimates and Crop Assessment Division, (202) 720–0888. Prepared or estimated on the basis of official statistics of foreign governments, other foreign source materials, reports of U.S. Agricultural Counselors, Attachés and Foreign Service Officers, results of office research, and related information.

Table 1-57.—Grains and grain products: Total and per capita civilian consumption as food, United States, 1995–2004

Calendar year [1]	Wheat			Rye		Rice (milled)	
	Total consumed [2]	Per capita consumption of food products		Total consumed [2]	Per capita consumption of rye flour	Total consumed [4]	Per capita consumption
		Flour [3]	Cereal				
	Million bushels	Pounds	Pounds	Million bushels	Pounds	Million cwt.	Pounds
1995	858	140	4.1	3.2	0.5	52.6	19.6
1996	896	146	4.0	3.5	0.6	53.5	19.7
1997	902	147	3.9	3.3	0.5	54.4	19.8
1998	911	143	3.9	3.6	0.6	57.2	20.6
1999	920	144	3.8	3.3	0.5	59.3	21.1
2000	951	146	3.8	3.3	0.5	61.1	21.5
2001	934	141	3.8	3.3	0.5	62.9	21.9
2002	913	137	3.7	3.3	0.4	64.7	22.3
2003	919	137	3.7	3.3	0.4	65.4	22.4
2004 [9]	905	134	3.7	3.3	0.4	67.1	22.7

Calendar year [1]	Corn						Oats		Barley	
	Total consumed [5]	Per capita consumption of food products					Total consumed [6]	Per capita consumption of oat food products	Total consumed [7]	Per capita consumption of food products [8]
		Flour and meal	Hominy and grits	Syrup	Dextrose	Starch				
	Million bushels	Pounds	Pounds	Pounds	Pounds	Pounds	Million bushels	Pounds	Million bushels	Pounds
1995	895	16.2	4.0	73.9	4.0	4.7	67	5.4	6.4	0.7
1996	933	16.5	4.5	74.2	4.0	4.9	63	5.0	6.5	0.7
1997	961	16.7	4.9	77.7	3.7	4.8	59	4.6	6.4	0.7
1998	970	17.0	5.3	79.0	3.6	4.8	57	4.4	6.4	0.7
1999	984	17.3	5.8	80.0	3.5	4.7	56.8	4.4	6.4	0.7
2000	970	17.5	6.2	78.5	3.4	4.7	56.7	4.3	6.4	0.7
2001	980	17.8	6.6	78.1	3.3	4.6	59.2	4.5	6.3	0.7
2002	976	18.1	7.0	78.3	3.3	4.6	60.2	4.5	6.4	0.7
2003	986	18.3	7.4	76.2	3.1	4.6	62.1	4.6	6.5	0.7
2004 [9]	973	18.6	7.8	75.0	3.3	4.5	63.0	4.7	6.6	0.7

[1] Data are in marketing year; for corn, September 1-August 31; for oats and barley, June 1-May 31; and rice, August 1-July 31. Wheat, rye, syrup, and sugar are in calendar year. [2] Excludes quantities used in alcoholic beverages. [3] Includes white, whole wheat, and semolina flour. [4] Does not include shipments to U.S. territories. Excludes rice used in alcoholic beverages. Includes imports and rice used in processed foods and pet foods. [5] Includes an allowance for the quantity used as hominy and grits. This series is not adjusted for trade. [6] Oats used in oatmeal, prepared breakfast foods, infant foods, and food products. [7] Malt for food, breakfast food uses, pearl barley, and flour. [8] Malt equivalent of barley food products. [9] Preliminary. Estimates of corn syrup and sugar are unofficial estimates; industry data were not reported after April 1968.

ERS, Market & Trade Economics Division, (202) 694-5290. All figures are estimates based on data from private industry sources, the U.S. Department of Commerce, the Internal Revenue Service, and other Government agencies.

Table 1-58.—Barley: Support operations, United States, 1996–2005

Marketing year beginning June 1	Income support payment rates per bushel [1]	Program price levels per bushel		Put under support [4]		Acquired by CCC under loan program [5]	Owned by CCC at end of marketing year
		Loan [2]	Target [3]	Quantity	Percentage of production		
	Dollars	Dollars	Dollars	Million bushels	Percent	Million bushels	Million bushels
1996/1997	0.33	1.55	NA	28.7	7.3	0.0	0.0
1997/1998	0.28	1.57	NA	33.3	9.3	1.5	0.0
1998/1999	0.43	1.56	NA	25.9	7.4	3.6	0.3
1999/2000	0.54	1.59	NA	13.6	4.9	1.3	0.1
2000/2001	0.52	1.62	NA	16.0	5.0	0.7	0.1
2001/2002	0.44	1.65	NA	10.6	4.2	0.2	0.0
2002/2003	0.24/0.00	1.88	2.21	10.4	4.6	0.0	0.0
2003/2004	0.24/0.00	1.88	2.21	17.9	6.4	0.1	0.0
2004/2005	0.24/0.15	1.85	2.24	8.3	3.0	0.1	0.0
2005/2006	0.24/0.15	1.85	2.24	....	....	....	....

[1] Payment rates for the 1995/96 and prior crop years were calculated according to the deficiency payment/production adjustment program provisions. Payment rates for the 1996/97 through 2001/02 crops were calculated according to the Production Flexibility Contract (PFC) program provisions of the Federal Agriculture Improvement and Reform Act of 1996 (1996 Act) and include supplemental PFC payment rates for 1998 through 2001. Payment rates for the 2002/03 and subsequent crops are calculated according to the Direct and Counter-cyclical program provisions, following enactment of the Farm Security and Rural Investment Act of 2002 (2002 Act). Payment rates are rounded to the nearest cent. Beginning with 2002/03, the first entry is the direct payment rate and the second entry is the counter-cylical payment rate. [2] The national average loan rate was also known as the price support rate prior to enactment of the 1996 Act. [3] Between the 1996/97 and 2001/02 marketing years, target prices were no longer applicable; however, target prices were reestablished under the 2002 Act. [4] Represents loans made, purchases, and purchase agreements entered into. Purchases and purchase agreements are no longer authorized for the 1996 and subsequent crops following enactment of the 1996 Act. Percentage of production is on a grain basis. Excludes quantity on which loan deficiency payments were made. [5] Acquisition of all loans forfeited during the marketing year. For 2004/05, as of October 25, 2005. NA-not applicable.

FSA, Feed Grains & Oilseeds Analysis Group, (202) 720–8838.

Table 1-59.—Sorghum: Area, yield, production, and value, United States, 1996–2005

Year	Area planted for all purposes [1]	Sorghum for grain [2]					Sorghum for silage		
		Area harvested	Yield per harvested acre	Production	Marketing year average price per cwt [3]	Value of production [3]	Area harvested	Yield per harvested acre	Production
	1,000 acres	*1,000 acres*	*Bushels*	*1,000 bushels*	*Dollars*	*1,000 dollars*	*1,000 acres*	*Tons*	*1,000 tons*
1996	13,097	11,811	67.3	795,274	4.17	1,986,316	423	11.8	4,976
1997	10,052	9,158	69.2	633,545	3.95	1,408,534	412	13.1	5,385
1998	9,626	7,723	67.3	519,933	2.97	904,123	308	11.4	3,526
1999	9,288	8,544	69.7	595,166	2.80	937,081	320	11.6	3,716
2000	9,195	7,726	60.9	470,526	3.37	845,755	278	10.5	2,932
2001	10,248	8,579	59.9	514,040	3.46	978,783	352	11.0	3,860
2002	9,589	7,125	50.6	360,713	4.14	855,140	408	9.6	3,913
2003	9,420	7,798	52.7	411,237	4.26	964,978	343	10.4	3,552
2004	7,486	6,517	69.6	453,654	3.19	843,464	352	13.6	4,776
2005 [4]	6,454	5,736	68.7	393,893	3.04	715,327	311	13.6	4,218

[1] Grain and sweet sorghum for all uses, including sirup.   [2] Includes both grain sorghum for grain, and sweet sorghum for grain or seed.   [3] Based on the reported price of grain sorghum.   [4] Preliminary.

NASS, Crops Branch, (202) 720–2127.

Table 1-60.—Sorghum grain: Stocks on and off farms, United States, 1997–2006

Year beginning previous Dec.	On farms				Off farms [1]			
	Dec. 1	Mar. 1	Jun. 1	Sep. 1	Dec. 1	Mar. 1	Jun. 1	Sep. 1
	1,000 bushels	*1,000 bushels*	*1,000 bushels*	*1,000 bushels*	*1,000 bushels*	*1,000 bushels*	*1,000 bushels*	*1,000 bushels*
1997	144,590	76,980	38,815	15,487	322,818	197,374	80,329	31,974
1998	99,625	56,760	27,200	13,700	274,244	177,916	68,944	35,203
1999	95,900	60,900	27,400	13,800	239,416	161,536	88,680	51,363
2000	90,300	51,700	27,300	12,200	259,136	173,932	99,606	53,175
2001	74,300	40,100	19,000	8,900	187,681	127,027	57,411	32,851
2002	72,400	38,100	17,300	7,400	241,477	156,007	88,178	53,573
2003	53,600	27,500	11,150	4,500	178,252	135,423	70,744	38,530
2004	45,200	21,000	7,650	3,700	190,736	137,652	72,944	29,849
2005	78,700	33,400	16,000	5,900	203,505	170,122	97,170	51,041
2006	55,000	NA	NA	NA	233,891	NA	NA	NA

[1] Includes stocks at mills, elevators, warehouses, terminals, and processors.   NA-not available.

NASS, Crops Branch, (202) 720–2127.

Table 1-61.—Sorghum: Supply and disappearance, United States, 1996–2005

Year beginning September 1	Supply			Disappearance						Ending stocks Aug. 31		
	Beginning stocks	Production	Total	Domestic use			Exports	Total disappearance	Privately owned [1]	Government	Total
				Feed and residual	Food, seed and industrial	Total					
	Million bushels	*Million bushels*	*Million bushels*	*Million bushels*	*Million bushels*	*Million bushels*	*Million bushels*	*Million bushels*	*Million bushels*	*Million bushels*	*Million bushels*
1996	18	795	814	516	45	561	205	766	47	0	47
1997	47	634	681	365	55	420	212	632	49	0	49
1998	49	520	569	262	45	307	197	504	65	0	65
1999	65	595	660	285	55	340	255	595	65	0	65
2000	65	471	536	222	35	258	237	494	41	1	42
2001	42	514	556	230	23	253	242	495	61	0	61
2002	61	361	422	170	24	194	184	379	43	0	43
2003	43	411	454	182	40	222	199	421	34	0	34
2004 [2]	34	454	487	191	55	246	184	430	57	0	57
2005 [3]	57	394	451	150	55	205	170	375	76	0	76

[1] Includes quantity under loan and farmer–owned reserve.   [2] Preliminary.   [3] Projected as of January 12, 2006, World Agricultural and Supply Demand Estimates. Totals may not add due to independent rounding.

ERS, Market and Trade Economics Division, (202) 694–5296.

Table 1-62.—Sorghum: Area, yield, and production, by States, 2003–2005

State	Area planted for all purposes			Sorghum for grain								
				Area harvested			Yield per harvested acre			Production		
	2003	2004	2005[1]	2003	2004	2005[1]	2003	2004	2005[1]	2003	2004	2005[1]
	1,000 acres	1,000 acres	1,000 acres	1,000 acres	1,000 acres	1,000 acres	Bushels	Bushels	Bushels	1,000 bushels	1,000 bushels	1,000 bushels
AL	10	10	10	6	6	6	45.0	43.0	53.0	270	258	318
AZ	17	20	23	6	6	7	90.0	95.0	95.0	540	570	665
AR	225	60	66	210	56	62	82.0	84.0	80.0	17,220	4,704	4,960
CA	18	28	26	10	12	10	90.0	90.0	90.0	900	1,080	900
CO	270	280	160	160	180	110	27.0	30.0	31.0	4,320	5,400	3,410
DE[2]	2	2	....	1	1	....	70.0	83.0	....	70	83	....
GA	55	45	40	38	25	27	47.0	47.0	50.0	1,786	1,175	1,350
IL	110	85	85	105	82	83	82.0	109.0	92.0	8,610	8,938	7,636
KS	3,550	3,200	2,750	2,900	2,900	2,600	45.0	76.0	75.0	130,500	220,400	195,000
KY	33	15	25	32	13	24	95.0	80.0	90.0	3,040	1,040	2,160
LA	170	85	90	165	80	88	85.0	65.0	99.0	14,025	5,200	8,712
MD[2]	6	5	....	3	4	....	65.0	84.0	....	195	336	....
MS	75	20	25	73	18	23	84.0	79.0	80.0	6,132	1,422	1,840
MO	215	150	135	210	145	130	77.0	108.0	76.0	16,170	15,660	9,880
NE	660	550	340	500	415	250	62.0	78.0	87.0	31,000	32,370	21,750
NM	140	140	120	62	92	97	27.0	46.0	45.0	1,674	4,232	4,365
NC	18	17	16	14	14	13	50.0	52.0	50.0	700	728	650
OK	300	270	270	250	240	240	37.0	60.0	52.0	9,250	14,400	12,480
PA	15	12	11	5	4	4	87.0	83.0	50.0	435	332	200
SC	7	7	10	5	5	7	52.0	52.0	51.0	260	260	357
SD	270	250	180	150	150	85	45.0	42.0	52.0	6,750	6,300	4,420
TN	45	20	22	40	17	20	82.0	90.0	92.0	3,280	1,530	1,840
TX	3,200	2,210	2,050	2,850	2,050	1,850	54.0	62.0	60.0	153,900	127,100	111,000
VA[2]	9	5	....	3	2	....	70.0	68.0	....	210	136	....
US	9,420	7,486	6,454	7,798	6,517	5,736	52.7	69.6	68.7	411,237	453,654	393,893

[1] Preliminary.　[2] Estimates discontinued in 2005.
NASS, Crops Branch, (202) 720–2127.

Table 1-63.—Sorghum: Utilization for silage, by States, 2003–2005

State	Silage								
	Area harvested			Yield per acre			Production		
	2003	2004	2005[1]	2003	2004	2005[1]	2003	2004	2005[1]
	1,000 acres	1,000 acres	1,000 acres	Tons	Tons	Tons	1,000 tons	1,000 tons	1,000 tons
AL	3	2	2	15.0	12.0	13.0	45	24	26
AZ	11	12	15	23.0	20.0	20.0	253	240	300
AR	3	2	2	10.0	10.0	10.0	30	20	20
CA	8	16	16	18.0	15.0	18.0	144	240	288
CO	15	19	22	14.0	14.0	13.0	210	266	286
DE[2]	1	1	....	14.0	8.0	....	14	8	....
GA	15	15	10	12.0	10.0	13.0	180	150	130
IL	3	2	1	7.0	10.0	9.0	21	20	9
KS	70	65	60	8.0	14.0	13.0	560	910	780
LA	1	1	0	11.0	10.0	....	11	10	....
MD[2]	3	1	....	10.0	8.0	....	30	8	....
MS	1	1	1	13.0	13.0	12.0	13	13	12
MO	5	4	3	8.0	10.0	6.0	40	40	18
NE	35	25	20	9.5	9.5	10.5	333	238	210
NM	10	35	14	15.0	17.0	15.0	150	595	210
NC	3	2	2	10.0	11.0	12.0	30	22	24
OK	18	15	14	10.0	8.0	7.0	180	120	98
PA	8	7	5	9.0	10.0	7.0	72	70	35
SC	2	2	3	13.0	10.0	9.0	26	20	27
SD	50	40	20	7.0	8.5	11.5	350	340	230
TN	2	2	1	18.0	16.0	15.0	36	32	15
TX	70	80	100	11.0	17.0	15.0	770	1,360	1,500
VA[2]	6	3	....	9.0	10.0	....	54	30	....
US	343	352	311	10.4	13.6	13.6	3,552	4,776	4,218

[1] Prelimary.　[2] Estimates discontinued in 2005.
NASS, Crops Branch, (202) 720–2127.

Table 1-64.—Sorghum grain: Marketing year average price and value of production, by States, crop of 2003, 2004, and 2005

State	Marketing year average price per cwt			Value of production		
	2003	2004	2005 [1]	2003	2004	2005 [1]
	Dollars	Dollars	Dollars	1,000 dollars	1,000 dollars	1,000 dollars
AL	4.10	4.00	3.25	620	578	579
AZ	4.80	4.80	5.00	1,452	1,532	1,862
AR	4.18	3.81	3.25	40,309	10,036	9,027
CA	4.65	4.20	3.80	2,344	2,540	1,915
CO	4.14	3.16	2.90	10,015	9,556	5,538
DE [2]	4.30	3.40		169	158	
GA	4.10	3.25	3.07	4,101	2,139	2,321
IL	4.64	3.37	3.20	22,372	16,868	13,684
KS	4.21	2.95	2.90	307,667	364,101	316,680
KY	4.55	3.48	3.20	7,746	2,027	3,871
LA	4.30	3.80	3.80	33,772	11,066	18,539
MD [2]	4.30	3.40		470	640	
MS	4.05	4.06	3.60	13,907	3,233	3,709
MO	4.29	3.21	3.25	38,847	28,150	17,982
NE	4.19	3.08	2.85	72,738	55,832	34,713
NM	4.45	3.30	3.70	4,172	7,821	9,044
NC	4.10	3.57	4.00	1,607	1,455	1,456
OK	4.16	3.14	3.30	21,549	25,321	23,063
PA	4.64	4.16	4.55	1,130	773	510
SC	4.11	3.25	3.20	598	473	640
SD	4.00	3.30	3.00	15,120	11,642	7,426
TN	4.30	3.87	3.35	7,898	3,316	3,452
TX	4.13	3.99	3.85	355,940	283,992	239,316
VA [2]	3.70	2.82		435	215	
US	4.26	3.19	3.04	964,978	843,464	715,327

[1] Preliminary.   [2] Estimates discontinued in 2005.
NASS, Crops Branch, (202) 720–2127.

Table 1-65.—Sorghum grain: Support operations, United States, 1996–2005

Marketing year beginning September 1	Income support payment rates per cwt [1]	Program price levels per cwt		Put under support [4]		Acquired by CCC under loan program [5]	Owned by CCC at end of marketing year
		Loan [2]	Target [3]	Quantity	Percentage of production		
	Dollars	Dollars	Dollars	Million cwt.	Percent	Millions cwt.	Million cwt.
1996/1997	0.58	3.23	NA	11.4	2.6	0.0	0.0
1997/1998	0.97	3.14	NA	9.8	2.8	0.1	0.1
1998/1999	1.21	3.11	NA	12.0	4.1	0.6	0.2
1999/2000	1.55	3.11	NA	9.6	2.9	0.5	0.0
2000/2001	1.49	3.05	NA	8.6	3.3	0.4	0.0
2001/2002	1.24	3.05	NA	9.6	3.3	0.1	0.1
2002/2003	0.63/0.00	3.54	4.54	3.7	1.8	0.0	0.0
2003/2004	0.63/0.00	3.54	4.54	3.5	1.5	0.0	0.0
2004/2005	0.63/0.48	3.48	4.59	5.5	2.1	0.1	0.0
2005/2006	0.63/0.48	3.48	4.59				

[1] Payment rates for the 1995/96 and prior crop years were calculated according to the deficiency payment/production adjustment program provisions. Payment rates for the 1996/97 through 2001/02 crops were calculated according to the Production Flexibility Contract (PFC) program provisions of the Federal Agriculture Improvement and Reform Act of 1996 (1996 Act) and include supplemental PFC payment rates for 1998 through 2001. Payment rates for the 2002/03 and subsequent crops are calculated according to the Direct and Counter-cyclical program provisions, following enactment of the Farm Security and Rural Investment Act of 2002 (2002 Act). Payment rates are rounded to the nearest cent. Beginning with 2002/03, the first entry is the direct payment rate and the second entry is the counter-cylical payment rate.   [2] The national average loan rate was also known as the price support rate prior to enactment of the 1996 Act.   [3] Between the 1996/97 and 2001/02 marketing years, target prices were no longer applicable; however, target prices were reestablished under the 2002 Act.   [4] Represents loans made, purchases, and purchase agreements entered into. Purchases and purchase agreements are no longer authorized for the 1996 and subsequent crops following enactment of the 1996 Act. Percentage of production is on a grain basis. Excludes quantity on which loan deficiency payments were made.   [5] Acquisition of all loans forfeited during the marketing year. For 2004/05, as of October 25, 2005.   NA-not applicable.
FSA, Feed Grains & Oilseeds Analysis Group, (202) 720–8838.

GRAIN AND FEED

Table 1-66.—Coarse grains: International trade, 2003–2005 [1]

Country	2002/2003	2003/2004	2004/2005 [2]
	1,000 metric tons	*1,000 Metric tons*	*1,000 metric tons*
Exports:			
Argentina	9,089	10,834	14,760
Australia	5,323	6,752	5,045
Brazil	3,897	6,191	1,456
Canada	2,443	3,474	3,250
China	8,630	7,723	7,614
South Africa	1,230	838	1,552
Russia	2,668	1,767	1,493
Ukraine	3,827	3,811	6,405
EU-25	7,659	2,342	5,105
Others	3,953	5,303	4,927
Subtotal	48,719	49,035	51,607
United States	53,964	54,077	50,495
Total	104,271	103,112	102,102
Imports:			
Algeria	1,696	1,788	2,105
Brazil	817	814	805
Canada	4,098	2,081	2,500
Chile	1,110	1,148	1,115
China	1,830	1,533	2,007
Colombia	2,402	2,214	2,360
Costa Rica	514	583	600
Dominican Republic	906	824	1,100
Ecuador	306	469	470
Egypt	4,867	3,747	5,320
Guatemala	513	513	650
Indonesia	1,633	1,436	500
Iran	2,157	2,233	3,450
Israel	1,382	2,012	1,850
Japan	20,321	19,982	19,766
Jordan	820	771	950
Korea, North	144	89	275
Korea, South	8,886	8,992	8,793
Libya	322	531	675
Malaysia	2,408	2,401	2,400
Mexico	8,766	8,873	9,045
Morocco	1,212	1,326	1,905
Peru	992	1,114	1,275
Russia	376	949	750
Saudi Arabia	8,926	7,522	7,700
South Africa	750	612	306
Syria	1,198	1,541	2,350
Taiwan	4,829	5,114	4,675
Thailand	7	6	10
Tunisia	1,046	883	1,150
Turkey	1,603	1,215	325
Venezuela	675	688	100
Zimbabwe	636	263	1,000
EU-25	6,721	7,616	3,160
United States	2,543	2,241	2,176
Subtotal	97,412	94,124	93,618
Other Countries	5,690	5,898	6,522
Unaccounted	1,169	3,090	1,962
World Total	102,683	104,271	102,102

[1] Year beginning Oct 1.   [2] Preliminary.

FAS, Grain and Feed Division, (202) 720–6219. www.fas.usda.gov/grain/default.html. Prepared or estimated on the basis of official statistics frpm foreign governments, other foreign source materials, reports of U.S. Agricultural Counselors, Attachés and Foreign Service Officers, results of office research, and related information.

Table 1-67.—Commercial feeds: Disappearance for feed, United States, 1995–2004

Year beginning October	Oilseed cake and meal						Animal protein			
	Soybean	Cotton-seed	Linseed	Peanut [1]	Sun-flower	Total	Tankage and meat meal	Fish meal	Dried milk [2]	Total
	1,000 tons	*1,000 tons*	*1,000 tons*	*1,000 tons*	*1,000 tons*	*1,000 tons*	*1,000 tons*	*1,000 tons*	*1,000 tons*	*1,000 tons*
1995	26,524	2,960	129	181	478	30,272	2,536	290	420	3,246
1996	27,203	3,113	149	141	462	31,068	2,803	308	428	3,538
1997	28,611	2,956	185	95	531	32,378	2,504	270	411	3,184
1998	30,099	2,533	169	94	635	33,530	2,694	326	273	3,293
1999	30,080	2,908	192	140	582	33,902	2,456	265	306	3,026
2000	31,264	2,855	196	113	496	34,924	2,168	246	303	2,717
2001	32,568	3,340	124	151	395	36,578	1,938	274	318	2,530
2002	32,074	2,691	178	178	256	35,377	1,878	252	340	2,470
2003	31,449	2,786	197	122	349	34,903	2,320	233	472	3,025
2004 [3]	33,563	3,473	206	95	147	37,484	2,101	124	306	2,530

Year beginning October	Mill products [4]					Total commercial feeds
	Wheat millfeeds	Gluten feed and meal [5]	Rice millfeeds	Alfalfa meal	Total	
	1,000 tons	*1,000 tons*	*1,000 tons*	*1,000 tons*	*1,000 tons*	*1,000 tons*
1995	7,373	881	664	255	9,173	42,690
1996	7,280	1,833	617	248	9,978	44,584
1997	7,030	1,914	603	276	9,823	45,385
1998	7,026	1,349	655	225	9,254	46,077
1999	7,324	1,199	716	NA	9,239	46,167
2000	7,303	1,432	689	NA	9,424	49,782
2001	6,895	1,475	678	NA	9,049	50,686
2002	6,948	2,275	688	NA	9,912	50,229
2003	6,755	2,421	594	NA	9,771	50,724
2004 [3]	6,684	2,318	614	NA	9,616	52,160

[1] Year beginning August 1. [2] Includes dried skim milk, and whey for feed, but does not include any milk products fed on farms. [3] Preliminary. [4] Other mill products that are not listed include screenings, hominy, and oats feed etc., for which no statistics are available. [5] Adjusted for export data. NA-not available.

ERS, Market and Trade Economics Division, (202) 694–5290.

Table 1-68.—High-protein feeds: Quantity for feeding, high-protein animal units, quantity per animal unit, and prices, United States, 1995–2004

Year beginning October	Quantity for feeding [1]						High-protein animal units	Quantity per animal unit	High protein feed prices
	Oilseed meal			Animal protein	Grain protein [3]	Total			
	Soybean meal	Other oilseed meals [2]	Total						
	1,000 tons	*1,000 tons*	*1,000 tons*	*1,000 tons*	*1,000 tons*	*1,000 tons*	*Million units*	*Pounds*	*Index numbers 1992=100*
1995	29,176	3,469	32,646	3,547	524	36,717	132.2	555.510	122
1996	29,923	3,573	33,496	3,893	1,092	38,481	133.1	578.171	140
1997	31,472	3,470	34,942	3,480	1,140	39,561	136.2	580.782	104
1998	33,109	3,154	36,263	3,737	803	40,803	137.1	595.376	73
1999	33,088	3,521	36,609	3,384	714	40,707	137.5	591.951	87
2000	34,390	3,371	37,762	3,013	853	41,627	138.4	601.593	91
2001	35,825	3,715	39,540	2,775	879	43,194	139.6	618.909	89
2002	35,281	3,058	38,339	2,683	1,355	42,377	138.4	612.533	95
2003	34,594	3,186	37,780	3,232	1,442	42,454	139.3	609.580	131
2004	36,919	3,628	40,547	2,779	1,381	44,707	141.2	633.355	96

[1] In terms of 44 percent protein soybean meal equivalent. [2] Includes cottonseed, linseed, peanut meal, and sunflower meal. [3] Beginning 1974, adjusted for exports of corn gluten feed and meal.

ERS, Market and Trade Economics Division (202) 694–5290.

Table 1-69.—Feed concentrates: Fed to livestock and poultry, 1995–2004

Year beginning October	Feed grains				Wheat [2]	Rye [2]	By-product feeds [3]	Total concentrates	Grain consuming animal units	Concentrates fed per grain-consuming animal unit
	Corn [1]	Sorghum [1]	Oats [2] and barley [2]	Total						
	Million tons	*Million tons*	*Million tons*	*Million tons*	*Million tons*	*Million tons*	*Million tons*	*Million tons*	*Millions*	*Tons*
1995	131.4	8.3	7.6	147.3	6.8	0.2	49.4	203.6	85.0	2.40
1996	147.8	14.4	6.9	169.1	8.5	0.1	52.2	229.9	85.3	2.69
1997	153.5	10.2	7.3	171.0	9.7	0.1	53.1	233.9	88.0	2.66
1998	153.1	7.3	6.8	167.3	7.0	0.1	53.6	228.0	88.1	2.59
1999	158.6	8.0	6.7	173.3	9.9	0.1	54.0	237.3	89.0	2.67
2000	163.6	6.2	6.1	175.9	6.6	0.1	57.7	240.3	89.4	2.69
2001	164.2	6.4	4.9	175.5	3.9	0.1	58.5	238.0	89.8	2.65
2002	155.8	4.8	5.2	165.7	7.4	0.1	57.9	231.0	88.3	2.62
2003	162.3	5.1	4.6	172.0	4.6	0.1	58.5	235.1	89.3	2.63
2004	172.5	5.4	4.6	182.5	5.7	0.1	60.4	248.6	90.2	2.76

[1] Marketing year beginning Sept. 1. [2] Marketing year beginning June 1. [3] Oilseed meals, animal protein feeds, mill by-products, and mineral supplements.

ERS, Market and Trade Economics Division (202) 694–5290.

Table 1-70.—Feed: Consumed per head and per unit of production, by class of livestock or poultry, with quantity expressed in equivalent feeding value of corn, 1995–2004

Year beginning October	Dairy cattle			Beef cattle				Sheep and lambs	
	Milk cows		Other dairy cattle per head	Cattle on feed per head Jan. 1 [1]	Other beef cattle per head	All beef cattle per head	Cattle and calves per 100 pounds produced [2]	Per head	Per 100 pounds produced [3]
	Per head	Per 100 pounds milk produced							
	Pounds	Pounds	Pounds	Pounds	Pounds	Pounds	Pounds	Pounds	Pounds
1995	12,408	76	6,444	8,929	5,284	5,808	1,292	1,266	1,610
1996	13,027	78	6,548	9,834	5,316	5,991	1,359	1,277	1,631
1997	12,957	76	6,537	9,732	5,312	6,007	1,328	1,276	1,522
1998	12,826	73	6,515	9,541	5,306	5,960	1,292	1,273	1,527
1999	13,004	71	6,544	9,801	5,315	6,057	1,271	1,277	1,539
2000	13,102	73	6,561	9,944	5,320	6,105	1,254	1,278	1,583
2001	13,040	70	6,551	9,853	5,317	6,079	1,255	1,277	1,556
2002	12,970	70	6,539	9,751	5,313	6,021	1,240	1,276	1,533
2003	13,014	68	6,546	9,816	5,315	6,074	1,241	1,277	1,533
2004 [4]	13,271	68	6,589	10,192	5,329	6,137	1,289	1,282	1,566

Year beginning October	Poultry								Hogs per 100 pounds produced	Horses and mules [2] years and over per head
	Hens and pullets		Chickens raised		Broilers produced		Turkeys raised			
	Per head Jan. 1	Per 100 eggs	Per head	Per 100 pounds live weight	Per head	Per 100 pounds produced	Per head	Per 100 pounds produced		
	Pounds	Pounds	Pounds	Pounds	Pounds	Pounds	Pounds	Pounds	Pounds	Pounds
1995	117	45	28	827	10.2	225	87	386	545	3,735
1996	129	51	31	976	11.4	242	96	401	619	3,856
1997	128	51	30	980	11.3	237	95	381	639	3,842
1998	125	50	30	945	11.0	232	93	361	590	3,816
1999	129	51	31	875	11.4	230	96	376	587	3,851
2000	131	52	31	867	11.6	233	97	380	594	3,870
2001	130	51	31	890	11.5	230	96	369	593	3,858
2002	128	50	31	847	11.3	219	95	349	574	3,844
2003	129	50	31	911	11.4	225	96	340	584	3,853
2004 [4]	135	52	32	932	11.9	233	100	352	603	3,903

[1] Feed consumed by all cattle divided by the number on feed Jan. 1.   [2] Feed for all cattle, except milk cows, divided by the net live-weight production of cattle and calves. It includes the growth on dairy heifers and calves as well as all beef cattle.   [3] Including wool produced.   [4] Preliminary.

ERS, Market and Trade Economics Division, (202) 694–5290.

Table 1-71.—Feed: Consumed by livestock and poultry, by type of feed, with quantity expressed in equivalent feeding value of corn, 1995–2004

Year beginning October	Concentrates	Harvested roughage	Pasture	Total
	Million tons	Million tons	Million	Million tons
1995	218	80	186	483
1996	244	81	179	504
1997	248	81	173	503
1998	243	81	173	497
1999	253	84	166	502
2000	259	87	159	505
2001	257	85	160	502
2002	250	83	162	495
2003	255	85	155	495
2004 [1]	268	86	157	512

[1] Preliminary.

ERS, Market and Trade Economics Division, (202) 694–5290.

Table 1-72.—Animal units fed: Grain-consuming, roughage-consuming, and grain-and-roughage-consuming, United States, 1996–2005 [1]

Year beginning October	Grain-consuming [2]	Roughage-consuming [3]	Grain and roughage-consuming [4]
	1,000 units	*1,000 units*	*1,000 units*
1996	85,340	76,420	79,363
1997	87,955	74,861	79,417
1998	88,145	74,493	79,248
1999	88,987	73,241	78,793
2000	89,436	72,442	78,462
2001	89,771	72,083	78,380
2002	88,291	72,048	77,791
2003	89,318	70,721	77,355
2004	90,236	71,551	78,188
2005 [5]	91,921	72,972	79,690

[1] Index series based on average feeding rates for years 1969–71. In calculations for the feeding years 1969 to date, cattle numbers used are the new categories shown in the Livestock and Poultry Inventory, published by NASS, USDA. [2] Livestock and poultry numbers weighted by all concentrates consumed. [3] Livestock and poultry numbers weighted by all roughage (including pasture) consumed. [4] Livestock and poultry numbers weighted by all feed (including pasture) fed to livestock. [5] Preliminary.
ERS, Market and Trade Economics Division, (202) 694–5290.

Table 1-73.—Feed grains: Average price, selected markets and grades, 1996–2005 [1]

Calendar year	Kansas City			Minneapolis			
	Corn, No. 2 Yellow (truck)	Corn, No. 2 White (truck)	Sorghum, No. 2 Yellow (truck)	Corn, No. 2 Yellow	Barley, No. 3 or Better malting	Duluth Barley, No. 2 Feed	Oats, No. 2 White
	Dollars per bushel	*Dollars per bushel*	*Dollars per cwt.*	*Dollars per bushel*	*Dollars per bushel*	*Dollars per bushel*	*Dollars per bushel*
1996	4.12	4.07	6.66	3.74	3.69	2.67	2.28
1997	2.84	3.09	4.54	2.65	3.18	2.32	2.03
1998	2.49	2.93	4.11	2.39	2.50	1.86	1.70
1999	2.01	2.42	3.29	1.88	2.30	1.86	1.33
2000	1.93	1.94	3.19	1.79	....	1.22	1.28
2001	1.85	....	4.03	....	....	....	....
2002	2.13	2.51	4.27	2.11	2.85	1.70	1.82
2003	2.36	2.58	4.07	2.22	3.34	1.91	1.82
2004	2.40	2.52	4.23	2.38	2.55	1.79	1.71
2005	1.87	2.19	3.34	1.79	2.53	NA	1.84

Calendar year	Omaha: Corn, No. 2 Yellow (truck)	Chicago: Corn, No. 2 Yellow	Texas High Plains: Sorghum, No. 2 Yellow	Memphis		St. Louis: Corn, No. 2 Yellow (truck)
				Corn, No. 2 Yellow	Barley, No. 2 Western	
	Dollars per bushel	*Dollars per bushel*	*Dollars per cwt.*	*Dollars per bushel*	*Dollars per bushel*	*Dollars per cwt.*
1996	3.87	3.97	7.30	4.02	7.45	4.06
1997	2.70	2.84	5.02	2.88	7.84	2.90
1998	2.36	2.56	4.72	2.59	6.68	2.62
1999	1.88	2.15	3.79	2.13	....	2.17
2000	1.82	1.97	3.51	2.01	....	2.00
2001	....	1.98	....	2.03	....	....
2002	2.13	2.24	4.27	2.29	....	2.33
2003	2.24	2.34	3.94	2.42	....	2.38
2004	2.36	2.48	4.70	2.55	....	2.64
2005	1.77	1.97	3.98	2.11	....	2.01

[1] Simple average of daily prices.
AMS, Livestock and Grain Market News Branch, (202) 720–6231.

Table 1-74.—Feedstuffs: Average price per ton bulk, in wholesale lots, at leading markets, 1996–2005

Year beginning October	Soybean meal 44% protein Decatur	Soybean meal 49–50% protein Decatur	Cottonseed meal 41% protein Kansas City	Cottonseed meal 41% protein Memphis	Linseed meal 34% protein Minneapolis	Meat meal 50% protein Kansas City	Fish meal 60% protein Gulf Coast	Wheat bran Kansas City	Wheat middlings Minneapolis
	Dollars per ton	Dollars per ton	Dollars per ton	Dollars per ton	Dollars per ton	Dollars per ton	Dollars per ton	Dollars per ton	Dollars per ton
1996	260.40	279.70	207.50	91.40	158.70	271.80	502.60	88.30	80.20
1997	175.00	196.10	162.60	144.00	117.50	180.90	545.60	73.30	61.20
1998	132.00	138.50	132.20	109.60	84.50	138.10	387.50	55.90	60.90
1999	131.99	138.55	130.20	109.60	84.50	138.10	387.80	55.88	49.58
2000	160.03	168.10	146.50	130.70	103.10	166.50	326.40	53.37	45.81
2001	165.21	173.60	165.00	142.70	121.90	166.50	358.20	62.93	50.81
2002	153.82	167.72	160.10	136.20	119.20	166.00	460.00	59.74	58.42
2003	115.60	208.95	172.52	152.24	134.31	196.30	487.50	65.07	56.05
2004	(1)	237.30	193.58	167.68	148.09	190.63	524.97	67.82	64.19
2005	(1)	188.17	156.59	128.89	115.70	169.19	....	54.34	44.53

Year beginning October	Wheat shorts or middlings Kansas City	Wheat millrun Portland	Gluten feed 21% protein Illinois Points	Hominy feed Illinois Points	Distillers' dried grains Lawrence-burg	Brewers' dried grains Columbus	Alfalfa meal Dehydrated, 17% protein Kansas City	Alfalfa meal Sun-cured Kansas City	Blackstrap molasses New Orleans
	Dollars per ton	Dollars per ton	Dollars per ton	Dollars per ton	Dollars per ton	Dollars per ton	Dollars per ton	Dollars per ton	Dollars per ton
1996	80.30	95.70	90.20	88.90	139.80	134.60	142.70	130.9	60.10
1997	73.30	74.30	67.20	78.90	105.80	107.20	126.80	115	51.50
1998	55.90	54.30	60.40	62.20	85.20	84.20	101.50	92.00	35.50
1999	55.88	54.34	60.41	67.19	85.15	84.19	101.55	91.99	35.48
2000	57.82	60.92	53.64	58.71	79.90	90.94	97.59	86.58	38.72
2001	62.88	63.25	60.55	55.02	80.62	94.00	139.06	130.38	63.16
2002	59.77	75.95	59.63	63.23	80.19	94.00	154.05	134.34	68.63
2003	65.27	85.49	70.15	72.66	93.13	94.95	138.61	122.48	58.00
2004	67.82	85.00	68.83	77.02	106.04	(1)	121.35	109.26	57.28
2005	54.23	74.72	68.17	50.50	75.47	(1)	135.83	110.57	NA

[1] Discontinued.

AMS, Livestock and Grain Market News Branch, (202) 720–6231.

GRAIN AND FEED

Table 1-75.—Proso millet: Area, yield, production, and value, United States, 1999-2005

Year	Area		Yield per harvested acre	Production	Marketing year average price per bushel received by farmers	Value of production
	Planted	Harvested				
	1,000 acres	1,000 acres	Bushels	1,000 bushels	Dollars	1,000 dollars
1999	600	540	33.2	17,910	2.12	38,033
2000	440	370	19.8	7,320	4.79	35,034
2001	650	585	33.2	19,405	2.02	39,109
2002	520	275	13.3	3,668	7.22	26,462
2003	730	620	18.5	11,450	2.95	33,730
2004	710	595	25.3	15,065	2.83	42,611
2005 [1]	565	515	26.3	13,545	3.35	45,117

[1] Preliminary.
NASS, Crops Branch, (202) 720-2127.

Table 1-76.—Proso millet: Area, yield, and production, by States, 2003-2005

State	Area planted			Area harvested		
	2003	2004	2005 [1]	2003	2004	2005 [1]
	1,000 acres	1,000 acres	1,000 acres	1,000 acres	1,000 acres	1,000 acres
CO	320	370	290	285	330	275
NE	200	160	135	170	135	125
SD	210	180	140	165	130	115
US	730	710	565	620	595	515

State	Yield per acre			Production		
	2003	2004	2005 [1]	2003	2004	2005 [1]
	Bushels	Bushels	Bushels	1,000 bushels	1,000 bushels	1,000 bushels
CO	19.0	24.0	20.0	5,415	7,920	5,500
NE	19.0	25.0	34.0	3,230	3,375	4,250
SD	17.0	29.0	33.0	2,805	3,770	3,795
US	18.5	25.3	26.3	11,450	15,065	13,545

[1] Preliminary.
NASS, Crops Branch, (202) 720-2127.

Table 1-77.—Proso millet: Marketing year average price and value, by States, crop of 2003, 2004, and 2005

State	Marketing year average price per bushel			Value of production		
	2003	2004	2005 [1]	2003	2004	2005 [1]
	Dollars	Dollars	Dollars	1,000 dollars	1,000 dollars	1,000 dollars
CO	2.70	2.70	3.10	14,621	21,384	17,050
NE	3.05	3.05	3.30	9,852	10,294	14,025
SD	3.30	2.90	3.70	9,257	10,933	14,042
US	2.95	2.83	3.35	33,730	42,611	45,117

[1] Preliminary.
NASS, Crops Branch, (202) 720-2127.

CHAPTER II

STATISTICS OF COTTON, TOBACCO, SUGAR CROPS, AND HONEY

In addition to tables on cotton, tobacco, sugar, and honey, this chapter includes tables on fibers other than cotton and syrups. Cottonseed data, however, are in the following chapter on oilseeds, fats, and oils.

Table 2-1.—Cotton: Area, yield, production, market year average price, and value, United States, 1996–2005

Year	Area		Yield per harvested acre	Production	Marketing year average price per pound received by farmers	Value of production
	Planted	Harvested				
	1,000 acres	1,000 acres	Pounds	1,000 bales [1]	Cents	1,000 dollars
1996	14,652.5	12,888.1	705	18,942.0	70.5	6,408,144
1997	13,898.0	13,406.0	673	18,793.0	66.2	5,975,585
1998	13,392.5	10,683.6	625	13,918.2	61.7	4,119,911
1999	14,873.5	13,424.9	607	16,968.0	46.8	3,809,560
2000	15,517.2	13,053.0	632	17,188.3	51.6	4,260,417
2001	15,768.5	13,827.7	705	20,302.8	32.0	3,121,848
2002	13,957.9	12,416.6	665	17,208.6	45.7	3,777,132
2003	13,479.6	12,003.4	730	18,255.2	63.0	5,516,761
2004	13,658.6	13,057.0	855	23,250.7	43.5	4,853,730
2005 [2]	14,195.4	13,702.6	831	23,719.0	49.0	5,574,119

[1] 480-pound net weight bales.   [2] Preliminary.
NASS, Crops Branch, (202) 720–2127.

Table 2-2.—Cotton: Area, yield, and production, by type, State, and United States, 2003–2005

State	Area planted			Area harvested			Yield per harvested acre			Production [1]		
	2003	2004	2005 [2]	2003	2004	2005 [2]	2003	2004	2005 [2]	2003	2004	2005 [2]
	1,000 acres	1,000 acres	1,000 acres	1,000 acres	1,000 acres	1,000 acres	Pounds	Pounds	Pounds	1,000 bales [3]	1,000 bales [3]	1,000 bales [3]
Upland:												
AL	525.0	550.0	550.0	510.0	540.0	545.0	772	724	749	820.0	814.0	850.0
AZ	215.0	240.0	230.0	213.0	238.0	229.0	1,239	1,458	1,300	550.0	723.0	620.0
AR	980.0	910.0	1,050.0	945.0	900.0	1,040.0	916	1,114	1,011	1,804.0	2,089.0	2,190.0
CA	550.0	560.0	430.0	545.0	557.0	428.0	1,317	1,543	1,178	1,495.0	1,790.0	1,050.0
FL	94.0	89.0	86.0	92.0	87.0	85.0	610	601	728	117.0	109.0	129.0
GA	1,300.0	1,290.0	1,220.0	1,290.0	1,280.0	1,210.0	785	674	853	2,110.0	1,797.0	2,150.0
KS	90.0	85.0	74.0	80.0	80.0	66.0	537	424	655	89.5	70.7	90.0
LA	525.0	500.0	610.0	510.0	490.0	600.0	967	867	896	1,027.0	885.0	1,120.0
MS	1,110.0	1,110.0	1,210.0	1,090.0	1,100.0	1,200.0	934	1,024	864	2,120.0	2,346.0	2,160.0
MO	400.0	380.0	440.0	390.0	378.0	438.0	862	1,054	970	700.0	830.0	885.0
NM	53.0	68.0	56.0	38.0	64.0	51.0	884	848	941	70.0	113.0	100.0
NC	810.0	730.0	815.0	770.0	725.0	810.0	646	900	847	1,037.0	1,360.0	1,430.0
OK	180.0	220.0	255.0	170.0	200.0	240.0	616	727	730	218.0	303.0	365.0
SC	220.0	215.0	266.0	218.0	214.0	265.0	718	875	761	326.0	390.0	420.0
TN	560.0	530.0	640.0	530.0	525.0	635.0	806	900	847	890.0	984.0	1,120.0
TX	5,600.0	5,850.0	5,900.0	4,350.0	5,350.0	5,500.0	478	694	716	4,330.0	7,740.0	8,200.0
VA	89.0	82.0	93.0	85.0	81.0	92.0	674	956	965	119.4	161.4	185.0
US	13,301.0	13,409.0	13,925.0	11,826.0	12,809.0	13,434.0	723	843	824	17,822.9	22,505.1	23,064.0
American-Pima:												
AZ	2.5	3.0	4.1	2.4	3.0	4.1	920	896	937	4.6	5.6	8.0
CA	150.0	215.0	230.0	149.0	214.0	229.0	1,194	1,532	1,216	370.5	683.0	580.0
NM	6.1	10.6	11.5	6.0	10.5	11.5	1,056	869	918	13.2	19.0	22.0
TX	20.0	21.0	24.8	20.0	20.5	24.0	1,056	890	900	44.0	38.0	45.0
US	178.6	249.6	270.4	177.4	248.0	268.6	1,170	1,443	1,171	432.3	745.6	655.0
US, all ..	13,479.6	13,658.6	14,195.4	12,003.4	13,057.0	13,702.6	730	855	831	18,255.2	23,250.7	23,719.0

[1] Production ginned and to be ginned.   [2] Preliminary.   [3] 480-pound net weight bale.
NASS, Crops Branch, (202) 720–2127.

Table 2-3.—Cotton: Marketing year average price per pound, and value, by State and United States, crop of 2003, 2004, and 2005

State	Marketing year average price per pound			Value of production		
	2003	2004	2005 [1]	2003	2004	2005 [1]
	Dollars	Dollars	Dollars	1,000 dollars	1,000 dollars	1,000 dollars
Upland:						
AL	0.596	0.406	0.495	234,586	158,632	201,960
AZ	0.664	0.444	0.531	175,296	154,086	158,026
AR	0.625	0.411	0.470	541,200	412,118	494,064
CA	0.745	0.516	0.600	534,612	443,347	302,400
FL	0.655	0.464	0.510	36,785	24,276	31,579
GA	0.612	0.428	0.491	619,834	369,176	506,712
KS	0.606	0.396	0.465	26,034	13,439	20,088
LA	0.609	0.414	0.469	300,213	175,867	252,134
MS	0.604	0.410	0.452	614,630	461,693	468,634
MO	0.600	0.410	0.440	201,600	163,344	186,912
NM	0.563	0.480	0.540	18,917	26,035	25,920
NC	0.647	0.437	0.454	322,051	285,274	311,626
OK	0.606	0.396	0.460	63,412	57,594	80,592
SC	0.623	0.430	0.490	97,487	80,496	98,784
TN	0.570	0.405	0.471	243,504	191,290	253,210
TX	0.577	0.402	0.450	1,199,237	1,493,510	1,771,200
VA	0.640	0.380	0.455	36,680	29,439	40,404
US	0.618	0.416	0.469	5,266,078	4,539,616	5,204,245
American-Pima:						
AZ	1.170	0.837	1.140	2,583	2,250	4,378
CA	1.230	0.882	1.190	218,743	289,155	331,296
NM	1.100	0.830	1.050	6,970	7,570	11,088
TX	1.060	0.830	1.070	22,387	15,139	23,112
US	1.210	0.878	1.180	250,683	314,114	369,874
US, all	0.630	0.435	0.490	5,516,761	4,853,730	5,574,119

[1] Preliminary.
NASS, Crops Branch, (202) 720-2127.

Table 2-4.—Cotton, American upland: Support operations, United States, 1996–2005

Marketing Year beginning August 1	Income support payment rates per pound [1]	Program price levels per pound		Put under Loan [4]		Acquired by CCC under loan program [5]	Owned by CCC at end of marketing year
		Loan [2]	Target [3]	Quantity	Percentage of production		
	Cents	Cents	Cents	1,000 bales [6]	Percent	1,000 bales [6]	1,000 bales [6]
1996/97	8.88	51.92	NA	3,340	18.1	0	0
1997/98	7.63	51.92	NA	4,281	23.5	1.3	0
1998/99	12.24	51.92	NA	4,724	36.8	31	3
1999/2000	15.76	51.92	NA	8,721	54.9	0	1
2000/2001	15.21	51.92	NA	8,837	52.6	89	5
2001/2002	12.66	51.92	NA	13,655	69.7	257	2
2002/2003	6.67/13.73	52.00	74.20	12,740	77.1	44	106
2003/2004	6.67/3.93	52.00	74.20	10,345	58.0	4	0
2004/2005	6.67/13.73	52.00	74.20	17,091	79.1	0	0
2005/2006	6.67/---	52.00	74.20				

[1] Payment rates for the 1995/96 & prior crop years were calculated according to the provisions of the now defunct deficiency payment/production adjustment program. Payment rates for the 1996/97 through 2001/02 crops were calculated according to the provisions of the Production Flexibility Contract (PFC) program, following enactment of the Federal Agriculture Improvement and Reform Act of 1996 (1996 Act) and included supplemental PFC payment rates for 1998 through 2001. Payment rates for the 2002/03 and subsequent crops are calculated according to the provisions of the Direct Payment program, following enactment of the Farm Security and Rural Investment Act of 2002 (2002 Act) and includes a PFC payment rate for 2002 only. Payment rates are rounded to the nearest thousandth of a cent. [2] The national average loan rate was also known as the price support rate prior to enactment of the 1996 Act. [3] Between the 1996/97 and 2001/02 marketing years, target prices were no longer applicable; however, with enactment of the 2002 Act, target prices were reestablished. [4] Represents loans made, purchases, and purchase agreements entered into. Purchases and purchase agreements are no longer authorized for the 1996 and subsequent crops following enactment of the 1996 Act. [5] Acquisitions from the crop harvested in the year indicated. [6] Running bales. [7] Less than 500 bales. NA-not applicable.

FSA, Fiber Analysis Group, (202) 720-7954.

Table 2-5.—Cotton: Area, yield, and production in specified countries, 2002/2003–2004/2005 [1]

Continent and country	Area [2]			Yield per hectare			Production		
	2002/ 2003	2003/ 2004	2004/ 2005 [3]	2002/ 2003	2003/ 2004	2004/ 2005 [3]	2002/ 2003	2003/ 2004	2004/ 2005 [3]
	1,000 hec- tares	1,000 hec- tares	1,000 hec- tares	Kilo- grams	Kilo- grams	Kilo- grams	1,000 metric tons[3]	1,000 metric tons[3]	1,000 metric tons[3]
North America:									
Mexico	42	62	105	1,071	1,258	1,295	45	78	136
United States	5,025	4,858	5,284	746	818	958	3,747	3,975	5,062
Total	5,067	4,920	5,389	748	824	965	3,792	4,053	5,198
South America:									
Argentina	148	250	374	426	448	393	63	112	147
Bolivia	3	10	20	667	700	500	2	7	10
Brazil	735	1,100	1,172	1,152	1,191	1,096	847	1,310	1,285
Colombia	51	64	65	647	781	800	33	50	52
Ecuador	1	3	3	500	667	667	1	2	2
Paraguay	170	280	215	353	393	302	60	110	65
Peru	65	83	100	738	723	700	48	60	70
Venezuela	13	10	10	1,154	1,300	1,300	15	13	13
Total	1,186	1,800	1,959	901	924	839	1,069	1,664	1,644
Central America:									
Costa Rica	1	1	1	218	218	218	0	0	0
El Salvador	1	1	1	218	218	218	0	0	0
Guatemala	2	2	2	327	327	327	1	1	1
Honduras	2	2	2	436	436	436	1	1	1
Nicaragua	4	4	4	544	544	544	2	2	2
Total	10	10	10	414	414	414	4	4	4
Carribean:									
Cuba	4	4	4	218	218	218	1	1	1
Dominican Rep	4	4	4	163	163	163	1	1	1
Haiti	7	7	7	156	156	156	1	1	1
Total	15	15	15	174	174	174	3	3	3
European Union:									
Cyprus	0	0	0	0	0	0	0	0	0
Greece	355	363	375	1,051	917	1,045	373	333	392
Italy	0	0	0	0	0	0	0	0	0
Portugal	1	1	1	218	218	218	0	0	0
Spain	85	92	90	1,127	1,017	1,209	96	94	109
Total	441	456	466	1,064	936	1,075	469	427	501
Other Europe:									
Albania	1	1	1	218	218	218	0	0	0
Bulgaria	6	2	2	733	1,100	1,100	4	2	2
Total	7	3	3	660	806	806	5	2	2
Former USSR:									
Azerbaijan, Rep	65	60	80	468	650	613	30	39	49
Kazakhstan, Rep	165	185	216	686	649	685	113	120	148
Kyrgyzstan, Rep	35	35	46	1,243	1,143	874	44	40	40
Tajikistan, Rep	265	285	290	595	596	600	158	170	174
Turkmenistan	490	480	550	307	427	376	150	205	207
Uzbekistan, Rep	1,421	1,431	1,456	705	624	777	1,002	893	1,132
Total	2,441	2,476	2,638	613	592	663	1,497	1,467	1,750
Middle East:									
Iran	160	145	167	626	779	802	100	113	134
Iraq	20	20	20	325	325	325	7	7	7
Israel	13	10	14	1,338	1,780	1,857	17	18	26
Syria	186	200	234	1,317	1,415	1,487	245	283	348
Turkey	700	710	700	1,300	1,258	1,291	910	893	904
Yemen	40	40	40	350	350	350	14	14	14
Total	1,119	1,125	1,175	1,155	1,180	1,219	1,293	1,327	1,433

See footnotes at end of table.

Table 2-5.—Cotton: Area, yield, and production in specified countries, 2002/2003–2004/2005 [1]—Continued

Continent and country	Area			Yield per hectare			Bales		
	2002/ 2003	2003/ 2004	2004/ 2005 [2]	2002/ 2003	2003/ 2004	2004/ 2005 [2]	2002/ 2003	2003/ 2004	2004/ 2005 [2]
	1,000 hec- tares	*1,000 hec- tares*	*1,000 hec- tares*	*Kilo- grams*	*Kilo- grams*	*Kilo- grams*	*1,000 metric tons[3]*	*1,000 metric tons[3]*	*1,000 metric tons[3]*
Africa:									
Angola	2	2	2	545	545	545	1	1	1
Benin	303	314	300	460	440	486	139	138	146
Burkina	405	460	550	403	457	467	163	210	257
Cameroon	200	220	220	463	495	495	93	109	109
Central Afric	50	35	35	152	187	187	8	7	7
Chad	425	300	425	166	163	192	71	49	82
Congo Dem.	30	30	30	109	109	109	3	3	3
Cote d'Ivoir	322	230	300	473	379	435	152	87	131
Egypt	302	218	307	944	919	945	285	200	290
Ethiopia	113	113	60	177	116	254	20	13	15
Ghana	30	20	25	232	272	261	7	5	7
Guinea	27	25	14	444	348	218	12	9	3
Kenya	50	50	37	87	87	118	4	4	4
Madagascar	14	20	30	280	403	435	4	8	13
Malawi	47	47	47	222	371	417	10	17	20
Morocco	1	1	1	218	218	218	0	0	0
Mozambique	155	190	155	133	175	155	21	33	24
Niger	5	5	5	218	218	218	1	1	1
Nigeria	340	375	380	250	241	241	85	90	91
Senegal	42	45	42	337	484	518	14	22	22
Somalia	12	12	12	127	127	127	2	2	2
South Africa	30	41	26	508	664	796	15	27	21
Sudan	155	179	210	527	377	544	82	67	114
Tanzania	387	387	500	158	132	229	61	51	114
Togo	160	185	200	435	382	354	70	71	71
Tunisia	2	2	2	1,089	1,089	1,089	2	2	2
Uganda	250	250	400	83	109	109	21	27	44
Zambia	240	300	375	218	236	203	52	71	76
Zimbabwe	330	360	330	257	302	231	85	109	76
Total	**4,429**	**4,416**	**5,020**	**335**	**325**	**348**	**1,483**	**1,435**	**1,745**
Asia:									
Afghanistan	50	50	50	370	370	370	19	19	19
Bangladesh	48	49	44	313	311	297	15	15	13
Burma	264	270	300	178	207	196	47	56	59
China, People	4,184	5,110	5,690	1,176	950	1,110	4,921	4,855	6,314
India	7,667	7,785	9,000	301	386	460	2,308	3,005	4,137
Indonesia	12	10	10	671	697	697	8	7	7
Korea, Democr	19	19	19	630	630	630	12	12	12
Korea, Republ	1	1	1	218	218	218	0	0	0
Mali	420	516	533	428	506	449	180	261	239
Pakistan	2,796	3,092	3,190	607	546	771	1,698	1,687	2,460
Philippines	2	5	5	436	392	479	1	2	2
Sri Lanka	5	5	5	218	218	218	1	1	1
Thailand	11	11	12	1,287	1,207	1,034	14	13	12
Vietnam	34	28	27	333	443	387	11	12	10
Total	**15,513**	**16,951**	**18,886**	**595**	**587**	**704**	**9,235**	**9,946**	**13,287**
Oceania:									
Australia	220	196	314	1,664	1,888	2,080	366	370	653
World total	**30,450**	**32,370**	**35,880**	**631**	**639**	**731**	**19,227**	**20,684**	**26,228**

[1] Harvest season beginning Aug. 1.   [2] Preliminary.   [3] 480-pound net weight.   [4] Less than 500 thousand.

FAS, Production Estimates and Crop Assessment Division, (202) 720–0888. Prepared or estimated on the basis of official statistics of foreign governments, other foreign source materials, reports of U.S. Agricultural Counselors, Attachés, and Foreign Service Officers, results of office research, and related information.

Table 2-6.—Cotton: Supply and distribution, United States, 1995–2004

Year beginning August 1	Supply			Distribution				
	Beginning of season total [2]	Ginnings in season [1]	Total supply [2]	Consumption [2]			Exports	Carryover, end of season [2]
				Upland	American Pima	Total		
	1,000 bales	*1,000 bales*	*1,000 bales*	*1,000 bales*	*1,000 bales*	*1,000 bales*	*1,000 bales*	*1,000 bales*
1995	2,608	17,500	20,499	10,089	127	10,216	7,277	2,695
1996	2,695	18,393	21,474	10,601	101	10,702	6,708	3,895
1997	3,985	18,445	22,444	10,793	109	10,902	7,279	4,079
1998	4,079	13,469	17,956	10,067	143	10,210	4,087	3,866
1999	3,866	16,692	20,542	9,665	132	9,797	6,557	4,056
2000	4,056	16,596	20,657	8,410	118	8,528	6,425	5,930
2001	5,930	19,729	25,650	7,289	99	7,388	10,649	7,305
2002	7,305	16,683	23,989	7,022	100	7,122	11,571	5,293
2003	5,193	17,729	22,921	6,076	61	6,137	13,330	3,381
2004 [3]	3,381	22,576	25,957	5,971	60	6,031	13,593	5,411

[1] Ginnings during the 12 months, Aug. 1–July 31. Includes an allowance for "city crop" which consists of rebaled samples and pickings from cotton damaged by fire and weather. [2] May include small volume of foreign growths. [3] Preliminary.

AMS, Cotton Program, (901) 384–3016. Compiled from reports of the Bureau of the Census.

Table 2-7.—Cotton, American Upland: Percentage distribution of fiber strength, United States, 2000–2004

Fiber strength [1]	Year				
	2000	2001	2002	2003	2004
17 and below	*	*	*	*	*
18	*	*	*	*	*
19	*	*	*	*	*
20	*	*	*	*	*
21	0.1	*	*	*	*
22	0.4	0.2	0.1	*	*
23	1.5	0.5	0.3	0.1	0.2
24	4.5	1.7	1.5	0.8	0.6
25	10.3	5.0	5.0	2.9	2.2
26	17.5	11.3	10.6	8.1	5.9
27	20.6	18.6	16.7	16.4	12.0
28	16.4	21.4	20.4	21.9	18.5
29	9.8	17.0	17.4	18.5	20.4
30	5.5	10.0	10.8	11.7	15.9
31	3.9	5.5	5.9	7.3	10.0
32	3.6	3.8	3.8	5.1	6.3
33	3.0	2.7	3.1	3.7	4.2
34	1.7	1.4	2.4	2.2	2.5
35	0.7	0.5	1.3	0.9	1.0
36 and above	0.4	0.2	0.7	0.3	0.3
Average	27.6	28.3	28.6	28.8	29.2

[1] Fiber strength expressed in terms of ⅛" gage (grams per tex). *Less than 0.05 percent.

AMS, Cotton Program, (901) 384–3016.

STATISTICS OF COTTON, TOBACCO, SUGAR CROPS, AND HONEY

Table 2-8.—Cotton, American Upland: Estimated percentage of the crop forward contracted by growers, by States, 1997–2004

State	Crop of—							
	1997	1998	1999	2000	2001	2002	2003	2004
	Percent	*Percent*	*Percent*	*Percent*	*Percent*	*Percent*	*Percent*	*Percent*
AL	19	21	8	29	19	17	17	19
AZ	11	16	3	2	1	*	33	1
AR	53	53	11	14	5	4	39	*
CA	41	59	15	6	*	*	*	*
FL	12	15	*	3	12	....	4	....
GA	29	37	9	21	8	8	21	19
LA	26	27	8	21	12	4	58	13
MS	18	21	8	11	6	2	7	2
MO	71	94	16	26	11	12	51	*
NM	7	14	9	13	2	3	3	....
NC	37	53	42	23	6	4	9	16
OK	(1)	*	....	*	....	....	....	....
SC	49	56	23	30	6	11	11	25
TN	14	29	11	20	10	4	71	*
TX	13	28	5	8	7	....	10	5
US	24	36	10	14	7	4	21	7

*Less than 0.5 percent.
AMS, Cotton Program, (901) 384–3016.

Table 2-9.—Cotton, American Upland: Carryover and crop, running bales, by grade groupings, United States, 1996–2005

Year beginning August 1	White color grades					Light spotted color grades				Other color grades [1]	All grades [2]
	21 and higher	31	41	51	61 and 71	22 and higher	32	42	52 and lower		
	1,000 bales	*1,000 bales*	*1,000 bales*	*1,000 bales*	*1,000 bales*	*1,000 bales*	*1,000 bales*	*1,000 bales*	*1,000 bales*	*1,000 bales*	*1,000 bales*
Carryover:											
1996	498	764	562	56	1	81	240	303	51	53	2,609
1997	689	1,463	750	43	1	57	423	273	33	57	3,789
1998	570	1,541	871	86	5	160	299	243	116	130	4,021
1999	407	1,094	978	147	3	134	271	471	145	131	3,781
2000	1,274	1,007	981	123	8	68	85	192	42	50	3,830
2001	1,392	1,712	1,464	218	4	63	148	601	150	67	5,819
2002	1,234	2,325	1,976	107	2	99	238	769	54	76	6,700
2003	596	988	1,804	502	8	37	193	475	251	115	4,972
2004	435	1,573	1,106	54	1	22	47	51	7	19	3,314
2005	975	1,042	1,609	530	18	42	154	505	186	339	5,402
Crop:											
1995	3,660	5,453	3,568	232	9	552	1,393	1,505	202	176	16,751
1996	3,219	6,629	3,413	233	16	481	1,602	1,513	339	224	17,669
1997	4,078	5,670	3,571	442	12	739	1,193	1,164	332	375	17,576
1998	1,821	3,893	3,486	251	5	385	849	1,669	419	234	13,012
1999	5,793	4,504	3,215	283	5	359	539	758	168	149	15,773
2000	3,459	5,100	4,454	504	13	292	624	1,408	302	192	16,348
2001	4,950	6,593	3,997	443	8	391	654	1,296	276	431	19,039
2002	2,248	3,389	5,610	1,086	29	122	594	1,627	859	488	16,053
2003	3,971	7,755	4,423	193	2	156	278	319	67	124	17,290
2004	4,063	5,228	7,079	1,955	45	180	605	1,328	567	782	21,832

[1] Includes all color grades of Spotted, Tinged, Yellow Stained, and Below Grade. [2] Carryover as reported by the Bureau of the Census, Crop as reported by AMS, Cotton Program.
AMS, Cotton Program, (901) 384–3016.

Table 2-10.—Cotton, American Upland: Carryover and crop, running bales, by staple groupings, United States, 1995–2005

Year beginning August 1	Staple										All staples [1]
	26 and shorter	28	29	30	31	32	33	34	35	36 and longer	
	1,000 bales	1,000 bales	1,000 bales	1,000 bales	1,000 bales	1,000 bales	1,000 bales	1,000 bales	1,000 bales	1,000 bales	1,000 bales
Carryover:											
1996	8	6	3	12	36	143	197	508	771	925	2,609
1997	(2)	(2)	(2)	2	6	30	235	947	1,471	1,099	3,789
1998	(2)	(2)	3	16	62	296	553	802	1,076	1,213	3,782
1999	(2)	1	2	18	81	439	765	902	785	788	3,781
2000	(2)	1	10	46	85	386	651	969	820	862	3,830
2001	1	4	22	88	241	558	1,209	1,385	1,341	970	5,819
2002			4	9	32	200	708	1,995	2,071	1,681	6,700
2003		(2)	15	35	69	214	708	1,495	1,357	1,079	4,972
2004		1	3	14	33	142	389	1,189	869	674	3,314
2005		1	4	17	77	213	543	1,128	1,615	1,803	5,402
Crop:											
1995	(2)	1	7	36	155	489	1,389	3,422	5,208	6,044	16,751
1996	(2)	(2)	2	13	72	277	1,035	3,226	5,668	7,376	17,669
1997		(2)	4	27	144	553	1,704	3,306	4,207	7,631	17,576
1998	(2)	2	15	81	307	887	2,282	3,692	3,202	2,544	13,012
1999	1	8	51	177	562	1,553	3,077	4,102	3,415	2,827	15,773
2000	2	20	86	229	558	1,408	2,915	4,196	3,661	3,273	16,348
2001	(2)	1	9	53	256	974	3,084	5,592	4,947	4,123	19,039
2002	(2)	2	22	123	457	1,259	2,840	4,324	3,596	3,429	16,053
2003	(2)	1	10	57	202	624	2,205	4,873	4,805	4,512	17,290
2004	(2)	1	9	56	196	723	2,175	4,630	6,543	7,499	21,832

[1] Carryover as reported by the Bureau of the Census, Crop as reported by AMS, Cotton Program.   [2] Fewer than 500 bales.

AMS, Cotton Program, (901) 384–3016.

Table 2-11.—Cotton, American Pima: Carryover and crop, running bales, by grade and staple, United States, 2000–2005

Year beginning August 1	Grade					Staple				All grades and staples [1]
	01 and 02	03	04	05	06 and 07	42 and shorter	44	46	48 and longer	
	1,000 bales	1,000 bales	1,000 bales	1,000 bales	1,000 bales	1,000 bales	1,000 bales	1,000 bales	1,000 bales	1,000 bales
Carryover:										
2001	87.5	22.8	2.2	0.1			28.9	68.8	14.9	112.6
2002	280.5	19.1	2.6		0.8	1.1	111.8	163.1	27.0	303.0
2003	163.5	40.7	8.8	5.1	3.2	1.1	113.3	76.8	30.2	221.3
2004	20.6	39.3	39.0	1.9	0.8	0.0	2.0	54.0	10.5	66.5
2005	7.9	1.9	0.7	0.4	0.4	0.0	0.3	9.2	1.2	10.7
Crop:										
2000	304.5	58.9	9.4	1.2	0.1	0.1	71.9	255.6	46.5	374.1
2001	618.9	25.1	9.0	8.1	3.6	0.6	160.5	424.1	79.6	664.8
2002	606.7	28.9	10.0	5.1	1.9	0.7	62.1	353.9	236.4	653.0
2003	390.0	15.1	2.8	0.7	0.1	0.1	16.3	190.2	202.8	409.3
2004	382.1	15.2	2.8	0.7	0.1	0.5	62.3	398.0	256.3	717.0

[1] Carryover as reported by the Bureau of the Census; Crop as reported by AMS, Cotton Program.

AMS, Cotton Program, (901) 384–3016.

Table 2-12.—Cotton, Upland: Average staple length of Upland cotton classed, by States, 1999–2004

State	Average staple length (32ds of an inch) [1]					
	1999	2000	2001	2002	2003	2004
AL	33.4	33.7	34.3	33.7	34.3	34.4
AZ	35.3	35.3	35.3	35.5	35.4	35.5
AR	34.8	34.6	34.8	34.8	34.8	35.3
CA	36.2	36.1	36.1	36.9	37.2	36.3
FL	34.2	34.4	34.4	33.7	34.3	34.8
GA	33.8	34.3	34.1	33.9	34.2	34.7
KS	(2)	(2)	(2)	33.1	32.2	31.7
LA	34.0	33.9	34.2	34.5	34.6	35.4
MS	34.2	34.0	34.3	34.5	34.5	35.3
MO	34.8	34.8	35.4	35.0	35.2	35.0
NM	35.8	36.5	36.3	35.9	36.1	36.1
NC	34.6	34.2	34.6	33.4	34.5	35.1
OK	33.1	34.1	35.4	33.9	34.6	34.4
SC	33.9	34.0	33.8	33.5	34.5	35.0
TN	33.8	33.5	33.9	34.2	34.2	34.0
TX	33.1	32.9	33.8	33.5	34.3	34.5
VA	35.0	34.6	34.9	34.1	35.3	35.0
Other States	(2)	(2)	(2)	(2)	(2)	(2)
US	34.1	34.2	34.5	34.3	34.7	34.9

[1] Average calculated on numerical equivalents of the staple-length designations. For example, 7/8-inch = 28, 29/32-inch = 29, etc. [2] Not available.
AMS, Cotton Program, (901) 384–3016.

Table 2-13.—Cotton: United States exports, by country of destination, 2002/2003–2004/2005

Country of destination	Year beginning August		
	2002/2003	2003/2004	2004/2005
	1,000 bales [1]	1,000 bales [1]	1,000 bales [1]
China, Peoples Republic	1,840	4,919	4,076
Turkey	1,639	1,416	1,998
Mexico	1,777	1,620	1,586
Indonesia	869	889	1,134
Thailand	592	527	846
Taiwan	556	396	713
Pakistan	461	598	658
Korea, Republic of	480	469	642
Canada	303	303	311
Japan	380	284	301
Hong Kong	364	169	272
Vietnam	137	192	187
Bangladesh	234	218	173
India	505	211	163
New Zealand	2	1	156
Colombia	179	224	155
Peru	180	123	144
Guatemala	83	87	135
Philippines	104	100	110
Brazil	312	315	88
Subtotal	10,996	13,060	13,848
El Salvador	98	82	85
Ecuador	67	93	83
Italy	81	63	75
United Kingdom	4	20	60
Belgium-Luxembourg	54	113	47
Switzerland	33	79	29
Germany	43	30	28
Venezuela	20	50	23
Malaysia	18	16	20
Chile	56	14	19
Ireland	28	19	15
Estonia	16	17	14
United Arab Emirates	5	8	8
Honduras	3	8	8
Cuba	0	7	8
Bahrain	8	14	6
Dominican Republic	3	4	5
Sri Lanka	9	9	4
Portugal	8	6	4
Spain	1	0	4
Other Countries	57	45	16
Total	11,607	13,758	14,409

[1] 480 pounds net.
FAS, Cotton, Oilseeds, Tobacco, and Seeds Division, (202) 720–9516. Compiled from reports of the U.S. Department of Commerce.

Table 2-14.—Cotton: International trade, 1999/2000-2004/2005 [1]

Country	1999/2000	2000/2001	2001/2002	2002/2003	2003/2004	2004/2005
	1,000 bales [2]	*1,000 bales* [2]	*1,000 bales* [2]	*1,000 bales* [2]	*1,000 bales* [2]	*1,000 bales* [2]
Principal exporting countries:						
United States	6,750	6,740	11,000	11,900	13,758	14,409
Uzbekistan, Republic of	4,200	3,450	3,500	3,400	3,100	3,950
Australia	3,211	3,903	3,130	2,655	2,157	2,002
Brazil	12	315	674	489	964	1,557
Greece	1,080	1,424	1,000	1,150	1,225	1,150
Burkina	520	520	650	725	950	975
Mali	900	575	925	850	1,175	950
India	70	94	60	56	625	800
Syria	950	1,050	1,000	750	700	700
Kazakhstan, Republic of	270	400	500	500	525	625
Tajikistan, Republic of	355	390	475	625	625	612
Egypt	425	375	410	700	400	600
Benin	650	625	650	725	675	575
Pakistan	415	575	160	231	200	550
Cote d'Ivoire	735	600	400	400	500	500
Cameroon	300	370	350	400	475	400
Tanzania, United Republic	142	146	154	215	155	375
Turkmenistan	800	675	450	400	550	375
Chad	330	300	275	300	250	300
Paraguay	325	400	190	225	451	297
Subtotal	22,440	22,927	25,953	26,696	29,460	31,702
Others	4,753	3,465	3,052	3,624	3,695	3,255
World total	27,193	26,392	29,005	30,320	33,155	34,957

Country	1999/2000	2000/2001	2001/2002	2002/2003	2003/2004	2004/2005
Principal importing countries:						
China, Peoples Republic of	117	230	449	3,127	8,832	6,385
Turkey	2,411	1,758	2,977	2,265	2,370	3,409
Indonesia	2,076	2,650	2,356	2,228	2,150	2,400
Thailand	1,696	1,573	1,882	1,945	1,678	2,282
Mexico	1,813	1,865	2,065	2,330	1,858	1,810
Bangladesh	775	1,000	1,200	1,600	1,540	1,700
Russian Federation	1,600	1,650	1,800	1,650	1,475	1,450
Pakistan	475	450	1,000	850	1,850	1,400
Korea, Republic of	1,525	1,421	1,616	1,492	1,274	1,344
Taiwan	1,438	1,040	1,531	1,219	1,011	1,337
Italy	1,456	1,349	1,288	1,216	925	990
Japan	1,280	1,138	1,063	1,013	778	815
India	1,600	1,567	2,388	1,216	800	800
Vietnam	330	400	435	400	540	680
Egypt	130	130	130	75	375	400
Germany	640	702	580	490	374	380
Canada	286	352	400	460	375	365
Portugal	649	623	557	484	389	325
Hong Kong	517	519	491	463	250	318
Malaysia	462	340	228	231	253	254
Subtotal	21,276	20,757	24,436	24,754	29,097	28,844
Others	6,781	5,536	5,116	5,447	4,918	4,174
World total	28,057	26,293	29,552	30,201	34,015	33,018

[1] Marketing year beginning Aug. 1.　　[2] 480-pound net weight.
FAS, Cotton, Oilseeds, Tobacco and Seeds Division, (202) 720–9516. Prepared or estimated on the basis of official statistics of foreign governments, other foreign source materials, reports of U.S. Agricultural Counselors, Attachés, and Foreign Service Officers, results of office research, and related information.

Table 2-15.—Cotton, American Upland: High, low, and season average spot prices for the base quality in the designated markets, cents per pound, 1996-2004

Season beginning August 1	Color 41, Leaf 4, Staple 34 [1]		
	Average	High	Low
	Cents	*Cents*	*Cents*
1996	71.59	78.11	68.00
1997	67.79	77.79	59.82
1998	60.12	74.19	47.21
1999	52.36	60.71	45.94
2000	51.56	63.57	35.39
2001	33.10	41.39	25.94
2002	47.46	55.86	36.56
2003	60.15	77.66	42.45
2004	45.61	52.30	40.39

[1] Prices are for mixed lots, net weight, compressed, FOB car/truck.
AMS, Cotton Program, (901) 384–3016.

Table 2-16.—Cotton and cotton linters: United States imports for consumption, by country of origin, 2000/2001–2004/2005

Country of origin	Year beginning August				
	2000/2001	2001/2002	2002/2003	2003/2004	2004/2005
	Bales¹	Bales¹	Bales¹	Bales¹	Bales¹
Cotton, raw:					
Egpyt	12,345	20,860	62,576	41,110	26,021
Israel	0	0	0	0	1,988
Pakistan	0	0	0	46	907
India	1,029	176	237	305	202
Italy	103	2	0	0	3
Australia	0	0	4,328	2,786	0
Barbados	114	0	0	0	0
Canada	1	0	0	0	0
China, Peoples Republic ...	0	2	0	0	0
Benin	0	0	0	200	0
Other countries	2,304	318	0	553	0
Total	15,897	21,359	67,142	45,001	29,121
Cotton linters:					
Syria	58,374	24,733	25,333	15,578	33,616
Mexico	2,940	9,885	6,586	334	4,278
Turkey	66,840	3,060	0	0	3,921
Brazil	37,563	69,666	1,086	179	3,321
Zimbabwe	0	0	0	0	362
Israel	621	700	184	440	179
China, People Republic	0	0	0	4	9
Argentina	3,083	2,267	0	0	0
Australia	458	0	0	0	0
Germany	219	0	0	0	0
Greece	4,119	2,888	0	0	0
India	810	375	0	0	0
Japan	0	0	11	0	0
Korea, Republic of	0	27	0	0	0
Latvia	25,214	0	0	0	0
Spain	0	0	0	12	0
Total	200,240	113,601	33,200	16,548	45,686

¹ 480 pounds net.

FAS, Cotton, Oilseeds, Tobacco and Seeds Division, (202) 720–9516. Compiled from reports of the U.S. Department of Commerce.

Table 2-17.—Cotton, American Upland: Percentage distribution of mike readings, by specified groups, United States, 1995–2004

Year beginning August 1	Mike groups						
	26 and below	27 to 29	30 to 32	33 to 34	35 to 49	50 to 52	53 and above
	Percent	Percent	Percent	Percent	Percent	Percent	Percent
1995	0.4	1.2	2.7	3.0	79.6	10.6	2.5
1996	0.4	1.2	3.1	3.5	80.8	8.7	2.5
1997	*	0.4	1.8	2.5	88.4	5.9	0.9
1998	*	0.4	1.2	1.5	83.4	11.1	2.3
1999	0.3	0.5	1.3	1.8	80.2	12.6	3.3
2000	0.1	0.7	2.0	2.8	85.8	7.1	1.3
2001	*	0.2	0.7	1.1	75.9	15.7	6.0
2002	*	0.3	0.7	1.1	74.2	17.7	5.8
2003	*	0.3	0.9	1.4	83.6	11.2	5.8
2004	0.4	1.5	3.4	3.7	83.8	6.4	0.8

(*) Less than 0.05 percent.

AMS, Cotton Program, (901) 384–3016.

Table 2-18.—Cotton, American Upland: Average spot prices for specified grades of staple 34 in the designated markets for mixed lots, net weight, FOB car/truck, compressed, cents per pound, 1995–2004

Year beginning August 1	White				Light Spotted			Spotted	
	Color 31 Leaf 3	Color 41 Leaf 4	Color 51 Leaf 5	Color 61 Leaf 6	Color 32 Leaf 3	Color 42 Leaf 4	Color 52 Leaf 5	Color 33 Leaf 3	Color 43 Leaf 4
	Cents	Cents	Cents	Cents	Cents	Cents	Cents	Cents	Cents
1995	84.82	83.03	77.61	73.25	82.49	78.38	72.97	76.31	71.30
1996	72.87	71.59	66.71	63.02	71.12	67.62	62.50	65.93	61.25
1997	69.08	67.79	63.31	60.38	67.08	64.44	60.06	64.47	59.43
1998	61.93	60.12	54.68	51.49	59.43	56.25	51.71	55.47	50.53
1999	53.99	52.36	46.52	42.52	51.50	48.03	43.15	46.50	41.99
2000	52.98	51.56	47.18	43.50	51.36	48.78	45.15	47.81	43.88
2001	34.66	33.10	29.32	26.87	33.26	31.04	28.12	30.42	27.50
2002	49.72	47.46	43.38	41.40	47.53	44.94	42.22	44.99	42.04
2003	62.24	60.15	56.05	53.89	60.03	57.42	54.89	57.15	54.58
2004	48.40	45.61	41.59	39.11	45.70	43.30	40.38	42.51	40.75

AMS, Cotton Program, (901) 384–3016.

Table 2-19.—Cotton, American Upland: Average spot prices for specified staple lengths of Grade 41 Leaf 4, in the designated markets for mixed lots, net weight, FOB car/truck, compressed, cents per pound, 1995–2004

Year beginning August 1	Staple							
	28	29	30	31	32	33	34	35
	Cents	Cents	Cents	Cents	Cents	Cents	Cents	Cents
1995	76.00	76.00	77.82	79.49	79.99	81.06	83.03	83.58
1996	64.61	64.61	66.38	67.18	67.93	69.81	71.59	72.20
1997	60.97	60.97	62.65	63.48	64.56	66.07	67.79	68.39
1998	47.05	47.05	48.93	51.08	53.75	57.27	60.12	61.05
1999	39.71	39.71	41.15	43.10	44.57	48.02	52.36	54.19
2000	43.14	43.14	44.59	45.90	46.10	48.24	51.56	52.82
2001	29.12	29.12	29.77	30.53	30.01	31.24	33.10	34.31
2002	43.07	43.07	43.57	44.60	44.40	45.64	47.46	49.13
2003	55.39	55.39	55.94	56.95	57.08	58.42	60.15	61.71
2004	41.54	41.54	42.13	43.28	43.32	44.07	45.61	47.02

AMS, Cotton Program, (901) 384–3016.

Table 2-20.—Cotton, American Upland: Season average spot prices for the base quality, by designated markets, cents per pound, 1999–2004 [1]

Market	Color 41, Leaf 4, Staple 34 [2]					
	1999	2000	2001	2002	2003	2004
	Cents	Cents	Cents	Cents	Cents	Cents
Southeast	53.81	52.63	33.02	48.28	60.80	45.91
North Delta	53.34	52.32	33.24	48.47	60.78	46.02
South Delta	53.34	52.32	33.24	48.46	60.85	46.02
East TX–OK	50.49	51.03	32.59	46.76	59.95	44.22
West Texas	50.12	50.71	32.39	46.51	59.71	44.08
Desert SW	48.79	49.47	32.60	46.27	59.23	45.66
SJ Valley	56.67	52.45	34.64	47.52	59.71	47.38
Average	52.36	51.56	33.10	47.46	60.15	45.61

[1] Year beginning August 1.   [2] Prices are for mixed lots, net weight, compressed, FOB car/truck.
AMS, Cotton Program, (901) 384–3016.

Table 2-21.—Cotton: Supply and distribution, by countries, 2004–05

Country	Beginning stocks Aug. 1	Produc-tion	Imports	Total supply	Consump-tion [1]	Exports	Ending stocks July 31	Total dis-tribution
	1,000 bales [2]	1,000 bales [2]	1,000 bales [2]	1,000 bales [2]	1,000 bales [2]	1,000 bales [2]	1,000 bales [2]	1,000 b ales [2]
Importing countries:								
China, Peoples Republic	12,808	29,000	6,385	48,193	39,400	30	10,563	49,993
Turkey	1,278	4,150	3,409	8,837	7,000	152	1,685	8,837
Indonesia	366	32	2,400	2,798	2,300	20	478	2,798
Thailand	419	57	2,282	2,758	2,175	0	583	2,758
Mexico	1,144	625	1,810	3,579	2,075	135	1,369	3,579
Bangladesh	316	60	1,700	2,076	1,730	0	346	2,076
Russian Federation	197	0	1,450	1,647	1,425	0	222	1,647
Pakistan	2,030	11,300	1,400	14,730	10,775	550	3,405	14,730
Korea, Republic of	409	1	1,344	1,754	1,325	9	420	1,754
Taiwan	351	0	1,337	1,688	1,200	0	488	1,688
Italy	143	0	990	1,133	945	25	163	1,133
Japan	141	0	815	956	815	0	141	956
India	4,061	19,000	800	23,861	15,000	800	8,061	23,861
Vietnam	135	48	680	863	675	0	188	863
Egypt	601	1,332	400	2,333	963	600	770	2,333
Germany	21	0	380	401	290	80	31	401
Canada	87	0	365	452	360	2	90	452
Portugal	99	1	325	425	325	2	98	425
Hong Kong	72	0	318	390	300	30	60	390
Malaysia	36	0	254	290	250	1	39	290
France	44	0	242	286	215	33	38	286
Poland	15	0	240	255	230	1	24	255
Czech Republic	36	0	230	266	220	10	36	266
Peru	226	320	221	767	440	8	319	767
Brazil	4,626	5,900	212	10,738	4,300	1,557	5,081	10,938
Belgium/Luxembourg	11	0	205	216	115	79	22	216
Morocco	29	1	150	180	150	0	30	180
Philippines	43	11	150	204	150	0	54	204
Colombia	103	240	131	474	405	11	58	474
South Africa, Republic of	98	95	124	317	249	0	68	317
Subtotal	29,945	72,173	30,749	132,867	95,802	4,135	34,930	134,867
Others	10,670	48,256	2,269	61,195	12,319	30,822	16,054	59,195
Total world	40,615	120,429	33,018	194,062	108,121	34,957	50,984	194,062
Exporting countries:								
United States	3,506	23,251	29	26,786	6,727	14,409	5,650	26,786
Uzbekistan, Republic of	923	5,200	0	6,123	875	3,950	1,298	6,123
Australia	894	3,000	1	3,895	190	2,002	1,953	4,145
Brazil	4,626	5,900	212	10,738	4,300	1,557	5,081	10,938
Greece	390	1,800	10	2,200	550	1,150	500	2,200
Burkina	196	1,180	0	1,376	4	975	397	1,376
Mali	257	1,100	0	1,357	20	950	387	1,357
India	4,061	19,000	800	23,861	15,000	800	8,061	23,861
Syria	299	1,600	0	1,899	700	700	499	1,899
Kazakhstan, Republic of	133	680	0	813	40	625	148	813
Tajikistan, Republic of	157	800	0	957	150	612	195	957
Egypt	601	1,332	400	2,333	963	600	770	2,333
Benin	186	670	0	856	10	575	271	856
Pakistan	2,030	11,300	1,400	14,730	10,775	550	3,405	14,730
Cote d'Ivoire	323	600	0	923	65	500	358	923
Cameroon	115	500	0	615	45	400	170	615
Tanzania, United Repub-lic	152	525	0	677	70	375	232	677
Turkmenistan	399	950	0	1,349	415	375	559	1,349
Chad	63	375	0	438	15	300	123	438
Paraguay	129	300	0	429	30	297	102	429
Togo	122	325	0	447	15	285	147	447
Sudan	78	525	0	603	9	284	310	603
Zimbabwe	176	350	0	526	115	250	161	526
Spain	119	500	75	694	280	235	179	694
Zambia	255	350	0	605	65	225	315	605
Kyrgyzstan, Republic of	32	185	0	217	5	175	37	217
Azerbaijan, Republic of	70	225	0	295	25	160	110	295
Turkey	1,278	4,150	3,409	8,837	7,000	152	1,685	8,837
Mexico	1,144	625	1,810	3,579	2,075	135	1,369	3,579
Uganda	80	200	0	280	35	110	135	280
Subtotal	22,794	87,498	8,146	118,438	50,568	33,713	34,607	118,888
Others	17,821	32,931	24,872	75,624	57,553	1,244	16,377	75,174
Total world	40,615	120,429	33,018	194,062	108,121	34,957	50,984	194,062

[1] Includes cotton destroyed or unaccounted for.   [2] Bales of 480 pounds net weight.

FAS, Cotton, Oilseeds, Tobacco and Seeds Division, (202) 720–9516. Prepared or estimated on the basis of official statis-tics of foreign governments, other foreign source materials, reports of U.S. Agricultural Counselors, Attaches, and Foreign Service Officers, results of office research, and related information.

Table 2-22.—Sugarbeets: Area, yield, production, and value, United States, 1996–2005 [1]

Year	Area Planted	Area Harvested	Yield per harvested acre	Production	Marketing year average price per ton received by farmers [2]	Value of production
	1,000 acres	*1,000 acres*	*Tons*	*1,000 tons*	*Dollars*	*1,000 dollars*
1996 ...	1,368.4	1,323.3	20.2	26,680	45.40	1,211,001
1997 ...	1,459.3	1,428.3	20.9	29,886	38.80	1,160,029
1998 ...	1,497.8	1,450.7	22.4	32,499	36.40	1,181,494
1999 ...	1,560.6	1,527.3	21.9	33,420	37.20	1,242,895
2000 ...	1,564.2	1,373.0	23.7	32,541	34.20	1,113,030
2001 ...	1,365.3	1,241.1	20.7	25,708	39.80	1,023,054
2002 ...	1,427.3	1,360.7	20.4	27,707	39.60	1,097,329
2003 ...	1,365.4	1,347.8	22.8	30,710	41.40	1,270,026
2004 ...	1,345.6	1,306.7	23.0	30,021	36.90	1,106,878
2005 [3]	1,294.8	1,238.9	22.3	27,654	NA	NA

[1] Relates to year of intended harvest except for overwintered spring planted beets in CA.　[2] Prices do not include Government payments under the Sugar Act.　[3] Preliminary.　NA-not available.
NASS, Crops Branch, (202) 720–2127.

Table 2-23.—Sugarbeets: Area, yield, and production, by States, 2003–2005 [1]

State	Area planted			Area harvested			Yield per harvested acre			Production		
	2003	2004	2005	2003	2004	2005	2003	2004	2005	2003	2004	2005
	1,000 acres	*1,000 acres*	*1,000 acres*	*1,000 acres*	*1,000 acres*	*1,000 acres*	*Tons*	*Tons*	*Tons*	*1,000 tons*	*1,000 tons*	*1,000 tons*
CA	50.8	49.1	44.4	50.1	48.9	44.1	39.1	40.8	38.7	1,959	1,995	1,707
CO	28.6	36.0	36.4	27.4	33.5	34.3	23.5	25.0	24.3	644	838	833
ID	208.0	195.0	169.0	207.0	192.0	167.0	29.2	28.7	28.3	6,044	5,510	4,726
MI	179.0	165.0	149.0	178.0	163.0	148.0	19.1	21.1	21.4	3,400	3,439	3,167
MN	492.0	486.0	491.0	487.0	470.0	460.0	20.6	20.9	20.4	10,032	9,823	9,384
MT	51.7	53.7	53.9	51.5	52.1	49.9	25.4	21.7	22.9	1,308	1,131	1,143
NE	45.3	49.8	48.4	42.4	47.5	45.3	20.3	22.1	20.4	861	1,050	924
ND	259.0	256.0	255.0	255.0	246.0	243.0	20.4	19.7	18.9	5,202	4,846	4,593
OH [2]	2.0	1.9	....	1.9	1.7	....	24.2	21.8	....	46	37	....
OR	10.0	12.9	9.8	9.8	12.6	9.7	30.7	31.4	31.6	301	396	307
WA	4.0	3.8	1.7	4.0	3.8	1.7	40.3	37.9	40.6	161	144	69
WY	35.0	36.4	36.2	33.7	35.6	35.9	22.3	22.8	22.3	752	812	801
US	1,365.4	1,345.6	1,294.8	1,347.8	1,306.7	1,238.9	22.8	23.0	22.3	30,710	30,021	27,654

[1] Relates to year of intended harvest except for overwintered spring planted beets in CA.　[2] No acreage reported in 2005.
NASS, Crops Branch, (202) 720–2127.

Table 2-24.—Sugarbeets: Production and value, by States, crop of 2003–2004 [1]

State	Production		Marketing year average price per ton received by farmers		Value of production	
	2003	2004	2003	2004	2003	2004
	1,000 tons	*1,000 tons*	*Dollars*	*Dollars*	*1,000 dollars*	*1,000 dollars*
CA	1,959	1,995	39.50	37.90	77,381	75,611
CO	644	838	38.40	36.30	24,730	30,419
ID	6,044	5,510	35.90	37.10	216,980	204,421
MI	3,400	3,439	36.70	26.40	124,780	90,790
MN	10,032	9,823	44.20	37.80	443,414	371,309
MT	1,308	1,131	43.00	40.80	56,244	46,145
NE	861	1,050	42.30	39.90	36,420	41,895
ND	5,202	4,846	46.30	39.50	240,853	191,417
OH	46	37	36.00	26.40	1,656	977
OR	301	396	35.90	37.10	10,806	14,692
WA	161	144	35.90	37.10	5,780	5,342
WY	752	812	41.20	41.70	30,982	33,860
US	30,710	30,021	41.40	36.90	1,270,026	1,106,878

[1] Relates to year of intended harvest in all States except CA. In CA, relates to year of intended harvest for fall planted beets in central CA and to year of planting for overwintered beets in central and southern CA.
NASS, Crops Branch, (202) 720–2127.

Table 2-25.—Sugarcane for sugar and seed: Area, yield, production, and value, United States, 1996–2005

Year[1]	Area harvested			Yield of cane per acre			Production		
	For sugar	For seed	Total	For sugar	For seed	For sugar and seed	For sugar	For seed	Total
	1,000 acres	*1,000 acres*	*1,000 acres*	*Tons*	*Tons*	*Tons*	*1,000 tons*	*1,000 tons*	*1,000 tons*
1996	829.5	59.4	888.9	33.4	29.9	33.1	27,687	1,777	29,464
1997	860.3	53.7	914.0	34.9	31.8	34.7	30,003	1,706	31,709
1998	888.3	58.8	947.1	36.9	33.4	36.6	32,743	1,964	34,707
1999	941.4	51.9	993.3	35.7	33.2	35.5	33,577	1,722	35,299
2000	976.7	55.6	1,032.3	35.1	32.8	35.0	34,291	1,823	36,114
2001	970.3	57.5	1,027.8	33.8	31.5	33.7	32,775	1,812	34,587
2002	971.9	51.3	1,023.2	34.9	32.2	34.7	33,903	1,650	35,553
2003	930.6	61.7	992.3	34.3	31.1	34.1	31,942	1,916	33,858
2004	879.5	58.7	938.2	31.0	30.2	30.9	27,243	1,770	29,013
2005[2]	866.4	57.5	923.9	29.4	28.7	29.4	25,485	1,649	27,134

Year[1]	Marketing year average price per ton received by farmers[3]	Value of production[4]	
		Of cane used for sugar	Of cane used for sugar and seed[4]
	Dollars	*1,000 dollars*	*1,000 dollars*
1996	28.30	784,113	833,297
1997	28.10	842,840	890,257
1998	27.30	893,049	944,562
1999	25.60	859,175	901,900
2000	26.10	895,917	941,791
2001	29.00	951,813	1,003,046
2002	28.40	961,896	1,007,142
2003	29.50	943,646	998,269
2004	28.30	771,734	821,118
2005	NA	NA	NA

[1] In Hawaii, harvest continues throughout the year and production statistics are on a calendar year basis. In other states, harvest is seasonal and the production statistics year relates to the year in which the season begins. [2] Preliminary. [3] Prices do not include Government payments under the Sugar Act. [4] Price per ton of cane for sugar used in evaluating value of production for seed. NA-not available.

NASS, Crops Branch, (202) 720–2127.

Table 2-26.—Sugarcane for sugar and seed: Production and value, by States, crop of 2003–2004

State	Sugarcane for sugar						Sugar and seed: Value of production	
	Production		Price per ton[1]		Value of production[1]			
	2003	2004	2003	2004	2003	2004	2003	2004
	1,000 tons	*1,000 tons*	*Dollars*	*Dollars*	*1,000 dollars*	*1,000 dollars*	*1,000 dollars*	*1,000 dollars*
FL	17,231	14,281	31.90	30.30	525,297	407,141	549,669	432,714
HI	2,082	2,026	31.70	31.10	64,351	61,547	65,999	63,009
LA	12,838	11,067	25.80	25.30	304,182	258,920	331,220	279,995
TX	1,707	1,639	30.10	27.70	49,816	44,126	51,381	45,400
Total ..	33,858	29,013	29.50	28.30	943,646	771,734	998,269	821,118

[1] Price per ton of cane for sugar used in evaluating value of production for seed.

NASS, Crops Branch, (202) 720–2127.

Table 2-27.—Sugarcane for sugar and seed: Area, yield, and production, by States, 2003–2005

State	Sugarcane for sugar and seed [1]								
	Area harvested			Yield of cane per acre			Cane production		
	2003	2004	2005	2003	2004	2005	2003	2004	2005
	1,000 acres	*1,000 acres*	*1,000 acres*	*Tons*	*Tons*	*Tons*	*1,000 tons*	*1,000 tons*	*1,000 tons*
For sugar:									
FL	419.0	385.0	383.0	39.3	34.9	34.0	16,467	13,437	13,022
HI	19.9	21.8	22.4	102.0	90.8	90.3	2,030	1,979	2,023
LA	450.0	430.0	420.0	26.2	23.8	23.0	11,790	10,234	9,660
TX	41.7	42.7	41.0	39.7	37.3	37.7	1,655	1,593	1,546
US	930.6	879.5	866.4	34.3	31.0	30.3	31,942	27,243	26,251
For seed:									
FL	19.0	21.0	18.0	40.2	40.2	41.0	764	844	738
HI	1.4	1.4	1.5	37.3	33.5	36.0	52	47	54
LA	40.0	35.0	35.0	26.2	23.8	23.0	1,048	833	805
TX	1.3	1.3	2.0	40.2	35.0	24.5	52	46	49
US	61.7	58.7	56.5	31.1	30.2	29.1	1,916	1,770	1,646
For sugar and seed:									
FL	438.0	406.0	401.0	39.3	35.2	34.3	17,231	14,281	13,760
HI	21.3	23.2	23.9	97.7	87.3	86.9	2,082	2,026	2,077
LA	490.0	465.0	455.0	26.2	23.8	23.0	12,838	11,067	10,465
TX	43.0	44.0	43.0	39.7	37.3	37.1	1,707	1,639	1,595
US	992.3	938.2	922.9	34.1	30.9	30.2	33,858	29,013	27,897

[1] In Hawaii, harvest continues throughout the year and production statistics are on a calendar year basis. In other states, harvest is seasonal and the production statistics year relates to the year in which the season begins.

NASS, Crops Branch, (202) 720–2127.

Table 2-28.—Sugar, cane (raw value [1]): Refiners' raw stocks, receipts, meltings, continental United States, 1995–2004

Year	Jan. 1 stocks	Receipts [2]	Meltings
	1,000 tons	*1,000 tons*	*1,000 tons*
1995	448	5,220	5,323
1996	334	6,071	6,072
1997	323	5,955	5,954
1998	322	5,636	5,704
1999	332	5,558	5,646
2000	356	5,543	5,575
2001	274	5,362	5,221
2002	351	5,607	5,681
2003	299	5,408	5,533
2004	286	5,127	5,164

[1] Raw value is the equivalent in terms of 96° sugar.   [2] Receipts include refiners' total offshore raw sugar receipts in continental U.S. ports, whether entered through the customs or held pending availability of quota and raw cane sugar produced from sugarcane in the continental United States.

FSA, Dairy and Sweeteners Analysis, (202) 720–6733.

Table 2-29.—Sugar, cane and beet: Domestic marketings, by source of supply, continental United States, 2002–2004 [1]

Area of supply	2002	2003	2004
	1,000 tons	*1,000 tons*	*1,000 tons*
Domestic areas:			
Mainland (beet)	4,596	4,476	5,153
Mainland and Hawaii (cane)	5,773	5,581	5,463
Puerto Rico	10	0	0
Total domestic areas	10,379	10,057	10,616

[1] Source: U.S. Census.

FSA, Dairy and Sweeteners Analysis Division, (202) 720–6733.

Table 2-30.—Sugar (Centrifugal Sugar, raw value): Production in specified countries, 2003/2004–2005/2006

Continent and country	2003/2004	2004/2005	2005/2006 [1] [2]
	1,000 metric tons	*1,000 metric tons*	*1,000 metric tons*
North America:			
United States	7,847	7,146	6,824
Canada	98	120	95
Mexico	5,330	6,149	6,000
Total	13,275	13,415	12,919
Caribbean:			
Cuba	2,450	1,300	1,450
Dominican Republic	553	485	520
Other	356	318	289
Total	3,359	2,103	2,259
Central America:			
Guatemala	1,850	1,982	2,000
Other	2,045	2,045	2,052
Total	3,895	4,027	4,052
South America:			
Brazil	26,400	28,175	28,700
Colombia	2,680	2,720	2,420
Argentina	1,925	1,815	2,050
Other	3,229	3,145	3,228
Total	34,234	35,855	36,398
Western Europe:			
EU [3]	17,132	21,825	21,233
Other Western Europe	201	234	235
Total	17,333	22,059	21,468
Eastern Europe:			
Poland [4]	2,116	NA	NA
Ukraine	1,580	1,900	1,830
Russian Federation	1,930	2,250	2,250
Other	2,055	931	927
Total	7,681	5,081	5,007
Africa:			
South Africa, Rep.	2,560	2,360	2,665
Other	5,741	5,905	5,850
Total	8,301	8,265	8,515
Middle East:			
Egypt	1,335	1,360	1,370
Turkey	1,915	2,109	2,175
Other	1,435	1,520	1,570
Total	4,685	4,989	5,115
Asia and Oceania:			
India	15,150	14,210	18,430
China, Peoples Republic	10,734	9,826	10,500
Thailand	7,010	5,187	4,330
Australia	5,178	5,388	5,200
Pakistan	4,047	2,937	2,890
Indonesia	1,730	2,050	1,800
Philippines	2,340	2,150	2,100
Japan	896	921	890
Other	2,552	2,348	2,278
Total	49,637	45,017	48,418
World total	142,400	140,811	144,151

[1] Crop years are on a September/August basis, but include the outturn of sugar from harvests of several Southern Hemisphere countries which begin in September.   [2] Preliminary.   [3] EU-15 for years prior to 2004/2005. EU-25 beginning in 2004/2005.   [4] No data available beyond 2003/2004 because Polish data is included in EU total beginning in 2004/2005.   NA-not available.

FAS, Horticultural and Tropical Products Division, (202) 720-3423. The division's Home Page is located at, www.fas.usda.gov/htp. You can also email the division at htp@fas.usda.gov. Note: Prepared or estimated on the basis of official statistics of foreign governments, other foreign source materials, reports of U.S. Agricultural Counselors, and Attachés, results of office research, and related information. This information can be obtained from the following web site: www.fas.usda.gov/psd.

Table 2-31.—Sugar, cane and beet (refined): Stocks, production or receipts, and deliveries, continental United States, 1996–2005

Item and year	Cane sugar refineries	Beet sugar factories	Importers of direct consumption sugar	Mainland cane sugar mills [1]	Total
JAN. 1 STOCKS [2]	*1,000 tons*	*1,000 tons*	*1,000 tons*	*1,000 tons*	*1,000 tons*
1996	195	1,383	0	12	1,590
1997	196	1,520	0	18	1,734
1998	212	1,535	0	22	1,769
1999	255	1,499	0	22	1,776
2000	208	1,554	0	22	1,784
2001	262	1,972	0	19	2,253
2002	288	1,812	0	19	2,119
2003	298	1,374	0	6	1,678
2004	326	1,853	0	5	2,184
2005	368	1,782	0	4	2,154
PRODUCTION OR RECEIPTS					
1995	5,366	4,471	44	14	9,895
1996	6,074	4,149	33	19	10,275
1997	5,968	4,117	27	20	10,132
1998	5,811	4,431	24	23	10,289
1999	5,840	4,767	40	24	10,671
2000	5,681	6,014	37	32	11,764
2001	5,467	4,839	58	26	10,390
2002	5,896	4,258	109	8	10,271
2003	5,761	4,817	58	8	10,644
2004	5,389	5,305	64	16	10,774
DELIVERIES [3]					
1995	5,397	4,645	44	15	10,101
1996	6,074	4,007	33	14	10,128
1997	5,940	4,060	27	16	10,043
1998	5,708	4,410	24	20	10,162
1999	5,777	4,678	40	24	10,519
2000	5,738	5,573	37	15	11,363
2001	5,538	4,961	58	13	10,570
2002	5,768	4,596	109	15	10,488
2003	5,573	4,476	58	8	10,115
2004	5,362	5,153	64	16	10,595

[1] Sugar for human consumption only.   [2] Stocks include sugar in bond and in Customs custody and control.   [3] Consists of all refined sugar.
FSA, Dairy and Sweeteners Analysis, (202) 720–7923.

Table 2-32.—Sugar (raw and refined): Average price per pound at specified markets, 1996–2005

Year	Cane sugar		Refined beet: Mid-west	Retail price, granulated: United States
	Raw, 96 centrifugal			
	Caribbean ports, f.o.b. and stowed, plus freight to Far East	New York, c.i.f. duty paid		
	Cents	*Cents*	*Cents*	*Cents*
1996	12.24	22.40	29.20	41.79
1997	12.06	21.96	27.09	43.26
1998	9.68	22.06	26.12	42.98
1999	6.54	21.16	26.71	43.27
2000	8.51	19.09	20.80	42.41
2001	9.12	21.11	23.31	43.42
2002	7.88	20.87	25.79	43.10
2003	7.51	21.42	26.21	42.68
2004	8.61	20.46	23.48	42.64
2005	11.35	21.28	29.54	43 54

ERS, Specialty Crops Branch, (202) 694–5247. Compiled from the following sources: (New York) Coffee, Sugar & Cocoa Exchange; the U.S. Department of Labor, Bureau of Labor Statistics; Milling and Baking News.

Table 2-33.—Refined sugar, centrifugal (raw value) [1]: United States exports, by country of destination, 2002/2003–2004/2005

Country of destination	2002/2003	2003/2004	2004/2005 [2]
	Metric tons	*Metric tons*	*Metric tons*
North America:			
Mexico	27,451	84,461	132,278
Canada	10,408	14,345	33,548
Total	37,859	98,806	165,826
Caribbean:			
Netherlands Antilles	2,462	4,605	3,445
Bahamas, The	2,579	1,295	2,145
Jamaica	4,983	473	1,341
Leeward-Windward Isl	601	723	717
Cayman Islands	461	589	427
Turks and caicos Isl	131	113	98
Barbados	634	47	86
Bermuda	0	0	48
Haiti	7	8	30
Dominican Republic	2	0	26
Trinidad and Tobago	180	270	26
Total	12,040	8,123	8,389
Central America:			
Panama	36	11	60
Belize	11	5	43
El Salvador	5	63	27
Costa Rica	75	52	16
Honduras	0	0	7
Guatemala	0	7	0
Nicaragua	0	1	0
Total	127	140	153
South America:			
Suriname	0	2	41
Uruguay	0	0	27
Chile	13	1	23
Guyana	0	0	22
Colombia	14	5	14
Brazil	0	21	12
Argentina	26	82	0
Bolivia	0	22	0
Ecuador	2	0	0
Venezuela	5	7	0
Total	60	142	139
European Union - 25:			
Germany	624	1,290	1,939
Netherlands	897	1,216	808
United Kingdom	695	542	403
Ireland	16	6	320
France	151	196	111
Spain	67	19	46
Sweden	0	0	38
Belgium-Luxembourg	53	26	20
Cyprus	0	0	14
Italy	0	0	12
Poland	22	12	11
Denmark	0	0	3
Greece	0	0	2
Finland	0	0	1
Austria	0	1	0
Slovenia	0	8	0
Total	2,524	3,316	3,730
Other Europe:			
Norway	12	13	13
Switzerland	4	25	2
Romania	15	0	0
Total	31	37	16

See footnotes at end of table.

Table 2-33.—Refined sugar, centrifugal (raw value)[1]: United States exports, by country of destination, 2002/2003–2004/2005—Continued

Country of destination	2002/2003	2003/2004	2004/2005[2]
	Metric tons	Metric tons	Metric tons
Former Soviet Union:			
Kazakhstan, Republic	0	0	187
Russian Federation	42	159	22
Total	42	159	209
East Asia:			
Japan	1,722	201	222
Korea, Republic of	331	380	207
China, Peoples Republic	69	28	70
Taiwan	6	2	54
Hong Kong	63	277	33
Total	2,191	888	587
Middle East:			
Kuwait	4	51	622
United Arab Emirates	17	44	576
Jordan	469	11	361
Yemen	0	0	346
Qatar	54	5	269
Saudi Arabia	46	147	255
Oman	0	0	221
Bahrain	0	289	97
Israel	69	2,268	52
Turkey	4	0	0
Total	664	2,816	2,798
North Africa:			
Morocco	0	0	169
Egypt	0	0	5
Total	0	0	174
Sub-Saharan Africa:			
Djibouti Afars-Issas	0	0	200
French Ind. Ocean Terr	0	1	8
Ghana	6	23	0
Nigeria	0	97	0
South Africa, Republic	11	8	0
Namibia	0	65	0
Total	17	194	208
South Asia:			
India	19	89	260
Afghanistan	0	0	29
Sri Lanka	0	0	13
Pakistan	0	0	5
Bangladesh	0	2	0
Total	19	91	307
Southeast Asia:			
Philippines	22	61	68
Indonesia	6	6	35
Thailand	15	8	19
Singapore	25	35	4
Vietnam	0	3	0
Total	68	113	127
Oceania:			
New Zealand	93	115	168
Australia	1	22	164
Micronesia, Federate	1	1	0
Marshal Islands	11	0	0
Palau	0	2	0
Total	106	139	332
World total[3]	55,746	114,964	182,995

[1] Refined sugar are tariff codes 1701.91 and 1701.99.　[2] October–September.　[3] Data are actual weight X 1.07.

FAS, Horticultural and Tropical Products Division, (202) 720-3423. The division's Home Page is located at, www.fas.usda.gov/htp. You can also email the division at htp@fas.usda.gov. Note: Compiled from reports of the U.S. Department of Commerce. U.S. trade can be obtained from the following web site: U.S. agricultural trade database, www.fas.usda.gov/ustrade.

Table 2-34.—Sugar, centrifugal (raw value): United States international trade in marketing years 2003/2004–2005/2006

Continent and country	2003/2004		2004/2005		2005/2006 [1]	
	Exports	Imports	Exports	Imports	Exports	Imports
	1,000 metric tons	*1,000 metric tons*	*1,000 metric tons*	*1,000 metric tons*	*1,000 metric tons*	*1,000 metric tons*
North America:						
United States	261	1,591	239	1,870	159	1,978
Canada	12	1,364	15	1,345	12	1,335
Mexico	14	327	145	226	344	101
Total	287	3,282	399	3,441	515	3,414
Caribbean:						
Cuba	1,900	100	700	100	1,032	300
Dominican Republic	185	0	185	0	206	0
Other	290	283	259	329	220	332
Total	2,375	383	1,144	429	1,458	632
Central America:						
Guatemala	1,335	0	1,386	0	1,391	0
Other	940	2	977	2	993	2
Total	2,275	2	2,363	2	2,384	2
South America:						
Brazil	15,240	0	18,020	0	18,250	0
Colombia	1,200	56	1,240	30	970	50
Argentina	201	10	225	15	530	10
Other	638	593	578	816	635	791
Total	17,279	659	20,063	861	20,385	851
Western Europe:						
EU	4,900	1,900	5,382	2,257	7,130	2,257
Other	10	514	10	407	11	400
Total	4,910	2,414	5,392	2,664	7,141	2,657
Eastern Europe:						
Poland	478	55	0	0	0	0
Ukraine	310	565	60	230	50	350
Russian Federation	110	3,670	110	4,300	110	4,200
Other	718	2,872	754	2,398	674	2,348
Total	1,616	7,162	924	6,928	834	6,898
Africa:						
South Africa	1,070	230	1,020	245	1,300	250
Other	2,210	5,446	2,321	5,476	2,326	5,507
Total	3,280	5,676	3,341	5,721	3,626	5,757
Middle East:						
Egypt	0	1,215	0	1,220	0	1,230
Turkey	232	1	50	0	50	0
Other	2,308	6,660	2,330	6,705	2,310	6,780
Total	2,540	7,876	2,380	7,925	2,360	8,010
Asia:						
India	250	550	50	2,000	200	1,000
China	67	1,235	182	1,250	110	1,300
Thailand	4,860	0	3,620	0	2,700	0
Australia	4,157	10	4,388	10	4,240	10
Pakistan	214	0	215	825	0	850
Indonesia	0	1,500	0	1,450	0	1,800
Philippines	202	0	306	0	156	0
Japan	18	1,364	10	1,431	10	1,402
Other Asia and Oceania	1,532	8,000	1,540	7,940	1,601	8,131
Total	11,300	12,659	10,311	14,906	9,017	14,493
World total [2]	45,862	40,113	46,317	42,877	47,720	42,714

[1] Preliminary.

FAS, Horticultural and Tropical Products Division, (202) 720-3423. The division's Home Page is located at, www.fas.usda.gov/htp. You can also email the division at htp@fas.usda.gov. Note: Compiled from reports of the U.S. Department of Commerce. U.S. trade can be obtained from the following web site: U.S. agricultural trade database, www.fas.usda.gov/ustrade.

Table 2-35.—Sugar, centrifugal (raw value): United States imports, by country of origin, 2002/2003–2004/2005[1][2][3]

Country of origin	2002/2003	2003/2004	2004/2005
	Metric tons	*Metric tons*	*Metric tons*
North America:			
Canada	5,753	9,917	66,394
Mexico	17	2	9
Total	5,770	9,918	66,402
Caribbean:			
Dominican Republic	184,525	184,361	184,452
Jamaica	0	11,489	2,921
French West Indies	0	10	9
Trinidad And Tobago	7,368	0	0
Total	191,892	195,860	187,382
Central America:			
Guatemala	196,246	228,363	164,824
El Salvador	54,363	65,610	122,971
Nicaragua	22,146	31,761	98,031
Costa Rica	40,834	105,712	55,019
Panama	40,403	32,642	53,543
Honduras	26,361	13,633	37,327
Belize	26,168	22,792	11,583
Total	406,520	500,512	543,297
South America:			
Brazil	120,462	154,782	145,935
Colombia	196,026	109,484	125,713
Peru	43,316	41,733	51,401
Argentina	45,310	44,106	50,618
Guyana	23,053	12,401	15,394
Paraguay	8,063	7,123	14,893
Ecuador	11,552	11,524	11,590
Bolivia	8,423	8,163	8,424
Uruguay	7,245	7,624	6,760
Total	463,451	396,940	430,728
European Union - 25:			
Belgium-Luxembourg	0	40	5
France	1	16	1
Germany	9	16	0
Netherlands	0	1	0
Total	10	72	6
Other Europe:			
Bosnia And Herzegovia	0	0	8
Total	0	0	8
Former Soviet Union:			
Russian Federation	7	4	0
Total	7	4	0
East Asia:			
Taiwan	12,636	12,595	12,362
China, Peoples Republic	20	71	10
Hong Kong	0	0	4
Total	12,656	12,666	12,376

See footnotes at end of table.

Table 2-35.—Sugar, centrifugal (raw value): United States imports, by country of origin, 2002/2003–2004/2005 [1] [2] [3]—Continued

Country of origin	2003	2004	2005
	Metric tons	Metric tons	Metric tons
Sub-Saharan Africa:			
South Africa, Republic	23,797	24,480	29,044
Swaziland	16,485	16,865	16,353
Mozambique	13,690	13,455	13,639
Zimbabwe	12,421	12,040	12,492
Malawi	9,149	10,350	10,113
Congo (Brazzaville)	7,271	7,271	7,255
Mauritius	2,387	20,887	3,870
Total	85,201	105,349	92,765
South Asia:			
India	8,427	8,235	0
Total	8,427	8,235	0
Southeast Asia:			
Philippines	142,002	141,265	142,160
Thailand	14,754	14,675	14,743
Malaysia	10	0	0
Total	156,767	155,940	156,903
Oceania:			
Australia	86,164	86,395	103,175
Other Pacific Island	9,351	9,378	9,321
Papua New Guinea	7,068	7,258	7,075
Total	102,583	103,031	119,571
World total	1,433,284	1,488,526	1,609,438

[1] Imports for consumption, imports in bonded warehouses (general imports) are tallied when customs as sugar whose content of sucrose, by weight in the dry state, corresponds to a polarimeter reading less than 99.5 degrees.   [2] Data are actual weight x 1.035.   [3] October/September.

FAS, Horticultural and Tropical Products Division, (202) 720-3423. The division's Home Page is located at, www.fas.usda.gov/htp. You can also email the division at htp@fas.usda.gov. Note: Compiled from reports of the U.S. Department of Commerce. U.S. trade can be obtained from the following web site: U.S. agricultural trade database, www.fas.usda.gov/ustrade.

Table 2-36.—Sugar, cane and beet (raw value): Production, stocks, trade, and supply available for consumption in continental United States includes Puerto Rico, 1996–2005

Year	Production	Visible stocks beginning of period	Imports	Exports	Total deliveries
	1,000 short tons	1,000 short tons	1,000 short tons	1,000 short tons	1,000 short tons
1996	7,268	2,908	2,927	331	9,619
1997	7,419	3,195	2,677	187	9,755
1998	7,881	3,377	2,148	203	9,854
1999	9,100	3,422	1,806	203	10,167
2000	8,955	3,855	1,639	109	10,091
2001	8,642	4,337	1,643	147	10,075
2002	7,504	4,525	1,574	136	9,994
2003	8,929	3,432	1,564	148	9,713
2004	8,366	4,088	1,652	241	9,901
2005	7,478	4,029	2,223	254	10,185

ERS, Specialty Crops Branch, (202) 694–5247.

Table 2-37.—Sugar, centrifugal (raw value): Beginning stocks in marketing years 2003/2004–2005/2006 [1]

Country	2003/2004	2004/2005	2005/2006
	1,000 metric tons	*1,000 metric tons*	*1,000 metric tons*
North America:			
United States	1,515	1,721	1,229
Canada	269	288	288
Mexico	1,194	1,237	2,043
Total	2,978	3,246	3,560
Caribbean:			
Cuba	182	132	132
Dominican Republic	22	60	28
Other	120	88	94
Total	324	280	254
Central America:			
Guatemala	173	188	223
Other	265	314	350
Total	438	502	573
South America:			
Brazil	270	1,030	585
Colombia	71	97	87
Argentina	76	250	245
Other	1,474	1,350	1,394
Total	1,891	2,727	2,311
Western Europe:			
EU [2]	3,581	4,699	5,773
Other Western Europe	333	346	346
Total	3,914	5,045	6,119
Eastern Europe:			
Poland [3]	979	NA	NA
Ukraine	496	181	151
Russian Federation	1,050	440	580
Other	1,856	1,534	1,377
Total	4,381	2,155	2,108
Africa:			
South Africa, Republic of	586	870	920
Other Africa	2,536	2,571	2,672
Total	3,122	3,441	3,592
Middle East:			
Egypt	233	363	459
Turkey	730	484	583
Other	1,918	1,920	1,945
Total	2,881	2,767	2,987
Asia and Oceana:			
India	12,430	9,070	5,730
China, Peoples Republic	2,021	2,323	1,617
Thailand	1,045	1,215	732
Australia	662	543	403
Pakistan	797	1,030	827
Indonesia	1,340	1,170	1,120
Philippines	277	405	239
Japan	359	354	433
Other	2,064	2,554	2,537
Total	20,995	18,664	13,638
World total	40,924	38,827	35,142

[1] 2003/04 - forecast.   [2] EU-15 for years prior to 2004/05. EU-25 beginning in 2004/2005.   [3] No data available beyond 2003/2004 due to Poland's accession to the EU.   NA-not available.

FAS, Horticultural and Tropical Products Division, (202) 720-3423. The division's Home Page is located at, www.fas.usda.gov/htp. You can also email the division at htp@fas.usda.gov. Note: Prepared or estimated on the basis of official statistics of foreign governments, other foreign source materials, reports of U.S. Agricultural Counselors and Attaches, results of office research, and related information. Note: This information can be obtained from the following web site: www.fas.usda.gov/psd.

Table 2-38.—Honey: United States imports for consumption, by country of origin, 2002–2004

Continent and country of origin	2002	2003	2004
	Metric tons	*Metric tons*	*Metric tons*
North America:			
Canada	19,617	11,607	10,172
Mexico	11,544	7,351	3,254
Total	31,161	18,958	13,426
Carribean:			
Dominican Republic	99	245	164
Bahamas	15	6	4
Jamaica	7	0	0
Total	121	251	168
Central America:			
Guatamala	149	0	24
Total	149	0	24
South America:			
Brazil	5,363	7,297	3,690
Argentina	8,692	4,425	3,620
Uruguay	5,968	5,308	3,137
Chile	2,665	4,550	844
Peru	363	935	529
Colombia	0	40	0
Venezuela	0	6	0
Total	23,051	22,561	11,820
European Union:			
Germany	541	380	406
Czech Republic	39	217	298
Poland	19	62	235
Hungary	252	292	148
Italy	100	51	70
France	47	48	60
Greece	49	59	43
Austria	21	15	30
Denmark	73	179	27
United Kingdom	10	15	27
Spain	247	31	26
Portugal	6	8	11
Lithuania	0	3	6
Sweden	1	1	1
Latvia	1	2	0
Netherlands	0	75	0
Total	1,406	1,438	1,388
Other European:			
Romania	526	1,003	877
Bulgaria	177	631	657
Switzerland	134	58	69
Macedonia	0	0	7
Croatia	4	0	5
Bosnia And Herzegovia	0	0	1
Total	841	1,692	1,616
Former Soviet Union:			
Ukraine	573	2,226	1,596
Russian Federation	27	135	1,075
Armenia	0	10	41
Moldova	173	164	19
Kazakhstan	0	20	0
Total	773	2,555	2,731
East Asia:			
China	7,583	22,827	26,916
Taiwan	44	81	759
Hong Kong	126	25	10
Korea	1	0	1
Total	7,754	22,933	27,686
Middle East:			
Turkey	1,763	2,223	622
Israel	22	13	17
Iran	0	0	3
Jordan	1	0	0
Total	1,786	2,236	642

See end of table.

Table 2-38.—Honey [1]: United States imports for consumption, by country of origin, 2002–2004—Continued

Continent and country of origin	2002	2003	2004 [2]
	Metric tons	*Metric tons*	*Metric tons*
North Africa:			
Egypt	40	226	295
Total	40	226	295
Sub-Saharan Africa:			
Kenya	0	0	1
Ethiopia	0	0	1
Djibouti Afars-Issas	0	0	1
South Africa	1	0	0
Burkina	0	19	0
Total	1	19	3
South Asia:			
India	2,465	4,645	6,948
Pakistan	57	345	472
Bangladesh	1	0	0
Total	2,523	4,990	7,420
Southeast Asia:			
Vietnam	14,356	7,979	9,792
Indonesia	0	384	1,876
Thailand	4,445	799	769
Malaysia	1,039	3,534	442
Burma	94	56	0
Total	19,934	12,752	12,879
Oceania:			
Australia	2,329	101	780
New Zealand	42	195	114
Total	2,371	296	894
Grand total [3]	91,907	90,906	80,994

[1] HS Code 040900. [2] Calendar year data. [3] Totals may not add due to rounding.

FAS, Horticultural and Tropical Products Division, (202) 720-3423. The division's Home Page is located at, www.fas.usda.gov/htp. You can also email the division at htp@fas.usda.gov. Note: Compiled from reports of the U.S. Department of Commerce. U.S. trade can be obtained from the following web site: U.S. agricultural trade database, www.fas.usda.gov/ustrade.

II-26   STATISTICS OF COTTON, TOBACCO, SUGAR CROPS, AND HONEY

Table 2-39.—Beeswax, crude: United States imports for consumption, by country of origin, 2002–2004

Continent and country of origin	2002	2003	2004[1]
	Metric tons	Metric tons	Metric tons
North America:			
Canada	320	596	449
Mexico	6	7	19
Total	326	603	468
Carribean:			
Dominican Republic	22	22	11
Total	22	22	11
South America:			
Argentina	235	235	191
Brazil	20	3	0
Uruguay	0	51	0
Total	255	289	191
European Union:			
Germany	185	279	112
United Kingdom	0	5	70
France	21	17	25
Belgium-Luxembourg	0	1	10
Poland	0	0	8
Netherlands	9	1	2
Italy	1	0	0
Total	216	303	227
Other Europe:			
Switzerland	2	1	1
Total	2	1	1
East Asia:			
China	305	382	279
Japan	7	4	59
Taiwan	13	0	0
Total	325	386	338
Sub-Saharan Africa:			
Ethiopia	70	140	88
Tanzania, United Rep	17	162	60
Central African Republic	54	31	54
Guinea	0	0	34
Djibouti Afars-Issas	0	36	34
Madagascar	0	0	24
Mali	0	22	20
Cameroon	33	34	17
South Africa	2	1	11
Ghana	5	0	0
Nigeria	6	0	0
Senegal	0	18	0
Zambia	1	1	0
Total	188	445	342
South Asia:			
India	0	0	1
Total	0	0	1
Southeast Asia:			
Vietnam	0	0	15
Malaysia	40	39	0
Philippines	0	1	0
Total	40	40	15
Oceana:			
Australia	107	86	97
New Zealand	9	21	21
Total	116	107	118
Grand total[2]	1,488	2,195	1,712

[1] Preliminary.   [2] Totals may not add due to rounding.

FAS, Horticultural and Tropical Products Division, (202) 720-3423. The division's Home Page is located at, www.fas.usda.gov/htp. You can also email the division at htp@fas.usda.gov. Note: Compiled from reports of the U.S. Department of Commerce. U.S. trade can be obtained from the following web site: U.S. agricultural trade database, www.fas.usda.gov/ustrade.

Table 2-40.—Honey: Number of colonies, yield, production, stocks, price and value, United States, 1996–2005 [1][2]

State	Honey producing colonies	Yield per colony	Production	Stocks Dec 15[3]	Average price per pound	Value of production
	1,000	Pounds	1,000 pounds	1,000 pounds	Cents	1,000 dollars
1996 ...	2,581	77.3	199,511	47,206	88.8	177,166
1997 ...	2,631	74.7	196,536	70,696	75.2	147,795
1998 ...	2,637	83.7	220,527	80,907	65.5	147,468
1999 ...	2,652	76.4	203,068	78,664	60.1	125,004
2000 ...	2,622	84.1	220,286	85,244	59.7	132,865
2001 ...	2,550	73.0	186,051	64,907	70.4	132,989
2002 ...	2,574	66.7	171,718	39,393	132.7	228,338
2003 ...	2,599	69.9	181,727	40,785	138.7	253,106
2004 ...	2,556	71.8	183,582	61,222	106.9	196,259
2005 ...	2,410	72.5	174,643	62,406	90.4	157,795

[1] For producers with 5 or more colonies.   [2] U.S. price weighted by survey expanded sales.   [3] Stocks held by producers.

NASS, Livestock Branch, (202) 720-6351.

Table 2-41.—Honey: Number of colonies, yield, production, stocks, price and value, by State and United States, 2005 [1]

State	Honey producing colonies	Yield per colony	Production	Stocks Dec 15[2]	Average price per pound[3]	Value of production
	1,000	Pounds	1,000 pounds	1,000 pounds	Cents	1,000 dollars
AL	13	66	858	266	128	1,098
AZ	36	50	1,800	720	97	1,746
AR	36	69	2,484	571	95	2,360
CA	400	75	30,000	9,300	84	25,200
CO	26	70	1,820	837	104	1,893
FL	160	86	13,760	2,477	87	11,971
GA	59	49	2,891	434	84	2,428
HI	9	131	1,179	283	134	1,580
ID	95	37	3,515	1,793	80	2,812
IL	8	85	680	408	162	1,102
IN	8	64	512	189	119	609
IA	28	88	2,464	1,232	98	2,415
KS	16	50	800	328	124	992
KY	5	50	250	40	212	530
LA	35	97	3,395	611	71	2,410
ME	8	26	208	193	187	389
MI	65	68	4,420	2,519	99	4,376
MN	120	74	8,880	1,598	83	7,370
MS	16	80	1,280	346	66	845
MO	15	50	750	180	121	908
MT	130	67	8,710	3,136	80	6,968
NE	40	68	2,720	2,530	89	2,421
NV	12	46	552	442	209	1,154
NJ	12	32	384	104	118	453
NM	7	49	343	113	102	350
NY	60	73	4,380	2,321	122	5,344
NC	10	54	540	146	194	1,048
ND	370	91	33,670	8,418	81	27,273
OH	15	69	1,035	580	141	1,459
OR	39	42	1,638	557	108	1,769
PA	28	56	1,568	800	105	1,646
SD	220	79	17,380	11,818	76	13,209
TN	7	55	385	92	164	631
TX	84	71	5,964	954	85	5,069
UT	23	45	1,035	331	103	1,066
VT	6	91	546	169	106	579
VA	8	37	296	59	221	654
WA	51	55	2,805	1,935	106	2,973
WV	8	51	408	102	124	506
WI	64	83	5,312	2,922	114	6,056
WY	40	56	2,240	291	87	1,949
Oth Sts[4] ..	18	44	786	261	278	2,184
US[5][6] ...	2,410	72.5	174,643	62,406	90.4	157,795

[1] For producers with 5 or more colonies. Colonies which produced honey in more than one State were counted in each State.   [2] Stocks held by producers.   [3] Price weighted by sales.   [4] CT, DE, MD, MA, NH, OK, RI, and SC not published separately to avoid disclosing data for individual operations.   [5] Total colonies multiplied by total yield may not exactly equal production.   [6] U.S. value of production is U.S. production multiplied by U.S. price per pound.

NASS, Livestock Branch, (202) 720-3570.

Table 2-42.—U.S. per capita caloric sweeteners estimated deliveries for domestic food and beverage, use by calendar year 1995–2004

Calendar year	U.S.population (July 1)	Refined sugar	Corn Sweetener				Pure honey	Edible syrups	Total caloric sweeteners
			HFCS	Glucose syrup	Dex-trose	Total			
									Millions
1995	266.6	64.9	57.6	16.3	4.0	77.9	0.9	0.4	144.1
1996	269.7	65.2	57.8	16.4	4.0	78.2	1.0	0.4	144.7
1997	272.9	64.9	60.4	17.3	3.7	81.5	0.9	0.4	147.8
1998	276.1	64.9	61.9	17.1	3.6	82.7	0.9	0.4	149.0
1999	279.3	66.3	63.7	16.3	3.5	83.5	1.1	0.4	151.4
2000	282.3	65.5	62.7	15.8	3.4	81.8	1.1	0.4	148.9
2001	285.0	64.5	62.6	15.5	3.3	81.4	0.9	0.4	147.3
2002	287.7	63.3	62.9	15.5	3.3	81.6	1.1	0.4	146.5
2003	290.3	61.0	61.0	15.2	3.1	79.3	1.0	0.4	141.7
2004	293.0	61.9	59.4	15.6	3.3	78.4	0.9	0.4	141.5

Note: Total may not add exactly, due to rounding.
ERS, Market and Trade Economics Division, Specialty Crops Branch, (202) 694–5247.

Table 2-43.—Tobacco: Area, yield, production, price, and value, United States, 1996–2005

Year	Area harvested	Yield per acre	Production [1]	Marketing year average price per pound received by farmers	Value of production
	Acres	*Pounds*	*1,000 pounds*	*Dollars*	*1,000 dollars*
1996	733,060	2,072	1,518,704	1.882	2,853,739
1997	836,230	2,137	1,787,399	1.802	3,217,176
1998	717,620	2,062	1,479,891	1.828	2,700,925
1999	647,160	1,997	1,292,692	1.828	2,356,304
2000	469,420	2,244	1,053,264	1.910	2,001,811
2001	432,490	2,292	991,293	1.956	1,938,892
2002	427,310	2,039	871,122	1.936	1,686,809
2003	411,150	1,952	802,560	1.964	1,576,436
2004	408,050	2,161	881,973	1.987	1,752,335
2005	298,020	2,147	639,709	1.647	1,053,430

[1] Production figures are on farm-sales-weight basis.
NASS, Crops Branch, (202) 720–2127.

Table 2-44.—Tobacco: Area, yield, and production, by States, 2003–2005

State	Area harvested			Yield per harvested acre			Production		
	2003	2004	2005 [1]	2003	2004	2005 [1]	2003	2004	2005 [1]
	Acres	*Acres*	*Acres*	*Pounds*	*Pounds*	*Pounds*	*1,000 pounds*	*1,000 pounds*	*1,000 pounds*
CT	2,180	2,360	2,430	1,321	1,574	1,674	2,880	3,714	4,067
FL	4,400	4,000	2,500	2,500	2,450	2,200	11,000	9,800	5,500
GA	27,000	23,000	16,000	2,200	2,030	1,735	59,400	46,690	27,760
IN [2]	4,200	4,200	....	1,950	2,050	....	8,190	8,610	....
KY	111,650	114,950	79,700	2,016	2,044	2,099	225,042	235,003	167,260
MD [2]	1,100	1,100	....	1,450	1,700	....	1,595	1,870	....
MA	1,250	1,240	1,200	1,392	1,587	1,500	1,740	1,968	1,800
MO	1,400	1,450	1,400	2,020	2,300	2,000	2,828	3,335	2,800
NC	159,700	156,100	126,000	1,878	2,246	2,213	299,995	350,560	278,900
OH	5,300	5,600	3,400	1,650	1,960	1,980	8,745	10,976	6,732
PA	3,700	4,000	5,000	2,130	2,025	2,140	7,880	8,100	10,700
SC	30,000	27,000	20,000	2,100	2,350	2,100	63,000	63,450	42,000
TN	31,140	30,260	22,950	2,108	2,161	2,251	65,632	65,381	51,670
VA	25,110	29,680	17,040	1,546	2,267	2,338	38,818	67,285	39,840
WV	1,200	1,300	400	1,300	1,300	1,700	1,560	1,690	680
WI [2]	1,820	1,810	....	2,338	1,956	....	4,255	3,541	....
US	411,150	408,050	298,020	1,952	2,161	2,147	802,560	881,973	639,709

[1] Preliminary.   [2] Estimates discontinued in 2005.
NASS, Crops Branch, (202) 720–2127.

Table 2-45.—Tobacco: Area, yield, production, stocks, supply, disappearance, and price, by types, United States including Puerto Rico, 1996–2005 (farm-sales-weight basis)

Type and crop year	Area	Yield per acre	Produc- tion	Stocks [1]	Supply	Disappearance			Average price per pound to growers
						Total	Exports	Domes- tic	
	Acres	Pounds	1,000 pounds	1,000 pounds	1,000 pounds	1,000 pounds	1,000 pounds	1,000 pounds	Cents
Total flue-cured, types 11–14: [2]									
1996	422,200	2,151	908,345	1,166,427	2,063,769	947,261	391,200	556,061	183.4
1997	458,300	2,285	1,047,438	1,116,508	2,130,041	876,878	335,900	540,978	172.0
1998	368,800	2,204	812,797	1,253,163	2,067,956	833,676	341,550	492,446	175.5
1999	303,800	2,162	656,752	1,234,280	1,888,172	698,684	261,818	436,866	173.6
2000	250,000	2,396	598,915	1,189,488	1,753,609	717,242	238,025	479,217	179.3
2001	238,100	2,432	579,091	1,036,367	1,580,790	664,912	276,007	388,905	185.8
2002	245,600	2,105	525,940	915,878	1,480,678	643,008	219,631	423,377	182.5
2003	233,400	1,957	525,941	837,670	1,345,326	522,478	215,520	306,958	184.9
2004	228,400	2,272	499,330	822,848	1,322,178	526,210	188,627	337,583	184.4
2005 [3]	178,800	2,176	389,120	795,968	1,185,088	....	....	....	....
Total fire-cured, types 21–23:									
1996	16,580	2,668	44,228	83,662	127,890	42,165	0	22,456	224.5
1997	16,550	2,554	42,262	85,725	127,987	40,839	18,235	22,604	225.6
1998	16,840	2,365	39,835	87,148	126,983	37,593	15,727	21,866	222.5
1999	16,420	2,319	38,075	89,390	127,465	36,246	12,979	14,312	226.4
2000	17,540	2,944	51,635	91,219	142,854	44,869	26,292	18,600	213.7
2001	14,620	3,096	45,299	97,962	143,261	38,955	16,379	22,576	213.0
2002	10,970	3,182	33,437	104,306	139,214	37,342	10,733	26,609	235.7
2003	11,250	3,067	34,508	101,872	136,380	33,788	20,259	13,259	245.5
2004	11,020	3,167	37,151	102,592	139,743	33,702	9,198	24,504	251.3
2005 [3]	10,700	3,186	39,725	106,041	145,766	....	....	....	....
Virginia fire-cured, type 21:									
1996	1,100	1,580	1,738	3,468	5,206	2,812	2,500	312	179.0
1997	1,200	1,640	1,968	2,394	4,362	1,989	554	1,435	212.5
1998	1,500	1,560	2,340	2,373	4,713	2,044	550	1,494	193.6
1999	1,600	1,670	2,672	2,669	5,341	1,897	979	918	181.9
2000	1,300	1,700	2,548	3,444	5,992	1,806	1,000	806	163.7
2001	1,200	1,835	2,202	4,168	6,388	1,567	150	1,417	175.8
2002	730	2,015	1,471	4,821	6,292	1,997	64	1,933	188.4
2003	550	1,525	839	4,295	5,134	1,359	63	1,296	160.3
2004	710	1,895	1,345	3,775	5,120	1,242	400	842	178.4
2005 [3]	550	2,300	805	3,878	4,683	....	....	....	....
Kentucky and Ten- nessee fire-cured, types 22–23:									
1996	15,480	2,745	42,490	80,194	122,684	39,353	17,209	22,144	224.5
1997	15,350	2,625	40,294	83,331	123,625	38,850	17,681	21,169	225.6
1998	15,340	2,444	37,495	84,775	122,270	35,549	15,177	20,372	222.5
1999	14,970	2,365	35,403	86,721	122,124	34,349	20,955	13,394	229.8
2000	16,240	3,023	49,087	87,775	136,862	43,086	25,292	17,794	216.3
2001	13,420	3,211	43,097	89,766	136,873	37,388	16,229	21,159	217.2
2002	10,240	3,265	33,437	90,787	132,922	35,345	10,669	24,676	214.9
2003	10,700	3,147	33,669	93,162	131,246	32,429	13,196	19,233	176.7
2004	11,020	3,249	35,806	94,315	134,623	32,460	8,798	23,662	161.6
2005 [3]	12,470	3,121	38,920	102,163	141,083	....	....	....	....
Burley, type 31: [2]									
1996	268,300	1,940	520,483	890,390	1,406,731	655,740	209,446	446,294	192.2
1997	335,300	1,943	648,633	750,991	1,379,222	547,616	168,395	379,221	188.5
1998	307,100	1,896	582,336	831,606	1,421,903	520,488	168,853	351,635	190.3
1999	300,600	1,829	555,185	901,415	1,452,573	412,531	139,262	273,269	189.9
2000	185,400	1,957	362,788	1,040,042	1,355,481	666,022	142,020	523,012	196.3
2001	164,400	2,032	334,066	689,459	1,033,119	385,238	139,802	245,436	197.3
2002	157,700	1,861	303,895	647,881	947,726	369,561	148,618	220,943	197.4
2003	152,300	1,850	303,896	578,165	849,861	309,832	173,650	136,182	197.7
2004	153,150	1,908	303,897	540,029	820,129	329,645	227,571	102,074	199.0
2005 [3]	105,300	1,826	192,285	490,484	682,769	....	....	....	....
Maryland, type 32: [2]									
1996	11,400	1,451	16,030	15,007	31,037	11,549	6,442	6,572	185.6
1997	11,200	1,629	17,700	19,488	37,188	14,645	6,515	8,130	158.5
1998	9,800	1,568	15,370	22,543	37,913	18,855	6,228	12,627	129.1
1999	9,500	1,511	14,350	16,003	30,353	14,353	10,157	4,196	134.5
2000	8,400	1,595	13,395	13,361	26,756	17,071	12,690	4,381	138.7
2001	3,300	1,620	5,346	9,685	15,031	6,817	4,126	2,691	155.4
2002	2,500	1,682	4,205	8,214	12,419	5,070	3,306	1,764	134.8
2003	2,400	1,748	4,195	7,349	11,544	5,594	3,802	1,792	146.3
2004	3,300	1,767	5,830	4,970	10,800	6,500	1,278	5,222	147.6
2005 [3]	1,500	2,000	3,000	4,300	7,300	....	....	....	....

See footnotes at end of table.

Table 2-45.—Tobacco: Area, yield, production, stocks, supply, disappearance, and price, by types, United States including Puerto Rico, 1996–2005 (farm-sales-weight basis)—Continued

Type and crop year	Area	Yield per acre	Production	Stocks [1]	Supply	Disappearance			Average price per pound to growers
						Total	Exports	Domestic	
	Acres	Pounds	1,000 pounds	1,000 pounds	1,000 pounds	1,000 pounds	1,000 pounds	1,000 pounds	Cents
Total dark air-cured, types 35–37:									
1996	3,850	2,250	8,662	25,472	34,134	10,277	61	10,216	195.0
1997	3,710	2,241	8,315	23,857	32,172	4,260	10	9,250	201.6
1998	4,435	2,206	9,785	22,512	32,297	8,203	42	8,161	195.1
1999	5,100	3,878	11,795	24,094	35,889	9,176	1,433	7,743	203.3
2000	5,580	4,551	16,061	26,713	42,774	9,896	1,022	8,874	196.4
2001	5,070	4,347	14,103	32,878	46,981	8,614	322	8,292	182.7
2002	3,830	4,466	10,686	38,367	48,956	11,679	230	11,449	209.8
2003	4,150	4,146	11,314	37,374	48,688	11,307	176	11,131	215.4
2004	4,260	4,586	11,924	37,381	49,305	11,767	58	11,709	218.7
2005 [3]	4,040	2,787	11,258	37,538	48,796	....	....	....	
One Sucker, Green River type 35-36:									
1996	3,780	2,262	8,550	25,424	33,974	10,141	....	10,141	191.1
1997	3,630	2,258	8,196	23,833	32,029	9,564	10	9,554	201.7
1998	4,335	2,229	9,663	22,465	32,128	8,107	10	8,097	195.5
1999	5,000	2,328	11,640	24,021	35,661	9,036	1,337	7,699	203.9
2000	5,480	2,901	15,896	26,625	42,521	9,824	1,000	8,824	197.1
2001	4,970	2,807	13,949	32,697	46,646	8,391	100	8,291	182.9
2002	3,760	2,811	10,570	38,255	48,825	11,548	100	11,448	210.1
2003	4,090	2,746	11,230	37,277	48,507	11,231	100	11,131	215.7
2004	4,190	2,816	11,798	37,276	49,074	11,561	58	11,503	219.4
2005 [3]	4,040	2,787	11,258	37,513	48,771	....	....	....	
Virginia sun-cured, type 37:									
1996	70	1,600	112	48	160	136	49	27	178.2
1997	80	1,490	119	24	143	96	51	45	190.8
1998	100	1,220	122	47	169	96	38	58	170.9
1999	100	1,550	155	73	228	140	44	96	159.0
2000	100	1,650	165	88	253	72	50	22	180.0
2001	100	1,540	154	181	335	223	1	222	168.6
2002	70	1,655	116	112	131	131	1	130	177.8
2003	60	1,400	84	97	181	76	0	76	142.9
2004	70	1,770	126	105	231	205	....	205	145.8
2005	11		....	26	26	....	....	....	
Total continental cigar filler, types 41–44:									
1996	4,600	2,040	9,384	17,939	27,323	14,124	*	12,700	155.0
1997	4,600	2,100	9,660	13,199	22,859	11,000	*	11,000	160.0
1998	4,500	2,100	9,450	12,969	22,419	11,039	*	11,039	130.0
1999	3,200	1,850	5,920	11,380	17,300	7,768	*	7,768	130.0
2000	2,400	2,100	5,040	9,532	14,572	2,453	*	2,453	NA
2001	2,000	2,060	4,120	12,119	16,239	3,968	*	3,968	150.0
2002	2,100	2,100	4,410	12,271	16,681	6,014	*	6,014	145.0
2003	2,400	2,200	5,280	10,667	15,947	6,218	*	6,218	140.0
2004	1,800	2,300	4,140	9,729	13,869	5,545	*	5,545	145.0
2005 [3]	1,300	2,200	2,860	1,000	11,184	....			
Pennsylvania seedleaf filler, type 41:									
1996	4,800	2,140	10,272	17,939	28,211	15,012	*	15,012	155
1997	4,900	2,200	10,780	13,199	23,979	11,010	*	11,010	160
1998	4,500	2,100	9,450	12,969	22,100	10,720	*	10,720	130
1999	3,200	1,850	5,920	11,380	17,300	7,768	*	7,768	130
2000	2,400	2,100	5,040	9,532	14,572	2,453	*	2,453	NA
2001	2,000	2,060	4,120	12,119	16,239	3,968	*	3,968	150
2002	2,100	2,100	4,410	12,271	16,681	6,014	*	6,014	145
2003	2,400	2,200	5,280	10,667	15,947	6,218	*	6,218	140
2004	1,800	2,300	4,140	9,729	13,869	5,545	*	5,545	145
2005 [3]	1,300	2,200	2,860	8,324	11,184	....			
Puerto Rican filler, type 46: [5]									
1996	*	*	*	*	*	*	*	*	*
1997	*	*	*	*	*	*	*	*	*
1998	*	*	*	*	*	*	*	*	*
1999	*	*	*	*	*	*	*	*	*
Total cigar binder, types 51–55:									
1996	4,500	1,792	8,063	23,317	31,380	12,398	11,832	12,198	321.1
1997	4,590	2,032	9,327	18,982	28,243	10,655	9,051	10,255	378.4
1998	4,410	1,783	7,863	18,253	26,116	8,335	6,440	1,895	334.7
1999	3,680	1,899	6,987	17,781	24,768	9,321	8,057	1,264	342.7
2000	1,860	1,787	3,325	15,447	18,772	6,735	346	5,389	263.3
2001	3,650	2,039	7,441	12,037	19,478	8,954	162	8,592	367.3
2002	3,650	2,147	7,838	10,524	18,362	8,149	1,379	6,770	356.5
2003	4,410	1,783	7,863	18,253	26,116	8,335	1,895	6,440	334.7
2004	3,680	1,899	6,987	17,781	24,768	9,321	1,046	8,275	342.7
2005 [3]	2,400	1,744	4,185	1,180	5,365	....	....		

See footnotes at end of table.

Table 2-45.—Tobacco: Area, yield, production, stocks, supply, disappearance, and price, by types, United States including Puerto Rico, 1996–2005 (farm-sales-weight basis)—Continued

Type and crop year	Area	Yield per acre	Production	Stocks[1]	Supply	Disappearance			Average price per pound to growers
						Total	Exports	Domestic	
	Acres	*Pounds*	*1,000 pounds*	*1,000 pounds*	*1,000 pounds*	*1,000 pounds*	*1,000 pounds*	*1,000 pounds*	*Cents*
Connecticut Valley binder, types 51–52:									
1996	1,630	1,780	2,901	1,577	4,478	3,298	605	2,932	628.0
1997	2,040	1,783	3,637	3,637	4,817	2,766	605	2,161	741.4
1998	2,360	1,539	3,633	2,051	5,684	2,199	775	1,424	549.9
1999	2,500	1,668	4,169	3,485	7,654	4,888	264	4,624	473.7
2000	900	1,189	1,070	2,766	3,836	1,522	346	1,176	491.6
2001	2,140	1,786	3,822	2,314	6,136	4,308	162	4,146	558.9
2002	2,200	1,828	4,021	1,828	5,849	3,779	1,379	2,400	536.9
2003	2,140	1,786	3,822	2,314	6,136	4,308	46	4,262	558.9
2004	2,420	1,557	3,767	1,576	5,343	4,163	600	3,563	530.9
2005[3]	2,400	1,744	4,185	1,180	5,365				
Wisconsin binder, types 54–55:									
1996	2,870	1,799	5,162	21,740	26,902	9,100	200	8,900	148.6
1997	2,550	2,231	5,690	17,802	23,492	7,290	400	6,890	150.5
1998	2,050	2,063	4,230	16,202	20,432	6,136	1,120	5,016	149.8
1999	1,180	2,388	2,818	14,296	17,114	4,433	1,000	3,433	149.0
2000	960	2,348	2,255	12,681	14,936	5,213	1,000	4,213	155.0
2001	1,510	2,397	3,619	9,723	13,342	4,646	200	4,646	165.0
2002	1,450	2,632	3,817	8,696	12,513	4,370	50	4,320	175.0
2003	1,180	2,388	2,818	14,296	17,114	4,433	1,000	3,433	149.0
2004	1,810	1,956	3,541	8,637	12,178	3,336	1,000	2,336	174.6
2005[3]	[12]			8,842	8,842				
Southern Wisconsin, type 54:									
1996	1,900	1,900	3,610						148.0
1997	1,800	2,330	4,194						151.0
1998	1,500	2,180	3,270						150.0
1999	890	2,530	2,252						149.0
2000	710	2,500	1,825						155.0
2001	1,200	2,535	3,042						165.0
2002	1,150	2,740	3,151						
2003	1,400	2,480	3,472						175.0
2004	1,400	1,960	2,744						175.0
2005[3]	[11]								
Northern Wisconsin, type 55:									
1996	970	1,600	1,552						
1997	750	1,995	1,496						
1998	550	1,745	960						
1999[2]	290	1,952	566						
2000	230	1,865	430						
2001	310	1,860	577						
2002	300	2,220	666						
2003	420	1,865	783						175.0
2004	410	1,945	797						175.0
2005[12]									

See footnotes at end of table.

Table 2-45.—Tobacco: Area, yield, production, stocks, supply, disappearance, and price, by types, United States including Puerto Rico, 1996–2005 (farm-sales-weight basis)—Continued

Type and crop year	Area	Yield per acre	Produc- tion	Stocks [1]	Supply	Disappearance			Aver- age price per pound to grow- ers
						Total	Exports	Domestic	
	Acres	*Pounds*	*1,000 pounds*	*1,000 pounds*	*1,000 pounds*	*1,000 pounds*	*1,000 pounds*	*1,000 pounds*	*Cents*
Total cigar wrapper, types 61:									
1996	1,430	1,473	2,106	2,256	4,362	2,410	2,000	210	[3]
1997	1,680	1,431	2,404	1,952	4,356	2,309	2,100	209	2,520
1998	1,720	1,413	2,431	2,047	4,478	3,202	2,722	480	NA
1999	1,860	1,951	3,628	1,276	4,904	4,127	3,021	1,106	[3]
2000	1,250	1,472	1,840	777	2,617	1,494	1,300	194	2,530
2001	1,270	1,605	1,757	1,123	2,880	1,093	800	293	NA
2002	960	1,201	1,153	1,787	2,940	2,232	750	1,482	2,250
2003	1,060	1,164	1,234	708	1,942	1,566	450	1,116	2,600
2004	1,170	1,597	1,869	376	2,245	1,631	800	831	NA
2005 [3]	1,200	1,538	1,845	614	2,459	....	....	....	NA
Total tobacco, types 11–72: [6]									
1996	733,060	2,072	1,518,704	2,225,443	3,744,147	1,713,445	631,036	1,082,409	[9] 188.2
1997	836,230	2,137	1,787,399	2,030,702	3,818,101	1,567,860	532,211	1,035,649	[9] 180.3
1998	717,605	2,062	1,479,867	2,250,241	3,730,108	1,429,440	537,023	892,417	[9] 182.8
1999	647,160	1,997	1,292,692	2,300,668	3,593,360	1,204,925	432,876	772,049	[9] 182.8
2000	472,410	2,229	1,052,999	2,388,435	3,441,434	1,548,344	414,414	1,133,930	186.9
2001	432,310	2,293	991,223	1,893,090	2,884,313	1,145,915	436,142	709,773	[9] 192.0
2002	427,310	2,039	871,122	1,738,398	2,609,520	1,025,502	382,976	642,526	[9] 193.6
2003	411,150	1,952	802,560	1,584,018	2,386,578	857,460	407,271	450,189	196.4
2004	408,040	2,155	879,227	1,529,118	2,408,345	944,345	515,767	428,578	198.8
2005 [3]	307,010	2,099	644,278	1,464,000	2,108,278	....	....	....	....

[1] July 1 for flue-cured types 11-14 and cigar types 61 and 62; Oct. 1 for all other types.   [2] Flue-cured (type11-14) and Burley (type 31) supply based on actual marketing. Maryland (type 32) based on October 1 stocks.   [3] Preliminary.   [4] Not available.   [5] No longer produced.   [6] Includes Perique.   [7] Does not include cigar wrapper type 61.   [8] Does not include cigar filler type 41.   [9] Does not include cigar wrapper type 61.   [10] Does not include cigar filler type 41.   [11] No longer produced.   [12] Data no longer collected.   NA-not applicable.   *Negligible.

ERS, Specialty Crops Branch, (202) 694–5311. Basic export data from the official reports of the Department of Commerce.

Table 2-46.—Tobacco: Price-support loan operations, United States, 1996–2005 [1]

	Flue-cured, types 11–14			Burley, type 31		
Year	Support price per pound	Placed under loan		Support price per pound	Placed under loan	
		Quantity	Percentage of production		Quantity	Percentage of production
	Cents	*Million pounds*	*Percent*	*Cents*	*Million pounds*	*Percent*
1996	160.1	1.8	0.2	173.7	0.0	0.0
1997	162.1	195.5	19.4	176	124.5	19.8
1998	162.8	82.5	10.1	177.8	73.2	12.4
1999	163.2	136.4	21.2	178.9	230.6	42.0
2000	164.0	27.4	4.4	180.5	19.3	4.8
2001	166.0	15.0	2.6	182.6	12.4	3.5
2002	165.6	24.8	4.8	183.5	24.3	31.0
2003	166.3	59.8	11.8	184.9	40.2	14.8
2004	169.0	94.9	18.5	187.3	48.0	16.1
2005	(2)	(2)	(2)	(2)	(2)	(2)

[1] Support operations for other kinds of tobacco not shown. Burley and flue-cured usually account for over 95 percent of tobacco loan placements.   [2] Price support and loans discontinued for 2005 and subsequent crops of tobacco by the Fair and Equitable Tobacco Return Act of 2004.

FSA, Tobacco and Peanuts Division, (202)–720–5291.

Table 2-47.—Tobacco: Stocks owned by dealers and manufacturers, by types, United States, 2000–2004 (farm-sales-weight basis) [1]

Type and year	Jan. 1	Apr. 1	July 1	Oct. 1
	1,000 pounds	*1,000 pounds*	*1,000 pounds*	*1,000 pounds*
Flue-cured, types 11–14:				
2000	1,416,141	1,202,551	1,036,367	1,218,789
2001	1,077,226	922,184	821,083	1,042,878
2002	1,109,625	997,234	837,670	1,069,352
2003	982,985	877,333	822,848	954,187
2004	982,985	877,333	822,848	954,187
Virginia fire-cured, type 21:				
2000	3,158	4,579	4,278	4,186
2001	3,593	4,223	4,020	3,906
2002	4,355	4,866	4,558	4,295
2003	4,447	4,507	4,178	3,775
2004	4,447	4,507	4,178	3,775
Kentucky and Tennessee fire-cured, types 22–23:				
2000	83,955	109,569	105,963	93,776
2001	91,383	106,293	100,652	92,723
2002	98,824	115,145	107,009	97,577
2003	99,611	116,437	107,124	98,817
2004	99,611	116,437	107,124	98,817
Burley, type 31:				
2000	1,159,848	1,057,925	786,926	689,459
2001	664,725	639,769	577,785	520,777
2002	734,675	724,140	640,887	578,165
2003	687,669	677,609	596,533	540,029
2004	687,669	677,609	596,533	540,029
Maryland, type 32:				
2000	13,361	36,306	35,339	32,697
2001	8,380	8,604	8,349	7,353
2002	8,214	7,768	7,252	7,349
2003	7,188	7,729	7,312	5,950
2004	7,188	7,729	7,312	5,950
One Sucker and Green River, types 35–36:[2]				
2000	28,222	36,306	35,339	32,697
2001	32,914	36,575	35,505	33,388
2002	39,993	43,074	40,138	37,277
2003	40,337	42,982	39,986	37,276
2004	40,337	42,982	39,986	37,276
Virginia sun-cured, type 37:				
2000	90	149	185	181
2001	151	104	100	96
2002	235	209	91	97
2003	139	350	586	105
2004	139	350	586	105
Pennsylvania seedleaf, type 41:				
2000	8,757	11,449	10,314	12,119
2001	9,620	12,603	11,701	10,092
2001	11,639	15,012	12,986	10,667
2002	....	....	....	....
2003	8,867	12,183	11,521	9,729
2004	8,867	12,183	11,521	9,729
Connecticut Valley, types 51–52:				
2000	2,747	2,356	2,545	2,314
2001	2,209	1,940	1,734	1,459
2002	2,238	1,710	2,137	2,070
2003	2,073	2,137	2,039	1,576
2004	2,073	2,137	2,039	1,576
Wisconsin binder, types 54–55:				
2000	11,384	12,498	10,871	9,723
2001	8,278	10,624	9,478	8,405
2002	7,647	10,289	9,271	8,143
2003	7,244	10,667	9,346	8,637
2004	7,244	10,667	9,346	8,637
Cigar Wrapper, type 61:				
2000	1,289	1,004	678	1,787
2001	1,169	1,131	770	553
2002	1,530	988	708	1,471
2003	914	838	376	130
2004	914	838	376	1,303

See footnotes at end of table.

Table 2-47.—Tobacco: Stocks owned by dealers and manufacturers, by types, United States, 2000–2004 (farm-sales-weight basis) [1]—Continued

Type and year	Jan. 1	Apr. 1	July 1	Oct. 1
	1,000 pounds	*1,000 pounds*	*1,000 pounds*	*1,000 pounds*
Perique, type 72:				
2000	31	29	21	71
2001	30	20	19	19
2002	28	27	25	23
2003	22	22	22	21
2004	22	22	22	21
Other miscellaneous domestic, type 73:				
2000	2,198	2,763	2,822	3,098
2001	2,477	2,165	1,753	2,557
2002	3,766	3,059	3,607	3,630
2003	9,945	2,088	573	529
2004	9,945	2,088	573	529
Foreign-grown cigar-leaf, types 81–89:				
2000	102,295	100,965	103,126	97,788
2001	74,555	71,520	69,128	72,430
2002	98,481	94,534	95,277	96,207
2003	101,083	94,384	93,066	98,272
2004	101,083	94,384	93,066	98,272
Foreign-grown cigarette and smoking, types 91–99:				
2000	790,892	789,941	753,082	766,382
2001	662,709	654,484	621,554	644,749
2002	744,609	734,865	737,292	755,748
2003	773,395	738,409	735,979	715,558
2004	773,395	738,409	735,979	715,558

[1] Stocks shown have been converted to a farm-sales-weight basis—the equivalent of weight at the time of sale by grower—thereby making these data of leaf-tobacco stocks comparable with the data of leaf-tobacco production. [2] One Sucker and Green River combined. [3] Stocks on the island of Puerto Rico are included.

AMS Market Informaiton and Program Analysis Branch, (202) 205–0489.

Table 2-48.—Tobacco products: Consumption, total and per capita (18 years of age and over) in the United States, 1996–2004 [1]

Year	Cigarettes			Large cigars [2]			Smoking, chewing, and snuff [3]		All tobacco products [3]	
	Total	Total	Per capita	Total	Total	Per capita	Total	Per capita	Total	Per capita
	Billion	*Million pounds*	*Number*	*Billion*	*Million pounds [4]*	*Number*	*Million pounds*	*Pounds*	*Million pounds*	*Pounds*
1996	487	814	2,482	3.1	52	16	90	0.68	960	4.70
1997	480	805	2,422	3.5	58	18	88	0.65	1,004	4.66
1998	465	781	2,320	3.7	60	18	87	0.64	962	4.49
1999	435	721	2,148	3.8	63	19	87	0.64	876	4.32
2000	430	711	2,056	3.9	63	19	92	0.62	866	4.21
2001	425	696	2,026	3.9	67	20	89	0.63	863	4.25
2002	415	719	1,979	3.8	68	19	89	0.62	881	4.15
2003	400	674	1,837	4.4	71	21	89	0.60	837	3.85
2004 [5]	388	650	1,770	4.5	73	24	89	0.60	819	3.76

[1] Includes consumption by overseas forces. [2] Weighing over 3 pounds per 1,000. [3] Unstemmed-processing weight equivalent. [4] Includes weight of small cigars. [5] Preliminary. NA-not available.

ERS, Market and Trade Economics Division, Specialty Crops Branch, (202) 694–5311. No adjustment made for quantities lost, destroyed, bartered, etc., under war and postwar conditions, but such adjustments probably would be small in relation to totals.

Table 2-49.—Tobacco products: Cigars, cigarettes, chewing and smoking tobacco, and snuff, manufactured in the United States, 1995–2004

Year	Cigars		Cigarettes		Chewing tobacco			
	Large	Small	Large [1]	Small	Firm	Moist	Twist	Looseleaf
	Millions	*Millions*	*Millions*	*Millions*	*1,000 pounds*	*1,000 pounds*	*1,000 pounds*	*1,000 pounds*
1995 ...	2,056.8	1,430.4	0.01	743,519.1	2,886	1,247	1,085	57,678
1996 ...	NA	NA	NA	NA	2,905	1,039	1,114	56,012
1997 ...	2,323.6	1,476.1	0.00	765,324.2	2,562	911	977	53,663
1998 ...	2,750.4	1,710.3	0.00	679,746.6	2,359	744	976	49,235
1999 ...	2,938.1	2,316.6	0.00	606,318.5	2,187	633	886	47,177
2000 [2]	2,824.5	2,468.9	0.00	593,173.0	2,048	543	829	45,978
2001 [2]	NA	NA	NA	NA	1,867	475	821	43,872
2002 [2]	3,815.8	2,478.3	0.00	484,332.1	1,782	376	787	41,515
2003 [2]	4,017.1	2,616.2	0.00	499,401.2	1,420	328	705	39,185
2004 [2]	3,359.8	4,341.7	0.00	492,749.4	1,403	271	651	37,012
	Taxable removals and domestic invoices [3]							
1995 ...	2,364.6	1,397.0	0.00	489,265.8	3,015	1,147	1,092	57,048
1996 ...	NA	NA	NA	NA	2,797	952	1,104	55,136
1997 [2]	3,031.2	1,587.3	0.00	495,237.5	2,517	806	1,010	52,480
1998 [2]	3,185.1	1,638.0	0.00	457,871.7	2,288	674	968	48,562
1999 [2]	3,348.7	2,195.9	0.00	429,556.4	2,119	581	894	46,916
2000 [2]	3,369.8	2,243.2	0.00	421,597.4	2,049	485	863	45,059
2001 [2]	NA	NA	NA	NA	1,828	429	803	43,532
2002 [2]	3,703.2	2,247.9	0.00	394,871.9	1,722	329	750	40,225
2003 [2]	4,018.5	2,298.2	0.00	376,682.4	1,417	289	714	38,020
2004 [2]	4,319.2	2,701.6	0.00	374,977.6	1,325	245	656	35,721
	Tax-free removals and exports							
1995 ...	84.6	16.8	0.2	250,765.6	91	33	0	83
1996 ...	NA	NA	NA	NA	73	32	0	100
1997 [2]	115.7	66.3	0.00	310,329.5	65	23	0	94
1998 [2]	134.3	1,323	0.00	212,364.9	48	30	0	73
1999 [2]	121.3	1,152.2	0.00	165,443.8	30	23	0	69
2000 [2]	113.7	228.6	0.00	153,633.8	31	34	0	85
2001 [2]	NA	NA	NA	NA	30	31	0	75
2002 [2]	79.6	270.5	0.00	136,582.4	28	26	0	68
2003 [2]	93.7	354.9	0.00	126,631.3	24	25	0	68
2004 [2]	114.5	658.6	0.00	107,210.9	28	19	0	55

See footnotes at end of table.

Table 2-49.—Tobacco products: Cigars, cigarettes, chewing and smoking tobacco, and snuff, manufactured in the United States, 1995–2004—Continued

Year	Smoking tobacco			Snuff	Total chewing, smoking, and snuff
	Pipe	Granulated	Cigarette cut		
	1,000 pounds	*1,000 pounds*	*1,000 pounds*	*1,000 pounds*	*1,000 pounds*
1995 ...	7,614	93	4,536	60,202	135,026
1996 ...	6,939	90	4,954	61,539	134,592
1997 ...	6,770	72	4,555	64,336	133,846
1998 ...	6,154	60	6,251	65,477	131,256
1999 ...	6,726	61	7,908	66,992	132,570
2000 ...	5,982	50	7,327	69,556	132,313
2001 ...	5,088	0	7,674	70,893	130,690
2002 ...	5,018	0	10,474	72,696	132,648
2003 ...	4,744	0	12,636	74,895	133,913
2004 [2]	4,512	0	11,626	79,333	134,808
	Taxable removals and domestic invoices [3]				
1995 ...	7,032	100	4,686	59,339	133,459
1996 ...	6,469	88	4,790	61,390	132,726
1997 [2]	5,999	70	4,746	62,481	130,109
1998 [2]	5,604	62	6,356	64,051	128,565
1999 [2]	5,701	59	7,688	65,518	129,476
2000 [2]	4,620	50	8,398	68,605	130,129
2001 [2]	4,815	0	10,094	69,661	131,162
2002 [2]	4,643	0	11,258	71,668	130,595
2003 [2]	4,125	0	12,610	73,841	131,016
2004 [2]	3,773	0	11,675	77,356	130,751
	Tax-free removals and exports				
1995 ...	524	0	0	821	1,552
1996 ...	429	0	0	292	926
1997 [2]	532	0	0	862	1,576
1998 [2]	531	0	0	682	1,364
1999 [2]	532	0	0	788	1,442
2000 [2]	546	0	0	742	1,438
2001 [2]	455	0	0	65	1,356
2002 [2]	598	0	0	704	1,424
2003 [2]	624	0	0	697	1,438
2004 [2]	652	0	0	726	1,480

[1] Weighing more than three pounds per thousand.   [2] Preliminary.   [3] Includes cigars and cigarettes imported or brought into the United States and Puerto Rico.   NA-not available.
AMS, Market Information and Program Analysis Branch, (202) 205–0489.

Table 2-50.—Cigarettes and cigars: Total output, domestic consumption, tax-exempt removals, and exports, United States, 1996–2004

Year	Cigarettes				Cigars [3]			
	Total output	Domestic consump-tion [1]	Tax-exempt removals [2]		Total output [4]	Domestic consump-tion [1]	Tax-exempt removals [2]	
			Total	Exports			Total	Exports
	Billion	*Billion*	*Billion*	*Billion*	*Million*	*Million*	*Million*	*Million*
1996	755	487	261	244	2,413	3,054	98	84
1997	720	480	232	217	2,324	3,517	110	136
1998	680	465	213	201	2,751	3,655	112	158
1999	607	435	166	151	2,938	3,845	121	84
2000	595	430	154	148	2,825	3,850	114	113
2001	562	425	145	134	3,743	3,941	130	124
2002	532	415	136	127	3,819	3,833	80	123
2003	500	400	124	122	4,017	4,527	94	130
2004 [5]	495	390	NA	119	4,428	4,253	191	171

[1] As indicated by taxable removals and imports, and estimated inventory changes.   [2] In addition to exports, tax-exempt removals include principally shipments to forces overseas, to United States possessions, and ships' stores.   [3] Includes cigarillos but excludes small (approximately cigarette-size) cigars.   [4] Includes cigars shipped to mainland United States from Puerto Rico.   [5] Preliminary.   NA-not applicable.
ERS, Market and Trade Economics Division, Specialty Crops Branch, (202) 694–5311. Compiled from annual and monthly reports of the Internal Revenue Service, U.S. Treasury Department, and the Commerce Department.

Table 2-51.—Tobacco, unmanufactured: United States imports for consumption, by country of origin, 2002–2004

Type and country of origin	2002	2003	2004 [1]
	Metric tons	Metric tons	Metric tons
Cigarette leaf:			
Brazil	60,286	80,522	61,650
Turkey	37,659	34,542	38,906
Malawi	14,503	15,842	13,818
Argentina	13,368	16,206	13,110
Italy	6,087	8,937	7,904
Bulgaria	6,056	4,197	7,623
Indonesia	6,500	7,821	7,448
Greece	7,551	8,928	7,426
Canada	9,068	8,826	7,416
Others	41,774	48,762	43,953
Total	202,852	234,581	209,253
Cigar wrapper:			
Ecuador	443	714	933
Dominican Republic	176	68	168
Indonesia	65	93	81
Cameroon	27	35	78
Honduras	518	197	61
Others	187	97	59
Total	1,415	1,205	1,380
Scrap:			
Turkey	6,686	2,682	2,603
Dominican Republic	695	1,371	1,848
Indonesia	109	27	1,392
Canada	1,977	1,000	941
Argentina	0	277	761
Greece	350	317	558
Others	1,429	3,532	1,417
Total	11,245	9,206	9,521
Stems:			
Brazil	28,188	25,533	19,266
Argentina	4,195	7,536	4,872
Turkey	2,200	694	1,954
Malawi	5,075	5,140	1,868
Tanzania	0	0	1,113
Macau	0	0	649
Others	5,060	4,161	3,957
Total	44,718	43,064	33,679
Grand total [2]	260,230	288,056	253,832

[1] Preliminary. [2] Includes unstemmed and stemmed cigar filler.
FAS, Cotton, Oilseeds, Tobacco and Seeds Division, (202) 720–9516. Compiled from U.S. Bureau of the Census records.

Table 2-52.—Tobacco, unmanufactured: United States exports (domestic), by country of destination, total and by types, 2002–2004

Type and country of destination	2002	2003	2004 [1]
	Metric tons	Metric tons	Metric tons
Total Leaf: [2]			
Germany	27,003	25,314	24,266
Japan	22,495	19,219	15,690
Russian Federation	3,423	2,283	15,442
Belgium/Luxembourg	13,341	27,993	12,433
France	4,659	3,974	7,384
Netherlands	4,679	4,465	7,189
Spain	2,862	2,120	5,254
Indonesia	2,281	2,456	4,664
Denmark	6,184	6,113	4,610
Dominican Republic	4,168	4,053	4,355
Switzerland	12,393	15,773	4,276
Romania	0	211	3,887
Thailand	5,730	1,616	3,859
Malaysia	6,860	2,884	3,597
Korea, Republic of	4,580	4,343	3,313
China, People's Republic of	93	649	3,088
Italy	3,912	3,428	2,981
Czech Republic	69	38	2,377
Nigeria	1,668	2,822	2,193
Portugal	2,653	2,809	2,101
Austria	218	410	1,905
Others	24,155	22,750	28,832
Total	153,427	155,723	163,694
Flue-cured:			
Germany	13,550	14,536	15,833
Japan	12,650	8,139	7,698
Belgium/Luxembourg	6,886	14,641	3,881
Spain	743	765	3,479
Malaysia	4,018	2,461	3,291
Netherlands	655	1,783	3,235
China, People's Republic of	40	305	2,864
Indonesia	844	1,377	2,473
Switzerland	7,391	4,095	2,469
Korea, Republic of	3,912	4,187	2,299
Denmark	3,084	3,101	2,210
Russian Federation	779	676	2,078
France	1,490	779	1,886
United Kingdom	1,809	1,424	1,574
Italy	2,010	1,225	1,492
Thailand	1,071	725	1,016
Egypt	0	0	979
Australia	1,773	2,113	938
Norway	731	801	653
Hong Kong	346	511	646
Portugal	1,165	1,821	605
Others	7,892	4,858	5,072
Total	72,839	70,323	66,669

See footnotes at end of table.

Table 2-52.—Tobacco, unmanufactured: United States exports (domestic), by country of destination, total and by types, 2002–2004—Continued

Type and country of destination	2002	2003	2004 [1]
	Metric tons	*Metric tons*	*Metric tons*
Burley:			
Russian Federation	0	91	12,559
Belgium/Luxembourg	5,623	9,298	7,146
Germany	9,212	5,052	5,506
Japan	5,571	5,912	5,090
Romania	0	0	3,517
Thailand	2,181	891	2,609
Czech Republic	0	0	2,012
Netherlands	1,973	335	1,814
Austria	0	0	1,709
Spain	0	1,097	1,600
Greece	0	37	1,551
Lithuania	0	0	1,426
Italy	1,067	1,922	1,250
Switzerland	2,948	10,776	1,205
Ukraine	0	0	1,125
France	1,110	746	1,113
Portugal	837	517	1,054
Kazakhstan, Republic	0	0	950
Denmark	1,466	1,436	949
Taiwan	1,278	824	888
Senegal	0	0	507
Others	5,817	2,623	2,529
Total	39,084	41,555	58,107
Dark-fired Kentucky and Tennessee:			
Netherlands	1,629	1,607	1,296
Sri Lanka	770	966	615
Belgium/Luxembourg	56	343	328
Indonesia	671	240	279
Nigeria	0	94	229
Egypt	267	413	132
Canada	19	17	26
Denmark	26	10	23
Sweden	154	87	22
United Arab Emirates	0	3	19
Germany	31	24	15
China, Peoples Republic of	22	0	10
Haiti	0	20	9
Japan	6	1,317	7
Thailand	0	0	3
Iraq	0	0	3
Australia	43	0	0
France	358	158	0
Others	1,152	205	0
Total	5,204	5,507	3,016
VA Fire and Sun-Cured:			
Ukraine	0	0	56
Dominican Republic	0	0	14
Germany	0	0	3
Haiti	15	1	3
Others	6	0	0
Total	21	1	75
Maryland:			
Switzerland	45	144	211
Indonesia	106	17	159
Germany	734	35	158
Israel	156	81	100
Belgium/Luxembourg	0	16	24
Others	45	28	6
Total	1,086	321	657

See footnotes at end of table.

Table 2-52.—Tobacco, unmanufactured: United States exports (domestic), by country of destination, total and by types, 2002–2004—Continued

Type and country of destination	2002	2003	2004 [1]
	Metric tons	*Metric tons*	*Metric tons*
Cigar Wrapper:			
Dominican Republic	907	941	803
Honduras	3	39	337
Nicaragua	47	2	10
Spain	0	0	9
Panama	0	0	5
Austria	0	0	2
Venezuela	0	2	1
El Salvador	0	0	1
Japan	0	0	1
Mexico	2	1	0
Chile	20	0	0
Others	37	2	0
Total	1,017	987	1,170
Stems Refuse:			
France	1,700	2,290	4,384
Germany	3,154	5,267	2,331
Japan	3,679	2,114	1,340
Denmark	1,502	1,399	1,313
Mexico	36	363	1,279
Russian Federation	1,551	1,515	805
Netherlands	311	643	714
Belgium/Luxembourg	629	3,450	593
Portugal	647	470	442
Australia	13	269	403
Switzerland	1,657	693	384
Czech Republic	0	0	307
Israel	386	104	301
Korea, Republic of	9	154	234
Thailand	2,478	0	230
Italy	393	257	211
Honduras	0	218	188
Peru	250	353	134
United Kingdom	772	638	134
Norway	127	119	116
Spain	344	193	114
Others	854	1,284	434
Total	20,492	21,794	16,391
Other unmanufactured, Tobacco:			
Dominican Republic	3,238	2,566	2,783
Nigeria	1,668	2,689	1,841
Cyprus	0	681	1,612
Indonesia	217	706	1,559
Japan	589	1,736	1,555
Jordan	0	1,403	1,263
Korea, Republic of	183	2	781
Brazil	526	136	600
Belgium/Luxembourg	139	245	461
Germany	320	399	421
Egypt	540	86	392
Honduras	696	350	391
Taiwan	4	255	324
Argentina	56	233	297
Turkey	0	8	289
Singapore	4	354	251
Austria	218	410	194
Canada	511	396	162
Netherlands	108	97	129
Nicaragua	19	24	126
Chile	0	179	117
Others	4,586	1,875	1,256
Total	13,622	14,829	16,803

[1] Preliminary.

FAS, Cotton, Oilseeds, Tobacco and Seeds Division, (202) 720–9516. Compiled from U.S. Bureau of the Census records.

CHAPTER III
STATISTICS OF OILSEEDS, FATS, AND OILS

This chapter includes information on cottonseed, flaxseed, olive oil, peanuts, soybeans, margarine, and fats and oils. Most butter statistics are included in the chapter on dairy and poultry statistics. Lard data are mostly in the chapter on livestock.

Table 3-1.—Cottonseed: All cotton harvested area and cottonseed production, farm disposition, marketing year average price per ton received by farmers, and value, United States, 1996–2005

Year	Harvested area of all cotton	Cottonseed					
		Production	Farm disposition		Marketing year average price	Value of production
			Sales to oil mills	Other [1]		
	1,000 acres	*1,000 tons*	*1,000 tons*	*1,000 tons*	*Dollars/tons*	*1,000 dollars*
1996	12,888.1	7,143.5	4,363.2	2,780.3	126.00	914,564
1997	13,406.0	6,934.6	4,182.4	2,752.2	121.00	835,371
1998	10,683.6	5,365.4	3,261.1	2,104.3	129.00	687,179
1999	13,424.9	6,353.5	3,340.2	3,013.3	89.00	559,157
2000	13,053.0	6,435.6	3,452.2	2,983.4	105.00	667,800
2001	13,827.7	7,452.2	3,860.9	3,591.3	90.50	667,348
2002	12,416.6	6,183.9	3,287.9	2,896.0	101.00	616,352
2003	12,003.4	6,664.6	3,383.6	3,281.0	117.00	778,994
2004	13,057.0	8,242.1	4,546.0	3,696.1	107.00	877,372
2005 [2]	13,702.6	8,501.0	NA	NA	95.50	808,598

[1] Includes planting seed, feed, exports, inter-farm sales, shrinkage, losses, and other uses.   [2] Preliminary.   NA-not available.

NASS, Crops Branch, (202) 720–2127.

Table 3-2.—Cottonseed: Production and farm disposition, by State and United States, 2003–2005

State	Production			Farm disposition [1]				Used for planting [2]	
	2003	2004	2005 [4]	Sales to oil mills		Other [3]		2004	2005 [4]
				2003	2004	2003	2004 [4]		
	1,000 tons	*1,000 tons*	*1,000 tons*	*1,000 tons*	*1,000 tons*	*1,000 tons*	*1,000 tons*	*1,000 tons*	*1,000 tons*
AL	327.0	282.0	306.0	49.0	16.0	278.0	266.0	6.3	6.4
AZ	216.8	301.6	243.0	4.1	3.3	212.7	298.3	2.2	2.1
AR	689.0	734.0	808.0	462.0	529.0	227.0	205.0	8.6	9.3
CA	680.0	902.0	588.0	93.5	116.0	586.5	786.0	6.6	6.1
FL	37.0	35.0	40.0	21.7	25.0	15.3	10.0	1.0	0.9
GA	732.0	560.0	718.0	405.0	343.0	327.0	217.0	15.0	14.0
KS	34.2	26.0	34.0	4.2	7.0	30.0	19.0	0.9	0.8
LA	365.0	295.0	393.0	191.0	138.0	174.0	157.0	4.5	5.6
MS	773.0	804.0	768.0	604.0	675.0	169.0	129.0	11.0	13.0
MO	274.0	268.0	315.0	200.0	186.0	74.0	82.0	4.0	4.3
NM	31.6	52.5	43.0	3.8	11.7	27.8	40.8	0.8	0.8
NC	349.0	447.0	478.0	52.0	79.0	297.0	368.0	6.6	6.8
OK	79.0	113.0	136.0	64.0	91.0	15.0	22.0	2.4	2.5
SC	109.0	94.0	137.0	58.0	54.0	51.0	40.0	1.5	1.6
TN	311.0	336.0	391.0	232.0	262.0	79.0	74.0	5.7	6.0
TX	1,616.0	2,939.0	3,040.0	939.3	2,010.0	676.7	929.0	53.2	51.5
VA	41.0	53.0	63.0	0.0	0.0	41.0	53.0	0.7	0.8
US ..	6,664.6	8,242.1	8,501.0	3,383.6	4,546.0	3,281.0	3,696.1	131.0	132.5

[1] 2003 farm disposition not available.   [2] Included in 'other' farm disposition. Seed for planting is produced in crop year shown, but used in the following year.   [3] Includes planting seed, feed, exports, inter-farm sales, shrinkage, losses, and other uses.   [4] Preliminary.

NASS, Crops Branch, (202) 720–2127.

OILSEEDS, FATS, AND OILS

Table 3-3.—Cottonseed: Marketing year average price per ton and value of production, by State and United States, crop of 2003–2005

State	Marketing year average price per ton			Value of production		
	2003	2004	2005 [1]	2003	2004	2005 [1]
	Dollars	*Dollars*	*Dollars*	*1,000 dollars*	*1,000 dollars*	*1,000 dollars*
AL	98.50	91.00	82.00	32,210	25,662	25,092
AZ	148.00	163.00	135.00	32,086	49,161	32,805
AR	110.00	99.50	90.00	75,790	73,033	72,720
CA	152.00	150.00	146.00	103,360	135,300	85,848
FL	99.00	86.00	75.00	3,663	3,010	3,000
GA	95.50	85.00	74.00	69,906	47,600	53,132
KS	118.00	85.50	83.50	4,036	2,223	2,839
LA	100.00	102.00	87.50	36,500	30,090	34,388
MS	100.00	93.50	83.50	77,300	75,174	64,128
MO	125.00	91.50	79.00	34,250	24,522	24,885
NM	145.00	118.00	108.00	4,582	6,195	4,644
NC	108.00	96.00	81.50	37,692	42,912	38,957
OK	125.00	76.00	72.00	9,875	8,588	9,792
SC	101.00	90.50	77.00	11,009	8,507	10,549
TN	130.00	99.50	85.00	40,430	33,432	33,235
TX	125.00	104.00	101.00	202,000	305,656	307,040
VA	105.00	119.00	88.00	4,305	6,307	5,544
US	117.00	107.00	95.50	778,994	877,372	808,598

[1] Preliminary.

NASS, Crops Branch, (202) 720–2127.

Table 3-4.—Cottonseed: Crushings, output of products and product prices, United States, 1995–2004

Year beginning August	Quantity crushed	Cottonseed products and prices			
		Oil		Cake and meal	
		Quantity	Price [1]	Quantity	Price [2]
	1,000 tons	*Million pounds*	*Cents per pound*	*1,000 tibs*	*Dollars per ton*
1995	3,882	1,229	26.5	1,748	190.74
1996	3,860	1,216	25.6	1,752	192.00
1997	3,889	1,224	28.8	1,769	145.00
1998	2,719	832	27.3	1,232	110.00
1999	3,064	939	21.6	1,390	127.33
2000	2,753	847	16.0	1,338	143.35
2001	2,791	876	18.0	1,294	136.16
2002	2,495	725	37.8	1,114	147.10
2003	2,643	874	32.0	1,244	187.00
2004 [3]	2,900	915	24.0	1,305	125.00

[1] Tanks, f.o.b. Valley Points. [2] 41 percent protein, solvent, Memphis. [3] Forecast

ERS, Field Crops Branch, (202) 694–5300. Compiled from annual reports of the U.S. Department of Commerce.

Table 3-5.—Cottonseed oil and cottonseed cake and meal: United States exports by country of destination 2002/2003–2004/2005 (Marketing year October–September)

Continent and country of destination	Cottonseed oil [1]			Cottonseed cake and meal		
	2002/ 2003	2003/ 2004	2004/ 2005 [2]	2002/ 2003	2003/ 2004	2004/ 2005 [2]
	Metric tons	*Metric tons*	*Metric tons*	*Metric tons*	*Metric tons*	*Metric tons*
North and Central America; incl. Caribbean:						
Barbados	77	94	109	0	0	0
Canada	36,801	29,278	14,378	508	599	196
El Salvador	407	0	0	0	0	0
Guatemala	0	0	0	0	0	0
Mexico	7,298	9,151	5,753	40,010	57,528	81,032
Total [3]	44,954	39,030	20,669	40,519	58,704	85,055
South America:						
Brazil	0	0	0	20	31	92
Total [3]	63	52	72	20	178	1,568
Europe:						
Austria	0	0	0	222	1,631	0
Belgium-Luxembourg	0	0	0	80	391	1,388
France	3	0	0	0	0	0
Germany	132	39	133	2,086	197	6,872
Ireland	0	0	0	0	0	0
Netherlands	0	0	0	0	0	0
United Kingdom	354	114	0	2,225	1,212	745
Total [3]	489	172	133	5,077	4,084	9,956
Africa:						
Egypt	0	0	0	0	0	0
Total [3]	5	0	0	0	0	0
Asia:						
China	3	207	271	0	0	0
Israel	571	0	0	0	23	0
Japan	1,521	10,737	4,450	304	286	289
Korea, Republic of	2,308	7	114	278	514	437
Total [3]	4,493	11,032	4,885	630	907	776
Oceania	0	0	16	0	0	0
Grand total [4]	50,004	50,286	25,775	46,246	63,872	97,355

[1] Crude and refined (includes shipments under P.L. 480). [2] Preliminary. [3] Includes quantities exported to countries not shown. [4] May not add due to rounding.

FAS, Cotton, Oilseeds, Tobacco and Seeds Division, (202) 720–9516. (Compiled from reports of the U.S. Department of Commerce.)

Table 3-6.—Cottonseed: Area, yield, and production in specified countries, 2002/2003–2004/2005 [1]

Continent and country	Area [2]			Yield per hectare			Production		
	2002/ 2003	2003/ 2004	2004/ 2005 [3]	2002/ 2003	2003/ 2004	2004/ 2005 [3]	2002/ 2003	2003/ 2004	2004/ 2005 [3]
	1,000 hectares	1,000 hectares	1,000 hectares	Metric tons	Metric tons	Metric tons	1,000 metric tons	1,000 metric tons	1,000 metric tons
North and Central America, and the Caribbean:									
Mexico	42	62	105	1.67	1.92	2.10	70	119	220
United States	5,025	4,858	5,284	1.12	1.24	1.42	5,610	6,046	7,477
Total	5,067	4,920	5,389	1.12	1.25	1.43	5,680	6,165	7,697
South America:									
Argentina	148	250	374	0.68	0.72	0.64	100	180	240
Brazil	735	1,100	1,172	1.97	2.04	1.96	1,448	2,240	2,300
Paraguay	170	280	215	0.57	0.64	0.49	97	178	105
Total	1,053	1,630	1,761	1.56	1.59	1.50	1,645	2,598	2,645
European Union	440	460	470	1.48	1.30	1.60	652	597	750
Former Soviet Union:									
Azerbaijan	65	60	80	0.85	1.17	1.11	55	70	89
Kazakhstan ..	165	185	216	1.21	1.16	1.23	200	215	265
Kyrgyzstan ...	35	35	46	2.29	2.00	1.52	80	70	70
Tajikistan	265	285	290	1.08	1.07	1.09	285	305	315
Turkmenistan	490	480	550	0.56	0.77	0.69	275	370	380
Uzbekistan ...	1,421	1,431	1,456	1.41	1.26	1.55	2,000	1,800	2,250
Total	2,441	2,476	2,638	1.19	1.14	1.28	2,895	2,830	3,369
Africa:									
Cameroon	200	220	220	1.06	1.14	1.14	212	250	250
Egypt	302	218	307	1.35	1.41	1.41	407	308	434
Mali	420	516	533	0.79	0.94	0.83	330	484	444
South Africa, Republic of	30	41	26	1.03	1.34	1.62	31	55	42
Sudan	155	179	210	1.23	0.89	1.28	190	159	269
Tanzania, United Republic of ...	387	387	500	0.32	0.27	0.46	123	103	230
Zimbabwe	330	360	330	0.48	0.53	0.40	160	190	133
Total	1,824	1,921	2,126	0.80	0.81	0.85	1,453	1,549	1,802
Asia and the Middle East:									
China, Peoples Republic of	4,184	5,110	5,690	2.12	1.74	2.02	8,850	8,870	11,500
India	7,667	7,785	9,000	0.57	0.75	0.90	4,400	5,859	8,070
Iran	160	145	175	0.84	1.05	1.05	135	152	183
Pakistan	2,796	3,092	3,190	1.21	1.06	1.50	3,396	3,290	4,797
Syria	186	200	234	2.66	2.85	3.01	495	570	705
Turkey	700	710	700	1.94	1.88	1.99	1,356	1,332	1,390
Total	15,693	17,042	18,989	1.19	1.18	1.40	18,632	20,073	26,645
Oceania:									
Australia	220	196	314	2.26	2.45	2.90	498	480	912
Total	220	196	314	2.26	2.45	2.90	498	480	912
World Total [4]	29,389	31,250	34,738	1.12	1.14	1.31	32,811	35,629	45,379

[1] Split year includes Northern Hemisphere crop harvested in the late months of the first year shown combined with Southern Hemisphere and certain Northern Hemisphere crops harvested in the early months of the following year.   [2] Harvested area.   [3] Preliminary.   [4] Includes all countries in USDA data base.

FAS, Cotton, Oilseeds, Tobacco and Seeds Division, (202) 720-9518. (Compiled from reports of the U.S. Department of Commerce.)

Table 3-7.—Flaxseed: Area, yield, production, disposition, and value, United States, 1996-2005

Year	Area planted	Area harvested	Yield per harvested acre	Production	Marketing year average price per bushel received by farmers	Value of production
	1,000 acres	*1,000 acres*	*Bushels*	*1,000 bushels*	*Dollars*	*1,000 dollars*
1996	96	92	17.4	1,602	6.37	10,197
1997	151	146	16.6	2,420	5.81	14,046
1998	336	329	20.4	6,708	5.05	33,809
1999	387	381	20.6	7,864	3.79	30,098
2000	536	517	20.8	10,730	3.30	35,569
2001	585	578	19.8	11,455	4.29	49,004
2002	784	703	16.9	11,863	5.77	68,564
2003	595	588	17.9	10,516	5.88	61,900
2004	523	511	20.3	10,368	8.07	83,767
2005 [1]	983	955	20.6	19,695	5.90	116,305

[1] Preliminary.
NASS, Crops Branch, (202) 720-2127.

Table 3-8.—Flaxseed: Supply and disappearance, United States, 1995-2004

Year beginning June	Supply				Disappearance			
	Stocks June 1	Production	Imports	Total	Total used for seed	Exports	Crushings [2]	Total domestic disappearance [3]
	1,000 bushels	*1,000 bushels*	*1,000 bushels*	*1,000 bushels*	*1,000 bushels*	*1,000 bushels*	*1,000 bushels*	*1,000 bushels*
1995	1,170	2,212	7,248	10,630	78	119	9,000	9,281
1996	1,230	1,602	8,390	11,222	122	144	10,000	10,625
1997	453	2,420	9,636	12,509	272	174	10,500	11,154
1998	1,181	6,708	5,992	13,881	313	476	10,600	11,247
1999	2,158	7,864	6,629	16,651	434	201	11,500	14,683
2000	1,767	10,730	2,849	15,346	474	1,017	12,000	13,021
2001	1,308	11,455	1,904	14,667	635	2,386	10,000	11,388
2002	893	11,863	2,901	15,657	482	3,181	10,500	11,398
2003	1,078	10,516	4,573	16,167	509	2,516	10,860	12,421
2004 [1]	1,288	10,471	3,537	15,296	529	2,016	10,410	12,363

[1] Preliminary. [2] From domestic and imported seed. [3] Total supply minus exports and stocks June 1 of following year.
ERS, Field Crops Branch, (202) 694-5300.

Table 3-9.—Flaxseed: Area, yield, and production, by States, 2003-2005

State	Area planted			Area harvested			Yield per harvested acre			Production		
	2003	2004	2005	2003	2004	2005	2003	2004	2005	2003	2004	2005
	1,000 acres	*1,000 acres*	*1,000 acres*	*1,000 acres*	*1,000 acres*	*1,000 acres*	*Bush- els*	*Bush- els*	*Bush- els*	*1,000 bush- els*	*1,000 bush- els*	*1,000 bush- els*
MN	8	3	13	7	3	12	23.0	17.0	11.0	161	51	132
MT	17	20	55	17	19	54	13.0	18.0	17.0	221	342	918
ND	560	490	890	555	480	865	18.0	20.5	21.0	9,990	9,840	18,165
SD	10	10	25	9	9	24	16.0	15.0	20.0	144	135	480
US	595	523	983	588	511	955	17.9	20.3	20.6	10,516	10,368	19,695

NASS, Crops Branch, (202) 720-2127.

Table 3-10.—Flaxseed: Marketing year average price and value of production, by States, crop of 2003, 2004, and 2005

State	Marketing year average price per bushel			Value of production		
	2003	2004	2005[1]	2003	2004	2005[1]
	Dollars	Dollars	Dollars	1,000 dollars	1,000 dollars	1,000 dollars
MN	6.70	10.30	6.05	1,079	525	799
MT	5.80	7.94	6.20	1,282	2,715	5,692
ND	5.88	8.05	5.90	58,741	79,212	107,174
SD	5.54	9.74	5.50	798	1,315	2,640
US	5.88	8.07	5.90	61,900	83,767	116,305

[1] Preliminary.

NASS, Crops Branch, (202) 720–2127.

Table 3-11.—Flaxseed: Support operations, United States, 1996–2005

Marketing year beginning June 1	Income support payment rates per bushels[1]	Program price levels per bushel		Put under loan[4]		Acquired by CCC under loan program[5]	Owned by CCC at end of marketing year
		Loan[2]	Target[3]	Quantity	Percentage of production		
	Dollars	Dollars	Dollars	1,000 bushels	Percent	1,000 bushels	1,000 bushels
1996/1997 ...		5.07		39.3	2.5	0.0	0.0
1997/1998 ...		5.21		105.4	4.4	0.0	0.0
1998/1999 ...		5.21		513.9	7.7	0.0	0.0
1999/2000 ...	0.12	5.21		432.4	5.5	225.8	4.5
2000/2001 ...	0.23	5.21		352.6	3.3	151.8	0.0
2001/2002 ...		5.21		107.6	0.9	35.7	1.3
2002/2003[6]	0.45/0.00	3.91	5.49	157.2	1.3	1.8	0.0
2003/2004 ...	0.45/---	5.39	5.49	276.8	2.6	0.0	0.0
2004/2005 ...	0.45/---	5.21	5.66	157.5	0.9	0.0	0.0
2005/2006 ...	0.45/---	5.21	5.66				

[1] Oilseeds producer payment rates for 1999/2000 were calculated according to the provisions of the Agriculture, Rural Development, Food and Drug Administration, and Related Agencies Appropriations Act, 2000. Rates for 2000/01 were calculated according to the provisions of the Agricultural Risk Protection Act of 2000, and included supplemental oilseeds payment rates. Payment rates for the 2002/2003 and subsequent crops are calculated according to the provisions of the Direct and Counter-Cylclical Payment program, following enactment of the Farm Security and Rural Investment Act of 2002 (2002 Act). Payment rates are rounded to the nearest tenth of a cent.   [2] The national average loan rate was also known as the price support rate prior to enactment of the Federal Agricultural Improvement and Reform Act of 1996.   [3] Target prices for the 2002/03 and subsequent crops were enacted with the 2002 Act.   [4] Does not include quantity on which loan deficiency payments were made.   [5] Acquisition of all loans forfeited during the marketing year including loans made in previous years.   [6] Beginning with 2002/2003, the first entry is the direct payment rate and the second entry is the counter-cyclical payment rate.

FSA, Feed Grains & Oilseeds Analysis Group, (202) 720–8838.

Table 3-12.—Flaxseed and linseed oil and meal: Average price Minneapolis, 1995–2004

Year	Average price received by farmers per bushel	Minneapolis	
		Oil, per pound[1]	Meal, per ton[2]
	Dollars	Cents	Dollars
1995	5.25	33.73	91.96
1996	6.21	36.54	133.54
1997	5.75	35.97	169.74
1998	5.25	36.33	131.40
1999	3.79	36.42	91.63
2000	3.30	35.83	93.77
2001	4.29	36.00	116.23
2002	5.77	33.10	119.62
2003	5.90	39.86	122.89
2004[3]	7.90	41.75	158.90

[1] Raw oil in tank cars.   [2] Bulk carlots, 34 percent protein.   [3] Preliminary.

ERS, Field Crops Branch, (202) 694–5300.

Table 3-13.—Flaxseed and products: Flaxseed crushed; production, imports, and exports of linseed oil, cake, and meal; and June 1 stocks of oil, United States, 1995-2004

Year beginning June	Total flaxseed crushed	Linseed oil			Linseed cake and meal		
		Stocks June 1	Production	Exports	Production	Imports for consumption	Exports
	1,000 bushels	*Million pounds*	*Million pounds*	*Million pounds*	*1,000 tons*	*1,000 tons*	*1,000 tons*
1995 ...	9,000	45	180	26	162	2	35
1996 ...	10,000	50	200	66	180	13	44
1997 ...	10,500	35	205	58	189	15	19
1998 ...	10,600	42	207	63	191	4	26
1999 ...	11,500	48	224	74	207	1	19
2000 ...	12,000	49	234	73	216	5	25
2001 ...	10,000	43	195	50	180	6	62
2002 ...	10,500	45	205	70	189	19	31
2003 ...	10,860	45	212	76	195	26	32
2004 [1]	10,410	45	203	78	187	15	32

[1] Preliminary.

ERS, Field Crops Branch, (202) 694-5300.

Table 3-14.—Sunflowerseed, sunflowerseed oil, and sunflowerseed cake and meal: United States exports by country of destination 2002/2003-2004/2005 [1]

Continent and country of destination	Sunflowerseed			Sunflowerseed oil [2]			Sunflowerseed cake and meal		
	2002/ 2003	2003/ 2004	2004/ 2005 [3]	2002/ 2003	2003/ 2004	2004/ 2005 [3]	2002/ 2003	2003/ 2004	2004/ 2005 [3]
	Metric tons	*Metric tons*	*Metric tons*	*Metric tons*	*Metric tons*	*Metric tons*	*Metric tons*	*Metric tons*	*Metric tons*
North and Central America; incl. Caribbean:									
Canada	21,828	16,173	32,660	16,754	19,537	40,523	1,711	231	303
Mexico	6,399	10,308	8,674	5,259	63,787	5,332	1,371	1,454	2,492
Total [4]	28,424	26,980	41,518	25,775	85,549	46,188	3,082	1,692	2,795
South America:									
Venezuela	4	8	0	0	0	0	0	0	0
Total [4]	259	764	109	0	39	0	10	501	0
Europe:									
France	2,703	2,921	2,932	0	0	0	0	0	129
Germany	40,051	29,528	22,534	0	0	0	0	39	0
Netherlands	13,508	17,979	6,219	16	30	0	0	0	0
Spain	29,159	34,326	27,176	0	2,461	2,061	0	0	194
Total [4]	113,156	119,756	84,517	98	3,234	4,782	0	9,783	322
Former Soviet Union [5]	167	665	486	0	0	0	0	0	0
Africa:									
Algeria	0	0	0	0	12,099	0	0	0	0
Egypt	40	0	0	3,000	0	0	0	0	0
Total [4]	1,237	155	375	5,638	12,099	0	0	0	0
Asia:									
Japan	2,083	2,035	496	10,227	3,570	3,214	21	0	0
Korea, Rep. of	1,393	1,393	1,393	296	354	158	0	0	0
Turkey	7,557	7,557	7,557	0	0	0	0	0	0
Total [4]	24,877	24,877	24,877	19,946	6,461	5,717	21	0	0
Oceania	1,513	1,513	1,513	0	2	2	0	0	0
Grand total [6]	234,689	234,689	234,689	51,458	107,384	56,689	3,112	11,976	3,117

[1] For sunflowerseed, year begins September 1; for sunflowerseed oil cake and meal, year begins October 1. [2] Crude and refined oil. [3] Preliminary. [4] Includes quantities exported to countries not shown. [5] Former Soviet Union; includes the 12 Republics of the USSR that are not members of the European Union. [6] May not add due to rounding.

FAS, Cotton, Oilseeds, Tobacco and Seeds Division, (202) 720-9516. Compiled from reports of the U.S. Department of Commerce.

Table 3-15.—Peanuts: Area, yield, production, disposition, marketing year average price per pound received by farmers, and value, United States, 1996–2005

Year	Area planted	Peanuts for nuts				
		Area harvested	Yield per acre	Production[1]	Marketing year average	Value of production
	1,000 acres	1,000 acres	Pounds	1,000 pounds	Cents	1,000 dollars
1996	1,401.5	1,380.0	2,653	3,661,205	28.1	1,029,774
1997	1,434.0	1,413.8	2,503	3,539,380	28.3	1,002,703
1998	1,521.0	1,467.0	2,702	3,963,440	28.4	1,125,919
1999	1,534.5	1,436.0	2,667	3,829,490	25.4	971,608
2000	1,536.8	1,336.0	2,444	3,265,505	27.4	896,097
2001	1,541.2	1,411.9	3,029	4,276,704	23.4	1,000,512
2002	1,353.0	1,291.7	2,571	3,321,040	18.2	599,624
2003	1,344.0	1,312.0	3,159	4,144,150	19.3	799,428
2004	1,430.0	1,394.0	3,076	4,288,200	18.9	813,551
2005[2]	1,657.0	1,629.0	2,960	4,821,250	17.4	845,873

[1] Estimates comprised of quota and non-quota peanuts. [2] Preliminary.
NASS, Crops Branch, (202) 720–2127.

Table 3-16.—Peanuts, farmer stock: Stocks, production, and quantity milled, United States, 1995–2004

Year beginning August	Stocks Aug. 1[1]	Production harvested for nuts[1]	Imports	Total supply	Milled[1][2]
	1,000 pounds	1,000 pounds	1,000 pounds	1,000 pounds	1,000 pounds
1995	48,574	3,461,475	8,628	3,518,677	2,558,954
1996	66,392	3,661,205	6,988	3,734,585	2,919,054
1997	22,714	3,539,380	5,907	3,568,001	2,899,138
1998	27,284	3,963,440	5,320	3,996,044	3,652,670
1999	158,646	3,829,490	5,341	3,993,477	3,703,266
2000	139,210	3,265,505	7,625	3,412,340	3,254,950
2001	116,994	4,276,704	0	4,393,698	3,663,304
2002	483,702	3,321,040	251	3,804,993	3,585,900
2003	123,428	4,144,150	321	4,267,899	4,014,994
2004	234,770	4,288,200	0	4,522,970	3,675,410

[1] Net weight basis. [2] Includes peanuts milled for seed.
NASS, Crops Branch, (202) 720–2127.

Table 3-17.—Peanuts: Crushings, and oil and meal stocks, production, and foreign trade, United States, 1995–2004

Year beginning August	Peanuts crushed (shelled basis)	Peanut oil				Peanut cake and meal	
		Stocks Aug. 1[1]	Production of crude	Imports	Exports[2]	Stocks Aug. 1[3]	Production
	1,000 pounds	1,000 pounds	1,000 pounds	1,000 pounds	1,000 pounds	1,000 pounds	1,000 pounds
1995	751,281	19,763	320,909	4,678	108,146	14,910	420,919
1996	520,413	26,992	220,877	14,445	20,625	7,212	294,590
1997	409,249	22,936	175,853	8,118	13,097	8,291	228,276
1998	345,825	29,297	145,254	72,534	10,516	24,004	192,425
1999	536,164	6,770	228,839	12,835	17,519	2,847	291,491
2000	411,558	10,881	178,523	79,119	13,824	4,721	230,099
2001	521,173	3,812	230,791	38,665	8,386	3,800	296,874
2002	644,194	21,986	285,685	69,995	41,868	1,292	356,888
2003	402,958	27,098	172,977	126,346	27,695	7,769	226,995
2004	295,769	13,368	126,249	32,639	9,985	5,732	172,668

[1] Crude plus refined. [2] Reported as edible peanut oil and crude peanut oil; in this tabulation added without converting. [3] Holding at producing mills only.
NASS, Crops Branch, (202) 720–2127, ERS, and Bureau of the Census.

AGRICULTURAL STATISTICS 2006

III–9

Table 3-18.—Cleaned peanuts (roasting stock): Supply and disposition, United States, 1995–2004

Year beginning August	Supply				Disposition		
	Stocks Aug. 1	Production	Imports	Total	Exports	Domestic disappearance	
						Total	Per capita
	1,000 pounds	*1,000 pounds*	*1,000 pounds*	*1,000 pounds*	*1,000 pounds*	*1,000 pounds*	*Pounds*
1995	70,620	188,954	8,628	268,202	73,937	163,283	0.62
1996	30,981	264,337	6,988	302,306	80,607	167,928	0.63
1997	53,771	229,912	5,907	289,590	75,154	183,868	0.68
1998	30,568	277,552	5,320	313,440	58,864	181,468	0.66
1999	73,108	235,756	5,341	314,205	53,406	200,877	0.72
2000	59,922	228,185	7,625	245,732	41,054	216,306	0.77
2001	38,372	245,783	0	284,155	39,099	179,907	0.63
2002	65,149	207,881	251	273,281	40,192	184,189	0.64
2003	48,900	254,048	321	303,269	32,202	211,104	0.73
2004	54,963	261,823	0	321,786	36,808	215,323	0.73

NASS, Crops Branch, (202) 720–2127, and ERS. Foreign trade from the Bureau of the Census.

Table 3-19.—Shelled peanuts (all grades): Supply, exports, and quantity crushed, United States, 1995–2004

Year beginning August	Supply						Exports	Crushed
	Stocks Aug. 1		Production		Imports	Total		
	Edible	Oil stock	Edible	Oil stock				
	1,000 pounds	*1,000 pounds*	*1,000 pounds*	*1,000 pounds*	*1,000 pounds*	*1,000 pounds*	*1,000 pounds*	*1,000 pounds*
1995	752,814	58,188	1,253,451	491,818	108,303	2,664,574	564,021	751,281
1996	370,431	126,318	1,692,581	305,674	95,041	2,590,045	440,438	520,413
1997	498,954	41,000	1,694,016	290,882	101,792	2,626,644	455,264	409,249
1998	580,370	14,091	2,227,037	310,459	112,643	3,249,600	377,171	345,825
1999	855,572	16,587	2,157,828	448,875	129,819	3,608,681	503,675	536,164
2000	707,554	70,103	1,939,736	337,324	147,103	3,201,820	354,419	411,558
2001	693,209	14,463	2,090,776	485,092	150,276	3,433,816	495,559	521,173
2002	680,850	16,648	1,983,016	611,627	54,117	3,346,258	337,332	644,194
2003	504,186	24,231	2,439,231	390,893	26,811	3,385,352	362,669	402,958
2004	603,504	17,686	2,357,314	246,663	25,261	3,250,428	340,667	295,769

NASS, Crops Branch, (202) 720–2127, and ERS. Foreign trade from the U.S. Bureau of the Census.

Table 3-20.—Peanuts: Shelled (raw basis) by types, used in primary products and apparent disappearance of peanuts (cleaned in shell), United States, 1995-2004

Type, and year beginning August	Shelled uses					Apparent disappearance (cleaned in shell)[2]
	Peanut butter[1]	Snack	Candy	Other	Total	
	1,000 pounds	*1,000 pounds*	*1,000 pounds*	*1,000 pounds*	*1,000 pounds*	*1,000 pounds*
Virginia:						
1995	71,310	93,041	25,176	13,656	203,183	
1996	64,274	91,882	24,158	12,852	193,166	
1997	59,228	80,309	28,428	14,135	182,100	
1998	57,864	99,401	36,178	3,492	196,935	
1999	73,926	100,384	23,173	3,321	200,804	
2000	102,050	100,650	19,101	3,271	225,072	
2001	106,573	97,046	26,640	3,097	233,356	
2002	77,018	75,100	26,930	4,178	183,226	
2003	88,053	68,257	23,580	1,669	181,559	
2004	112,027	70,216	25,466	1,702	209,411	
Runner:						
1995	634,350	169,142	304,285	15,942	1,123,719	
1996	634,387	176,851	318,924	19,185	1,149,347	
1997	676,839	206,718	302,791	20,598	1,206,946	
1998	670,705	234,486	321,838	17,719	1,244,748	
1999	690,564	278,440	315,467	15,922	1,300,393	
2000	643,229	247,739	320,304	15,884	1,227,156	
2001	702,454	250,079	303,668	13,575	1,269,776	
2002	734,844	257,258	312,192	19,552	1,323,846	
2003	805,852	333,198	328,560	13,847	1,481,457	
2004	824,876	367,671	349,437	20,708	1,562,692	
Spanish:						
1995	22,416	14,906	21,202	2,417	60,941	
1996	28,870	21,369	17,764	1,788	69,791	
1997	24,163	19,881	19,798	738	64,580	
1998	16,137	15,919	22,161	920	55,137	
1999	7,614	15,297	16,313	984	40,208	
2000	7,960	13,127	16,205	843	38,135	
2001	9,900	13,791	19,421	612	43,724	
2002	16,667	12,555	15,110	649	44,981	
2003	7,732	13,133	13,843	414	35,122	
2004	1,611	12,894	14,793	137	29,435	
All types:						
1995	728,076	277,089	350,663	32,015	1,387,843	2,059,522
1996	727,531	290,102	360,846	33,825	1,412,304	1,998,270
1997	760,230	306,908	351,017	35,471	1,453,626	1,930,391
1998	744,706	349,806	380,177	22,131	1,496,820	2,259,798
1999	772,104	394,121	354,953	20,227	1,541,405	2,701,205
2000	753,239	361,516	355,610	19,998	1,490,363	2,347,528
2001	818,927	360,916	349,729	17,284	1,546,856	2,586,042
2002	828,525	344,913	354,232	24,379	1,552,053	2,763,724
2003	901,637	414,588	365,983	15,930	1,698,138	2,737,351
2004	938,514	450,781	389,696	22,547	1,801,538	2,723,299

[1] Excludes peanut butter made by manufacturers for own use in candy. Includes peanut butter used in spreads, sandwiches, and cookies. [2] Apparent disappearance represents stocks beginning of year plus production, minus stocks at end of year.

NASS, Crops Branch, (202) 720-2127, and ERS.

Table 3-21.—Peanuts: Area, yield, and production, by States, 2003-2005

State	Area planted			Peanuts for nuts								
	2003	2004	2005[1]	Area harvested			Yield per harvested acre			Production[2]		
				2003	2004	2005[1]	2003	2004	2005[1]	2003	2004	2005[1]
	1,000 acres	*1,000 acres*	*1,000 acres*	*1,000 acres*	*1,000 acres*	*1,000 acres*	*Pounds*	*Pounds*	*Pounds*	*1,000 pounds*	*1,000 pounds*	*1,000 pounds*
AL	190.0	200.0	225.0	185.0	199.0	223.0	2,750	2,800	2,750	508,750	557,200	613,250
FL	125.0	145.0	160.0	115.0	130.0	152.0	3,000	2,800	2,700	345,000	364,000	410,400
GA	545.0	620.0	755.0	540.0	610.0	750.0	3,450	2,980	2,870	1,863,000	1,817,800	2,152,500
MS[3]			15.0			14.0			3,200			44,800
NM	18.0	17.0	19.0	17.0	17.0	19.0	2,700	3,500	3,300	45,900	59,500	62,700
NC	101.0	105.0	97.0	100.0	105.0	96.0	3,200	3,500	3,000	320,000	367,500	288,000
OK	37.0	35.0	35.0	35.0	33.0	33.0	2,800	3,100	3,200	98,000	102,300	105,600
SC	19.0	35.0	63.0	17.0	33.0	60.0	3,400	3,400	2,800	57,800	112,200	168,000
TX	275.0	240.0	265.0	270.0	235.0	260.0	3,000	3,420	3,500	810,000	803,700	910,000
VA	34.0	33.0	23.0	33.0	32.0	22.0	2,900	3,250	3,000	95,700	104,000	66,000
US	1,344.0	1,430.0	1,657.0	1,312.0	1,394.0	1,629.0	3,159	3,076	2,960	4,144,150	4,288,200	4,821,250

[1] Preliminary. [2] Estimates comprised of quota and non-quota peanuts. [3] Estimates began in 2005.
NASS, Crops Branch, (202) 720-2127.

Table 3-22.—Peanuts: Marketing year average price, and value of production, by States, crop of 2003, 2004, and 2005

State	Marketing year average price per pound			Value of production		
	2003	2004	2005[1]	2003	2004	2005[1]
	Dollars	*Dollars*	*Dollars*	*1,000 dollars*	*1,000 dollars*	*1,000 dollars*
AL	0.183	0.178	0.169	93,101	99,182	103,639
FL	0.185	0.181	0.170	63,825	65,884	69,768
GA	0.187	0.185	0.171	348,381	336,293	368,078
MS[2]	....	....	0.170	....	....	7,616
NM	0.230	0.240	0.250	10,557	14,280	15,675
NC	0.229	0.216	0.197	73,280	79,380	56,736
OK	0.183	0.186	0.177	17,934	19,028	18,691
SC	0.216	0.210	0.180	12,485	23,562	30,240
TX	0.195	0.192	0.179	157,950	154,310	162,890
VA	0.229	0.208	0.190	21,915	21,632	12,540
US	0.193	0.189	0.174	799,428	813,551	845,873

[1] Preliminary. [2] Estimates began in 2005.

NASS, Crops Branch, (202) 720–2127.

Table 3-23.—Peanuts, farmers' stock: Price-support operations, United States, 1996–2005

Marketing year beginning August 1	Income support payment rates per pound[1]	Price support level per pound		Put under support[4] [5]			Owned by CCC at end of marketing year[5]
		Quota[2]	Additional[3]	Quantity	Quantity	Percentage of production	
	Cents	*Cents*	*Cents*	*Million pounds*	*Million pounds*	*Percent*	*Million pounds*
1996/97		30.5	6.6	....	320	8.7	0
1997/98		30.5	6.6	....	417	11.8	0
1998/99		30.5	6.6	....	802	20.2	0
1999/2000		30.5	6.6	....	459	12.0	0
2000/01		30.5	6.6	....	450	13.9	0
2001/02		30.5	6.6	....	....	21.9	0
				MAL total	LDP total		
		Loan[6]	Target[6]	1,000 s.t.	1,000 s.t.		
2002/03	1.80	17.75	24.75	668	904	94.6	0
2003/04	1.8	17.75	24.75	1,657	0	80.0	0
2004/05	1.8	17.75	24.75	1,948	0	91.4	9.1
2005/06	1.8	17.75	24.75	....	....	....	....

[1] Enactment of the Farm Security and Rural Investment Act of 2002 (2002 Act) repealed the peanut quota marketing program; and established payment rates for the 2002/03 and subsequent crops according to the provisions of the Direct Payment program. [2] Quota peanuts are those peanuts grown within the farm poundage quota. [3] Additional peanuts are those peanuts grown in excess of the quota. [4] Includes loans made and direct purchases. [5] Includes shelled peanuts converted to farmers' stock basis. [6] Loan rates and target prices for the 2002/03 and subsequent crops were enacted with t0he 2002 Act.

FSA, Tobacco and Peanuts Division, (202) 720–5291.

Table 3-24.—Peanuts: Simple average of monthly f.o.b. price per pound of cleaned and shelled peanuts by approximate crop years, 1995–2004[1]

Classification	1995	1996	1997	1998	1999	2000	2001	2002	2003	2004
Southeastern area:	*Cents*	*Cents*	*Cents*	*Cents*	*Cents*	*Cents*	*Cents*	*Cents*	*Cents*	*Cents*
Georgia, Alabama, and Florida:										
Shelled:.										
Runner Jumbo	67½	60¾	62	59	60¾	62	59½	41½	43	41¾
Runner Medium	66	59½	60⅝	56	58¼	59½	57½	39¾	40¼	39¾
Runner U.S. Splits	65½	57	57¾	56¼	56¾	58½	56¾	38¾	39	39¼
Virginia-North Carolina:										
Clean unshelled Virginias:										
Fancy	57¼	49¾	54⅛	45¾	49⅛	50⅞	51	55	41⅝	38¼
Shelled Virginias:										
Extra large	68½	66½	69⅛	62	66	70½	66½	72¾	53¾	54½
Medium	67½	60¼	64⅝	59¼	60¾	60⅝	59¼	42½	42¼	38
No. 2 with 70% splits	60½	55½	56⅜	53¾	53⅞	55⅛	55	37¾	38½	35¾
Southwestern area:										
Texas and Oklahoma:										
Shelled:										
Spanish No. 1	65½	60¼	60⅝	72	79½	70⅞	62¾	41½	41¼	41¾
Spanish U.S. Splits	66¼	58	58¼	60½	57¼	59	NA	37¾	NA	NA

[1] Crop year begins about Oct. 1 in the Virginia-North Carolina area and about Sept. 1 in the Southeastern and the Southwestern States. Prices are for shipment within 6 months. NA-not available.

AMS, Fruit and Vegetable Division, Market News Branch (229) 228–1208.

Table 3-25.—Peanuts:[1] Area, yield, and production in specified countries and the world, 2001–2002/2003–2004

Continent and country	Area[2]			Yield per hectare			Production		
	2001/ 2002	2002/ 2003	2003/ 2004[3]	2001/ 2002	2002/ 2003	2003/ 2004[3]	2001/ 2002	2002/ 2003	2003/ 2004[3]
	1,000 hec-tares	*1,000 hec-tares*	*1,000 hec-tares*	*Metric tons*	*Metric tons*	*Metric tons*	*1,000 metric tons*	*1,000 metric tons*	*1,000 metric tons*
North America and Caribbean:									
Mexico	62	63	60	1.21	1.43	1.52	75	90	91
United States	523	531	564	2.88	3.54	3.43	1,506	1,880	1,933
Total[4]	585	594	624	2.70	3.32	3.24	1,581	1,970	2,024
South America:									
Argentina	156	167	210	2.03	2.51	2.79	316	420	585
Brazil	85	100	129	2.06	2.17	2.40	175	217	310
Total[4]	241	267	339	2.04	2.39	2.64	491	637	895
Central America:									
Nicaragua	16	20	21	3.19	3.20	3.19	51	64	67
Middle East:									
Turkey	29	30	30	3.10	2.83	2.67	90	85	80
Africa:									
Benin	164	170	160	0.89	0.74	0.81	146	125	130
Burkina	343	345	345	0.94	0.93	0.93	324	320	320
Cameroon	205	205	205	0.98	0.98	0.98	200	200	200
Central African Republic	116	120	125	1.10	1.12	1.12	128	134	140
Chad	480	480	480	0.94	0.94	0.94	450	450	450
Congo, Democratic	457	458	460	0.78	0.79	0.79	355	360	364
Cote d'Ivoire	150	150	150	1.00	1.00	1.00	150	150	150
Egypt	59	60	60	3.24	3.17	3.17	191	190	190
Gambia, The	90	95	105	0.80	0.98	1.00	72	93	105
Ghana	384	465	470	1.02	0.94	0.94	390	439	440
Guinea	204	210	210	1.22	1.19	1.19	248	250	250
Malawi	220	210	210	0.86	0.77	0.76	190	161	160
Morocco	20	24	22	2.00	1.21	2.05	40	29	45
Mozambique	293	290	290	0.38	0.38	0.38	110	110	110
Niger	230	260	260	0.43	0.42	0.42	100	110	110
Nigeria	1,230	1,230	1,240	1.23	1.23	1.23	1,510	1,510	1,520
Senegal	750	525	747	0.35	0.85	0.77	260	445	573
South Africa, Rep.	50	72	43	1.20	1.60	1.77	60	115	76
Sudan	550	550	550	0.67	0.67	0.67	370	370	370
Tanzania, United Rep.	117	117	117	0.64	0.64	0.64	75	75	75
Togo	64	66	66	0.56	0.56	0.56	36	37	37
Uganda	211	211	211	0.70	0.70	0.70	148	148	148
Zambia	135	135	135	0.41	0.41	0.41	55	55	55
Zimbabwe	240	220	220	0.21	0.20	0.20	50	45	45
Total[4]	6,762	6,668	6,881	0.84	0.89	0.88	5,658	5,921	6,063
Asia:									
Bangladesh	27	27	27	1.26	1.19	1.19	34	32	32
Burma	567	575	580	1.33	1.23	1.23	756	710	715
China, Peoples Rep.	4,920	5,057	4,745	3.01	2.65	3.02	14,818	13,420	14,340
India	6,800	8,000	8,000	0.79	0.96	0.85	5,400	7,700	6,800
Indonesia	670	700	720	1.62	1.61	1.60	1,086	1,130	1,150
Japan	10	9	9	2.40	2.44	2.22	24	22	20
Korea, Republic of	5	4	3	1.40	1.75	2.33	7	7	7
Malaysia	1	1	1	2.00	2.00	2.00	2	2	2
Mali	205	210	210	0.59	0.74	0.86	121	156	180
Pakistan	87	92	92	0.72	1.03	1.03	63	95	95
Philippines	27	27	27	1.00	1.00	1.00	27	27	27
Taiwan	25	25	25	3.08	2.92	3.00	77	73	75
Thailand	69	87	85	1.62	1.38	1.53	112	120	130
Vietnam	247	250	250	1.61	1.60	1.60	397	400	400
Total[4]	13,660	15,064	14,774	1.68	1.59	1.62	22,924	23,894	23,973
Oceania:									
Australia	18	20	20	1.17	2.00	2.00	21	40	40
World total[4]	21,311	22,660	22,690	1.45	1.44	1.46	30,816	32,610	33,140

[1] Peanuts in the shell. Split year includes Northern Hemisphere crop harvested in the late months of the first year shown combined with Southern Hemisphere and certain Northern Hemisphere crops harvested in the early months of the following year. [2] Harvested area as far as possible. [3] Preliminary. [4] Regional totals include other countries not shown. World total for all countries not in USDA data base.

FAS, Production Estimates and Crop Assessment Division, (202) 720–0888. Prepared or estimated on the basis of official statistics of foreign governments, other foreign source materials, reports of U.S. Agricultural Counselors, Attachés, and Foreign Service Officers, results of office research, and related information.

Table 3-26.—Soybeans: Area, yield, production, and value, United States, 1996–2005

Year	Area planted	Soybeans for beans				
		Area harvested	Yield per acre	Production	Marketing year average price per bushel received by farmers	Value of production
	1,000 acres	*1,000 acres*	*Bushels*	*1,000 bushels*	*Dollars*	*1,000 dollars*
1996	64,195	63,349	37.6	2,380,274	7.35	17,439,971
1997	70,005	69,110	38.9	2,688,750	6.47	17,372,628
1998	72,025	70,441	38.9	2,741,014	4.93	13,493,891
1999	73,730	72,446	36.6	2,653,758	4.63	12,205,352
2000	74,266	72,408	38.1	2,757,810	4.54	12,466,572
2001	74,075	72,975	39.6	2,890,682	4.38	12,605,717
2002	73,963	72,497	38.0	2,756,147	5.53	15,252,691
2003	73,404	72,476	33.9	2,453,665	7.34	18,013,753
2004	75,208	73,958	42.2	3,123,686	5.74	17,894,948
2005	72,142	71,361	43.3	3,086,432	5.50	16,927,898

NASS, Crops Branch, (202) 720–2127.

Table 3-27.—Soybeans: Stocks on and off farms, United States, 1996–2005

Year	On farms				Off farms [1]			
	Dec. 1	Mar. 1	June 1	Sep. 1 [2]	Dec. 1	Mar. 1	June 1	Sep. 1 [2]
	1,000 bushels	*1,000 bushels*	*1,000 bushels*	*1,000 bushels*	*1,000 bushels*	*1,000 bushels*	*1,000 bushels*	*1,000 bushels*
1996	935,100	514,000	216,000	43,600	889,984	541,754	283,890	88,233
1997	1,048,000	637,000	318,000	84,300	951,417	565,922	275,654	115,499
1998	1,187,000	815,000	458,000	145,000	999,440	642,338	390,573	203,482
1999	1,150,000	730,000	370,000	112,500	1,032,666	665,986	404,425	177,662
2000	1,217,000	780,000	365,000	83,500	1,022,991	623,908	343,180	164,247
2001	1,240,000	687,000	301,200	62,700	1,035,618	648,987	383,721	145,361
2002	1,172,000	636,500	272,500	58,000	943,373	565,528	329,862	120,329
2003	820,000	355,900	110,000	29,400	868,653	549,947	300,604	83,014
2004	1,300,000	795,000	356,100	99,700	1,004,640	586,364	343,174	156,038
2005 [3]	1,345,000	NA	NA	NA	1,157,389	NA	NA	NA

[1] Includes stocks at mills, elevators, warehouses, terminals, and processors.　[2] Old crop only.　[3] Preliminary.　NA-not available.
NASS, Crops Branch, (202) 720–2127.

Table 3-28.—Soybeans, soybean meal, and oil: Average price at specified markets, 1995–2004

Year [1]	Soybeans per bushel: No. 1 Yellow Chicago	Soybean oil per pound crude, tanks, f.o.b. Decatur	Soybean meal per short ton: 48 percent protein Decatur
	Dollars	*Cents*	*Dollars*
1995 ...	5.73	24.70	235.92
1996 ...	7.39	22.51	270.90
1997 ...	7.80	25.83	185.28
1998 ...	6.64	19.80	138.55
1999 ...	5.00	15.59	167.70
2000 ...	4.90	14.15	173.60
2001 ...	4.77	16.46	167.70
2002 ...	4.79	22.04	181.60
2003 ...	5.90	29.97	256.05
2004 [2]	8.22	22.00	165.00

[1] Year beginning September for soybeans and October for oil and meal.　[2] Preliminary.
ERS, Field Crops Branch, (202) 694–5300.

Table 3-29.—Soybeans: Supply and disappearance, United States, 1995–2004

Year beginning September	Supply				
	Stocks by position			Production	Total [1]
	Farm	Terminal market, interior mill, elevator, and warehouse	Total		
	1,000 bushels	*1,000 bushels*	*1,000 bushels*	*1,000 bushels*	*1,000 bushels*
1995	105,130	229,684	334,814	2,174,254	2,513,524
1996	59,523	123,935	183,458	2,380,274	2,572,636
1997	43,600	88,233	131,833	2,688,750	2,825,589
1998	84,300	115,499	199,799	2,741,014	2,944,334
1999	145,000	203,482	348,482	2,653,758	3,006,411
2000	112,500	177,662	290,162	2,757,810	3,051,540
2001	83,500	164,247	247,747	2,890,682	3,140,749
2002	62,700	145,361	208,061	2,756,147	2,968,869
2003	58,000	120,329	178,329	2,453,665	2,637,556
2004 [2]	29,400	83,014	112,414	3,123,686	3,241,676

Year beginning September	Disappearance			
	Crushed [3]	Seed, feed and residual	Exports	Total
	1,000 bushels	*1,000 bushels*	*1,000 bushels*	*1,000 bushels*
1995	1,369,541	111,441	849,084	2,330,066
1996	1,436,961	118,954	885,888	2,440,803
1997	1,596,983	154,476	874,334	2,625,793
1998	1,589,787	201,414	804,651	2,595,852
1999	1,577,650	165,194	973,405	2,716,249
2000	1,639,670	168,252	995,871	2,803,793
2001	1,699,741	169,296	1,063,651	2,932,688
2002	1,615,464	131,380	1,044,372	2,790,540
2003	1,529,699	108,892	886,551	2,525,142
2004 [2]	1,696,088	187,386	1,102,695	2,986,169

[1] Includes imports, beginning with 1988.   [2] Preliminary.   [3] Reported by the U.S. Department of Commerce.
ERS, Field Crops Branch, (202) 694–5300.

Table 3-30.—Soybeans: Support operations, United States, 1996–2005

Marketin year beginning September 1	Income support payment rates per bushels [1]	Program price levels per bushel		Put under loan [4]		Acquired by CCC under loan program [5]	Owned by CCC at end of marketing year
		Loan [2]	Target [3]	Quantity	Percentage of production		
	Dollars	*Dollars*	*Dollars*	*Million bushels*	*Percent*	*Million bushels*	*Million bushels*
1996/1997		4.97		195.9	8.2	0.0	0.0
1997/1998		5.26		266.3	9.9	0.5	0.6
1998/1999		5.26		340.9	12.4	7.7	3.8
1999/2000	0.14	5.26		286.9	10.8	13.7	7.0
2000/2001	0.26	5.26		313.0	11.3	10.0	2.0
2001/2002		5.26		311.8	10.8	3.8	2.7
2002/2003	0.44/---	5.00	5.80	384.3	13.9	0.2	0.7
2003/2004	0.44/---	5.00	5.80	156.6	6.4	0.0	0.0
2004/2005	0.44/---	5.00	5.80	426.0	13.6	0.0	0.0
2005/2006	0.44/---	5.00	5.80				

[1] Oilseeds payment rates for 1999/2000 were calculated according to the provisions of the Agriculture, Rural Development, Food and Drug Administration, and Related Agencies Appropriations Act, 2000. Rates for 2000/01 were calculated according to the provisions of the Agricultural Risk Protection Act of 2000, and included supplemental oilseeds payment rates. Payment rates for the 2002/2003 and subsequent crops are calculated according to the provisions of the Direct and Counter-Cylical Payment program, following enactment of the Farm Security and Rural Investment Act of 2002 (2002 Act). Payment rates are rounded to the nearest tenth of a cent. Beginning with 2002/03, the first entry is the direct payment rate and the second entry is the counter-cylical payment rate. Counter-cyclical payment rate for 2004/05 is preliminary.   [2] The national average loan rate was also known as the price support rate prior to enactment of the Federal Agricultural Improvement and Reform Act of 1996 (1996 Act).   [3] Target prices for the 2002/03 and subsequent crops were enacted with the 2002 Act.   [4] Does not include quantity on which loan deficiency payments were made.   [5] Acquisition of all loans forfeited during the marketing year including loans made in previous years.
FSA, Feed Grains & Oilseeds Analysis Group, (202) 720–8838.

Table 3-31.—Soybeans: Area, yield, and production, by State and United States, 2003–2005

State	Area planted			Soybeans for beans								
				Area harvested			Yield per harvested acre			Production		
	2003	2004	2005	2003	2004	2005	2003	2004	2005	2003	2004	2005
	1,000 acres	1,000 acres	1,000 acres	1,000 acres	1,000 acres	1,000 acres	Bushels	Bushels	Bushels	1,000 bushels	1,000 bushels	1,000 bushels
AL	170	210	150	160	190	145	36.0	35.0	33.0	5,760	6,650	4,785
AR	2,920	3,200	3,030	2,890	3,150	3,000	38.5	39.0	34.0	111,265	122,850	102,000
DE	180	210	185	178	208	182	36.0	42.0	26.0	6,408	8,736	4,732
FL	13	19	9	12	17	8	30.0	34.0	32.0	360	578	256
GA	190	280	180	180	270	175	33.0	31.0	26.0	5,940	8,370	4,550
IL	10,300	9,950	9,500	10,260	9,900	9,450	37.0	50.0	47.0	379,620	495,000	444,150
IN	5,450	5,550	5,400	5,370	5,520	5,380	38.0	51.5	49.0	204,060	284,280	263,620
IA	10,600	10,200	10,100	10,550	10,150	10,050	32.5	49.0	53.0	342,875	497,350	532,650
KS	2,600	2,800	2,900	2,480	2,710	2,850	23.0	41.0	37.0	57,040	111,110	105,450
KY	1,250	1,310	1,260	1,240	1,300	1,250	43.5	44.0	43.0	53,940	57,200	53,750
LA	760	1,100	880	740	990	850	34.0	33.0	34.0	25,160	32,670	28,900
MD	435	500	480	430	495	470	37.0	43.0	34.0	15,910	21,285	15,980
MI	2,000	2,000	2,000	1,990	1,980	1,990	27.5	38.0	39.0	54,725	75,240	77,610
MN	7,500	7,300	6,900	7,450	7,050	6,800	32.0	33.0	45.0	238,400	232,650	306,000
MS	1,440	1,670	1,610	1,430	1,640	1,590	39.0	37.5	37.0	55,770	61,500	58,830
MO	5,000	5,000	5,000	4,950	4,960	4,960	29.5	45.0	37.0	146,025	223,200	183,520
NE	4,550	4,800	4,700	4,500	4,750	4,660	40.5	46.0	50.5	182,250	218,500	235,330
NJ	90	105	95	88	103	91	34.0	42.0	28.0	2,992	4,326	2,548
NY	140	175	190	138	172	188	35.0	39.0	42.0	4,830	6,708	7,896
NC	1,450	1,530	1,490	1,400	1,500	1,460	30.0	34.0	27.0	42,000	51,000	39,420
ND	3,150	3,750	2,950	3,050	3,570	2,900	29.0	23.0	37.0	88,450	82,110	107,300
OH	4,300	4,450	4,500	4,280	4,420	4,480	38.5	47.0	45.0	164,780	207,740	201,600
OK	270	320	325	245	290	305	26.0	30.0	26.0	6,370	8,700	7,930
PA	380	430	430	375	425	420	41.0	46.0	41.0	15,375	19,550	17,220
SC	430	540	430	420	530	420	28.0	27.0	20.5	11,760	14,310	8,610
SD	4,250	4,150	3,900	4,200	4,120	3,850	27.5	34.0	36.0	115,500	140,080	138,600
TN	1,150	1,210	1,130	1,120	1,180	1,100	42.0	41.0	38.0	47,040	48,380	41,800
TX	200	290	260	185	270	230	29.0	32.0	26.0	5,365	8,640	5,980
VA	500	540	530	480	530	510	34.0	39.0	30.0	16,320	20,670	15,300
WV	16	19	18	15	18	17	41.0	46.0	35.0	615	828	595
WI	1,720	1,600	1,610	1,670	1,550	1,580	28.0	34.5	44.0	46,760	53,475	69,520
US	73,404	75,208	72,142	72,476	73,958	71,361	33.9	42.2	43.3	2,453,665	3,123,686	3,086,432

NASS, Crops Branch, (202) 720–2127.

Table 3-32.—Soybeans: Crushings, and oil and meal stocks, production, and foreign trade, United States, 1995–2003

Year beginning October	Soybeans crushed					Soybean oil			Soybean cake and meal		
	Oct.-Dec.	Jan.-Mar.	Apr.-Jun.	Jul.-Sep.	Total	Stocks Oct. 1	Production	Exports	Stocks Oct. 1	Production	Exports
	1,000 bushels	1,000 bushels	1,000 bushels	1,000 bushels	1,000 bushels	Million pounds	Million pounds	Million pounds	1,000 tons	1,000 tons	1,000 tons
1995	369,123	349,480	325,929	318,490	1,363,022	1,137	15,240	992	223	32,527	6,004
1996	398,225	392,509	334,449	320,681	1,445,864	2,015	15,752	2,033	212	34,211	6,994
1997	438,067	436,997	375,370	359,643	1,610,077	1,520	18,143	3,079	210	38,176	9,330
1998	430,007	403,987	377,640	388,027	1,599,661	1,382	18,078	2,372	218	37,797	7,122
1999	435,943	395,117	360,423	381,273	1,572,756	1,520	17,825	1,375	330	37,591	7,331
2000	434,530	417,420	391,733	395,327	1,639,010	1,993	18,420	1,401	293	39,385	7,703
2001	452,756	443,946	414,407	382,738	1,693,847	2,767	18,898	2,519	383	40,292	7,508
2002	445,333	414,609	378,150	381,990	1,620,082	2,359	18,430	2,261	240	38,194	6,019
2003 [1]	437,588	406,939	339,334	339,213	1,523,047	1,491	17,080	935	220	36,324	4,344

[1] Preliminary.
ERS, Field Crops Branch, (202) 694–5300. Data from the U.S. Department of Commerce.

Table 3-33.—Soybeans for beans: Marketing year average price and value, by State and United States, crop of 2003, 2004, and 2005

State	Marketing year average price per bushel			Value of production		
	2003	2004	2005[1]	2003	2004	2005[1]
	Dollars	Dollars	Dollars	1,000 dollars	1,000 dollars	1,000 dollars
AL	7.25	6.25	5.85	41,760	41,563	27,992
AR	7.11	5.88	5.80	791,094	722,358	591,600
DE	7.65	5.40	5.45	49,021	47,174	25,789
FL	6.90	5.60	5.40	2,484	3,237	1,382
GA	7.47	5.70	5.50	44,372	47,709	25,025
IL	7.51	5.84	5.50	2,850,946	2,890,800	2,442,825
IN	7.67	5.66	5.50	1,565,140	1,609,025	1,449,910
IA	7.70	5.76	5.45	2,640,138	2,864,736	2,902,943
KS	7.68	5.39	5.30	438,067	598,883	558,885
KY	7.40	5.87	5.55	399,156	335,764	298,313
LA	6.80	6.29	5.85	171,088	205,494	169,065
MD	7.65	5.35	5.45	121,712	113,875	87,091
MI	7.30	5.72	5.55	399,493	430,373	430,736
MN	7.26	5.90	5.45	1,730,784	1,372,635	1,667,700
MS	6.61	6.20	5.80	368,640	381,300	341,214
MO	7.52	5.62	5.45	1,098,108	1,254,384	1,000,184
NE	7.02	5.54	5.50	1,279,395	1,210,490	1,294,315
NJ	7.35	5.45	5.65	21,991	23,577	14,396
NY	7.80	5.40	5.20	37,674	36,223	41,059
NC	7.29	5.56	5.45	306,180	283,560	214,839
ND	6.62	5.75	5.30	585,539	472,133	568,690
OH	7.20	5.74	5.55	1,186,416	1,192,428	1,118,880
OK	7.40	5.70	5.30	47,138	49,590	42,029
PA	7.33	5.43	5.55	112,699	106,157	95,571
SC	7.60	5.60	5.55	89,376	80,136	47,786
SD	6.96	5.58	5.30	803,880	781,646	734,580
TN	7.05	5.58	5.55	331,632	269,960	231,990
TX	7.00	5.85	5.45	37,555	50,544	32,591
VA	7.67	5.32	5.55	125,174	109,964	84,915
WV	7.54	5.34	5.45	4,637	4,422	3,243
WI	7.11	5.70	5.50	332,464	304,808	382,360
US	7.34	5.74	5.50	18,013,753	17,894,948	16,927,898

[1] Preliminary.
NASS, Crops Branch, (202) 720–2127.

Table 3-34.—Soybeans: Area, yield, and production in specified countries and the world, 2002–2003/2004–2005[1]

Continent and country	Area[2]			Yield per hectare			Production		
	2002/ 2003	2003/ 2004	2004/ 2005[3]	2002/ 2003	2003/ 2004	2004/ 2005[3]	2002/ 2003	2003/ 2004	2004/ 2005[3]
	1,000 hec- tares	1,000 hec- tares	1,000 hec- tares	Metric tons	Metric tons	Metric tons	1,000 metric tons	1,000 metric tons	1,000 metric tons
North America:									
Canada	1,024	1,044	1,174	2.28	2.17	2.59	2,336	2,263	3,042
Mexico	56	68	80	1.59	1.84	1.56	89	125	125
United States	29,339	29,330	29,930	2.56	2.28	2.86	75,010	66,778	85,484
Total	30,419	30,442	31,184	2.55	2.27	2.84	77,435	69,166	88,651
South America:									
Argentina	12,600	14,000	14,400	2.82	2.36	2.71	35,500	33,000	39,000
Bolivia	710	863	920	2.32	2.14	2.21	1,650	1,850	2,030
Brazil	18,448	21,520	22,840	2.82	2.35	2.23	52,000	50,500	51,000
Colombia	25	28	31	2.44	2.25	2.06	61	63	64
Ecuador	58	53	56	1.67	1.62	1.61	97	86	90
Paraguay	1,550	1,750	2,000	2.90	2.23	1.90	4,500	3,911	3,800
Peru	1	1	1	2.00	2.00	2.00	2	2	2
Uruguay	77	247	280	2.38	1.53	1.79	183	377	500
Venezuela	2	3	3	1.00	1.00	1.00	2	3	3
Total	33,471	38,465	40,531	2.81	2.33	2.38	93,995	89,792	96,489

See footnotes at end of table.

Table 3-34.—Soybeans: Area, yield, and production in specified countries and the world, 2002–2003/2004–2005 [1]—Continued

Continent and country	Area [2]			Yield per hectare			Production		
	2002/ 2003	2003/ 2004	2004/ 2005 [3]	2002/ 2003	2003/ 2004	2004/ 2005 [3]	2002/ 2003	2003/ 2004	2004/ 2005 [3]
	1,000 hec- tares	1,000 hec- tares	1,000 hec- tares	Metric tons	Metric tons	Metric tons	1,000 metric tons	1,000 metric tons	1,000 metric tons
Central America:									
Guatemala	11	11	11	2.91	2.91	2.91	32	32	32
Nicaragua	2	2	2	1.50	1.50	1.50	3	3	3
Total	13	13	13	2.69	2.69	2.69	35	35	35
European Union:									
Austria	14	15	17	2.50	2.60	2.65	35	39	45
Czech Republic	3	8	9	2.00	1.50	1.44	6	12	13
France	75	81	61	2.80	1.60	2.52	210	130	154
Germany	1	1	1	2.00	2.00	2.00	2	2	2
Hungary	26	30	27	2.04	1.67	2.44	53	50	66
Italy	152	152	151	3.72	2.56	3.32	566	389	501
Slovakia	9	11	9	1.67	1.09	1.56	15	12	14
Spain	1	1	0	1.00	3.00	0.00	1	3	0
Total	281	299	275	3.16	2.13	2.89	888	637	795
Other Europe:									
Bosnia-Hercegovina	3	4	4	2.33	1.25	2.25	7	5	9
Croatia	48	50	50	2.69	1.66	2.20	129	83	110
Romania	62	110	122	1.69	1.73	2.46	105	190	300
Serbia and Montenegro	120	140	130	2.00	2.14	2.31	240	300	300
Switzerland	2	3	5	2.50	2.33	2.80	5	7	14
Total	235	307	311	2.07	1.91	2.36	486	585	733
Former Soviet Union:									
Russian Fed.	362	401	555	1.17	0.98	1.00	423	393	555
Ukraine	98	188	256	1.28	1.23	1.42	125	232	363
Total	460	589	811	1.19	1.06	1.13	548	625	918
Middle East:									
Iran	90	90	90	1.50	1.28	1.50	135	115	135
Syria	3	3	3	1.67	2.00	1.67	5	6	5
Turkey	35	15	10	2.71	3.00	2.50	95	45	25
Total	128	108	103	1.84	1.54	1.60	235	166	165
Africa:									
Egypt	4	8	14	2.00	2.25	2.43	8	18	34
Morocco	1	1	0	1.00	0.00	0.00	1	0	0
Nigeria	390	400	410	0.97	1.00	1.00	380	400	410
South Africa, Republic of	100	135	153	1.37	1.63	1.81	137	220	277
Uganda	151	151	151	1.10	1.10	1.10	166	166	166
Zambia	13	13	13	2.31	2.31	2.31	30	30	30
Zimbabwe	40	40	40	1.38	1.25	1.25	55	50	50
Total	699	748	781	1.11	1.18	1.24	777	884	967
Asia:									
Burma	118	125	125	1.05	1.04	1.04	124	130	130
China, Peop.	8,720	9,313	9,590	1.89	1.65	1.81	16,510	15,394	17,400
India	5,670	6,450	7,200	0.71	1.05	0.76	4,000	6,800	5,500
Indonesia	550	630	640	1.42	1.30	1.29	780	820	825
Japan	150	152	137	1.80	1.53	1.20	270	232	165
Korea, Dem.	315	315	315	1.14	1.14	1.14	360	360	360
Korea, Rep.	81	80	85	1.42	1.31	1.64	115	105	139
Pakistan	2	2	2	1.00	1.00	1.00	2	2	2
Philippines	1	1	1	1.00	1.00	1.00	1	1	1
Taiwan	0	0	0	0.00	0.00	0.00	0	0	0
Thailand	180	180	165	1.39	1.22	1.45	250	220	240
Vietnam	158	182	190	1.27	1.24	1.26	201	225	240
Total	15,945	17,430	18,450	1.42	1.39	1.36	22,613	24,289	25,002
Oceania:									
Australia	10	33	33	1.80	2.24	1.82	18	74	60
World total	81,695	88,430	92,490	2.41	2.11	2.31	197,079	186,260	213,350

[1] Split year includes Northern Hemisphere crop harvested in the late months of the first year shown combined with Southern Hemisphere and certain Northern Hemisphere crops harvested in the early months of the following year. [2] Harvested area as far as possible. [3] Preliminary.

FAS, Production Estimates and Crop Assessment Division, (202) 720–0888. Prepared or estimated on the basis of official statistics of foreign governments, other foreign source materials, reports of U.S. Agricultural Counselors, Attachés, and Foreign Service Officers, results of office research, and related information.

Table 3-35.—Soybeans, soybean oil, and soybean cake and meal: United States exports by country of destination, 2002/2003–2004/2005 [1]

Continent and country of destination	Soybeans			Soybean oil [2]			Soybean cake and meal		
	2002/2003	2003/2004	2004/2005 [3]	2002/2003	2003/2004	2004/2005 [3]	2002/2003	2003/2004	2004/2005 [3]
	Metric tons	Metric tons	Metric tons	Metric tons	Metric tons	Metric tons	Metric tons	Metric tons	Metric tons
North and Central America; incl. Caribbean:									
Canada	652,646	572,110	388,630	124,667	96,450	68,134	1,071,936	1,072,753	1,173,621
Costa Rica	220,524	152,094	216,330	696	1,422	5	6,629	6,482	13,664
Jamaica & Dep	17	78	17	18,990	12,610	11,648	88,388	86,786	69,279
Mexico	4,108,915	3,005,396	3,363,617	188,992	97,099	162,665	616,495	806,054	1,177,716
Total [4]	5,193,220	3,970,655	4,185,132	506,500	304,707	382,175	2,824,065	2,761,205	3,460,542
South America:									
Brazil	0	0	0	123	3	0	425	118	91
Columbia	163,779	124,613	151,901	1,665	563	4,253	59,955	141,613	207,836
Peru	0	0	0	20,349	25,097	15,459	22,721	20,076	25,873
Venezuela	2,698	8,000	9,031	311	169	6,063	217,086	128,364	157,763
Total [4]	166,885	132,981	161,003	22,720	26,058	26,077	337,926	298,609	493,276
Europe:									
Belgium and Luxembourg	401,631	195,833	412,819	0	58	0	945	2,544	2,775
Denmark	74,428	103,729	73,291	0	0	0	8,908	0	0
France	121,163	54,741	113,169	2	39	85	89	153	369
Germany	1,557,940	717,388	2,004,025	3,927	664	942	239	2,859	261
Greece	134,231	21,000	50,093	0	0	0	7,521	121	140
Italy	241,650	40,473	152,192	11	0	0	62	522	199
Netherlands	582,321	570,621	472,307	4,121	1,098	432	42,455	61,992	26,725
Norway	0	0	0	0	0	0	18	0	188
Spain	1,247,833	825,659	780,149	2,000	0	39	5,759	31,684	30,005
United Kingdom	127,157	72,631	163,635	22	75	8	17,706	5,408	20,768
Total [4]	5,094,368	2,988,381	4,487,820	10,446	2,255	1,580	167,333	154,307	205,437
Former Soviet Union: [5]									
Russia	0	0	0	16	0	3,400	57,026	41,940	17,530
Ukraine	142	155	0	0	14	0	80	19	0
Total [4]	56,142	10,155	10,002	5,151	13,173	7,170	57,106	48,140	17,796
Africa:									
Egypt	40,640	71,025	298,791	54,079	322	0	36,139	15,401	187,910
Morocco	267,965	124,200	179,020	26,517	15,518	6,579	0	0	0
Total [4]	314,927	195,225	477,924	155,709	46,036	73,740	270,574	255,704	343,995
Asia:									
China	7,731,935	8,315,837	11,822,824	94,053	195	159	120	155	290
India	0	0	1,049	42,727	14,561	29,385	26	74	2,636
Israel	306,255	183,880	246,084	1,529	1,289	714	46,364	30,801	7,641
Japan	3,533,379	3,329,867	3,006,789	26,810	8,332	12,363	287,885	214,316	506,929
Korea, Rep. of	1,006,754	1,069,875	726,522	44,158	1,196	14,143	102,561	2,799	5,095
Philippines	196,376	179,276	196,301	114	138	78	289,610	254,483	514,306
Taiwan	1,517,350	1,264,877	1,589,705	10,059	140	136	716	14,901	60,908
Total [4]	16,987,609	16,392,123	20,179,421	325,868	32,277	109,628	1,693,418	879,128	1,863,934
Oceania:									
Australia	43,635	8,233	3	46	12	19	299,489	225,447	198,777
Total [4]	43,729	8,233	59,737	203	48	23	377,518	293,049	273,753
Grand Total [6][7]	28,423,161	24,127,961	30,010,450	1,026,638	424,554	600,393	5,727,939	4,690,142	6,658,733

[1] For soybeans, year begins September 1; for soybean oil and cake and meal, year begins October 1.　[2] Crude and refined oil (includes shipments under P.L. 480).　[3] Preliminary.　[4] Includes quantities exported to countries not shown.　[5] Former Soviet Union; includes the 12 Republics of the USSR that are not members of the European Union.　[6] May not add due to rounding.　[7] Includes quantities transshipped via Canada to unidentified countries.

FAS, Cotton, Oilseeds, Tobacco and Seeds Division, (202) 720–9518. Compiled from reports of the U.S. Department of Commerce.

Table 3-36.—Soybeans: International trade, 2000/2001–2004/2005 [1]

Country	2000/2001	2001/2002	2002/2003	2003/2004	2004/2005
	1,000 metric tons	*1,000 metric tons*	*1,000 metric tons*	*1,000 metric tons*	*1,000 metric tons*
Principal exporting countries:					
United States	28,948	28,423	24,128	30,011	29,257
Brazil	15,000	19,734	19,816	20,538	24,000
Argentina	6,005	8,713	6,710	9,600	9,700
Paraguay	2,285	2,806	2,776	2,600	3,000
Canada	495	700	914	1,025	1,050
Subtotal	52,733	60,376	54,344	63,774	67,007
Others	673	801	1,517	1,475	1,468
World total	53,406	61,177	55,861	65,249	68,475

Country	2000/2001	2001/2002	2002/2003	2003/2004	2004/2005
Principal importing countries:					
EU-25	17,525	18,539	16,872	14,638	15,800
China, Peoples Republic	13,245	10,385	21,417	16,933	25,802
Indonesia	1,127	1,414	1,235	1,316	1,175
Japan	4,767	5,023	5,087	4,688	4,295
Korea, Republic of	1,389	1,434	1,516	1,368	1,240
Mexico	4,381	4,510	4,230	3,797	3,500
Taiwan	2,330	2,578	2,351	2,218	2,300
Thailand	1,290	1,560	1,708	1,407	1,517
Subtotal	46,054	45,443	54,416	46,365	55,629
Others	7,107	9,010	8,508	7,881	9,318
World total	53,161	54,453	62,924	54,246	64,947

[1] Marketing year beginning Aug. 1.

FAS, Cotton, Oilseeds, Tobacco and Seeds Division, (202) 720–9516. Prepared or estimated on the basis of official statistics of foreign governments, other foreign source materials, reports of U.S. Agricultural Counselors, Attachés, and Foreign Service Officers, results of office research, and related information.

Table 3-37.—Sunflower: Area, yield, production, and value, United States, 1996–2005 [1]

Year	Area planted	Area harvested	Yield per harvested acre	Production	Price per cwt.	Value of production
	1,000 acres	*1,000 acres*	*Pounds*	*1,000 pounds*	*Dollars*	*1,000 dollars*
1996	2,536	2,479	1,436	3,559,343	11.70	414,842
1997	2,888	2,792	1,317	3,676,952	11.60	426,766
1998	3,568	3,492	1,510	5,273,162	10.60	536,971
1999	3,553	3,441	1,262	4,341,862	7.53	339,985
2000	2,840	2,647	1,339	3,544,428	6.89	246,869
2001	2,633	2,555	1,338	3,418,759	9.62	325,950
2002	2,581	2,167	1,131	2,451,247	12.10	294,595
2003	2,344	2,197	1,213	2,665,226	12.10	316,214
2004	1,873	1,711	1,198	2,049,613	13.70	272,732
2005	2,709	2,610	1,540	4,018,355	11.50	472,470

[1] Estimates include all States except AK and HI.
NASS, Crops Branch, (202) 720–2127.

Table 3-38.—Sunflower, Oil Varieties: Area, yield, production, and value, United States, 1996–2005 [1]

Year	Area planted	Area harvested	Yield per harvested acre	Production	Price per cwt.	Value of production
	1,000 acres	*1,000 acres*	*Pounds*	*1,000 pounds*	*Dollars*	*1,000 dollars*
1996	1,967	1,934	1,470	2,843,763	10.80	309,057
1997	2,284	2,212	1,350	2,985,700	11.00	329,858
1998	2,953	2,897	1,549	4,486,360	9.37	423,775
1999	2,757	2,695	1,298	3,497,820	6.33	229,593
2000	2,248	2,116	1,375	2,909,844	5.89	175,306
2001	2,117	2,060	1,361	2,803,704	9.07	254,705
2002	2,126	1,806	1,144	2,065,899	11.70	241,851
2003	1,998	1,874	1,206	2,259,666	11.30	254,076
2004	1,533	1,424	1,238	1,763,378	12.80	223,836
2005	2,104	2,032	1,564	3,177,635	10.00	329,424

[1] Estimates include all States except AK and HI.
NASS, Crops Branch, (202) 720–2127.

Table 3-39.—Sunflower, non-oil varieties: Area, yield, production, and value, United States, 1996–2005 [1]

Year	Area planted	Area harvested	Yield per harvested acre	Production	Price per cwt.	Value of production
	1,000 acres	*1,000 acres*	*Pounds*	*1,000 pounds*	*Dollars*	*1,000 dollars*
1996	569	545	1,313	715,580	13.80	105,785
1997	604	580	1,192	691,252	14.30	96,908
1998	615	595	1,322	786,802	14.60	113,196
1999	796	746	1,131	844,042	13.40	110,392
2000	592	531	1,195	634,584	11.20	71,563
2001	516	495	1,243	615,055	11.60	71,245
2002	455	361	1,067	385,348	13.70	52,744
2003	346	323	1,256	405,560	15.20	62,138
2004	340	287	997	286,235	17.20	48,896
2005	605	578	1,455	840,720	16.30	143,046

[1] Estimates include all States except AK and HI.
NASS, Crops Branch, (202) 720–2127.

Table 3-40.—Sunflower: Area, yield, production, and value by type, State and United States, 2004–2005

Type and State	Area planted		Area harvested		Yield per harvested acre	
	2004	2005[1]	2004	2005[1]	2004	2005[1]
	1,000 acres	1,000 acres	1,000 acres	1,000 acres	Pounds	Pounds
Oil:						
CO	90	150	80	145	1,350	1,250
KS	150	255	140	245	1,460	1,540
MN	30	75	28	72	1,200	1,600
NE	36	60	35	58	1,000	1,400
ND	720	910	660	885	1,040	1,610
SD	410	500	394	481	1,460	1,650
TX	18	50	16	48	1,300	1,600
Other States[2] ...	79	104	71	98	1,408	1,300
US	1,533	2,104	1,424	2,032	1,238	1,564
Non-oil:						
CO	45	65	43	60	900	1,350
KS	21	45	18	44	1,220	1,700
MN	30	60	25	55	920	1,250
NE	20	39	18	38	1,050	1,600
ND	160	230	130	220	810	1,490
SD	25	50	21	49	1,500	1,700
TX	23	95	22	92	1,600	1,300
Other States[2] ...	16	21	10	20	1,168	1,234
US	340	605	287	578	997	1,455
Total:						
CO	135	215	123	205	1,193	1,279
KS	171	300	158	289	1,433	1,564
MN	60	135	53	127	1,068	1,448
NE	56	99	53	96	1,017	1,479
ND	880	1,140	790	1,105	1,002	1,586
SD	435	550	415	530	1,462	1,655
TX	41	145	38	140	1,474	1,403
Other States[2] ...	95	125	81	118	1,378	1,289
US	1,873	2,709	1,711	2,610	1,198	1,540

Type and State	Production		Marketing year average price per cwt.		Value of production	
	2004	2005[1]	2004	2005[1]	2004	2005[1]
	1,000 pounds	1,000 pounds	Dollars	Dollars	1,000 dollars	1,000 dollars
Oil:						
CO	108,000	181,250	11.00	10.20	11,880	18,488
KS	204,400	377,300	11.60	8.65	23,710	32,636
MN	33,600	115,200	14.40	12.20	4,838	14,054
NE	35,000	81,200	13.00	11.60	4,550	9,419
ND	686,400	1,424,850	13.00	10.30	89,232	146,760
SD	575,240	793,650	12.80	10.30	73,631	81,746
TX	20,800	76,800	13.20	13.30	2,746	10,214
Other States[2] ...	99,938	127,385	13.30	12.60	13,249	16,107
US	1,763,378	3,177,635	12.80	10.00	223,836	329,424
Non-oil:						
CO	38,700	81,000	16.30	18.40	6,308	14,904
KS	21,960	74,800	17.20	18.10	3,777	13,539
MN	23,000	68,750	18.60	17.80	4,278	12,238
NE	18,900	60,800	17.00	17.40	3,213	10,579
ND	105,300	327,800	18.50	16.70	19,481	54,743
SD	31,500	83,300	14.70	11.70	4,631	9,746
TX	35,200	119,600	14.70	19.00	5,174	22,724
Other States[2] ...	11,675	24,670	17.40	18.50	2,034	4,573
US	286,235	840,720	17.20	16.30	48,896	143,046
Total:						
CO	146,700	262,250	12.40	12.70	18,188	33,392
KS	226,360	452,100	12.70	12.10	27,487	46,175
MN	56,600	183,950	16.70	15.90	9,116	26,292
NE	53,900	142,000	14.40	14.10	7,763	19,998
ND	791,700	1,752,650	14.10	11.40	108,713	201,503
SD	606,740	876,950	13.10	10.60	78,262	91,492
TX	56,000	196,400	14.20	16.80	7,920	32,938
Other States[2] ...	111,613	152,055	13.70	13.60	15,283	20,680
US	2,049,613	4,018,355	13.70	11.50	272,732	472,470

[1] Preliminary.   [2] For 2004, Other States include CA, GA, IL, LA, MI, MO, MT, NM, NY, OH, OK, PA, SC, UT, WA, WI, and WY. For 2005, Other States include CA, IL, MI, MO, MT, OK, WI, and WY.

NASS, Crops Branch, (202) 720–2127.

Table 3-41.—Sunflowerseeds: Area, yield and production in specified countries, 2002/2003–2004/2005 [1]

Continent and country	Area [2]			Yield per hectare			Production		
	2002/ 2003	2003/ 2004	2004/ 2005 [3]	2002/ 2003	2003/ 2004	2004/ 2005 [3]	2002/ 2003	2003/ 2004	2004/ 2005 [3]
	1,000 hectares	1,000 hectares	1,000 hectares	Metric tons	Metric tons	Metric tons	1,000 metric tons	1,000 metric tons	1,000 metric tons
North America:									
Canada	95	115	60	1.65	1.30	1.00	157	150	60
Mexico	1	1	1	1.00	1.00	1.00	1	1	1
United States	877	889	692	1.27	1.36	1.34	1,112	1,209	929
Total	973	1,005	753	1.31	1.35	1.31	1,270	1,360	990
South America:									
Argentina	2,350	1,830	1,890	1.57	1.77	1.90	3,700	3,240	3,600
Bolivia	134	83	160	0.58	1.11	1.06	78	92	170
Brazil	43	55	45	1.30	1.56	1.67	56	86	75
Paraguay	30	30	30	1.33	1.67	1.67	40	50	50
Uruguay	170	110	129	1.38	1.61	1.28	234	177	165
Total	2,727	2,108	2,254	1.51	1.73	1.80	4,108	3,645	4,060
European Union:									
Austria	21	26	29	2.76	2.73	2.69	58	71	78
Czech Republic	24	49	39	2.29	2.33	2.18	55	114	85
France	616	691	614	2.43	2.16	2.37	1,497	1,492	1,455
Germany	26	37	32	2.00	1.97	2.19	52	73	70
Greece	10	18	19	2.00	1.22	1.21	20	22	23
Hungary	415	511	479	1.88	1.94	2.50	779	992	1,198
Italy	166	151	124	2.13	1.57	2.21	354	237	274
Portugal	38	38	36	0.55	0.47	0.47	21	18	17
Slovakia	64	131	90	1.88	1.93	2.18	120	253	196
Spain	754	790	780	1.00	0.97	1.01	757	763	785
Total	2,134	2,442	2,242	1.74	1.65	1.86	3,713	4,035	4,181
Other Europe:									
Bulgaria	430	610	490	1.35	1.18	1.73	580	720	850
Croatia	27	28	28	2.33	2.46	2.21	63	69	62
Macedonia (Skopje)	7	5	5	1.29	1.40	1.00	9	7	5
Romania	880	1,100	950	1.01	1.27	1.50	890	1,400	1,425
Serbia and Montenegro	170	220	210	1.65	1.82	2.10	280	400	440
Total	1,514	1,963	1,683	1.20	1.32	1.65	1,822	2,596	2,782
Frm. USSR (non-Baltics):									
Kazakhstan	320	431	400	0.59	0.68	0.66	190	292	265
Moldova	256	280	250	1.55	1.43	1.44	397	400	360
Russian Federation	3,798	4,850	4,650	0.97	1.00	1.02	3,685	4,850	4,750
Ukraine	2,720	3,807	3,400	1.20	1.12	0.90	3,270	4,252	3,050
Total [4]	7,094	9,368	8,700	1.06	1.05	0.97	7,542	9,794	8,425
Middle East:									
Iran	79	80	80	0.52	0.53	0.56	41	42	45
Israel	15	7	6	0.80	1.71	2.33	12	12	14
Turkey	550	500	480	1.49	1.20	1.35	820	600	650
Total	644	587	566	1.36	1.11	1.25	873	654	709
Africa:									
Egypt	2	1	2	1.50	2.00	2.00	3	2	4
Morocco	50	105	145	0.32	0.53	0.34	16	56	50
South Africa, Republic of	606	530	497	1.06	1.23	1.34	642	651	665
Total	658	636	644	1.00	1.11	1.12	661	709	719
Asia and the Middle East:									
Burma	486	490	490	0.57	0.58	0.58	279	285	285
China,Peoples Republic of	1,131	1,173	1,100	1.72	1.49	1.54	1,946	1,743	1,690
India	2,700	2,800	2,850	0.60	0.61	0.61	1,625	1,700	1,750
Pakistan	65	110	184	1.18	1.21	1.24	77	133	228
Total	4,382	4,573	4,624	0.90	0.84	0.85	3,927	3,861	3,953
Oceania:									
Australia	40	46	46	0.63	1.26	1.35	25	58	62
World total [4]	20,166	22,728	21,512	1.19	1.18	1.20	23,941	26,712	25,882

[1] Split year includes Northern Hemisphere crop harvested in the late months of the first year shown combined with Southern Hemisphere and certain Northern Hemisphere crops harvested in the early months of the following year. [2] Harvested area as far as possible. [3] Preliminary. [4] Regional totals include other countries not shown. World total for all countries in USDA data base.

FAS, Production Estimates and Crop Assessment Division, (202) 720–0888. Prepared or estimated on the basis of official statistics of foreign governments, other foreign source materials, reports of U.S. Agricultural Counselors, Attaches, and Foreign Service Officers, results of office research, and related information.

Table 3-42.—Peppermint oil: Area, yield, production, and value, United States, 1996–2005

Year	Area harvested	Yield per harvested acre	Production	Price per pound	Value of production
	1,000 acres	Pounds	1,000 pounds	Dollars	1,000 dollars
1996	132.0	72	9,446	13.60	128,778
1997	135.6	74	9,971	12.90	128,846
1998	124.0	78	9,727	11.90	116,037
1999	106.3	71	7,537	10.70	80,951
2000	88.5	78	6,877	10.80	74,320
2001	79.5	82	6,512	10.90	70,860
2002	78.5	89	6,958	11.90	82,560
2003	79.4	88	6,996	12.00	84,218
2004	78.7	92	7,236	11.90	86,421
2005 [1]	76.0	92	6,980	12.00	83,791

[1] Preliminary.
NASS, Crops Branch (202), 720–2127.

Table 3-43.—Spearmint oil: Area, yield, production, and value, United States, 1996–2005

Year	Area harvested	Yield per harvested acre	Production	Price per pound	Value of production
	1,000 acres	Pounds	1,000 pounds	Dollars	1,000 dollars
1996	23.1	94	2,167	12.00	26,094
1997	25.5	96	2,441	11.90	29,128
1998	27.4	109	2,987	11.00	32,731
1999	24.4	101	2,454	9.75	23,925
2000	21.7	101	2,199	9.06	19,919
2001	19.5	105	2,052	9.09	18,645
2002	18.4	109	2,010	9.11	18,308
2003	15.8	113	1,778	9.29	16,521
2004	15.8	116	1,839	9.62	17,700
2005 [1]	17.7	109	1,933	10.30	19,966

[1] Preliminary.
NASS, Crops Branch, (202) 720–2127.

Table 3-44.—Mint oil: Production and value, by States, 2003–2005

State	Production			Price per pound			Value of production		
	2003	2004	2005 [1]	2003	2004	2005 [1]	2003	2004	2005 [1]
	1,000 pounds	1,000 pounds	1,000 pounds	Dollars	Dollars	Dollars	1,000 dollars	1,000 dollars	1,000 dollars
Peppermint:									
ID	1,330	1,260	1,400	11.60	11.50	11.80	15,428	14,490	16,520
IN	495	594	495	11.20	11.50	11.60	5,544	6,831	5,742
MI	44	45	35	11.00	10.90	12.00	484	491	420
OR	2,375	2,205	2,185	13.10	13.20	13.10	31,113	29,106	28,624
WA	2,524	2,880	2,645	11.60	11.40	11.40	29,278	32,832	30,153
WI	228	252	220	10.40	10.60	10.60	2,371	2,671	2,332
US	6,996	7,236	6,980	12.00	11.90	12.00	84,218	86,421	83,791
Spearmint:									
ID	84	72	75	8.60	9.50	11.40	722	684	855
IN	76	64	72	9.60	9.80	10.60	730	627	763
MI	64	72	56	9.50	9.30	9.50	608	670	532
OR	126	203	252	9.50	10.00	10.80	1,197	2,030	2,722
WA	1,343	1,378	1,418	9.30	9.60	10.20	12,490	13,229	14,464
WI	85	50	60	9.10	9.20	10.50	774	460	630
US	1,778	1,839	1,933	9.29	9.62	10.30	16,521	17,700	19,966

[1] Preliminary.
NASS, Crops Branch, (202) 720–2127.

Table 3-45.—Olive oil: World production, 2001–2003 [1] [2]

Continent and country	2001	2002	2003 [3]
	1,000 metric tons	1,000 metric tons	1,000 metric tons
European Union	1,944	2,351	2,098
Middle East:			
Israel	9	3	7
Jordan	28	11	27
Lebanon	6	4	6
Syria	185	120	200
Turkey	170	65	175
Total [4]	398	203	415
Africa:			
Algeria	33	60	30
Morocco	17	17	17
Tunisia	75	290	130
Libya	7	7	7
Total [4]	165	442	227
World total [4]	2,508	2,997	2,741

[1]Marketing year begins November 1. [2]Production excludes residue oil. [3]Preliminary. [4]Includes other countries not listed separately.

FAS, Cotton, Oilseeds, Tobacco and Seeds Division, (202) 720–9516. Prepared or estimated on the basis of official statistics of foreign governments, other foreign source materials, reports of U.S. Agricultural Counselors, AttacheAE1s, and Foreign Service Officers, results of office research, and related information.

Table 3-46.—Margarine, actual weight: Supply and disposition, United States, 1994–2003

Year	Supply			Disposition		
	Production	Stocks, Jan. 1	Total supply	Exports	Domestic disappearance	
					Total	Per capita
	Million pounds	Million pounds	Million pounds	Million pounds	Million pounds	Pounds
1994	2,623	66	2,693	21	2,610	9.9
1995	2,490	62	2,557	36	2,463	9.3
1996	2,480	58	2,544	29	2,471	9.2
1997	2,367	44	2,417	29	2,344	8.6
1998	2,311	44	2,363	32	2,297	8.3
1999	2,274	35	2,319	36	2,241	8.0
2000	NA	42	NA	31	2,353	8.3
2001	NA	69	NA	31	NA	NA
2002	NA	34	NA	28	NA	NA
2003	NA	30	NA	29	NA	NA

NA-not available.

ERS, Field Crops Branch. (202) 694–5300. Totals and per capita estimates computed from unrounded numbers.

Table 3-47.—Margarine: Selected reported fats and oils used in manufacture, United States, 1994–2003

Year	Vegetable oils			Animal fats [1]	Total [2]
	Soybean oil	Cottonseed oil	Corn oil		
	Million pounds	Million pounds	Million pounds	Million pounds	Million pounds
1994	1,793	NA	NA	42	2,003
1995	1,684	NA	NA	41	1,847
1996	1,694	NA	77	28	1,816
1997	1,650	NA	61	14	1,733
1998	1,606	NA	55	22	1,692
1999	1,574	NA	NA	21	1,664
2000	1,465	NA	56	12	1,547
2001	1,298	NA	NA	7	1,394
2002	1,212	NA	NA	16	1,300
2003 [3]	1,138	NA	NA	10	1,214

[1]Lard and edible tallow. [2]Includes small quantities of nuts, coconut, palm, and sunflower oil. NA-not available. [3]Preliminary

ERS, Field Crops Branch, (202) 694–5300. Compiled from reports of the U.S. Department of Commerce. Totals computed from unrounded numbers.

Table 3-48.—Shortening: Supply and disposition, United States, 1994–2003

| Year | Supply | | | Disposition | | |
| | Factory and warehouse stocks, Jan. 1 | Production | Total supply | Exports and shipments | Domestic disappearance | |
					Total	Per capita
	Million pounds	Million pounds	Million pounds	Million pounds	Million pounds	Pounds
1994	94	6,334	6,427	44	6,305	23.9
1995	90	5,975	6,065	45	5,926	22.2
1996	106	5,929	6,035	43	5,914	21.9
1997	81	5,656	5,737	42	5,606	20.5
1998	91	5,724	5,815	54	5,670	20.5
1999	92	5,945	6,037	65	5,886	21.1
2000	86	9,043	9,130	69	8,932	31.6
2001	129	9,420	9,549	83	9,315	32.6
2002	151	9,685	9,836	89	9,607	33.3
2003	140	9,622	9,762	91	9,549	32.8

ERS, Market and Trade Economics Division, Field Crops Branch, (202) 694–5300. Compiled from reports of the Commerce and Agriculture Departments.

Table 3-49.—Shortening: Fats and oils used in manufacture, United States, 1994–2003

| Year | Vegetable oils | | | | Animal fats | | Total primary and secondary fats and oils [1] |
	Cottonseed oil	Soybean oil	Coconut oil	Palm oil	Lard	Edible tallow	
	Million pounds	Million pounds	Million pounds	Million pounds	Million pounds	Million pounds	Million pounds
1994	216	4,929	([2])	([2])	287	405	6,365
1995	212	4,673	([2])	([2])	325	374	6,031
1996	237	4,690	([2])	([2])	284	320	5,935
1997	256	4,517	([2])	([2])	272	312	5,679
1998	200	4,748	([2])	([2])	280	259	5,749
1999	167	5,069	([2])	([2])	241	262	5,968
2000	188	7,908	([2])	([2])	([2])	283	9,023
2001	185	8,234	([2])	([2])	([2])	([2])	9,405
2002	195	8,566	([2])	([2])	([2])	([2])	9,685
2003 [3]	165	8,296	([2])	([2])	([2])	([2])	9,333

[1] Includes small quantities of corn, peanut, safflower, and sunflower oil. [2] Not included to avoid disclosure. [3] Preliminary.
ERS, Market and Trade Economics Division, Field Crops Branch, (202) 694–5300. Compiled from reports of the U.S. Department of Commerce. Totals computed from unrounded numbers.

Table 3-50.—Inedible tallow and grease: Supply and disposition, United States, and price per pound at Chicago, 1994–2003

| Year | Supply | | | | Disposition | | | Price of inedible tallow No. 1 at Chicago, per pound [1] |
| | Stocks Jan. 1 | Production | Total | Exports | Factory consumption | | | |
					Total	Use in soap	Use in feed	
	Million pounds	Million pounds	Million pounds	Million pounds	Million pounds	Million pounds	Million pounds	Cents
1994	320	6,712	7,032	2,176	3,190	301	2,102	17.4
1995	348	6,745	7,093	2,683	3,223	264	2,167	19.2
1996	373	6,376	6,749	2,004	3,289	245	2,253	21.7
1997	266	6,249	6,516	1,689	3,399	245	2,401	20.7
1998	339	6,575	6,914	2,300	3,442	228	2,452	17.7
1999	437	7,076	7,513	1,940	3,728	229	2,751	13.0
2000	405	7,149	7,554	1,745	3,662	148	2,756	10.2
2001	331	5,931	6,261	1,335	3,030	([2])	2,187	12.4
2002	316	6,462	6,777	1,747	3,131	([2])	2,314	13.1
2003	242	6,245	6,487	1,555	3,170	([2])	2,405	18.3

[1] Includes small quantities of corn, peanut, safflower, and sunflower. [2] Not included to avoid disclosure.
ERS, Market and Trade Economics Division, Field Crops Branch, (202) 694–5300.

Table 3-51.—Fats, oils, and oilseeds (fat or oil equivalent): World production, 2001-2002/2004-2005

Commodity	World production [1]			
	2001-2002	2002-2003	2003-2004	2004-2005 [2]
	1,000 metric tons	1,000 metric tons	1,000 metric tons	1,000 metric tons
Edible vegetable oils:				
Cottonseed	3,840	3,521	3,852	4,772
Olive [3]	2,745	2,508	2,997	2,741
Peanut	5,117	4,560	4,946	4,908
Rapeseed	13,062	12,261	14,150	15,816
Soybean	28,846	30,435	29,879	32,448
Sunflower	7,481	8,185	9,230	9,154
Total	61,091	61,470	65,054	69,839
Tropical oils:				
Coconut	3,213	3,171	3,239	3,271
Palm	25,435	27,784	29,676	33,169
Palm kernel	3,124	3,356	3,664	4,006
Total	31,772	34,311	36,579	40,446
Animal fats:				
Butter (fat content)	6,145	6,555	6,599	6,638
Total	6,145	6,555	6,599	6,638
Grand total [4]	99,008	102,336	108,232	116,923

[1] Split year includes Northern Hemisphere crop harvested in the late months of the first year shown combined with Southern Hemisphere and certain Northern Hemisphere crops harvested in the early months of the following year.　[2] Preliminary.　[3] Excludes olive residue oil.　[4] Excludes linseed oil.

FAS, Cotton, Oilseeds, Tobacco and Seeds Division, (202) 720-9516. Prepared or estimated on the basis of official statistics of foreign governments, other foreign source materials, reports of U.S. Agricultural Counselors, Attachés, and Foreign Service Officers, results of office research, and related information.

Table 3-52.—Fats, oils, oilseeds, and oilseed cake and meal: Exports of selected items, United States, 1995-2004

Year beginning January	Lard	Inedible animal tallow, greases, and oils [1]	Oilseeds				
			Cottonseed	Flaxseed	Peanuts unshelled	Peanuts shelled	Soybeans
	Metric tons	Metric tons	Metric tons	Metric tons	Metric tons	Metric tons	Metric tons
1995	56,260	1,616,550	147,600	2,317	44,529	285,530	22,757,454
1996	45,690	1,192,335	83,078	2,890	30,005	208,733	25,565,559
1997	53,729	1,032,246	114,804	5,319	41,245	225,640	26,206,065
1998	59,412	1,363,415	127,554	10,987	27,718	182,324	20,302,984
1999	66,867	1,238,001	117,406	3,167	22,708	184,388	23,141,059
2000	78,930	1,108,129	194,780	27,282	23,963	229,715	26,985,945
2001	46,869	964,642	245,418	60,733	13,843	144,563	28,723,781
2002	38,202	1,286,452	301,984	75,461	17,394	235,593	27,674,537
2003	53,215	1,189,855	281,013	66,953	17,062	130,961	30,834,700
2004 [2]	131,183	1,074,132	345,512	31,599	16,617	170,813	25,137,971

Year beginning January	Vegetable oils							Oilseed cake and meal	
	Cocoa butter	Coconut oil	Cotton-seed oil [3]	Linseed oil	Mar-garine	Peanut oil	Soybean oil [3]	Soybean	Other [4]
	Metric tons	Metric tons	Metric tons	Metric tons	Metric tons	Metric tons	Metric tons	Metric tons	Metric tons
1995	2,972	9,090	137,693	15,422	17,080	47,741	1,037,306	5,858,626	511,125
1996	2,862	3,987	96,014	14,925	13,015	37,300	571,209	5,861,575	195,306
1997	3,886	5,170	110,575	34,691	13,210	8,818	1,015,356	6,992,801	187,401
1998	5,606	3,799	86,902	29,237	14,326	4,183	1,463,885	8,411,679	167,932
1999	5,566	4,276	56,973	28,373	16,090	5,806	845,206	6,979,286	218,166
2000	9,313	5,636	58,484	33,390	14,059	5,515	595,409	6,560,697	160,176
2001	17,506	4,394	64,573	28,575	14,171	6,490	688,336	7,498,102	260,923
2002	15,095	3,309	63,210	44,445	12,471	3,653	1,134,299	6,941,766	190,521
2003	13,497	5,441	46,194	29,414	13,201	27,032	971,442	5,694,100	254,922
2004 [2]	13,734	5,130	47,851	49,428	15,038	4,781	464,006	5,112,103	411,293

[1] Includes edible and inedible tallow, choice white grease, wool grease, yellow grease, edible and inedible oleo and oleo stearine, and animal oils, fats and oils, n.e.c.　[2] Preliminary.　[3] Includes shipments under PL480.　[4] Includes corn meal.

FAS, Cotton, Oilseeds, Tobacco and Seeds Division, (202) 720-9518. Compiled from reports of the U.S. Department of Commerce.

Table 3-53.—Oilseeds, oils, and oilseed cake and meal: Imports of selected items, United States, 1995–2004

Year beginning January	Oilseeds					
	Castor beans	Copra	Flaxseed	Peanuts unshelled	Peanuts shelled [1]	Poppy seed
	Metric tons	Metric tons	Metric tons	Metric tons	Metric tons	Metric tons
1995	0	1,387	186,153	1,628	32,901	6,122
1996	2	1,023	202,313	2,986	38,051	6,243
1997	0	1,170	223,519	3,176	41,272	5,238
1998	0	1,367	171,093	2,686	46,792	5,865
1999	0	605	182,859	2,413	50,225	6,192
2000	0	1,008	122,330	2,423	59,682	5,300
2001	0	879	49,667	3,458	66,022	4,742
2002	0	19	60,231	26	67,919	5,431
2003	133	1	98,798	88	26,212	5,021
2004 [2]	0	60	104,363	146	11,983	4,383

Year beginning January	Oilseeds—continued			Vegetable oils			
	Rapeseed	Sesame seed	Soybeans	Cocoa butter	Castor oil	Coconut oil	Linseed oil
	Metric tons	Metric tons	Metric tons	Metric tons	Metric tons	Metric tons	Metric tons
1995	200,054	39,356	130,141	57,158	41,417	491,151	1,744
1996	261,348	46,563	86,981	68,762	39,938	423,189	2,699
1997	318,249	42,629	258,602	87,689	41,025	589,192	3,102
1998	350,469	47,437	148,780	65,307	48,477	587,062	4,306
1999	210,262	42,214	84,263	80,587	46,671	308,902	5,635
2000	241,585	49,042	110,451	94,629	40,739	477,466	6,102
2001	237,426	49,072	93,686	80,746	45,395	477,278	4,478
2002	151,022	46,299	82,750	54,788	32,339	475,010	5,809
2003	110,732	37,348	162,415	78,315	26,702	365,281	7,029
2004 [2]	451,297	42,859	113,064	94,890	40,674	394,413	3,881

Year beginning January	Vegetable oils—continued						Total oilseed cake and meal
	Olive oil	Palm oil	Palm kernel oil	Peanut oil	Rapeseed oil	Tung oil	
	Metric tons	Metric tons	Metric tons	Metric tons	Metric tons	Metric tons	Metric tons
1995	122,270	101,621	121,949	3,165	430,648	4,427	831,047
1996	112,778	125,383	148,719	1,639	522,183	3,943	999,513
1997	163,469	134,519	161,706	6,648	491,083	6,265	1,071,926
1998	164,972	115,871	149,305	30,336	499,574	3,879	1,293,373
1999	162,733	138,864	208,145	9,633	523,219	5,822	1,130,363
2000	203,960	165,606	167,756	19,548	533,485	3,554	1,252,473
2001	211,893	174,054	150,366	32,789	539,796	11,430	1,074,434
2002	221,517	215,541	171,723	31,123	485,579	4,165	1,008,824
2003	214,550	199,821	222,968	6,724	451,846	4,288	1,312,264
2004 [2]	245,890	271,185	247,123	66,392	556,929	2,974	1,757,313

[1] Includes blanched or roasted peanuts. [2] Preliminary.
FAS, Cotton, Oilseeds, Tobacco and Seeds Division, (202) 720–9518. Compiled from reports of the U.S. Department of Commerce.

Table 3-54.—Animal tallow, greases, and oils:[1] United States exports by region and country of destination 2000–2004

Continent and country	2000	2001	2002	2003	2004[2]
	Metric tons	Metric tons	Metric tons	Metric tons	Metric tons
Mexico	329,855	348,330	405,864	427,867	533,741
Turkey	117,073	88,399	136,265	119,304	130,938
Venezuela	52,949	46,969	86,871	109,060	85,408
Canada	68,062	72,136	70,324	69,416	63,861
Guatemala	55,990	56,827	60,204	53,001	43,027
Dominican Republic	50,947	57,137	55,027	34,700	36,926
Honduras	21,820	17,721	39,578	41,521	36,197
China	32,193	15,540	32,885	64,790	34,044
Colombia	21,885	33,321	40,710	32,556	33,465
Nigeria	30,500	25,539	47,283	39,698	29,604
Peru	602	12,290	40,516	29,878	25,285
Korea, South	75,484	25,302	54,877	25,009	19,964
Japan	34,898	32,001	27,665	16,414	19,964
Nicaragua	10,445	13,149	15,451	16,320	17,862
El Salvador	40,671	22,988	46,247	42,852	16,872
South Africa	2,710	2,761	14,657	16,794	16,841
Pakistan	17,047	0	18,262	20,999	15,500
Haiti	11,540	10,450	10,452	12,315	10,551
Saudi Arabia	8,480	5,834	8,812	7,367	7,247
Ukraine	0	0	2,000	3,249	6,499
Leeward-Windward	6,587	3,104	3,741	2,122	5,442
Egypt	13,294	4,557	5,849	2,555	5,125
Morocco	10,023	4,000	12,068	9,079	4,901
Jamaica	5,544	7,344	4,505	8,041	4,851
Taiwan	19,940	6,914	7,386	12,720	4,167
Others	152,789	127,354	92,260	55,170	23,186
Grand total	1,191,328	1,039,967	1,339,759	1,272,797	1,231,468

[1] This category includes edible tallow; inedible tallow; choice white grease; pig and poultry fat, yellow grease; sheep or goat, other fat, raw or rendered; lard stearin, lard oil, oleo-oil and tallow oil; other animal fat, and baking and frying fat.   [2] Preliminary.

FAS, Dairy, Livestock and Poultry Division, (202) 720–8031. Updated data available at http://www.fas.usda.gov/ustrade.

Table 3-55.—Fats and oils: Use in products for civilian consumption, total and per capita, United States, 1994–2003

Calendar year	Food products[1]													
	Butter (actual weight)		Lard and tallow (direct use)[2]		Margarine (actual weight)		Baking and frying fats (shortening)		Salad and cooking oils		Other edible use		All food products (fat content)	
	Total	Per capita	Total	Per capita	Total	Per capita	Total	Per capita	Total	Per capita	Total	Per capita	Total	Per capita
	Million lbs	*Lbs*	*Million lbs*	*Lbs*	*Million lbs*	*Lbs*	*Million lbs*	*Lbs*	*Million lbs*	*Lbs*	*Million lbs*	*Lbs*	*Million lbs*	*Lbs*
1994	1,255	4.8	1,110	4.2	2,610	9.9	6,305	24.0	6,845	26.0	426	1.6	17,778	67.6
1995	1,187	4.4	963	3.6	2,463	9.3	5,926	22.3	7,057	26.5	434	1.6	17,300	65.0
1996	1,148	4.3	1,059	3.9	2,471	9.2	5,914	22.0	6,924	25.7	361	1.4	17,153	63.7
1997	1,116	4.1	1,102	4.0	2,344	8.6	5,606	20.6	7,652	28.1	297	1.1	17,426	64.0
1998	1,208	4.4	1,409	5.1	2,297	8.3	5,670	20.6	7,532	27.3	365	1.3	17,780	64.5
1999	1,307	4.7	1,545	5.6	2,241	8.0	5,886	21.1	8,030	28.8	431	1.6	18,731	67.2
2000	1,277	4.5	1,686	6.0	2,153	7.6	8,838	31.3	9,522	33.7	429	1.5	23,218	82.2
2001	1,264	4.4	1,528	5.4	NA	NA	NA	NA	NA	NA	408	1.4	NA	NA
2002	1,281	4.4	1,683	5.8	NA	NA	NA	NA	NA	NA	402	1.4	NA	NA
2003[3]	1,208	4.5	1,817	6.2	NA	NA	NA	NA	NA	NA	391	1.3	NA	NA

Calendar year	Industrial products											
	Soap		Fatty acids		Animal feeds		Other inedible products		All inedible products[4]		All products[5]	
	Total	Per capita	Total	Per capita	Total	Per capita	Total	Per capita	Total	Per capita	Total	Per capita
	Million pounds	*Pounds*	*Million pounds*	*Pounds*	*Million pounds*	*Pounds*	*Million pounds*	*Pounds*	*Million pounds*	*Pounds*	*Million pounds*	*Pounds*
1994	687	3.0	1,959	8.8	2,340	8.7	654	2.9	6,103	25.2	23,881	92.2
1995	594	2.8	1,964	8.6	2,341	8.8	747	2.5	6,101	23.0	23,401	89.5
1996	469	1.8	1,921	7.2	2,430	9.1	782	2.9	6,018	22.7	23,171	87.8
1997	567	2.1	2,342	8.7	2,646	9.9	557	2.1	6,535	24.4	23,961	89.9
1998	561	2.1	2,187	8.1	2,878	10.6	578	2.1	6,573	24.2	24,353	90.5
1999	565	2.1	2,028	7.4	3,200	11.7	553	2.0	6,733	24.7	25,464	93.8
2000	423	1.5	2,108	7.5	2,602	9.2	426	1.5	5,954	21.2	27,472	100.3
2001	366	1.3	2,060	7.4	2,651	9.6	476	1.7	6,344	22.2	NA	NA
2002	374	1.3	2,178	7.6	2,670	9.3	489	3.0	6,637	23.0	NA	NA
2003[3]	304	1.0	2,235	7.7	2,782	9.6	445	1.5	6,572	22.6	NA	NA

[1] Domestic disappearance data are computed by ERS. [2] Includes edible tallow direct use beginning in 1979. [3] Preliminary. [4] Including paint, varnish, resin, plastic, and lubricants. [5] Including only fat content of butter and margarine. [6] N. A.=Not Available.

ERS, Market and Trade Economics Division, Field Crops Branch, (202) 694–5300.

Table 3-56.—Fats and oils: Index numbers of wholesale prices, leading markets, United States, 1995–2004

[1982=100]

Year	All fats and oils excluding butter	Seventeen major fats and oils								
			Classified by origin		Classified by use					
		All fats and oils	Animal	Vegetable	Edible			Industrial		
				Domestic origin	Butter	Lard	All edible	Soap fats	Drying oils	All industrial
1995	91.9	115.5	59.6	174.4	NA	151.1	66.1	242.8	92.9	124.0
1996	87.8	655.3	68.1	159.3	NA	162.5	67.1	1,346.9	10.8	129.0
1997	90.2	67.8	74.2	156.8	NA	157.1	67.6	150.2	11.0	141.4
1998	94.3	78.4	88.5	174.3	NA	146.4	84.5	143.6	10.8	135.1
1999	70.1	58.1	67.0	128.0	NA	140.3	60.4	106.6	10.8	102.7
2000	57.7	51.8	63.3	107.7	NA	128.9	55.4	82.9	10.2	81.4
2001	60.8	59.9	85.5	104.6	NA	(¹)	64.7	92.0	9.3	88.6
2002	74.7	60.1	69.7	127.2	NA	(¹)	62.5	112.0	9.6	106.1
2003	112.5	82.8	80.4	204.9	NA	(¹)	88.2	144.5	9.4	133.8
2004	140.1	113.8	107.1	27.9	NA	(¹)	129.5	148.5	8.7	136.7

NA-not available.   ¹ Discontinued.
ERS, Market and Trade Economics Division, Field Crops Branch, (202) 694–5300.

Table 3-57.—Fats and oils: Wholesale price per pound, 1999–2004 ¹

Item and market	1999	2000	2001	2002	2003	2004
	Cents	Cents	Cents	Cents	Cents	Cents
Castor oil, No. 1, Brazilian, tanks, imported, New York	48.00	47.42	47.92	47.25	47.04	47.08
Coconut oil, crude, tanks, f.o.b. New York	39.89	23.34	24.15	21.93	25.86	38.05
Corn oil, crude, tank cars, f.o.b. Decatur	23.31	20.50	15.75	20.78	28.64	27.66
Cottonseed oil, crude, tank cars, f.o.b. Valley	23.95	20.86	15.41	23.33	36.73	28.87
Linseed oil, raw, tank cars, Minneapolis	36.00	35.83	36.79	39.06	41.80	48.50
Palm oil, U.S. ports, refined	22.86	16.28	15.73	23.31	32.02	34.09
Canola oil, Midwest	20.23	16.38	18.86	27.17	28.61	33.21
Safflower oil, tanks, New York	59.00	59.00	78.75	79.00	77.75	69.00
Soybean oil, crude, tank cars, f.o.b. Decatur	17.72	15.01	14.49	18.25	23.57	28.57
Sunflower oil, crude, Minneapolis	19.09	15.85	17.34	26.91	33.03	34.13
Tallow, inedible, number delivered Chicago	12.99	10.18	11.50	14.80	20.34	19.74
Tung oil, imported, drums, f.o.b. New York	84.83	79.33	61.63	43.77	75.63	85.42

¹ All prices are calendar year basis.
ERS, Market and Trade Economics Division, Field Crops Branch, (202) 694–5300. Compiled from the Chemical Marketing Reporter, the National Provisioner, the Wall Street Journal, and the U.S. Department of Labor.

CHAPTER IV
STATISTICS OF VEGETABLES AND MELONS

This chapter contains statistics on potatoes, sweet potatoes, and commercial vegetables and melons. For potatoes and sweet potatoes, the estimates of area, production, value, and farm disposition pertain to the total crop and include quantities produced both for sale and for use on farms where grown. Potato statistics are shown on a within-year seasonal grouping of winter, spring, summer, and fall crops, by States. Some States have production in more than one seasonal group.

For processing vegetables, the estimates of area, production, and value for each of 10 crops relate to production used by commercial canners, freezers, and other processors, except dehydrators. These estimates include raw products grown by processors themselves and those grown under contract or purchased on the open market. This production and the actual area harvested are not duplicated in the fresh market estimates for the same commodities. The production of those vegetables used for processing for which regular processing estimates are not made is included in the fresh market estimates. The processed segment of production for asparagus, broccoli, and cauliflower, combined with fresh market production during the year, is published at the end of the season, separately. In 2000, estimates were added for collard greens, kale, mustard greens, turnip greens, okra, chili peppers, pumpkins, radishes, and squash. In 2002, estimates for fresh market lima beans, beets for canning, Brussels sprouts, cabbage for kraut, eggplant, escarole/endive, collard greens, kale, mustard greens, turnip greens, okra, and radishes were discontinued. Additionally, States were removed from the program for certain commodities. For details on the 2002 program changes see the following website: http:/www.usda.gov/nass/events/programchg/vegprogchgs.htm.

Seasonal Groups and Marketing Period

Prospective Area For Harvest

Winter: January, February, March   Summer: July, August, September

Spring: April, May, June   Fall: October, November, December

Annual Acreage, Yield, Production, and Value

The seasonal patterns of harvest do not correspond precisely in all States to the estimating period or periods designated. In some cases, only one seasonal group is shown for a State, but marketing may be active in earlier or later months. Because of the small volume from this earlier or later period, the crop estimate has been placed in the seasonal group where the largest portion is harvested.

In 2002, commercial vegetables for fresh market include 24 principal vegetable and melon crops in the major producing States. These estimates relate to crops which are grown primarily for sale, and they do not include vegetables and melons produced in farm and nonfarm gardens. The bulk of the production of the principal vegetable and melon crops is for consumption in the fresh state. However, quantities used by processors of artichokes, celery, garlic, onions, bell peppers, chile peppers, pumpkins, and squash are included, and separate estimates of commercial processing are not made for these crops. The commercial estimates of the principal crops include local market production from areas near consuming centers as well as production from well recognized commercial areas which specialize in producing supplies for shipment to distant markets.

For fresh market vegetables and melons, value per unit and total value are on a f.o.b. basis. For processed vegetables, value per unit and total value are at processing plant door.

Aggregate data for the years 2000, 2000 and 2001, and 2001 and 2002 lack comparability with data from other years because of program changes altering the crops included.

STATISTICS OF VEGETABLES AND MELONS

Table 4-1.—Vegetables, commercial: Area, production, and value of principal crops, United States, 1996–2005

Year	Area [1]		
	For fresh market [2]	For processing [3]	Total
	Acres	Acres	Acres
1996	1,886,780	1,485,020	3,371,800
1997	1,849,730	1,423,000	3,272,730
1998	1,840,650	1,443,510	3,284,160
1999	1,890,450	1,512,750	3,403,200
2000	2,068,870	1,449,930	3,488,800
2001	2,020,220	1,333,310	3,353,530
2002	1,930,650	1,339,520	3,270,170
2003	1,928,420	1,336,870	3,265,290
2004	1,939,820	1,297,070	3,236,890
2005	1,935,650	1,285,660	3,221,310

Year	Production [4]		
	For fresh market [2]	For processing [3]	Total
	Tons	Tons	Tons
1996	20,600,500	17,547,062	38,147,562
1997	21,828,450	16,229,609	38,058,059
1998	20,683,250	15,476,230	36,159,480
1999	22,349,350	19,063,030	41,412,380
2000	23,811,150	17,031,310	40,842,460
2001	23,477,150	14,988,950	38,466,100
2002	23,148,800	17,074,350	40,223,150
2003	23,432,700	15,559,380	38,992,080
2004	24,226,750	17,675,590	41,902,340
2005	23,635,250	15,718,470	39,353,720

Year	Value [5]		
	For fresh market [2]	For processing [3]	Total
	1,000 dollars	1,000 dollars	1,000 dollars
1996	6,883,050	1,470,784	8,353,834
1997	8,070,906	1,372,269	9,443,175
1998	7,971,765	1,354,576	9,326,341
1999	7,518,948	1,660,051	9,178,999
2000	9,089,706	1,415,628	10,505,334
2001	8,877,326	1,255,589	10,132,915
2002	9,416,299	1,334,583	10,750,882
2003	9,769,278	1,289,353	11,058,631
2004	9,701,288	1,395,774	11,097,062
2005	9,819,240	1,267,265	11,086,505

[1] Area for fresh market is area for harvest, including any partially harvested or not harvested because of low prices or other economic factors. Area for processing is area harvested.   [2] Area, production, and farm value of the following crops for which regular seasonal estimates are prepared in major producing States: Artichokes, asparagus, snap beans, lima beans, broccoli, brussels sprouts, cabbage, cantaloups, carrots, cauliflower, celery, sweet corn, cucumbers, eggplant, escarole/endive, garlic, honeydew melons, head lettuce, leaf lettuce, romaine lettuce, onions, green peppers, spinach, tomatoes, and watermelons. In 2000, collard greens, kale, mustard greens, turnip greens, okra, chile peppers, pumpkins, radishes, and squash were added. In 2002, fresh market lima beans, beets for canning, Brussels sprouts, cabbage for kraut, eggplant, escarole/endive, collard greens, kale, mustard greens, turnip greens, okra, and radishes were discontinued. Additionally in 2002, States were removed from the program for certain commodities. See table footnotes when comparing years.   [3] Area, production, and farm value of the following 10 crops in all States: Lima beans, snap beans, beets, cabbage (sauerkraut), carrots, sweet corn, cucumbers (pickles), green peas, spinach, and tomatoes. Production of other vegetables processed included in fresh market series of estimates.   [4] Production for fresh market excludes some quantities not marketed because of low prices or other economic factors.   [5] Value for all fresh market vegetables. For processing vegetables, value at processing plant door.

NASS, Crops Branch, (202) 720–2127.

Table 4-2.—Vegetables, commercial: Area of principal crops, by States, 2003–2005 [1] [2]

State	For fresh market [3]			For processing [5]			Total		
	2003	2004	2005 [4]	2003	2004	2005 [4]	2003	2004	2005 [4]
	Acres	Acres	Acres	Acres	Acres	Acres	Acres	Acres	Acres
AL	6,400	6,400	5,100				6,400	6,400	5,100
AZ	132,300	133,900	137,600				132,300	133,900	137,600
AR	3,500	3,300	3,200				3,500	3,300	3,200
CA	828,200	836,900	847,000	312,300	310,800	288,000	1,140,500	1,147,700	1,135,000
CO	27,200	30,700	29,500				27,200	30,700	29,500
CT	4,100	4,300	4,000				4,100	4,300	4,000
DE	4,800	5,400	5,800	38,400	37,200	36,800	43,200	42,600	42,600
FL	184,200	185,800	181,100				184,200	185,800	181,100
GA	122,800	134,400	139,800	10,900	8,000	6,900	133,700	142,400	146,700
ID	9,800	10,400	9,500				9,800	10,400	9,500
IL	14,150	18,450	19,850	43,600	36,900	44,700	57,750	55,350	64,550
IN	16,500	17,000	16,900				16,500	17,000	16,900
IA				1,430	1,600	2,900	1,430	1,600	2,900
LA	1,800	1,300					1,800	1,300	
ME	2,000	2,000	2,000				2,000	2,000	2,000
MD	10,180	11,820	11,630	17,820	18,020	18,540	28,000	29,840	30,170
MA	5,600	5,800	5,700				5,600	5,800	5,700
MI	64,200	63,800	61,800	52,700	56,600	64,800	116,900	120,400	126,600
MN	210	150		227,750	212,450	219,740	227,960	212,600	219,740
MS	3,000	2,700	2,900				3,000	2,700	2,900
MO	4,700	4,400	3,500	16,300	16,300	17,500	21,000	20,700	21,000
NV	4,300	4,600	3,000				4,300	4,600	3,000
NH	1,900	1,800	1,700				1,900	1,800	1,700
NJ	26,700	27,500	26,300	5,900	7,900	6,850	32,600	35,400	33,150
NM	22,400	22,500	22,600				22,400	22,500	22,600
NY	84,100	77,600	76,600	51,100	57,400	58,700	135,200	135,000	135,300
NC	47,500	43,300	39,200				47,500	43,300	39,200
OH	33,480	33,300	35,720	7,900	11,200	9,200	41,380	44,500	44,920
OK	6,000	5,000	5,000				6,000	5,000	5,000
OR	31,300	30,000	30,200	70,400	65,200	55,420	101,700	95,200	85,620
PA	33,700	34,100	31,300	9,970	15,200	10,800	43,670	49,300	42,100
RI	1,000	1,100	1,000				1,000	1,100	1,000
SC	15,800	15,900	14,900				15,800	15,900	14,900
TN	15,200	14,500	16,700	5,800	5,800	5,800	21,000	20,300	22,500
TX	82,600	74,600	68,100	20,700	18,900	19,800	103,300	93,500	87,900
UT	1,800	1,500	1,500				1,800	1,500	1,500
VT	1,100	1,000	1,100				1,100	1,000	1,100
VA	17,500	17,800	19,150	1,000	2,800	510	18,500	20,600	19,660
WA	42,400	41,800	41,700	154,200	142,000	127,000	196,600	183,800	168,700
WI	14,000	13,000	13,000	209,300	193,300	219,300	223,300	206,300	232,300
Other States [6]				79,400	79,500	72,400	79,400	79,500	72,400
US	1,928,420	1,939,820	1,935,650	1,336,870	1,297,070	1,285,660	3,265,290	3,236,890	3,221,310

[1] Area for fresh market and for processing is area harvested.   [2] Commodity estimates for 2002 and 2003 are comparable. Estimates for 2001 are not comparable due to vegetable estimation program changes. These changes are documented in footnotes found under each individual commodity table.   [3] Area of the following crops for which regular seasonal estimates are prepared in major producing States: Artichokes, asparagus, snap beans, lima beans, broccoli, brussels sprouts, cabbage, cantaloups, carrots, cauliflower, celery, sweet corn, cucumbers, eggplant, escarole/endive, garlic, honeydew melons, head lettuce, leaf lettuce, romaine lettuce, onions, green peppers, spinach, tomatoes, and watermelons. In 2000, collard greens, kale, mustard greens, turnip greens, okra, chile peppers, pumpkins, radishes, and squash were added. In 2002, fresh market lima beans, beets for canning, Brussels sprouts, cabbage for kraut, eggplant, escarole/endive, collard greens, kale, mustard greens, turnip greens, okra, and radishes were discontinued. Additionally in 2002, States were removed from the program for certain commodities. See table footnotes when comparing years.   [4] Preliminary.   [5] Includes Lima beans, snap beans, beets, cabbage (sauerkraut), carrots, sweet corn, cucumbers (pickles), green peas, spinach, and tomatoes. Other vegetables processed (dual purpose) included in fresh market series of estimates.   [6] Processing, 2002 - AL, FL, ID, IN, MA, NC, and SC. 2003 - AL, FL, ID, IN, MA, NC, and SC. 2004 - AL, FL, ID, IN, MA, NC, and SC.

NASS, Crops Branch, (202) 720–2127.

Table 4-3.—Vegetables, commercial: Production of principal crops, by States, 2003-2005 [1]

State	For fresh market [2]			For processing [4]			Total		
	2003	2004	2005 [3]	2003	2004	2005 [3]	2003	2004	2005 [3]
	1,000 Cwt	1,000 Cwt	1,000 Cwt	Tons	Tons	Tons	Tons	Tons	Tons
AL	1,000	708	566				50,000	35,400	28,300
AZ	41,139	41,401	41,430				2,056,950	2,070,050	2,071,500
AR	729	487	574				36,450	24,350	28,700
CA	229,089	234,405	229,033	9,534,270	11,957,760	9,874,980	20,988,720	23,678,010	21,326,630
CO	8,212	10,676	9,493				410,600	533,800	474,650
CT	246	344	300				12,300	17,200	15,000
DE	711	1,161	1,184	107,430	92,450	97,420	142,980	150,500	156,620
FL	40,979	44,280	42,462				2,048,950	2,214,000	2,123,100
GA	20,601	21,270	23,319	53,220	32,880	28,550	1,083,270	1,096,380	1,194,500
ID	5,880	8,008	6,080				294,000	400,400	304,000
IL	3,776	5,169	5,687	189,710	180,000	190,330	378,510	438,450	474,680
IN	3,770	3,706	3,777				188,500	185,300	188,850
IA				10,790	9,600	18,930	10,790	9,600	18,930
LA	234	130					11,700	6,500	
ME	120	120	120				6,000	6,000	6,000
MD	906	1,139	1,249	67,080	77,250	79,990	112,380	134,200	142,440
MA	420	522	456				21,000	26,100	22,800
MI	9,854	9,553	9,742	389,710	374,780	400,460	882,410	852,430	887,560
MN	65	48		1,133,870	1,019,920	1,132,340	1,137,120	1,022,320	1,132,340
MS	435	378	435				21,750	18,900	21,750
MO	1,387	924	963	42,520	51,060	46,550	111,870	97,260	94,700
NV	2,042	2,320	2,220				102,100	116,000	111,000
NH	133	126	128				6,650	6,300	6,400
NJ	3,931	4,174	3,561	41,750	56,440	49,380	238,300	265,140	227,430
NM	6,889	6,874	6,402				344,450	343,700	320,100
NY	14,638	14,921	13,941	214,180	210,760	214,790	946,080	956,810	911,840
NC	6,218	5,105	5,470				310,900	255,250	273,500
OH	5,062	4,880	6,052	201,220	222,320	204,520	454,320	466,320	507,120
OK	840	700	725				42,000	35,000	36,250
OR	12,011	14,160	12,876	416,080	421,380	379,070	1,016,630	1,129,380	1,022,870
PA	2,822	3,682	3,267	68,250	54,750	34,010	209,350	238,850	197,360
RI	90	99	70				4,500	4,950	3,500
SC	2,750	3,116	1,465				137,500	155,800	73,250
TN	2,098	1,640	1,993	9,120	9,120	9,120	114,020	91,120	108,770
TX	19,592	17,878	16,429	134,150	116,560	99,030	1,113,750	1,010,460	920,480
UT	828	780	743				41,400	39,000	37,150
VT	88	55	77				4,400	2,750	3,850
VA	2,955	3,117	3,381	2,850	8,030	3,540	150,600	163,880	172,590
WA	13,806	14,799	15,160	1,181,240	1,111,620	1,025,180	1,871,540	1,851,570	1,783,180
WI	2,308	1,680	1,875	1,173,250	1,007,130	1,185,570	1,288,650	1,091,130	1,279,320
Other States [5]				588,690	661,780	644,710	588,690	661,780	644,710
US	468,654	484,535	472,705	15,559,380	17,675,590	15,718,470	38,992,080	41,902,340	39,353,720

[1] Commodity estimates for 2002 and 2003 are comparable. Estimates for 2001 are not comparable due to vegetable estimation program changes. These changes are documented in footnotes found under each individual commodity table. [2] Production of the following crops for which regular seasonal estimates are prepared in major producing States: Artichokes, asparagus, snap beans, lima beans, broccoli, brussels sprouts, cabbage, cantaloups, carrots, cauliflower, celery, sweet corn, cucumbers, eggplant, escarole/endive, garlic, honeydew melons, head lettuce, leaf lettuce, romaine lettuce, onions, green peppers, spinach, tomatoes, and watermelons.In 2002, collard greens, kale, mustard greens, turnip greens, okra, chile peppers, pumpkins, radishes, and squash were added. In 2002, fresh market lima beans, beets for canning, Brussels sprouts, cabbage for kraut, eggplant, escarole/endive, collard greens, kale, mustard greens, turnip greens, okra, and radishes were discontinued. Additionally in 2002, states were removed from the program for certain commodities. See table footnotes when comparing years. [3] Preliminary. [4] Includes Lima beans, snap beans, beets, cabbage (sauerkraut), carrots, sweet corn, cucumbers (pickles), green peas, spinach, and tomatoes. Other vegetables processed (dual purpose) included in fresh market series of estimates. [5] 2002 - AL, FL, ID, IN, MA, NC, and SC. 2003 - AL, FL, ID, IN, MA, NC, and SC. 2004 - AL, FL, ID, IN, MA, NC, and SC.

NASS, Crops Branch, (202) 720-2127.

Table 4-4.—Vegetables, commercial: Value of principal crops, by States, 2003–2005 [1]

State	For fresh market [2]			For processing [4]			Total		
	2003	2004	2005 [3]	2003	2004	2005 [3]	2003	2004	2005 [3]
	1,000 dollars	*1,000 dollars*	*1,000 dollars*	*1,000 dollars*	*1,000 dollars*	*1,000 dollars*	*1,000 dollars*	*1,000 dollars*	*1,000 dollars*
AL	15,687	15,489	12,663	....	....	....	15,687	15,489	12,663
AZ	615,393	861,370	859,573	....	....	....	615,393	861,370	859,573
AR	19,941	6,134	19,242	....	....	....	19,941	6,134	19,242
CA	5,345,001	5,191,700	4,569,275	574,763	712,646	612,526	5,919,764	5,904,346	5,181,801
CO	97,058	110,179	127,258	....	....	....	97,058	110,179	127,258
CT	6,765	9,804	8,400	....	....	....	6,765	9,804	8,400
DE	8,901	12,846	14,713	20,762	17,075	18,695	29,663	29,921	33,408
FL	1,162,254	1,156,539	1,549,286	....	....	....	1,162,254	1,156,539	1,549,286
GA	380,138	375,330	472,397	20,320	8,883	8,176	400,458	384,213	480,573
ID	55,709	42,486	62,055	....	....	....	55,709	42,486	62,055
IL	23,644	27,762	30,154	25,405	23,405	16,026	49,049	51,167	46,180
IN	60,734	62,187	48,731	....	....	....	60,734	62,187	48,731
IA	....	....	....	863	768	1,457	863	768	1,457
LA	1,732	819	....	....	....	....	1,732	819	....
ME	3,900	3,960	4,080	....	....	....	3,900	3,960	4,080
MD	16,622	24,686	25,980	12,079	11,955	12,255	28,701	36,641	38,235
MA	13,230	16,965	15,732	....	....	....	13,230	16,965	15,732
MI	170,366	175,402	163,334	56,446	56,502	53,206	226,812	231,904	216,540
MN	509	243	....	109,941	96,858	115,129	110,450	97,101	115,129
MS	3,263	3,213	4,785	....	....	....	3,263	3,213	4,785
MO	7,629	3,788	6,356	7,832	9,356	8,475	15,461	13,144	14,831
NV	33,582	37,408	33,462	....	....	....	33,582	37,408	33,462
NH	5,586	5,292	5,312	....	....	....	5,586	5,292	5,312
NJ	105,124	107,786	97,181	4,593	6,815	5,550	109,717	114,601	102,731
NM	98,672	96,409	100,494	....	....	....	98,672	96,409	100,494
NY	267,279	281,527	287,756	32,117	33,263	33,555	299,396	314,790	321,311
NC	110,081	86,055	93,075	....	....	....	110,081	86,055	93,075
OH	120,795	109,229	161,002	22,863	28,887	25,274	143,658	138,116	186,276
OK	5,628	5,880	6,815	....	....	....	5,628	5,880	6,815
OR	109,283	94,722	157,164	49,963	51,327	45,816	159,246	146,049	202,980
PA	59,568	78,898	82,203	8,649	10,911	6,695	68,217	89,809	88,898
RI	2,790	3,465	2,450	....	....	....	2,790	3,465	2,450
SC	53,873	49,238	27,664	....	....	....	53,873	49,238	27,664
TN	72,823	54,281	67,780	1,898	1,921	1,884	74,721	56,202	69,664
TX	433,136	341,915	357,638	27,074	24,337	13,637	460,210	366,252	371,275
UT	7,259	4,092	7,116	....	....	....	7,259	4,092	7,116
VT	3,168	2,145	3,157	....	....	....	3,168	2,145	3,157
VA	78,229	112,139	111,749	690	1,408	423	78,919	113,547	112,172
WA	164,454	107,597	196,675	108,122	94,998	87,057	272,576	202,595	283,732
WI	29,472	22,308	26,533	104,321	96,900	98,776	133,793	119,208	125,309
Other States [5]	....	....	....	100,652	107,559	102,653	100,652	107,559	102,653
US	9,769,278	9,701,288	9,819,240	1,289,353	1,395,774	1,267,265	11,058,631	11,097,062	11,086,505

[1] Commodity estimates for 2002 and 2003 are comparable. Estimates for 2001 are not comparable due to vegetable estimation program changes. These changes are documented in footnotes found under each individual commodity table. [2] Value of the following crops for which regular seasonal estimates are prepared in major producing States: Artichokes, asparagus, snap beans, lima beans, broccoli, brussels sprouts, cabbage, cantaloups, carrots, cauliflower, celery, sweet corn, cucumbers, eggplant, escarole/endive, garlic, honeydew melons, head lettuce, leaf lettuce, romaine lettuce, onions, green peppers, spinach, tomatoes, and watermelons. In 2002, collard greens, kale, mustard greens, turnip greens, okra, chile peppers, pumpkins, radishes, and squash were added. In 2002, fresh market lima beans, beets for canning, Brussels sprouts, cabbage for kraut, eggplant, escarole/endive, collard greens, kale, mustard greens, turnip greens, okra, and radishes were discontinued. Additionally in 2002, states were removed from the program for certain commodities. See table footnotes when comparing years. [3] Preliminary. [4] Includes Lima beans, snap beans, beets, cabbage (sauerkraut), carrots, sweet corn, cucumbers (pickles), green peas, spinach, and tomatoes. Other vegetables processed (dual purpose) included in fresh market series of estimates. [5] 2002 - AL, FL, ID, IN, MA, NC, and SC. 2003 - AL, FL, ID, IN, MA, NC, and SC. 2004 - AL, FL, ID, IN, MA, NC, and SC.

NASS, Crops Branch, (202) 720–2127.

Table 4-5.—Artichokes for fresh market and processing: Area, production, and value per hundredweight, California, 2003-2005

Crop	Area harvested			Production			Value per unit		
	2003	2004	2005	2003	2004	2005	2003	2004	2005
	Acres	Acres	Acres	1,000 cwt.	1,000 cwt.	1,000 cwt.	Dollars per cwt.	Dollars per cwt.	Dollars per cwt.
CA	7,200	7,500	7,300	1,008	825	840	75.10	88.10	45.10

NASS, Crops Branch, (202) 720-2127.

Table 4-6.—Asparagus, commercial crop: Area, yield, production, value per hundredweight and per ton, and total value, United States, 1996-2005

Year	Total crop					For fresh market			For processing		
	Area for harvest	Yield per acre	Produc-tion	Value [1]		Produc-tion	Value [1]		Produc-tion	Value [2]	
				Per cwt.	Total		Per cwt.	Total		Per ton	Total
	Acres	Cwt.	1,000 cwt.	Dollars	1,000 dollars	1,000 cwt.	Dollars	1,000 dollars	Tons	Dollars	1,000 dollars
1996	73,560	27	1,989	78.70	156,623	1,114	92.90	103,480	43,780	1,210.00	53,143
1997	74,030	27	2,026	90.10	182,531	1,248	108.00	134,860	38,920	1,220.00	47,571
1998	74,430	27	1,979	101.00	199,482	1,264	124.00	156,734	35,720	1,200.00	42,748
1999	75,890	29	2,176	107.00	233,170	1,455	131.00	190,719	36,070	1,180.00	42,451
2000	77,400	29	2,272	97.40	221,299	1,504	117.00	176,017	38,400	1,180.00	45,282
2001	70,150	30	2,078	110.00	228,925	1,372	140.00	192,346	35,290	1,040.00	36,579
2002	66,000	28	1,868	92.50	172,876	1,267	110.00	139,609	30,050	1,110.00	33,267
2003	58,000	32	1,843	88.40	162,901	1,194	105.00	125,086	32,450	1,170.00	37,815
2004	61,500	34	2,062	105.00	217,060	1,524	122.00	185,468	26,900	1,170.00	31,592
2005 [3] ...	54,000	33	1,804	87.80	158,350	1,414	97.50	137,902	19,500	1,050.00	20,448

[1] Price and value on F.O.B. basis. [2] Price and value at processing plant door. [3] Preliminary.
NASS, Crops Branch, (202) 720-2127.

Table 4-7.—Asparagus, commercial crop: Area, production, and value per hundredweight and per ton, by States, 2003-2005

State	Area harvested [1]			Production			Value per unit		
	2003	2004	2005 [2]	2003	2004	2005 [2]	2003	2004	2005 [2]
	Acres	Acres	Acres	1,000 cwt.	1,000 cwt.	1,000 cwt.	Dollars per cwt.	Dollars per cwt.	Dollars per cwt.
CA [3]	27,000	34,000	29,000	918	1,190	1,044	122.00	141.00	114.00
MI	15,000	13,500	12,000	317	270	228	60.80	64.70	51.60
WA	16,000	14,000	13,000	608	602	532	52.00	52.80	51.80
US	58,000	61,500	54,000	1,843	2,062	1,804	88.40	105.00	87.80

State	For fresh market						For processing					
	Production			Value per unit			Production			Value per unit		
	2003	2004	2005 [2]	2003	2004	2005 [2]	2003	2004	2005 [2]	2003	2004	2005 [2]
	1,000 cwt.	1,000 cwt.	1,000 cwt.	Dollars per cwt.	Dollars per cwt.	Dollars per cwt.	Tons	Tons	Tons	Dollars per ton	Dollars per ton	Dollars per ton
CA [3]	918	1,190	1,044	122.00	141.00	114.00	....	....	....	....	....	....
MI	43	26	54	66.00	90.00	63.00	13,700	12,200	8,700	1,200.00	1,240.00	960.00
WA	233	308	316	44.00	49.80	49.00	18,750	14,700	10,800	1,140.00	1,120.00	1,120.00
US	1,194	1,524	1,414	105.00	122.00	97.50	32,450	26,900	19,500	1,170.00	1,170.00	1,050.00

[1] Asparagus for fresh market and for processing is frequently harvested from the same area; therefore it is not practical to make individual area estimates for these segments. [2] Preliminary. [3] Includes a small amount of processing asparagus.
NASS, Crops Branch, (202) 720-2127.

Table 4-8.—Lima beans for processing: Area, production, and value per ton, by States, 2003–2005 [1]

State	Area harvested			Production			Value per unit		
	2003	2004	2005 [2]	2003	2004	2005 [2]	2003	2004	2005 [2]
	Acres	Acres	Acres	Tons	Tons	Tons	Dollars per ton	Dollars per ton	Dollars per ton
US	45,800	41,600	39,220	60,180	53,550	53,510	442.00	425.00	410.00

[1] 2003 - 2005 - CA, DE, IL, MD, OR, TN, WA, and WI.   [2] Preliminary.
NASS, Crops Branch, (202) 720–2127.

Table 4-9.—Snap beans for fresh market: Area, production, and value per hundredweight, by States, 2003–2005

State	Area harvested			Production			Value per unit		
	2003	2004	2005 [1]	2003	2004	2005 [1]	2003	2004	2005 [1]
	Acres	Acres	Acres	1,000 cwt.	1,000 cwt.	1,000 cwt.	Dollars per cwt.	Dollars per cwt.	Dollars per cwt.
CA	5,800	6,800	6,500	580	680	650	55.70	66.80	63.60
FL	31,800	33,200	34,000	2,639	2,822	2,210	57.20	47.20	64.40
GA	16,000	17,000	17,500	800	901	700	40.00	28.00	35.20
MD	1,400	1,700	1,900	42	34	67	35.00	41.00	40.00
MI	4,000	4,100	4,200	160	185	210	25.00	45.00	25.00
NJ	2,300	3,100	2,900	81	124	116	33.00	52.00	47.00
NY	9,800	7,600	8,100	392	190	300	68.00	73.70	76.80
NC	6,200	6,000	5,000	310	270	250	35.00	32.00	30.00
SC	1,300	1,000	1,000	52	60	30	49.00	45.00	40.00
TN	9,500	7,600	10,500	409	365	662	29.00	33.00	36.00
VA	4,800	4,600	5,100	230	138	260	23.00	26.00	37.00
US	92,900	92,700	96,700	5,695	5,769	5,455	49.30	45.20	52.60

[1] Preliminary.
NASS, Crops Branch, (202) 720–2127.

Table 4-10.—Snap beans for processing, commercial crop: Area, production, and value per ton, by States, 2003–2005

State	Area harvested			Production			Value per unit		
	2003	2004	2005 [1]	2003	2004	2005 [1]	2003	2004	2005 [1]
	Acres	Acres	Acres	Tons	Tons	Tons	Dollars per ton	Dollars per ton	Dollars per ton
DE	2,900	....	....	9,250	....	....	192.00	....	....
FL	....	....	2,100	....	....	9,300	....	....	245.00
IL	16,600	11,200	16,700	56,040	52,320	68,180	176.00	163.00	104.00
IN	6,200	5,700	5,500	17,340	17,630	17,200	169.00	182.00	183.00
MD	2,700	....	....	6,350	....	....	203.00	....	....
MI	14,300	17,300	22,200	45,010	61,280	62,460	160.00	169.00	168.00
NY	21,900	20,400	21,200	77,380	66,310	68,970	178.00	195.00	186.00
OR	16,000	17,800	18,500	100,200	115,320	116,530	178.00	179.00	188.00
PA	7,800	13,500	9,500	23,190	46,760	27,380	198.00	217.00	222.00
VA	1,000	....	....	2,850	....	....	242.00	....	....
WI	66,200	73,100	75,500	270,840	322,640	311,280	112.00	115.00	92.00
Other States [2][3]	34,000	41,990	39,420	119,190	153,620	140,470	203.00	188.00	165.00
Total	189,600	200,990	210,620	727,640	835,880	821,770	157.00	158.00	141.00

[1] Preliminary.   [2] 2003 - AR, CA, FL, GA, MN, MO, NJ, NC, and TX. 2004 - AR, CA, DE, FL, GA, MD, MN, MO, NJ, NC, TX, VA, and WA. 2005 - AR, CA, DE, GA, MD, MN, MO, NJ, NC, TX, and VA.   [3] WA estimates discontinued in 2005.
NASS, Crops Branch, (202) 720–2127.

Table 4-11.—Snap beans for processing, commercial crop: Area, yield, production, value per ton, and total value, United States, 1996–2005

Year	Area harvested	Yield per acre	Production	Value [1] Per ton	Value [1] Total
	Acres	*Tons*	*Tons*	*Dollars*	*1,000 dollars*
1996	207,050	3.79	784,920	178.00	139,755
1997	195,080	3.74	729,250	176.00	128,032
1998	198,700	3.68	730,990	172.00	125,373
1999	212,150	3.67	778,430	173.00	134,501
2000	218,380	3.82	833,490	171.00	142,502
2001	193,980	3.55	688,140	161.00	111,114
2002	201,800	3.93	793,710	151.00	120,190
2003	189,600	3.84	727,640	157.00	114,520
2004	200,400	4.16	835,880	158.00	131,865
2005 [2] ...	210,620	3.90	821,770	141.00	115,545

[1] Price and value at processing plant door.   [2] Preliminary.

NASS, Crops Branch, (202) 720–2127.

Table 4-12.—Broccoli, commercial crop: Area, yield, production, value per hundredweight and per ton, and total value, United States, 1996–2005 [1]

Year	Total crop Area for harvest	Total crop Yield per acre	Total crop Production	Total crop Value [2] Per cwt.	Total crop Value [2] Total	For fresh market Production	For fresh market Value [2] Per cwt.	For fresh market Value [2] Total	For processing Production	For processing Value [3] Per ton	For processing Value [3] Total
	Acres	*Cwt.*	*1,000 cwt.*	*Dollars*	*1,000 dollars*	*1,000 cwt.*	*Dollars*	*1,000 dollars*	*Tons*	*Dollars*	*1,000 dollars*
1996	133,500	118	15,693	26.50	415,695	14,428	27.10	391,194	63,250	387.00	24,501
1997	130,800	129	16,880	28.50	481,459	15,744	29.10	457,423	56,810	423.00	24,036
1998	133,500	129	17,251	29.50	508,101	16,128	30.20	486,332	56,148	388.00	21,769
1999	148,000	140	20,664	23.90	493,087	19,491	24.10	468,882	58,656	413.00	24,205
2000	144,300	141	20,315	30.50	620,606	19,502	31.20	607,958	40,670	311.00	12,648
2001	133,100	140	18,690	25.90	484,467	17,755	26.50	469,694	46,750	316.00	14,773
2002	130,400	141	18,375	30.90	567,767	17,595	31.40	552,713	39,000	386.00	15,054
2003	131,600	148	19,450	31.60	615,334	17,861	32.70	583,514	79,454	403.00	32,020
2004	133,800	148	19,835	32.20	638,079	18,081	33.20	600,289	87,680	431.00	37,790
2005 [4]	133,900	148	19,790	28.50	563,673	(5)	(5)	(5)	(5)	(5)	(5)

[1] Sprouting broccoli only. Does not include broccoli rabe nor heading (cauliflower) broccoli.   [2] Price and value on f.o.b. basis.   [3] Price and value at processing plant door.   [4] Preliminary.   [5] Not published to avoid disclosure of individual operations.

NASS, Crops Branch, (202) 720–2127.

Table 4-13.—Broccoli, commercial crop: Area, production, and value per hundredweight, and per ton, by States, 2003–2005 [1]

State	Area harvested 2003	Area harvested 2004	Area harvested 2005 [2]	Production 2003	Production 2004	Production 2005 [2]	Value per unit 2003	Value per unit 2004	Value per unit 2005 [2]
	Acres	*Acres*	*Acres*	*1,000 cwt.*	*1,000 cwt.*	*1,000 cwt.*	*Dollars per cwt.*	*Dollars per cwt.*	*Dollars per cwt.*
AZ	11,600	11,800	11,900	1,450	1,535	1,490	27.80	33.20	33.50
CA	120,000	122,000	122,000	18,000	18,300	18,300	32.00	32.10	28.10
US	131,600	133,800	133,900	19,450	19,835	19,790	31.60	32.20	28.50

State	For fresh market Production 2003	For fresh market Production 2004	For fresh market Production 2005 [2]	For fresh market Value per unit 2003	For fresh market Value per unit 2004	For fresh market Value per unit 2005 [2]	For processing Production 2003	For processing Production 2004	For processing Production 2005 [2]	For processing Value per unit 2003	For processing Value per unit 2004	For processing Value per unit 2005 [2]
	1,000 cwt.	*1,000 cwt.*	*1,000 cwt.*	*Dollars per cwt.*	*Dollars per cwt.*	*Dollars per cwt.*	*Tons*	*Tons*	*Tons*	*Dollars per ton*	*Dollars per ton*	*Dollars per ton*
AZ	1,450	1,535	(3)	27.80	33.20	(3)	....	....	(3)	....	....	(5)
CA	16,411	16,546	(3)	33.10	33.20	(3)	79,454	87,680	(3)	403.00	431.00	(5)
US	17,861	18,081	(3)	32.70	33.20	(3)	79,454	87,680	(3)	403.00	431.00	(5)

[1] Sprouting broccoli only. Does not include broccoli rabe nor heading (cauliflower) broccoli.   [2] Preliminary.   [3] Not published to avoid disclosure of individual operations.

NASS, Crops Branch, (202) 720–2127.

Table 4-14.—Cabbage for fresh market: Area, production, and value per hundredweight, by States, 2003–2005

State	Area harvested			Production			Value per unit		
	2003	2004	2005[1]	2003	2004	2005[1][2]	2003	2004	2005[1]
	Acres	*Acres*	*Acres*	*1,000 cwt.*	*1,000 cwt.*	*1,000 cwt.*	*Dollars per cwt.*	*Dollars per cwt.*	*Dollars per cwt.*
AZ	3,800	3,400	3,100	1,520	1,666	1,395	18,392	26,989	22,739
CA	13,500	13,600	13,300	5,265	5,576	4,655	84,240	100,368	70,291
CO	2,800	3,300	3,400	1,120	1,584	1,632	10,640	15,365	15,504
FL	7,600	7,600	7,800	2,356	2,812	2,652	23,089	30,932	31,294
GA	9,800	10,000	10,000	1,274	2,500	2,800	15,288	27,500	30,800
IL	650	650	750	104	88	236	1,248	950	2,053
MI	1,800	1,600	1,400	576	432	630	5,760	5,184	3,969
NJ	1,400	1,500	1,500	455	563	390	4,778	6,475	6,942
NY	9,800	10,600	9,700	3,822	3,710	4,559	38,351	42,317	67,289
NC	7,700	7,000	6,500	1,656	1,120	1,430	18,216	11,200	15,730
OH	1,500	1,400	1,400	465	595	252	6,138	5,932	4,360
PA	1,500	1,600	1,300	293	336	221	3,135	3,998	4,022
TX	7,700	8,300	8,700	2,541	3,237	2,610	53,869	60,208	41,499
VA	700	800	750	203	124	169	1,827	1,190	2,451
WI	4,600	4,200	4,100	989	630	615	9,593	6,111	6,519
US	74,850	75,550	73,700	22,639	24,973	24,246	294,564	344,719	325,462

[1] Preliminary.　[2] Includes some quantities of fall storage in NY harvested but not sold because of shrinkage and loss: 2003, 367,000 cwt; and 2004, 404,000 cwt and 2005, 456,000 cwt.
NASS, Crops Branch, (202) 720–2127.

Table 4-15.—Cantaloups for fresh market: Area, production, and value per hundredweight, by States, 2003–2005

State	Area harvested			Production			Value per unit		
	2003	2004	2005[1]	2003	2004	2005[1]	2003	2004	2005[1]
	Acres	*Acres*	*Acres*	*1,000 cwt.*	*1,000 cwt.*	*1,000 cwt.*	*Dollars per cwt.*	*Dollars per cwt.*	*Dollars per cwt.*
AZ	15,200	17,700	19,400	5,624	5,400	5,820	15.50	13.90	17.10
CA	49,000	48,000	51,000	12,005	13,200	13,260	15.40	13.80	10.00
CO	1,600	1,700	1,600	304	272	304	15.90	14.50	13.60
GA	6,000	6,600	6,300	1,290	990	851	12.50	19.00	17.90
IN	2,800	2,700	2,900	560	500	450	18.70	19.80	15.70
MD	500	550	660	34	61	63	22.00	16.00	32.00
PA	1,200	1,100	1,000	106	99	140	21.90	26.80	29.40
SC	1,200	1,100	1,100	144	154	88	12.00	12.50	17.00
TX	8,500	7,500	5,200	2,040	1,200	1,144	31.10	22.30	29.90
US	86,000	86,950	89,160	22,107	21,876	22,120	16.80	14.70	13.60

[1] Preliminary.
NASS, Crops Branch, (202) 720–2127.

IV-10 STATISTICS OF VEGETABLES AND MELONS

Table 4-16.—Carrots for fresh market, commercial crop: Area, production, and value per hundredweight, by States, 2003–2005

State	Area harvested			Production			Value per unit		
	2003	2004	2005[1]	2003	2004	2005[1]	2003	2004	2005[1]
	Acres	Acres	Acres	1,000 cwt.	1,000 cwt.	1,000 cwt.	Dollars per cwt.	Dollars per cwt.	Dollars per cwt.
AZ	2,600	2,000	....	858	680	....	15.60	21.60	....
CA	68,000	66,500	67,000	20,400	20,283	20,435	20.40	21.50	21.70
CO	2,400	1,700	....	1,152	1,071	....	10.30	10.40	....
MI	4,200	4,200	4,200	1,470	1,302	1,050	13.10	12.20	14.00
TX	2,100	2,100	2,300	609	704	690	20.80	26.00	29.00
Other States[2]	6,500	6,100	10,200	2,625	2,590	4,384	15.90	16.30	17.80
US	85,800	82,600	83,700	27,114	26,630	26,559	19.00	20.20	20.90

[1] Preliminary. [2] 2003-2004 - GA and WA. 2005 - AZ, CO, GA, and WA.
NASS, Crops Branch, (202) 720–2127.

Table 4-17.—Carrots for processing, commercial crop: Area, production, and value per ton, by States, 2003–2005

State	Area harvested			Production			Value per unit		
	2003	2004	2005[1]	2003	2004	2005[1]	2003	2004	2005[1]
	Acres	Acres	Acres	Tons	Tons	Tons	Dollars per tons	Dollars per tons	Dollars per tons
CA	3,500	4,300	4,100	113,050	137,600	143,500	100.00	111.00	82.00
MI	1,600	1,300	1,400	38,400	32,500	30,800	69.00	62.00	63.00
MN	850	160	770	21,100	3,010	27,220	53.40	55.50	78.10
TX	1,400	800	800	23,050	12,360	12,400	70.60	65.90	69.00
WA	5,300	5,400	4,200	157,940	162,000	122,220	68.00	70.00	70.00
WI	3,300	3,800	3,900	96,030	84,930	86,390	65.60	59.90	62.20
US	15,950	15,760	15,170	449,570	432,400	422,530	75.10	80.20	72.50

[1] Preliminary.
NASS, Crops Branch, (202) 720–2127.

Table 4-18.—Cauliflower, commercial crop: Area, yield, production, value per hundredweight and per ton, and total value, 1996–2005[1]

Year	Total crop					For fresh market			For processing		
	Area for harvest	Yield per acre	Produc- tion	Value[2]		Produc- tion	Value[2]		Produc- tion	Value[3]	
				Per cwt.	Total		Per cwt.	Total		Per ton	Total
	Acres	Cwt.	1,000 cwt.	Dollars	1,000 dollars	1,000 cwt.	Dollars	1,000 dollars	Tons	Dollars	1,000 dollars
1996	48,200	153	7,354	32.30	237,342	6,801	33.00	224,168	27,640	477.00	13,174
1997	43,500	158	6,889	31.60	217,534	6,323	32.30	203,957	28,300	480.00	13,577
1998	44,200	156	6,897	32.80	226,560	5,468	34.50	188,477	71,450	533.00	38,083
1999	46,400	161	7,450	28.70	213,833	6,666	29.70	197,767	39,186	410.00	16,066
2000	43,160	165	7,120	31.00	220,817	6,350	32.10	203,770	38,480	443.00	17,047
2001	42,050	160	6,708	28.30	190,085	5,920	29.20	172,690	39,410	441.00	17,395
2002	41,000	152	6,220	31.80	197,568	5,842	32.20	188,340	18,910	488.00	9,228
2003	39,000	168	6,546	34.60	226,202	6,216	35.10	217,952	16,500	500.00	8,250
2004	37,700	170	6,425	30.50	195,889	6,097	30.80	188,040	16,420	500.00	7,849
2005[4]	37,500	174	6,510	30.30	197,419	[5]	[5]	[5]	[5]	[5]	[5]

[1] Includes heading (cauliflower) broccoli. [2] Price and value on f.o.b. basis. [3] Price and value at processing plant door. [4] Preliminary. [5] Not published to avoid disclosure of individual operations.
NASS, Crops Branch, (202) 720–2127.

Table 4-19.—Cauliflower, commercial crop: Area, production, and value per hundredweight and per ton, by States, 2003–2005 [1]

State	Area harvested			Production			Value per unit		
	2003	2004	2005[2]	2003	2004	2005[2]	2003	2004	2005[2]
	Acres	Acres	Acres	1,000 cwt.	1,000 cwt.	1,000 cwt.	Dollars per cwt.	Dollars per cwt.	Dollars per cwt.
AZ	4,200	4,700	4,700	1,050	800	1,010	30.40	36.20	39.30
CA	34,000	32,000	31,900	5,440	5,600	5,396	35.30	29.60	28.60
NY	800	1,000	900	56	25	104	37.00	36.80	32.20
US	39,000	37,700	37,500	6,546	6,425	6,510	34.60	30.50	30.30

State	For fresh market						For processing					
	Production			Value per unit			Production			Value per unit		
	2003	2004	2005[2]	2003	2004	2005[2]	2003	2004	2005[2]	2003	2004	2005[2]
	1,000 cwt.	1,000 cwt.	1,000 cwt.	Dollars per cwt.	Dollars per cwt.	Dollars per cwt.	Tons	Tons	Tons	Dollars per ton	Dollars per ton	Dollars per ton
AZ	1,050	800	(3)	30.40	36.20	(3)	(3)	(3)	(3)	(3)	(3)	(3)
CA	5,110	5,272	(3)	36.00	30.00	(3)	16,500	16,420	(3)	500.00	478.00	(3)
NY	56	25	104	37.00	36.80	32.20	(3)	(3)	(3)	(3)	(3)	(3)
US	6,216	6,097	(3)	35.10	30.80	(3)	16,500	16,420	(3)	500.00	478.00	(3)

[1] Includes heading (cauliflower) broccoli.   [2] Preliminary.   [3] Not published to avoid disclosure of individual operations.
NASS, Crops Branch, (202) 720–2127.

Table 4-20.—Celery, commercial crop: Area, production, and value per hundredweight, by States, 2003–2005 [1]

State	Area harvested			Production			Value per unit		
	2003	2004	2005[2]	2003	2004	2005[2]	2003	2004	2005[2]
	Acres	Acres	Acres	1,000 cwt.	1,000 cwt.	1,000 cwt.	Dollars per cwt.	Dollars per cwt.	Dollars per cwt.
CA	25,300	25,700	25,400	18,090	18,247	18,034	13.30	15.00	14.20
MI	2,200	2,200	2,200	1,166	1,232	1,144	15.10	12.30	15.90
US	27,500	27,900	27,600	19,256	19,479	19,178	13.40	14.80	14.30

[1] Mostly for fresh market use, but includes some quantities used for processing.   [2] Preliminary.
NASS, Crops Branch, (202) 720–2127.

Table 4-21.—Celery, commercial crop: Area, yield, production, value per hundredweight, and total value, United States, 1996–2005 [1]

Year	Area for harvest	Yield per acre	Production	Value[2]	
				Per cwt.	Total
	Acres	Cwt.	1,000 cwt.	Dollars	1,000 dollars
1996 ...	27,840	683	19,015	10.50	199,877
1997 ...	26,910	673	18,119	14.70	266,321
1998 ...	27,200	662	18,000	11.70	210,753
1999 ...	27,500	681	18,727	12.00	224,702
2000 ...	26,200	703	18,425	18.50	341,391
2001 ...	27,800	678	18,856	14.40	272,391
2002 ...	27,100	691	18,737	12.80	239,846
2003 ...	27,500	700	19,256	13.40	258,965
2004 ...	27,900	698	19,479	14.80	288,791
2005[3]	27,600	695	19,178	14.30	274,331

[1] Mostly for fresh market use, but includes quantities used for processing.   [2] Price and value on f.o.b. basis.   [3] Preliminary.
NASS, Crops Branch, (202) 720–2127.

STATISTICS OF VEGETABLES AND MELONS

Table 4-22.—Corn, sweet, commercial crop: Area, production, and value per hundredweight and per ton, by States, 2003–2005

Utilization and State	Area harvested			Production			Value per unit		
	2003	2004	2005[1]	2003	2004	2005[1]	2003	2004	2005[1]
FOR FRESH MARKET	Acres	Acres	Acres	1,000 cwt.	1,000 cwt.	1,000 cwt.	Dollars per cwt.	Dollars per cwt.	Dollars per cwt.
AL	1,900	2,000	1,100	125	78	44	17.20	19.40	19.20
CA	28,500	24,500	25,300	5,415	4,655	4,934	21.80	29.50	22.10
CO	7,600	9,300	9,700	1,292	1,395	1,455	9.10	11.70	13.70
CT	4,100	4,300	4,000	246	344	300	27.50	28.50	28.00
DE	3,000	3,300	3,000	189	363	330	25.00	20.00	20.00
FL	38,800	38,700	33,600	5,626	5,999	5,376	16.00	18.40	20.10
GA	20,000	27,000	29,000	2,800	3,645	3,625	16.50	12.80	21.90
IL	5,600	5,300	6,200	414	509	484	17.90	21.60	24.90
IN	5,100	5,400	5,200	372	486	328	23.60	23.30	23.40
ME	2,000	2,000	2,000	120	120	120	32.50	33.00	34.00
MD	4,200	4,500	4,700	269	315	353	19.00	20.00	26.00
MA	5,600	5,800	5,700	420	522	456	31.50	32.50	34.50
MI	9,500	9,500	9,500	855	713	808	16.60	19.50	20.00
NH	1,900	1,800	1,700	133	126	128	42.00	42.00	41.50
NJ	7,800	7,500	7,100	507	525	568	23.90	20.80	21.50
NY	35,600	28,000	28,200	4,094	2,800	2,679	20.60	21.40	22.60
NC	8,200	7,500	7,000	820	675	644	16.00	17.00	17.50
OH	15,200	15,300	16,100	1,474	1,285	1,385	19.70	21.10	22.10
OR	5,000	4,700	4,400	480	423	528	17.50	17.00	21.00
PA	18,800	19,600	17,700	1,166	1,392	1,080	24.90	22.10	29.40
RI	1,000	1,100	1,000	90	99	70	31.00	35.00	35.00
TX	2,900	1,900	1,700	261	152	119	20.60	18.00	22.20
VT	1,100	1,000	1,100	88	55	77	36.00	39.00	41.00
VA	3,000	3,000	3,200	198	375	205	16.00	14.00	16.90
WA	3,000	2,800	3,800	420	392	570	17.00	24.00	26.20
WI	7,400	6,900	6,900	629	442	600	23.00	27.00	24.90
US	246,800	242,700	238,900	28,503	27,885	27,266	19.30	20.80	22.10
FOR PROCESSING	Acres	Acres	Acres	Tons[2]	Tons[2]	Tons[2]	Dollars per ton	Dollars per ton	Dollars per ton
DE	9,400	7,300	7,300	56,400	49,020	49,020	98.00	78.00	78.00
MD	5,500	6,500	6,700	31,000	46,200	41,840	93.10	98.00	67.00
MN	139,400	135,400	136,400	938,800	894,590	977,090	64.50	72.50	71.60
NY	14,400	19,000	17,600	108,780	110,200	116,160	78.40	77.70	80.40
OR	30,100	28,500	23,200	271,680	259,910	233,810	81.00	80.70	79.20
PA	770	1,700	1,300	3,970	7,990	6,630	85.90	94.20	94.70
WA	98,300	94,800	81,200	900,540	826,140	792,160	76.90	70.10	69.00
WI	92,100	78,600	92,000	681,420	511,220	680,230	62.30	67.20	61.70
Other States[3]	36,630	34,000	38,210	273,460	262,910	277,180	66.60	69.30	55.50
US	426,600	405,800	403,910	3,266,050	2,968,180	3,174,120	70.40	72.10	68.40
Grand total	673,400	648,500	642,810	4,691,200	4,362,430	4,537,420			

[1] Preliminary. [2] Tonnage in husk. [3] 2003 - ID, IL, IA, NJ, and TN. 2004 - ID, IL, IA, NJ, TN, and VA. 2005 - ID, IL, IA, NJ, TN, and VA.

NASS, Crops Branch, (202) 720–2127.

Table 4-23.—Corn, sweet, commercial crop: Area, yield, production, value per hundredweight and per ton, and total value, United States, 1996–2005

Year	For fresh market					For processing				
	Area for harvest	Yield per acre	Production	Value[1]		Area for harvest	Yield per acre	Production	Value[2]	
				Per cwt.	Total				Per ton	Total
	Acres	Cwt.	1,000 cwt.	Dollars	1,000 dollars	Acres	Tons	Tons	Dollars	1,000 dollars
1996	227,800	102	23,127	16.90	390,737	474,200	6.95	3,296,330	78.50	258,840
1997	236,400	100	23,641	17.70	418,617	465,800	7.18	3,342,330	74.90	250,329
1998	237,350	111	26,307	17.20	452,278	467,300	6.97	3,255,560	73.30	238,748
1999	236,950	109	25,762	16.90	436,094	466,300	7.07	3,297,390	71.10	234,418
2000	239,200	109	26,027	18.50	481,016	460,400	6.86	3,160,020	73.40	232,021
2001	244,930	109	26,815	19.50	523,567	447,150	7.04	3,147,530	73.00	229,678
2002	245,730	108	26,480	19.20	509,421	417,100	7.35	3,067,690	68.00	208,703
2003	246,800	115	28,503	19.30	550,024	426,600	7.66	3,266,050	70.40	229,788
2004	242,700	115	27,885	20.80	580,320	405,800	7.31	2,968,180	72.10	213,993
2005[3]	238,900	114	27,266	22.10	601,519	403,910	7.86	3,174,120	68.40	217,096

[1] Price and value on f.o.b. basis. [2] Price and value at processing plant door. [3] Preliminary.

NASS, Crops Branch, (202) 720–2127.

Table 4-24.—Cucumbers for fresh market: Area, production, and value per hundredweight, by States, 2003–2005

State	Area harvested			Production			Value per unit		
	2003	2004	2005[1]	2003	2004	2005[1]	2003	2004	2005[1]
	Acres	*Acres*	*Acres*	*1,000 cwt.*	*1,000 cwt.*	*1,000 cwt.*	*Dollars per cwt.*	*Dollars per cwt.*	*Dollars per cwt.*
CA	4,000	4,400	4,600	800	1,078	1,242	33.30	41.80	23.90
FL	11,300	10,700	10,500	2,712	2,515	2,835	22.20	20.10	26.00
GA	12,500	14,000	16,000	2,125	1,960	2,800	10.60	16.60	23.90
MD	500	770	770	28	62	39	25.00	26.00	30.00
MI	6,400	7,400	7,200	1,024	1,295	1,224	20.40	17.20	16.50
NJ	3,000	3,100	3,200	600	682	480	20.00	22.70	20.20
NY	4,800	4,600	4,500	528	874	540	23.60	27.60	28.30
NC	6,500	6,300	5,000	780	630	525	17.00	18.00	16.00
SC	1,900	2,000	1,700	209	240	119	25.90	23.00	21.00
TX	1,400	1,500	700	322	525	98	21.00	21.00	26.90
VA	2,700	2,400	3,000	297	240	330	22.00	17.00	13.10
US	55,000	57,170	57,170	9,425	10,101	10,232	19.90	22.10	22.90

[1] Preliminary.
NASS, Crops Branch, (202) 720–2127.

Table 4-25.—Cucumbers (for pickles), commercial crop: Area, yield, production, value per ton, total value, and pickle stocks, United States, 1996–2005

Year	For processing					Pickle stocks on hand Dec. 1 [2]
	Area harvested	Yield per acre	Production	Value [1]		
				Per ton	Total	
	Acres	*Tons*	*Tons*	*Dollars*	*1,000 dollars*	*Tons*
1996	105,200	5.36	563,689	248.00	139,985	392,970
1997	103,370	6.00	620,100	234.00	145,371	282,190
1998	102,870	5.77	593,720	237.00	140,553	359,512
1999	105,300	5.97	628,360	238.00	149,839	452,445
2000	104,710	5.86	613,160	269.00	164,956	387,544
2001	108,260	5.37	581,540	291.00	168,958	552,303
2002	117,800	5.26	619,310	273.00	169,006	300,580
2003	118,800	5.46	648,430	275.00	178,328	353,573
2004	113,000	5.23	591,380	269.00	158,793	240,644
2005[3]	113,700	5.02	570,720	260.00	148,324	299,273

[1] Price and value at processing plant door. [2] Stocks in hands of original salters of both salt and dill pickles, sold and unsold, in tanks and barrels, on Dec. 1. Includes stocks of fresh-pack pickles. [3] Preliminary.
NASS, Crops Branch, (202) 720–2127.

Table 4-26.—Cucumbers (for pickles), commercial crop: Area, production, and value per ton, by States, 2003–2005

State	Area harvested			Production			Value per unit		
	2003	2004	2005[1]	2003	2004	2005[1]	2003	2004	2005[1]
	Acres	*Acres*	*Acres*	*Tons*	*Tons*	*Tons*	*Dollars per ton*	*Dollars per ton*	*Dollars per ton*
FL	6,500	6,500	6,500	70,850	70,850	66,950	465.00	464.00	476.00
IN	1,700	1,700	1,600	10,100	7,410	6,560	187.00	160.00	163.00
MD	4,300	4,300	3,000	19,350	8,600	11,700	280.00	240.00	264.00
MI	33,500	34,500	38,000	180,900	172,500	182,400	200.00	205.00	168.00
NC	17,400	16,500	16,000	80,040	69,300	68,800	295.00	280.00	290.00
OH	2,200	5,000	3,400	27,940	45,000	29,240	337.00	333.00	374.00
SC	4,000	4,000	4,100	18,400	22,800	14,350	239.00	220.00	222.00
TX	8,000	6,600	7,800	42,400	33,000	35,880	412.00	448.00	235.00
WI	5,500	4,600	4,700	36,080	30,180	29,700	178.00	157.00	195.00
Oth Sts[2]	35,700	29,300	28,600	162,370	131,740	125,140	250.00	215.00	267.00
US	118,800	113,000	113,700	648,430	591,380	570,720	275.00	269.00	260.00

[1] Preliminary. [2] 2003-2005 - AL, CA, DE, GA, MA, MO, and WA.
NASS, Crops Branch, (202) 720–2127.

Table 4-27.—Garlic for fresh market and processing: Area, production, and value per hundredweight, by States, 2003–2005

State	Area harvested			Production			Value per unit		
	2003	2004	2005[1]	2003	2004	2005[1]	2003	2004	2005[1]
	Acres	Acres	Acres	1,000 cwt.	1,000 cwt.	1,000 cwt.	Dollars per cwt.	Dollars per cwt.	Dollars per cwt.
CA	29,000	26,000	24,500	5,365	4,680	4,000	27.20	27.70	44.10
NV	1,200	1,200	600	182	144	108	21.00	18.00	16.50
OR	4,800	4,400	4,300	694	400	538	14.90	15.90	22.00
US	35,000	31,600	29,400	6,241	5,224	4,646	25.70	26.50	40.90

[1] Preliminary.
NASS, Crops Branch, (202) 720-2127.

Table 4-28.—Honeydew melons, commercial crop: Area, yield, production, value per hundredweight, and total value, United States, 1996–2005

Year	Area for harvest	Yield per acre	Production	Value[1]	
				Per cwt.	Total
	Acres	Cwt.	1,000 cwt.	Dollars	1,000 dollars
1996	27,300	174	4,737	17.00	80,405
1997	26,600	182	4,828	18.90	91,040
1998	25,500	197	5,013	21.60	108,155
1999	27,500	188	5,160	21.10	109,082
2000	26,000	197	5,116	19.20	98,244
2001	24,200	195	4,720	21.10	99,500
2002	24,400	208	5,065	18.10	91,453
2003	23,200	227	5,275	18.80	98,961
2004	21,900	238	5,221	17.60	92,133
2005[2]	21,200	213	4,505	15.30	69,010

[1] Price and value on f.o.b. basis. [2] Preliminary.
NASS, Crops Branch, (202) 720–2127.

Table 4-29.—Honeydew melons, commercial crop: Area, production, and value per hundredweight, by States, 2003–2005

State	Area harvested			Production			Value per unit		
	2003	2004	2005 [1]	2003	2004	2005 [1]	2003	2004	2005 [1]
	Acres	*Acres*	*Acres*	*1,000 cwt.*	*1,000 cwt.*	*1,000 cwt.*	*Dollars per cwt.*	*Dollars per cwt.*	*Dollars per cwt.*
AZ	3,000	2,700	2,600	1,005	945	950	14.70	14.40	23.80
CA	18,500	17,900	17,900	3,700	3,938	3,401	17.50	17.70	13.0 0
TX	1,700	1,300	700	570	338	154	34.10	26.10	14.20
US	23,200	21,900	21,200	5,275	5,221	4,505	18.80	17.60	15.30

[1] Preliminary.
NASS, Crops Branch, (202) 720–2127.

Table 4-30.—Head lettuce, commercial crop: Area, production, and value per hundredweight, by States, 2003–2005

State	Area harvested			Production			Value per unit		
	2003	2004	2005 [1]	2003	2004	2005 [1]	2003	2004	2005 [1]
	Acres	*Acres*	*Acres*	*1,000 cwt.*	*1,000 cwt.*	*1,000 cwt.*	*Dollars per cwt.*	*Dollars per cwt.*	*Dollars per cwt.*
AZ:									
Western	49,600	46,500	45,600	17,856	16,740	15,510	10.30	22.20	14.60
Other	800	500	500	208	150	145	12.30	18.50	22.20
CA	132,000	131,000	131,000	49,500	48,470	47,160	21.00	15.10	15.90
CO	1,800	2,200	1,900	522	704	684	11.00	12.50	11.80
NJ	900	800	500	158	164	95	22.00	23.60	35.00
US	185,100	181,000	179,500	68,244	66,228	63,594	18.10	16.90	15.60

[1] Preliminary.
NASS, Crops Branch, (202) 720–2127.

Table 4-31.—Head lettuce, commercial crop: Area, yield, production, value per hundredweight, and total value, United States, 1996–2005

Year	Area for harvest	Yield per acre	Production	Value [1]	
				Per cwt.	Total
	Acres	*Cwt.*	*1,000 cwt.*	*Dollars*	*1,000 dollars*
1996	217,600	285	62,072	14.70	912,586
1997	203,000	339	68,794	17.50	1,201,899
1998	198,500	330	65,461	16.20	1,060,070
1999	192,800	380	73,181	13.30	972,917
2000	185,200	377	69,673	17.30	1,208,140
2001	184,300	374	68,917	17.90	1,234,981
2002	184,500	369	68,140	21.10	1,435,296
2003	185,100	369	68,244	18.10	1,235,193
2004	181,000	366	66,228	16.90	1,118,970
2005 [2]	179,500	354	63,594	15.60	990,905

[1] Price and value on f.o.b. basis. [2] Preliminary.
NASS, Crops Branch, (202) 720–2127.

STATISTICS OF VEGETABLES AND MELONS

Table 4-32.—Leaf lettuce for fresh market: Area, production, and value per hundredweight, by States, 2003–2005

State	Area harvested			Production			Value per unit		
	2003	2004	2005 [1]	2003	2004	2005 [1]	2003	2004	2005 [1]
	Acres	*Acres*	*Acres*	*1,000 cwt.*	*1,000 cwt.*	*1,000 cwt.*	*Dollars per cwt.*	*Dollars per cwt.*	*Dollars per cwt.*
AZ	7,400	7,500	7,600	2,220	2,100	2,205	29.60	45.50	52.70
CA	49,000	54,000	55,000	11,270	12,690	13,200	31.80	28.30	31.60
US	56,400	61,500	62,600	13,490	14,790	15,405	31.40	30.70	34.60

[1] Preliminary.
NASS, Crops Branch, (202) 720–2127.

Table 4-33.—Romaine lettuce for fresh market: Area, production, and value per hundredweight, by States, 2003–2005

State	Area harvested			Production			Value per unit		
	2003	2004	2005 [1]	2003	2004	2005 [1]	2003	2004	2005 [1]
	Acres	*Acres*	*Acres*	*1,000 cwt.*	*1,000 cwt.*	*1,000 cwt.*	*Dollars per cwt.*	*Dollars per cwt.*	*Dollars per cwt.*
AZ	16,500	17,200	19,400	4,703	5,755	6,400	19.20	18.60	24.20
CA	60,000	58,000	63,000	18,000	17,400	17,325	29.70	19.30	17.50
US	76,500	75,200	82,400	22,703	23,155	23,725	27.50	19.10	19.30

[1] Preliminary.
NASS, Crops Branch, (202) 720–2127.

Table 4-34.—Onions (fresh market): Foreign trade, United States, 1995–2004 [1]

Year beginning July	Imports	Domestic exports
	1,000 cwt.	*1,000 cwt.*
1995	5,725	5,942
1996	5,875	5,814
1997	5,537	6,589
1998	5,312	5,666
1999	5,005	7,040
2000	5,671	7,964
2001	5,925	6,788
2002	6,322	6,838
2003	6,563	6,174
2004	6,713	6,924

[1] Includes onion sets and pearl onions.
ERS, Specialty Crops Branch, (202) 694–5253. Compiled from reports of the U.S. Department of Commerce.

Table 4-35.—Onions, commercial crop: Area, yield, production, shrinkage and loss, value per hundredweight, and total value, United States, 1996–2005 [1]

Year	Area for harvest	Yield per acre	Production [2]	Shrinkage and loss	Value [3] Per cwt.	Value [3] Total
	Acres	Cwt.	1,000 cwt.	1,000 cwt.	Dollars	1,000 dollars
1996 ...	166,210	386	64,106	6,678	10.50	604,789
1997 ...	165,910	414	68,769	7,540	12.60	769,974
1998 ...	172,140	394	67,747	6,428	13.00	798,227
1999 ...	175,500	428	75,032	9,173	9.74	641,278
2000 ...	167,070	437	72,948	7,131	11.20	735,939
2001 ...	164,990	424	69,961	6,564	10.70	680,350
2002 ...	162,720	429	69,844	6,425	12.10	764,994
2003 ...	166,090	442	73,363	5,593	13.70	929,274
2004 ...	168,950	491	83,007	8,909	10.50	777,339
2005 [4]	161,520	457	73,769	6,468	13.70	922,369

[1] Mostly for fresh market use, but includes some quantities used for processing. [2] Includes storage crop onions harvested but not sold because of shrinkage and waste. [3] Price and value on f.o.b. basis. [4] Preliminary.
NASS, Crops Branch, (202) 720–2127.

Table 4-36.—Onions, commercial crop: Area, production, shrinkage and loss, and value per hundredweight, by States, 2003–2005 [1]

Season and State	Area harvested			Production			Shrinkage and loss			Value per unit		
	2003	2004	2005 [2]	2003	2004	2005 [2]	2003	2004	2005 [2]	2003	2004	2005 [2]
	Acres	Acres	Acres	1,000 cwt.	1,000 cwt.	1,000 cwt.	1,000 cwt.	1,000 cwt.	1,000 cwt.	Dollars per cwt.	Dollars per cwt.	Dollars per cwt.
Spring:												
AZ	1,500	1,600	2,000	750	800	920	....	....	....	9.89	8.80	10.20
CA	7,500	7,100	7,300	3,675	3,586	3,468	....	....	....	22.90	15.10	12.00
GA	12,500	14,500	10,500	2,188	3,770	2,205	....	....	....	34.30	23.50	29.70
TX	11,000	12,500	15,500	3,520	3,875	4,650	....	....	....	38.10	22.60	29.70
Total	32,500	35,700	35,300	10,133	12,031	11,243	....	....	....	29.70	19.70	22.60
Summer:												
Non-storage:												
CA	7,500	8,400	8,700	3,975	4,704	4,785	....	....	....	13.70	13.20	9.50
NV	3,100	3,400	2,400	1,860	2,176	2,112	....	....	....	16.00	16.00	15.00
NM	7,700	7,100	6,400	4,235	3,657	3,392	....	....	....	13.60	12.60	15.80
TX	2,500	2,800	900	1,000	1,036	333	....	....	....	24.60	24.10	33.70
WA	1,400	1,500	1,400	518	525	518	....	....	....	26.80	16.00	23.60
Non-storage total	22,200	23,200	19,800	11,588	12,098	11,140	....	....	....	15.60	14.60	13.80
Storage: [3]												
CA [4]	35,000	30,000	27,800	14,700	13,200	11,815	250	250	250	7.54	7.46	8.32
CO	9,600	11,000	9,500	3,696	5,500	4,180	460	1,400	630	15.00	12.20	17.70
ID	9,800	10,400	9,500	5,880	8,008	6,080	950	1,760	1,155	11.30	6.80	12.60
MI	3,600	3,200	3,700	1,152	928	962	230	185	190	14.50	10.80	12.50
MN [5] [6]	210			65			10			9.25		
NY	11,900	13,000	13,600	3,808	5,200	3,808	500	730	495	13.30	12.10	13.80
OH [7]	380			122			8			14.60		
OR (Malheur)	12,200	11,100	11,200	7,198	8,658	7,616	1,370	1,620	1,600	11.10	6.90	14.40
OR (Other)	6,900	7,400	7,700	3,243	4,218	3,696	480	630	520	7.45	7.20	11.90
UT	1,800	1,500		828	780		130	160		10.40	6.60	
WA	18,000	20,000	19,500	10,260	11,600	11,700	1,130	2,090	1,400	8.50	2.90	9.60
WI	2,000	1,900	2,000	690	608	660	75	65	70	8.80	7.85	8.60
Other states [8]		550	1,920		178	869		19	158		12.90	12.30
Storage total	111,390	110,050	106,420	51,642	58,878	51,386	5,593	8,909	6,468	9.73	7.28	11.40
Total summer	133,590	133,250	126,220	63,230	70,976	62,526	....	....	....	10.90	8.70	11.90
US	166,090	168,950	161,520	73,363	83,007	73,769	....	....	....	13.70	10.50	13.70

[1] Mostly for fresh market use, but includes some quantities used for processing. [2] Preliminary. [3] Includes some quantities of storage crop onions harvested but not sold because of shrinkage and loss. [4] Includes fresh and processed. [5] 2004 data not published to avoid disclosure of individual operations. [6] Estimates discontinued in 2005. [7] 2004 and 2005 data not published to avoid disclosure of individual operations. [8] 2004 - MN and OH. 2005 - OH and UT.
NASS, Crops Branch, (202) 720–2127.

Table 4-37.—Peas, green (for processing), commercial crop: Area, yield, production, value per ton, and total value, United States, 1996–2005

Year	Area harvested	Yield per acre	Production	Value[1]	
				Per ton	Total
	Acres	Tons[2]	Tons	Dollars	1,000 dollars
1996 ...	249,800	1.67	417,672	285.00	118,910
1997 ...	271,200	1.77	480,000	288.00	138,482
1998 ...	273,900	1.77	483,900	282.00	136,584
1999 ...	271,640	1.70	461,590	275.00	126,925
2000 ...	277,240	1.91	530,550	248.00	131,817
2001 ...	211,640	1.85	390,980	264.00	103,313
2002 ...	212,200	1.65	349,860	253.00	88,439
2003 ...	232,100	2.01	467,670	250.00	117,087
2004 ...	206,900	1.92	397,570	250.00	99,280
2005[3]	211,500	1.79	378,830	267.00	101,080

[1] Price and value at processing plant door. [2] Shelled basis: 2½ pounds of peas in the shell produce approximately 1 pound of shelled peas. [3] Preliminary.
NASS, Crops Branch, (202) 720–2127.

Table 4-38.—Peas, green (for processing), commercial crop: Area, production, and value per ton, States, 2003–2005[1]

State	Area harvested			Production			Value per unit		
	2003	2004	2005[2]	2003	2004	2005[2]	2003	2004	2005[2]
	Acres	Acres	Acres	Tons	Tons	Tons	Dollars per ton	Dollars per ton	Dollars per ton
DE	5,900	6,000	6,000	10,620	9,000	9,000	355.00	350.00	350.00
MN	81,800	72,100	76,200	158,110	107,350	105,100	288.00	285.00	384.00
NY	14,800	18,000	19,900	28,020	34,250	29,660	350.00	343.00	385.00
OR	22,200	16,700	12,800	39,260	41,400	26,750	208.00	188.00	175.00
WA	44,300	35,300	36,700	98,340	89,950	89,140	208.00	196.00	193.00
WI	37,700	29,600	39,400	84,310	54,500	73,650	206.00	269.00	216.00
Other States[3]	25,400	29,200	20,500	49,010	61,120	45,530	245.00	225.00	183.00
US	232,100	206,900	211,500	467,670	397,570	378,830	250.00	250.00	267.00

[1] Shelled basis; 2½ pounds of peas in the shell produce approximately 1 pound of shelled peas. [2] Preliminary. [3] 2003-2005 - CA, ID, IL, MD, and NJ.
NASS, Crops Branch, (202) 720–2127.

Table 4-39.—Bell peppers for fresh market: Area, production, and value per hundredweight, by States, 2003–2005

State	Area harvested			Production			Value per unit		
	2003	2004	2005[1]	2003	2004	2005[1]	2003	2004	2005[1]
	Acres	Acres	Acres	1,000 cwt.	1,000 cwt.	1,000 cwt.	Dollars per cwt.	Dollars per cwt.	Dollars per cwt.
CA	18,500	19,000	23,000	7,215	7,600	6,900	28.60	34.20	26.90
FL	17,700	18,300	19,000	4,956	5,673	4,580	35.90	38.50	46.60
GA	4,500	3,900	3,600	1,350	663	684	30.00	30.00	30.00
MI	1,800	1,800	1,600	450	522	480	22.00	26.00	23.00
NJ	3,600	3,500	3,200	882	928	832	29.00	25.00	24.70
NC	5,000	4,000	3,700	500	440	444	25.00	23.00	24.00
OH	2,000	1,900	2,200	590	494	484	27.10	20.40	35.70
TX	700	500	700	175	80	105	32.40	49.80	33.70
US	53,800	52,900	57,000	16,118	16,400	14,509	30.70	34.10	33.30

[1] Preliminary.
NASS, Crops Branch, (202) 720–2127.

Table 4-40.—Potatoes: Area, yield, production, season average price, and value, United States, 1996-2005

Year	Area planted	Area harvested	Yield per harvested acre	Production	Season average price per cwt. received by farmers[1]	Value of production
	1,000 acres	1,000 acres	Cwt.	1,000 cwt.	Dollars	1,000 dollars
1996	1,454.7	1,425.9	350	499,254	4.91	2,423,476
1997	1,383.5	1,353.6	345	467,091	5.64	2,622,621
1998	1,415.8	1,386.9	343	475,667	5.56	2,633,941
1999	1,376.1	1,331.8	359	478,093	5.76	2,742,428
2000	1,383.1	1,347.5	381	513,544	5.08	2,590,053
2001	1,246.9	1,220.9	358	437,673	6.99	3,055,876
2002	1,299.6	1,265.9	362	458,171	6.67	3,045,310
2003	1,272.6	1,248.6	367	457,814	5.89	2,685,822
2004	1,193.3	1,166.9	391	456,041	5.67	2,575,204
2005	1,107.2	1,084.6	388	420,879	6.90	2,903,137

[1] 1996-2004 obtained by weighting State prices by quantity sold. 2005 obtained by weighting State prices by production.
NASS, Crops Branch, (202) 720-2127.

Table 4-41.—Potatoes: Production, seed used, and disposition, United States, 1995-2004

Year	Production	Total used for seed	Used on farms where produced			Sold
			For seed, feed, and household use	Shrinkage and loss		
	1,000 cwt.	1,000 cwt.	1,000 cwt.	1,000 cwt.		1,000 cwt.
1995	445,099	30,561	5,755	29,630		409,714
1996	499,254	29,138	6,221	41,238		451,795
1997	467,091	29,975	5,475	32,183		429,433
1998	475,667	29,206	5,764	35,449		434,454
1999	478,093	29,580	5,545	35,550		436,998
2000	513,544	27,137	5,287	43,685		464,572
2001	437,673	28,625	5,386	31,227		401,060
2002	458,171	28,149	5,622	30,905		421,644
2003	457,814	26,687	5,543	35,294		416,977
2004	456,041	24,695	4,796	37,408		413,837

NASS, Crops Branch, (202) 720-2127.

Table 4-42.—Fall potatoes: Production and total stocks held by growers and local dealers, 15 Major States, 1995-2004

Crop year	Production	Total stocks							
		Dec. 1	Following year						
			Jan. 1	Feb. 1	Mar. 1	Apr. 1	May 1	June 1[1]	
	1,000 cwt.	1,000 cwt.	1,000 cwt.	1,000 cwt.	1,000 cwt.	1,000 cwt.	1,000 cwt.	1,000 cwt.	
1995	394,785	256,710	223,550	189,360	156,020	115,855	75,870	....	
1996	443,704	295,100	261,320	226,080	189,210	147,635	103,210	....	
1997	413,513	278,830	246,550	212,562	175,870	134,190	92,840	....	
1998	423,170	280,910	246,230	209,640	173,650	131,220	87,895	50,270	
1999	420,567	275,100	239,910	207,150	169,620	128,410	86,915	47,220	
2000	458,827	310,300	275,270	234,260	197,670	153,520	109,160	61,270	
2001	387,033	258,750	224,680	192,090	158,590	119,950	81,200	42,990	
2002	407,085	264,485	231,490	199,020	165,210	125,770	83,040	45,880	
2003	403,181	267,900	233,590	200,230	166,280	126,110	85,000	46,020	
2004	403,587	271,100	236,700	203,490	168,020	128,900	88,550	51,700	

[1] Estimates begun in 1998.
NASS, Crops Branch, (202) 720-2127.

STATISTICS OF VEGETABLES AND MELONS

Table 4-43.—Potatoes: Area, production, and marketing year price per hundredweight received by farmers, by States, 2003–2005

Season and State	Area harvested			Yield			Production		
	2003	2004	2005	2003	2004	2005	2003	2004	2005
	1,000 acres	*1,000 acres*	*1,000 acres*	*Cwt.*	*Cwt.*	*Cwt.*	*1,000 cwt.*	*1,000 cwt.*	*1,000 cwt.*
Winter:									
CA	8.5	13.0	14.0	310	250	250	2,635	3,250	3,500
FL	5.8	5.5	5.8	240	285	240	1,392	1,568	1,392
Total	14.3	18.5	19.8	282	260	247	4,027	4,818	4,892
Spring:									
AZ	7.6	6.2	4.3	275	285	275	2,090	1,767	1,183
CA	19.0	17.5	15.1	440	475	405	8,360	8,313	6,116
FL	28.6	24.5	23.2	280	313	281	8,008	7,678	6,527
Hastings	20.3	18.0	17.0	280	320	280	5,684	5,760	4,760
Other	8.3	6.5	6.2	280	295	285	2,324	1,918	1,767
NC	17.0	13.5	15.0	175	200	190	2,975	2,700	2,850
TX	12.5	10.5	9.1	240	210	225	3,000	2,205	2,048
Total	84.7	72.2	66.7	288	314	281	24,433	22,663	18,724
Summer:									
AL	1.8	1.3	1.3	185	175	155	333	228	202
CA	7.2	7.0	6.2	385	350	340	2,772	2,450	2,108
CO	6.4	5.7	4.8	360	350	365	2,304	1,995	1,752
DE	3.6	3.1	3.1	240	260	260	864	806	806
IL	6.1	4.8	4.3	360	415	340	2,196	1,992	1,462
KS	2.7	3.4	4.0	380	400	360	1,026	1,360	1,440
MD	4.6	4.6	3.4	240	260	260	1,104	1,196	884
MO	7.1	6.2	5.8	265	310	340	1,882	1,922	1,972
NJ	2.7	2.2	2.1	250	270	255	675	594	536
NM [1]	1.9	1.0	....	280	340	....	532	340	....
TX	8.4	9.6	8.7	420	440	465	3,528	4,224	4,046
VA	6.2	5.0	4.9	250	240	210	1,550	1,200	1,029
Total	58.7	53.9	48.6	320	340	334	18,766	18,307	16,237
Fall:									
CA	8.3	7.6	7.2	425	480	450	3,528	3,648	3,240
CO	65.7	64.3	57.9	360	370	385	23,652	23,791	22,292
ID	358.0	353.0	323.0	344	374	362	123,180	131,970	116,975
10 S.W. counties	25.0	25.0	21.0	465	490	465	11,625	12,250	9,765
Other counties	333.0	328.0	302.0	335	365	355	111,555	119,720	107,210
IN [2]	3.7	3.2	....	250	350	....	925	1,120	....
ME	65.5	61.5	56.2	260	310	280	17,030	19,065	15,736
MA	2.7	2.5	2.4	265	320	260	716	800	624
MI	45.5	42.0	43.5	330	325	320	15,015	13,650	13,920
MN	58.0	44.0	43.0	385	430	410	22,330	18,920	17,630
MT	10.6	10.6	10.9	315	335	315	3,339	3,551	3,434
NE	23.2	21.6	19.4	420	430	425	9,744	9,288	8,245
NV	8.0	6.7	5.5	415	430	425	3,320	2,881	2,338
NM	4.0	4.0	4.2	400	430	420	1,600	1,720	1,764
NY	21.7	19.2	20.1	300	270	260	6,510	5,184	5,226
ND	112.0	101.0	82.0	245	265	250	27,440	26,765	20,500
OH	4.3	3.6	3.6	255	300	240	1,097	1,080	864
OR	42.6	37.0	37.1	493	534	594	20,991	19,775	22,023
Malheur County	5.8	5.2	3.8	415	470	450	2,407	2,444	1,710
Other counties	36.8	31.8	33.3	505	545	610	18,584	17,331	20,313
PA	12.5	11.0	11.0	270	240	250	3,375	2,640	2,750
RI	0.6	0.5	0.5	285	290	210	171	145	105
SD [3]	1.0	....	....	340	....	....	340	....	....
UT [3]	1.0	....	....	335	....	....	335	....	....
WA	162.0	159.0	154.0	575	590	620	93,150	93,810	95,480
WI	80.0	70.0	68.0	410	435	410	32,800	30,450	27,880
Total	1,090.9	1,022.3	949.5	376	401	401	410,588	410,253	381,026
US	1,248.6	1,166.9	1,084.6	367	391	388	457,814	456,041	420,879

[1] Summer potatoes combined with fall potatoes in 2005. [2] Estimates discontinued in 2005. [3] Estimates discontinued in 2004.

NASS, Crops Branch, (202) 720–2127.

Table 4-44.—Fall potatoes: Total stocks held by growers and local dealers, 15 States, crop of 2003 and 2004[1]

State	Crop of 2003						
	Dec. 1, 2003	Jan. 1, 2004	Feb. 1, 2004	Mar. 1, 2004	Apr. 1, 2004	May 1, 2004	June 1, 2004
	1,000 cwt.	1,000 cwt.	1,000 cwt.	1,000 cwt.	1,000 cwt.	1,000 cwt.	1,000 cwt.
CA	2,700	2,100	1,600	1,300	1,000	700	....
CO	17,500	15,300	12,600	10,700	7,500	4,300	2,600
ID	86,000	76,500	67,000	58,000	46,000	33,000	19,500
ME	13,500	12,100	10,500	8,900	6,500	4,100	2,300
MI	9,200	7,700	6,200	5,100	3,200	1,500	....
MN	14,000	12,400	11,100	9,600	7,600	5,500	3,500
MT	3,100	3,000	2,900	2,800	2,300	1,000	....
NE	6,500	5,600	4,600	....	....	....	....
NY	2,700	1,600	1,100	700	300	....	....
ND	19,400	17,300	15,000	13,000	11,200	8,300	5,000
OH	300	190	130	....	....	....	....
OR	18,000	15,800	13,500	11,000	8,400	5,500	2,900
PA	1,900	1,500	1,100	700	400	....	....
WA	51,000	44,000	38,000	29,500	21,500	15,000	7,000
WI	22,100	18,500	14,900	11,200	7,600	4,200	1,500
Other	....	....	....	3,780	2,610	1,900	1,720
15 State total	267,900	233,590	200,230	166,280	126,110	85,000	46,020

State	Crop of 2004						
	Dec. 1, 2004	Jan. 1, 2005	Feb. 1, 2005	Mar. 1, 2005	Apr. 1, 2005	May 1, 2005	June 1, 2005
	1,000 cwt.	1,000 cwt.	1,000 cwt.	1,000 cwt.	1,000 cwt.	1,000 cwt.	1,000 cwt.
CA	3,000	2,300	1,800	1,500	900	500	....
CO	18,000	15,800	13,700	11,900	9,000	6,000	3,600
ID	93,500	84,500	75,000	64,000	52,000	38,500	24,000
ME	15,000	12,800	11,100	9,400	7,500	5,000	2,900
MI	8,000	6,300	4,800	3,600	2,200	900	....
MN	13,000	11,600	10,200	8,800	6,500	5,100	3,500
MT	3,500	3,400	3,350	2,400	1,900	500	....
NE	6,300	5,300	....	....	....	....	....
NY	2,300	1,600	1,200	700	300	....	....
ND	19,600	17,500	15,300	13,100	10,300	7,200	4,800
OH	200	100	....	....	....	....	....
OR	17,000	14,500	11,800	9,000	6,200	3,900	2,100
PA	1,600	1,100	900	700	500	....	....
WA	50,000	43,000	36,500	29,000	22,000	15,500	8,000
WI	20,100	16,900	13,600	10,700	7,200	3,800	1,600
Other	....	....	4,240	3,220	2,400	1,650	1,200
15 State total	271,100	236,700	203,490	168,020	128,900	88,550	51,700

[1] Blank States combined into Other.
NASS, Crops Branch, (202) 720-2127.

Table 4-45.—Frozen French fries: U.S. imports from principal suppliers for marketing years, 2000/2001–2004/2005 [1]

Countries	2000/2001	2001/2002	2002/2003	2003/2004	2004/2005
	Metric tons	Metric tons	Metric tons	Metric tons	Metric tons
Canada	493,680	605,912	623,328	775,508	724,015
Other	1,387	1,664	3,667	3,556	3,248
Total [2]	495,067	607,576	626,995	779,065	727,263

[1] July through June. [2] Totals may not add due to rounding

FAS, Horticultural and Tropical Products Division, (202) 720-3423. The division's Home Page is located at, www.fas.usda.gov/htp. You can also email the division at htp@fas.usda.gov. Note: Compiled from reports of the U.S. Department of Commerce. U.S. trade can be obtained from the following web site: U.S. agricultural trade database, www.fas.usda.gov/ustrade.

STATISTICS OF VEGETABLES AND MELONS

Table 4-46.—Potatoes: Utilization, United States, crop years 1997–2004

Item	1997	1998	1999	2000
SALES	1,000 cwt.	1,000 cwt.	1,000 cwt.	1,000 cwt.
Table stock	131,670	125,099	133,947	139,590
For processing:				
Chips and shoestring	48,130	51,504	52,941	52,405
Dehydration	48,389	55,521	50,830	54,332
Frozen french fries	131,628	142,933	140,196	146,869
Other frozen products	33,397	25,162	23,592	26,723
Canned potatoes	2,822	2,730	3,311	2,368
Other canned products (hash, stews, soups)	2,675	1,964	2,394	2,709
Starch and flour	1,311	1,585	1,310	1,966
Total	268,352	281,399	274,574	287,372
Other sales:				
Livestock feed [1]	3,603	3,111	3,141	14,265
Seed	25,808	24,845	25,336	23,345
Total	29,411	27,956	28,477	37,610
Total sales	429,433	434,454	436,998	464,572
NON-SALES				
Seed used on farms where grown	4,167	4,361	4,244	3,792
Household use	1,308	1,403	1,301	1,495
Shrinkage and loss	32,183	35,449	35,550	43,685
Total non-sales	37,658	41,213	41,095	48,972
Total production	467,091	475,667	478,093	513,544

Item	2001	2002	2003	2004
SALES	1,000 cwt.	1,000 cwt.	1,000 cwt.	1,000 cwt.
Table stock	122,552	131,889	133,143	130,418
For processing:				
Chips and shoestring	54,080	51,640	52,790	50,068
Dehydration	40,759	51,357	48,418	48,527
Frozen french fries	126,711	124,875	126,515	131,592
Other frozen products	23,598	28,951	23,870	23,003
Canned potatoes	2,590	2,744	3,086	2,843
Other canned products (hash, stews, soups)	1,722	2,089	1,168	998
Starch and flour	1,015	1,050	1,379	1,531
Total	250,475	262,706	257,226	258,562
Other sales:				
Livestock feed	3,496	3,044	2,005	1,942
Seed	24,537	24,005	24,603	22,915
Total	28,033	27,049	26,608	24,857
Total sales	401,060	421,644	416,977	413,837
NON-SALES				
Seed used on farms where grown	4,088	4,144	4,000	3,614
Household use	1,298	1,478	1,543	1,182
Shrinkage and loss [2]	31,227	30,905	35,294	37,408
Total non-sales	36,613	36,527	40,837	42,204
Total production	437,673	458,171	457,814	456,041

[1] Includes 6,872 thousand cwt sold for livestock feed under Government Diversion Program for 2000.   [2] Includes potatoes disposed of by the United Fresh Potato Growers of Idaho for 2004.

NASS, Crops Branch, (202) 720–2127.

Table 4-47.—Potatoes: Production, seed used, and disposition, by seasonal groups, crop of 2004

Season and State	Production	Total used for seed	Used on farms where produced		Sold
			For seed, feed, and household use	Shrinkage and loss	
	1,000 cwt.	*1,000 cwt.*	*1,000 cwt.*	*1,000 cwt.*	*1,000 cwt.*
Winter:					
CA	3,250	462	20	34	3,196
FL	1,568	156		9	1,559
Total	4,818	618	20	43	4,755
Spring:					
AZ	1,767	86	2	14	1,751
CA	8,313	400	47	756	7,510
FL	7,678	614	1	38	7,639
Hastings	5,760	450	1	23	5,736
Other	1,918	164		15	1,903
NC	2,700	203	24	108	2,568
TX	2,205	152	30	40	2,135
Total	22,663	1,455	104	956	21,603
Summer:					
AL	228	24	4	17	207
CA	2,450	186	24	25	2,401
CO	1,995	118	5	140	1,850
DE	806	45	4	23	779
IL	1,992	99	35	42	1,915
KS	1,360	98		54	1,306
MD	1,196	53	35	37	1,124
MO	1,922	120	1	19	1,902
NJ	594	29	3	18	573
NM [1]	340		1	11	328
TX	4,224	141	10	58	4,156
VA	1,200	70	1	42	1,157
Total	18,307	983	123	486	17,698
Fall:					
CA	3,648	202	1	292	3,355
CO	23,791	1,513	1,300	2,286	20,205
ID [2]	131,970	7,260	1,250	10,902	119,818
IN [3]	1,120		5	50	1,065
ME	19,065	1,188	190	4,900	13,975
MA	800	61	5	6	789
MI	13,650	860	194	1,656	11,800
MN	18,920	837	100	1,750	17,070
MT	3,551	264	156	276	3,119
NE	9,288	532	212	906	8,170
NV	2,881	110		350	2,531
NM [1]	1,720	106	10	110	1,600
NY	5,184	451	100	513	4,571
ND	26,765	1,755	240	2,825	23,700
OH	1,080	71	6	33	1,041
OR	19,775	892	241	1,264	18,270
Malheur County	2,444	99	5	64	2,375
Other counties	17,331	793	236	1,200	15,895
PA	2,640	253	39	216	2,385
RI	145	14		3	142
SD [4]					
UT [4]					
WA	93,810	3,850	240	5,600	87,970
WI	30,450	1,420	260	1,985	28,205
Total	410,253	21,639	4,549	35,923	369,781
US	456,041	24,695	4,796	37,408	413,837

[1] Summer potatoes combined with fall potatoes in 2005. "Total Used for Seed" for summer potatoes is included with fall potatoes. [2] Shrink and loss includes potatoes disposed of by the United Fresh Potato Growers in Idaho. [3] Estimates discontinued in 2005. "Total Used for Seed" not available. [4] Estimates discontinued in 2004.

NASS, Crops Branch, (202) 720–2127.

STATISTICS OF VEGETABLES AND MELONS

Table 4-48.—Potatoes,[1] white: United States exports by country of destination and imports by country of origin, 2001/2002–2004/2005

Item and country	Year beginning July			
	2001/2002	2002/2003	2003/2004	2004/2005[2]
	Metric tons	Metric tons	Metric tons	Metric tons
Exports				
Canada	267,206	259,296	175,137	173,598
Mexico	7,553	10,238	26,245	38,416
Other	23,278	15,658	29,535	33,711
Total[3]	298,037	285,192	230,917	245,725
Imports				
Certified seed:				
Canada	116,274	113,291	82,742	71,802
Others	76	39	17	2
Total	116,350	113,330	82,759	71,804

[1] Includes seed potatoes.   [2] July-June.   [3] Totals may not add due to rounding.
FAS, Horticultural and Tropical Products Division, (202) 720-3423. The division's Home Page is located at, www.fas.usda.gov/htp. You can also email the division at htp@fas.usda.gov. Note: Compiled from reports of the U.S. Department of Commerce. U.S. trade can be obtained from the following web site: U.S. agricultural trade database, www.fas.usda.gov/ustrade.

Table 4-49.—Potatoes (fresh): Foreign trade, United States, 1995–2004 [1]

Year beginning July	Imports for consumption	Domestic exports
	1,000 cwt.	*1,000 cwt.*
1995	10,741	5,501
1996	7,129	6,402
1997	10,481	7,172
1998	9,090	5,718
1999	9,094	6,541
2000	5,986	6,695
2001	8,644	6,571
2002	9,265	6,287
2003	7,611	5,091
2004	7,604	5,417

[1] Includes seed.
ERS, Specialty Crops Branch, (202) 694–5253. Compiled from reports of the U.S. Department of Commerce.

Table 4-50.—Spinach for fresh market: Area, production, and value per hundredweight, by States, 2003–2005

State	Area harvested			Production			Value per unit		
	2003	2004	2005[1]	2003	2004	2005[1]	2003	2004	2005[1]
	Acres	*Acres*	*Acres*	*1,000 cwt.*	*1,000 cwt.*	*1,000 cwt.*	*Dollars per cwt.*	*Dollars per cwt.*	*Dollars per cwt.*
AZ	5,200	6,000	6,400	780	1,050	1,090	22.10	23.90	30.50
CA	26,000	27,000	31,000	4,160	4,590	5,270	40.70	40.90	20.20
CO	1,400			126			29.20		
MD	780			47			35.00		
NJ	1,800	1,900	1,900	252	171	200	30.00	22.20	30.90
TX	1,700	2,000	2,100	204	250	210	38.30	38.50	15.40
Oth Sts[2]		2,700	2,800		205	231		33.10	36.20
US	36,880	39,600	44,200	5,569	6,266	7,001	37.20	37.20	22.50

[1] Preliminary.   [2] 2004 - CO and MD.
NASS, Crops Branch, (202) 720–2127.

Table 4-51.—Spinach for processing: Area, production, and value per ton, by States, 2003–2005

State	Area harvested			Production			Value per unit		
	2003	2004	2005[1]	2003	2004	2005[1]	2003	2004	2005[1]
	Acres	Acres	Acres	Tons	Tons	Tons	Dollars per ton	Dollars per ton	Dollars per ton
CA	10,600	7,800	6,000	84,410	87,360	77,300	110.00	118.00	108.00
Other States[2]	3,500	4,600	3,500	35,720	42,860	19,570	99.40	112.00	113.00
US	14,100	12,400	9,500	120,130	130,220	96,870	107.00	116.00	109.00

[1] Preliminary. [2] 2003-2005 - NJ and TX.
NASS, Crops Branch, (202) 720–2127.

Table 4-52.—Sweet Potatoes: Area, yield, production, season average price per hundredweight received by farmers, and value, United States, 1996–2005

Year	Area harvested	Yield per acre	Production	Price[1]	Value of production
	1,000 acres	Cwt.	1,000 cwt.	Dollars	1,000 dollars
1996	83.7	158	13,216	14.40	190,529
1997	82.1	162	13,327	15.80	211,177
1998	83.7	148	12,365	15.30	189,532
1999	83.0	147	12,221	17.60	214,754
2000	94.8	145	13,780	15.30	210,351
2001	93.6	155	14,515	15.30	222,658
2002	82.3	156	12,799	16.80	214,650
2003	92.6	172	15,891	19.20	305,448
2004	92.8	174	16,112	17.50	281,559
2005	87.8	179	15,747	19.60	309,090

[1] Obtained by weighting State prices by production.
NASS, Crops Branch, (202) 720–2127.

Table 4-53.—Sweet Potatoes: Area, production, and season average price per hundredweight received by farmers, by States, 2003–2005

State	Area harvested			Production			Price for crop of-		
	2003	2004	2005	2003	2004	2005	2003	2004	2005
	1,000 acres	1,000 acres	1,000 acres	1,000 cwt.	1,000 cwt.	1,000 cwt.	Dollars	Dollars	Dollars
AL	2.5	2.3	2.5	475	380	375	25.50	20.80	16.20
CA	10.7	11.5	11.7	3,210	3,220	3,510	25.20	25.00	29.50
LA	18.0	15.5	17.0	3,150	2,325	2,465	20.00	17.70	18.70
MS	13.6	15.3	16.6	2,380	2,601	2,988	20.80	17.70	20.90
NJ	1.1	1.2	1.2	138	168	156	25.80	26.30	23.70
NC	42.0	43.0	35.0	5,880	6,880	5,950	14.50	13.50	14.00
SC	1.0	0.8	0.9	150	96	135	12.50	15.00	15.90
TX	3.2	2.8	2.6	448	392	130	19.00	16.80	10.40
VA	0.5	0.4	0.3	60	50	38	12.20	12.80	11.30
US	92.6	92.8	87.8	15,891	16,112	15,747	19.20	17.50	19.60

NASS, Crops Branch, (202) 720–2127.

Table 4-54.—Taro: Area, yield, total production, price, and value, Hawaii, 1996–2005

Year	Total area	Yield per acre [1]	Production	Price per pound	Value of production
	Acres	1,000 pounds	1,000 pounds	Dollars	1,000 dollars
1996	530		5,700	0.490	2,793
1997	450		5,500	0.510	2,805
1998	490		6,000	0.530	3,180
1999	500		6,800	0.530	3,604
2000	470		7,000	0.530	3,710
2001	440		6,400	0.530	3,392
2002	430		6,100	0.540	3,294
2003	420		5,000	0.540	2,700
2004	370		5,200	0.540	2,808
2005 [2]	360		4,000	0.540	2,160

[1] Yield not estimated.   [2] Preliminary.
NASS, Crops Branch, (202) 720–2127.

Table 4-55.—Tomatoes: Foreign trade, United States,1995–2004

Year beginning July	Imports			Domestic exports [2]				
	Fresh	Canned [1]	Paste	Fresh	Canned whole	Catsup and sauces	Paste	Juice [3]
	1,000 pounds	1,000 pounds	1,000 pounds	1,000 pounds	1,000 pounds	1,000 pounds	1,000 pounds	1,000 pounds
1995	1,702,019	221,894	33,590	288,021	59,312	265,503	193,215	51,002
1996	1,678,128	230,685	31,818	307,555	82,467	279,845	284,377	24,455
1997	1,794,808	281,251	23,601	332,416	55,165	319,960	298,407	23,050
1998	1,618,343	305,598	167,627	311,056	69,571	329,251	173,369	27,290
1999	1,596,470	189,526	46,790	356,676	85,641	334,125	198,029	29,417
2000	1,885,424	238,248	32,717	398,458	77,988	355,414	215,569	39,324
2001	1,708,004	457,088	40,729	375,744	78,828	367,393	206,113	14,480
2002	2,114,478	398,645	24,482	324,097	78,082	389,279	250,924	2,956
2003	1,984,044	386,956	15,681	333,895	82,375	417,271	280,244	3,588
2004	1,985,717	437,732	9,536	365,022	96,458	393,949	302,342	3,739

[1] Includes all canned tomato and tomato product imports except paste, and is on a product-weight-basis.   [2] Includes exports for military-civilian feeding abroad.   [3] Converted to pounds from liters.
ERS, Specialty Crops Branch, (202) 694–5253. Compiled from reports of the U.S. Department of Commerce.

Table 4-56.—Tomatoes, commercial crop: Area, yield, production, value per hundredweight and per ton, and total value, United States, 1996–2005

Year	For fresh market					For processing				
	Area harvested	Yield per acre	Production	Value [1]		Area harvested	Yield per acre	Production	Value [2]	
				Per cwt.	Total				Per ton	Total
	Acres	Cwt.	1,000 cwt.	Dollars	1,000 dollars	Acres	Tons	Tons	Dollars	1,000 dollars
1996	120,640	279	33,634	28.20	947,031	339,140	33.64	11,407,301	62.30	711,043
1997	115,190	285	32,777	31.70	1,040,382	283,390	35.19	9,973,259	60.70	604,905
1998	121,710	268	32,628	35.20	1,149,713	299,960	31.34	9,402,010	65.30	613,954
1999	132,880	276	36,735	25.90	951,046	350,410	36.63	12,836,020	71.10	912,988
2000	123,170	306	37,665	30.80	1,159,590	289,600	37.49	10,858,240	59.80	649,066
2001	124,250	286	35,527	30.40	1,080,166	274,860	33.65	9,248,720	59.20	547,473
2002	129,020	307	39,588	31.60	1,252,801	312,200	37.38	11,670,820	58.20	679,823
2003	121,700	292	35,578	37.40	1,332,361	293,920	33.41	9,819,710	58.70	576,441
2004	131,100	292	38,346	37.50	1,439,197	300,620	40.80	12,266,410	58.60	719,285
2005 [3] ..	129,800	304	39,462	41.50	1,637,394	282,040	36.17	10,200,120	61.00	622,143

[1] Price and value of f.o.b. basis.   [2] Price and value at processing plant door.   [3] Preliminary.
NASS, Crops Branch, (202) 720–2127.

Table 4-57.—Tomatoes, commercial crop: Area, production, and value per hundredweight and per ton, by States, 2003–2005 [1]

Utilization and State	Area harvested			Production			Value per unit		
	2003	2004	2005 [2]	2003	2004	2005 [2]	2003	2004	2005 [2]
FOR FRESH MARKET	Acres	Acres	Acres	1,000 cwt.	1,000 cwt.	1,000 cwt.	Dollars per cwt.	Dollars per cwt.	Dollars per cwt.
AL	1,200	1,200	1,100	330	342	319	28.80	34.80	31.00
AR	1,200	1,300	1,200	384	137	414	46.00	32.00	43.00
CA	34,000	42,000	42,600	10,200	13,020	11,928	35.90	39.20	31.00
FL	43,000	42,000	42,000	14,190	15,120	15,540	38.80	33.10	51.80
GA	4,500	5,800	6,300	1,530	986	2,142	31.50	45.00	35.00
IN	1,600	1,700	1,500	248	272	225	69.10	77.70	62.30
MD	1,000	1,200	....	90	156	....	42.00	49.00	....
MI	2,200	2,100	2,200	484	546	616	34.00	48.00	37.00
NJ	3,100	3,000	3,000	682	690	600	41.00	37.00	41.50
NY	2,300	2,400	2,000	322	360	360	80.60	63.50	59.60
NC	2,800	2,000	2,500	896	620	800	30.00	29.00	28.00
OH	6,600	6,700	6,600	1,155	1,106	2,145	41.50	44.80	34.90
PA	4,200	3,700	3,800	441	555	513	33.00	45.70	51.00
SC	3,300	3,500	3,000	1,023	1,050	390	31.50	26.50	36.10
TN	4,600	5,900	5,200	1,610	1,180	1,248	37.00	34.00	34.00
TX	1,300	1,100	1,200	169	116	150	45.00	65.00	57.50
VA	4,800	5,500	5,600	1,824	2,090	2,072	33.00	45.90	42.60
US	121,700	131,100	129,800	35,578	38,346	39,462	37.40	37.50	41.50
FOR PROCESSING	Acres	Acres	Acres	Tons	Tons	Tons	Dollars per ton	Dollars per ton	Dollars per ton
CA	274,000	281,000	264,000	9,252,000	11,672,000	9,600,000	57.20	57.40	59.60
IN	8,200	8,300	7,900	202,290	274,810	266,470	86.80	85.80	84.80
MI	3,300	3,500	....	125,400	108,500	....	83.00	81.00	81.00
OH	5,700	6,200	5,800	173,280	177,320	175,280	77.60	78.40	81.80
Other states [3]	2,720	1,620	4,340	66,740	33,780	158,370	87.10	90.10	82.40
US	293,920	300,620	282,040	9,819,710	12,266,410	10,200,120	58.70	58.60	61.00
Grand total	415,620	431,720	411,840	11,598,610	14,183,710	12,173,220	....	....	....

[1] Cherry, grape, tomatillo, and greenhouse tomatoes are exclued.   [2] Preliminary.   [3] 2003 - MD, NJ, and PA. 2004 - MD and NJ. 2005 - MI, MD, and NJ. PA estimates discontinued in 2004.

NASS, Crops Branch, (202) 720–2127.

STATISTICS OF VEGETABLES AND MELONS

Table 4-58.—Vegetables and melons, fresh: Total reported domestic rail, truck, and air shipments, 2004

Commodity	Jan.	Feb.	Mar.	Apr.	May	June	July	Aug.	Sep.	Oct.	Nov.	Dec.	Total
	1,000 cwt.	1,000 cwt.	1,000 cwt.	1,000 cwt.	1,000 cwt.	1,000 cwt.	1,000 cwt.	1,000 cwt.	1,000 cwt.	1,000 cwt.	1,000 cwt.	1,000 cwt.	1,000 cwt.
Vegetables:													
Artichokes	12	43	115	107	72	40	52	29	30	20	22	37	579
Asparagus	4	20	354	480	276	116	5	....	....	....	....	....	1,255
Beans	145	245	333	312	360	186	122	144	67	137	262	126	2,439
Broccoli	775	640	687	650	641	621	570	535	470	542	611	730	7,472
Cabbage	1,319	1,160	1,583	1,071	1,160	605	502	765	828	920	1,030	1,293	12,236
Carrots	753	721	879	811	746	788	667	639	688	725	770	779	8,966
Cauliflower	525	347	514	472	402	309	320	299	318	345	386	370	4,607
Celery	1,574	1,421	1,547	1,310	1,387	1,430	1,249	1,264	1,225	1,350	1,942	1,593	17,292
Chinese cabbage	104	90	111	90	93	104	99	91	84	88	83	82	1,119
Corn, sweet	261	324	676	1,700	2,800	2,357	455	182	174	240	188	354	9,711
Cucumbers	94	31	197	501	882	583	514	646	584	461	498	222	5,213
Eggplant	25	9	7	56	136	205	44	....	17	81	70	51	701
Endive	25	22	22	19	14	11	10	9	8	9	10	20	179
Escarole	27	22	24	23	13	11	10	8	7	9	9	24	187
Greens	365	313	307	301	214	79	73	53	59	98	296	358	2,516
Lettuce, iceberg	2,640	2,358	3,117	3,261	3,256	3,152	3,199	3,260	3,313	3,187	3,310	3,026	37,079
Lettuce, other	304	289	350	352	330	324	299	294	299	313	340	354	3,848
Lettuce, Processed	1,348	1,140	1,358	1,986	170	205	289	210	70	793	1,482	1,029	10,080
Lettuce, Romaine	1,127	1,043	1,288	1,212	1,052	1,051	1,022	1,093	1,124	1,115	1,087	1,266	13,480
Misc Asian Vegetables	73	59	66	52	56	64	59	50	53	54	56	53	695
Misc Herbs	53	56	62	46	43	50	47	42	47	50	56	57	609
Onions, dry	3,702	3,047	2,938	2,985	3,916	3,935	4,216	4,214	4,259	3,892	3,769	3,864	44,737
Onions, green	11	11	12	15	47	75	83	70	72	39	22	15	472
Okra	....	....	....	8	16	14	11	6	5	10	8	1	79
Parsley	37	38	41	38	27	24	21	19	18	18	22	39	342
Peppers, bell	522	570	929	1,212	1,306	1,446	760	739	757	649	641	677	10,208
Peppers, other	38	33	49	38	50	4	....	....	....	3	21	40	276
Potatoes, table	9,611	7,917	9,245	8,916	9,404	9,117	7,938	8,474	8,767	9,247	9,414	9,413	107,463
Potatoes, chipper	3,353	2,453	2,892	2,982	4,239	4,311	4,665	3,271	3,129	3,920	3,388	3,636	42,239
Potatoes, seed	563	756	2,768	6,353	4,101	126	....	....	17	123	89	232	15,128
Radishes	62	50	60	57	39	13	12	12	12	12	40	81	450
Rhubarb	....	....	2	12	12	4	5	4	....	....	....	....	39
Spinach	71	65	68	47	45	49	37	45	43	43	54	74	641
Squash	97	148	167	211	87	66	196	183	84	19	97	74	1,429
Sweet potatoes	310	303	336	337	268	236	246	230	281	392	751	390	4,080
Tomatoes	2,350	1,405	1,611	1,956	2,569	2,657	2,021	2,125	1,933	2,028	1,133	2,402	24,190
Tomatoes, cherry	68	36	50	48	77	52	62	40	36	25	26	57	577
Tomatoes, plum	299	171	217	219	331	377	303	186	270	245	133	440	3,191
Tomatoes,Grapes Type	205	184	179	146	209	172	90	40	35	44	79	186	1,569
Total	32,852	27,540	35,161	40,392	40,846	34,969	30,273	29,271	29,183	31,246	32,195	33,445	397,373
Melons:													
Cantaloups	....	....	....	29	3,105	4,050	2,827	2,584	1,748	1,661	992	15	17,011
Honeydews	....	....	....	1	494	658	625	622	454	315	151	3	3,323
Mixed and miscellaneous	....	....	....	....	50	175	84	80	11	....	....	1	401
Watermelons, seeded	2	....	....	167	1,905	2,892	1,654	871	170	18	32	1	7,712
Watermelons, seedless	4	....	....	234	3,791	6,442	4,040	2,912	585	63	30	1	18,102
Total	6	....	....	431	9,345	14,217	9,230	7,069	2,968	2,057	1,205	21	46,549
Grand total	32,858	27,540	35,161	40,823	50,191	49,186	39,503	36,340	32,151	33,303	33,400	33,466	443,922

AMS, Fruit and Vegetable Division, Market News Branch, (202) 720–3343.

Table 4-59.—Vegetables (fresh), melons, potatoes, sweet potatoes: Per capita civilian utilization (farm-weight basis), United States, 1996–2005[1]

Year	Cabbage	Cucumbers	Tomatoes[2]	Asparagus	Broccoli	Carrots	Head Lettuce	Leaf/romaine
	Pounds	Pounds	Pounds	Pounds	Pounds	Pounds	Pounds	Pounds
1996	8.3	5.9	17.4	0.6	4.5	12.4	21.6	5.8
1997	9.0	6.4	17.3	0.7	5.0	14.1	23.9	6.6
1998	8.4	6.5	18.5	0.7	5.0	9.5	22.3	6.6
1999	7.6	6.7	19.1	0.9	6.2	9.3	24.9	7.6
2000	8.9	6.4	19.0	1.0	5.9	9.2	23.5	8.4
2001	8.8	6.3	19.2	0.9	5.4	9.4	23.0	8.0
2002	8.3	6.5	20.3	1.0	5.4	8.4	22.5	9.6
2003	7.6	6.1	19.5	1.0	5.5	8.8	22.2	11.2
2004	8.3	6.3	19.3	1.0	5.9	8.9	22.5	12.0
2005[4]	8.3	6.5	19.4	1.1	5.9	8.7	22.1	12.2

Year	Snap beans	Garlic	Cauliflower	Celery	Sweet Corn	Onions	Spinach	Bell peppers
	Pounds	Pounds	Pounds	Pounds	Pounds	Pounds	Pounds	Pounds
1996	1.5	2.3	1.7	7.0	8.3	18.3	0.6	7.1
1997	1.3	2.0	1.8	6.5	8.3	18.8	1.1	6.4
1998	1.6	2.6	1.5	6.5	9.3	18.4	1.0	6.4
1999	1.9	3.3	1.8	6.5	9.1	18.5	1.0	6.7
2000	2.0	2.2	1.7	6.3	9.0	18.9	1.4	7.0
2001	2.2	2.4	1.5	6.4	9.2	18.5	1.1	6.9
2002	2.1	2.5	1.4	6.3	9.0	19.3	1.4	6.8
2003	2.0	2.8	1.6	6.3	9.5	19.5	1.8	6.9
2004	1.9	2.6	1.8	6.0	9.6	21.7	2.1	7.1
2005[4]	2.0	2.7	1.8	6.0	9.7	20.7	2.2	7.1

Year	Watermelon	Cantaloupe	Honeydew melons	Others[3]	Total vegetables and melons	Potatoes	Sweet potatoes
	Pounds	Pounds	Pounds	Pounds	Pounds	Pounds	Pounds
1996	16.6	10.3	2.0	5.9	158.1	49.9	4.3
1997	15.5	10.5	2.2	6.1	163.5	47.3	4.3
1998	14.3	10.6	2.3	6.6	158.6	46.9	3.8
1999	15.2	11.4	2.5	5.9	166.1	47.7	3.7
2000	13.8	11.1	2.3	14.0	172.0	47.1	4.2
2001	15.0	11.2	2.0	12.9	170.3	46.6	4.4
2002	14.0	11.1	2.2	13.3	171.4	44.3	3.8
2003	13.6	10.8	2.2	13.1	172.0	46.8	4.7
2004	13.0	9.5	2.2	13.7	175.4	45.8	4.7
2005[4]	13.0	10.4	2.2	13.5	175.5	42.8	4.4

[1] Fresh vegetable consumption computed for total commercial production for fresh market. Does not include production for home use. Consumption obtained by dividing the total apparent consumption by total July 1 population as reported by the Bureau of the Census. All data for calendar year.   [2] After 1996, includes an ERS estimate of domestically produced hothouse tomatoes. Hothouse imports included in all years.   [3] Includes artichokes, eggplant, radishes, brussels sprouts, squash, green limas, and escarole/endive. Beginning in 2000, also includes collards, mustard greens, turnip greens, kale, okra, and pumpkins.   [4] Preliminary.

ERS, Market and Trade Economics Division, Specialty Crops Branch, (202)694–5253.

STATISTICS OF VEGETABLES AND MELONS

Table 4-60.—Vegetables, frozen: Commercial pack, United States, 1995–2004

Commodity	1995	1996	1997	1998	1999
	1,000 pounds	1,000 pounds	1,000 pounds	1,000 pounds	1,000 pounds
Artichokes					
Asparagus	13,985	10,364	10,103	9,792	12,439
Beans, butter	7,664	4,577	5,170	7,298	8,515
Beans, green, regular cut	208,246	232,711	231,625	214,871	207,345
Beans, green, French cut	70,144	65,041	75,086	73,613	54,955
Beans, green, wax	7,120	7,592	7,324	8,927	7,132
Beans, green, Italian	17,267	17,711	18,364	18,531	23,537
Beans, whole	53,538	50,517	58,605	71,481	63,714
Beans, baby lima	101,295	85,053	97,598	101,757	83,381
Beans, lima, Fordhook	20,489	22,645	25,872	22,962	14,369
Broccoli	205,471	138,387	109,013	144,361	174,283
Brussels sprouts	25,519	30,154	26,436	33,062	35,264
Carrots	418,816	397,967	409,044	388,094	424,691
Cauliflower	64,207	48,541	44,590	44,607	42,445
Celery	32,279	40,343	32,985	29,001	32,966
Collards	17,647	17,709	22,244	26,915	17,685
Corn, cut	694,110	646,389	736,545	753,429	665,209
Corn-on-cob	414,662	383,840	464,942	435,271	404,831
Kale	3,664	2,671	4,213	2,936	2,319
Mushrooms	26,778	23,219	13,139	10,349	11,152
Mustard greens	10,302	11,294	57,725	85,762	90,004
Okra	64,810	60,847	115,980	122,289	123,414
Onions	99,639	111,021	25,363	27,684	16,328
Peas, blackeye	29,352	22,901	468,074	491,377	444,050
Peas, green	505,350	369,390	40,746	42,599	53,876
Peppers, green and red	54,155	44,858	8,381,370	8,723,707	8,765,173
Potato products	8,364,909	8,419,203	24,570	19,329	17,889
Pumpkin and cooked squash	28,097	20,058	7,505	6,558	6,353
Rhubarb	8,228	6,910	198,715	190,028	174,149
Spinach	185,640	182,999	46,579	42,976	34,842
Squash, summer	54,957	42,387	20,992	¹	¹
Sweet potatoes and yams	14,127	14,532	20,490	21,719	32,499
Turnip greens	18,203	19,112	25,119	19,485	²
Turnip greens with turnips	17,041	19,568	22,537	20,415	25,117
Miscellaneous vegetables	41,004	36,715	46,363	42,849	51,341
Total	11,898,715	11,607,226	11,895,026	12,253,884	12,121,317

Commodity	2000	2001	2002	2003	2004 ³
	1,000 pounds	1,000 pounds	1,000 pounds	1,000 pounds	1,000 pounds
Artichokes					
Asparagus	10,224	9,348	7,221	7,033	
Beans, butter	7,771	7,688	5,639	10,722	
Beans, green, regular cut	235,279	240,514	177,248	143,496	
Beans, green, French cut	52,454	55,503	44,836	37,832	
Beans, green, wax	8,512	8,341	6,867	7,576	
Beans, green, Italian	15,979	15,818	19,473	17,255	
Beans, whole	77,619	74,005	75,118	60,361	
Beans, baby lima	71,637	67,000	72,966	89,488	
Beans, lima, Fordhook	15,443	11,920	53,501	8,540	
Broccoli	138,944	137,705	135,219	97,739	
Brussels sprouts	22,583	19,951	22,258	17,388	
Carrots	412,744	206,085	261,789	246,396	
Cauliflower	40,405	35,550	22,403	33,066	
Celery	48,402	45,683	25,464	32,324	
Collards	26,949	23,420	27,959	26,020	
Corn, cut	666,617	622,984	610,664	557,800	
Corn-on-cob	372,794	343,516	343,367	333,783	
Kale	2,850	2,169	3,458	2,221	
Mustard greens	13,799	12,841	9,776	8,457	
Okra	75,770	70,735	66,781	35,447	
Onions ¹	108,891	107,592	110,022	59,627	
Peas, blackeye	26,652	24,973	10,090	12,446	
Peas, green	451,995	446,389	281,493	287,868	
Peppers, green and red	46,165	45,712	47,508	55,901	
Potato products	7,622,629	7,847,656	7,799,687	8,644,989	
Pumpkin and cooked squash	20,346	19,133	9,924	7,076	
Rhubarb	3,276	2,594	2,594	2,097	
Spinach	172,543	173,850	211,272	183,751	
Squash, summer	31,117	28,700	40,728	36,734	
Sweet potatoes and yams	¹	12,736	¹	¹	
Turnip greens	22,933	15,049	24,879	25,081	
Turnip greens with turnips	15,780	0	7,641	0	
Mushrooms	36,928	32,332	35,531	36,017	
Miscellaneous vegetables	39,467	55,383	55,383	94,739	
Total	10,915,497	10,822,875	10,763,430	11,219,270	

¹ Included with miscellaneous vegetables.   ² Included with turnip greens.   ³ Data not available.
ERS, Specialty Crops Branch, (202) 694–5253. Data from American Frozen Food Institute.

Table 4-61.—Vegetables, canning: Per capita utilization (farm weight), United States, 1996–2005

Year	Cabbage for kraut	Asparagus	Snap beans	Carrots	Green peas
	Pounds	Pounds	Pounds	Pounds	Pounds
1996	1.0	0.2	3.8	1.7	1.5
1997	1.4	0.2	3.6	1.5	1.5
1998	1.4	0.2	3.8	1.4	1.4
1999	1.2	0.2	3.7	1.4	1.4
2000	1.4	0.2	4.0	1.1	1.5
2001	1.3	0.2	3.8	1.1	1.4
2002	1.2	0.2	3.4	1.9	1.1
2003	1.1	0.2	3.7	1.2	1.3
2004	1.1	0.2	3.7	1.6	1.2
2005 [1]	1.1	0.2	3.8	1.4	1.2

Year	Tomatoes	Corn	Pickles	Other [2]	Total [3]
	Pounds	Pounds	Pounds	Pounds	Pounds
1996	73.4	10.4	4.1	6.8	102.9
1997	72.6	9.1	5.2	6.8	101.9
1998	74.0	9.2	4.0	6.8	102.2
1999	71.2	9.1	4.2	7.2	99.6
2000	70.1	9.0	4.9	7.8	100.0
2001	65.5	8.7	3.7	7.8	93.5
2002	69.3	7.8	5.4	8.1	98.4
2003	69.8	8.3	4.4	8.1	98.1
2004	70.4	8.2	4.6	8.8	99.8
2005 [1]	70.8	8.2	4.3	8.4	99.4

[1] Preliminary. [2] Includes beets, chile peppers (all uses), green lima beans and spinach. [3] Totals may not add due to rounding.
ERS, Specialty Crops Branch, (202) 694–5253.

Table 4-62.—Watermelon for fresh market: Area, production, and value per hundredweight, by States, 2003–2005

State	Area harvested			Production			Value per unit		
	2003	2004	2005 [1]	2003	2004	2005 [1]	2003	2004	2005 [1]
	Acres	Acres	Acres	1,000 cwt.	1,000 cwt.	1,000 cwt.	Dollars per cwt.	Dollars per cwt.	Dollars per cwt.
AL	3,300	3,200	2,900	545	288	203	7.40	7.20	9.50
AZ	5,900	6,900	6,700	2,655	3,280	3,350	11.80	7.80	15.20
AR	2,300	2,000	2,000	345	350	160	6.60	5.00	9.00
CA	11,700	13,500	13,000	5,616	6,615	6,370	12.10	10.60	6.40
DE	1,800	2,100	2,800	522	798	854	8.00	7.00	9.50
FL	24,000	25,000	26,000	7,200	8,000	8,190	8.60	8.40	15.50
GA	25,000	23,000	25,000	5,375	3,795	5,250	7.80	7.00	7.90
IN	7,000	7,200	7,300	2,590	2,448	2,774	9.40	8.10	7.20
LA [2]	1,800	1,300		234	130		7.40	6.30	
MD	1,800	1,900	2,400	396	456	672	8.00	10.00	13.00
MS	3,000	2,700	2,900	435	378	435	7.50	8.50	11.00
MO	4,700	4,400	3,500	1,387	924	963	5.50	4.10	6.60
NC	7,800	7,500	6,100	975	1,050	1,037	7.00	6.00	7.00
OK	6,000	5,000	5,000	840	700	725	6.70	8.40	9.40
SC	7,000	7,000	7,000	1,190	1,470	770	6.50	5.00	9.00
TX	35,000	27,500	22,300	7,700	6,050	5,798	8.80	10.00	12.80
VA	1,500	1,500	1,500	203	150	345	6.00	14.00	10.50
US	149,600	141,700	136,400	38,208	36,882	37,896	8.98	8.49	10.80

[1] Preliminary. [2] Estimates discontinued in 2005.
NASS, Crops Branch, (202) 720–2127.

Table 4-63.—Vegetables, freezing: Per capita utilization (farm weight basis), United States, 1996–2005

Year	Leafy, green, and yellow vegetables				
	Asparagus	Snap beans	Carrots	Peas	Broccoli
	Pounds	Pounds	Pounds	Pounds	Pounds
1996	0.1	1.9	2.8	1.9	2.5
1997	0.1	1.8	2.6	2.0	2.3
1998	0.1	2.0	2.8	1.9	2.1
1999	0.1	2.0	2.4	2.0	2.1
2000	0.1	1.8	2.7	2.1	2.3
2001	0.1	1.9	1.5	2.0	2.0
2002	0.1	1.8	1.9	1.7	2.1
2003	0.1	1.9	1.5	1.9	2.6
2004	0.1	1.9	1.6	1.6	2.7
2005 [1]	0.1	1.9	1.6	1.8	2.6

Year	Cauliflower	Sweet Corn	Other [2]	Total vegetables excluding potatoes	Potato products	Grand total
	Pounds	Pounds	Pounds	Pounds	Pounds	Pounds
1996	0.5	10.4	3.1	23.2	60.2	83.4
1997	0.4	10.1	3.0	22.3	57.8	80.1
1998	0.8	9.8	2.9	22.3	58.1	80.4
1999	0.5	10.1	3.2	22.5	58.5	81.0
2000	0.6	9.0	3.2	21.8	57.5	79.3
2001	0.5	9.3	3.2	20.4	58.2	78.6
2002	0.3	9.3	4.3	21.5	55.2	76.7
2003	0.4	9.0	3.8	21.1	57.2	78.3
2004	0.4	9.1	4.3	21.6	57.3	78.9
2005 [1]	0.4	8.9	4.6	21.8	55.7	77.5

[1] Preliminary.   [2] Includes green lima beans, spinach, and miscellaneous freezing vegetables.   [3] Totals may not sum due to rounding.
ERS, Specialty Crops Branch, (202) 694–5253.

Table 4-64.—Commercially produced vegetables: Per capita utilization, United States, 1996–2005 [1]

Year	Total fresh and processed	Farm weight equivalent				Percentage of annual total			
		Fresh [2]	Processed [3]			Fresh	Processed		
			Total	Canning	Freezing		Total	Canning	Freezing
	Pounds	Pounds	Pounds	Pounds	Pounds	Percent	Percent	Percent	Percent
1996	284.2	158.1	126.1	102.9	23.2	55.6	44.4	36.2	8.2
1997	287.7	163.5	124.2	101.9	22.3	56.8	43.2	35.4	7.8
1998	283.1	158.6	124.5	102.2	22.3	56.0	44.0	36.1	7.9
1999	288.2	166.1	122.1	99.6	22.5	57.6	42.4	34.6	7.8
2000	293.8	172.0	121.8	100.0	21.8	58.5	41.5	34.0	7.4
2001	284.2	170.3	113.9	93.5	20.4	59.9	40.1	32.9	7.2
2002	291.3	171.4	119.9	98.4	21.5	58.8	41.2	33.8	7.4
2003	291.2	172.0	119.2	98.1	21.1	59.1	40.9	33.7	7.2
2004	296.8	175.4	121.4	99.8	21.6	59.1	40.9	33.6	7.3
2005 [4]	296.7	175.5	121.2	99.4	21.8	59.2	40.8	33.5	7.3

[1] Excludes potatoes, sweet potatoes, pulses, dehydrating onions, and mushrooms.   [2] See table 4-59 for items included. Includes melons.   [3] See table 4-60 for items included.   [4] Preliminary
ERS, Market and Trade Economics Division, Specialty Crops Branch, (202) 694–5253.

Table 4-65.—Frozen Vegetables and potato products: Cold storage holdings, end of month, United States, 2004 and 2005

Month	Asparagus		Green beans, regular cut		Green beans, French cut		Green beans, total	
	2004	2005	2004	2005	2004	2005	2004	2005
	1,000 pounds	*1,000 pounds*	*1,000 pounds*	*1,000 pounds*	*1,000 pounds*	*1,000 pounds*	*1,000 pounds*	*1,000 pounds*
January	7,600	8,774	113,562	150,711	21,898	20,915	135,460	171,626
February	6,933	8,206	102,991	143,724	19,252	17,634	122,243	161,358
March	6,435	7,799	80,039	118,892	15,680	16,416	95,719	135,308
April	7,385	6,968	61,740	100,815	12,383	14,193	74,123	115,008
May	10,825	8,454	58,805	86,226	9,295	10,714	68,100	96,940
June	14,059	13,943	51,146	76,655	6,595	8,157	57,741	84,812
July	15,485	13,670	112,861	138,773	17,745	17,026	130,606	155,799
August	13,811	13,501	189,833	203,449	27,046	26,560	216,879	230,009
September ...	12,007	11,714	248,212	256,015	35,662	33,938	283,874	289,953
October	10,933	10,072	227,307	244,160	30,875	32,844	258,182	277,004
November	9,562	9,629	195,822	209,597	26,595	26,884	222,417	236,481
December	9,280	9,029	151,229	171,319	22,925	22,341	174,154	193,660

Month	Broccoli spears		Broccoli, chopped & cut		Broccoli, total		Brussels sprouts	
	2004	2005	2004	2005	2004	2005	2004	2005
	1,000 pounds	*1,000 pounds*	*1,000 pounds*	*1,000 pounds*	*1,000 pounds*	*1,000 pounds*	*1,000 pounds*	*1,000 pounds*
January	32,890	44,576	46,069	57,590	78,959	102,166	20,655	23,246
February	35,524	45,268	46,841	65,779	82,365	111,047	18,095	21,964
March	40,269	42,961	52,822	63,998	93,091	106,959	15,410	21,014
April	45,533	49,265	55,759	66,456	101,292	115,721	14,160	18,952
May	49,910	49,535	59,857	66,942	109,767	116,477	13,037	17,198
June	50,313	45,581	58,844	62,382	109,157	107,963	11,588	16,027
July	50,950	46,889	60,576	61,798	111,526	108,687	10,036	13,388
August	46,901	46,279	59,249	59,579	106,150	105,858	8,701	10,819
September ...	44,367	43,342	55,808	60,840	100,175	104,182	7,219	9,768
October	39,441	39,762	58,362	62,046	97,803	101,808	11,397	11,802
November	31,257	38,217	53,911	58,716	85,168	96,933	17,542	15,982
December	35,882	35,814	57,607	62,561	93,489	98,375	23,519	21,660

Month	Fordhook lima beans		Baby lima beans		Mixed vegetables		Okra	
	2004	2005	2004	2005	2004	2005	2004	2005
	1,000 pounds	*1,000 pounds*	*1,000 pounds*	*1,000 pounds*	*1,000 pounds*	*1,000 pounds*	*1,000 pounds*	*1,000 pounds*
January	5,790	7,325	40,789	47,593	42,643	49,052	31,709	27,900
February	4,959	6,718	37,051	46,818	43,720	49,373	27,687	24,880
March	3,041	6,379	33,405	38,601	46,802	48,271	21,607	16,146
April	2,755	4,720	29,669	34,319	46,368	50,890	18,306	11,687
May	2,483	4,018	27,102	30,842	51,760	46,163	20,104	16,134
June	1,960	3,502	21,715	27,360	56,156	49,493	33,110	28,097
July	1,475	2,773	18,634	24,823	55,263	49,049	33,630	35,936
August	6,633	7,575	30,687	27,593	53,226	52,889	42,175	36,894
September ...	12,293	10,494	53,479	41,422	52,368	56,989	45,452	40,637
October	10,986	9,142	56,332	50,118	53,409	55,774	42,607	39,889
November	9,701	8,161	51,497	45,908	46,466	54,291	38,327	34,506
December	8,450	7,687	48,955	40,867	46,892	55,236	33,875	31,734

Month	Carrots, diced		Carrots, other		Carrots, total		Cauliflower	
	2004	2005	2004	2005	2004	2005	2004	2005
	1,000 pounds	*1,000 pounds*	*1,000 pounds*	*1,000 pounds*	*1,000 pounds*	*1,000 pounds*	*1,000 pounds*	*1,000 pounds*
January	85,435	118,508	139,613	129,732	225,048	248,240	27,295	37,000
February	81,668	107,143	129,110	123,374	210,778	230,517	24,605	36,703
March	71,007	93,928	109,696	112,779	180,703	206,707	19,182	30,410
April	65,481	86,375	98,657	107,773	164,138	194,148	15,871	24,709
May	57,651	76,614	90,010	102,926	147,661	179,540	14,254	21,656
June	49,477	63,871	75,561	93,702	125,038	157,573	12,518	19,916
July	36,316	52,669	72,278	89,532	108,594	142,201	12,417	17,539
August	32,919	50,575	69,482	82,673	102,401	133,248	13,956	18,824
September ...	44,421	45,283	76,685	87,458	121,106	132,741	14,365	23,935
October	89,750	92,931	137,590	150,160	227,340	243,091	25,736	33,915
November	128,713	138,986	146,433	162,871	275,146	301,857	31,162	44,002
December	117,745	128,877	140,440	150,317	258,185	279,194	37,123	43,836

See end of table.

Table 4-65.—Frozen Vegetables and potato products: Cold storage holdings, end of month, United States, 2004 and 2005—Continued

Month	Corn, cut		Corn, cob		Corn, total		Onion rings	
	2004	2005	2004	2005	2004	2005	2004	2005
	1,000 pounds	1,000 pounds	1,000 pounds	1,000 pounds	1,000 pounds	1,000 pounds	1,000 pounds	1,000 pounds
January	458,556	448,809	260,167	226,424	718,723	675,233	6,395	6,486
February	408,974	414,544	223,367	199,547	632,341	614,091	6,831	5,689
March	362,149	362,385	194,069	161,320	556,218	523,705	6,420	5,656
April	314,684	312,820	163,456	130,932	478,140	443,752	6,730	6,372
May	268,781	268,295	133,870	104,806	402,651	373,101	9,525	6,135
June	227,385	234,965	109,612	78,525	336,997	313,490	8,426	6,181
July	233,685	224,782	108,509	78,715	342,194	303,497	7,841	5,171
August	368,960	366,371	190,176	155,597	559,136	521,968	7,465	4,422
September ...	595,374	521,330	295,913	256,009	891,287	777,339	6,739	4,555
October	620,274	580,345	303,479	321,361	923,753	901,706	7,120	4,752
November	576,596	519,738	275,419	295,810	852,015	815,548	6,651	3,734
December	502,025	464,177	242,271	263,818	744,296	727,995	6,587	4,757

Month	Onions, other		Blackeye peas		Green peas		Peas & carrots mixed	
	2004	2005	2004	2005	2004	2005	2004	2005
	1,000 pounds	1,000 pounds	1,000 pounds	1,000 pounds	1,000 pounds	1,000 pounds	1,000 pounds	1,000 pounds
January	26,120	33,942	3,414	2,508	156,435	205,775	6,698	6,120
February	27,249	38,502	3,448	1,845	130,864	182,590	6,007	5,629
March	30,304	43,039	3,367	2,253	104,346	159,511	6,189	5,147
April	32,634	44,207	3,521	2,325	78,373	136,455	5,655	4,948
May	32,000	41,283	2,920	2,115	65,225	115,936	5,262	4,998
June	28,982	39,424	3,004	1,878	206,182	227,162	5,674	5,416
July	27,004	42,060	2,702	2,567	361,946	354,835	6,726	5,778
August	27,423	40,412	3,629	2,712	349,150	338,130	6,596	6,003
September ...	29,087	36,163	3,095	3,376	326,667	306,054	5,753	5,926
October	30,978	38,999	2,713	3,143	300,471	295,896	5,758	5,756
November	33,138	41,956	3,296	2,954	263,533	251,401	5,088	4,745
December	33,278	43,349	2,693	2,348	230,326	214,851	5,641	5,227

Month	Spinach		Squash		Southern greens		Other vegetables	
	2004	2005	2004	2005	2004	2005	2004	2005
	1,000 pounds	1,000 pounds	1,000 pounds	1,000 pounds	1,000 pounds	1,000 pounds	1,000 pounds	1,000 pounds
January	40,399	34,984	39,251	50,968	19,649	17,120	306,091	345,554
February	36,922	36,922	36,041	50,022	19,251	17,718	283,641	343,449
March	55,321	42,511	31,602	44,709	18,710	20,479	263,862	303,141
April	81,836	50,742	26,928	40,575	19,289	18,886	260,748	278,099
May	88,861	74,009	29,224	39,832	16,397	17,935	259,857	272,386
June	83,316	87,983	31,435	39,164	16,330	18,905	248,596	267,346
July	75,642	77,762	36,240	39,708	17,834	17,538	294,332	290,025
August	67,454	63,870	40,215	41,600	17,322	15,315	323,971	332,690
September ...	56,783	56,152	45,306	51,596	14,739	14,446	357,852	379,018
October	47,681	54,077	51,971	62,783	14,048	14,005	393,788	393,021
November	38,776	45,803	50,078	62,949	16,125	12,706	368,056	371,003
December	34,724	42,464	45,724	61,518	17,328	13,750	363,579	353,315

Month	Total vegetables		French fries		Other frozen potatoes		Total frozen potatoes	
	2004	2005	2004	2005	2004	2005	2004	2005
	1,000 pounds	1,000 pounds	1,000 pounds	1,000 pounds	1,000 pounds	1,000 pounds	1,000 pounds	1,000 pounds
January	1,939,123	2,101,612	906,579	920,765	260,680	248,037	1,167,259	1,168,802
February	1,761,031	1,994,041	930,547	903,956	276,811	248,862	1,207,358	1,152,818
March	1,591,734	1,767,745	919,539	855,761	272,912	237,905	1,192,451	1,093,666
April	1,467,921	1,603,483	874,233	926,504	284,469	247,822	1,158,702	1,174,326
May	1,377,015	1,485,152	894,667	940,430	291,270	237,642	1,185,937	1,178,072
June	1,411,984	1,515,635	864,384	947,971	264,270	242,493	1,128,654	1,190,464
July	1,670,127	1,702,806	858,246	905,112	258,824	249,766	1,117,070	1,154,878
August	1,996,980	2,004,332	877,320	893,612	249,711	227,719	1,127,031	1,121,331
September ...	2,439,646	2,356,460	915,922	932,179	262,726	248,179	1,178,648	1,180,358
October	2,573,006	2,606,753	1,000,376	962,477	274,491	237,494	1,274,867	1,199,971
November	2,423,744	2,460,549	959,360	905,274	259,985	217,183	1,219,345	1,122,457
December	2,218,098	2,250,852	838,438	848,595	236,381	202,504	1,074,819	1,051,099

NASS, Livestock Branch, (202) 720–3570.

CHAPTER V
STATISTICS OF FRUITS, TREE NUTS, AND HORTICULTURAL SPECIALTIES

For most fruits, production is estimated at two levels—total and utilized. Total production is the quantity of fruit harvested plus quantities which would have been acceptable for fresh market or processing but were not harvested or utilized because of economic and other reasons. Utilized production is the amount sold plus the quantities used on farms where grown and quantities held in storage. The difference between total and utilized production is the quantity of marketable fruit not harvested and fruit harvested but not sold or utilized because of economic and other reasons. Production relates to the crop produced on all farms, except for apples and strawberries. In accordance with Congressional enactment, the Department's estimates of apple production since 1938 have related only to commercial production. The estimates for strawberries cover production on area grown primarily for sale. Statistics on utilization of fruit by commercial processors refer to first utilization, not necessarily final utilization. For example, frozen fruit includes fruit which may later be used for preserves.

The price shown for each crop is a marketing year average price for all methods of sales. Prices for most fresh fruit are the average prices producers received at the point of first sale, commonly referred to as the "average price as sold." Since the point of first sale is not the same for all producers, prices for the various methods of sale are weighted by the proportionate quantity sold. For example, if in a given State part of the fruit crop is sold f.o.b. packed by growers, part sold as bulk fruit at the packinghouse door, and some sold retail at roadside stands, the fresh fruit average price as sold is a weighted average of the average price for each method of sale.

The annual estimates are checked and adjusted at the end of each marketing season on the basis of shipment and processing records from transportation agencies, processors, cooperative marketing associations, and other industry organizations. The estimates are reviewed (and revised if necessary) at 5-year intervals, when the Census of Agriculture data become available. The Department's available statistics are limited to the major tree fruits and nuts and to grapes, cranberries, and strawberries, and exclude some States where census data indicate production is of only minor importance.

Table 5-1.—Fruits and planted nuts: Bearing area, United States, 1996–2005

Year	Citrus fruits [1]	Major deciduous fruits [2]	Miscellaneous fruits [3]	Planted nuts [4]	Fruits and planted nuts
	1,000 acres	*1,000 acres*	*1,000 acres*	*1,000 acres*	*1,000 acres*
1996	1,104.5	1,796.1	287.8	732.1	3,919.2
1997	1,152.5	1,810.9	292.2	748.6	4,004.2
1998	1,125.5	1,833.3	295.8	774.7	4,029.2
1999	1,114.3	1,866.2	298.0	801.1	4,079.6
2000	1,094.8	1,889.0	300.2	831.0	4,115.0
2001	1,086.9	1,843.9	293.8	858.8	4,083.4
2002	1,053.9	1,833.5	299.0	885.0	4,071.4
2003	1,031.9	1,829.5	297.0	896.8	4,055.1
2004	983.6	1,808.3	296.9	926.2	4,015.1
2005	939.0	1,796.1	232.7	943.3	3,911.8

[1] Oranges, tangerines, Temples, grapefruit, lemons, limes, and tangelos. Area is for the year of harvest. [2] Commercial apples, apricots, cherries, grapes, nectarines, peaches, pears, plums, and prunes. [3] Avocados, bananas, berries, cranberries, dates, figs, guavas, kiwifruit, olives, papayas, pineapples, and strawberries. [4] Almonds, hazelnuts, macadamia nuts, pistachios, and walnuts.

NASS, Crops Branch, (202) 720–2127.

Table 5-2.—Fruits: Total production in tons, United States, 1996-2005 [1]

Year	Apples, commercial crop [2]	Peaches	Pears	Grapes (fresh basis)	Sweet cherries	Tart cherries	Apricots	Figs (fresh basis)
	1,000 tons	1,000 tons	1,000 tons	1,000 tons	1,000 tons	1,000 tons	1,000 tons	1,000 tons
1996	5,191	1,052	821	5,554	154	136	79	46
1997	5,162	1,312	1,043	7,291	226	147	139	58
1998	5,823	1,190	990	5,820	197	174	119	52
1999	5,316	1,252	1,044	6,236	216	128	91	47
2000	5,291	1,276	993	7,688	208	144	97	56
2001	4,712	1,204	1,027	6,569	230	185	83	41
2002	4,262	1,268	890	7,339	181	31	90	53
2003	4,397	1,260	934	6,664	246	113	98	49
2004	5,225	1,307	877	6,240	283	107	101	51
2005 [3]	4,935	1,183	813	6,975	251	135	81	51

Year	Plums (CA)	Prunes (CA)	Prunes & Plums (ID,MI,OR,WA)	Olives	Strawberries [4]	Pineapples [4]	Avocados [5]	Nectarines
	1,000 tons	1,000 tons	1,000 tons	1,000 tons	1,000 tons	1,000 tons	1,000 tons	1,000 tons
1996	228	704	20	166	813	347	191	247
1997	246	655	26	104	814	324	178	264
1998	188	346	26	90	819	332	159	224
1999	196	516	23	142	916	352	183	274
2000	197	681	24	53	950	354	239	267
2001	210	420	21	134	826	323	223	275
2002	201	519	16	103	942	320	199	300
2003	209	578	16	118	1,078	300	233	273
2004	156	144	25	104	1,107	220	179	269
2005 [3]	171	274	9	139	1,161	212	NA	249

Year	Oranges [6]	Tangerines [6]	Grapefruit [6][7]	Lemons [6]	K-Early Citrus [6]	Limes [6]	Tangelos [6]	Temples [6]
	1,000 tons	1,000 tons	1,000 tons	1,000 tons	1,000 tons	1,000 tons	1,000 tons	1,000 tons
1996	11,426	349	2,718	992	7	14	110	97
1997	12,692	425	2,885	962	7	14	178	108
1998	13,670	360	2,593	897	2	19	128	101
1999	9,824	327	2,513	747	4	22	115	81
2000	12,997	458	2,763	840	5	26	99	88
2001	12,221	373	2,462	996	2	11	95	56
2002	12,374	420	2,424	801	1	7	97	70
2003	11,545	382	2,063	1,026	NA	NA	105	59
2004	12,872	417	2,165	798	NA	NA	45	63
2005 [3]	9,112	331	1,008	813	NA	NA	70	29

Year	Cranberries	Bananas [4]	Kiwifruit	Dates	Papayas [4]	Berries [8]	Guavas	Total
	1,000 tons	1,000 tons	1,000 tons	1,000 tons	1,000 tons	1,000 tons	1,000 tons	1,000 tons
1996	234	7	32	23	21	119	8	31,906
1997	275	7	35	21	19	199	8	35,785
1998	272	11	37	25	20	178	7	34,869
1999	318	12	27	22	21	200	5	31,170
2000	286	15	34	17	27	229	8	36,410
2001	267	14	26	20	28	216	5	33,278
2002	285	10	26	24	23	210	5	33,491
2003	310	11	25	18	21	227	3	26,341
2004	309	8	27	17	18	243	4	33,361
2005 [3]	311	(9)	42	17	16	253	NA	28,641

[1] For some crops in certain years, production includes some quantities unharvested for economic reasons or excess cullage fruit.　[2] Estimates of the commercial crop refer to production in orchards of 100 or more bearing-age trees.　[3] Preliminary.　[4] Utilized production only.　[5] Year of bloom.　[6] Year harvest was complete.　[7] Excludes economic abandonment in 1996 of 127,500 tons; in 1997 of 255,000 tons; and in 1998 of 255,000 tons.　[8] Wild Blueberries added in 1998.　[9] Estimates for 2005 not published to avoid disclosure of individual operations.　NA-not available.

NASS, Crops Branch, (202) 720-2127.

Table 5-3.—Apples, commercial crop:[1] Production and season average price per pound, by States, 2003–2005

State	Total production			Utilized production			Price[3] for crop of—		
	2003	2004	2005[2]	2003	2004	2005[2]	2003	2004	2005[2]
	Million pounds	Million pounds	Million pounds	Million pounds	Million pounds	Million pounds	Dollars	Dollars	Dollars
AZ	7.0	37.0	22.2	7.0	37.0	22.0	0.078	0.153	0.240
AR[4]	2.4	1.9	....	2.3	1.1	....	0.235	0.355	....
CA	450.0	355.0	370.0	440.0	355.0	370.0	0.178	0.149	0.220
CO	22.0	28.0	31.0	21.0	27.0	27.0	0.185	0.154	0.235
CT	21.5	19.5	15.0	20.0	18.5	14.5	0.371	0.395	0.391
GA	13.0	12.0	14.0	12.9	12.0	14.0	0.106	0.228	0.236
ID	70.0	90.0	80.0	70.0	90.0	80.0	0.202	0.118	0.187
IL	52.5	56.5	49.0	45.9	51.3	38.5	0.291	0.237	0.390
IN	51.0	60.0	50.0	48.0	58.0	37.5	0.263	0.219	0.220
IA	6.0	5.3	2.1	4.7	4.8	1.9	0.424	0.466	0.462
KS[4]	3.4	2.8	....	2.6	2.2	....	0.273	0.276	....
KY	7.5	7.7	5.5	7.1	7.0	4.7	0.327	0.368	0.346
ME	44.0	47.0	31.5	40.0	43.0	30.0	0.298	0.320	0.308
MD	40.0	34.1	41.0	38.4	33.0	40.8	0.156	0.136	0.151
MA	42.5	42.0	28.5	37.0	37.0	26.0	0.346	0.381	0.383
MI	890.0	730.0	790.0	890.0	730.0	790.0	0.117	0.123	0.127
MN	27.0	25.0	22.0	19.8	19.9	16.0	0.436	0.468	0.535
MO	40.0	48.0	49.0	40.0	47.0	47.0	0.208	0.164	0.165
NH	26.0	30.5	18.0	24.5	28.0	17.0	0.279	0.301	0.332
NJ	40.0	40.0	45.0	40.0	38.0	45.0	0.146	0.151	0.170
NM[4]	2.0	4.6	....	1.8	2.5	....	0.307	0.418	....
NY	1,070.0	1,280.0	1,040.0	1,060.0	1,280.0	1,020.0	0.145	0.151	0.183
NC	135.0	155.0	145.0	130.0	132.0	135.0	0.132	0.132	0.126
OH	90.0	90.0	102.0	88.0	90.0	101.0	0.274	0.276	0.282
OR	133.0	163.0	140.0	132.0	160.0	140.0	0.175	0.163	0.141
PA	442.0	405.0	455.0	442.0	400.0	450.0	0.103	0.101	0.113
RI	2.3	2.2	1.8	2.0	2.1	1.6	0.393	0.480	0.445
SC	6.0	6.0	4.0	4.5	3.0	2.5	0.219	0.108	0.172
TN	12.0	11.0	8.5	11.5	10.5	7.5	0.252	0.263	0.268
UT	28.0	32.0	38.0	27.5	31.4	37.1	0.230	0.268	0.280
VT	42.0	41.5	32.5	37.5	38.0	30.0	0.266	0.225	0.264
VA	270.0	300.0	300.0	262.0	297.0	297.0	0.096	0.149	0.113
WA	4,550.0	6,150.0	5,800.0	4,550.0	6,150.0	5,800.0	0.259	0.160	0.190
WV	87.0	81.0	87.0	85.0	80.0	86.0	0.097	0.091	0.091
WI	68.0	57.0	52.0	58.0	55.0	49.5	0.334	0.336	0.397
US	8,793.1	10,450.6	9,869.6	8,703.0	10,371.3	9,779.1	0.209	0.159	0.183

[1] In orchards of 100 or more bearing-age trees. [2] Preliminary. [3] Fresh fruit prices are equivalent packinghouse-door returns for CA, MI, NY, and WA; prices at point of first sale for other States. Processing prices are equivalent at processing plant door. [4] Estimates discontinued in 2005.
NASS, Crops Branch, (202) 720–2127.

Table 5-4.—Apples: Production and value, United States, 1996–2005

Year	Apples, commercial crop[1]			
	Total production	Utilized production	Marketing year average price[2]	Value
	Million pounds	Million pounds	Cents per pound	1,000 dollars
1996	10,381.9	10,330.0	15.9	1,641,462
1997	10,323.8	10,254.3	15.4	1,575,403
1998	11,646.4	10,762.5	12.2	1,317,322
1999	10,631.6	10,447.4	15.0	1,563,814
2000	10,580.9	10,319.8	12.8	1,320,618
2001	9,423.0	9,209.2	15.8	1,452,344
2002	8,523.9	8,374.1	18.9	1,581,260
2003	8,793.1	8,703.0	20.9	1,817,240
2004	10,450.6	10,371.3	15.9	1,647,983
2005[3]	9,869.6	9,779.1	18.3	1,786,674

[1] In orchards of 100 or more bearing-age trees. [2] Fresh fruit prices are equivalent packinghouse-door returns for CA, NY, MI, and WA; prices at point of first sale for other States. Processing prices are equivalent at processing plant door. [3] Preliminary.
NASS, Crops Branch, (202) 720–2127.

Table 5-5.—Apples, fresh: Production in specified countries, 2003/2004–2005/2006

Continent and country	2003/2004	2004/2005	2005/2006
	1,000 metric tons	*1,000 metric tons*	*1,000 metric tons*
North America:			
Canada	379	370	360
Mexico	494	573	495
United States	3,988	4,726	4,254
Total	4,862	5,670	5,109
South America:			
Argentina [1]	900	1,300	NA
Chile [1]	1,252	1,160	NA
Total	2,152	2,460	NA
European Union:			
Belgium-Luxembourg [2]	322	NA	NA
France [2]	2,080	NA	NA
Germany	1,518	1,945	1,415
Greece	203	275	280
Hungary	500	666	450
Italy	1,878	2,110	2,159
Netherlands [2]	385	NA	NA
Poland	2,428	2,440	2,300
Slovakia	61	42	39
Spain	791	577	720
Sweden [2]	52	46	NA
United Kingdom	144	205	199
Total EU	10,360	8,305	7,562
Other Europe:			
Russia Federation	1,489	1,796	1,660
Turkey	2,600	2,100	2,500
Total	4,089	3,896	4,160
Total Europe	14,449	12,201	11,722
Africa:			
South Africa, Rep [1]	725	706	NA
Total	725	706	NA
Asia:			
China	21,100	23,675	21,300
Japan	842	755	870
Taiwan	3	7	7
Total	21,946	24,436	22,177
Oceania:			
Australia [1]	250	300	NA
New Zealand [1]	550	504	NA
Total	800	804	NA
Total selected countries	44,932	46,277	39,007

[1] It is too early to make reliable forecast for the Southern Hemisphere countries for the 2005/06 season. [2] Belgium-Luxembourg, Sweden, France and the Netherlands are no longer reporting. NA-not available.

FAS, Horticultural and Tropical Products Division, (202) 720-3423. The division's Home Page is located at, www.fas.usda.gov/htp. You can also email the division at htp@fas.usda.gov. Note: Prepared or estimated on the basis of official statistics of foreign governments, other foreign source materials, reports of U.S. Agricultural Counselors and Attachés, results of office research, and related information. Note: This information can be obtained from the following web site: www.fas.usda.gov/psd.

Table 5-6.—Apples, commercial crop: Production and utilization, United States, 1996–2005

Crop of—	Total production	Utilized production	Utilization of quantities sold				
			Fresh[1]	Processed (fresh basis)			
				Canned	Dried	Frozen	Other[2][3]
	Million pounds	*Million pounds*	*Million pounds*	*Million pounds*	*Million pounds*	*Million pounds*	*Million pounds*
1996	10,381.9	10,330.0	6,206.9	1,294.2	316.6	267.8	2,244.5
1997	10,323.8	10,254.3	5,814.5	1,498.8	267.0	349.0	2,325.0
1998	11,646.4	10,762.5	6,412.5	1,173.8	329.9	266.0	2,580.3
1999	10,631.6	10,447.4	5,995.7	1,318.6	263.2	271.3	2,598.6
2000	10,580.9	10,319.8	6,265.5	1,183.6	248.2	195.9	2,426.6
2001	9,423.0	9,209.2	5,467.5	1,257.2	221.0	248.5	2,015.0
2002	8,523.9	8,374.1	5,366.0	1,078.7	207.9	191.7	1,529.8
2003	8,793.1	8,703.0	5,461.8	1,235.6	182.2	282.8	1,540.6
2004	10,450.6	10,371.3	6,643.0	1,257.9	201.8	256.0	2,012.6
2005[3]	9,869.6	9,779.1	NA	NA	NA	NA	NA

[1] Includes "Home use." [2] Mostly crushed for vinegar, cider, and juice. For some States, small quantities canned, dried, and frozen are included. Beginning in 2004, "fresh slices" included. [3] Preliminary. NA-not available.
NASS, Crops Branch, (202) 720–2127.

Table 5-7.—Apples, commercial crop: Production and utilization, by States, crop of 2004

State	Total production	Utilized production	Utilization					
			Fresh[1]	Processed (fresh basis)				
				Canned	Dried	Frozen	Juice and cider	Other[2]
	Million pounds	*Million pounds*	*Million pounds*	*Million pounds*	*Million pounds*	*Million pounds*	*Million pounds*	*Million pounds*
CA	355.0	355.0	165.0	....	....	....	170.0	....
MI	730.0	730.0	240.0	210.0	....	....	115.0	8.0
NY	1,280.0	1,280.0	660.0	340.0	....	50.0	200.0	30.0
OR	163.0	160.0	110.0	....	....	....	27.0	....
PA	405.0	400.0	110.0	230.0	....	....	52.0	....
VA	300.0	297.0	132.0	125.0	....	....	24.0	....
WA	6,150.0	6,150.0	4,600.0	190.0	....	....	1,110.0	....
WV	81.0	80.0	15.0	47.0	....	....	13.0	....
Other States[3]	986.6	919.3	611.0	115.9	....	206.0	168.6	41.0
US	10,450.6	10,371.3	6,643.0	1,257.9	201.8	256.0	1,879.6	79.0

[1] Includes "Home use." [2] Mostly vinegar, wine, and fresh slices for pie making. [3] AZ, AR, CO, CT, GA, ID, IL, IN, IA, KS, KY, ME, MD, MA, MN, MO, NH, NJ, NM, NC, OH, RI, SC, TN, UT, VT, and WI.
NASS, Crops Branch, (202) 720–2127.

Table 5-8.—Apples, fresh: United States exports by country of destination and imports by country of origin, 2002/2003–2004/2005

Country	Year beginning July		
	2002/2003	2003/2004	2004/2005
	Metric tons	Metric tons	Metric tons
Exports			
Europe:			
Finland	764	968	1,548
France	0	2	18
Iceland	1,349	1,309	865
Ireland	611	1,261	957
Netherlands	447	728	2,525
Norway	795	1,271	840
Sweden	880	646	1,382
United Kingdom	22,380	26,501	34,454
Other countries	1,918	2,782	2,400
Total	29,142	35,468	44,990
Latin America:			
Brazil	25	227	0
Colombia	4,374	3,423	7,785
Costa Rica	4,457	3,515	4,104
El Salvador	2,691	3,001	2,104
Guatemala	2,694	3,183	6,060
Nicaragua	344	93	230
Mexico	112,346	86,512	117,307
Honduras	3,211	2,980	3,098
Panama	2,644	2,170	3,041
Ecuador	944	225	488
Venezuela	1,614	2,797	5,740
Other countries	809	477	489
Total	136,152	108,602	150,446
Caribbean:			
Bahamas	632	548	540
Barbados	389	306	283
Bermuda	152	136	188
Dominican Republic	6,094	2,133	5,518
French West Indies	39	0	83
Haiti	65	7	15
Leeward and Windward Islands	135	316	262
Netherlands Antilles	82	215	94
Trinidad and Tobago	3,555	2,505	2,840
Other countries	1,572	1,929	1,161
Total	12,715	8,095	10,983
Asia:			
Taiwan	46,130	47,822	43,076
Hong Kong	36,840	32,392	33,870
Indonesia	41,632	39,287	33,089
Malaysia	25,195	25,357	24,744
Japan	63	4	8
Philippines	5,312	3,311	1,865
Singapore	8,433	6,871	6,150
Thailand	10,193	7,019	8,105
Other countries	21,432	15,633	39,855
Total	195,232	177,696	190,761
Other countries:			
Canada	112,806	96,386	113,557
French Pacific Islands	86	182	209
Kuwait	3,498	2,245	5,150
New Zealand	331	724	165
Saudi Arabia	6,818	6,358	19,942
United Arab Emirates	17,332	10,756	27,312
Total	140,871	116,651	166,335
Grand total	514,111	446,512	563,515
Imports			
Canada	45,707	30,811	30,359
Chile	79,460	107,174	61,388
Japan	58	61	407
Mexico	0	19	0
New Zealand	45,787	67,938	31,043
South Africa	1,927	2,977	1,132
Other countries	4,876	4,589	1,704
Total	177,815	213,568	126,034

FAS, Horticultural and Tropical Products Division, (202) 720-3423. The division's Home Page is located at, www.fas.usda.gov/htp. You can also email the division at htp@fas.usda.gov. Note: Compiled from reports of the U.S. Department of Commerce. U.S. trade can be obtained from the following web site: U.S. agricultural trade database, www.fas.usda.gov/ustrade.

Table 5-9.—Apples: [1] Foreign trade, United States, 1995–2004

Year beginning October	Imports, fresh and dried, in terms of fresh	Domestic exports	
		Fresh	Dried, in terms of fresh[1]
	Metric tons	*Metric tons*	*Metric tons*
1995	196,069	564,953	24,621
1996	197,341	689,749	20,366
1997	173,565	539,081	18,042
1998	171,778	660,251	15,729
1999	195,255	571,860	21,521
2000	180,616	743,644	33,308
2001	193,893	592,955	21,232
2002	231,504	522,525	26,250
2003	243,293	438,300	32,960
2004	155,095	634,445	32,443

[1] Dried converted to terms of fresh apples on following basis; 1 pound dried is equivalent to 8 pounds fresh. No re-exports reported.

ERS, Specialty Crops Branch, (202) 694–5260.

Table 5-10.—Apricots: Production and value, United States, 1996–2005 [1]

Year	Total production	Utilized production	Market year average price per ton [2]	Value
	Tons	*Tons*	*Dollars*	*1,000 dollars*
1996	79,300	79,290	444.00	35,171
1997	139,230	129,630	332.00	43,072
1998	118,490	108,080	327.00	35,358
1999	90,500	90,500	391.00	35,377
2000	96,900	87,760	369.00	32,346
2001	82,460	75,430	353.00	26,598
2002	90,040	80,030	357.00	28,565
2003	97,580	97,560	356.00	34,702
2004	101,130	92,590	378.00	35,012
2005 [3]	81,350	76,345	533.00	40,723

[1] Production, price, and value for CA, UT, and WA.   [2] Fresh fruit prices are equivalent packinghouse-door returns for CA and WA. Quantities processed are priced at the equivalent processing plant door level.   [3] Preliminary.

NASS, Crops Branch, (202) 720–2127.

Table 5-11.—Apricots: Production and marketing year average price per ton, by States, 2003–2005

State	Total production			Utilized production			Price [2] for crop of—		
	2003	2004	2005 [1]	2003	2004	2005 [1]	2003	2004	2005 [1]
	Tons	*Tons*	*Tons*	*Tons*	*Tons*	*Tons*	*Dollars*	*Dollars*	*Dollars*
CA	92,500	94,000	75,500	92,500	85,500	70,500	316.00	334.00	495.00
UT	180	330	250	160	290	245	588.00	610.00	959.00
WA	4,900	6,800	5,600	4,900	6,800	5,600	1,100.00	921.00	997.00
Total	97,580	101,130	81,350	97,560	92,590	76,345	356.00	378.00	533.00

[1] Preliminary.   [2] Fresh fruit prices are equivalent packinghouse-door returns for CA and WA. Quantities processed are priced at the equivalent processing plant door level.

NASS, Crops Branch, (202) 720–2127.

Table 5-12.—Apricots: Production and utilization, United States,[1] 1996–2005

Crop of—	Total production	Utilized production	Utilization of quantities sold			
			Fresh[2]	Processed[3]		
				Canned[4]	Dried (fresh basis)	Frozen
	Tons	*Tons*	*Tons*	*Tons*	*Tons*	*Tons*
1996	79,300	79,290	13,490	20,000	15,000	9,000
1997	139,230	129,630	26,830	46,700	12,000	15,100
1998	118,490	108,080	22,880	40,700	9,000	10,400
1999[5]	90,500	90,500	25,800			
2000	96,900	87,760	26,580	32,000	8,000	10,000
2001	82,460	75,430	18,230	31,000	6,000	9,000
2002	90,040	80,030	18,290	30,500	8,000	10,500
2003	97,580	97,560	26,250	30,000	6,800	11,000
2004	101,130	92,590	23,650	(7)	11,800	9,700
2005[6]	81,350	76,345	23,545	23,500	11,500	(7)

[1] CA, UT, and WA. [2] Includes "Home use." [3] CA only. [4] Includes some quantities frozen or otherwise processed. [5] Breakdown of processed utilization for 1999 unpublished to avoid disclosure of individual operations. [6] Preliminary. [7] Missing data not published to avoid disclose of individual operations.

NASS, Crops Branch, (202) 720–2127.

Table 5-13.—Apricots: Production and utilization, by States, crop of 2005 (preliminary)

State	Total production	Utilized production	Utilization			
			Fresh	Processed[1]		
				Canned[2]	Dried (fresh basis)	Frozen
	Tons	*Tons*	*Tons*	*Tons*	*Tons*	*Tons*
CA	75,500	70,500	18,500	23,500	11,500	(4)
UT[3]	250	245				(4)
WA[3]	5,600	5,600				(4)
US	81,350	76,345	23,545	23,500	11,500	(4)

[1] CA only. [2] Some quantities used for juice are included in "Canned" to avoid disclosure of individual operations. [3] Missing data not published to avoid disclosure of individual operations, but included in U.S. total. [4] Missing data not published to avoid disclose of individual operations.

NASS, Crops Branch, (202) 720–2127.

Table 5-14.—Apricots: Foreign trade, United States, 1995–2004

Year beginning October	Domestic exports				
	Fresh	Canned[1]	Dried[1]	Dried, in fruit salad[2]	Total, in terms of fresh[3]
	Metric tons	*Metric tons*	*Metric tons*	*Metric tons*	*Metric tons*
1995	3,492	1,338	1,376	519	14,196
1996	6,604	1,086	1,034	396	14,737
1997	6,980	968	927	428	14,639
1998	9,197	980	1,797	340	20,886
1999	6,204	1,695	1,349	176	15,260
2000	7,663	769	2,251	312	21,393
2001	7,732	1,600	2,004	202	20,218
2002	7,914	1,554	3,310	402	28,113
2003	7,534	1,588	1,073	486	16,683
2004	5,516	1,581	706	719	13,974

[1] Net processed weight. [2] Dried apricots are 12⅓ percent of total dried fruit for salad. [3] Dried fruit converted to unprocessed dry weight by dividing by 1.07. Unprocessed dry weight converted to terms of fresh fruit on the basis that 1 pound dried equals 5.5 pounds fresh. Canned apricots converted to terms of fresh on the basis that 1 pound canned equals 0.717 pounds fresh.

ERS, Specialty Crops Branch, (202) 694–5260.

Table 5-15.—Avocados: Foreign trade, United States, 1995–2004

Year beginning October	Imports
	Metric tons
1995	23,118
1996	27,667
1997	39,847
1998	59,637
1999	63,944
2000	73,070
2001	103,339
2002	136,708
2003	132,644
2004	248,356

ERS, Specialty Crops Branch, (202) 694–5260.

Table 5-16.—Avocados: Production, marketing year average price per ton, and value, United States, 1994–95 to 2004–2005

Season	California [1]			Florida [1]		
	Production [2]	Price [3]	Value	Production [2]	Price [3]	Value
	Tons	*Dollars*	*1,000 dollars*	*Tons*	*Dollars*	*1,000 dollars*
1995–96	171,000	1,370	234,831	19,000	596	11,324
1996–97	167,000	1,560	260,162	23,500	528	12,408
1997–98	154,000	1,710	263,473	24,000	584	14,016
1998–99	136,000	2,400	327,002	23,000	716	16,468
1999–2000	161,000	2,110	339,594	22,000	748	16,456
2000–2001	213,000	1,480	315,842	26,000	584	15,184
2001–2002	200,000	1,790	358,000	23,000	676	15,548
2002–2003	168,000	2,170	364,560	31,000	556	17,236
2003–2004	216,000	1,760	380,160	17,000	808	13,736
2004–2005 [4]	151,000	1,840	277,840	28,000	516	14,448

Season	Hawaii			United States		
	Production [2]	Price [3]	Value	Production [2]	Price [3]	Value
	Tons	*Dollars*	*1,000 dollars*	*Tons*	*Dollars*	*1,000 dollars*
1995–96	250	1,090.00	273	190,250	1,300.00	246,428
1996–97	200	1,070.00	214	190,700	1,430.00	272,784
1997–98	250	1,060.00	265	178,250	1,560.00	277,754
1998–99	250	1,040.00	260	159,250	2,160.00	343,730
1999–2000	300	1,200.00	360	183,300	1,950.00	356,410
2000–2001	320	1,160.00	371	239,320	1,400.00	331,397
2001–2002	300	1,140.00	342	223,300	1,670.00	373,890
2002–2003	350	1,120.00	392	199,350	1,920.00	382,188
2003–2004	380	1,240.00	471	233,380	1,690.00	394,367
2004–2005 [4]	370	1,260.00	466	179,370	1,630.00	292,754

[1] Season from Nov. 1 to Nov. 30 (following year) for California and June 20 to Feb. 28 for Florida.   [2] Production is the quantity sold or utilized.   [3] Quantities processed are priced at the equivalent processing plant door level.   [4] Preliminary.
NASS, Crops Branch, (202) 720–2127.

Table 5-17.—Bananas: Area, yield, utilized production, marketing year average price, and value, Hawaii, 1996–2005

Year	Area harvested	Yield per acre	Production	Price per pound	Value
	Acres	1,000 pounds	1,000 pounds	Cents	1,000 dollars
1996	960	13.5	13,000	40.0	5,200
1997	950	14.4	13,700	38.0	5,206
1998	1,420	14.8	21,000	35.0	7,350
1999	1,420	17.3	24,500	35.0	8,575
2000	1,460	19.9	29,000	36.0	10,440
2001	1,490	18.8	28,000	38.0	10,640
2002	1,330	15.0	20,000	43.0	8,600
2003	1,350	16.7	22,500	41.0	9,225
2004	1,000	16.5	16,500	49.0	8,085
2005 [1]	(2)	(2)	(2)	(2)	(2)

[1] Preliminary.   [2] Missing data not shown.
NASS, Crops Branch, (202) 720–2127.

Table 5-18.—Kiwifruit: Area, yield, utilized production, marketing year average price, and value, California, 1996–2005

Year	Bearing acreage	Yield [1]	Production	Price per ton	Value
	Acres	Tons	Tons	Dollars	1,000 dollars
1996	5,700	5.53	28,000	470	13,157
1997	5,300	6.60	31,800	518	16,483
1998	5,300	6.91	33,000	744	24,544
1999	5,300	5.09	24,000	634	15,215
2000	5,300	6.42	30,500	455	13,888
2001	4,900	5.27	23,000	667	15,340
2002	4,500	5.80	23,100	783	18,097
2003	4,500	5.64	25,400	853	20,472
2004	4,500	5.93	26,700	809	19,977
2005	4,500	9.24	41,600	NA	NA

[1] Yield based on total production.
NASS, Crops Branch, (202) 720-2127.

Table 5-19.—Cherries: Foreign trade, United States, 1995–2004

Year beginning October	Imports		Domestic exports	
	Fresh	Dried and preserved	Fresh	Canned
	Metric tons	Metric tons	Metric tons	Metric tons
1995	1,906	1,345	34,702	18,975
1996	1,909	1,884	39,400	15,044
1997	968	2,054	37,591	13,357
1998	2,088	1,750	42,655	14,500
1999	2,815	2,184	43,289	14,970
2000	3,858	2,561	42,880	20,515
2001	6,680	3,023	36,232	19,355
2002	8,548	3,062	47,829	12,519
2003	5,170	3,652	43,079	13,144
2004	7,233	3,712	43,063	14,280

ERS, Specialty Crops Branch, (202) 694–5260.

Table 5-20.—Sweet cherries: Production and value, United States, 1996–2005

Year	Total production	Utilized production	Marketing year average price per ton [1]	Value
	Tons	Tons	Dollars	1,000 dollars
1996	154,100	151,700	1,470.00	223,022
1997	225,770	223,490	1,250.00	278,511
1998	196,900	193,910	1,100.00	213,109
1999	216,120	213,260	1,100.00	234,879
2000	207,900	205,420	1,340.00	274,995
2001	230,380	219,620	1,230.00	270,914
2002	181,355	177,305	1,550.00	274,471
2003	245,700	243,580	1,400.00	342,113
2004	283,060	279,160	1,570.00	437,133
2005 [2]	251,170	243,910	1,980.00	483,504

[1] Fresh fruit prices are equivalent packinghouse-door returns for Western States, and the average price as sold for other States. Quantities processed are priced at the equivalent processing plant door level. [2] Preliminary.
NASS, Crops Branch, (202) 720–2127.

Table 5-21.—Tart cherries: Production and value, United States, 1996–2005

Year	Total production	Utilized production	Marketing year average price per ton [1]	Value
	Million pounds	Million pounds	Dollars	1,000 dollars
1996	271.8	260.1	0.161	41,747
1997	292.9	283.3	0.159	44,911
1998	348.1	305.6	0.145	44,186
1999	256.1	254.1	0.218	55,505
2000	288.5	281.4	0.187	52,488
2001	370.1	307.9	0.186	57,150
2002	62.5	62.2	0.448	27,879
2003	226.3	226.3	0.354	80,210
2004	213.0	213.0	0.326	69,501
2005 [2]	270.4	268.4	0.243	65,296

[1] Fresh fruit prices are equivalent packinghouse-door returns for Western States, and the average price as sold for other States. Quantities processed are priced at the equivalent processing plant door level. [2] Preliminary.
NASS, Crops Branch, (202) 720–2127.

Table 5-22.—Sweet cherries: Production and season average price, by States, 2003–2005

State	Total production			Utilized production			Price [2]		
	2003	2004	2005 [1]	2003	2004	2005 [1]	2003	2004	2005 [1]
	Tons	Tons	Tons	Tons	Tons	Tons	Dollars per ton	Dollars per ton	Dollars per ton
CA	65,600	73,000	52,700	63,900	70,300	48,600	1,670.00	1,750.00	1,740.00
ID	2,900	3,100	1,700	2,900	3,100	1,700	1,400.00	1,390.00	1,950.00
MI	13,000	24,700	27,000	13,000	24,700	27,000	830.00	660.00	620.00
MT	2,060	2,360	1,170	1,850	2,220	1,120	1,710.00	2,010.00	3,660.00
NY	600	900	800	590	890	740	1,770.00	1,400.00	1,710.00
OR	41,000	43,000	28,000	41,000	42,000	25,000	1,080.00	1,150.00	1,460.00
PA [3]	340	400			340	350	2,360.00	2,980.00	
UT	2,200	1,600	1,800	2,000	1,600	1,750	900.00	996.00	1,100.00
WA	118,000	134,000	138,000	118,000	134,000	138,000	1,430.00	1,770.00	2,430.00
Total 9 States	245,700	283,060	251,170	243,580	279,160	243,910	1,400.00	1,570.00	1,980.00

[1] Preliminary. [2] Fresh fruit prices are equivalent packinghouse-door returns for CA, OR, and WA, and the average price as sold for other States. Quantities processed are priced at the equivalent processing plant door level. [3] Estimates discontinued in 2005.
NASS, Crops Branch, (202) 720–2127.

Table 5-23.—Tart cherries: Production and season average price, by States, 2003–2005

State	Total production			Utilized production			Price [2]		
	2003	2004	2005 [1]	2003	2004	2005 [1]	2003	2004	2005 [1]
	Million pounds	Million pounds	Million pounds	Million pounds	Million pounds	Million pounds	Dollars per ton	Dollars per ton	Dollars per ton
CO [3]	0.4	0.2		0.4	0.2		0.380	0.210	
MI	154.0	149.0	208.0	154.0	149.0	208.0	0.376	0.335	0.230
NY	7.2	10.7	7.5	7.2	10.7	7.5	0.314	0.409	0.432
OR	1.4	3.9	0.3	1.4	3.9	0.3	0.361	0.369	0.380
PA	3.9	3.0	2.6	3.9	3.0	2.6	0.434	0.353	0.430
UT	26.0	22.0	28.0	26.0	22.0	26.0	0.228	0.218	0.250
WA	20.1	17.5	16.5	20.1	17.5	16.5	0.323	0.309	0.239
WI	13.3	6.7	7.5	13.3	6.7	7.5	0.394	0.375	0.332
Total 8 States	226.3	213.0	270.4	226.3	213.0	268.4	0.354	0.326	0.243

[1] Preliminary. [2] Fresh fruit prices are equivalent packinghouse-door returns for OR and WA, and the average price as sold for other States. Quantities processed are priced at the equivalent processing plant door level. [3] Estimates discontinued in 2005.
NASS, Crops Branch, (202) 720–2127.

Table 5-24.—Sweet cherries: Production and utilization, by States, crop of 2005 (preliminary)

State	Total production	Utilized production	Utilization			
			Fresh[1]	Processed		
				Canned and otherwise processed	Brined	Other[2]
	Tons	Tons	Tons	Tons	Tons	Tons
CA	52,700	48,600	35,600	...	...	...
MI	27,000	27,000	600	4,350	17,800	4,250
OR	28,000	25,000	14,000	1,300	6,900	2,800
WA	138,000	138,000	113,000	4,000	13,000	8,000
Other States[3]	5,470	5,310	3,950	...	7,530	6,830
US	251,170	243,910	167,150	9,650	45,230	21,880

[1]Includes "Home use." [2]Includes California canned utilization and other processed utilizations from all States. [3]ID, MT, NY, and UT.
NASS, Crops Branch, (202) 720-2127.

Table 5-25.—Tart cherries: Production and utilization, by States, crop of 2005 (preliminary)

State	Total production	Utilized production	Utilization			
			Fresh[2]	Processed		
				Canned and otherwise processed[3]	Frozen	Other[3]
	Million pounds	Million pounds	Million pounds	Million pounds	Million pounds	Million pounds
MI	208.0	208.0	0.5	51.0	146.0	10.5
Other States[4]	62.4	60.4	0.7	6.9	41.9	10.9
US	270.4	268.4	1.2	57.9	187.9	21.4

[1]Preliminary. [2]Includes "Home use." [3]Some quantities used for juice, wine, brined, and dried. [4]NY, OR, PA, UT, WA, and WI.
NASS, Crops Branch, (202) 720-2127.

Table 5-26.—Sweet cherries: Production and utilization, United States,[1] 1996-2005

Crop of—	Total production	Utilized production	Utilization of quantities sold		
			Fresh[2]	Processed	
				Other[3]	Brined
	Tons	Tons	Tons	Tons	Tons
1996	154,100	151,700	80,670	22,070	48,960
1997	225,770	223,490	115,440	30,400	77,650
1998	196,900	193,910	101,960	31,200	60,750
1999	216,120	213,260	123,410	26,065	63,785
2000	207,900	205,420	120,760	27,710	56,950
2001	230,380	219,620	145,710	25,730	48,180
2002	181,355	177,305	126,595	18,570	32,140
2003	245,700	243,580	175,570	25,960	42,050
2004	283,060	279,160	185,050	33,380	60,730
2005[4][5]	251,170	243,910	167,150	31,530	45,230

[1]CA, ID, MI, NY, OR, PA, UT, and WA. [2]Includes "Home use." [3]Includes canned utilization and other processed utilizations from all States. [4]Preliminary. [5]Estimates discontinued for PA in 2005.
NASS, Crops Branch, (202) 720-2177.

Table 5-27.—Tart cherries: Production and utilization, United States,[1] 1996-2005

Crop of—	Total production	Utilized production	Utilization of quantities sold		
			Fresh[2]	Processed	
				Other[3]	Frozen
	Million pounds	Million pounds	Million pounds	Million pounds	Million pounds
1996	271.8	260.1	2.5	87.4	170.2
1997	292.9	283.3	2.6	107.1	173.6
1998	348.1	305.6	2.3	103.5	199.8
1999	256.1	254.1	1.8	114.4	137.9
2000	288.5	281.4	1.8	135.3	144.3
2001	370.1	307.9	1.9	129.2	176.8
2002	62.5	62.2	0.8	32.5	28.9
2003	226.3	226.3	1.0	76.6	148.7
2004	213.0	213.0	1.3	61.6	150.1
2005[4][5]	270.4	268.4	1.2	79.3	187.9

[1]CO, MI, NY, OR, PA, UT, WA, and WI. [2]Includes "Home use." [3]Includes canned utilization and other processed utilizations from all states. [4]Preliminary. [5]Estimates discontinued for CO in 2005.
NASS, Crops Branch, (202) 720-2177.

Table 5-28.—Citrus fruit: Utilized production and value, United States, for season of 1995–96 to 2004–2005

Season [1]	Production	Marketing year average returns per box [2]	Value	Quantities processed [3]	Production	Marketing year average returns per box [2]	Value	Quantities processed [3]
	Oranges [4]				Grapefruit			
	1,000 boxes	Dollars	1,000 dollars	1,000 boxes	1,000 boxes [5]	Dollars	1,000 dollars	1,000 boxes
1995–96	263,890	6.85	1,821,579	207,365	66,200	4.33	290,152	33,582
1996–97	293,020	6.16	1,836,662	228,565	70,100	4.00	284,749	36,665
1997–98	315,525	6.13	1,965,358	247,004	63,150	4.13	268,598	32,460
1998–99	224,580	7.41	1,687,928	192,194	61,200	5.33	334,626	30,027
1999–2000	299,760	5.56	1,666,100	244,582	66,980	6.07	409,716	38,509
2000–2001	280,935	5.88	1,682,790	223,232	59,750	4.69	285,065	32,600
2001–2002 [6]	283,760	6.37	1,846,199	228,276	58,660	4.92	292,156	32,113
2002–2003 [6]	267,040	5.80	1,564,658	206,000	50,080	5.24	269,381	26,150
2003–2004 [6]	294,620	5.90	1,782,157	238,690	52,540	5.91	317,218	27,225
2004–2005 [6]	212,800	6.87	1,498,063	155,452	25,340	15.59	397,909	9,465
	Lemons				Temples (FL)			
1995–96	26,100	10.01	261,281	12,533	2,150	6.52	14,024	1,457
1996–97	25,300	12.00	303,476	12,206	2,400	5.23	12,541	1,845
1997–98	23,600	10.21	240,846	12,250	2,250	5.12	11,510	1,684
1998–99	19,650	12.79	251,397	7,523	1,800	7.25	13,050	1,207
1999–2000	22,100	13.51	298,677	8,476	1,950	4.70	9,173	1,510
2000–2001	26,200	9.06	237,362	12,793	1,250	4.23	5,282	907
2001–2002	21,100	15.54	327,964	6,678	1,550	4.46	6,919	1,132
2002–2003 [6]	27,000	10.79	291,425	12,354	1,300	4.30	5,591	995
2003–2004 [6]	21,000	12.85	269,753	6,792	1,400	3.51	4,915	1,058
2004–2005 [6]	21,400	16.44	351,897	6,605	650	5.10	3,314	437
	Tangerines [7]				Tangelos (FL)			
1995–96	8,100	13.94	110,573	2,390	2,450	6.16	15,100	1,432
1996–97	9,650	12.47	122,172	3,096	3,950	4.75	18,759	2,918
1997–98	8,200	11.78	96,524	2,642	2,850	4.19	11,950	1,937
1998–99	7,400	15.74	116,632	2,047	2,550	7.17	18,277	1,712
1999–2000	10,350	10.43	108,192	3,640	2,200	5.11	11,232	1,464
2000–2001	8,450	11.26	96,789	2,517	2,100	3.90	8,193	1,358
2001–2002	9,420	12.97	124,718	2,665	2,150	5.00	10,758	1,454
2002–2003 [6]	8,730	13.23	117,432	1,989	2,350	4.89	11,489	1,742
2003–2004 [6]	9,390	12.19	116,475	2,545	1,000	10.02	10,021	455
2004–2005 [6]	7,650	16.79	130,068	1,614	1,550	5.16	8,004	1,055
	Limes (FL)				K-Early Citrus (FL)			
1995–96	300	13.05	3,914	60	160	4.82	771	98
1996–97	320	11.93	3,816	65	150	3.95	592	93
1997–98	440	11.90	5,235	110	40	1.13	45	10
1998–99	500	17.83	8,913	90	80	4.45	356	56
1999–2000	600	16.21	9,728	100	110	3.24	356	95
2000–2001	250	17.00	4,249	30	40	4.68	187	19
2001–2002	150	11.55	1,732	25	30	3.77	113	24
2002–2003 [6]	(8)	(8)	(8)	(8)	(8)	(8)	(8)	(8)
2003–2004 [6]	(8)	(8)	(8)	(8)	(8)	(8)	(8)	(8)
2004–2005 [6]	(8)	(8)	(8)	(8)	(8)	(8)	(8)	(8)

[1] See footnote 1, table 5-29. [2] Equivalent packing-house door returns. [3] Includes quantities used for juice, concentrates, grapefruit segments, and other citrus products. In some seasons, includes appreciable quantities of oranges and lemons in CA delivered to processing plants which were not utilized, but for which growers received payment. [4] Includes small quantities of tangerines in TX. Excludes FL Temples. [5] Excludes FL economic abandonment in 1995–96 of 3 million boxes of Colored Seedless; in 1996–97 of 3 million boxes of White Seedless and 3 million boxes of Colored Seedless; in 1997–98 of 5 million boxes of White Seedless and 1 million boxes of Colored Seedless. [6] Preliminary. [7] AZ and CA tangelos and tangors included. [8] Estimates discontinued.

NASS, Crops Branch, (202) 720–2127.

Table 5-29.—Citrus fruit: Utilized production and marketing year average returns per box, by States, 2003–2004 to 2004–2005 [1]

Crop and State	Utilized production		Market year average price [2]	
	2003–2004	2004–2005	2003–2004	2004–2005
ORANGES	*1,000*	*1,000*		
Early, midseason, and Navel varieties: [3]	*boxes*	*boxes*	*Dollars*	*Dollars*
AZ	300	240	6.55	7.88
CA	39,500	43,000	10.66	9.83
FL	126,000	79,100	4.26	4.77
TX	1,420	1,500	4.21	7.01
Total early, midseason, and Navel varieties	167,220	123,840	5.58	6.36
Valencia:				
AZ	170	190	4.74	3.70
CA	11,000	18,000	12.95	12.12
FL	116,000	70,500	5.79	6.60
TX	230	270	4.82	5.45
Total Valencia	127,400	88,960	6.31	7.56
All oranges:				
AZ	470	430	5.90	6.03
CA	50,500	61,000	11.16	10.51
FL	242,000	149,600	4.99	5.63
TX	1,650	1,770	4.29	6.77
US, all oranges	294,620	212,800	5.90	6.87
GRAPEFRUIT				
AZ	140	140	9.69	15.16
CA	5,800	5,800	12.10	19.12
FL, all	40,900	12,800	5.45	16.26
Colored seedless	25,000	9,400	6.34	16.91
White seedless	15,900	3,400	4.06	14.48
TX	5,700	6,600	3.98	11.62
US, all grapefruit	52,540	25,340	5.91	15.59
LEMONS				
AZ	3,000	2,400	9.70	11.51
CA	18,000	19,000	13.37	17.07
US, lemons	21,000	21,400	12.85	16.44
TANGELOS				
FL	1,000	1,550	10.02	5.16
TANGERINES				
AZ [4]	690	400	11.63	14.84
CA [5]	2,200	2,800	16.68	19.90
FL	6,500	4,450	11.04	15.38
US, tangerines	9,390	7,650	12.19	16.79
TEMPLES				
FL	1,400	650	3.51	5.10

[1] The crop year begins with the bloom of the first year shown and ends with completion of harvest the following year.   [2] Equivalent packinghouse-door returns.   [3] Includes small quantities of tangerines in TX. Excludes FL Temples.   [4] Net lbs. per box: oranges—AZ and CA, 75; FL, 90; and TX, 85; grapefruit—AZ and CA, 67; FL, 85; TX, 80; lemons—76; tangelos, K-Early Citrus and Temples—90; tangerines—AZ and CA, 75.   [5] Includes tangelos and tangors.

NASS, Crops Branch, (202) 720–2127.

Table 5-30.—Citrus fruits: Production in specified countries, 2002–2003 to 2004–2005 [1]

Commodity and country	2002–2003	2003–2004	2004–2005 [2]
	1,000 metric tons	1,000 metric tons	1,000 metric tons
Oranges:			
Argentina	700	750	770
Australia	407	461	500
Brazil	15,382	19,054	16,606
China, Peoples Republic of	3,600	4,036	4,200
Cuba	480	398	200
Cyprus	89	92	90
Egypt	1,734	1,740	1,750
Greece	1,145	950	820
Israel	143	133	160
Italy	1,723	1,835	1,997
Japan	17	16	15
Mexico	3,734	3,901	4,120
Morocco	800	705	813
South Africa, Republic of	1,148	1,113	1,120
Spain	2,950	3,052	2,700
Turkey	1,250	1,250	1,280
United States [3]	10,527	11,734	8,293
Total	45,829	51,220	45,434
Tangerines:			
Argentina	380	420	430
China; Peoples Republic of	6,545	6,870	6,950
Cuba	3	4	2
Egypt	504	506	506
Greece	79	53	60
Israel	71	88	119
Italy	540	528	578
Japan	1,332	1,330	1,235
Korea; Republic of	690	630	593
Morocco	478	408	443
Spain	2,025	2,060	2,100
Turkey	590	550	565
United States [4]	442	419	364
Total	13,679	13,866	13,945
Grapefruit:			
Argentina	185	160	170
China, Peoples Republic of	1,526	1,642	1,724
Cuba	227	20	5
Cyprus	38	38	38
Israel	255	237	247
Italy	7	7	7
Mexico	281	288	310
South Africa, Republic of	256	264	270
Turkey	125	135	110
United States	1,872	1,964	914
Total	4,772	4,755	3,795
Lemons:			
Argentina	1,200	1,220	1,300
Cyprus	21	20	20
Greece	108	70	63
Israel	20	20	23
Italy	528	534	542
Japan	2	2	2
Morocco	20	15	20
South Africa, Republic of	182	183	180
Spain	920	1,130	900
Turkey	525	550	535
United States	931	724	738
Total	4,457	4,468	4,323

[1] Split years refer to harvest periods which usually begin in the fall and extend through the following spring. This corresponds roughly with October–June in the Northern Hemisphere and April–December of the second year shown in the Southern Hemisphere. [2] Preliminary. [3] Includes temple oranges. [4] Includes tangelos.

FAS, Horticultural and Tropical Products Division, (202) 720-3423. The division's Home Page is located at, www.fas.usda.gov/htp. You can also email the division at htp@fas.usda.gov. Note: Prepared or estimated on the basis of official statistics of foreign governments, other foreign source materials, reports of U.S. Agricultural Counselors and Attachés, results of office research, and related information. Note: This information can be obtained from the following web site: www.fas.usda.gov/psd.

FRUITS, TREE NUTS, AND HORTICULTURAL SPECIALTIES

Table 5-31.—Oranges, fresh:[1] United States exports by country of destination, 2001/2002–2003/2004

Country of destination	Year beginning November		
	2001/2002	2002/2003	2003/2004
	Metric tons	Metric tons	Metric tons
Caribbean:			
Bahamas, The	257	214	224
Barbados	26	10	59
Bermuda	3	0	46
Haiti	21	142	12
Netherlands Antilles	6	6	13
Total	313	372	354
Central America:			
Costa Rica	87	0	0
El Salvador	0	28	0
Guatemala	8	61	131
Honduras	11	7	0
Panama	17	7	20
Total	124	103	151
East Asia:			
China, Peoples Republic of	21,673	36,645	30,951
Hong Kong	71,775	78,360	77,796
Japan	88,311	93,310	80,131
Korea, Republic of	88,336	134,809	149,981
Taiwan	8,975	10,823	7,251
Total	279,069	353,948	346,111
European Union - 25:			
Belgium-Luxembourg	86	105	533
Denmark	0	0	14
France	0	0	588
Germany	17	19	37
Ireland	0	0	132
Italy	0	0	6
Netherlands	710	3,185	1,673
Poland	0	0	165
Slovenia	0	55	220
United Kingdom	712	400	170
Total	1,527	3,765	3,537
Former Soviet Union - 12:			
Russian Federation	19	32	0
Total	19	32	0
Middle East:			
Bahrain	18	0	0
Oman	22	22	627
Saudi Arabia	30	0	0
United Arab Emirates	126	523	1,389
Total	195	545	2,016
North America:			
Canada	153,082	187,609	177,406
Mexico	14,567	28,439	14,462
Total	167,649	216,048	191,869
Oceania:			
Australia	7,667	7,571	10,920
French Pacific Islands	40	10	222
New Zealand	2,959	6,343	5,999
Other Pacific Islands, NEC	0	0	28
Total	10,666	13,924	17,170
Other Europe:			
Croatia	0	77	0
Iceland	0	0	19
Switzerland	2	156	78
Total	2	233	97

See footnotes at end of table.

Table 5-31.—Oranges, fresh:[1] United States exports by country of destination, 2001/2002–2003/2004—Continued

Country of destination	Year beginning November		
	2001/2002	2002/2003	2003/2004
	Metric tons	Metric tons	Metric tons
South America:			
Brazil	106	0	0
Chile	0	288	171
Colombia	127	26	102
Ecuador	36	151	251
Peru	106	0	0
Total	376	465	524
South Asia:			
Bangladesh	327	949	924
India	17	305	343
Sri Lanka	91	334	548
Total	435	1,587	1,815
Southeast Asia:			
Brunei	33	49	17
Cambodia	19	0	0
Indonesia	2,222	1,494	2,122
Malaysia	18,325	29,588	27,481
Philippines	2,777	4,121	2,944
Singapore	15,363	19,053	18,163
Thailand	253	220	129
Vietnam	434	230	208
Total	39,426	54,757	51,064
Sub-Saharan Africa:			
Namibia	0	0	10
Senegal	0	0	68
Total	0	0	78
World Total	499,802	645,779	614,784

[1] Includes Temple oranges.

FAS, Horticultural and Tropical Products Division, (202) 720-3423. The division's Home Page is located at, www.fas.usda.gov/htp. You can also email the division at htp@fas.usda.gov. Note: Compiled from reports of the U.S. Department of Commerce. U.S. trade can be obtained from the following web site: U.S. agricultural trade database, www.fas.usda.gov/ustrade.

Table 5-32.—Fresh citrus fruits: Foreign trade, United States, 1995–2004

Year [1]	Oranges		Grapefruit		Lemons		Limes		Tangerines	
	Imports	Domes-tic ex-ports	Imports	Domes-tic ex-ports	Imports	Domes-tic ex-ports	Imports	Domes-tic ex-ports	Imports	Domes-tic ex-ports
	Metric tons	Metric tons	Metric tons	Metric tons	Metric tons	Metric tons	Metric tons	Metric t ons	Metric tons	Metric tons
1995	23,393	513,630	14,797	497,339	11,042	132,269	130,073	3,363	3,786	17,178
1996	29,653	590,428	12,807	484,403	21,736	120,279	147,065	3,517	3,877	15,285
1997	39,961	642,010	5,171	387,215	22,132	113,282	164,577	3,886	3,712	13,282
1998	101,923	263,199	15,521	428,618	22,991	114,109	151,613	3,921	2,576	11,543
1999	48,885	511,852	5,769	390,958	25,160	106,249	179,394	3,752	5,674	10,983
2000	52,785	570,162	19,409	389,629	34,127	110,373	182,412	3,846	4,117	11,786
2001	56,789	499,988	27,327	396,400	36,351	99,906	179,101	3,659	4,324	12,678
2002	55,590	638,079	17,781	350,953	27,901	99,566	251,973	2,236	4,545	14,406
2003	58,041	626,060	18,983	396,229	34,461	101,603	267,027	2,364	3,593	17,030
2004	69,986	572,122	14,562	225,968	35,377	97,842	306,090	3,377	4,780	13,121

[1] Year beginning October for all commodities.

ERS, Specialty Crops Branch, (202) 694-5260.

Table 5-33.—Concentrated citrus juices: Annual packs, Florida, 1994–2003

Season beginning December	Frozen concentrated juice [1]		
	Orange [2]	Grapefruit [2]	Tangerine
	1,000 gallons	1,000 gallons	1,000 gallons
1994	216,502	31,344	1,192
1995	202,353	26,930	1,102
1996	241,800	30,032	2,386
1997	253,734	24,223	1,461
1998	158,884	24,512	1,191
1999	207,708	28,642	1,646
2000	235,933	27,481	1,065
2001	241,609	27,552	1,853
2002	195,362	20,416	1,596
2003	241,171	21,019	1,542

[1] Net pack. [2] Frozen orange juice reported in 42.0° Brix; Grapefruit 40.0° Brix. Includes concentrated juice for manufacture.

ERS, Specialty Crops Branch, (202) 694–5260.

Table 5-34.—Dates: Area, yield, total production, marketing year average price per ton, and value, California, 1996–2005

Year	Bearing acreage	Yield per acre	Production	Price per ton	Value
	Acres	Tons	Tons	Dollars	1,000 dollars
1996	4,680	4.91	23,000	1,090	25,070
1997	4,800	4.38	21,000	1,100	23,100
1998	5,000	4.98	24,900	1,220	30,378
1999	5,100	4.35	22,200	1,240	27,528
2000	5,000	3.48	17,400	1,230	21,402
2001	4,900	4.02	19,700	1,360	26,792
2002	4,800	5.04	24,200	1,550	37,510
2003	4,700	3.81	17,900	2,300	41,170
2004	4,700	3.64	17,100	2,260	38,646
2005 [1]	4,600	3.61	16,600	2,000	33,200

[1] Preliminary.

NASS, Crops Branch, (202) 720–2127.

Table 5-35.—Dates: Foreign trade, United States, 1995–2004

Year beginning October	Imports
	Metric tons
1995	4,683
1996	2,587
1997	3,207
1998	5,179
1999	5,006
2000	2,996
2001	4,347
2002	5,253
2003	5,536
2004	5,195

ERS, Specialty Crops Branch, (202) 694–5260.

Table 5-36.—Cranberries: Area, yield, production, season average price per barrel, value and quantities processed, United States, 1996–2005 [1]

Year	Area harvested	Yield per acre [2]	Total production [3]	Utilized production	Price [4]	Value	Quantities processed [5]
	Acres	Barrels [6]	Barrels [6]	Barrels [6]	Dollars	1,000 dollars	Barrels [6]
1996	34,000	137.4	4,671,000	4,671,000	65.90	307,827	4,330,000
1997	35,700	154.0	5,497,000	5,497,000	63.70	350,147	5,072,000
1998	36,600	148.7	5,444,000	5,444,000	36.60	199,114	5,200,000
1999	38,200	166.4	6,357,000	6,357,000	17.20	109,072	6,000,000
2000	37,200	153.5	5,712,000	5,579,000	18.10	100,851	5,137,000
2001	35,600	149.7	5,329,000	4,783,000	23.80	113,646	4,357,000
2002	39,400	144.4	5,689,000	5,682,000	32.20	182,783	5,312,000
2003	39,600	156.4	6,193,000	6,193,000	33.90	209,834	5,842,000
2004	39,200	157.5	6,175,000	6,167,000	32.30	199,296	5,770,000
2005 [7]	39,100	159.2	6,225,000	6,225,000	34.00	211,527	5,865,000

[1] Estimates relate to MA, NJ, OR, WA, and WI. [2] Derived from total production. [3] Differences between utilized and total production are quantities unharvested for economic reasons or excess cullage and/or set-aside production under provisions of the Cranberry Marketing Order. [4] Average price of utilized production. Equivalent returns at first delivery point, screened basis of utilized production. [5] Mainly for canning. [6] Barrels of 100 pounds. [7] Preliminary.

NASS, Crops Branch, (202) 720–2127.

Table 5-37.—Cranberries: Area, yield, production, and season average price per barrel, by States, 2003–2005

State	Area harvested			Yield per acre			Total production			Price per barrel[2]		
	2003	2004	2005[1]	2003	2004	2005[1]	2003	2004	2005[1]	2003	2004	2005[1]
	Acres	*Acres*	*Acres*	*Bbl.[3]*	*Bbl.[3]*	*B bl.[3]*	*Bbl.[3]*	*Bbl.[3]*	*Bbl.[3]*	*Dollars*	*Dollars*	*Dollars*
MA ..	14,400	14,100	14,200	97.6	128.2	98.9	1,406,000	1,808,000	1,405,000	34.10	32.10	33.70
NJ ...	3,200	3,100	3,100	150.0	129.7	171.9	480,000	402,000	533,000	31.90	30.40	33.70
OR ..	2,900	2,900	2,700	175.9	170.7	163.0	510,000	495,000	440,000	34.10	32.50	34.30
WA	1,700	1,700	1,700	111.8	100.0	110.0	190,000	170,000	187,000	35.20	34.70	36.30
WI ...	17,400	17,400	17,400	207.3	189.7	210.3	3,607,000	3,300,000	3,660,000	34.00	32.50	34.00
US ..	39,600	39,200	39,100	156.4	157.5	159.2	6,193,000	6,175,000	6,225,000	33.90	32.30	34.00

[1] Preliminary. [2] Average price of utilized production. Equivalent returns at first delivery point, screened basis of utilized production. [3] Barrels of 100 pounds.
NASS, Crops Branch, (202) 720–2127.

Table 5-38.—Figs: Total production, marketing year average price per ton, and value, California, 1996–2005

Year	Dried (dry basis)				Total		
	Production			Price per ton	Production (fresh basis)[1]	Price per ton	Value
	Total	Standard	Substandard				
	Tons	*1,000 tons*	*Tons*	*Dollars*	*Tons*	*Dollars*	*1,000 dollars*
1996	14,500	13,100	1,400	774	45,500	283	12,894
1997	18,500	15,900	2,600	699	57,500	265	15,209
1998	16,600	13,300	3,300	594	51,600	222	11,445
1999	15,100	13,800	1,300	681	47,300	268	12,685
2000	17,300	15,400	1,900	672	55,900	272	15,226
2001	13,000	11,700	1,300	....	41,000	366	15,012
2002	16,900	15,000	1,900	....	53,200	340	18,087
2003	15,200	13,300	1,900	....	48,500	317	15,373
2004	15,600	13,700	1,900	....	51,100	396	20,214
2005[2]	15,400	13,500	1,900	....	50,900	NA	NA

[1] Dried figs converted to fresh basis at ratio of 3 pounds fresh to 1 pound dried. [2] Preliminary. NA-not available.
NASS, Crops Branch, (202) 720–2127.

Table 5-39.—Figs, dried: Foreign trade, United States, 1995–2004

Year beginning October	Imports for consumption	Domestic exports
	Metric tons	*Metric tons*
1995	4,834	4,850
1996	3,780	2,027
1997	4,414	1,824
1998	3,425	2,010
1999	3,900	2,763
2000	3,070	2,506
2001	6,788	2,399
2002	7,627	2,962
2003	4,477	3,607
2004	6,247	4,182

ERS, Specialty Crops Branch, (202) 694–5260.

Table 5-40.—Ginger Root: Area, yield, production, marketing year average price, and value, Hawaii, 1995/96–2004/2005

Year	Area harvested	Yield per acre	Total production	Price per pound	Value
	Acres	1,000 pounds	1,000 pounds	Cents	1,000 dollars
1995–96 ...	200	47.0	9,400	75.0	7,050
1996–97 ...	275	44.0	12,100	67.0	8,107
1997–98 ...	360	50.0	18,000	40.0	7,200
1998–99 ...	350	46.0	16,100	50.0	8,050
1999–2000	270	50.0	13,500	66.0	8,910
2000–2001	360	50.0	18,000	45.0	8,100
2001–2002	320	45.0	14,400	30.0	4,320
2002–2003	160	37.5	6,000	60.0	3,600
2003–2004	150	40.0	6,000	90.0	5,400
2004–2005	120	42.5	5,100	80.0	4,080

NASS, Crops Branch, (202) 720–2127.

Table 5-41.—Grapes: Production, price, and value, United States, 1996–2005

Year	Grapes			
	Production (fresh basis)		Market year average price per ton [1]	Value
	Total	Utilized		
	Tons	Tons	Dollars	1,000 dollars
1996	5,553,600	5,537,325	429.00	2,376,111
1997	7,290,900	7,287,365	429.00	3,126,537
1998	5,819,950	5,816,405	454.00	2,640,470
1999	6,235,910	6,234,380	469.00	2,926,745
2000	7,687,970	7,687,330	403.00	3,098,427
2001	6,569,250	6,568,100	449.00	2,947,867
2002	7,338,900	7,336,810	387.00	2,841,569
2003	6,643,530	6,489,630	402.00	2,609,289
2004	6,240,030	6,229,930	483.00	3,010,958
2005 [2]	6,974,900	6,971,650	432.00	3,013,418

[1] Fresh fruit prices are equivalent packinghouse-door returns for California and Washington, and the average price as sold for other States. Quantities processed are priced at the equivalent processing plant door level.   [2] Preliminary.
NASS, Crops Branch, (202) 720–2127.

Table 5-42.—Grapes: Production and marketing year average price per ton, by States, 2003–2005

State	Total production			Utilized production			Price per ton [1]		
	2003	2004	2005 [2]	2003	2004	2005 [2]	2003	2004	2005 [2]
	Tons	Tons	Tons	Tons	Tons	Tons	Dollars	Dollars	Dollars
AZ	8,000	4,000	1,000	8,000	4,000	1,000	1,030.00	334.00	550.00
AR	2,400	3,000	1,900	2,300	2,700	1,900	485.00	502.00	539.00
CA:									
All types	5,861,000	5,623,000	6,130,000	5,786,000	5,623,000	6,130,000	402.00	492.00	445.00
Wine	2,909,000	2,815,000	3,200,000	2,909,000	2,815,000	3,200,000	530.00	570.00	570.00
Table [3]	732,000	770,000	830,000	678,000	770,000	830,000	601.00	695.00	373.00
Raisin [3]	2,220,000	2,038,000	2,100,000	2,199,000	2,038,000	2,100,000	170.00	306.00	283.00
GA	3,100	3,300	3,500	2,800	3,200	3,500	978.00	1,160.00	1,390.00
MI	94,500	62,500	100,000	80,500	58,000	100,000	262.00	242.00	214.00
MO	3,030	3,630	3,900	3,030	3,630	3,900	610.00	720.00	774.00
NY	198,000	142,000	178,000	152,000	142,000	178,000	252.00	229.00	193.00
NC	2,800	3,500	3,900	2,800	3,500	3,900	1,070.00	962.00	937.00
OH	8,100	4,800	8,500	7,000	4,800	8,500	350.00	417.00	318.00
OR	24,000	24,000	24,600	24,000	19,400	22,800	1,510.00	1,660.00	1,580.00
PA	85,000	86,800	90,000	68,000	86,800	90,000	231.00	221.00	218.00
TX	6,000	8,800	9,700	5,800	8,500	8,500	900.00	919.00	1,250.00
VA	3,600	3,700	4,900	3,400	3,400	4,650	1,300.00	1,300.00	1,370.00
WA:.									
All types	344,000	267,000	415,000	344,000	267,000	415,000	420.00	456.00	342.00
Wine	112,000	107,000	110,000	112,000	107,000	110,000	920.00	925.00	930.00
Juice [4]	232,000	160,000	305,000	232,000	160,000	305,000	178.00	143.00	....
US	6,643,530	6,240,030	6,974,900	6,489,630	6,229,930	6,971,650	402.00	483.00	432.00

[1] Fresh fruit prices are equivalent packinghouse-door returns for CA and WA, and the average price as sold for other States. Quantities processed are priced at the equivalent processing plant door level.   [2] Preliminary.   [3] Fresh equivalent of dried and not dried.   [4] Official estimate of price for 2005 is not published.
NASS, Crops Branch, (202) 720–2127.

Table 5-43.—Grapes: Production and utilization, United States, 1996–2005

| Crop of— | Total production[1] | Utilized production | Fresh | Utilization of quantities sold |
| | | | | Processed |
				Canned	Dried (fresh basis)	Crushed for wine	Crushed for juice, etc.[2]
	Tons	Tons	Tons	Tons	Tons	Tons	Tons
1996	5,553,600	5,537,325	767,025	36,000	1,329,000	3,042,850	362,450
1997	7,290,900	7,287,365	937,115	44,000	1,806,500	4,034,400	465,350
1998	5,819,950	5,816,405	780,795	36,000	1,331,600	3,314,760	353,250
1999	6,235,910	6,234,380	887,161	35,000	1,459,900	3,350,419	501,900
2000	7,687,970	7,687,330	906,825	32,000	2,194,600	4,129,655	424,250
2001	6,569,250	6,568,100	864,330	29,000	1,736,800	3,568,190	369,780
2002	7,338,900	7,336,810	982,340	31,000	1,907,000	3,998,970	417,500
2003	6,643,530	6,489,630	805,460	27,000	1,597,000	3,581,420	478,750
2004	6,240,030	6,229,930	882,580	25,000	1,107,000	3,818,130	397,220
2005[3]	6,974,900	6,971,650	947,370	16,000	1,325,000	4,068,070	615,210

[1] Total production includes utilized production plus production not harvested and harvested not sold: 1996—16,275 tons fresh equivalent; 1997—3,535 tons fresh equivalent; 1998—3,545 tons fresh equivalent; 1999—1,530 tons fresh equivalent; 2000—640 tons fresh equivalent; 2001—1,150 tons fresh equivalent; 2002—2,090 tons fresh equivalent; 2003—153,900 tons fresh equivalent; 2004—10,100 tons fresh equivalent; and 2005—3,250 tons fresh equivalent.　[2] Mostly juice, but includes some quantities used for jam, jelly, etc.　[3] Preliminary.

NASS, Crops Branch, (202) 720–2127.

Table 5-44.—Grapes: Production and utilization, by States, crop of 2005 (preliminary)

State	Total production	Utilized production	Fresh	Utilization
				Processed
				Canned	Dried (fresh basis)[1]	Crushed for—
						Wine	Juice, etc.[2]
	Tons	Tons	Tons	Tons	Tons	Tons	Tons
AZ	1,000	1,000					
AR	1,900	1,900					
CA:							
All types	6,130,000	6,130,000	940,000		1,325,000	3,849,000	
Wine	3,200,000	3,200,000	50,000			3,150,000	
Table	830,000	830,000	682,000		15,000	133,000	
Raisin	2,100,000	2,100,000	208,000	16,000	1,310,000	566,000	
GA	3,500	3,500					
MI	100,000	100,000	700			4,600	94,700
MO	3,900	3,900	60			3,830	
NY	178,000	178,000	3,000			40,000	135,000
NC	3,900	3,900	200			3,700	
OH	8,500	8,500	100			1,400	7,000
OR	24,600	22,800				22,800	
PA	90,000	90,000	500			16,300	73,200
TX	9,700	8,500				8,500	
VA	4,900	4,650				4,650	
WA:.							
All types	415,000	415,000				110,000	305,000
Wine	110,000	110,000				110,000	
Juice	305,000	305,000					305,000
Other States			2,810			3,290	310
US	6,974,900	6,971,650	947,370	16,000	1,325,000	4,068,070	615,210

[1] Equivalent raisins produced (dried basis): 276,100 tons.　[2] Mostly juice, but includes some quantities used for jam, jelly, etc.

NASS, Crops Branch, (202) 720–2127.

Table 5-45.—Raisins and currants: United States exports by country of destination, 2002/2003–2004/2005

Country of destination	Year beginning August		
	2002/2003	2003/2004	2004/2005
	Metric tons	Metric tons	Metric tons
North America:			
Canada	13,305.9	12,799.5	12,160.8
Mexico	1,261.9	2,342.0	3,318.3
Total	14,567.9	15,141.5	15,479.1
Carribean:			
Trinidad And Tobago	235.0	404.5	407.4
Dominican Republic	406.1	346.2	367.1
Jamaica	41.2	111.5	176.9
Bermuda	37.0	18.8	86.0
Netherlands Antilles	54.9	58.1	76.9
Bahamas, The	80.4	67.9	60.1
Haiti	24.8	43.7	14.1
Cuba	10.8	115.7	11.0
Leeward-Windward Islands	0.7	0.7	2.1
Barbados	0.0	17.6	0.6
Total	890.9	1,184.7	1,202.3
Central America:			
Panama	280.8	400.5	412.0
Costa Rica	83.0	127.2	126.6
El Salvador	118.0	141.0	113.9
Guatemala	118.8	301.4	96.3
Nicaragua	77.4	66.8	81.4
Honduras	47.0	78.5	77.3
Belize	0.0	2.8	0.0
Total	725.0	1,118.1	907.5
South America:			
Brazil	481.9	592.1	421.1
Venezuela	282.9	389.5	350.2
Colombia	90.7	62.2	132.7
Ecuador	36.6	27.4	60.6
Chile	33.9	110.3	38.1
Guyana	0.0	12.8	16.3
Peru	4.0	0.0	0.0
Total	930.0	1,194.2	1,019.0
European Union:			
United Kingdom	22,787.8	23,541.0	19,571.3
Germany	7,389.3	6,958.3	7,308.9
Sweden	3,747.1	4,093.3	5,210.6
Netherlands	3,391.2	2,885.4	2,758.7
Denmark	3,298.1	2,915.2	2,180.6
Finland	1,964.9	1,571.8	1,901.5
France	1,121.8	1,409.6	681.8
Greece	100.0	118.9	669.0
Poland	679.6	588.1	607.3
Spain	716.2	777.4	513.3
Belgium-Luxembourg	1,123.0	678.7	461.1
Czech Republic	527.9	403.8	423.2
Ireland	142.5	392.8	299.7
Slovakia	39.5	203.0	248.2
Italy	160.4	181.8	214.1
Latvia	85.2	99.3	172.4
Hungary	37.0	55.4	150.2
Austria	20.4	78.5	80.0
Lithuania	0.0	41.5	48.4
Estonia	0.0	0.0	41.5
Portugal	0.0	0.0	20.0
Total	47,331.8	46,993.8	43,561.7
Other Europe:			
Switzerland	2,118.6	2,300.8	2,487.2
Norway	2,121.6	2,276.5	1,876.3
Iceland	105.8	99.0	149.8
Romania	0.0	20.8	0.0
Total	4,346.0	4,697.1	4,513.4
Former Soviet Union:			
Ukraine	270.5	210.3	207.1
Russian Federation	23.7	70.7	38.7
Armenia, Republic Of	18.6	0.0	0.0
Total	312.8	281.0	245.8

See end of table.

Table 5-45.—Raisins and currants: United States exports by country of destination, 2002/2003–2004/2005—Continued

Country of destination	Year beginning August		
	2002/2003	2003/2004	2004/2005
	Metric tons	Metric tons	Metric tons
East Asia:			
Japan	20,374.9	20,024.7	17,398.7
China, Peoples Repub	4,133.8	5,611.6	8,549.5
Taiwan	4,324.4	3,837.1	2,578.7
Korea, Republic Of	2,691.4	2,622.0	2,447.7
Hong Kong	1,589.6	1,059.5	1,987.8
Mongolia	18.5	0.0	0.0
Total	33,132.7	33,154.9	32,962.4
Middle East:			
Saudi Arabia	1,751.4	2,831.0	2,374.5
Israel	1,722.0	2,171.8	1,732.0
United Arab Emirates	259.4	447.6	465.5
Kuwait	340.4	324.1	403.0
Bahrain	0.0	37.4	38.9
Lebanon	3.0	9.6	17.0
Qatar	18.7	0.0	0.0
Turkey	13.0	0.0	0.0
Total	4,107.9	5,821.6	5,030.7
North Africa:			
Egypt	87.5	9.8	19.6
Algeria	51.0	429.4	0.0
Morocco	54.0	0.0	0.0
Tunisia	0.0	120.4	0.0
Total	192.6	559.6	19.6
Sub-Saharan Africa:			
South Africa, Republic	0.0	0.0	42.5
Mauritius	0.0	0.0	2.0
Mali	17.6	0.0	0.0
Swaziland	17.9	0.0	0.0
French Ind. Ocean Terr.	94.1	0.0	0.0
Total	129.6	0.0	44.5
South Asia:			
Pakistan	0.0	25.5	214.0
Afghanistan	16.5	0.0	34.0
Sri Lanka	13.2	0.0	19.1
India	40.8	8.4	16.1
Total	70.5	33.8	283.1
Southeast Asia:			
Malaysia	1,984.5	2,111.5	1,942.3
Singapore	1,892.8	1,722.1	1,434.2
Philippines	1,306.4	1,202.4	1,358.9
Thailand	366.9	584.4	693.1
Vietnam	452.6	416.2	304.2
Indonesia	235.8	335.5	222.7
Brunei	11.7	19.0	11.0
Total	6,250.6	6,391.1	5,966.4
Oceania:			
Australia	1,880.0	1,924.2	1,232.5
New Zealand	1,008.2	1,287.4	1,129.9
French Pacific Island	6.4	0.0	2.3
Total	2,894.6	3,211.6	2,364.6
Grand total	115,882.9	119,783.1	113,600.0

FAS, Horticultural and Tropical Products Division, (202) 720-3423. The division's Home Page is located at, www.fas.usda.gov/htp. You can also email the division at htp@fas.usda.gov. Note: Compiled from reports of the U.S. Department of Commerce. U.S. trade can be obtained from the following web site: U.S. agricultural trade database, www.fas.usda.gov/ustrade.

Table 5-46.—Grapes and raisins: Foreign trade, United States, 1995–2004

Year beginning October	Grapes		Raisins[1]	
	Imports, fresh	Domestic exports, fresh	Imports for consumption	Domestic exports
	Metric tons	*Metric tons*	*Metric tons*	*Metric tons*
1995	341,098	240,097	13,528	128,330
1996	351,567	233,558	12,319	124,217
1997	419,956	217,467	11,793	130,597
1998	387,165	222,317	28,328	112,650
1999	452,182	272,901	18,283	90,539
2000	418,012	303,396	12,571	118,838
2001	501,055	293,754	16,421	125,319
2002	564,512	307,602	15,416	121,438
2003	532,746	321,079	11,955	134,329
2004	614,633	301,206	23,301	119,454

Raisins converted to sweatbox or production basis by multiplying by 1.08.
ERS, Specialty Crops Branch, (202) 694–5260.

Table 5-47.—Guavas: Area, yield, utilized production, marketing year average price, and value, Hawaii, 1996–2005

Year	Area harvested	Yield per acre	Production	Price per pound	Value
	Acres	*1,000 pounds*	*1,000 pounds*	*Cents*	*1,000 dollars*
1996	750	21.7	16,300	13.8	2,249
1997	730	21.8	15,900	12.2	1,940
1998	710	20.6	14,600	12.2	1,781
1999	630	17.0	10,700	12.0	1,284
2000	680	23.4	15,900	12.9	2,051
2001	610	25.1	15,300	14.1	2,157
2002	550	17.6	9,700	15.0	1,455
2003	530	12.6	6,700	13.8	925
2004	500	16.2	8,100	14.4	1,166
2005	NA	NA	NA	NA	NA

NA-not available.
NASS, Crops Branch, (202) 720–2127.

Table 5-48.—Nectarines: Production, utilization, and value, United States, 1996–2005 [1]

Crop of—	Production	Utilization		Marketing year average price per ton[3]	Value
		Fresh[2]	Processed (fresh basis)		
	Tons	*Tons*	*Tons*	*Dollars*	*1,000 dollars*
1996	247,000	239,800	7,200	474.00	116,977
1997	264,000	258,500	5,500	375.00	98,895
1998	224,000	207,600	16,400	471.00	105,466
1999	274,000	256,300	17,700	411.00	112,497
2000	267,000	260,700	6,300	398.00	106,256
2001	275,000	265,400	9,600	464.00	127,642
2002	300,000	300,000	0	382.00	114,600
2003	273,000	273,000	0	436.00	119,028
2004	269,000	252,000	(5)	342.00	86,184
2005[4]	249,400	249,400	(5)	521.00	129,969

[1] Washington added in 2005, prior years are California only.   [2] Includes "Home use."   [3] Processing fruit prices are equivalent returns at processing plant door.   [4] Preliminary.   [5] Small quantities of processed nectarines are included in fresh to avoid disclosure of individual operations.
NASS, Crops Branch, (202) 720–2127.

Table 5-49.—Olives: Total production, marketing year average price, value, and processed utilization, California, 1996–2005

Year	Production	Marketing year average price per ton	Value	Processed utilization			
				Crushed for oil	Canned	Limited	Undersized
	Tons	*Dollars*	*1,000 dollars*	*Tons*	*Tons*	*Tons*	*Tons*
1996	166,000	617	102,364	7,000	123,000	29,000	6,500
1997	104,000	642	66,801	3,600	82,200	10,200	7,500
1998	90,000	459	41,331	4,100	64,200	12,800	8,400
1999	142,000	387	55,011	5,000	86,000	36,500	14,000
2000	53,000	656	34,743	3,000	41,400	5,100	3,000
2001	134,000	672	90,096	3,000	109,700	15,300	5,500
2002	103,000	573	58,983	6,000	82,800	9,900	3,800
2003	118,000	409	48,289	7,500	96,000	10,500	3,500
2004	104,000	571	59,379	8,000	74,400	16,100	5,000
2005 [1]	139,000	548	76,126	11,000	100,000	21,200	6,30 0

[1] Preliminary.
NASS, Crops Branch, (202) 720–2127.

Table 5-50.—Olives and olive oil: Foreign trade, United States, 1995–2004

Year beginning October	Imports			
	Olives		Olive oil	
	In brine	Dried	Edible	Inedible
	Metric tons	*Metric tons*	*Metric tons*	*Metric tons*
1995	66,725	396	113,588	126
1996	74,759	271	148,052	43
1997	82,513	402	161,014	0
1998	87,594	438	170,087	1
1999	89,920	314	189,302	0
2000	102,652	415	212,341	0
2001	105,076	367	217,649	276
2002	111,925	464	219,883	97
2003	108,734	504	244,976	26
2004	112,473	726	248,299	78

ERS, Specialty Crops Branch, (202) 694–5260.

Table 5-51.—Peaches: Production and value, United States, 1996–2005

Year	Total production	Utilized production	Marketing year average price [1]	Value
	Million pounds	Million pounds	Cents per pound	1,000 dollars
1996	2,104.6	2,043.8	19.1	389,894
1997	2,624.6	2,508.4	17.7	444,137
1998	2,379.2	2,304.2	18.9	434,889
1999	2,503.3	2,411.3	18.7	451,728
2000	2,551.4	2,460.9	19.1	470,399
2001	2,407.8	2,309.9	20.9	483,043
2002	2,535.0	2,435.4	20.0	488,011
2003	2,519.0	2,410.3	18.9	454,286
2004	2,614.2	2,459.6	18.8	461,629
2005 [2]	2,365.2	2,286.3	18.8	509,745

[1] Fresh fruit prices are equivalent packinghouse-door returns for CA and WA except equivalent returns for bulk fruit at the first delivery point for CA Clingstone, and the average price as sold for other States. Quantities processed are priced at the equivalent processing plant door level.   [2] Preliminary.

NASS, Crops Branch, (202) 720–2127.

Table 5-52.—Peaches: Foreign trade, United States, 1995–2004

Year beginning October	Domestic exports				
	Fresh	Canned	Canned, in fruit salad [2]	Dried, in fruit salad [1][3]	Total, in terms of fresh [4]
	Metric tons	Metric tons	Metric tons	Metric tons	Metric tons
1995	74,822	17,777	10,696	884	108,208
1996	103,345	16,873	8,990	675	132,958
1997	79,987	21,580	9,189	728	114,801
1998	98,123	27,170	10,225	579	138,732
1999	113,098	16,875	7,869	299	139,506
2000	129,292	13,008	4,677	532	149,931
2001	127,434	10,922	3,885	344	144,152
2002	120,802	29,850	3,478	685	157,937
2003	112,506	42,418	4,438	827	163,955
2004	103,578	32,902	7,237	1,224	150,518

[1] Net processed weight.   [2] Canned peaches are 40 percent of total canned fruit for salad.   [3] Dried peaches are 21 percent of total dried fruit for salad.   [4] Dried fruit converted to unprocessed dry weight by dividing by 1.08. Unprocessed dry weight converted to terms of fresh fruit on the basis that 1 pound dried equals 6.0 pounds fresh. Canned peaches converted to terms of fresh on basis that 1 pound canned equals about 1 pound fresh.

ERS, Specialty Crops Branch, (202) 694–5260.

Table 5-53.—Peaches: Production and utilization, United States, 1996–2005

Crop of—	Total production [1]	Utilized production	Utilization of quantities sold				
			Fresh [2]	Processed (fresh basis)			
				Canned	Dried	Frozen	Other [3]
	Million pounds	Million pounds	Million pounds	Million pounds	Million pounds	Million pounds	Million pounds
1996	2,104.6	2,043.8	769.8	994.3	32.7	183.2	63.8
1997	2,624.6	2,508.4	1,126.8	1,107.8	34.1	201.0	38.7
1998	2,379.2	2,304.2	979.2	985.1	25.0	185.8	129.1
1999	2,503.3	2,411.3	1,077.8	995.9	31.4	204.1	102.1
2000	2,551.4	2,460.9	1,133.4	1,026.6	25.2	219.5	56.2
2001	2,407.8	2,309.9	1,129.3	906.4	29.2	200.7	44.3
2002	2,535.0	2,435.4	1,074.5	1,061.0	28.4	204.3	67.2
2003	2,519.0	2,410.3	1,085.5	997.0	20.3	222.9	84.6
2004	2,614.2	2,459.6	1,071.2	1,047.9	20.8	211.5	108.2
2005 [4]	2,365.2	2,286.3	1,007.1	953.1	25.4	198.8	101.9

[1] Includes harvested not sold and unharvested production for California Clingstone peaches.   [2] Includes "Home use."   [3] Used for jams, preserves, pickles, wine, brandy, baby food, etc. Includes small quantities frozen for some years.   [4] Preliminary.

NASS, Crops Branch, (202) 720–2127.

Table 5-54.—Peaches: Production and season average price per pound, 2003–2005

State	Total production			Utilized production			Price[2] for crop of—		
	2003	2004	2005[1]	2003	2004	2005[1]	2003	2004	2005[1]
	Tons	Tons	Tons	Tons	Tons	Tons	Dollars	Dollars	Dollars
AL	4,500	14,000	12,000	3,850	10,000	10,000	940.00	661.00	808.00
AR	4,450	4,500	4,950	4,000	3,700	4,650	942.00	842.00	1,100.00
CA:									
Freestone	413,000	436,000	385,000	413,000	390,000	385,000	336.00	281.00	408.00
CO	10,500	13,000	12,000	10,000	12,000	11,000	1,220.00	944.00	1,080.00
CT	750	850	700	750	850	700	1,400.00	1,600.00	1,600.00
GA	55,000	52,500	40,000	53,000	49,500	37,000	559.00	667.00	743.00
ID	6,500	9,000	8,000	6,300	8,500	8,000	713.00	752.00	967.00
IL	10,250	10,600	11,200	9,900	9,500	11,000	1,020.00	770.00	1,250.00
IN[3]	1,700	1,200	....	1,650	1,200	....	1,430.00	1,380.00	.
KY	900	800	750	900	750	650	1,110.00	1,290.00	1,000.00
LA	800	850	650	750	800	650	1,630.00	1,420.00	1,730.00
MD	4,250	4,100	4,200	4,250	4,100	4,200	947.00	558.00	919.00
MA	1,500	960	1,000	1,350	950	990	1,600.00	1,500.00	1,500.00
MI	23,500	18,700	11,200	21,500	18,700	11,200	362.00	549.00	607.00
MO	5,000	4,500	5,800	5,000	4,500	5,800	860.00	780.00	980.00
NJ	35,000	32,500	35,000	31,000	30,500	33,700	780.00	760.00	916.00
NY	6,500	6,000	4,000	6,000	5,900	3,850	703.00	717.00	709.00
NC	3,000	3,500	6,000	3,000	3,500	6,000	800.00	840.00	850.00
OH	5,650	5,100	2,100	5,500	5,000	2,100	996.00	1,000.00	1,100.00
OK	1,500	2,000	2,000	1,400	2,000	1,860	1,240.00	1,030.00	1,010.00
OR	2,250	3,300	2,800	1,750	3,200	2,700	1,120.00	867.00	966.00
PA	36,500	23,000	26,600	35,500	22,700	26,600	659.00	710.00	734.00
SC	50,000	70,000	75,000	40,000	55,000	45,000	723.00	557.00	703.00
TN	1,750	1,950	2,000	1,600	1,800	1,800	1,030.00	1,070.00	1,280.00
TX	3,500	12,200	8,750	3,350	9,900	8,350	1,460.00	1,520.00	1,680.00
UT	4,500	5,000	4,700	4,350	4,550	4,420	789.00	627.00	775.00
VA	5,000	4,500	4,700	4,600	4,400	4,630	756.00	660.00	800.00
WA	19,500	21,500	22,000	19,500	21,500	22,000	473.00	349.00	535.00
WV	6,250	6,000	5,500	5,900	5,800	5,300	740.00	478.00	724.00
Total above	723,500	768,110	698,600	699,650	690,800	659,150	494.00	463.00	588.00
CA:									
Clingstone	536,000	539,000	484,000	505,500	539,000	484,000	215.00	263.00	252.00
US	1,259,500	1,307,110	1,182,600	1,205,150	1,229,800	1,143,150	377.00	375.00	446.00

[1] Preliminary.   [2] Fresh fruit prices are equivalent packinghouse-door returns for CA and WA except equivalent returns for bulk fruit at the first delivery point for CA Clingstone, and the average price as sold for other States. Quantities processed are priced at the equivalent processing plant door level.   [3] Estimates discontinued in 2005.
NASS, Crops Branch, (202) 720–2127.

Table 5-55.—Peaches: Production and utilization, by States, crop of 2005 (preliminary)

State	Total production	Utilized production[1]	Utilization				
			Fresh[2]	Processed (fresh basis)			
				Canned	Dried	Frozen	Other[3]
	Tons	Tons	Tons	Tons	Tons	Tons	Tons
CA, all[3]	869,000	869,000	252,000	....	....	....	....
Clingstone[3]	484,000	484,000	....	459,000	....	....	....
Freestone	385,000	385,000	252,000	....	12,700	82,800	....
GA	40,000	37,000	....	....	....	....	....
NJ	35,000	33,700	....	....	....	....	....
PA	26,600	26,600	....	....	....	....	....
SC	75,000	45,000	42,000	....	....	....	....
WA	22,000	22,000	....	....	....	....	....
Other States	115,000	109,850	209,550	17,560	....	16,600	50,940
US	1,182,600	1,143,150	503,550	476,560	12,700	99,400	50,940

[1] Difference between total and utilized production is harvested not sold and unharvested production.   [2] Includes "Home use."   [3] Used for jams, preserves, brandy, etc.
NASS, Crops Branch, (202) 720–2127.

Table 5-56.—Peaches, canned: United States exports by country of destination, 2002/2003–2004/2005 [1]

Country of destination	Year beginning June		
	2002/2003	2003/2004	2004/2005
	Metric tons	Metric tons	Metric tons
Mexico	9,097	12,032	12,691
Canada	4,496	7,260	9,693
Thailand	2,737	577	5,041
Taiwan	173	487	1,939
Australia	0	559	1,797
Guatemala	57	29	1,211
China	60	4,452	954
Japan	171	277	257
Philippines	1,375	9	228
United Kingdom	158	0	206
New Zealand	0	52	178
Panama	444	409	173
Singapore	126	148	144
Honduras	15	15	137
Israel	54	215	134
Costa Rica	243	86	132
Bahamas, The	19	38	80
Netherlands	0	8,199	24
Germany	0	4,738	0
Greece	589	1,008	0
Italy	0	3,024	0
Other	586	638	316
World total [2]	20,398	44,254	35,334

[1] One metric ton equals 48.99 standard cases of 24 x 2 1/2 cans.   [2] Totals may not add due to rounding.

FAS, Horticultural and Tropical Products Division, (202) 720-3423. The division's Home Page is located at, www.fas.usda.gov/htp. You can also email the division at htp@fas.usda.gov. Note: Compiled from reports of the U.S. Department of Commerce. U.S. trade can be obtained from the following web site: U.S. agricultural trade database, www.fas.usda.gov/ustrade.

Table 5-57.—Pineapples: Total area, utilized production, utilization, marketing year average price, and value, Hawaii, 1996–2005

Year	Total area	Utilized production	Utilization		Price per ton	Value
			Fresh	Processed		
	Acres	Tons	Tons	Tons	Dollars	1,000 dollars
1996	20,000	347,000	115,000	232,000	276	95,914
1997	19,900	324,000	103,000	221,000	283	91,721
1998	21,000	332,000	111,000	221,000	279	92,776
1999	21,000	352,000	122,000	230,000	288	101,448
2000	20,700	354,000	122,000	232,000	287	101,530
2001	20,100	323,000	110,000	213,000	298	96,337
2002	19,100	320,000	117,000	203,000	314	100,616
2003	16,000	300,000	130,000	170,000	338	101,470
2004	13,000	220,000	104,000	116,000	378	83,104
2005 [1]	14,000	212,000	106,000	106,000	374	79,288

[1] Preliminary.

NASS, Crops Branch, (202) 720–2127.

Table 5-58.—Pears: Production and value, United States 1996–2005

Year	Total production	Utilized production	Marketing year average price [1]	Value
	Tons	Tons	Dollars per ton	1,000 dollars
1996	820,550	820,250	376.00	308,367
1997	1,042,500	1,041,930	276.00	287,822
1998	990,140	987,795	294.00	290,331
1999	1,044,250	1,042,235	294.00	306,505
2000	993,250	975,270	267.00	260,626
2001	1,026,930	989,430	266.00	263,431
2002	890,020	888,570	297.00	264,334
2003	934,050	928,450	294.00	273,142
2004	877,260	872,400	340.00	296,291
2005 [2] ...	812,325	811,670	388.00	315,240

[1] Fresh fruit prices are equivalent packinghouse-door returns for CA, OR, and WA, and the average price as sold for other States. Quantities processed are priced at the equivalent processing plant door level. [2] Preliminary.
NASS, Crops Branch, (202) 720–2127.

Table 5-59.—Pears: Production and season average price per ton, by States, 2003–2005

Variety and State	Total production			Utilized production			Price [2] for crop of—		
	2003	2004	2005 [1]	2003	2004	2005 [1]	2003	2004	2005 [1]
	Tons	Tons	Tons	Tons	Tons	Tons	Dollars	Dollars	Dollars
CA, all	272,000	271,000	200,000	272,000	271,000	200,000	233.00	297.00	399.00
Bartlett	217,000	223,000	164,000	217,000	223,000	164,000	216.00	252.00	301.00
Other	55,000	48,000	36,000	55,000	48,000	36,000	300.00	503.00	846.00
CO	2,800	2,600	2,500	2,800	2,500	2,200	600.00	561.00	455.00
CT	1,300	900	1,000	1,270	900	1,000	1,000.00	800.00	952.00
MI	4,800	3,460	2,000	4,300	3,400	1,970	259.00	311.00	423.00
NY	15,500	16,500	8,500	14,800	13,900	8,200	373.00	386.00	499.00
OR, all	210,000	212,000	196,000	206,000	210,000	196,000	327.00	365.00	376.00
Bartlett	54,000	63,000	58,000	54,000	61,000	58,000	334.00	345.00	344.00
Other	156,000	149,000	138,000	152,000	149,000	138,000	324.00	374.00	390.00
PA	5,200	4,500	2,100	4,900	4,400	2,100	697.00	564.00	597.00
UT	450	300	225	380	300	200	784.00	393.00	645.00
WA, all	422,000	366,000	400,000	422,000	366,000	400,000	306.00	350.00	384.00
Bartlett	185,000	171,000	170,000	185,000	171,000	170,000	321.00	296.00	340.00
Other	237,000	195,000	230,000	237,000	195,000	230,000	295.00	397.00	416.00
US	934,050	877,260	812,325	928,450	872,400	811,670	294.00	340.00	388.00

[1] Preliminary. [2] Fresh fruit prices are equivalent packinghouse-door returns for CA, OR, and WA, and the average price as sold for other States. Quantities processed are priced at the equivalent processing plant door level.
NASS, Crops Branch, (202) 720–2127.

Table 5-60.—Pears: Foreign trade, United States, 1995–2004

Year beginning October	Imports for consumption, fresh	Domestic exports				
		Fresh [1]	Canned	Dried, in fruit salad [1 2]	Canned, in fruit salad [3]	Total, in terms of fresh fruit [4]
	Metric tons	Metric tons	Metric tons	Metric tons	Metric tons	Metric tons
1995	57,371	144,426	7,245	702	9,359	165,460
1996	78,521	126,409	3,494	536	7,866	141,150
1997	67,636	156,996	5,228	578	8,041	173,911
1998	87,422	145,725	4,327	459	8,947	161,897
1999	89,827	162,629	4,655	238	6,885	175,669
2000	85,219	158,333	5,887	422	4,092	170,976
2001	79,967	175,346	6,181	273	3,400	186,649
2002	86,328	160,240	4,944	544	3,043	171,659
2003	66,923	167,084	4,952	656	3,883	180,061
2004	76,834	141,976	10,129	972	6,332	164,567

[1] Net processed weight. [2] Dried pears are 16⅔ percent of total dried fruit for salad. [3] Canned pears are 35 percent of total canned fruit for salad. [4] Dried converted to unprocessed dry weight by dividing by 1.03. Unprocessed dry weight converted to terms of fresh on the basis that 1 pound dried equals about 6.5 pounds fresh. Canned converted to terms of fresh on basis that 1 pound of canned equals about 1 pound fresh.
ERS, Specialty Crops Branch, (202) 694–5260.

Table 5-61.—Pears, fresh: Production in specified countries, 2003/2004–2005/2006

Continent and country	2003/2004	2004/2005	2005/2006
	1,000 metric tons	*1,000 metric tons*	*1,000 metric tons*
North America:			
Canada	15	14	15
Mexico	29	33	30
United States	847	808	774
Total	892	855	819
South America:			
Argentina [1]	525	640	NA
Chile [1]	257	257	NA
Total	782	897	NA
European Union:			
Belgium-Luxembourg [2]	176	NA	NA
France [2]	254	NA	NA
Germany	49	61	43
Greece	30	41	44
Italy	872	895	882
Netherlands [2]	160	NA	NA
Spain	704	660	639
Sweden [2]	8	9	NA
United Kingdom	30	23	17
Total EU	2,282	1,689	1,624
Other Europe:			
Russian Federation	278	336	320
Turkey	370	320	350
Total	648	656	670
Total Europe	2,930	2,345	2,294
Africa:			
South Africa	308	303	NA
Total	308	303	NA
Asia:			
China	9,798	10,640	11,200
Japan	366	352	387
Total	10,164	10,992	11,587
Oceania:			
Australia	141	150	NA
New Zealand	12	7	NA
Total	153	157	NA
Total selected countries	15,229	15,549	14,700

[1] It is too early to make reliable forecast for the Southern Hemisphere countries for the 2005/06 season.   [2] Belgium-Luxembourg, Sweden, France and the Netherlands are no longer reporting.   NA-not available.

FAS, Horticultural and Tropical Products Division, (202) 720-3423. The division's Home Page is located at, www.fas.usda.gov/htp. You can also email the division at htp@fas.usda.gov. Note: Prepared or estimated on the basis of official statistics of foreign governments, other foreign source materials, reports of U.S. Agricultural Counselors and Attachés, results of office research, and related information. Note: This information can be obtained from the following web site: www.fas.usda.gov/psd.

Table 5-62.—Pears: Production and utilization, by States, crop of 2005 (preliminary)

State and variety	Total production	Utilized production	Utilization Fresh [1]	Utilization Processed [2]
	Tons	Tons	Tons	Tons
CA, all	200,000	200,000	([3])	([3])
Bartlett	164,000	164,000	46,000	118,000
Other	36,000	36,000	([3])	([3])
CO	2,500	2,200	([3])	([3])
CT	1,000	1,000	([3])	([3])
MI	2,000	1,970	([3])	([3])
NY	8,500	8,200	([3])	([3])
OR, all	196,000	196,000	([3])	([3])
Bartlett	58,000	58,000	28,000	30,000
Other	138,000	138,000	([3])	([3])
PA	2,100	2,100	([3])	([3])
UT	225	200	([3])	([3])
WA, all	400,000	400,000	([3])	([3])
Bartlett	170,000	170,000	64,000	106,000
Other	230,000	230,000	([3])	([3])
US	812,325	811,670	512,400	299,270

[1] Includes "Home use." [2] Mostly canned, but includes small quantities dried, juiced, and other uses. [3] Data not published to avoid disclosure of individual operations, but included in U.S. totals.
NASS, Crops Branch, (202) 720–2127.

Table 5-63.—Pears: Production and utilization, United States, 1996–2005

Crop of—	Total production	Utilized production	Utilization of quantities sold—Fresh [1]
	Tons	Tons	Tons
1996	820,550	820,250	459,550
1997	1,042,500	1,041,930	572,310
1998	990,140	987,795	533,795
1999	1,044,250	1,042,235	564,975
2000	993,250	975,270	573,230
2001	1,026,930	989,430	568,320
2002	890,020	888,570	524,440
2003	934,050	928,450	559,950
2004	877,260	872,400	513,270
2005 [2]	812,325	811,670	512,400

[1] Includes "Home use." [2] Preliminary.
NASS, Crops Branch, (202) 720–2127.

Table 5-64.—Papayas: Area, utilized production, utilization, marketing year average price, and value, Hawaii, 1996–2005

Year	Area harvested	Utilized production	Utilization Fresh	Utilization Processed	Price per pound	Value
	Acres	1,000 pounds	1,000 pounds	1,000 pounds	Cents	dollars
1996	1,835	41,800	37,800	4,000	40.8	17,054
1997	1,985	38,800	35,700	3,100	48.9	18,978
1998	2,120	39,900	35,600	4,300	31.6	12,589
1999	1,940	42,400	39,400	3,000	37.6	15,929
2000	1,650	54,500	50,250	4,250	29.4	16,007
2001	1,950	55,000	52,000	3,000	26.5	14,598
2002	1,720	45,900	42,700	3,200	26.0	11,924
2003	1,565	42,600	40,800	1,800	30.7	13,069
2004	1,235	35,800	34,100	1,700	34.5	12,361
2005 [1]	1,450	32,500	30,200	2,300	33.8	10,971

[1] Preliminary.
NASS, Crops Branch, (202) 720–2127.

Table 5-65.—Plums, California: Production, value, and utilization, 1996–2005

Season	Total production	Utilized production	Marketing year average price per ton [1]	Value
	Tons	Tons	Dollars	1,000 dollars
1996	228,000	228,000	420.00	95,831
1997	246,000	246,000	312.00	76,825
1998	188,000	188,000	529.00	99,388
1999	196,000	196,000	419.00	82,041
2000	197,000	197,000	442.00	87,115
2001	210,000	210,000	306.00	64,362
2002	201,000	201,000	386.00	77,586
2003	209,000	209,000	418.00	87,362
2004	156,000	144,000	516.00	74,347
2005 [2]	171,000	171,000	551.00	94,163

[1] Fresh fruit prices are equivalent returns at point of first sale. Processing fruit prices are equivalent returns at processing plant door.   [2] Preliminary.
NASS, Crops Branch, (202) 720–2127.

Table 5-66.—Prunes (dried basis): Production, price and value, California, 1996–2005 [1]

Season	Total production	Utilized production	Marketing year average price per ton [2]	Value
	Tons	Tons	Dollars	1,000 dollars
1996	223,000	223,000	839.00	187,097
1997	214,000	205,000	883.00	181,015
1998	108,000	103,000	764.00	78,692
1999	178,000	165,000	861.00	142,065
2000	219,000	201,000	770.00	154,770
2001	150,000	135,000	726.00	98,010
2002	172,000	163,000	810.00	132,030
2003	181,000	168,000	772.00	129,696
2004	49,000	48,000	1,500.00	72,000
2005 [3]	90,000	87,000	1,500.00	130,500

[1] The drying ratio is approximately 3 pounds of fresh fruit to 1 pound of dried fruit.   [2] Equivalent returns at the processing plant door.   [3] Preliminary.
NASS, Crops Branch, (202) 720–2127.

Table 5-67.—Prunes and plums: [1] Production, value, and utilization, 4-States, 1996–2005

Year	Total production	Utilized production	Marketing year average price per ton	Value	Utilization of quantities sold			
					Fresh [2]	Processed (fresh basis)		
						Dried and other	Canned	Frozen
	Tons	Tons	Dollars	1,000 dollars	Tons	Tons	Tons	Tons
1996	19,500	18,700	442.00	8,272	10,650	1,900	5,700	450
1997	25,500	23,700	273.00	6,481	10,500	2,800	8,700	1,700
1998	25,600	24,800	311.00	7,707	11,750	4,150	7,250	1,650
1999	22,900	21,620	208.00	4,500	11,150	4,120	5,400	950
2000	23,900	21,950	239.00	5,247	9,400	5,650	5,400	1,500
2001	21,200	20,000	273.00	5,459	11,000	3,250	4,470	1,280
2002	15,650	14,790	286.00	4,237	6,360	3,930	3,340	1,160
2003	16,300	14,880	353.00	5,260	7,700	2,780	3,100	1,300
2004	25,000	18,920	360.00	6,802	10,350	4,390	3,140	1,040
2005 [3]	8,700	8,650	577.00	4,993	5,500	580	2,150	420

[1] ID, MI, OR, and WA. Mostly prunes; however, estimates include small quantities of plums in all States.   [2] Includes "Home use."   [3] Preliminary.
NASS, Crops Branch, (202) 720–2127.

Table 5-68.—Prunes and plums (fresh basis): Production and season average price per ton, by States, 2003–2005

State	Total production			Utilized production			Price per ton[2]		
	2003	2004	2005[1]	2003	2004	2005[1]	2003	2004	2005[1]
	Tons	*Tons*	*Tons*	*Tons*	*Tons*	*Tons*	*Dollars*	*Dollars*	*Dollars*
ID	2,500	4,000	2,000	2,480	3,920	1,950	518.00	613.00	999.00
MI	3,600	2,500	2,000	3,600	2,000	2,000	355.00	353.00	365.00
OR	5,500	13,000	1,500	4,100	7,500	1,500	272.00	352.00	416.00
WA	4,700	5,500	3,200	4,700	5,500	3,200	337.00	193.00	528.00
Total, 4 States	16,300	25,000	8,700	14,880	18,920	8,650	353.00	360.00	577.00

[1] Preliminary. [2] Fresh fruit prices are equivalent packinghouse-door returns for OR and WA, and the average price as sold for other States. Quantities processed are priced at the equivalent processing plant door level.

NASS, Crops Branch, (202) 720–2127.

Table 5-69.—Prunes and plums: Utilization and marketing year average price per ton, by State, 1999–2005

State and season	Quantity				Price [3]			
	Fresh[1]	Dried and other[2]	Canned	Frozen	Fresh	Dried and other	Canned	Frozen
	Tons	*Tons*	*Tons*	*Tons*	*Dollars*	*Dollars*	*Dollars*	*Dollars*
MI:								
1999	1,100	[4]	[4]	[4]	440.00	[4]	[4]	[4]
2000	1,250	[4]	[4]	[4]	270.00	[4]	[4]	[4]
2001	1,800	[4]	[4]	[4]	442.00	[4]	[4]	[4]
2002	60	[4]	[4]	[4]	600.00	[4]	[4]	[4]
2003	1,100	[4]	[4]	[4]	480.00	[4]	[4]	[4]
2004	350	[4]	[4]	[4]	769.00	[4]	[4]	[4]
2005	450	[4]	[4]	[4]	760.00	[4]	[4]	[4]
WA:								
1999	[4]	[4]	[4]	[4]	[4]	[4]	[4]	[4]
2000	[4]	[4]	[4]	[4]	[4]	[4]	[4]	[4]
2001	[4]	[4]	[4]	[4]	[4]	[4]	[4]	[4]
2002	[4]	[4]	[4]	[4]	[4]	[4]	[4]	[4]
2003	[4]	[4]	[4]	[4]	[4]	[4]	[4]	[4]
2004	[4]	[4]	[4]	[4]	[4]	[4]	[4]	[4]
2005	[4]	[4]	[4]	[4]	[4]	[4]	[4]	[4]
Total 4 States:[5]								
1999	11,150	4,120	5,400	950	232.00	218.00	135.00	297.00
2000	9,400	5,650	5,400	1,500	321.00	145.00	182.00	287.00
2001	11,000	3,250	4,470	1,280	345.00	157.00	190.00	241.00
2002	6,360	3,930	3,340	1,160	396.00	245.00	175.00	147.00
2003	7,700	2,780	3,100	1,300	446.00	214.00	255.00	339.00
2004	10,350	4,390	3,140	1,040	468.00	237.00	196.00	289.00
2005	5,500	580	2,150	420	759.00	279.00	254.00	269.00

[1] Includes "Home use." [2] Some quantities otherwise processed are included to avoid disclosure of individual operations. [3] Prices for fresh sales are average prices as sold for ID and MI; equivalent packinghouse door returns for OR and WA. Quantities processed are priced at the equivalent processing plant door level. [4] Not published to avoid disclosure of individual operations, but is included in total. [5] ID, MI, OR, and WA.

NASS, Crops Branch, (202) 720–2127.

Table 5-70.—Prunes, dried: United States exports by country of destination, 2002/2003–2004/2005

Country of destination	Year beginning August		
	2002/2003	2003/2004	2004/2005
	Metric tons	*Metric tons*	*Metric tons*
North America:			
Canada	4,939.5	5,092.7	3,461.5
Mexico	1,310.0	1,118.9	656.7
Total	6,249.5	6,211.6	4,118.2
Carribean:			
Dominican Republic	158.5	73.4	65.1
Netherlands Antilles	50.1	56.2	60.9
Jamaica	34.9	39.3	28.4
Barbados	19.4	32.4	27.2
Bahamas, The	18.7	26.7	14.8
Trinidad And Tobago	3.5	14.3	14.5
Haiti	8.2	6.5	4.8
Bermuda	0.0	0.0	0.8
Total	293.4	248.9	216.4
Central America:			
Panama	205.0	121.1	98.4
Costa Rica	23.5	35.9	52.6
Guatemala	5.4	66.1	2.2
El Salvador	13.5	0.0	1.4
Belize	0.0	0.0	1.3
Nicaragua	0.0	11.8	0.7
Honduras	4.2	5.0	0.0
Total	251.5	239.9	156.7
South America:			
Brazil	5.3	8.7	22.2
Venezuela	5.0	0.0	19.7
Ecuador	31.4	36.2	17.7
Colombia	0.0	1.2	0.0
Total	41.7	46.1	59.7
European Union:			
Germany	6,622.9	9,135.1	6,155.8
United Kingdom	4,570.1	5,868.8	3,620.5
Italy	6,448.8	7,434.6	3,517.6
Netherlands	3,649.8	2,923.7	2,450.0
Belgium-Luxembourg	2,400.7	2,111.3	1,879.4
Sweden	1,039.7	995.7	1,357.7
Finland	1,459.4	1,709.7	1,356.0
Denmark	1,187.3	1,297.4	945.6
Latvia	35.8	1,322.9	539.5
Poland	572.6	950.7	498.5
Spain	669.7	1,458.2	418.4
Greece	694.7	591.5	196.5
France	1.6	160.0	56.3
Ireland	66.1	93.3	48.4
Czech Republic	99.1	253.2	39.0
Austria	0.0	37.5	18.0
Cyprus	9.1	8.5	9.5
Hungary	18.3	31.8	0.0
Lithuania	46.5	175.1	0.0
Malta	4.0	4.3	0.0
Portugal	18.3	0.0	0.0
Slovenia	0.0	20.0	0.0
Total	29,614.5	36,583.3	23,106.8
Other Europe:			
Norway	956.4	872.0	941.4
Switzerland	45.2	10.1	124.2
Iceland	79.6	75.3	36.5
Total	1,081.1	957.4	1,102.1
Former Soviet Union:			
Russian Federation	69.1	1,175.2	145.0
Georgia, Republic of	0.0	18.1	19.0
Belarus	69.9	0.0	0.0
Kazakhstan, Republic	18.3	18.3	0.0
Ukraine	19.0	93.5	0.0
Total	176.3	1,305.2	164.0

See end of table.

Table 5-70.—Prunes, dried: United States exports by country of destination, 2002/2003–2004/2005—Continued

Country of destination	Year beginning August		
	2002/2003	2003/2004	2004/2005
	Metric tons	Metric tons	Metric tons
East Asia:			
Japan	16,207.6	16,799.8	13,489.1
Hong Kong	2,511.2	3,065.9	1,433.2
China, Peoples Republic	63.6	647.7	181.4
Taiwan	544.5	413.0	168.2
Korea, Republic of	103.4	148.8	122.4
Total	19,430.5	21,075.1	15,394.3
Middle East:			
Israel	1,236.9	2,258.2	299.1
Saudi Arabia	173.6	93.0	47.1
Lebanon	6.3	24.9	27.1
Kuwait	32.7	45.9	17.8
Syria	0.0	150.2	0.0
United Arab Emirates	0.0	33.4	0.0
Turkey	55.2	73.3	0.0
Total	1,504.7	2,678.8	391.0
North Africa:			
Algeria	45.2	214.3	0.0
Egypt	90.4	160.7	0.0
Total	135.6	375.0	0.0
Sub-Saharan Africa:			
South Africa, Republic	68.3	8.4	0.0
Burkina	0.0	57.9	0.0
Total	68.3	66.3	0.0
South Asia:			
India	23.6	28.0	12.0
Sri Lanka	0.0	0.0	9.1
Afghanistan	0.0	0.0	5.2
Pakistan	1.0	0.0	0.0
Total	24.6	28.0	26.3
Southeast Asia:			
Malaysia	694.2	988.7	1,256.6
Singapore	1,238.4	1,187.3	651.4
Thailand	87.2	107.6	219.6
Philippines	217.2	298.6	129.2
Indonesia	39.4	34.2	47.2
Cambodia	0.0	0.0	2.9
Vietnam	0.0	17.7	1.7
Total	2,276.5	2,634.1	2,308.6
Oceania:			
New Zealand	869.2	896.0	831.8
Australia	2,347.3	1,933.5	481.5
French Pacific Island	0.0	0.0	7.2
Total	3,216.5	2,829.5	1,320.6
Grand total	64,364.8	75,279.2	48,364.7

FAS, Horticultural and Tropical Products Division, (202) 720-3423. The division's Home Page is located at, www.fas.usda.gov/htp. You can also email the division at htp@fas.usda.gov. Note: Compiled from reports of the U.S. Department of Commerce. U.S. trade can be obtained from the following web site: U.S. agricultural trade database, www.fas.usda.gov/ustrade.

Table 5-71.—Prunes: Foreign trade, United States, 1995-2004

Year beginning October	Imports				Domestic exports			
	Fresh prunes and plums	Other-wise pre-pared or pre-served	Dried prunes [1]	Total, in terms of fresh [2]	Fresh prunes and plums	Dried prunes [1]	Dried, in fruit salad [1][3]	Total, in terms of fresh [2]
	Metric tons	Metric tons	Metric tons	Metric tons	Metric tons	Metric tons	Metric tons	Metric tons
1995	20,218	826	250	21,610	67,193	62,548	1,811	234,279
1996	22,923	746	450	24,763	71,906	67,535	1,382	250,826
1997	19,509	811	214	20,792	48,855	70,831	1,491	236,613
1998	27,736	735	712	30,245	60,078	68,450	1,191	240,859
1999	22,893	778	510	24,915	61,354	66,304	613	235,080
2000	33,400	792	431	35,231	62,926	83,746	1,089	283,170
2001	32,459	811	969	35,704	62,802	69,660	704	245,478
2002	32,336	921	570	34,643	60,028	66,624	1,403	236,637
2003	35,959	1,039	677	38,650	45,105	73,976	1,693	241,554
2004	40,065	1,089	9,871	66,671	48,233	45,397	2,507	172,598

[1] Net processed weight. [2] Exports and imports of dried prunes converted to unprocessed dry weight by dividing by 1.04. Unprocessed dry weight converted to terms of fresh fruit on the basis that 1 pound dried equals 2.7 pounds fresh. "Other-wise prepared or preserved" converted to terms of fresh fruit on the basis that 1 pound equals 0.899 pound fresh. [3] Dried prunes in salad estimated at 43 percent of total dried fruit for salad.
ERS, Specialty Crops Branch, (202) 694-5260.

Table 5-72.—Strawberries, commercial crop: Production and value per hundredweight, by States, 2003-2005

Utilization, season, and State	Production			Value per unit		
	2003	2004	2005 [1]	2003	2004	2005 [1]
	1,000 cwt.	1,000 cwt.	1,000 cwt.	Dollars per cwt.	Dollars per cwt.	Dollars per cwt.
FOR FRESH MARKET:						
CA	14,377	14,830	15,825	72.80	73.30	61.80
FL	1,562	1,633	1,789	82.70	109.00	110.00
MI	58	36	47	105.00	105.00	99.00
NY	50	65	52	155.00	160.00	155.00
NC	170	176	195	90.00	90.00	95.00
OH	38	38	42	120.00	134.00	137.00
OR	20	29	24	115.00	124.00	118.00
PA	83	79	70	134.00	164.00	183.00
WA	21	17	20	113.00	110.00	113.00
WI	45	41	41	117.00	125.00	133.00
US	16,424	16,944	18,105	74.90	78.10	68.20
PROCESSING:						
CA	4,715	4,758	4,755	26.70	24.90	27.80
MI	5	5	5	46.00	45.00	45.00
OR	275	295	226	45.00	41.50	48.00
WA	141	136	130	43.00	40.00	36.00
US	5,136	5,194	5,116	28.10	26.30	28.90

[1] Preliminary. [2] Mostly for fresh market, but includes some quantities used for processing in States for which processing estimates are not prepared.
NASS, Crops Branch, (202) 720-2127.

Table 5-73.—Strawberries, commercial crop: Area, yield, production, value per hundred weight, and total value, United States, 1996–2005

Year	Fresh market and processing					Fresh market [3]			Processing		
	Area for harvest	Yield per acre	Produc-tion [1]	Value [2]		Produc-tion	Value [2]		Produc-tion	Value [2]	
				Per cwt	Total		Per cwt	Total		Per cwt	Total
	Acres	*Cwt.*	*1,000 cwt.*	*Dollars per cwt.*	*1,000 dollars*	*1,000 cwt.*	*Dollars per cwt.*	*1,000 dollars*	*1,000 cwt.*	*Dollars per cwt.*	*1,000 dollars*
1996	47,670	341	16,259	47.30	768,943	12,126	56.50	684 ,661	4,133	20.40	84,282
1997	44,260	368	16,278	55.50	903,350	12,018	65.60	787,974	4,260	27.10	115,376
1998	44,930	365	16,381	61.10	1,000,254	11,332	65.60	838,803	5,059	31.90	161,451
1999	46,460	394	18,314	62.50	1,144,876	13,052	74.40	971,114	5,262	33.00	173,762
2000	47,350	401	19,008	55.00	1,044,594	14,333	64.90	930,125	4,675	24.50	114,469
2001	45,700	361	16,509	64.70	1,068,582	12,597	75.80	954,413	3,912	29.20	114,169
2002	47,600	396	18,845	61.60	1,161,630	14,063	71.30	1,003,145	4,782	33.10	158,485
2003	48,400	445	21,560	63.80	1,375,142	16,424	74.90	1,230,583	5,136	28.10	144,559
2004	51,400	431	22,138	66.00	1,460,077	16,944	78.10	1,323,695	5,194	26.30	136,382
2005 [4]	52,200	445	23,221	59.60	1,383,064	18,105	68.20	1,235,122	5,116	28.90	147,942

[1] Excludes the following quantities not harvested and not marked because of economic conditions (1,000 hundredweight): 50 in 1995.   [2] Fresh market price and value on f.o.b. basis. Processing price and value at processing plant door.   [3] Mostly for fresh market, but includes some quantities used for processing in States for which processing estimates are not prepared.   [4] Preliminary.
NASS, Crops Branch, (202) 720–2127.

Table 5-74.—Strawberries, commercial crop: Area harvested, production, value per hundred weight, by States, 2003–2005 [1]

Season and State	Area harvested			Production			Value per unit		
	2003	2004	2005 [2]	2003	2004	2005 [2]	2003	2004	2005 [2]
	Acres	*Acres*	*Acres*	*1,000 cwt.*	*1,000 cwt.*	*1,000 cwt.*	*Dollars per cwt.*	*Dollars per cwt.*	*Dollars per cwt.*
CA	29,600	33,200	34,300	19,092	19,588	20,580	61.40	61.50	53.90
FL	7,100	7,100	7,300	1,562	1,633	1,789	82.70	109.00	110.00
MI	1,100	900	1,000	63	41	52	100.00	97.70	93.80
NY	1,500	1,500	1,500	50	65	52	155.00	160.00	155.00
NC	1,700	1,600	1,500	170	176	195	90.00	90.00	95.00
OH	800	800	800	38	38	42	120.00	134.00	137.00
OR	2,600	2,400	2,200	295	324	250	49.70	48.90	54.70
PA	1,300	1,300	1,300	83	79	70	134.00	164.00	183.00
WA	1,800	1,700	1,500	162	153	150	52.10	47.80	46.30
WI	900	900	800	45	41	41	117.00	125.00	133.00
US	48,400	51,400	52,200	21,560	22,138	23,221	63.80	66.00	59.60

[1] Includes quantities used for fresh market and processing.   [2] Preliminary.
NASS, Crops Branch, (202) 720–2127.

Table 5-75.—Fruits, noncitrus: Production, utilization, and value, United States, 1996–2005 [1]

Year	Utilized produc-tion	Fresh [2]	Processed						Value of utilized produc-tion
			Canned	Dried	Juice	Frozen	Wine	Other	
	1,000 tons	*1,000 tons*	*1,000 tons*	*1,000 tons*	*1,000 tons*	*1,000 tons*	*1,000 tons*	*1,000 tons*	*1,000 dollars*
1996	16,103	6,313	1,873	2,275	1,582	604	3,043	180	7,265,788
1997	18,400	6,642	2,130	2,660	1,666	699	4,035	293	8,189,821
1998	16,552	6,514	1,845	1,911	1,786	711	3,315	198	7,251,032
1999	17,347	6,691	1,986	2,154	1,887	717	3,351	244	8,077,404
2000	18,854	7,015	1,812	3,023	1,712	691	4,130	191	7,882,036
2001	16,740	6,488	1,859	2,290	1,462	665	3,568	169	7,918,636
2002	17,122	6,549	1,727	2,582	1,251	591	3,999	138	8,138,348
2003	16,853	6,676	1,762	2,293	1,295	716	3,582	219	8,633,919
2004	16,837	7,179	1,711	1,425	1,423	686	3,819	286	8,975,718
2005 [3]	17,163	NA	NA	NA	NA	NA	NA	NA	9,341,497

[1] Includes the following crops: Apples, apricots, avocados, bananas, berries, cherries, cranberries, dates, figs, grapes, guavas, kiwifruit, nectarines, olives, papayas, peaches, pears, pineapples, plums, prunes, and strawberries. [2] Includes "Home Use," local and roadside sales. [3] Preliminary. NA-not available.

NASS, Crops Branch, (202) 720–2127.

Table 5-76.—Fruits, fresh: Total reported domestic rail, truck, and air shipments, 2004

Commodity	Jan.	Feb.	Mar.	Apr.	May	Jun.	Jul.	Aug.	Sep.	Oct.	Nov.	Dec.	Total
	1,000 cwt.	*1,000 cwt.*	*1,000 cwt.*	*1,000 cwt.*	*1,000 cwt.*	*1,000 cwt.*	*1,000 cwt.*	*1,000 cwt.*	*1,000 cwt.*	*1,000 cwt.*	*1,000 cwt.*	*1,000 cwt.*	*1,000 cwt.*
Citrus:													
Grapefruit	2,550	2,804	2,597	1,157	313	68	30	22	25	420	1,131	1,407	12,524
Misc Citrus	1										1		2
Lemons	97	82	88	100	117	84	63	59	59	49	76	91	965
Oranges	959	910	1,083	972	727	506	135	81	98	429	912	1,290	8,102
Tangelos	41	5								1	51	151	249
Temples	111	84	25									1	221
Total	3,759	3,885	3,793	2,229	1,157	658	228	162	182	899	2,171	2,940	22,063
Noncitrus:													
Apples	4,534	4,415	3,712	3,357	3,594	1,986	1,772	2,025	3,347	5,691	4,574	4,853	43,860
Apricots					219	111	101	16		2			449
Avocados	199	261	384	422	685	571	803	687	413	272	120	90	4,907
Blueberries				23	127	307	367	141	48	12			1,025
Cherries				18	846	1,241	827	66					2,998
Cranberries									4	51	104	12	171
Fruit, Other	34	8					5	17	26	31	21	15	157
Grapes	36				614	1,090	2,379	3,253	3,055	2,462	1,868	541	15,298
Kiwifruit	57	24	8	2	1				48	39	47		226
Nectarines					451	1,344	1,381	872	446	18			4,512
Papaya	32	22	20	19	26	23	26	17	21	30	15	20	271
Peaches				11	758	1,652	1,914	1,676	699	135	2		6,847
Pears	1,202	888	869	690	491	234	472	649	913	1,470	1,173	1,119	10,170
Persimmons										25	28	14	67
Pineapples	344	391	197	126	151	148	221	125	109	194	144	149	2,299
Plums					167	557	863	696	371	40			2,694
Pomegranates		3	2	1				5	43	78	23	1	156
Raspberries	6	2	3	8	74	71	65	92	56	25	18	19	439
Prunes							2	42	28	6	1		79
Strawberries	402	505	1,574	2,035	1,948	1,626	1,318	880	556	446	257	173	11,720
Total	6,846	6,519	6,769	6,712	10,152	10,961	12,516	11,259	10,135	11,036	8,387	7,053	108,345
Grand total	10,605	10,404	10,562	8,941	11,309	11,619	12,744	11,421	10,317	11,935	10,558	9,993	130,408

AMS, Fruit and Vegetable Division, Market News Branch, (202) 720–3343.

Table 5-77.—Fruits, dried: Production (dry basis), California, 1996–2005

Year	Apples	Apricots	Dates	Figs[1]	Peaches[2]	Pears[3]	Prunes	Grapes[4]	Total
	Tons	Tons	Tons	Tons	Tons	Tons	Tons	Tons	Tons
1996	NA	2,170	23,000	14,500	1,700	700	223,000	313,900	578,970
1997	NA	1,740	21,000	18,500	1,894	800	205,000	432,800	681,734
1998	NA	1,250	24,900	16,600	1,551	1,100	103,000	281,200	429,601
1999	NA	NA	22,200	15,100	1,849	1,010	165,000	348,300	553,459
2000	NA	1,120	17,400	17,300	1,350	600	201,000	493,700	732,470
2001	NA	820	19,700	13,000	1,450	500	135,000	417,100	587,570
2002	NA	1,120	24,200	16,900	1,525	460	163,000	443,400	650,605
2003	NA	900	17,900	15,200	1,070	610	168,000	351,900	555,580
2004[5]	NA	1,630	16,400	15,900	870	620	48,000	266,100	349,520

[1] Standard and substandard.　[2] Freestone only.　[3] Bartlett only.　[4] Raisin and table type.　[5] Preliminary.　NA-not available.

ERS, Specialty Crops Branch, (202) 694–5260.

Table 5-78.—Raisins: Commercial production in specified countries,
2002/2003–2004/2005

Commodity/country	2002/2003	2003/2004	2004/2005[1]
	Metric tons	Metric tons	Metric tons
Australia	16,118	29,000	30,000
Chile	51,000	48,700	48,000
Greece	10,000	9,000	28,000
Mexico	7,140	7,440	7,500
South Africa	34,953	40,000	36,720
Turkey	230,000	215,000	280,000
United States	398,163	297,285	227,159
Total	747,374	646,425	657,379

[1] Preliminary.

FAS, Horticultural and Tropical Products Division, (202) 720-3423. The division's Home Page is located at, www.fas.usda.gov/htp. You can also email the division at htp@fas.usda.gov. Note: Prepared or estimated on the basis of official statistics of foreign governments, other foreign source materials, reports of U.S. Agricultural Counselors and Attaches, results of office research, and related information. Note: This information can be obtained from the following web site: www.fas.usda.gov/psd.

Table 5-79.—Fruits, frozen: Commercial pack, United States, 1995–2004

Commodity	1995	1996	1997	1998	1999	2000	2001	2002	2003	2004
	1,000 pounds	1,000 pounds	1,000 pounds	1,000 pounds	1,000 pounds	1,000 pounds	1,000 pounds	1,000 pounds	1,000 pounds	1,000 pounds
Apples	113,278	114,332	119,180	124,866	111,944	141,820	146,145	123,232	113,836	80,506
Apricots	10,046	17,759	24,267	20,929	18,492	22,786	30,638	20,591	14,767	4,804
Cherries, RSP	NA	NA	NA	NA	NA	NA	NA	6,912	40,709	40,332
Cherries, sweet	27,032	14,945	24,515	21,628	13,640	15,901	13,101	9,062	8,175	11,010
Peaches	89,743	109,598	124,220	110,491	123,942	148,083	131,694	135,884	136,204	123,378
Plums and prunes	1,042	1,736	789	1,518	986	1,331	1,380	680	1,732	1,359
Purees, noncitrus	71,862	78,457	85,333	100,239	85,535	74,663	58,924	36,052	31,359	31,253
Berries:										
Blackberries	26,823	20,404	26,272	24,734	23,895	26,857	22,884	25,074	23,938	12,962
Blueberries	118,064	90,085	122,767	90,850	96,567	102,185	98,369	39,887	52,750	38,122
Boysenberries	3,459	5,288	4,983	3,338	4,703	3,597	3,537	3,174	1,808	1,407
Loganberries	(¹)	(¹)	(¹)	(¹)	(¹)	(¹)	(¹)	(¹)	(¹)	(¹)
Raspberries	40,109	25,118	27,504	23,851	23,324	23,902	21,736	12,220	30,554	4,888
Strawberries	371,138	330,139	328,150	373,824	419,768	439,749	422,371	415,865	246,202	215,481
Miscellaneous fruits and berries	140,140	136,526	110,644	107,716	101,907	135,066	54,799	5,197	4,682	5,421
Total	1,012,736	944,387	998,624	1,003,984	1,024,703	1,135,940	1,005,578	826,918	706,716	570,918

[1] Included in miscellaneous. NA=not available

ERS, Specialty Crops Branch, (202) 694-5260. Data from American Frozen Food Institute.

Table 5-80.—Fruits: Per capita consumption, United States, 1995–2004[1]

Year	Fruits used fresh		
	Citrus fruit[2]	Noncitrus fruits[3]	Canned fruits[4]
	Per capita	Per capita	Per capita
	Pounds	Pounds	Pounds
1995	23.8	73.2	14.7
1996	24.6	73.0	16.1
1997	26.5	75.0	17.6
1998	26.6	75.1	15.1
1999	20.4	80.7	16.7
2000	23.5	77.6	15.3
2001	23.9	73.6	15.5
2002	23.3	76.0	14.8
2003	23.9	77.4	15.0
2004[8]	22.7	79.3	14.8

Year	Juice[5]	Frozen fruit[6]	Dried fruits[7]
	Per capita	Per capita	Per capita
	Gallons	Pounds	Pounds
1995	8.5	3.8	2.7
1996	8.7	4.1	2.8
1997	9.0	3.2	2.7
1998	9.0	3.7	2.8
1999	9.1	3.4	2.5
2000	8.8	3.9	2.5
2001	8.3	4.2	2.4
2002	8.2	3.4	2.5
2003	8.3	4.2	2.4
2004[8]	8.2	3.9	2.3

[1] Fresh citrus fruits, canned fruit, and fruit juices are on a crop-year basis. Dried fruits are on a pack-year basis. The per capita consumption was obtained by dividing the total consumption by total population. [2] Oranges and temples, tangerines and tangelos, lemons, limes, and grapefruit. [3] Apples, apricots, avocados, bananas, cherries, cranberries, grapes, kiwifruit, mangoes, peaches and nectarines, pears, pineapples, papayas, plums and prunes, and strawberries. [4] Apples, apricots, cherries, olives, peaches, pears, pineapples, and plums and prunes. [5] Orange, grapefruit, lemon, lime, apple, grape, pineapple, prune, and cranberry. [6] Blackberries, blueberries, raspberries, strawberries, other berries, apples, apricots, cherries, and peaches. [7] Apples, apricots, dates, figs, peaches, pears, prunes, and raisins. Dried data in terms of processed weight. [8] Preliminary.

ERS, Specialty Crops Branch, (202) 694–5260.

Table 5-81.—All tree nuts: Supply and utilization, United States, 1995/96–2004/05

Market year[1]	Beginning stocks	Marketable produc- tion[2]	Imports	Total supply	Exports	Ending stocks	Domestic consumption	
							Total	Per capita Pounds
	—Million pounds (shelled)—							
1995/96	334.1	770.1	204.0	1,308.2	542.4	251.9	513.9	1.92
1996/97	251.9	816.9	218.3	1,287.1	580.4	156.4	550.3	2.03
1997/98	156.4	1,214.4	243.8	1,614.7	667.1	348.7	598.8	2.18
1998/99	348.7	850.9	252.1	1,451.7	603.3	223.9	624.6	2.25
1999/2000	223.9	1,297.5	285.4	1,806.8	667.6	365.6	774.0	2.75
2000/2001	365.6	1,116.6	297.1	1,779.2	782.0	274.3	723.0	2.55
2001/2002	274.3	1,334.5	335.5	1,944.3	844.3	280.7	819.3	2.86
2002/2003	280.7	1,561.9	361.2	2,203.7	912.5	362.5	928.8	3.21
2003/2004	362.5	1,504.4	428.2	2,295.1	945.8	349.6	999.7	3.42
2004/2005[3]	349.6	1,506.4	462.4	2,372.3	998.9	315.0	1,058.4	3.58

[1] Marketing season begins July 1 for almonds, hazelnuts, macadamias, pecans, and other nuts; August 1 for walnuts; and September 1 for pistachios. [2] Utilized production (NASS data) minus inedibles and noncommercial useage. [3] Preliminary.

ERS, Specialty Crops Branch, (202) 694–5260.

Table 5-82.—Tree nuts: Commercial production in specified countries, 2002/2003–2004/2005

Commodity and country	2002/2003	2003/2004	2004/2005 [1]
	1,000 metric tons	*1,000 metric tons*	*1,000 metric tons*
Almonds (shelled basis):			
Greece	17.0	10.0	17.0
India	1.1	1.0	1.2
Italy	9.0	5.0	12.0
Spain	66.0	44.0	23.0
Turkey	14.0	13.7	13.0
United States	494.4	471.7	489.9
Total	601.5	545.4	556.1
Hazelnuts (in-shell basis):			
Italy	120.0	75.0	135.0
Spain	22.0	12.0	9.0
Turkey	625.0	480.0	425.0
United States	17.7	34.3	48.5
Total	784.7	601.3	617.5
Walnuts:			
Chile	13.8	14.0	14.5
China	340.2	320.0	350.0
France	33.3	23.4	25.0
India	32.0	31.0	34.0
Italy	20.0	18.0	12.0
Turkey	60.0	69.0	68.0
United States	255.8	295.7	294.8
Total	755.1	771.1	798.3
Pistachios:			
Greece	9.9	9.0	9.5
Syria	42.0	50.0	40.0
Turkey	35.0	90.0	30.0
United States	137.6	54.0	157.7
Total	224.4	203.0	237.2

[1] Preliminary.
FAS, Horticultural and Tropical Products Division, (202) 720-3423. The division's Home Page is located at, www.fas.usda.gov/htp. You can also email the division at htp@fas.usda.gov. Note: Prepared or estimated on the basis of official statistics of foreign governments, other foreign source materials, reports of U.S. Agricultural Counselors and Attachés, results of office research, and related information. Note: This information can be obtained from the following web site: www.fas.usda.gov/psd.

Table 5-83.—Almonds (shelled basis): Bearing acreage, yield, production, price, and value, California, 1996–2005 [1]

Year	Bearing Acreage	Yield per acre	Utilized production	Price per pound	Value
	Acres	*Pounds*	*1,000 pounds*	*Dollars*	*1,000 dollars*
1996	428,000	1,190	510,000	2.08	1,018,368
1997	442,000	1,720	759,000	1.56	1,160,640
1998	460,000	1,130	520,000	1.41	703,590
1999	485,000	1,720	833,000	0.86	687,742
2000	510,000	1,380	703,000	0.97	666,487
2001	530,000	1,570	830,000	0.91	740,012
2002	545,000	2,000	1,090,000	1.11	1,200,687
2003	550,000	1,890	1,040,000	1.57	1,600,144
2004	570,000	1,760	1,005,000	2.21	2,189,005
2005 [2]	580,000	1,550	900,000	3.08	2,724,876

[1] Price and value are based on edible portion of the crop only. Included in production are inedible quantities of no value as follows (million pounds): 1996-20.4; 1997-15.0; 1998-21.0; 1999-33.2; 2000-15.9; 2001-16.8; 2002-8.3; 2003-20.8; 2004-14.5; 2005-15.3. [2] Preliminary.
NASS, Crops Branch, (202) 720–2127.

Table 5-84.—Almonds (shelled basis [1]): Foreign trade, United States, 1995–2004

Year beginning October	Imports	Domestic exports
	Metric tons	*Metric tons*
1995	68	279,338
1996	24	162,629
1997	27	182,156
1998	29	184,501
1999	39	197,271
2000	173	225,550
2001	319	261,563
2002	750	289,589
2003	830	308,041
2004	1,233	302,986

[1] Imports of unshelled nuts converted to shelled basis at ratio of 1.67 to 1. Exports of unshelled nuts converted to shelled basis at ratio of 1.67 to 1.0.
ERS, Market and Trade Economics Division, Specialty Crops Branch, (202) 694–5260.

Table 5-85.—Hazelnuts (in-shell basis): Bearing acreage, yield, production, price, and value, Oregon, Washington, and United States, 1996–2005

Year	Bearing Acreage	Yield per acre	Utilized production	Price per ton	Value
	Acres	Tons	Tons	Dollars	1,000 dollars
			Oregon		
1996	28,200	0.66	18,750	859	16,106
1997	28,600	1.63	46,650	899	41,938
1998	29,100	0.53	15,400	964	14,846
1999	28,800	1.38	39,700	890	35,333
2000	28,300	0.79	22,300	890	19,847
2001	29,000	1.71	49,500	701	34,700
2002	29,200	0.67	19,500	1,000	19,500
2003	28,000	1.35	37,900	1,030	39,037
2004	28,400	1.32	37,500	1,440	54,000
2005 [1]	28,300	0.99	28,000	2,040	57,120
			Washington		
1996	400	0.63	250	940	235
1997	400	0.88	350	940	329
1998	430	0.23	100	960	96
1999	400	0.75	300	900	270
2000	350	0.57	200	960	192
2001 [2]					
2002					
2003					
2004					
2005					
			United States		
1996	28,600	0.66	19,000	860	16,341
1997	29,000	1.62	47,000	899	42,267
1998	29,530	0.52	15,500	964	14,942
1999	29,200	1.37	40,000	890	35,603
2000	28,650	0.79	22,500	891	20,039
2001	29,000	1.71	49,500	701	34,700
2002	29,200	0.67	19,500	1,000	19,500
2003	28,000	1.35	37,900	1,030	39,037
2004	28,400	1.32	37,500	1,440	54,000
2005 [1]	28,300	0.99	28,000	2,040	57,120

[1] Preliminary. [2] WA discontinued.
NASS, Crops Branch, (202) 720–2127.

Table 5-86.—Hazelnuts (shelled basis [1]): Foreign trade, United States, 1995–2004

Year beginning October	Imports	Domestic exports
	Metric tons	Metric tons
1995	3,525	6,848
1996	3,717	5,826
1997	4,265	10,119
1998	5,484	3,999
1999	5,425	6,563
2000	5,129	5,706
2001	6,736	11,110
2002	6,441	4,524
2003	4,916	11,142
2004	4,068	10,443

[1] Imports of unshelled nuts converted to shelled basis at ratio of 2.22 to 1. Exports of unshelled nuts converted to shelled basis at ratio of 2.50 to 1.
ERS, Specialty Crops Branch, (202) 694–5260.

Table 5-87.—Macadamia nuts (in-shell basis): Bearing acreage, yield, production, price, and value, Hawaii, 1996–2005

Year	Bearing Acreage	Yield per acre	Utilized production	Price per pound	Value
	Acres	Pounds	1,000 pounds	Cents	1,000 dollars
1996	19,200	2,940	56,500	78.0	44,070
1997	19,200	3,020	58,000	75.0	43,500
1998	19,200	2,990	57,500	65.0	37,375
1999	18,900	2,990	56,500	67.0	37,855
2000	17,700	2,820	50,000	59.0	29,500
2001	17,800	3,150	56,000	59.0	33,040
2002	17,800	2,980	53,000	57.0	30,210
2003	17,800	2,980	53,000	61.0	32,330
2004	17,800	3,170	56,500	73.0	41,245
2005 [1]	18,000	3,330	60,000	78.0	46,800

[1] Preliminary.
NASS, Crops Branch, (202) 720–2127.

Table 5-88.—Pecans (in-shell basis): Production, price per pound, and value, United States, 1996-2005

Year	Improved varieties			Native and seedling			All pecans		
	Utilized production	Price per pound	Value	Utilized production	Price per pound	Value	Utilized production	Price per pound	Value
	1,000 pounds	Cents	1,000 dollars	1,000 pounds	Cents	1,000 dollars	1,000 pounds	Cents	1,000 dollars
1996[1]	165,125	68.9	113,749	44,375	46.4	20,606	209,500	64.1	134,355
1997	202,900	93.3	189,226	132,100	53.0	69,994	335,000	77.4	259,220
1998	112,000	135.0	150,908	34,400	77.2	26,544	146,400	121.0	177,452
1999	219,400	101.0	222,647	186,700	57.7	107,751	406,100	81.4	330,398
2000	160,550	126.0	201,575	49,300	75.4	37,193	209,850	114.0	238,768
2001	246,550	66.2	163,204	91,950	41.2	37,897	338,500	59.4	201,101
2002	130,720	107.0	139,597	42,180	60.3	25,436	172,900	95.5	165,033
2003	202,900	110.0	223,547	79,200	68.3	54,082	282,100	98.4	277,629
2004	138,970	192.0	267,215	46,830	128.0	59,709	185,800	176.00	326,924
2005[2][3]	209,400	165.0	345,492	50,200	109.0	54,949	259,600	154.00	400,441

[1] MO and TN discontinued. [2] Preliminary. [3] MO added.
NASS, Crops Branch, (202) 720-2127.

Table 5-89.—Pecans (in-shell basis): Production and marketing year average price per pound, by States, 2003-2005

Item and State	Utilized production			Price per pound		
	2003	2004	2005[1]	2003	2004	2005[1]
IMPROVED VARIETIES[2]	1,000 pounds	1,000 pounds	1,000 pounds	Dollars	Dollars	Dollars
AL	7,000	1,000	3,200	0.940	1.210	1.040
AZ	22,500	14,000	21,000	1.040	1.850	1.600
AR	1,400	1,000	1,500	1.100	1.400	1.400
CA	3,700	3,500	3,700	1.420	2.210	1.620
FL	500	400	700	1.000	1.500	1.500
GA	60,000	42,000	60,000	1.000	1.770	1.300
LA	4,000	2,500	1,000	1.080	1.400	1.700
MS	4,800	700	700	0.860	1.300	1.500
MO[3]	....	....	200	....	....	1.550
NM	55,000	39,000	62,000	1.280	2.280	1.800
NC	2,200	70	1,400	0.850	2.000	0.900
OK	1,500	6,000	2,000	1.120	1.600	1.950
SC	3,300	800	2,000	0.850	1.800	1.300
TX	37,000	28,000	50,000	1.110	1.840	1.980
US	202,900	138,970	209,400	1.100	1.920	1.650
NATIVE AND SEEDLING						
AL	1,000	100	800	0.690	0.840	0.790
AR	2,400	700	1,200	0.720	1.200	1.100
FL	1,600	100	300	0.600	0.950	1.400
GA	15,000	3,000	10,000	0.640	1.240	0.800
KS	2,000	1,800	3,200	0.870	1.750	1.300
LA	16,000	6,500	3,000	0.680	0.950	0.900
MS	2,200	300	100	0.500	0.800	1.000
MO[3]	....	....	1,800	....	....	1.190
NC	300	30	300	0.500	1.500	0.750
OK	4,500	22,000	14,000	0.800	1.350	1.650
SC	1,200	300	500	0.720	1.400	0.900
TX	33,000	12,000	15,000	0.690	1.270	0.780
US	79,200	46,830	50,200	0.683	1.280	1.090
ALL PECANS						
AL	8,000	1,100	4,000	0.909	1.180	0.990
AZ	22,500	14,000	21,000	1.040	1.850	1.600
AR	3,800	1,700	2,700	0.860	1.320	1.270
CA	3,700	3,500	3,700	1.420	2.210	1.620
FL	2,100	500	1,000	0.695	1.390	1.470
GA	75,000	45,000	70,000	0.928	1.730	1.230
KS	2,000	1,800	3,200	0.870	1.750	1.300
LA	20,000	9,000	4,000	0.760	1.080	1.100
MS	7,000	1,000	800	0.747	1.150	1.440
MO[3]	....	....	2,000	....	....	1.230
NM	55,000	39,000	62,000	1.280	2.280	1.800
NC	2,500	100	1,700	0.808	1.850	0.874
OK	6,000	28,000	16,000	0.880	1.400	1.690
SC	4,500	1,100	2,500	0.815	1.690	1.220
TX	70,000	40,000	65,000	0.912	1.670	1.700
US	282,100	185,800	259,600	0.984	1.760	1.540

[1] Preliminary. [2] Budded, grafted or topworked varieties. [3] Estimates began in 2005.
NASS, Crops Branch, (202) 720-2127.

Table 5-90.—Pecans (shelled basis[1]): Foreign trade, United States, 1995-2004

Year beginning October	Imports	Domestic exports
	Metric tons	Metric tons
1995	12,782	7,917
1996	10,743	8,819
1997	14,577	9,796
1998	16,071	7,887
1999	12,152	9,238
2000	12,902	8,963
2001	14,323	11,115
2002	14,555	13,243
2003	20,953	15,275
2004	28,672	13,505

[1] Imports of unshelled nuts converted to shelled basis at ratio of 2.50 to 1. Exports of unshelled nuts converted to shelled basis at ratio of 2.50 to 1.

ERS, Specialty Crops Branch, (202) 694-5260.

Table 5-91.—Pistachios (in-shell basis): Bearing acreage, yield, production, price, and value, California, 1996-2005

Year	Bearing Acreage	Yield per acre	Utilized production	Price per pound	Value
	Acres	Pounds	1,000 pounds	Dollars	1,000 dollars
1996	64,300	1,630	105,000	1.16	121,800
1997	65,400	2,750	180,000	1.13	203,400
1998	68,000	2,760	188,000	1.03	193,640
1999	71,000	1,730	123,000	1.33	163,590
2000	74,600	3,260	243,000	1.01	245,430
2001	78,000	2,060	161,000	1.01	162,610
2002	83,000	3,650	303,000	1.10	333,300
2003	88,000	1,350	119,000	1.22	145,180
2004	93,000	3,730	347,000	1.34	464,980
2005 [1]	98,000	2,890	283,000	2.03	574,490

[1] Preliminary.
NASS, Crops Branch, (202) 720-2127.

Table 5-92.—Walnuts (English): Bearing acreage, yield, production, price, and value, California, 1996-2005

Year	Bearing Acreage	Yield per acre	Utilized production	Price per ton	Value
	Acres	Tons	Tons	Dollars	1,000 dollars
1996	192,000	1.08	208,000	1,580	328,640
1997	193,000	1.39	269,000	1,430	384,670
1998	198,000	1.15	227,000	1,050	238,350
1999	197,000	1.44	283,000	886	250,738
2000	200,000	1.20	239,000	1,240	296,360
2001	204,000	1.50	305,000	1,120	341,600
2002	210,000	1.34	282,000	1,170	329,940
2003	213,000	1.53	326,000	1,160	378,160
2004	217,000	1.50	325,000	1,390	451,750
2005 [1]	219,000	1.62	355,000	NA	NA

[1] Preliminary. NA-not available.
NASS, Crops Branch, (202) 720-2127.

Table 5-93.—Walnuts (shelled basis[1]): Foreign trade, United States, 1995-2004

Year beginning October	Imports	Domestic exports
	Metric tons	Metric tons
1995	1,050	45,997
1996	2,662	46,811
1997	97	39,341
1998	111	41,000
1999	76	41,428
2000	523	41,918
2001	49	46,937
2002	99	49,925
2003	170	56,608
2004	331	60,213

[1] Imports of unshelled nuts converted to shelled basis at ratio of 2.50 to 1. Exports of unshelled nuts converted to shelled basis at ratio of 2.50 to 1.

ERS, Specialty Crops Branch, (202) 694-5260.

Table 5-94.—Cacao beans: United States imports by country of origin, 2002-2004

Continent and country	2002	2003	2004 [1]
	Metric tons	Metric tons	Metric tons
Caribbean:			
Dominican Republic	24,248	28,111	23,746
Haiti	2,585	2,558	2,346
Jamaica	13	25	37
Leeward-Windward Islands	14	8	18
Trinidad and Tobago	63	74	58
Total	26,923	30,776	26,205
Central America:			
Costa Rica	0	25	14
El Salvador	3	7	6
Nicaragua	0	2	0
Panama	16	16	0
Total	19	50	19
European Union - 25:			
Belgium-Luxembourg	0	1	0
France	0	48	0
Italy	0	0	0
Netherlands	26	0	0
Spain	9	4	21
United Kingdom	500	0	0
Total	536	53	21
North America:			
Canada	23	49	45
Mexico	1,106	1,342	0
Total	1,129	1,391	45
Oceania:			
New Zealand	40	0	0
Other Pacific Islands, NEC	0	1	1
Papua New Guinea	12,934	2,489	13,180
Total	12,974	2,489	13,181
Other Europe:			
Switzerland	0	0	910
Total	0	0	910
South America:			
Brazil	40	0	0
Colombia	70	80	120
Ecuador	19,264	24,376	33,100
Peru	0	0	40
Venezuela	344	537	257
Total	19,718	24,993	33,517
South Asia			
India	13	0	1
Total	13	0	1
Southeast Asia:			
Indonesia	111,710	114,277	114,339
Malaysia	180	0	0
Singapore	0	9,889	3,000
Total	111,890	124,167	117,339
Sub-Saharan Africa:			
Cameroon	20	0	0
Cote d'Ivoire	138,092	180,556	254,162
Ghana	7,100	200	4,761
Guinea	0	0	19
Madagascar	38	38	25
Nigeria	4,807	22,502	1,036
South Africa, Republic of	0	0	101
Total	150,057	203,297	260,104
Grand total	323,257	387,217	451,344

[1] Preliminary.

FAS, Horticultural and Tropical Products Division, (202) 720-3423. The division's Home Page is located at, www.fas.usda.gov/htp. You can also email the division at htp@fas.usda.gov. Note: Compiled from reports of the U.S. Department of Commerce. U.S. trade can be obtained from the following web site: U.S. agricultural trade database, www.fas.usda.gov/ustrade.

Table 5-95.—Coffee: United States imports by country of origin, 2002-2004 [1]

Continent and country	2002	2003	2004 [2]
	Metric tons	*Metric tons*	*Metric tons*
Caribbean:			
Bahamas, The	17	0	0
Dominican Republic	66	75	56
Haiti	13	43	27
Jamaica	171	73	149
Leeward-Windward Islands	223	41	0
Total	490	232	232
Central America:			
Costa Rica	57,751	54,471	57,225
El Salvador	28,140	34,071	31,377
Guatemala	97,453	121,280	97,379
Honduras	24,406	21,771	28,594
Nicaragua	21,222	26,307	30,110
Panama	4,174	6,096	4,488
Total	233,146	263,995	249,173
East Asia:			
China, Peoples Republic of	816	385	104
Japan	0	67	34
Total	816	452	139
European Union - 25			
Austria	5	1	0
Belgium-Luxembourg	129	756	31
Denmark	0	138	0
France	6,126	7,225	2,265
Germany	33,889	35,123	47,412
Greece	36	1	2
Hungary	0	114	0
Italy	219	180	668
Lithuania	1	0	0
Netherlands	673	490	295
Poland	1	-55	11
Spain	0	61	0
Sweden	1	5	2
United Kingdom	77	41	1
Total	41,156	44,191	50,687
Former Soviet Union - 12:			
Armenia, Republic of	0	0	1
Georgia, Republic of	0	20	0
Russian Federation	0	2	0
Total	0	22	1
Middle East:			
Israel	3	1	0
Jordan	1	0	0
Lebanon	4	3	0
Turkey	39	38	4
Yemen	496	694	571
Total	542	737	575
North America:			
Canada	1,204	1,275	1,028
Mexica	112,844	79,752	72,533
Total	114,047	81,027	73,560
Oceania			
Australia	19	3	79
New Zealand	0	367	0
Papua New Guinea	8,106	11,953	9,331
Total	8,125	12,323	9,410
Other Europe:			
Bulgaria	1	0	0
Iceland	17	0	0
Norway	1	1	0
Romania	0	17	0
Switzerland	382	743	947
Total	400	762	947

See footnotes at end of table.

Table 5-95.—Coffee: United States imports by country of origin, 2002–2004[1]—Continued

Continent and country	2002	2003	2004[2]
	Metric tons	Metric tons	Metric tons
South America:			
Argentina	1	0	102
Bolivia	1,322	1,576	1,079
Brazil	292,175	303,488	255,340
Chile	46	0	158
Colombia	213,046	230,147	219,644
Ecuador	8,424	7,688	9,140
Peru	50,470	48,409	47,075
Suriname	0	1	0
Uruguay	0	19	1
Venezuela	10,154	10,813	5,839
Total	575,638	602,141	538,379
South Asia:			
India	4,461	3,118	6,401
Nepal	0	0	8
Total	4,461	3,118	6,409
Southeast Asia:			
Brunei	0	36	0
Cambodia	5	0	0
East Timor	0	1,687	0
Indonesia	45,496	57,871	98,005
Laos	18	33	506
Malaysia	0	0	6
Philippines	0	15	40
Singapore	3	256	18
Thailand	2,239	4,605	12,957
Vietnam	113,254	109,190	167,683
Total	161,015	173,694	279,215
Sub-Saharan Africa:			
Burkina	0	126	0
Burundi	833	5,453	166
Cameroon	1,046	241	987
Central African Republic	2	0	0
Chad	0	0	216
Congo (Brazzaville)	77	345	57
Congo, Democratic Republic of	0	519	489
Cote d'Ivoire	1,358	1,709	1,473
Djibouti Afars-Issas	8	0	18
Equatorial Guinea	38	76	0
Ethiopia	4,408	6,017	7,050
Ghana	107	27	0
Guinea	835	139	946
Kenya	5,018	5,862	4,810
Madagascar	3	437	779
Malawi	60	39	7
Nigeria	20	19	38
Rwanda	2,274	2,162	3,213
Sierra Leone	18	0	0
South Africa Republic of	30	53	46
St. Helena (Br W Afr)	0	0	7
Tanzania United Republic of	671	2,300	1,880
Uganda	5,634	10,497	6,975
Zambia	227	477	834
Zimbabwe	271	462	329
Total	22,939	36,960	30,320
Grand total	1,162,776	1,219,652	1,239,047

[1] Green coffee only; does not include roasted or soluble. Bags of 60 kilograms each (132.276 lbs.). [2] Preliminary.

FAS, Horticultural and Tropical Products Division, (202) 720-3423. The division's Home Page is located at, www.fas.usda.gov/htp. You can also email the division at htp@fas.usda.gov. Note: Compiled from reports of the U.S. Department of Commerce. U.S. trade can be obtained from the following web site: U.S. agricultural trade database, www.fas.usda.gov/ustrade.

Table 5-96.—Coffee: Exports from principal producing countries, 2001/02–2003/04 [1]

Continent and country of origin	2001/2002	2002/2003	2003/2004 [2]
	1,000 bags [3]	*1,000 bags* [3]	*1,000 bags* [3]
Colombian Milds:			
Colombia	10,625	10,478	10,154
Kenya	793	878	820
Tanzania	579	841	558
Total	11,997	12,197	11,532
Other Milds:			
Bolivia	67	82	76
Burundi	250	552	222
Costa Rica	1,881	1,676	1,505
Cuba	87	58	29
Dominican Republic	114	141	52
Ecuador	568	667	642
El Salvador	1,473	1,320	1,347
Guatemala	3,330	3,965	3,306
Haiti	49	37	16
Honduras	2,617	2,439	2,794
India	3,441	3,560	3,776
Jamaica	28	25	27
Malawi	53	39	41
Mexico	2,893	2,561	2,422
Nicaragua	920	978	1,270
Panama	78	84	92
Papua New Guinea	1,026	1,143	1,120
Peru	2,689	2,664	2,447
Rwanda	346	237	414
Venezuela	131	271	149
Zambia	105	127	121
Zimbabwe	116	100	75
Total	22,262	22,726	21,943
Brazil/Other Arabicas:			
Brazil	26,158	27,633	25,403
Ethiopia	1,939	2,277	2,374
Paraguay	5	5	7
Total	28,102	29,915	27,784
Robustas:			
Angola	10	15	7
Benin	0	0	0
Cameroon	617	732	831
Central African Rep	100	38	68
Congo	0	0	0
Congo, Dem. Rep. of	171	192	249
Cote d'Ivoire	3,388	2,608	2,604
Equatorial Guinea	0	0	0
Gabon	1	1	0
Ghana	24	23	13
Guinea	201	251	344
Indonesia	4,307	4,517	4,558
Liberia	0	0	NA
Madagascar	107	165	155
Nigeria	3	10	1
Philippines	6	9	25
Sierra Leone	48	21	9
Sri Lanka	2	2	2
Thailand	93	225	368
Togo	114	78	152
Trinidad and Tobago	1	1	0
Uganda	3,153	2,810	2,523
Vietnam	11,966	11,555	14,497
Total	24,312	23,253	26,406
Grand total	86,673	88,091	87,665

[1] October-September.   [2] Preliminary.   [3] One bag = 132.276 pounds, green bean equivalent.   NA-not available.

FAS, Horticultural and Tropical Products Division, (202) 720–3423. The division's home page is located at www.fas.usda.gov/htp. You can e-mail the division at htp@fas.usda.gov. Note: This information can be obtained from www.ico.org. International Coffee Organization.

Table 5-97.—Coffee: Area, yield, total production, marketing year average price, and value, Hawaii, 1994–95 to 2005–06

Season	Area harvested	Yield per harvested acre	Production [1]	Price per pound	Value
	Acres	*1,000 pounds*	*1,000 pounds*	*Dollars*	*1,000 dollars*
1996–97	5,400	1.2	6,400	3.25	20,800
1997–98	5,800	1.6	9,400	3.00	28,200
1998–99	6,100	1.6	9,500	2.60	24,700
1999–2000	6,400	1.6	10,000	2.10	21,000
2000–2001	6,800	1.3	8,700	2.65	23,055
2001–2002	6,300	1.3	8,000	2.45	19,600
2002–2003	5,900	1.3	7,500	3.10	23,250
2003–2004	5,900	1.4	8,300	2.90	24,070
2004–2005	5,800	1.0	5,600	3.55	19,880
2005–2006 [2]	6,100	1.1	6,400	3.80	24,320

[1] Parchment basis.   [2] Preliminary.
NASS, Crops Branch, (202) 720–2127.

Table 5-98.—Tea: United States imports by country of origin, 2002–2004

Continent and country	2002	2003	2004 [1]
	Metric tons	*Metric tons*	*Metric tons*
Caribbean:			
Bahamas, The	18	0	0
Haiti	0	0	6
Jamaica	25	106	56
Total	43	106	61
Central America:			
El Salvador	0	0	8
Guatemala	8	0	33
Total	8	0	41
East Asia:			
China, Peoples Republic of	15,780	18,354	18,073
Hong Kong	154	221	119
Japan	548	968	1,040
Korea, Republic of	88	85	117
Macau	0	0	0
Taiwan	576	622	712
Total	17,146	20,250	20,061
European Union - 25			
Austria	2	6	12
Belgium-Luxembourg	60	27	1
Czech Republic	0	0	0
France	61	28	38
Germany	3,569	4,722	5,440
Greece	2	2	2
Hungary	0	2	0
Ireland	180	157	172
Italy	10	26	7
Latvia	0	4	17
Lithuania	1	6	1
Netherlands	192	452	754
Poland	48	117	93
Portugal	4	3	0
Slovakia	0	1	0
Slovenia	19	26	12
Spain	38	7	14
Sweden	0	0	0
United Kingdom	1,384	1,575	1,203
Total	5,569	7,160	7,765
Former Soviet Union - 12:			
Georgia, Republic of	736	935	898
Russian Federation	9	8	8
Ukraine	0	9	8
Uzbekistan, Republic of	0	12	9
Total	745	964	923
Middle East			
Iran	0	1	6
Israel	64	106	171
Jordan	15	57	18
Lebanon	2	0	0
Saudi Arabia	0	0	0
Turkey	689	1,103	870
United Arab Emirates	8	0	0
Total	780	1,267	1,066

See footnote at end of table.

Table 5-98.—Tea: United States imports by country of origin, 2002-2004—Continued

Continent and country	2002	2003	2004 [1]
	Metric tons	Metric tons	Metric tons
North Africa:			
Egypt	3	4	6
Morocco	30	60	75
Tunisia	0	4	0
Total	34	68	81
North America:			
Canada	662	1,238	1,964
Mexico	8	10	0
Total	670	1,248	1,964
Oceania:			
Australia	21	18	14
French Pacific Islands	0	0	0
New Zealand	0	0	0
Papua New Guinea	975	729	727
Total	996	748	740
Other Europe:			
Bosnia and Herzegovina	2	46	1
Bulgaria	1	2	0
Croatia	0	6	1
Romania	2	4	0
Serbia and Montenegro	1	0	16
Switzerland	0	20	0
Total	5	78	18
South America:			
Argentina	37,852	34,034	38,487
Brazil	1,610	1,243	1,183
Chile	47	5	13
Colombia	0	23	37
Ecuador	497	512	455
Paraguay	5	6	22
Peru	0	20	0
Suriname	20	0	0
Uruguay	0	12	0
Venezuela	3	1	0
Total	40,035	35,856	40,197
South Asia:			
Afghanistan	0	1	0
Bangladesh	0	9	13
India	5,762	7,467	7,082
Nepal	2	1	1
Pakistan	13	26	51
Sri Lanka	3,414	3,731	3,771
Total	9,190	11,235	10,919
Southeast Asia:			
Indonesia	5,532	5,516	6,299
Malaysia	6	20	16
Philippines	0	0	3
Singapore	12	179	67
Thailand	127	176	177
Vietnam	1,886	1,621	2,629
Total	7,564	7,511	9,191
Sub-Saharan Africa:			
Burundi	0	0	8
Ethiopia	0	19	19
Kenya	5,114	2,965	2,450
Malawi	5,068	4,056	3,449
Mali	18	0	0
Mauritius	0	22	0
Mozambique	0	11	39
Niger	0	0	5
Rwanda	28	21	0
Sierra Leone	1	0	20
South Africa, Republic of	234	385	249
Tanzania, United Republic of	126	44	41
Uganda	0	9	0
Zimbabwe	89	161	156
Total	10,680	7,694	6,438
Grand total	93,465	94,186	99,467

[1] Preliminary.

FAS, Horticultural and Tropical Products Division, (202) 720-3423. The division's Home Page is located at, www.fas.usda.gov/htp. You can also email the division at htp@fas.usda.gov. Note: Compiled from reports of the U.S. Department of Commerce. U.S. trade can be obtained from the following web site: U.S. agricultural trade database, www.fas.usda.gov/ustrade.

Table 5-99.—Mushrooms, canned: United States imports by country of origin, 2000–2004

Country	2000	2001	2002	2003	2004
	Metric tons	Metric tons	Metric tons	Metric tons	Metric tons
Canada	1,579	1,559	1,463	1,832	1,670
China, People's Republic	3,900	9,312	10,057	22,319	28,504
Colombia	1,770	1,601	728	1,351	1,050
France	3,175	2,491	2,810	1,278	379
India	15,677	13,377	10,840	12,252	15,282
Indonesia	14,968	11,667	11,435	10,705	11,099
Italy	86	88	81	204	109
Malaysia	1,025	1,286	1,374	1,770	2,263
Mexico	2,986	2,285	2,122	2,167	635
Netherlands	12,958	8,232	9,548	5,629	2,500
Oman	365	343	358	433	471
Poland	86	108	86	234	554
Spain	1,032	1,446	1,722	2,724	988
Taiwan	8,483	6,001	4,349	4,599	4,031
Thailand	238	223	202	171	181
Vietnam	0	0	776	1,200	1,840
Total	68,327	60,018	57,951	68,869	71,558

FAS, Horticultural and Tropical Products Division, (202) 720-3423. The division's Home Page is located at, www.fas.usda.gov/htp. You can also email the division at htp@fas.usda.gov. Note: Compiled from reports of the U.S. Department of Commerce. U.S. trade can be obtained from the following web site: U.S. agricultural trade database, www.fas.usda.gov/ustrade.

Table 5-100.—Specialty mushrooms: Number of growers, total production, volume of sales, price per pound, and value of sales, July 1–June 30, 2002–2003/2004–2005 [1]

Year and variety	Growers [2]	All sales			
		Total production [3]	Volume of sales [4]	Price per pound [5]	Value of sales
	Number	1,000 pounds	1,000 pounds	Dollars	1,000 dollars
2002–2003					
Shiitake	134	7,476	7,059	3.08	21,718
Oyster	62	3,997	3,562	1.91	6,820
Other	23	1,431	1,287	4.56	5,873
US [6]	153	12,904	11,908	2.89	34,411
2003–2004					
Shiitake	146	7,731	7,517	3.24	24,391
Oyster	70	4,428	4,185	2.08	8,714
Other	25	1,691	1,650	4.49	7,404
US [6]	170	13,850	13,352	3.03	40,509
2004–2005					
Shiitake	146	9,081	8,613	3.21	27,677
Oyster	74	5,409	5,109	2.33	11,889
Other	26	1,435	1,327	4.90	6,501
US [6]	172	15,925	15,049	3.06	46,067

[1] Specialty mushroom estimates represent growers who have at least 200 natural wood logs in production or commercial indoor growing areas. [2] Growers counted only once for US total if growing more than one specialty type mushroom. Growers growing Agaricus and Specialty are included. [3] Total production includes all fresh market and processing sales plus amount harvested but not sold (shrinkage, cullage, dumped, etc.). [4] Virtually all specialty mushroom sales are for fresh market. [5] Prices for mushrooms are the average prices producers receive at the point of first sale, commonly referred to as the average price as sold. For example, if in a given State, part of the fresh mushrooms are sold F.O.B. packed by growers, part are sold bulk to brokers or repackers, and some are sold retail at roadside stands, the mushroom average price as sold is a weighted average of the average price for each method of sale. [6] 2002-03: AR, CA, CT, DE, FL, GA, HI, IA, KS, KY, ME, MA, MN, MO, MT, NH, NY, NC, OH, OK, OR, PA, SC, VT, VA, WA, WV, and WI. 2003-04: AR, CA, CT, DE, FL, HI, IN, KS, KY, ME, MD, MA, MI, MN, MO, MT, NH, NY, NC, OH, OR, PA, SC, TN, TX, VT, VA, WA, WV, and WI. 2004-05: AR, CA, CT, DE, FL, HI, IL, IN, KS, KY, ME, MD, MA, MI, MN, MO, MT, NH, NY, NC, OH, OR, PA, SC, TN, TX, VA, WA, WV, and WI.

NASS, Crops Branch, (202) 720–2127.

Table 5-101.—Agaricus mushrooms: Area, volume of sales, marketing year average price, and value of sales, United States, 1995/96–2004/2005 [1]

Year	Area in production	Volume of sales	Price per pound	Value of sales		
				Total	Fresh market	Processing
	1,000 sq. ft.	1,000 pounds	Cents	1,000 dollars	1,000 dollars	1,000 dollars
1995–96	135,320	777,870	93.5	727,578	588,126	139,452
1996–97	136,461	776,677	94.0	730,296	605,728	124,568
1997–98	145,094	808,678	95.7	773,617	670,168	103,449
1998–99	150,017	847,760	97.7	828,098	712,000	116,098
1999–2000	151,487	854,394	97.0	828,551	715,943	112,608
2000–2001	143,873	846,209	97.6	825,500	736,543	88,957
2001–2002	140,822	831,107	105.0	870,573	796,522	74,051
2002–2003	141,844	836,398	102.0	855,983	778,307	77,676
2003–2004	146,510	841,162	104.0	878,405	805,200	73,205
2004–2005	143,093	838,083	103.0	862,303	796,604	65,699

[1] Marketing year begins July 1 and ends June 30 the following year.
NASS, Crops Branch, (202) 720–2127.

Table 5-102.—Cut flowers: Sales and wholesale value for operations with $100,000+ sales, Surveyed States, 1995–2004

Year	Quantity sold	Wholesale price	Value of sales at wholesale [1]	Quantity sold	Wholesale price	Value of sales at wholesale [1]
	Standard carnations			Miniature carnations		
	1,000 blooms	Cents	1,000 dollars	1,000 bunches	Dollars	1,000 dollars
1995	112,067	15.3	17,199	9,152	1.21	11,060
1996	92,160	14.5	13,345	5,562	1.32	7,330
1997	74,368	15.8	11,739	4,907	1.52	7,441
1998	63,171	15.7	9,891	4,211	1.43	6,036
1999	38,456	18.7	7,201	2,631	1.46	3,829
2000	40,206	16.0	6,430	[2]	[2]	[2]
2001	24,760	15.6	3,870	[2]	[2]	[2]
2002	21,643	15.8	3,416	[2]	[2]	[2]
2003	13,491	17.6	2,374	[2]	[2]	[2]
2004	9,636	18.2	1,752	[2]	[2]	[2]
	Standard chrysanthemums			Pompon chrysanthemums		
	1,000 blooms	Cents	1,000 dollars	1,000 bunches	Dollars	1,000 dollars
1995	14,877	53.8	7,997	13,542	1.26	17,079
1996	15,184	55.6	8,438	12,003	1.26	15,145
1997	11,643	47.7	5,550	12,562	1.30	16,341
1998	14,265	51.6	7,362	12,427	1.35	16,828
1999	12,786	55.4	7,083	16,315	0.93	15,181
2000	[2]	[2]	[2]	13,159	1.31	17,214
2001	[2]	[2]	[2]	12,933	1.30	16,831
2002	[2]	[2]	[2]	14,766	1.31	19,351
2003	[2]	[2]	[2]	14,002	1.30	18,196
2004	[2]	[2]	[2]	15,066	1.33	20,030
	Hybrid Tea roses			Sweetheart roses		
	1,000 blooms	Cents	1,000 dollars	1,000 blooms	Cents	1,000 dollars
1995	352,154	32.5	114,594	45,350	24.5	11,133
1996	309,663	34.2	105,823	43,251	26.6	11,523
1997	343,879	32.0	109,914	44,706	28.2	12,601
1998	268,302	33.6	90,174	48,806	25.0	12,191
1999	223,562	34.3	76,709	39,377	26.1	10,294
2000	[2]	[2]	[2]	[2]	[2]	[2]
2001	[2]	[2]	[2]	[2]	[2]	[2]
2002	[2]	[2]	[2]	[2]	[2]	[2]
2003	[2]	[2]	[2]	[2]	[2]	[2]
2004	[2]	[2]	[2]	[2]	[2]	[2]
	Gladioli			Other cut flowers		
	1,000 spikes	Cents	1,000 dollars			1,000 dollars
1995	149,132	24.2	36,110			208,458
1996	144,941	22.9	33,260			217,836
1997	154,650	22.5	34,861			273,122
1998	129,297	25.6	33,138			235,975
1999	100,457	25.4	25,535			285,792
2000	127,109	22.3	28,339			133,648
2001	112,948	21.5	24,284			133,343
2002	126,001	21.3	26,853			136,515
2003	121,465	23.3	28,325			129,663
2004	113,610	22.8	25,957			120,565

See footnotes at end of table.

Table 5-102.—Cut flowers: Sales and wholesale value for operations with $100,000+ sales, Surveyed States, 1994–2003—Continued

Year	Quantity sold	Wholesale price	Value of sales at wholesale [1]	Quantity sold	Wholesale price	Value of sales at wholesale [1]
		Alstromeria			Delphinium & Larkspur	
	1,000 blooms	Cents	1,000 dollars	1,000 bunches	Dollars	1,000 dollars
1995						
1996						
1997						
1998						
1999						
2000	25,056	23.7	5,939	43,840	25.0	10,955
2001	21,253	24.2	5,137	45,515	24.2	11,008
2002	17,153	27.2	4,674	47,023	23.3	10,971
2003	13,402	29.7	3,978	40,945	23.9	9,797
2004	12,847	28.9	3,710	38,795	25.5	9,891
		Gerbera Daisy			Iris	
	1,000 blooms	Cents	1,000 dollars	1,000 bunches	Dollars	1,000 dollars
1995						
1996						
1997						
1998						
1999						
2000	67,520	30.9	20,886	79,012	25.8	20,395
2001	72,916	30.6	22,317	83,594	23.4	19,549
2002	84,917	29.8	25,343	81,837	22.4	18,344
2003	94,046	29.9	28,164	89,976	22.6	20,367
2004	101,598	30.8	31,280	92,285	23.5	21,708
		All Lilies			Lisianthus	
	1,000 blooms	Cents	1,000 dollars	1,000 blooms	Cents	1,000 dollars
1995						
1996						
1997						
1998						
1999						
2000	81,301	68.8	55,975	15,150	45.5	6,891
2001	91,267	66.7	60,876	19,040	44.7	8,505
2002	101,748	61.3	62,347	14,530	45.1	6,551
2003	112,946	65.0	73,400	14,410	45.0	6,491
2004	116,312	67.2	78,169	12,816	47.2	6,045
		All orchids			All roses	
	1,000 spikes	Cents	1,000 dollars			1,000 dollars
1995						
1996						
1997						
1998						
1999						
2000	11,719	68.9	8,071	185,975	37.3	69,294
2001	11,571	74.0	8,563	160,301	37.4	59,976
2002	11,113	70.2	7,796	157,253	37.4	58,878
2003	12,237	69.8	8,536	123,483	38.1	46,997
2004	11,702	70.0	8,195	108,341	39.8	43,111
		Snapdragons			Tulips	
	1,000 blooms	Cents	1,000 dollars	1,000 blooms	Cents	1,000 dollars
1995						
1996						
1997						
1998						
1999						
2000	63,711	30.1	19,166	77,039	34.7	26,760
2001	60,939	27.9	16,980	75,769	35.5	26,864
2002	60,860	28.0	17,041	90,625	32.0	29,001
2003	55,392	28.2	15,639	92,551	33.6	31,055
2004	50,452	29.7	15,001	100,379	36.1	36,217

[1] Equivalent wholesale value of all sales.   [2] This data series discontinued.

NASS, Crops Branch, (202) 720–2127.

Table 5-103.—Cut Greens: Sales and wholesale value for operations with $100,000+ sales, Surveyed States, 1995–2004

Year	Quantity sold	Wholesale price	Value of sales at wholesale[1]	Quantity sold	Wholesale price	Value of sales at wholesale[1]
	Leatherleaf Ferns			Other cut cultivated greens		
	1,000 bunches	*Dollars*	*1,000 dollars*	*1,000 bunches*	*Dollars*	*1,000 dollars*
1995	74,968	0.85	63,485			49,639
1996	77,982	0.87	67,993			50,192
1997	78,707	0.82	64,373			51,811
1998	72,981	0.83	60,498			57,191
1999	71,928	0.90	64,547			62,128
2000	75,611	0.88	66,245			59,923
2001	63,002	0.88	55,310			57,048
2002	61,907	0.87	53,634			60,139
2003	58,305	0.84	48,868			53,197
2004	53,873	0.88	47,283			45,162

[1] Equivalent wholesale value of all sales.
NASS, Crops Branch, (202) 720–2127.

Table 5-104.—Potted flowering and foliage plants: Sales and wholesale value for operations with $100,000+ sales, Surveyed States, 1995–2004

Year	Quantity sold		Wholesale Price		Value of sales at wholesale[1]
	Less than 5 inches	5 inches or more	Less than 5 inches	5 inches or more	
	African violets				
	1,000 pots	*1,000 pots*	*Dollars*	*Dollars*	*1,000 dollars*
1995	21,366	707	1.08	2.74	25,086
1996	20,683	269	1.06	3.79	23,029
1997	19,736	338	1.09	3.23	22,584
1998	19,277	216	1.14	3.20	22,672
1999	20,678	226	1.15	3.18	24,455
2000	16,043	257	1.12	3.57	18,909
2001	15,834	260	1.21	3.52	20,034
2002	15,513	621	1.24	2.52	20,816
2003	14,365	663	1.18	2.33	18,540
2004	12,635	637	1.17	2.38	16,308
	Florist chrysanthemums				
	1,000 pots	*1,000 pots*	*Dollars*	*Dollars*	*1,000 dollars*
1995	10,358	21,642	1.47	3.56	92,219
1996	8,905	19,825	1.42	3.64	84,726
1997	7,779	19,328	1.49	3.68	82,632
1998	7,207	17,727	1.53	3.52	73,408
1999	6,895	17,218	1.51	3.51	70,923
2000	8,439	19,936	1.50	3.47	81,869
2001	6,585	18,592	1.56	3.49	75,225
2002	7,096	23,948	1.58	3.09	85,128
2003	8,721	17,982	1.78	3.01	69,641
2004	12,080	17,553	1.76	3.06	75,018
	Potted florist roses[2]				
	1,000 pots	*1,000 pots*	*Dollars*	*Dollars*	*1,000 dollars*
1995					
1996					
1997					
1998					
1999					
2000	8,784	2,844	1.77	4.19	27,499
2001	7,257	3,072	1.91	3.83	25,645
2002	6,662	2,483	2.19	5.18	27,492
2003	6,863	945	2.44	3.87	20,394
2004	6,270	648	2.43	4.26	17,977
	Potted spring flowering bulbs[2]				
	1,000 pots	*1,000 pots*	*Dollars*	*Dollars*	*1,000 dollars*
1995					
1996					
1997					
1998					
1999					
2000	6,408	8,775	1.59	3.33	39,392
2001	7,517	10,360	1.50	3.36	46,075
2002	7,590	12,347	1.65	3.44	55,012
2003	7,206	12,181	1.66	3.52	54,927
2004	6,341	12,341	1.53	3.53	53,268

See footnotes at end of table.

Table 5-104.—Potted flowering and foliage plants: Sales and wholesale value for operations with $100,000+ sales, Surveyed States, 1995–2004—Continued

Year	Quantity sold		Wholesale Price		Value of sales at wholesale [1]
	Less than 5 inches	5 inches or more	Less than 5 inches	5 inches or more	
	Cyclamen				
	1,000 pots	*1,000 pots*	*Dollars*	*Dollars*	*1,000 dollars*
1995	2,568	2,537	2.01	3.93	15,145
1996	3,087	2,556	2.12	4.02	16,824
1997	3,129	2,767	2.17	4.02	17,902
1998	2,649	2,608	2.25	3.95	16,259
1999	2,683	3,125	2.23	3.68	17,491
2000	[3]	[3]	[3]	[3]	[3]
2001	[3]	[3]	[3]	[3]	[3]
2002	[3]	[3]	[3]	[3]	[3]
2003	[3]	[3]	[3]	[3]	[3]
2004	[3]	[3]	[3]	[3]	[3]
	Florist azaleas				
	1,000 pots	*1,000 pots*	*Dollars*	*Dollars*	*1,000 dollars*
1995	6,371	9,673	1.65	4.88	57,770
1996	3,251	8,567	1.77	4.45	43,843
1997	3,350	9,584	1.82	3.77	42,186
1998	2,714	7,224	2.04	4.72	39,635
1999	2,718	7,207	1.95	5.26	43,185
2000	4,880	10,032	2.08	5.14	61,719
2001	3,987	9,974	2.58	5.32	63,333
2002	3,035	7,679	2.64	5.29	48,603
2003	2,330	6,281	3.04	5.09	39,048
2004	2,448	7,154	2.31	4.96	41,130
	Kalanchoes				
	1,000 pots	*1,000 pots*	*Dollars*	*Dollars*	*1,000 dollars*
1995	3,279	1,982	1.47	3.53	11,832
1996	3,410	2,692	1.40	3.35	13,800
1997	3,615	2,863	1.44	3.51	15,251
1998	3,944	2,993	1.51	3.66	16,907
1999	3,930	2,853	1.50	3.74	16,550
2000	[3]	[3]	[3]	[3]	[3]
2001	[3]	[3]	[3]	[3]	[3]
2002	[3]	[3]	[3]	[3]	[3]
2003	[3]	[3]	[3]	[3]	[3]
2004	[3]	[3]	[3]	[3]	[3]
	Easter lilies				
	1,000 pots	*1,000 pots*	*Dollars*	*Dollars*	*1,000 dollars*
1995	37	9,193	2.97	4.00	36,925
1996	182	9,399	2.88	3.99	38,012
1997	177	12,175	2.72	3.76	46,278
1998	16	9,414	2.79	3.95	37,273
1999	13	9,096	2.31	3.95	35,990
2000	141	9,002	3.22	4.09	37,246
2001	214	9,236	3.07	4.01	37,735
2002	241	8,853	2.86	4.10	37,014
2003	244	8,580	2.33	4.19	36,434
2004	133	9,198	2.12	4.15	38,488

See footnotes at end of table.

Table 5-104.—Potted flowering and foliage plants: Sales and wholesale value for operations with $100,000+ sales, Surveyed States, 1995–2004—Continued

Year	Quantity sold		Wholesale Price		Value of sales at wholesale [1]
	Less than 5 inches	5 inches or more	Less than 5 inches	5 inches or more	

			Other lilies [3]		
	1,000 pots	*1,000 pots*	*Dollars*	*Dollars*	*1,000 dollars*
1995	286	1,971	1.72	3.76	7,895
1996					
1997					
1998					
1999					
2000					
2001					
2002					
2003					
2004					

			Poinsettias		
	1,000 pots	*1,000 pots*	*Dollars*	*Dollars*	*1,000 dollars*
1995	12,407	45,770	1.90	4.10	211,437
1996	12,819	46,873	1.92	4.07	215,248
1997	13,446	48,999	1.85	4.14	227,729
1998	14,915	46,993	1.87	4.18	224,441
1999	12,839	48,848	1.86	4.15	226,816
2000	15,457	50,931	1.81	4.28	246,263
2001	14,682	52,284	1.84	4.37	255,323
2002	14,837	51,707	1.86	4.36	252,983
2003	13,092	48,432	1.91	4.54	244,973
2004	11,760	49,197	1.94	4.57	247,845

			Other flowering [4]		
	1,000 pots	*1,000 pots*	*Dollars*	*Dollars*	*1,000 dollars*
1995	45,951	37,580	1.61	3.97	222,798
1996	52,440	39,266	1.30	3.41	201,865
1997	47,538	37,930	1.32	3.58	198,451
1998	53,219	40,238	1.71	3.74	241,357
1999	53,252	44,032	1.60	3.61	244,030
2000	33,585	35,766	1.73	3.90	197,684
2001	34,770	32,178	1.89	4.16	199,331
2002	37,033	31,103	1.89	4.35	205,157
2003	37,521	30,550	1.89	4.14	197,597
2004	37,192	31,541	1.84	4.09	197,494

			Potted Orchids [5]		
	1,000 pots	*1,000 pots*	*Dollars*	*Dollars*	*1,000 dollars*
1995					
1996	5,212	3,380	4.00	7.74	46,993
1997	4,134	5,447	5.93	8.32	69,856
1998	3,510	5,096	6.00	8.60	64,885
1999	4,805	5,695	5.51	9.29	79,398
2000	4,782	4,912	6.58	11.72	89,018
2001	6,992	5,208	6.17	11.31	102,049
2002	7,835	5,430	6.62	11.02	111,735
2003	8,871	6,209	6.15	10.85	121,908
2004	11,229	6,002	5.90	10.22	127,608

			Foliage		
	1,000 pots	*1,000 pots*	*Dollars*	*Dollars*	*1,000 dollars*
1995					413,566
1996					432,976
1997					421,398
1998					436,243
1999					432,532
2000					472,079
2001					568,668
2002					538,837
2003					566,984
2004					558,311

[1] Equivalent wholesale value of all sales except for potted foliage which is value of sales less cost of plant material purchased from other growers for growing on.   [2] Estimates began in 2000.   [3] This data series discontinued.   [4] Orchids included 1994-1995, cyclamen and kalanchoes included 2000-2003, potted florist roses and potted spring flowering bulbs included 1994-1999.   [5] Potted orchids reported separately for first time in 1996. Previously orchids were included in other flowering category.

NASS, Crops Branch, (202) 720–2127.

Table 5-105.—Flowering & foliage hanging baskets: Sales and wholesale value for operations with $100,000+ sales, Surveyed States, 1995–2004

Year	Quantity sold	Wholesale price	Value of sales at wholesale [1]	Quantity sold	Wholesale price	Value of sales at wholesale [1]
		Geraniums			Impatiens	
	1,000 baskets	*Dollars*	*1,000 dollars*	*1,000 baskets*	*Dollars*	*1,000 dollars*
1995	4,010	6.34	25,407	3,668	5.22	19,163
1996	4,691	6.43	30,142	4,004	5.22	20,891
1997	5,066	6.38	32,333	4,339	5.05	21,912
1998	4,734	6.56	31,058	4,140	4.96	20,536
1999	4,961	6.55	32,502	4,274	5.04	21,559
2000	(2)	(2)	(2)	4,054	5.15	20,859
2001	(2)	(2)	(2)	3,414	5.42	18,492
2002	(2)	(2)	(2)	4,096	5.12	20,972
2003	(2)	(2)	(2)	3,638	5.30	19,267
2004	(2)	(2)	(2)	3,684	5.39	19,865
		New Guinea Impatiens			Petunias	
	1,000 baskets	*Dollars*	*1,000 dollars*	*1,000 baskets*	*Dollars*	*1,000 dollars*
1995	3,812	6.04	23,014	1,345	5.36	7,205
1996	4,210	6.32	26,604	1,664	5.40	8,984
1997	4,674	6.25	29,198	2,013	5.40	10,876
1998	4,522	6.34	28,668	2,203	5.37	11,822
1999	4,911	6.35	31,196	2,823	5.59	15,774
2000	4,635	6.39	29,615	2,941	5.30	15,595
2001	4,663	6.34	29,572	3,102	5.89	18,269
2002	5,140	6.34	32,584	3,558	5.89	20,950
2003	4,540	6.44	29,247	3,933	5.93	23,325
2004	5,001	6.44	32,193	4,608	5.93	27,346
		Other flowering hanging baskets [3]			Foliage	
	1,000 baskets	*Dollars*	*1,000 dollars*	*1,000 baskets*	*Dollars*	*1,000 dollars*
1995	15,857	5.64	89,420	22,725	3.76	85,403
1996	15,595	5.76	89,874	19,736	3.85	75,971
1997	18,093	5.70	103,183	20,066	3.92	78,566
1998	19,404	5.95	115,437	15,993	4.14	66,258
1999	19,210	6.19	118,900	20,630	3.85	79,467
2000	14,760	6.01	88,656	20,983	4.20	88,113
2001	15,979	6.24	99,761	21,292	3.85	81,922
2002	17,679	6.25	110,492	19,984	4.19	83,723
2003	17,836	6.62	118,125	19,452	4.25	82,697
2004	20,618	6.48	133,528	18,297	4.41	80,668
		Begonias [4]			Marigolds [4]	
	1,000 baskets	*Dollars*	*1,000 dollars*	*1,000 baskets*	*Dollars*	*1,000 dollars*
1995						
1996						
1997						
1998						
1999						
2000	2,855	5.51	15,733	59	7.71	455
2001	3,335	5.72	19,062	50	6.90	345
2002	2,536	5.88	14,919	41	7.07	290
2003	3,352	5.14	17,229	23	5.96	137
2004	3,476	5.26	18,301	51	6.41	327

See footnotes at end of table.

Table 5-105.—Flowering & foliage hanging baskets: Sales and wholesale value for operations with $100,000+ sales, Surveyed States, 1995–2004—Continued

Year	Quantity sold	Wholesale price	Value of sales at wholesale [1]	Quantity sold	Wholesale price	Value of sales at wholesale [1]
	Geraniums from seed [4]			Geraniums from cuttings [4]		
	1,000 baskets	*Dollars*	*1,000 dollars*	*1,000 baskets*	*Dollars*	*1,000 dollars*
1995						
1996						
1997						
1998						
1999						
2000	684	5.83	3,991	4,146	7.00	29,024
2001	647	5.98	3,869	4,121	6.84	28,200
2002	567	6.47	3,666	4,431	7.00	30,997
2003	688	6.11	4,201	4,900	6.91	33,848
2004	701	6.12	4,291	5,404	6.88	37,161
	Pansy/Viola [4]					
	1,000 baskets	*Dollars*	*1,000 dollars*	*1,000 baskets*	*Dollars*	*1,000 dollars*
1995						
1996						
1997						
1998						
1999						
2000	303	6.38	1,932			
2001	466	5.97	2,784			
2002	600	6.09	3,651			
2003	747	5.87	4,383			
2004	729	5.90	4,304			

[1] Equivalent wholesale value of all sales.   [2] This data series discontinued.   [3] 1994–2000 data include Geraniums, Impatiens, New Guinea Impatiens, Petunias, and Pansy/Violas.   [4] Estimates began in 2000.

NASS, Crops Branch, (202) 720–2127.

Table 5-106.—Bedding plant flats: Sales and wholesale value for operations with $100,000+ sales, Surveyed States, 1995–2004

Year	Quantity sold	Wholesale price	Value of sales at wholesale [1]	Quantity sold	Wholesale price	Value of sales at wholesale [1]
		Geraniums			Impatiens	
	1,000 flats	*Dollars*	*1,000 dollars*	*1,000 flats*	*Dollars*	*1,000 dollars*
1995	3,978	9.01	35,859	15,501	6.90	107,014
1996	4,462	9.55	42,598	15,722	6.96	109,475
1997	3,940	9.46	37,281	18,147	6.97	126,394
1998	3,143	8.97	28,189	16,715	6.71	112,105
1999	3,205	9.77	31,325	16,331	7.04	114,939
2000	(2)	(2)	(2)	15,380	7.70	118,381
2001	(2)	(2)	(2)	14,904	7.81	116,331
2002	(2)	(2)	(2)	14,650	8.20	120,133
2003	(2)	(2)	(2)	13,418	7.80	104,689
2004	(2)	(2)	(2)	12,533	7.86	98,476
		New Guinea Impatiens			Petunias	
	1,000 flats	*Dollars*	*1,000 dollars*	*1,000 flats*	*Dollars*	*1,000 dollars*
1995	1,055	9.10	9,601	10,754	7.21	77,487
1996	877	8.98	7,873	10,761	7.10	76,359
1997	861	9.86	8,487	11,818	7.36	87,014
1998	589	10.66	6,277	11,783	7.23	85,189
1999	1,342	9.24	12,403	11,645	7.46	86,848
2000	657	9.71	6,381	12,093	7.90	95,488
2001	589	11.12	6,547	11,542	8.03	92,669
2002	793	9.93	7,872	11,635	8.47	98,595
2003	628	9.81	6,160	11,583	8.22	95,161
2004	422	10.60	4,474	10,990	8.38	92,083
		Other Flowering and foliar type bedding plant flats [3]			Vegetable type bedding plant flats [4]	
	1,000 flats	*Dollars*	*1,000 dollars*	*1,000 flats*	*Dollars*	*1,000 dollars*
1995	55,652	7.09	394,419	10,327	7.23	74,676
1996	57,654	7.20	414,932	10,666	7.46	79,578
1997	70,958	7.46	529,375	12,624	7.82	98,755
1998	63,123	7.58	478,431	12,044	7.70	92,723
1999	69,832	8.02	560,067	11,976	8.12	97,288
2000	47,709	8.04	383,686	8,604	7.97	68,604
2001	43,226	7.84	339,064	8,480	8.37	70,946
2002	40,978	8.35	342,326	8,121	8.74	70,991
2003	39,880	8.49	338,557	7,594	8.64	65,629
2004	38,535	8.61	331,783	7,459	8.42	62,824
		Begonias [3]			Marigolds [3]	
	1,000 flats	*Dollars*	*1,000 dollars*	*1,000 flats*	*Dollars*	*1,000 dollars*
1995	....	....	....	....	....	....
1996	....	....	....	....	....	....
1997	....	....	....	....	....	....
1998	....	....	....	....	....	....
1999	....	....	....	....	....	....
2000	6,645	7.83	52,004	5,443	7.75	42,169
2001	8,272	7.76	64,193	6,623	8.09	53,600
2002	7,906	7.69	60,817	6,311	8.50	53,616
2003	7,424	7.63	56,633	6,386	8.19	52,298
2004	6,717	7.84	52,650	6,144	8.30	50,965
		Geraniums from seed [3]			Geraniums from vegetable cuttings [3]	
	1,000 flats	*Dollars*	*1,000 dollars*	*1,000 flats*	*Dollars*	*1,000 dollars*
1995	....	....	....	....	....	....
1996	....	....	....	....	....	....
1997	....	....	....	....	....	....
1998	....	....	....	....	....	....
1999	....	....	....	....	....	....
2000	861	9.33	8,035	1,574	9.47	14,906
2001	766	10.27	7,868	1,003	10.82	10,849
2002	837	10.30	8,623	1,126	10.87	12,242
2003	749	10.31	7,725	914	10.11	9,239
2004	774	10.35	8,014	1,099	10.24	11,255
		Pansy/Viola [3]				
	1,000 flats	*Dollars*	*1,000 dollars*	*1,000 flats*	*Dollars*	*1,000 dollars*
1995	....	....	....	....	....	....
1996	....	....	....	....	....	....
1997	....	....	....	....	....	....
1998	....	....	....	....	....	....
1999	....	....	....	....	....	....
2000	10,153	8.23	83,521			
2001	13,109	7.87	103,151			
2002	14,201	8.55	121,452			
2003	14,179	8.35	118,358			
2004	13,042	8.39	109,396			

[1] Equivalent wholesale value of all sales. [2] This data series discontinued. [3] Begonias, Marigolds, Geraniums from seed/cuttings, and Pansy/Violas included in other flowering and foliar type flats prior to 2000. [4] Does not include vegetable transplants grown for use in commercial vegetable production.
NASS, Crops Branch, (202) 720–2127.

Table 5-107.—Potted flowering and foliar type bedding plants: Sales and wholesale value for operations with $100,000+ sales, Surveyed States, 1995–2004

Year	Quantity sold		Wholesale Price		Value of sales at wholesale [1]
	Less than 5 inches	5 inches or more	Less than 5 inches	5 inches or more	
			Hardy/Garden mums		
	1,000 pots	*1,000 pots*	*Dollars*	*Dollars*	*1,000 dollars*
1995	13,761	29,545	0.80	1.83	65,001
1996	13,001	32,588	0.82	1.79	69,078
1997	12,686	36,067	0.89	1.86	78,316
1998	14,965	42,505	0.89	1.89	93,787
1999	14,606	46,755	0.89	1.94	103,564
2000	17,813	48,534	1.11	1.78	106,385
2001	15,109	45,442	1.15	1.88	102,907
2002	12,705	50,295	1.00	2.03	114,524
2003	9,513	55,798	1.07	1.98	120,927
2004	13,006	55,577	1.04	2.16	133,641
			Geraniums (cuttings)		
	1,000 pots	*1,000 pots*	*Dollars*	*Dollars*	*1,000 dollars*
1995	49,726	15,902	1.24	2.64	103,569
1996	40,770	15,287	1.30	2.57	92,457
1997	40,620	15,190	1.41	2.61	97,123
1998	47,855	17,374	1.37	2.57	110,425
1999	45,221	17,099	1.39	2.76	110,291
2000	44,004	18,423	1.41	2.62	110,223
2001	42,033	18,126	1.47	2.79	112,417
2002	41,293	18,116	1.50	2.76	111,819
2003	41,245	19,287	1.56	2.88	119,921
2004	39,775	21,048	1.62	2.93	126,022
			Geraniums (seed)		
	1,000 pots	*1,000 pots*	*Dollars*	*Dollars*	*1,000 dollars*
1995	46,913	2,377	0.80	1.84	42,035
1996	42,339	2,184	0.82	1.91	38,841
1997	44,198	3,781	0.80	1.90	42,675
1998	42,564	2,173	0.80	1.82	37,889
1999	42,550	2,149	0.79	1.87	37,837
2000	46,834	1,295	0.83	2.08	41,756
2001	43,675	379	0.84	3.58	37,879
2002	40,451	1,365	0.84	1.65	36,273
2003	34,196	1,072	0.86	2.05	31,697
2004	37,497	902	0.90	2.15	35,508
			Impatiens		
	1,000 pots	*1,000 pots*	*Dollars*	*Dollars*	*1,000 dollars*
1995	21,916	2,334	0.64	1.57	17,585
1996	20,098	2,513	0.64	1.54	16,828
1997	20,791	3,211	0.63	1.53	18,102
1998	21,033	4,137	0.65	1.57	20,078
1999	24,730	4,926	0.64	1.41	22,837
2000	23,903	5,791	0.67	1.46	24,473
2001	26,839	4,340	0.71	1.76	26,736
2002	24,002	4,237	0.72	1.66	24,382
2003	26,557	4,788	0.71	1.71	26,989
2004	27,374	5,203	0.74	1.84	29,734

See footnotes at end of table.

Table 5-107.—Potted flowering and foliar type bedding plants: Sales and wholesale value for operations with $100,000+ sales, Surveyed States, 1995–2004—Continued

Year	Quantity sold		Wholesale Price		Value of sales at wholesale [1]
	Less than 5 inches	5 inches or more	Less than 5 inches	5 inches or more	
			New Guinea Impatiens		
	1,000 pots	*1,000 pots*	*Dollars*	*Dollars*	*1,000 dollars*
1995	8,452	3,072	1.24	2.45	18,003
1996	10,499	3,732	1.20	2.35	21,364
1997	11,960	4,427	1.31	2.38	26,222
1998	14,851	4,078	1.33	2.47	29,789
1999	15,860	5,090	1.33	2.48	33,802
2000	18,148	5,653	1.35	2.60	39,223
2001	16,382	6,678	1.40	2.35	38,601
2002	18,829	6,952	1.38	2.31	42,073
2003	18,135	7,051	1.43	2.54	43,790
2004	18,777	6,839	1.46	2.55	44,811
			Petunias		
	1,000 pots	*1,000 pots*	*Dollars*	*Dollars*	*1,000 dollars*
1995	10,656	1,129	0.66	1.78	9,030
1996	10,520	1,813	0.66	1.58	9,797
1997	9,061	1,985	0.68	1.60	9,321
1998	10,504	2,726	0.70	1.72	12,042
1999	11,491	3,115	0.75	1.75	14,072
2000	13,340	3,784	0.83	1.73	17,580
2001	14,724	5,280	0.90	2.11	24,389
2002	17,373	6,198	0.90	2.00	28,035
2003	17,268	7,237	0.92	2.12	31,190
2004	19,475	7,975	0.93	2.13	35,200
			Other flowering/foliar type [2]		
	1,000 pots	*1,000 pots*	*Dollars*	*Dollars*	*1,000 dollars*
1995	159,627	54,664	0.78	1.79	221,883
1996	134,216	83,260	0.82	1.69	250,656
1997	170,526	78,583	0.94	2.58	363,277
1998	232,674	111,416	0.88	2.82	520,239
1999	236,079	93,870	0.91	2.67	466,422
2000	159,357	53,236	0.96	2.35	277,692
2001	157,399	53,412	0.97	2.25	272,608
2002	169,081	56,477	1.02	2.29	301,859
2003	163,041	56,181	1.10	2.45	316,867
2004	169,581	64,871	1.10	2.44	343,953
			Vegetable type [3]		
	1,000 pots	*1,000 pots*	*Dollars*	*Dollars*	*1,000 dollars*
1995	16,185	2,910	0.74	1.56	16,596
1996	19,127	4,827	0.74	1.59	21,802
1997	23,545	6,101	0.76	1.50	27,115
1998	29,293	7,956	0.80	1.82	37,926
1999	26,454	5,721	0.79	1.87	31,513
2000	25,430	4,452	0.86	1.80	29,768
2001	24,930	4,625	0.91	1.85	31,309
2002	33,774	6,050	1.09	1.70	47,142
2003	42,492	6,305	0.93	1.82	51,028
2004	40,515	7,905	1.06	1.84	57,704

See footnotes at end of table.

Table 5-107.—Potted flowering and foliar type bedding plants: Sales and wholesale value for operations with $100,000+ sales, Surveyed States, 1995–2004—Continued

Year	Quantity sold		Wholesale Price		Value of sales at wholesale [1]
	Less than 5 inches	5 inches or more	Less than 5 inches	5 inches or more	
			Begonias [2]		
	1,000 pots	*1,000 pots*	*Dollars*	*Dollars*	*1,000 dollars*
1995	....	....	....	....	....
1996	....	....	....	....	....
1997	....	....	....	....	....
1998	....	....	....	....	....
1999	....	....	....	....	....
2000	12,559	2,321	0.89	1.81	15,427
2001	13,890	4,275	0.92	2.23	22,260
2002	15,969	4,493	0.89	1.98	23,142
2003	14,489	4,371	1.02	1.88	22,946
2004	18,262	4,949	0.91	1.91	26,175
			Marigolds [2]		
	1,000 pots	*1,000 pots*	*Dollars*	*Dollars*	*1,000 dollars*
1995	....	....	....	....	....
1996	....	....	....	....	....
1997	....	....	....	....	....
1998	....	....	....	....	....
1999	....	....	....	....	....
2000	4,994	1,111	0.78	1.33	5,368
2001	5,472	1,685	0.72	1.43	6,351
2002	7,760	1,497	0.76	1.35	7,895
2003	7,118	1,708	0.66	1.47	7,189
2004	8,003	2,750	0.67	1.53	9,599
			Pansy/viola [2]		
	1,000 pots	*1,000 pots*	*Dollars*	*Dollars*	*1,000 dollars*
1995	....	....	....	....	....
1996	....	....	....	....	....
1997	....	....	....	....	....
1998	....	....	....	....	....
1999	....	....	....	....	....
2000	19,985	3,932	0.74	1.55	20,882
2001	18,756	5,175	0.78	1.55	22,648
2002	25,244	7,906	0.70	1.67	31,053
2003	27,291	7,885	0.70	1.76	33,026
2004	29,162	8,690	0.73	1.86	37,553

See footnotes at end of table.

Table 5-107.—Potted flowering and foliar type bedding plants: Sales and wholesale value for operations with $100,000+ sales, Surveyed States, 1995–2004—Continued

Year	Quantity sold			Wholesale Price			Value of sales at wholesale [1]
	Less than 1 gallon	1-2 gallons	2 gallons or more	Less than 1 gallon	1-2 gallons	2 gallons or more	

				Potted Hosta [5]			
	1,000 pots	1,000 pots	1,000 pots	Dollars	Dollars	Dollars	1,000 dollars
1995							
1996							
1997							
1998							
1999							
2000	3,358	7,269	296	2.02	3.67	8.28	35,874
2001	2,889	7,341	1,161	2.71	3.48	5.49	39,755
2002	3,827	7,834	439	2.33	3.76	7.68	41,771
2003	4,148	8,533	433	2.35	3.69	7.51	44,498
2004	3,289	10,213	403	2.59	3.52	7.51	47,493

				Other Potted Herbaceous Perennials [5]			
	1,000 pots	1,000 pots	1,000 pots	Dollars	Dollars	Dollars	1,000 dollars
1995							
1996							
1997							
1998							
1999							
2000	66,995	56,181	3,904	1.47	3.04	5.84	291,734
2001	74,100	80,278	4,868	1.48	2.65	6.28	353,070
2002	98,314	90,170	8,527	1.44	2.92	5.96	455,793
2003	98,844	84,819	7,598	1.63	3.09	6.01	469,447
2004	98,819	96,222	8,358	1.69	2.98	6.18	505,916

[1] Equivalent wholesale value of all sales.   [2] Begonias, Marigolds, Pansy/Violas, Hosta, and Other herbaceous perennials included in other flowering/foliar type pots prior to 2000.   [3] Does not include vegetable transplants grown for use in commercial vegetable production.
NASS, Crops Branch, (202) 720–2127.

Table 5-108.—Floriculture: Growing area by type of cover, all operations with $10,000+ sales, 36 Surveyed States, 2003–2004

State	Glass greenhouses		Fiberglass and other rigid greenhouses		Film plastic (single/multi) greenhouses	
	2003	2004	2003	2004	2003	2004
	1,000 square feet	1,000 square feet	1,000 square feet	1,000 square feet	1,000 square feet	1,000 square feet
AL					8,346	7,954
AZ					1,567	1,793
AR					3,021	2,687
CA	14,291	14,161	29,301	29,624	57,229	52,539
CO	726	653	6,646	6,494	3,058	3,133
CT	1,405	1,360	1,134	977	6,590	6,072
FL	5,412	4,857	11,875	9,038	43,815	49,119
GA	637		358		6,834	6,143
HI	17		2,312		2,393	2,153
IL	2,702	2,738	1,660	1,734	8,960	8,661
IN	1,417	1,376	1,206	1,155	5,911	5,756
IA	1,065	1,022	1,508	1,124	3,968	3,861
KS	101	194	393	418	4,071	3,836
KY	586	713	305	293	4,234	4,309
LA	131	131	723	585	2,421	2,309
MD	1,994	1,880	649	609	4,016	3,829
MA	1,520	1,496	930	939	7,266	6,966
MI	4,657	4,549	4,191	4,559	37,424	38,217
MN	833	864	2,510	2,508	7,966	7,686
MS					2,347	2,241
MO	573	584	664	529	5,132	5,101
NJ	4,912	4,638	479	564	14,099	13,538
NM					2,590	2,601
NY	4,754	4,896	1,078	1,385	18,322	18,177
NC	3,253	3,125	225	236	15,477	16,173
OH	8,779	8,640	1,599	1,591	18,084	17,190
OK	157		342		3,506	3,529
OR	1,695	1,724	2,155	2,172	12,029	11,456
PA	4,136	4,104	2,594	2,608	15,935	15,987
SC	829	829	304	413	3,446	3,572
TN	1,001	1,008	313	329	5,950	5,906
TX	691	575	3,539	3,568	26,928	27,259
UT	35		1,938		4,017	3,758
VA	1,643	1,593	737	671	7,546	8,029
WA	2,173	2,364	606	672	7,130	7,242
WI	1,597	1,579	1,027	905	7,365	7,344
Oth Sts	1,122	1,651	2,507	7,232		
36 States	74,844	73,304	85,808	82,932	388,993	386,126

See end of table.

Table 5-108.—Floriculture: Growing area by type of cover, all operations with $10,000+ sales, 36 Surveyed States, 2003–2004—Continued

State	Shade and temporary cover		Total covered area		Open ground	
	2003	2004	2003	2004	2003	2004
	1,000 square feet	*1,000 square feet*	*1,000 square feet*	*1,000 square feet*	*Acres*	*Acres*
AL	113	146	9,366	8,934	143	397
AZ	900	837	3,498	3,342	311	296
AR	142	70	3,187	2,778	85	89
CA	26,258	25,062	127,079	121,386	9,648	9,020
CO	191	179	10,621	10,459	59	47
CT	470	406	9,599	8,815	663	547
FL	312,552	316,488	373,654	379,502	9,572	9,239
GA	501	316	8,330	7,377	380	193
HI	21,794	20,652	26,516	25,240	2,199	1,993
IL	699	906	14,021	14,039	509	511
IN	136	154	8,670	8,441	272	275
IA	166	147	6,707	6,154	131	127
KS	98	114	4,663	4,562	76	81
KY	116	162	5,241	5,477	113	112
LA	180	192	3,455	3,217	149	145
MD	163	170	6,822	6,488	219	214
MA	121	116	9,837	9,517	633	518
MI	1,569	1,425	47,841	48,750	3,237	2,996
MN	218	235	11,527	11,293	196	163
MS	130	181	2,594	2,519	92	95
MO	235	198	6,604	6,412	271	283
NJ	442	452	19,932	19,192	3,018	2,887
NM	120	132	4,260	4,313	16	19
NY	696	687	24,850	25,145	1,058	963
NC	566	699	19,521	20,233	452	448
OH	626	537	29,088	27,958	312	357
OK	314	270	4,319	4,234	549	525
OR	1,360	1,286	17,239	16,638	1,605	1,575
PA	386	290	23,051	22,989	582	579
SC	2,427	2,557	7,006	7,371	617	552
TN	96	241	7,360	7,484	326	262
TX	5,988	6,352	37,146	37,754	594	605
UT	150	155	6,140	5,764	183	214
VA	501	464	10,427	10,757	360	334
WA	940	623	10,849	10,901	1,721	1,340
WI	476	382	10,465	10,210	565	593
Oth Sts	....	....	....	....	....	....
36 States	381,840	383,283	931,485	925,645	40,916	38,594

NASS, Crops Branch, (202) 720-2127.

Table 5-109.—Floriculture: Total operations of $10,000+ sales and expanded wholesale value, 36 Surveyed States, 2003–2004

State	Total operations		Expanded wholesale value [1]	
	2003	2004	2003	2004
	Number	*Number*	*1,000 dollars*	*1,000 dollars*
AL	173	176	73,010	78,303
AZ	42	38	32,919	31,291
AR	78	66	14,229	12,923
CA	910	851	997,396	1,001,882
CO	132	123	71,122	74,072
CT	284	250	72,812	83,146
FL	1,106	1,001	809,959	806,337
GA	224	186	63,243	60,136
HI	436	425	62,314	60,174
IL	326	296	100,619	110,711
IN	299	281	45,935	46,917
IA	161	144	47,451	46,839
KS	130	115	27,726	22,048
KY	235	212	27,761	29,925
LA	129	113	18,316	17,540
MD	185	170	83,319	100,640
MA	431	395	63,446	65,490
MI	743	711	322,980	353,051
MN	256	235	78,735	75,736
MS	75	68	11,358	11,338
MO	256	221	37,513	32,385
NJ	419	392	129,090	143,196
NM	47	42	22,216	21,973
NY	886	837	168,633	183,437
NC	381	349	147,591	152,541
OH	636	576	180,381	186,246
OK	136	115	31,990	32,567
OR	302	279	84,419	86,048
PA	892	865	157,326	157,457
SC	121	111	103,038	104,878
TN	221	187	35,230	36,042
TX	357	340	285,241	279,551
UT	82	74	46,342	42,778
VA	243	235	74,167	78,397
WA	253	232	103,575	117,706
WI	409	388	70,485	67,704
Oth Sts	....	....	68,000	75,641
36 States	11,996	11,099	4,769,887	4,887,046

[1] Wholesale value of sales as reported by growers with $100,000 or more in sales of floriculture crops plus a calculated wholesale value of sales for growers with sales below $100,000. The value of sales for growers below the $100,000 level was estimated by multiplying the number of growers in each size group by the mid-point of each dollar value range.

NASS, Crops Branch, (202) 720-2127.

Table 5-110.—Floriculture Crops: Wholesale value of sales by category for operations with $100,000+ saies, 36 Surveyed States, 2003–2004

State	Total cut flowers		Total potted flowering plants		Total foliage for indoor or patio use	
	2003	2004	2003	2004	2003	2004
	1,000 dollars	*1,000 dollars*	*1,000 dollars*	*1,000 dollars*	*1,000 dollars*	*1,000 dollars*
AL			10,937	7,508	4,190	4,008
AZ			1,914	1,775	3,313	3,939
AR			1,483	1,336	401	287
CA	304,152	303,562	188,289	192,214	102,532	96,607
CO	2,817	2,845	10,554	8,874	481	523
CT			11,795	11,198	644	955
FL	19,807	19,519	91,887	92,617	423,882	415,967
GA			8,116	7,293	1,597	2,591
HI	17,331	15,510	21,771	22,501	14,607	15,443
IL	675		21,595	21,692	2,988	4,759
IN			6,338	6,349	1,182	1,259
IA			10,470	9,797	1,479	1,053
KS			6,693	3,212	328	551
KY			3,531	4,348	1,657	805
LA			2,839	2,398	4,145	4,500
MD			8,300	8,457	1,368	2,112
MA	2,433	2,453	9,992	8,619	3,957	1,163
MI	8,797	8,711	32,400	31,991	3,375	4,152
MN	5,245	5,018	10,281	10,519	1,265	1,251
MS			2,286	2,976	601	451
MO			9,905	8,719	3,421	
NJ	8,954	8,779	30,792	29,450	4,139	3,878
NM			17,105	17,138		
NY	4,984	4,406	43,110	56,213	4,057	6,740
NC	3,165	3,040	34,796	36,369	7,384	5,201
OH	1,046		22,077	23,672	7,045	10,542
OK			4,203	3,969	1,163	1,257
OR	10,209	10,828	19,422	19,017		1,700
PA	2,606	2,482	40,839	41,081	4,024	3,753
SC	851		14,681	15,256	1,056	1,103
TN			10,330	9,097	531	1,158
TX			42,142	43,546	29,074	28,622
UT			13,783	12,652	3,128	1,832
VA		1,250	20,606	23,824	1,649	1,920
WA	18,371	19,720	7,557	9,506	1,946	1,796
WI	1,929		10,643	9,953	1,453	1,137
Oth Sts	9,610	13,508			5,619	5,964
36 States	422,982	421,631	803,462	815,136	649,681	638,979

See end of table.

Table 5-110.—Floriculture Crops: Wholesale value of sales by category for operations with $100,000+ sales, 36 Surveyed States, 2003–2004[1]—Continued

State	Total bedding/garden plants[2]		Total cut cultivated greens		Propagative materials		Total reported wholesale value of floriculture crops	
	2003	2004	2003	2004	2003	2004	2003	2004
	1,000 dollars	1,000 dollars	1,000 dollars	1,000 dollars	1,000 dollars	1,000 dollars	1,000 dollars	1,000 dollars
AL	57,883	66,787					73,010	78,303
AZ	27,312	25,197			380	380	32,919	31,291
AR	12,345	11,300					14,229	12,923
CA	317,209	325,554	12,124	9,260	73,090	74,685	997,396	1,001,882
CO	57,270	61,830					71,122	74,072
CT	60,373	70,389				604	72,812	83,146
FL	111,513	117,004	81,072	76,495	81,798	84,735	809,959	806,337
GA	47,197	50,252			6,333		63,243	60,136
HI	3,584	3,493	634	664	4,387	2,563	62,314	60,174
IL	69,930	80,700			5,431	3,560	100,619	110,711
IN	35,984	36,914			2,431	2,395	45,935	46,917
IA	35,502	35,989					47,451	46,839
KS	20,705	18,285					27,726	22,048
KY	22,573	24,772					27,761	29,925
LA	11,332	10,642					18,316	17,540
MD	68,873	85,018			4,778	5,053	83,319	100,640
MA	42,916	48,481			4,148	4,774	63,446	65,490
MI	230,322	249,753			48,086	58,444	322,980	353,051
MN	61,944	58,948					78,735	75,736
MS	8,471	7,911					11,358	11,338
MO	24,187	23,666					37,513	32,385
NJ	85,205	89,358				11,731	129,090	143,196
NM	5,111	4,835					22,216	21,973
NY	107,510	107,874			8,972	8,204	168,633	183,437
NC	100,084	106,294			2,162	1,637	147,591	152,541
OH	132,311	134,068			17,902	17,964	180,381	186,246
OK	26,624	27,341					31,990	32,567
OR	43,982	46,251	6,576	4,287	4,230	3,965	84,419	86,048
PA	85,259	84,944			24,598	25,197	157,326	157,457
SC	86,450	88,519					103,038	104,878
TN	24,369	25,787					35,230	36,042
TX	206,535	199,670			7,490	7,713	285,241	279,551
UT	26,260	28,294			3,171		46,342	42,778
VA	51,912	51,403					74,167	78,397
WA	60,165	70,005			15,536	16,679	103,575	117,706
WI	52,875	53,822		9	3,585	2,783	70,485	67,704
Oth Sts	1,649	1,195	1,659	1,730	49,463	53,244	68,000	75,641
36 States	2,423,726	2,532,545	102,065	92,445	367,971	386,310	4,769,887	4,887,046

[1] Missing data were included in "Other States" to avoid disclosure of individual operations. [2] Includes Annual Bedding Plants and Herbaceous Perennials.
NASS, Crops Branch, (202) 720-2127.

Table 5-111.—Fruit and orange juice: Cold storage holdings, end of month, United States, 2004 and 2005

Month	Fresh					
	Apples, regular storage		Apples, CA storage		Apples, total	
	2004	2005	2004	2005	2004	2005
	1,000 pounds	1,000 pounds	1,000 pounds	1,000 pounds	1,000 pounds	1,000 pounds
January	395,445	547,180	2,987,516	3,829,865	3,382,961	4,377,045
February	312,963	316,524	2,533,968	3,336,128	2,846,931	3,652,652
March	183,302	207,320	1,795,704	2,588,717	1,979,006	2,796,037
April	89,851	93,610	1,358,010	1,957,788	1,447,861	2,051,398
May	21,812	34,507	902,068	1,388,216	923,880	1,422,723
June	8,030	13,494	549,974	845,302	558,004	858,796
July	3,120	14,912	271,174	433,693	274,294	448,605
August	17,991	24,990	65,804	91,812	83,795	116,802
September	1,052,119	1,009,941	1,953,788	1,968,279	3,005,907	2,978,220
October	1,595,369	1,563,343	4,695,285	4,258,308	6,290,654	5,821,651
November	1,183,155	1,028,602	4,710,686	4,415,177	5,893,841	5,443,779
December	858,184	676,437	4,328,804	4,077,970	5,186,988	4,754,407

Month	Fresh					
	Pears, Bartlett		Pears, other		Pears, total	
	2004	2005	2004	2005	2004	2005
	1,000 pounds	1,000 pounds	1,000 pounds	1,000 pounds	1,000 pounds	1,000 pounds
January	4,488	286	245,655	209,994	250,143	210,280
February	3,404	273	186,158	148,642	189,562	148,915
March	2,395	255	136,617	104,889	139,012	105,144
April	2,052	226	79,305	62,043	81,357	62,269
May	6	168	44,273	35,605	44,279	35,773
June	1,284	238	13,708	14,130	14,992	14,368
July	27,470	8,576	8,275	601	35,745	9,177
August	64,183	72,173	2,722	1,408	66,905	73,581
September	103,666	109,403	394,052	330,978	497,718	440,381
October	53,753	56,156	421,380	466,034	475,133	522,190
November	23,506	30,038	343,681	409,921	367,187	439,959
December	6,574	12,135	276,382	322,022	282,956	334,157

Month	Frozen							
	Apples		Apricots		Blackberries, IQF		Blackberries, pails & tubs	
	2004	2005	2004	2005	2004	2005	2004	2005
	1,000 pounds	1,000 pounds	1,000 pounds	1,000 pounds	1,000 pounds	1,000 pounds	1,000 pounds	1,000 pounds
January	48,596	73,380	6,135	5,862	13,754	18,121	1,574	3,262
February	50,778	89,863	4,680	5,025	12,596	14,810	1,360	2,989
March	55,604	89,088	4,285	4,175	11,126	12,707	1,756	2,746
April	76,921	93,840	4,367	4,016	8,604	9,102	1,230	2,683
May	73,897	89,904	2,762	3,620	7,858	7,376	983	1,926
June	66,704	85,146	14,891	8,069	12,712	7,137	1,595	1,506
July	58,519	76,475	11,224	6,789	27,713	23,160	4,081	4,780
August	50,023	62,193	10,043	6,841	27,392	22,214	4,949	5,436
September	40,485	50,321	9,194	6,358	25,635	21,099	4,614	5,319
October	50,749	54,953	8,033	5,744	24,310	20,024	4,303	5,158
November	59,468	59,732	7,319	5,027	23,122	17,857	4,153	4,706
December	65,909	69,932	6,645	4,630	20,834	16,457	3,760	4,148

Month	Frozen							
	Blackberries, barrels		Blackberries, concentrate		Blackberries, total		Blueberries	
	2004	2005	2004	2005	2004	2005	2004	2005
	1,000 pounds	1,000 pounds	1,000 pounds	1,000 pounds	1,000 pounds	1,000 pounds	1,000 pounds	1,000 pounds
January	2,889	3,611	1,247	471	19,464	25,465	69,985	77,105
February	2,540	3,318	1,072	333	17,568	21,450	63,981	68,419
March	2,201	2,854	766	273	15,849	18,580	51,158	53,737
April	1,674	1,946	727	219	12,235	13,950	43,190	41,516
May	944	1,311	575	211	10,360	10,824	37,504	34,673
June	907	1,101	408	297	15,622	10,041	34,073	28,198
July	9,900	5,546	332	397	42,026	33,883	76,874	62,282
August	9,584	5,713	222	306	42,147	33,669	113,322	111,198
September	9,010	5,724	317	299	39,576	32,441	110,057	109,229
October	7,078	5,926	279	263	35,970	31,371	103,575	97,204
November	5,619	5,031	559	280	33,453	27,874	93,845	88,008
December	4,391	4,805	649	267	29,634	25,677	85,109	77,360

See end of table.

Table 5-111.—Fruit and orange juice: Cold storage holdings, end of month, United States, 2004 and 2005—Continued

Month	Boysenberries		Cherries, Tart (RSP)		Cherries, Sweet		Grapes	
	2004	2005	2004	2005	2004	2005	2004	2005
	1,000 pounds	*1,000 pounds*	*1,000 pounds*	*1,000 pounds*	*1,000 pounds*	*1,000 pounds*	*1,000 pounds*	*1,000 pounds*
January	1,307	1,303	60,825	74,505	6,039	11,070	7,186	6,587
February	698	997	50,575	69,829	6,286	10,152	6,178	3,186
March	546	662	41,893	56,106	5,534	8,924	5,487	2,599
April	563	671	32,281	47,832	4,985	7,123	4,172	2,473
May	800	1,027	23,971	39,172	5,290	6,707	4,042	2,451
June	2,288	1,656	17,357	27,701	5,129	5,520	3,175	3,324
July	3,674	2,397	80,107	136,042	9,572	15,048	3,064	2,961
August	2,703	2,496	93,985	150,216	9,141	13,972	2,727	2,133
September ...	2,245	2,124	99,862	139,969	9,816	12,736	6,915	3,659
October	1,798	1,836	92,953	131,846	9,327	11,498	9,794	8,421
November	1,531	1,972	81,816	117,828	9,075	10,298	8,625	7,417
December	1,276	1,949	76,570	110,359	8,237	9,607	7,293	6,784

Month	Peaches		Raspberries, Black		Red Raspberries, IQF		Red Raspberries, pails & tubs	
	2004	2005	2004	2005	2004	2005	2004	2005
	1,000 pounds	*1,000 pounds*	*1,000 pounds*	*1,000 pounds*	*1,000 pounds*	*1,000 pounds*	*1,000 pounds*	*1,000 pounds*
January	62,887	58,340	945	800	16,356	12,306	8,344	8,402
February	52,296	51,616	765	748	15,237	11,915	7,350	7,693
March	43,404	41,581	743	448	13,602	11,218	5,735	6,483
April	34,341	33,336	622	320	11,830	11,579	5,307	5,395
May	26,214	24,672	486	243	11,110	11,432	3,595	4,935
June	22,968	17,559	1,764	1,138	14,729	13,323	5,113	5,178
July	38,381	17,771	1,814	2,992	30,806	23,780	15,092	15,518
August	65,100	45,220	1,410	2,354	26,481	21,652	15,042	14,845
September ...	87,117	72,567	1,259	2,003	25,741	19,681	13,651	13,992
October	86,544	73,327	1,311	1,647	22,520	17,661	11,835	13,288
November	77,899	61,957	1,282	1,173	19,245	15,538	11,639	11,786
December	69,665	61,937	983	829	17,000	14,867	9,767	11,131

Month	Red Raspberries, barrels		Red Raspberries, concentrate		Red Raspberries, total		Strawberries, IQF & Poly	
	2004	2005	2004	2005	2004	2005	2004	2005
	1,000 pounds	*1,000 pounds*	*1,000 pounds*	*1,000 pounds*	*1,000 pounds*	*1,000 pounds*	*1,000 pounds*	*1,000 pounds*
January	8,873	10,676	1,085	549	34,658	31,933	102,277	107,158
February	7,304	9,198	823	553	30,714	29,359	90,723	90,498
March	5,744	8,368	732	992	25,813	27,061	79,438	80,554
April	4,597	5,579	774	860	22,508	23,413	99,398	105,697
May	2,990	4,573	659	776	18,354	21,716	150,858	119,175
June	5,805	8,698	1,068	1,014	26,715	28,213	209,583	165,736
July	29,912	42,483	1,743	1,689	77,553	83,470	250,348	191,146
August	24,727	37,874	1,532	1,948	67,782	76,319	230,865	178,739
September ...	20,570	34,567	737	1,907	60,699	70,147	217,831	155,502
October	18,380	32,435	767	2,022	53,502	65,406	184,856	135,678
November	16,847	29,703	901	2,100	48,632	59,127	160,766	116,219
December	14,791	27,399	694	2,404	42,252	55,801	144,204	106,508

Month	Strawberries, pails & tubs		Strawberries, barrels & drums		Strawberries, juice stock		Strawberries, total	
	2004	2005	2004	2005	2004	2005	2004	2005
	1,000 pounds	*1,000 pounds*	*1,000 pounds*	*1,000 pounds*	*1,000 pounds*	*1,000 pounds*		
January	79,092	88,154	31,649	48,933	7,311	7,735	220,329	251,980
February	69,159	73,482	28,958	43,446	7,248	5,980	196,088	213,406
March	59,605	57,162	31,794	27,724	7,112	8,170	177,949	173,610
April	67,245	57,787	48,062	25,055	12,712	9,502	227,417	198,041
May	84,367	69,317	70,225	33,530	13,968	15,128	319,418	237,150
June	130,556	103,816	59,113	71,595	25,345	18,067	424,597	359,214
July	134,937	134,220	74,340	75,085	22,952	17,687	482,577	418,138
August	126,214	119,898	68,463	70,202	19,655	15,754	445,197	384,593
September ...	119,338	103,435	59,108	52,671	19,977	13,201	416,254	324,809
October	106,447	89,121	50,222	44,192	14,678	10,238	356,203	279,229
November	101,087	78,277	46,414	41,103	12,766	9,960	321,033	245,559
December	95,193	69,319	43,029	34,096	11,134	8,839	293,560	218,762

See end of table.

Table 5-111.—Fruit and orange juice: Cold storage holdings, end of month, United States, 2004 and 2005—Continued

Month	Other fruit		Total frozen fruit		Orange juice	
	2004	2005	2004	2005	2004	2005
	1,000 pounds	1,000 pounds	1,000 pounds	1,000 pounds	1,000 pounds	1,000 pounds
January	355,842	365,236	895,506	984,882	1,613,011	1,553,755
February	321,469	308,868	803,321	874,252	1,646,142	1,578,921
March	286,429	270,956	715,709	748,834	1,790,524	1,578,173
April	260,202	240,512	724,805	708,061	1,987,553	1,652,358
May	235,406	211,448	759,481	684,513	2,128,721	1,668,135
June	203,559	188,746	839,835	765,766	2,075,713	1,548,778
July	183,093	181,378	1,070,294	1,045,598	1,953,408	1,501,593
August	154,395	164,671	1,059,966	1,060,239	1,823,271	1,397,309
September ...	150,535	158,923	1,036,710	988,081	1,644,222	1,243,274
October	398,744	439,479	1,210,340	1,204,538	1,516,773	1,139,934
November	436,056	435,833	1,181,257	1,124,369	1,457,953	1,027,540
December	405,492	421,491	1,093,884	1,067,127	1,468,844	1,044,708

NASS, Livestock Branch, (202) 720–3570.

Table 5-112.—Nuts: Cold storage holdings, end of month, United States, 2004 and 2005

Month	Peanuts					
	Shelled		In-shell		Total	
	2004	2005	2004	2005	2004	2005
	1,000 pounds	1,000 pounds	1,000 pounds	1,000 pounds	1,000 pounds	1,000 pounds
January	267,504	357,520	13,365	14,585	280,869	372,105
February	307,766	395,582	12,674	20,796	320,440	416,378
March	333,671	473,962	20,575	33,140	354,246	507,102
April	375,658	491,138	22,215	44,478	397,873	535,616
May	395,852	454,718	27,286	48,551	423,138	503,269
June	391,829	387,006	36,971	43,701	428,800	430,707
July	399,079	294,224	28,304	28,153	427,383	322,377
August	306,668	272,468	21,028	19,939	327,696	292,407
September ...	225,378	219,375	11,945	14,944	237,323	234,319
October	230,565	251,291	11,176	16,679	241,741	267,970
November	249,983	236,963	15,136	19,294	265,119	256,257
December	300,480	263,923	15,700	21,670	316,180	258,633

Month	Pecans					
	Shelled		In-shell		Total	
	2004	2005	2004	2005	2004	2005
	1,000 pounds	1,000 pounds	1,000 pounds	1,000 pounds	1,000 pounds	1,000 pounds
January	22,567	21,494	186,878	137,743	209,445	159,237
February	31,230	25,282	192,749	145,263	223,979	170,545
March	32,692	27,980	169,860	127,509	202,552	155,489
April	32,480	30,412	145,237	105,487	177,717	135,899
May	36,349	35,909	121,339	82,509	157,688	118,418
June	37,341	34,336	102,426	61,713	139,767	96,049
July	37,660	34,207	79,895	42,461	117,555	76,668
August	35,125	31,241	53,773	24,026	88,898	55,267
September ...	25,272	23,010	38,359	13,910	63,631	36,920
October	20,077	17,123	27,460	7,741	47,537	24,864
November	18,515	13,303	48,476	21,066	66,991	34,369
December	16,971	17,020	95,581	98,287	112,552	115,307

Month	Other nuts					
	Shelled		In-shell		Total	
	2004	2005	2004	2005	2004	2005
	1,000 pounds	1,000 pounds	1,000 pounds	1,000 pounds	1,000 pounds	1,000 pounds
January	144,082	115,318	10,029	8,510	154,111	123,828
February	145,832	121,097	7,613	6,409	153,445	127,506
March	154,331	134,984	9,127	8,043	163,458	143,027
April	150,783	124,726	5,831	7,525	156,614	132,251
May	145,048	124,672	6,858	5,676	151,906	130,348
June	122,970	122,179	6,815	5,398	129,785	127,577
July	102,951	108,144	8,319	7,436	111,270	115,580
August	77,750	84,355	11,242	7,656	88,992	92,011
September ...	88,311	89,024	9,592	7,729	97,903	96,753
October	90,438	89,965	11,099	8,818	101,537	98,783
November	93,319	84,684	8,475	9,038	101,794	93,722
December	102,498	88,931	8,524	8,183	111,022	97,114

NASS, Livestock Branch, (202) 720-3570.

CHAPTER VI
STATISTICS OF HAY, SEEDS, AND MINOR FIELD CROPS

Chapter VI deals with hay, pasture, seeds, and various minor field crops.

Table 6-1.—Hay, all: Area, yield, production, and value, United States, 1996–2005

Year	Area harvested	Yield per acre	Production	Marketing year average price per ton received by farmers	Value of production
	1,000 acres	*Tons*	*1,000 tons*	*Dollars*	*1,000 dollars*
1996	61,169	2.45	149,779	95.80	12,726,992
1997	61,084	2.50	152,536	100.00	13,249,825
1998	60,006	2.52	151,387	84.60	11,575,791
1999	63,181	2.53	159,582	76.90	11,007,327
2000	60,355	2.54	153,603	84.60	11,556,882
2001	63,516	2.46	156,416	96.50	12,589,493
2002	63,942	2.34	149,467	92.40	12,338,010
2003	63,383	2.49	157,585	85.50	12,006,783
2004	61,966	2.55	158,247	92.00	12,211,868
2005 [1]	61,649	2.44	150,590	98.00	12,491,263

[1] Preliminary.
NASS, Crops Branch, (202) 720–2127.

Table 6-2.—Hay, all: Area, yield, and production, by States, 2003–2005

State	Area harvested			Yield per harvested acre			Production		
	2003	2004	2005 [1]	2003	2004	2005 [1]	2003	2004	2005 [1]
	1,000 acres	*1,000 acres*	*1,000 acres*	*Tons*	*Tons*	*Tons*	*1,000 tons*	*1,000 tons*	*1,000 tons*
AL	780	850	730	2.60	2.70	2.70	2,028	2,295	1,971
AZ	275	275	300	7.86	7.71	7.75	2,162	2,119	2,324
AR	1,340	1,420	1,310	2.22	2.51	1.71	2,974	3,570	2,239
CA	1,620	1,600	1,550	5.85	5.76	5.76	9,485	9,220	8,935
CO	1,500	1,520	1,550	2.41	2.41	2.64	3,610	3,666	4,085
CT	63	66	63	2.21	2.17	1.87	139	143	118
DE	13	14	14	2.92	2.93	2.79	38	41	39
FL	255	260	290	2.50	2.50	2.45	638	650	711
GA	600	600	550	3.00	2.70	3.00	1,800	1,620	1,650
ID	1,500	1,480	1,410	3.30	3.61	3.82	4,950	5,350	5,382
IL	775	750	730	3.51	3.41	2.96	2,723	2,560	2,159
IN	650	660	650	3.25	3.49	3.18	2,110	2,303	2,067
IA	1,600	1,600	1,600	3.45	3.90	3.66	5,515	6,240	5,860
KS	3,250	3,350	2,900	2.15	2.35	2.30	7,000	7,880	6,680
KY	2,450	2,340	2,410	2.60	2.53	2.40	6,375	5,928	5,777
LA	380	370	350	2.90	3.00	2.30	1,102	1,110	805
ME	144	155	151	1.83	1.91	1.59	264	296	240
MD	195	215	190	2.76	2.65	2.79	539	570	531
MA	79	88	89	1.91	2.06	2.12	151	181	189
MI	1,050	1,100	1,150	2.97	2.97	2.86	3,120	3,270	3,290
MN	2,075	2,000	2,050	2.53	2.95	2.95	5,245	5,895	6,055
MS	750	720	730	2.50	2.30	2.90	1,875	1,656	2,117
MO	4,250	4,350	4,000	1.91	2.17	1.68	8,122	9,420	6,718
MT	2,450	2,500	3,000	1.89	1.90	1.95	4,635	4,760	5,850
NE	3,150	2,800	2,850	2.41	2.29	2.44	7,600	6,423	6,945
NV	440	420	450	3.25	3.53	3.58	1,429	1,481	1,609
NH	52	57	57	2.06	1.84	1.84	107	105	105
NJ	120	120	115	2.23	2.35	1.84	267	282	212
NM	300	330	330	4.27	4.14	4.28	1,281	1,365	1,413
NY	1,850	1,270	1,650	1.99	2.30	1.59	3,680	2,916	2,625
NC	778	712	691	2.61	2.49	2.40	2,030	1,776	1,660
ND	2,950	2,730	3,030	1.56	1.34	1.86	4,598	3,666	5,646
OH	1,350	1,190	1,200	2.94	2.72	3.03	3,974	3,232	3,630
OK	2,810	3,060	2,920	1.89	1.95	1.74	5,304	5,958	5,084
OR	1,100	1,130	1,000	3.25	3.21	3.14	3,572	3,624	3,140
PA	1,650	1,700	1,600	2.47	2.53	2.12	4,070	4,296	3,397
RI	9	9	9	2.11	2.22	2.22	19	20	20
SC	340	330	290	2.60	2.40	2.70	884	792	783
SD	4,300	3,900	4,000	1.68	1.76	1.89	7,210	6,870	7,560
TN	2,030	1,935	1,885	2.33	2.52	2.32	4,726	4,883	4,367
TX	5,240	5,350	5,050	2.36	2.30	1.81	12,388	12,295	9,140
UT	700	715	690	3.56	3.45	3.76	2,490	2,469	2,594
VT	235	230	240	2.00	1.67	1.56	470	384	374
VA	1,280	1,290	1,320	2.69	2.54	2.68	3,445	3,272	3,542
WA	810	790	740	4.45	4.29	4.34	3,603	3,392	3,210
WV	545	575	575	1.95	1.85	1.86	1,063	1,062	1,070
WI	2,100	2,050	2,050	2.09	2.38	2.18	4,380	4,880	4,470
WY	1,200	990	1,140	2.00	2.08	1.93	2,395	2,061	2,202
US	63,383	61,966	61,649	2.49	2.55	2.44	157,585	158,247	150,590

[1] Preliminary.
NASS, Crops Branch, (202) 720–2127.

HAY, SEEDS, AND MINOR FIELD CROPS

Table 6-3.—Hay, alfalfa and alfalfa mixtures: Area, yield, and production, by States, 2003–2005

State	Area harvested			Yield per harvested acre			Production		
	2003	2004	2005 [1]	2003	2004	2005 [1]	2003	2004	2005 [1]
	1,000 acres	*1,000 acres*	*1,000 acres*	*Tons*	*Tons*	*Tons*	*1,000 tons*	*1,000 tons*	*1,000 tons*
AZ	235	240	260	8.50	8.20	8.40	1,998	1,968	2,184
AR	20	20	20	3.50	3.50	2.30	70	70	46
CA	1,090	1,050	1,000	7.00	7.00	6.90	7,630	7,350	6,900
CO	800	770	800	3.20	3.30	3.70	2,560	2,541	2,960
CT	8	7	8	2.90	2.70	2.40	23	19	19
DE	5	6	5	2.70	3.90	3.60	14	23	18
ID	1,200	1,180	1,140	3.70	4.00	4.20	4,440	4,720	4,788
IL	425	400	400	4.10	4.30	3.50	1,743	1,720	1,400
IN	350	350	340	3.80	4.10	3.80	1,330	1,435	1,292
IA	1,330	1,300	1,250	3.70	4.20	4.10	4,921	5,460	5,125
KS	1,000	950	850	3.40	4.00	4.00	3,400	3,800	3,400
KY	250	240	260	3.50	3.70	3.20	875	888	832
ME	9	10	11	2.30	2.00	2.70	21	20	30
MD	45	40	40	3.30	3.30	3.90	149	132	156
MA	14	13	14	2.40	2.40	2.20	34	31	31
MI	850	850	900	3.20	3.20	3.10	2,720	2,720	2,790
MN	1,375	1,350	1,350	3.00	3.50	3.50	4,125	4,725	4,725
MO	410	400	450	2.95	3.80	2.70	1,210	1,520	1,215
MT	1,600	1,400	1,750	2.10	2.30	2.20	3,360	3,220	3,850
NE	1,450	1,250	1,250	3.60	3.65	3.70	5,220	4,563	4,625
NV	265	250	260	4.40	4.70	4.80	1,166	1,175	1,248
NH	8	7	8	2.40	2.10	2.10	19	15	17
NJ	30	30	25	3.50	3.70	2.70	105	111	68
NM	230	240	240	4.90	4.90	5.10	1,127	1,176	1,224
NY	600	470	450	2.80	2.80	2.10	1,680	1,316	945
NC	18	12	11	3.00	2.20	2.50	54	26	28
ND	1,600	1,300	1,650	1.65	1.50	2.00	2,640	1,950	3,300
OH	580	470	510	3.40	3.20	3.60	1,972	1,504	1,836
OK	310	360	320	3.40	3.80	3.70	1,054	1,368	1,184
OR	480	480	400	4.60	4.30	4.40	2,208	2,064	1,760
PA	550	540	510	3.00	2.80	2.60	1,650	1,512	1,326
RI	2	2	2	2.50	2.30	3.00	5	5	6
SD	2,700	2,250	2,400	1.90	2.10	2.15	5,130	4,725	5,160
TN	30	35	35	4.20	3.80	3.20	126	133	112
TX	140	150	150	4.70	5.70	5.40	658	855	810
UT	545	560	530	4.00	3.80	4.20	2,180	2,128	2,226
VT	40	40	45	2.00	2.00	1.80	80	80	81
VA	130	110	110	3.50	4.00	3.60	455	440	396
WA	510	480	450	5.30	5.00	5.20	2,703	2,400	2,340
WV	45	45	35	2.50	2.40	2.80	113	108	98
WI	1,600	1,600	1,550	2.30	2.60	2.40	3,680	4,160	3,720
WY	650	450	600	2.50	2.90	2.50	1,625	1,305	1,500
US	23,529	21,707	22,389	3.24	3.48	3.38	76,273	75,481	75,771

[1] Preliminary.

NASS, Crops Branch, (202) 720–2127.

Table 6-4.—Hay, all other: Area, yield, and production, by States, 2003–2005

State	Area harvested			Yield per harvested acre			Production		
	2003	2004	2005[1]	2003	2004	2005[1]	2003	2004	2005[1]
	1,000 acres	*1,000 acres*	*1,000 acres*	*Tons*	*Tons*	*Tons*	*1,000 tons*	*1,000 tons*	*1,000 tons*
AL	780	850	730	2.60	2.70	2.70	2,028	2,295	1,971
AZ	40	35	40	4.10	4.30	3.50	164	151	140
AR	1,320	1,400	1,290	2.20	2.50	1.70	2,904	3,500	2,193
CA	530	550	550	3.50	3.40	3.70	1,855	1,870	2,035
CO	700	750	750	1.50	1.50	1.50	1,050	1,125	1,125
CT	55	59	55	2.10	2.10	1.80	116	124	99
DE	8	8	9	3.00	2.30	2.30	24	18	21
FL	255	260	290	2.50	2.50	2.45	638	650	711
GA	600	600	550	3.00	2.70	3.00	1,800	1,620	1,650
ID	300	300	270	1.70	2.10	2.20	510	630	594
IL	350	350	330	2.80	2.40	2.30	980	840	759
IN	300	310	310	2.60	2.80	2.50	780	868	775
IA	270	300	350	2.20	2.60	2.10	594	780	735
KS	2,250	2,400	2,050	1.60	1.70	1.60	3,600	4,080	3,280
KY	2,200	2,100	2,150	2.50	2.40	2.30	5,500	5,040	4,945
LA	380	370	350	2.90	3.00	2.30	1,102	1,110	805
ME	135	145	140	1.80	1.90	1.50	243	276	210
MD	150	175	150	2.60	2.50	2.50	390	438	375
MA	65	75	75	1.80	2.00	2.10	117	150	158
MI	200	250	250	2.00	2.20	2.00	400	550	500
MN	700	650	700	1.60	1.80	1.90	1,120	1,170	1,330
MS	750	720	730	2.50	2.30	2.90	1,875	1,656	2,117
MO	3,840	3,950	3,550	1.80	2.00	1.55	6,912	7,900	5,503
MT	850	1,100	1,250	1.50	1.40	1.60	1,275	1,540	2,000
NE	1,700	1,550	1,600	1.40	1.20	1.45	2,380	1,860	2,320
NV	175	170	190	1.50	1.80	1.90	263	306	361
NH	44	50	49	2.00	1.80	1.80	88	90	88
NJ	90	90	90	1.80	1.90	1.60	162	171	144
NM	70	90	90	2.20	2.10	2.10	154	189	189
NY	1,250	800	1,200	1.60	2.00	1.40	2,000	1,600	1,680
NC	760	700	680	2.60	2.50	2.40	1,976	1,750	1,632
ND	1,350	1,430	1,380	1.45	1.20	1.70	1,958	1,716	2,346
OH	770	720	690	2.60	2.40	2.60	2,002	1,728	1,794
OK	2,500	2,700	2,600	1.70	1.70	1.50	4,250	4,590	3,900
OR	620	650	600	2.20	2.40	2.30	1,364	1,560	1,380
PA	1,100	1,160	1,090	2.20	2.40	1.90	2,420	2,784	2,071
RI	7	7	7	2.00	2.20	2.00	14	15	14
SC	340	330	290	2.60	2.40	2.70	884	792	783
SD	1,600	1,650	1,600	1.30	1.30	1.50	2,080	2,145	2,400
TN	2,000	1,900	1,850	2.30	2.50	2.30	4,600	4,750	4,255
TX	5,100	5,200	4,900	2.30	2.20	1.70	11,730	11,440	8,330
UT	155	155	160	2.00	2.20	2.30	310	341	368
VT	195	190	195	2.00	1.60	1.50	390	304	293
VA	1,150	1,180	1,210	2.60	2.40	2.60	2,990	2,832	3,146
WA	300	310	290	3.00	3.20	3.00	900	992	870
WV	500	530	540	1.90	1.80	1.80	950	954	972
WI	500	450	500	1.40	1.60	1.50	700	720	750
WY	550	540	540	1.40	1.40	1.30	770	756	702
US	39,854	40,259	39,260	2.04	2.06	1.91	81,312	82,766	74,819

[1] Preliminary.
NASS, Crops Branch, (202) 720–2127.

Table 6-5.—Hay, all: Stocks on farms, United States, 1996–2005

Crop year	Dec. 1	May 1[1]
	1,000 tons	*1,000 tons*
1996	105,179	17,424
1997	103,044	21,827
1998	111,809	24,662
1999	109,115	28,848
2000	106,412	21,248
2001	110,384	22,458
2002	102,978	22,013
2003	111,027	25,947
2004	114,516	27,758
2005[2]	105,056	NA

[1] Following year.   [2] Preliminary.   NA-not available.
NASS, Crops Branch, (202) 720–2127.

Table 6-6.—Hay, all: Marketing year average price and value of production, by States, crop of 2003, 2004, and 2005

State	Marketing year average price per ton, baled			Value of production		
	2003	2004	2005 [1]	2003	2004	2005 [1]
	Dollars	Dollars	Dollars	1,000 dollars	1,000 dollars	1,000 dollars
AL	58.00	57.00	57.00	117,624	130,815	112,347
AZ	89.00	99.50	123.00	192,843	210,161	284,732
AR	55.50	54.00	63.50	148,632	169,820	142,544
CA	90.50	115.00	132.00	852,425	1,045,885	1,150,613
CO	86.00	84.00	100.00	318,630	305,985	409,210
CT	145.00	156.00	162.00	20,150	22,248	19,153
DE	158.00	127.00	150.00	5,940	5,192	5,850
FL	90.00	93.00	95.00	57,420	60,450	67,545
GA	62.00	55.00	59.00	111,600	89,100	97,350
ID	87.50	106.00	111.00	426,855	552,600	586,782
IL	87.00	86.50	91.50	229,828	215,760	193,046
IN	105.00	97.00	95.50	207,780	210,000	197,016
IA	79.50	83.00	78.00	438,865	513,240	453,356
KS	68.50	68.00	66.50	459,900	489,560	423,620
KY	73.00	71.00	81.00	437,375	388,824	436,909
LA	51.00	55.00	57.00	56,202	61,050	45,885
ME	110.00	122.00	139.00	28,803	36,044	33,330
MD	139.00	129.00	150.00	75,232	73,308	79,650
MA	147.00	152.00	157.00	22,330	27,485	29,596
MI	93.00	94.50	88.50	295,240	304,525	290,430
MN	66.50	68.50	66.00	353,355	411,683	416,150
MS	42.00	45.40	52.00	78,750	75,182	110,084
MO	64.00	58.50	59.50	487,684	505,980	384,108
MT	73.50	76.00	71.50	339,338	355,740	414,200
NE	54.00	50.50	47.50	407,290	323,853	326,400
NV	93.00	102.00	121.00	135,882	154,477	195,246
NH	140.00	151.00	167.00	15,110	15,825	17,554
NJ	125.00	122.00	121.00	33,855	34,845	25,648
NM	142.00	121.00	125.00	180,460	165,102	176,328
NY	113.00	115.00	115.00	417,760	325,324	282,450
NC	61.50	61.00	66.00	124,500	108,120	109,720
ND	52.50	58.50	50.00	233,398	202,761	271,821
OH	121.00	101.00	116.00	424,989	291,680	386,166
OK	67.00	70.50	78.50	319,549	339,975	344,008
OR	88.50	105.00	116.00	313,262	371,892	355,100
PA	137.00	122.00	133.00	555,060	528,024	451,562
RI	145.00	159.00	165.00	2,835	3,175	3,294
SC	65.00	63.00	63.00	57,460	49,896	49,329
SD	60.50	62.50	61.50	423,515	414,375	454,020
TN	55.00	52.00	54.50	262,260	256,215	238,619
TX	74.00	77.00	84.50	844,213	832,725	730,165
UT	81.50	88.50	95.00	199,840	216,672	244,240
VT	111.00	122.00	131.00	52,150	46,896	48,837
VA	88.50	88.50	90.50	304,655	289,368	320,452
WA	93.50	111.00	114.00	343,610	379,648	365,610
WV	57.50	61.50	61.00	61,039	65,178	65,498
WI	79.00	79.00	109.00	345,080	385,440	484,050
WY	79.00	73.50	74.00	186,210	149,765	161,640
US	85.50	92.00	98.00	12,006,783	12,211,868	12,491,263

[1] Preliminary.
NASS, Crops Branch, (202) 720–2127.

Table 6-7.—Hay: Area and production, by kinds, United States, 1996–2005

Year	Area harvested			Production		
	Alfalfa	All other hay	All hay	Alfalfa	All other hay	All hay
	1,000 acres	1,000 acres	1,000 acres	1,000 acres	1,000 acres	1,000 acres
1996	24,206	36,963	61,169	79,139	70,640	149,779
1997	23,551	37,533	61,084	78,535	74,001	152,536
1998	23,592	36,414	60,006	81,992	69,395	151,387
1999	24,066	39,115	63,181	84,405	75,177	159,582
2000	23,463	36,892	60,355	81,520	72,083	153,603
2001	23,952	39,564	63,516	80,354	76,062	156,416
2002	22,923	41,019	63,942	73,014	76,453	149,467
2003	23,529	39,854	63,383	76,273	81,312	157,585
2004	21,707	40,259	61,966	75,481	82,766	158,247
2005 [1]	22,389	39,260	61,649	75,771	74,819	150,590

[1] Preliminary.
NASS, Crops Branch, (202) 720–2127.

Table 6-8.—Hay: Supply and disappearance, prices, and number of animal units fed annually, United States, 1996–2005[1]

Year beginning May	Farm carryover May 1	Production	Total supply	Disappearance	Roughage-consuming animal units	Supply per animal unit	Disappearance per animal unit	Price received per ton
	Million tons	Million tons	Million tons	Million tons	Million units	Tons	Tons	Dollars
1996	20.8	149.8	170.5	152.8	76.4	2.23	2.00	95.8
1997	17.4	152.5	170.0	148.1	74.9	2.27	1.98	100.0
1998	21.8	151.4	173.2	148.6	74.5	2.33	1.99	84.6
1999	24.7	159.6	184.2	155.4	73.2	2.52	2.12	76.9
2000	28.8	153.6	182.5	161.2	72.4	2.52	2.23	84.6
2001	21.2	156.4	177.7	155.2	72.1	2.46	2.15	96.5
2002	22.5	149.5	171.9	149.9	72.0	2.39	2.08	92.4
2003	22.0	157.6	179.6	153.7	70.7	2.54	2.17	35.5
2004	25.9	158.2	184.2	156.4	71.6	2.57	2.19	89.7
2005[2]	27.8	150.6	178.3	NA	73.0	2.44	NA	NA

[1] Excludes trade.　[2] Preliminary.　NA-not available.
ERS, Market and Trade Economics Division, (202) 694-5296.

Table 6-9.—Field seeds: Average retail price paid by farmers for seed, Apr. 15, United States, 1996–2005

Kind of seed	1996	1997	1998	1999	2000	2001	2002	2003	2004	2005
					Price per 100 pounds					
	Dollars	Dollars	Dollars	Dollars	Dollars	Dollars	Dollars	Dollars	Dollars	Dollars
Alfalfa, uncertified varieties	185.00	185.00	205.00	184.00	165.00	158.00	157.00	178.00	163.00	177.00
Alfalfa, certified varieties	277.00	282.00	288.00	287.00	277.00	278.00	280.00	286.00	291.00	281.00
Clover, ladino	318.00	307.00	308.00	298.00	285.00	285.00	280.00	305.00	291.00	280.00
Clover, red	172.00	184.00	194.00	178.00	143.00	132.00	130.00	144.00	145.00	174.00
Lespedeza, Korean	99.00	90.00	89.00	76.15	77.50	160.00	98.00	102.00	81.50	79.30
Lespedeza, Striate, Kobe	125.00	112.00	108.00	96.00	90.00	180.00	104.00	108.00	93.60	83.10
Lespedeza, Sericea	291.00	220.00	290.00	294.00	310.00	330.00	300.00	281.00	230.00	220.00
Timothy	76.00	73.00	71.20	78.80	115.00	105.00	90.00	107.00	110.00	105.00
Orchardgrass	141.00	119.00	116.00	107.00	108.00	135.00	143.00	147.00	140.00	137.00
Blue Grass, Kentucky: Public and common	172.00	153.00	152.00	129.00	158.00	140.00	155.00	159.00	180.00	180.00
Proprietary, including Merion	243.00	224.00	216.00	204.00	214.00	220.00	225.00	228.00	217.00	235.00
Ryegrass, annual	58.80	57.90	65.30	64.20	60.50	55.50	58.00	51.30	52.60	59.30
Tall fescue	109.00	148.00	101	99.50	91.00	114.00	106.00	92.60	93.70	100.00
Sudangrass	51.90	51.40	53.70	52.20	53.00	53.00	56.00	55.30	55.60	57.40
Potatoes	10.30	7.60	9.10	8.50	10.45	8.50	10.90	10.80	9.69	9.30
Peanuts	82.00	81.75	83.60	80.90	81.70	82.60	82.10	55.90	56.90	56.40
Sunflower	313.00	355.00	380.00	400.00	395.00	407.00	407.00	417.00	425.00	476.00
Cottonseed, all	73.00	74.90	79.30	82.40	128.00	154.00	213.00	218.00	270.00	309.00
Biotech[1]	....	....	....	....	....	217.00	271.00	293.00	340.00	390.00
Non-biotech	....	....	....	....	....	87.00	94.00	107.00	108.00	110.00
Grain sorghum, hybrid ..	84.00	92.00	96.00	97.60	93.00	93.00	96.00	100.00	105.00	114.00
					Price per bushel					
	Dollars	Dollars	Dollars	Dollars	Dollars	Dollars	Dollars	Dollars	Dollars	Dollars
Corn, hybrid, all[2]	77.70	83.50	86.90	88.10	87.50	92.20	92.00	102.00	105.00	111.00
Biotech[1]	....	....	....	....	....	110.00	113.00	115.00	122.00	131.00
Non-biotec	....	....	....	....	....	85.30	85.80	90.90	91.10	93.40
Wheat (spring)	8.10	7.30	6.85	6.10	6.10	6.20	6.50	8.77	7.00	7.30
Wheat (winter)	8.50	10.00	8.25	7.35	7.05	7.20	7.70	8.01	8.26	9.06
Oats (spring)	5.19	5.32	5.02	4.60	4.50	4.70	5.35	7.05	5.88	5.54
Rice	17.50	19.00	19.50	19.10	17.25	15.70	14.90	14.90	19.60	20.80
Barley (spring)	6.49	6.13	6.04	5.80	5.80	5.80	5.80	6.90	6.39	6.72
Soybeans for seed, all ..	14.80	16.10	17.15	17.00	17.10	20.70	22.50	24.20	24.10	27.60
Biotech[1]	....	....	....	....	....	23.90	27.00	28.80	30.50	34.60
Non-biotec	....	....	....	....	....	17.90	15.00	19.60	17.40	19.10
Flaxseed	8.14	9.31	10.00	8.50	7.90	7.60	7.60	9.96	9.60	14.40

[1] Biotech varities are made to be resistant to herbicides, insects, or both. A technology fee is included within the price.　[2] Price per 80,000 kernels.
NASS, Environmental, Economics, and Demographics Branch, (202) 720–6146.

Table 6-10.—Beans, dry edible (cleaned basis): Production, by classes, United States, 2003–2005 [1]

Class	2003	2004	2005
	1,000 cwt.	*1,000 cwt.*	*1,000 cwt.*
Navy (pea beans)	2,514	2,142	3,942
Great northern	2,216	951	1,585
Small white	55	66	47
Pinto	10,453	7,814	13,101
Red kidney, light	1,095	806	1,111
Red kidney, dark	845	682	919
Pink	612	521	662
Small red	581	601	903
Cranberry	190	180	162
Black	1,263	1,870	1,802
Large lima (CA)	369	307	352
Baby lima (CA)	325	267	389
Blackeye	785	384	406
Small chickpeas (Garbanzo)	60	76	149
Large chickpeas (Garbanzo)	357	517	922
Chickpeas, all (Garbanzo)	417	593	1,071
Other	772	604	770
Total	22,492	17,788	27,222

[1] Excludes beans grown for garden seed.
NASS, Crops Branch, (202) 720–2127.

Table 6-11.—Beans, dry edible: Area, yield, and production, by States, 2003–2005 [1]

State	Area planted			Area harvested			Yield per acre (cleaned basis)			Production (cleaned basis)		
	2003	2004	2005	2003	2004	2005	2003	2004	2005	2003	2004	2005
	1,000 acres	*1,000 acres*	*1,000 acres*	*1,000 acres*	*1,000 acres*	*1,000 acres*	*Pounds*	*Pounds*	*Pounds*	*1,000 cwt.*	*1,000 cwt.*	*1,000 cwt.*
CA	77.0	60.0	66.0	75.0	57.0	65.0	1,840	2,020	2,130	1,380	1,152	1,385
CO	80.0	75.0	125.0	73.0	67.0	115.0	1,600	1,550	1,650	1,168	1,039	1,898
ID	75.0	80.0	100.0	73.0	78.0	98.0	2,050	2,100	1,900	1,497	1,638	1,862
KS	12.0	9.0	13.0	11.0	8.5	12.5	2,100	1,800	2,200	231	153	275
MI	170.0	190.0	235.0	165.0	185.0	230.0	1,500	1,700	1,700	2,475	3,145	3,910
MN	115.0	115.0	145.0	110.0	100.0	135.0	1,700	1,150	1,800	1,870	1,150	2,430
MT	13.0	13.0	18.0	12.8	12.7	14.1	1,820	2,240	2,000	233	285	282
NE	155.0	120.0	175.0	148.0	110.0	172.0	2,130	2,160	2,250	3,151	2,376	3,870
NM	10.0	6.0	6.3	10.0	6.0	6.3	1,860	2,600	2,200	186	156	139
NY	25.0	24.0	25.0	24.0	23.5	23.0	1,860	1,050	1,230	446	247	282
ND	540.0	560.0	620.0	520.0	475.0	565.0	1,500	1,000	1,520	7,800	4,750	8,588
OR	7.0	8.0	9.0	6.0	7.5	8.8	1,650	1,550	2,000	99	116	176
SD	8.0	9.0	17.5	7.5	8.9	17.4	1,770	1,840	1,730	133	164	301
TX	50.0	20.0	17.0	44.0	17.5	15.3	1,170	800	1,520	513	140	233
UT	5.6	5.3	4.5	5.2	4.8	4.5	310	300	500	16	14	23
WA	27.5	30.0	49.0	27.5	29.0	48.0	1,910	2,100	1,650	525	609	792
WI [2] ...	6.0	5.0	....	5.9	4.9	....	2,100	2,310	....	124	113	....
WY	30.0	25.0	34.0	29.0	24.0	33.0	2,220	2,250	2,350	645	541	776
US	1,406.1	1,354.3	1,659.3	1,346.9	1,219.3	1,562.9	1,670	1,459	1,742	22,492	17,788	27,222

[1] Excludes beans grown for garden seed.　[2] Estimates discontinued in 2005.
NASS, Crops Branch, (202) 720–2127.

Table 6-12.—Beans, dry edible: Area, yield, production, price, and value, United States, 1996–2005[1]

Year	Area planted	Area harvested	Yield per acre[2]	Production[2]	Marketing year average price per 100 pounds received by farmers	Value of production
	1,000 acres	1,000 acres	Pounds	1,000 cwt.	Dollars	1,000 dollars
1996	1,839.0	1,750.7	1,594	27,912	23.50	652,240
1997	1,869.8	1,758.8	1,670	29,370	19.30	576,658
1998	2,014.1	1,917.7	1,586	30,418	19.00	567,243
1999	2,027.5	1,881.0	1,762	33,146	16.40	548,784
2000	1,767.7	1,616.5	1,642	26,543	15.50	416,462
2001	1,437.4	1,250.0	1,569	19,610	22.10	427,055
2002	1,929.7	1,738.9	1,743	30,312	17.10	519,341
2003	1,406.1	1,346.9	1,670	22,492	18.40	422,793
2004	1,354.3	1,219.3	1,459	17,788	25.70	452,871
2005	1,659.3	1,562.9	1,742	27,222	18.40	526,044

[1] Excludes beans grown for garden seed.   [2] Cleaned basis.
NASS, Crops Branch, (202) 720–2127.

Table 6-13.—Beans, dry edible (cleaned basis): Marketing year average price and value of production, by States, crop of 2003, 2004, and 2005[1]

State	Marketing year average price per cwt.			Value of production		
	2003	2004	2005	2003	2004	2005
	Dollars	Dollars	Dollars	1,000 dollars	1,000 dollars	1,000 dollars
CA	35.30	36.90	40.40	48,714	42,509	55,954
CO	18.20	28.00	18.80	21,258	29,092	35,682
ID	19.20	25.10	21.80	28,742	41,114	40,592
KS	17.60	30.20	17.80	4,066	4,621	4,895
MI	19.30	22.50	19.20	47,768	70,763	75,072
MN	18.60	26.50	20.00	34,782	30,475	48,600
MT	18.20	28.70	17.00	4,241	8,180	4,794
NE	17.30	22.80	17.40	54,512	54,173	67,338
NM	16.00	28.00	32.00	2,976	4,368	4,448
NY	22.60	27.90	22.00	10,080	6,891	6,204
ND	16.10	24.80	15.80	125,580	117,800	135,690
OR	19.10	26.50	22.90	1,891	3,074	4,030
SD	15.80	21.90	18.80	2,101	3,592	5,659
TX	20.00	22.00	20.10	10,260	3,080	4,683
UT	18.00	30.00	17.40	288	420	400
WA	21.00	24.50	21.40	11,025	14,921	16,949
WI[2]	26.50	33.50	....	3,286	3,786	....
WY	17.40	25.90	19.40	11,223	14,012	15,054
US	18.40	25.70	18.40	422,793	452,871	526,044

[1] Excludes beans grown for garden seed.   [2] Estimates discontinued in 2005.
NASS, Crops Branch, (202) 720–2127.

Table 6-14.—Beans, dry edible: Season average wholesale price per 100 pounds,
selected markets, 1995–2004

Year beginning September	F.o.b. California points			F.o.b. Northern Colorado points: Pinto	F.o.b. Western Ne-braska points: Great northern	F.o.b. Southern Idaho points: Small red	F.o.b. Michigan points:		
	Baby lima	Large lima	Blackeye				Pea bean (Navy)	Black	Light red kidney
	Dollars	Dollars	Dollars	Dollars	Dollars	Dollars	Dollars	Dollars	Dollars
1995	39.66	49.97	29.36	25.74	38.55	28.87	23.88	22.86	29.02
1996	45.58	57.09	32.57	27.56	26.61	39.31	23.11	27.08	37.76
1997	27.51	39.81	31.00	26.48	27.07	28.16	20.82	33.19	28.44
1998	41.27	46.80	37.52	20.89	25.84	27.54	26.33	28.56	32.40
1999	28.57	35.90	23.40	19.75	24.25	21.92	19.16	18.37	26.79
2000	26.26	34.56	25.95	21.02	23.20	24.33	16.43	18.33	25.32
2001	33.73	41.65	29.93	31.39	23.52	33.36	25.65	37.44	34.04
2002	32.28	42.33	34.48	22.87	26.47	28.81	18.00	19.24	29.68
2003	32.34	42.40	30.12	22.19	22.22	28.53	23.53	24.64	29.71
2004 [1]	41.61	43.49	31.36	35.23	24.78	32.01	29.62	26.50	36.19

[1] Preliminary.

ERS, Specialty Crops Branch, (202) 694–5253. Compiled from the Bean Market Summary, Agricultural Marketing Service, U.S. Department of Agriculture, Greeley, Colorado.

Table 6-15.—Beans, dry edible: United States exports to specified countries,
2002/2003–2004/2005 [1] [2]

Country	2002/2003	2003/2004	2004/2005
	1,000 metric tons	1,000 metric tons	1,000 metric tons
Mexico	62	50	48
United Kingdom	36	24	29
Haiti	16	22	12
Canada	24	15	12
Dominican Republic	7	18	11
Japan	13	14	10
Korea, Democratic People	0	0	7
France	8	4	5
Australia	6	4	4
Guatemala	4	4	3
Angola	5	5	3
Netherlands	5	4	3
Korea, Republic Of	1	4	3
Spain	4	2	3
Uganda	0	10	3
Honduras	1	2	3
Belgium-Luxembourg	3	2	2
Zambia	11	4	2
Greece	3	2	2
Malawi	2	0	2
Other	72	82	32
Total	283	272	202

[1] Year beginning September 1.   [2] Excluding seed bean exports. Compiled from U.S. Census data.
FAS, Grain and Feed Division, (202) 720–8398.www.fas.usda.gov/grain/default.html

Table 6-16.—Beans, dry edible: United States exports by class and quantity, 1995/96–2004/2005

Year[1]	Navy or pea	Great northern	Other white	Pinto	Red kidney	Other[2]	Total
	Metric tons	Metric tons	Metric tons	Metric tons	Metric tons	Metric tons	Metric tons
1995/96	86,464	28,291	1,126	82,062	32,380	98,920	329,244
1996/97	95,279	40,773	551	74,960	35,335	107,092	353,992
1997/98	107,669	50,201	1,338	96,256	33,605	150,203	439,271
1998/99	90,679	42,011	1,561	94,991	27,680	117,212	374,135
1999/2000	67,222	38,204	1,388	64,337	28,662	127,581	327,394
2000/2001	89,997	50,742	1,565	93,037	31,939	103,492	370,771
2001/2002	63,088	48,179	882	71,198	20,163	77,124	280,634
2002/2003	66,361	24,210	11,746	56,908	33,067	90,521	282,813
2003/2004	54,937	19,347	1,687	92,165	11,396	92,421	271,953
2004/2005	45,580	16,793	1,611	53,873	10,099	73,962	201,918

[1] Year beginning September 1. [2] Includes other colored, black, blackeye, and limas.
FAS, Grain and Feed Division, (202) 720–8398. www.fas.usda.gov/grain/default/html. Compiled from reports of the U.S. Department of Commerce.

Table 6-17.—Peas, dry field: United States exports to specified countries, 2002/2003–2004/2005 [1][2]

Country	2002/2003	2003/2004	2004/2005
	Metric tons	Metric tons	Metric tons
Canada	36	17	53
Cuba	0	10	31
India	2	1	27
Sudan	3	4	19
Philippines	13	15	14
Kenya	8	7	14
Angola	2	6	14
Uganda	3	2	11
Burundi	2	4	5
Korea, Republic Of	4	4	5
Belgium-Luxembourg	1	0	4
South Africa, Republic	1	0	3
Taiwan	1	2	3
Mexico	1	2	3
Liberia	0	1	3
Peru	2	2	3
Congo, Dem Rep Of	1	16	2
Ethiopia	0	0	2
United Arab Emirates	0	0	2
Bolivia	1	1	2
Other	19	30	18
Total	102	128	240

[1] Year beginning September 1. [2] Excluding seed pea exports.
FAS, Grain and Feed Division, (202) 720–8398. www.fas.usda.gov/grain/default.html.

Table 6-18.—Peas, dry field and Chickpeas: United States exports to specified countries, 2002/2003–2004/2005 [1][2][3]

Country	2002/2003	2003/2004	2004/2005
	Metric tons	Metric tons	Metric tons
Canada	42	19	57
Cuba	0	10	33
India	2	1	27
Sudan	3	5	19
Philippines	13	15	14
Kenya	8	7	14
Angola	2	6	14
Uganda	3	2	11
Burundi	2	4	5
Korea, Republic of	4	4	5
Belgium-Luxembourg	1	0	4
South Africa, Republic	1	0	3
Taiwan	1	2	3
Mexico	1	2	3
Liberia	0	1	3
Peru	2	2	3
Congo, Dem Rep of	0	2	3
Ethiopia	0	16	2
United Arab Emirates	0	0	2
Bolivia	1	1	2
Other	27	34	22
Total	118	135	250

[1] Year beginning September 1. [2] Excluding seed pea exports.
FAS, Grain and Feed Division, (202) 720–8398. www.fas.usda.gov.

HAY, SEEDS, AND MINOR FIELD CROPS

Table 6-19.—Hops: Area, yield, production, price, value, and Sept. 1 stocks, United States, 1996–2005

Year	Area harvested	Yield per acre	Production	Marketing year average price per pound received by farmers	Value of production	Stocks Sept. 1
	1,000 acres	Pounds	1,000 pounds	Cents	1,000 dollars	1,000 pounds
1996	44.2	1,698	74,971	165.0	123,530	58,700
1997	43.3	1,729	74,872	160.0	119,840	62,000
1998	36.6	1,625	59,548	169.0	100,728	55,000
1999	34.3	1,881	64,456	169.0	109,099	54,000
2000	36.1	1,871	67,577	187.0	126,217	48,000
2001	35.9	1,861	66,832	185.0	123,843	54,000
2002	29.3	1,990	58,337	191.0	111,546	65,000
2003	28.7	1,903	54,565	186.0	101,637	69,000
2004	27.7	1,990	55,204	190.0	104,798	61,000
2005 [1]	29.5	1,791	52,915	195.0	103,294	....

[1] Preliminary.
NASS, Crops Branch, (202) 720–2127.

Table 6-20.—Hops: Area, yield, and production, by States, 2003–2005

State	Area harvested			Yield per acre			Production		
	2003	2004	2005 [1]	2003	2004	2005 [1]	2003	2004	2005 [1]
	1,000 acres	1,000 acres	1,000 acres	Pounds	Pounds	Pounds	1,000 pounds	1,000 pounds	1,000 pounds
ID	3,429	3,253	3,287	1,536	1,588	1,640	5,266.3	5,165.0	5,390.9
OR	5,748	5,107	5,163	1,626	1,686	1,560	9,347.6	8,612.0	8,054.0
WA	19,492	19,382	21,094	2,050	2,137	1,871	39,951.2	41,426.9	39,469.6
US	28,669	27,742	29,544	1,903	1,990	1,791	54,565.1	55,203.9	52,914.5

[1] Preliminary.
NASS, Crops Branch, (202) 720–2127.

Table 6-21.—Hops: Marketing year average price and value of production, by States, crop of 2003, 2004, and 2005

State	Marketing year average price per pound			Value of production		
	2003	2004	2005 [1]	2003	2004	2005 [1]
	Dollars	Dollars	Dollars	1,000 dollars	1,000 dollars	1,000 dollars
ID	1.62	1.60	1.63	8,531	8,264	8,787
OR	2.31	2.31	2.57	21,593	19,894	20,699
WA	1.79	1.83	1.87	71,513	75,811	73,808
Total	1.86	1.88	1.95	101,637	103,969	103,294

[1] Preliminary.
NASS, Crops Branch, (202) 720–2127.

Table 6-22.—Hops: United States exports by country of destination and imports by country of origin, 2002/2003–2004/2005

Item and country	Year beginning September		
	2002/2003	2003/2004	2004/2005
	Metric tons	Metric tons	Metric tons
Exports			
North America:			
Canada	1,069.9	1,106.4	998.5
Mexico	1,045.1	1,185.4	904.5
Total	2,115.0	2,291.9	1,902.9
Caribbean:			
Dominican Republic	173.1	158.7	235.1
Trinidad And Tobago	3.4	4.5	34.1
Jamaica	16.6	26.3	32.7
Haiti	6.0	10.9	12.0
Leeward-Windward Island	6.4	8.5	8.7
Barbados	5.4	1.5	3.0
Bahamas, The	1.3	1.5	0.5
Netherlands Antilles	0.9	0.4	0.4
Total	213.2	212.4	326.6
Central America:			
Guatemala	21.5	18.8	29.1
Costa Rica	14.8	18.7	21.5
Panama	7.8	10.6	21.0
El Salvador	15.7	10.2	12.0
Honduras	9.1	6.9	8.5
Nicaragua	1.2	4.0	5.5
Belize	3.7	2.3	1.4
Total	73.8	71.4	99.0
South America:			
Brazil	1,094.7	1,133.5	1,203.3
Colombia	394.2	415.5	284.0
Venezuela	159.4	245.4	211.3
Argentina	112.5	196.2	192.3
Peru	22.1	68.5	173.9
Ecuador	71.8	59.2	87.4
Bolivia	66.0	36.2	72.5
Chile	41.1	109.5	71.7
Paraguay	23.9	55.5	62.8
Suriname	0.9	0.8	1.7
Uruguay	5.5	9.9	1.2
Total	1,992.0	2,330.2	2,362.2
European Union-25:			
Belgium-Luxembourg	1,239.7	2,080.1	1,924.4
Germany	658.8	1,922.6	827.7
United Kingdom	680.7	781.5	537.0
Netherlands	251.0	214.5	143.2
Ireland	65.4	35.9	73.7
Italy	39.7	20.9	19.7
Hungary	0.0	47.7	18.0
Denmark	0.1	30.8	16.7
France	11.3	51.6	13.7
Spain	18.6	2.8	9.3
Finland	21.4	22.7	7.5
Poland	0.0	114.5	1.5
Portugal	3.4	0.2	0.4
Slovenia	0.0	0.0	0.2
Greece	36.0	31.4	0.1
Czech Republic	0.0	0.1	0.0
Sweden	0.0	1.8	0.0
Total	3,026.0	5,359.0	3,593.1
Other Europe:			
Switzerland	1.6	8.7	5.0
Croatia	0.9	0.6	1.1
Bulgaria	0.1	0.0	0.0
Macedonia, Republic	0.0	5.5	0.0
Serbia And Montenegro	0.1	0.2	0.0
Total	2.7	15.1	6.1
Former Soviet Union:			
Russian Federation	124.9	410.1	371.7
Ukraine	20.4	0.0	1.8
Total	145.3	410.1	373.5
East Asia:			
Japan	467.4	334.3	487.4
China, Peoples Republic	92.8	213.3	409.8
Korea, Republic of	386.7	234.8	358.0
Hong Kong	286.4	277.9	201.3
Taiwan	0.0	0.1	0.0
Total	1,233.2	1,060.5	1,456.5

See end of table.

HAY, SEEDS, AND MINOR FIELD CROPS

Table 6-22.—Hops: United States exports by country of destination and imports by country of origin, 2002/2003–2004/2005—Continued

Item and country	Year beginning September		
	2002/2003	2003/2004	2004/2005
	Metric tons	*Metric tons*	*Metric tons*
Middle East:			
Turkey	3.8	3.6	48.1
Israel	0.0	9.0	0.1
Bahrain	0.0	0.6	0.0
Lebanon	0.0	0.4	0.0
Total	3.8	13.6	48.1
Sub-Saharan Africa:			
Nigeria	87.9	151.1	81.3
Cameroon	24.0	32.1	38.1
Ghana	0.0	0.0	9.8
South Africa, Republic	17.7	12.4	2.4
Seychelles	0.0	0.0	0.5
Kenya	0.0	0.0	0.5
Uganda	0.0	0.0	0.5
Total	129.6	195.5	132.9
South Asia:			
Sri Lanka	0.5	3.7	5.0
India	3.3	53.1	3.4
Pakistan	0.0	0.0	2.5
Nepal	1.0	0.5	0.6
Bangladesh	4.6	3.5	0.0
Total	9.5	60.8	11.5
Southeast Asia:			
Thailand	303.9	363.3	355.0
Philippines	94.7	114.0	118.4
Vietnam	43.5	57.2	72.2
Malaysia	27.5	24.7	21.8
Singapore	13.6	27.7	16.7
Indonesia	1.9	1.4	9.9
Cambodia	1.0	2.2	0.0
Total	486.1	590.4	594.0
Oceania:			
Australia	107.4	183.5	143.0
New Zealand	118.1	37.1	36.2
Papua New Guinea	4.5	1.0	6.5
French Pacific Island	2.1	3.2	2.3
Total	232.0	224.7	188.0
Grand total	9,662.1	12,835.6	11,094.4
Imports			
North America:			
Canada	0.0	0.2	0.0
Total	0.0	0.2	0.0
European Union - 25:			
Germany	2,536.6	1,677.1	3,588.4
United Kingdom	156.4	139.9	148.2
Czech Republic	42.0	36.3	73.0
France	1,002.8	832.1	43.0
Slovenia	14.5	5.5	7.4
Poland	0.1	0.4	5.2
Belgium-Luxembourg	36.5	45.5	4.3
Austria	0.0	0.1	0.0
Spain	235.7	0.0	0.0
Total	4,024.6	2,737.0	3,869.6
Other Europe:			
Serbia And Montenegro	0.0	0.0	0.4
Total	0.0	0.0	0.4
East Asia:			
China, Peoples Republic	415.0	0.0	1.0
Total	415.0	0.0	1.0
Oceania:			
Australia	433.5	370.2	230.0
New Zealand	93.3	22.4	86.3
Total	526.9	392.7	316.3
Grand total	4,966.4	3,129.9	4,187.3

FAS, Horticultural and Tropical Products Division, (202) 720–3423. The division's Home Page is located at, www.fas.usda.gov/htp. You can also email the division at htp@fas.usda.gov. Note: Compiled from reports of the U.S. Department of Commerce. U.S. trade can be obtained from the following web site: U.S. agricultural trade database, www.fas.usda.gov/ustrade.

CHAPTER VII

STATISTICS OF CATTLE, HOGS, AND SHEEP

This chapter contains information about most kinds of farm livestock and livestock products, with the exception of dairy and poultry. The information relates to inventories, production, disposition, prices, and income for farm animals, and to livestock slaughter (including horse slaughter), meat production, and market statistics for meat animals.

Table 7-1.—All cattle and calves: Number and value, United States, Jan. 1, 1997–2006

Year	Total number	Value	
		Per head	Total
	Thousands	*Dollars*	*1,000 dollars*
1997	101,656	525	53,383,392
1998	99,744	603	60,193,070
1999	99,115	594	58,833,650
2000	98,199	683	67,099,440
2001	97,298	725	70,495,030
2002	96,723	747	72,283,865
2003	96,100	728	69,948,620
2004	94,888	818	77,594,700
2005	95,438	916	87,385,945
2006 [1]	97,102	1,008	97,872,945

[1] Preliminary.

NASS, Livestock Branch, (202) 720–3570.

Table 7-2.—All cattle and calves: Number, by classes, United States, Jan. 1, 1997–2006

Year	All cattle and calves [1]	Cows and heifers that have calved		500 pounds and over						Calves under 500 pounds
		Beef cows	Milk cows	Heifers			Steers	Bulls		
				Beef cow replacements	Milk cow replacements	Other				
	Thousands	*Thousands*	*Thousands*	*Thousands*	*Thousands*	*Thousands*	*Thousands*	*Thousands*		*Thousands*
1997	101,656	34,458	9,318	6,042	4,058	10,212	17,392	2,350		17,826
1998	99,744	33,885	9,199	5,764	3,986	10,051	17,189	2,270		17,401
1999	99,115	33,750	9,128	5,535	4,069	10,170	16,891	2,281		17,290
2000	98,199	33,575	9,183	5,503	4,000	10,147	16,682	2,293		16,816
2001	97,298	33,398	9,172	5,588	4,057	10,131	16,461	2,274		16,216
2002	96,723	33,134	9,106	5,571	4,055	10,057	16,804	2,244		15,753
2003	96,100	32,983	9,142	5,624	4,114	9,891	16,554	2,248		15,545
2004	94,888	32,861	8,990	5,518	4,020	9,806	16,277	2,206		15,210
2005	95,838	32,915	9,005	5,691	4,118	9,763	16,476	2,219		15,250
2006 [2]	97,102	33,253	9,058	5,905	4,278	9,795	16,923	2,263		15,626

[1] Totals may not add due to rounding. [2] Preliminary.

NASS, Livestock Branch, (202) 720–3570.

STATISTICS OF CATTLE, HOGS, AND SHEEP

Table 7-3.—All cattle and calves: Number and value, by States, Jan. 1, 2005–2006

State	Number		Value			
	2005	2006 [1]	Value per head		Total value	
			2005	2006 [1]	2005	2006 [1]
	Thousands	Thousands	Dollars	Dollars	1,000 dollars	1,000 dollars
AL	1,320	1,280	750	820	990,000	1,049,600
AK	14.5	15.5	980	1,030	14,210	15,965
AZ	910	940	1,020	1,090	928,200	1,024,600
AR	1,860	1,750	800	850	1,488,000	1,487,500
CA	5,400	5,500	1,130	1,280	6,102,000	7,040,000
CO	2,500	2,650	1,000	1,100	2,500,000	2,915,000
CT	56	52	1,070	1,210	59,920	62,920
DE	23	23	990	1,050	22,770	24,150
FL	1,710	1,690	810	900	1,385,100	1,521,000
GA	1,210	1,180	770	820	931,700	967,600
HI	155	161	640	630	99,200	101,430
ID	2,060	2,120	1,080	1,240	2,224,800	2,628,800
IL	1,370	1,340	860	980	1,178,200	1,313,200
IN	850	900	930	1,060	790,500	954,000
IA	3,600	3,800	900	980	3,240,000	3,724,000
KS	6,600	6,650	830	920	5,478,000	6,118,000
KY	2,250	2,400	810	860	1,822,500	2,064,000
LA	860	820	780	850	670,800	697,000
ME	92	92	1,100	1,170	101,200	107,640
MD	235	230	1,010	1,120	237,350	257,600
MA	48	47	1,100	1,190	52,800	55,930
MI	1,000	1,040	1,060	1,220	1,060,000	1,268,800
MN	2,400	2,350	950	1,030	2,280,000	2,420,500
MS	1,070	1,000	780	840	834,600	840,000
MO	4,400	4,550	850	950	3,740,000	4,322,500
MT	2,350	2,400	1,080	1,220	2,538,000	2,928,000
NE	6,350	6,550	910	980	5,778,500	6,419,000
NV	500	500	980	1,050	490,000	525,000
NH	40	39	1,170	1,280	46,800	49,920
NJ	44	42	1,110	1,220	48,840	51,240
NM	1,500	1,550	1,090	1,160	1,635,000	1,798,000
NY	1,410	1,410	1,120	1,280	1,579,200	1,804,800
NC	870	860	760	820	661,200	705,200
ND	1,710	1,720	1,070	1,230	1,829,700	2,115,600
OH	1,300	1,280	950	1,030	1,235,000	1,318,400
OK	5,350	5,450	820	870	4,387,000	4,741,500
OR	1,430	1,440	970	1,050	1,387,100	1,512,000
PA	1,620	1,610	1,080	1,240	1,749,600	1,996,400
RI	5.5	5.0	1,010	1,090	5,555	5,450
SC	435	415	780	860	339,300	356,900
SD	3,700	3,750	1,020	1,160	3,774,000	4,350,000
TN	2,170	2,240	770	830	1,670,900	1,859,200
TX	13,700	14,100	780	840	10,686,000	11,844,000
UT	860	820	940	1,020	808,400	836,400
VT	275	280	1,320	1,420	363,000	397,600
VA	1,640	1,690	770	840	1,262,800	1,419,600
WA	1,080	1,120	1,110	1,210	1,198,800	1,355,200
WV	405	410	780	860	315,900	352,600
WI	3,350	3,400	1,190	1,330	3,986,500	4,522,000
WY	1,350	1,440	1,020	1,130	1,377,000	1,627,200
US	95,438.0	97,101.5	916	1,008	87,385,945	97,872,945

[1] Preliminary.
NASS, Livestock Branch, (202) 720–3570.

Table 7-4.—Cattle and calves, Jan. 1: Number, by sex and weight classes, by States, 2005 and 2006

| State | Cows and heifers that have calved | | | | Heifers, 500 pounds and over | | | | | |
| | Beef cows | | Milk cows | | Beef cow replacements | | Milk cow replacements | | Other | |
	2005	2006 [1]	2005	2006 [1]	2005	2006 [1]	2005	2006 [1]	2005	2006 [1]
	Thousands	Thousands	Thousands	Thousands	Thousands	Thousands	Thousands	Thousands	Thousands	Thousands
AL	724	696	16	14	100	85	7	6	33	29
AK	5.4	6.1	1.2	0.8	1.2	1.3	0.5	0.3	0.2	0.2
AZ	175	190	165	165	28	30	42	44	22	19
AR	964	919	26	21	170	175	10	8	65	52
CA	720	700	1,740	1,770	130	120	760	790	190	180
CO	639	685	101	105	130	115	50	55	570	570
CT	7	5	20	20	2.0	1.5	11.0	11.0	1.0	0.5
DE	4	4	8	7	0.4	0.9	2.5	3.0	0.6	0.6
FL	932	926	138	134	145	140	40	40	20	20
GA	596	592	84	78	82	82	24	23	35	30
HI	81.3	87.4	5.7	4.6	15	15	2	2	5	5
ID	475	472	435	473	100	95	230	250	205	200
IL	460	446	105	104	65	70	55	52	150	158
IN	230	222	155	158	40	45	56	65	59	65
IA	1,013	1,053	187	187	120	140	100	120	580	630
KS	1,530	1,560	110	110	245	255	45	55	1,700	1,670
KY	1,100	1,128	110	102	170	180	45	50	100	100
LA	494	468	36	32	82	82	9	9	18	18
ME	12	12	33	32	4.5	4.0	19.0	18.0	1.5	2.0
MD	43	49	73	70	12	13	32	30	11	10
MA	7	8	17	16	2.0	2.0	8.5	8.5	0.5	0.5
MI	93	108	307	312	35	31	120	137	47	45
MN	395	390	460	445	95	95	270	265	190	170
MS	564	536	26	24	99	89	15	11	36	21
MO	2,121	2,236	119	114	290	340	65	70	275	260
MT	1,432	1,451	18	19	400	445	9	8	146	157
NE	1,909	1,930	61	60	300	295	20	15	1,470	1,490
NV	240	238	25	27	42	42	10	12	36	35
NH	4.0	4.0	16.0	16.0	1.5	1.5	9.0	8.5	0.5	0.5
NJ	10	9	12	12	3.0	2.0	6.0	6.0	2.0	2.0
NM	472	460	318	340	90	90	100	105	95	105
NY	80	78	650	652	27	25	305	325	48	40
NC	400	384	55	52	75	72	25	22	20	26
ND	947	937	33	33	147	157	18	19	210	190
OH	294	297	266	273	75	70	125	120	70	70
OK	2,015	2,075	75	75	375	410	15	20	550	530
OR	630	619	120	121	120	125	60	75	120	110
PA	154	152	566	558	40	40	275	275	55	50
RI	1.7	1.5	1.1	1.0	0.3	0.3	0.8	0.7	0.1	0.1
SC	222	213	18	17	38	40	7	6	15	10
SD	1,710	1,719	80	81	295	295	40	45	520	500
TN	1,078	1,110	72	70	195	185	40	40	75	85
TX	5,432	5,475	318	325	800	870	130	130	1,510	1,650
UT	347	335	88	85	65	60	45	45	70	65
VT	10	10	143	143	4	5	58	61	3	4
VA	705	747	105	103	125	125	43	40	72	63
WA	240	293	235	237	50	60	102	98	118	107
WV	197	204	13	13	45	38	4	4	21	25
WI	245	250	1,235	1,240	70	70	650	670	80	70
WY	756	763	4	7	145	175	3	5	142	155
US	32,915.4	33,253.0	9,005.0	9,058.4	5,690.9	5,904.5	4,118.3	4,278.0	9,763.4	9,795.4

See footnote at end of table.

Table 7-4.—Cattle and calves, Jan. 1: Number, by sex and weight classes, by States, 2005 and 2006—Continued

State	Steers, 500 pounds and over		Bulls, 500 pounds and over		Calves under 500 pounds	
	2005	2006 [1]	2005	2006 [1]	2005	2006 [1]
	Thousands	Thousands	Thousands	Thousands	Thousands	Thousands
AL	45	60	45	50	350	340
AK	1.2	1.1	1.9	2.4	2.9	3.3
AZ	330	340	23	22	125	130
AR	110	95	55	55	460	425
CA	670	645	70	75	1,120	1,220
CO	840	910	40	45	130	165
CT	2.5	2.0	1.0	1.0	11.5	11.0
DE	3.5	3.0	0.3	0.3	3.7	4.2
FL	20	20	60	60	355	350
GA	49	41	35	34	305	300
HI	7	7	5	5	34	35
ID	325	310	35	35	255	285
IL	275	255	25	25	235	230
IN	105	120	20	19	185	206
IA	1,020	1,100	70	70	510	500
KS	2,140	2,180	95	95	735	725
KY	190	205	75	75	460	560
LA	22	20	32	32	167	159
ME	3.5	4.0	1.5	1.5	17.0	18.5
MD	18	18	4	3	42	37
MA	2.0	2.0	1.0	1.0	10.0	9.0
MI	200	195	18	17	180	195
MN	440	450	40	35	510	500
MS	53	51	42	38	235	230
MO	460	460	120	130	950	940
MT	205	160	90	110	50	50
NE	2,180	2,340	100	100	310	320
NV	49	49	15	15	83	82
NH	1.5	1.5	0.5	0.5	7.0	6.5
NJ	3	3	1	1	7	7
NM	150	190	45	40	230	220
NY	34	28	16	17	250	245
NC	41	45	29	29	225	230
ND	230	220	55	59	70	105
OH	183	170	32	30	255	250
OK	1,180	1,180	130	140	1,010	1,020
OR	165	180	40	40	175	170
PA	170	160	25	25	335	350
RI	0.6	0.5	0.1	0.1	0.8	0.8
SC	13	10	18	17	104	102
SD	710	700	85	90	260	320
TN	115	125	75	75	520	550
TX	2,720	2,850	370	370	2,420	2,430
UT	110	105	22	20	113	105
VT	4.0	4.0	3.0	3.0	50.0	50.0
VA	150	170	40	42	400	400
WA	170	150	23	24	142	151
WV	35	38	15	14	75	74
WI	360	350	30	30	680	720
WY	165	200	45	45	90	90
US	16,475.8	16,923.1	2,219.3	2,262.8	15,249.9	15,626.3

[1] Preliminary.

NASS, Livestock Branch, (202) 720–3570.

Table 7-5.—Cattle and buffalo: Number in specified countries, 2001–2004

Country	2001	2002	2003	2004[1]
	Thousands	Thousands	Thousands	Thousands
India	313,774	317,000	323,000	327,250
Brazil	150,382	156,314	161,463	165,492
China, Peoples Republic	128,663	128,242	130,848	134,672
United States	97,298	96,723	96,100	94,888
European Union - 25	91,419	90,339	88,719	87,478
Argentina	50,167	50,369	50,869	50,768
Australia	27,720	27,870	27,870	26,640
Russian Federation	25,500	24,510	23,500	22,285
Mexico	28,449	28,481	29,224	28,437
South Africa, Republic	13,460	13,505	13,635	13,540
Canada	13,608	13,762	13,488	14,660
Uruguay	10,423	11,667	12,257	12,609
Turkey	11,350	11,200	10,950	10,700
New Zealand	9,390	9,656	9,760	9,540
Ukraine	9,424	9,433	9,108	7,712
Egypt	6,300	6,390	6,400	6,500
Philippines	5,472	5,512	5,567	5,567
Japan	4,530	4,563	4,524	4,478
Romania	2,870	2,895	2,878	2,897
Korea, Republic of	2,134	1,954	1,954	1,999
Colombia	22,676	23,757	NA	NA
Dominican Republic	1,918	1,918	NA	NA
Nicaragua	2,280	NA	NA	NA
Poland	5,722	5,499	NA	NA
Venezuela	13,400	13,500	NA	NA
Others	7,954	9,870	NA	NA
Total	900,039	906,693	1,024,117	1,030,116

[1] Preliminary.   NA-not available.
FAS, Dairy Livestock and Poultry Division, (202) 720–8031. Updated data available at http://www.fas.usda.gov/psd.

Table 7-6.—Cows and calf crop: Cows and heifers that have calved, Jan. 1, 2005–2006, and calves born, by States, 2004 and 2005

State	Cows and heifers that have calved Jan. 1		Calves born	
	2005	2006 [1]	2004	2005 [1]
	Thousands	Thousands	Thousands	Thousands
AL	740	710	670	630
AK	6.6	6.9	4.4	4.7
AZ	340	355	275	285
AR	990	940	850	820
CA	2,460	2,470	2,070	2,080
CO	740	790	740	770
CT	27	25	23	21
DE	12	11	8.5	9.0
FL	1,070	1,060	900	880
GA	680	670	560	550
HI	87	92	63	66
ID	910	945	880	900
IL	565	550	510	500
IN	385	380	340	330
IA	1,200	1,240	1,100	1,170
KS	1,640	1,670	1,480	1,500
KY	1,210	1,230	1,040	1,110
LA	530	500	415	400
ME	45	44	38	37
MD	116	119	98	85
MA	24	24	19	18
MI	400	420	335	355
MN	855	835	820	800
MS	590	560	490	470
MO	2,240	2,350	2,040	2,080
MT	1,450	1,470	1,520	1,480
NE	1,970	1,990	1,800	1,800
NV	265	265	210	210
NH	20.0	20.0	18	18
NJ	22	21	16	14
NM	790	800	600	600
NY	730	730	600	590
NC	455	436	410	400
ND	980	970	960	930
OH	560	570	470	470
OK	2,090	2,150	1,900	1,980
OR	750	740	700	670
PA	720	710	650	620
RI	2.8	2.5	2.5	2.4
SC	240	230	195	185
SD	1,790	1,800	1,720	1,720
TN	1,150	1,180	1,020	1,060
TX	5,750	5,800	5,000	5,150
UT	435	420	390	380
VT	153	153	125	135
VA	810	850	720	760
WA	475	530	430	450
WV	210	217	190	195
WI	1,480	1,490	1,340	1,350
WY	760	770	750	740
US	41,920.4	42,311.4	37,505.4	37,780.1

[1] Preliminary.
NASS, Livestock Branch, (202) 720–3570.

Table 7-7.—Cattle and calves: All cattle on feed, United States, Jan. 1, 1997–2006 [1]

Year	Number
	Thousands
1997	13,181
1998	13,608
1999	13,284
2000	14,073
2001	14,276
2002	14,050
2003	13,220
2004	13,812.9
2005	13,744.7
2006 [2]	14,131.9

[1] Cattle and calves on feed are animals for slaughter market being fed a ration of grain or other concentrates and are expected to produce a carcass that will grade select or better. [2] Preliminary.
NASS, Livestock Branch, (202) 720–3570.

Table 7-8.—Cattle and calves: Number on feed, 1,000+ capacity feedlots, by States, Jan. 1, 2005–2006[1]

State	2005	2006[2]
	1,000 Head	*1,000 Head*
AZ	331	334
AR	10	5
CA	535	550
CO	1,100	1,120
ID	300	280
IL	210	205
IN	125	115
IA	920	920
KS	2,460	2,550
KY	10	15
MD	12	12
MI	190	190
MN	290	290
MO	70	70
MT	60	55
NE	2,470	2,600
NV	10	8
NM	126	143
NY	23	18
NC	4	4
ND	60	60
OH	200	180
OK	355	375
OR	80	65
PA	75	75
SD	400	400
TN	5	8
TX	2,720	2,930
UT	35	30
VA	30	28
WA	195	155
WV	7	9
WI	225	230
WY	80	80
Other States[3]	21.7	22.9
US	13,744.7	14,131.9

[1] Inventory is the number on hand January 1 of the following year.   [2] Preliminary.   [3] AL, AK, CT, DE, FL, GA, HI, LA, ME, MA, MS, NH, NJ, RI, SC, and VT.

NASS, Livestock Branch, (202) 720–3570.

Table 7-9.—Cattle: Average price per 100 pounds, by grades, at Nebraska Direct, Sioux Falls, SD and South St. Paul, MN, 1996–2005

Year	Steers[1]		Heifers[2]		Sioux Falls, SD		South St. Paul, MN	
					Cows[3]		Cows[3]	
	Choice	Select	Choice	Select	Commer-cial	Breaking utility	Commer-cial	Boning Utility
	Dollars	*Dollars*	*Dollars*	*Dollars*	*Dollars*	*Dollars*	*Dollars*	*Dollars*
1996	74.50	61.83	64.18	61.22	35.24	33.64	37.69	35.22
1997	65.92	63.85	65.66	63.36	36.91	35.64	42.7	38.62
1998	60.07	56.17	59.23	55.17	43.22	39.23	40.15	37.02
1999 65-80%	65.64	....	65.68	....	45.04	40.29	43.52	38.55
2000 65-80%	69.52	....	69.55	....	49.25	44.51	49.26	41.77
2001 65-80%	67.68	....	67.81	....	52.35	46.67	50.35	47.91
2002 65-80%	66.39	....	67.39	....	44.99	40.97	45.16	42.50
2003 65-80%	82.37	....	82.06	....	53.49	49.50	58.50	51.75
2004 65-80%	84.78	....	84.40	....	60.64	57.22	59.60	54.18
2005 65-80%	86.54	....	87.35	....	61.89	57.82	60.67	55.64

[1] 1,100 to 1,500 pound weight range; weighted average of price range.   [2] 1,000 to 1,300 pound weight range; simple average of price range.   [3] All weights; simple average of price range.

AMS, Livestock and Grain Market News, (202) 720–7316.

Table 7-10.—Cattle and calves: Production, disposition, cash receipts, and gross income, United States, 1995–2004

Year	Calf crop [1]	Death loss		Marketings [2]		Cattle shipped in for feeding and breeding [3]	Farm slaughter
		Cattle	Calves	Cattle	Calves		Cattle and calves
	1,000 head	1,000 head	1,000 head	1,000 head	1,000 head	1,000 head	1,000 head
1995	40,264	1,645	2,739	48,741	9,656	23,507	227
1996	39,823	1,761	2,811	48,722	10,295	22,098	225
1997	38,961	1,847	2,829	49,647	10,154	23,828	223
1998	38,812	1,668	2,541	47,440	9,506	21,928	214
1999	38,796	1,658	2,455	48,683	9,540	22,836	213
2000	38,631	1,711	2,387	48,986	9,693	23,448	203
2001	38,300	1,722	2,487	47,102	9,183	21,813	194
2002	38,224	1,710	2,366	46,804	9,296	21,522	193
2003	37,903	1,710	2,320	47,686	9,613	22,405	191
2004 [4]	37,625	1,711	2,292	44,808	9,065	21,397	185

Year	Quantity produced (live weight) [5]	Value of production	Cash receipts from sales of cattle, calves, beef and veal [6]	Value of cattle and calves slaughtered for home consumption	Gross income [7]	Average price per 100 pounds received by farmers	
						Cattle	Calves
	1,000 pounds	1,000 dollars	1,000 dollars	1,000 dollars	1,000 dollars	Dollars	Dollars
1995	42,533,734	24,699,740	34,044,038	304,751	34,348,789	61.80	73.10
1996	40,883,614	22,034,934	30,976,861	274,011	31,250,872	58.70	58.40
1997	41,110,640	24,941,882	35,999,620	322,053	36,321,673	63.10	78.90
1998	41,698,894	24,187,549	33,442,843	304,406	33,747,249	59.60	78.80
1999	42,578,262	26,097,222	36,568,554	330,862	36,899,416	63.40	87.70
2000	43,040,893	28,498,670	40,783,472	366,744	41,150,216	68.60	104.00
2001	42,581,294	29,403,098	40,540,645	362,317	40,902,962	71.30	106.00
2002	42,409,258	27,097,532	38,095,116	333,768	38,428,884	66.50	96.40
2003	42,242,705	32,112,931	45,092,283	384,290	45,476,573	79.70	102.00
2004 [4]	41,501,303	34,887,821	47,295,574	427,865	47,723,439	85.90	119.00

[1] Calves born during the year. [2] Includes custom slaughter for use on farms where produced and State outshipments, but excludes interfarm sales within the State. [3] Includes cattle shipped in from other States and from central markets, but excludes cattle for immediate slaughter. [4] Preliminary. [5] Adjustments made for changes in inventory and for inshipments. [6] Receipts from marketings and sale of farm slaughter. [7] Cash receipts from sales of cattle, calves, beef, and veal plus value of cattle and calves slaughtered for home consumption.

NASS, Livestock Branch, (202) 720–3570.

Table 7-11.—Cattle: Weighted average weight and price per 100 pounds, Texas-Oklahoma, Kansas, Colorado, Nebraska, Iowa-So. Minnesota Feedlots, 1998–2005 [1]

Year	Steers SE/CH 65-80%			Steers SE/CH 35-65%		
	Price	Average Weight	Number of Head	Price	Average Weight	Number of Head
	Dollars	Pounds		Dollars	Pounds	
1998	61.05	1,282	408,859	61.79	1,211	3,135,109
1999	65.66	1,287	586,293	65.63	1,249	3,057,112
2000	69.82	1,294	584,809	70.15	1,253	2,631,692
2001	NA	NA	NA	NA	NA	NA
2002	66.74	1,327	270,924	67.40	1,263	1,965,036
2003	82.81	1,294	372,429	82.79	1,245	2,391,746
2004	84.65	1,319	389,144	85.03	1,242	2,336,418
2005	86.28	1,336	326,751	87.66	1,265	2,492,108

Year	Heifers SE/CH 65-80%			Heifers SE/CH 35-65%		
	Price	Average Weight	Number of Head	Price	Average Weight	Number of Head
	Dollars	Pounds		Dollars	Pounds	
1998	61.31	1,175	278,275	61.83	1,102	2,369,994
1999	65.75	1,182	493,893	65.81	1,135	2,410,684
2000	69.88	1,190	492,881	70.21	1,131	2,353,612
2001	68.22	1,208	377,415	69.01	1,134	2,065,438
2002	67.53	1,229	324,078	67.42	1,142	1,692,785
2003	82.70	1,192	358,900	83.59	1,126	2,077,258
2004	84.38	1,210	403,193	84.92	1,128	2,193,273
2005	87.23	1,219	313,240	87.90	1,145	1,901,730

[1] Sales FOB feedlots and delivered. Estimated net weights after 3-4 % shrink. NA-not available.
AMS, Livestock and Grain Market News, (202) 720–7316.

Table 7-12.—Cattle and calves: Receipts at selected public stockyards, 1996–2005 [1]

Year	Oklahoma City	Greeley	Amarillo	South St. Joseph	South St. Paul	All others reporting	Total markets reporting [2][3]
				Cattle			
	Thousands	*Thousands*	*Thousands*	*Thousands*	*Thousands*	*Thousands*	*Thousands*
1996	554			118	173	4,236	4,949
1997	526	71	136	121	180	4,214	4,953
1998	546	41	117	102	155	3,058	3,761
1999	516	48	105	110	157	4,598	5,018
2000	497	53	95	101	145	4,454	4,847
2001	512	49	89	113	135	2,698	3,593
2002	516	43	144	30	143	2,293	3,169
2003	574	37	130	110	147	2,530	3,528
2004	456	24	99	101	141	2,121	2,942
2005	491	51	87	97	124	2,124	2,974
				Calves			
	Thousands	*Thousands*	*Thousands*	*Thousands*	*Thousands*	*Thousands*	*Thousands*
1996						4	90	93
1997						3	92	96
1998						3	89	92
1999						2	92	94
2000						1	89	90
2001						1	86	87
2002		3				1	106	113
2003		7				1	486	494
2004		6		1			574	581
2005	0	13	0	0	0	216	229

[1] Total rail and truck receipts unloaded at public stockyards. Saleable receipts 1978 on. [2] Rounded totals of the complete figures. [3] The number of stockyards varies from 23 to 46.
AMS, Livestock & Grain Market News, (202) 720–7316. Compiled from reports received from stockyard companies.

Table 7-13.—Cattle and calves: Number slaughtered, United States, 1996–2005

Year	Cattle slaughter					Calf slaughter				
	Commercial			Farm	Total	Commercial			Farm	Total
	Federally in-spected	Other	Total [1]			Federally in-spected	Other	Total [1]		
	Thou-sands	*Thou-sands*	*Thou-sands*	*Thou-sands*	*Thou-sands*	*Thou-sands*	*Thou-sands*	*Thou-sands*	*Thou-sands*	*Thou-sands*
1996	35,721	862	36,583	177	36,760	1,714	55	1,768	47	1,815
1997	35,567	751	36,318	174	36,492	1,534	41	1,575	44	1,619
1998	34,787	678	35,465	172	35,637	1,422	36	1,458	43	1,501
1999	35,486	664	36,150	170	36,320	1,251	31	1,282	40	1,322
2000	35,631	615	36,246	170	36,416	1,089	43	1,132	40	1,172
2001	34,771	599	35,370	160	35,530	981	26	1,007	40	1,047
2002	35,120	614	35,735	153	35,888	1,019	26	1,045	37	1,082
2003	34,907	587	35,493	154	35,647	976	25	1,001	38	1,039
2004	32,156	573	32,728	152	32,880	823	20	842	37	879
2005	31,832	556	32,388	148	32,536	718	17	734	36	770

[1] Totals are based on unrounded numbers.
NASS, Iowa Agricultural Statistics Service, (515) 284–4340.

Table 7-14.—Cattle and calves: Number slaughtered commercially, total and average live weight, by States, 2005[1]

State	Cattle			Calves		
	Number slaughtered	Total live weight	Average live weight	Number slaughtered	Total live weight	Average live weight
	Thousands	*1,000 pounds*	*Pounds*	*Thousands*	*1,000 pounds*	*Pounds*
AL	4.7	4,195	884			
AK	0.9	967	1,135			
AZ						
AR	9.5	8,826	933	0.4	220	512
CA	1,362.4	1,813,221	1,331	85.7	13,225	154
CO	2,086.7	2,655,335	1,272			
DE-MD	42.3	55,201	1,304	1.4	429	296
FL						
GA	238.1	203,625	855	1.8	852	469
HI	9.2	9,957	1,083			
ID	376.0	482,976	1,284			
IL				72.9	32,086	440
IN				4.0	508	128
IA						
KS	7,321.4	9,097,379	1,243			
KY	16.4	14,820	904			
LA	13.0	11,514	884	5.8	2,851	490
MI	441.4	599,360	1,358	23.7	9,663	407
MN	644.0	901,493	1,400			
MS						
MO						
MT	20.7	23,918	1,157			
NE	7,028.9	9,078,200	1,292			
NV						
N ENG[2]	13.7	13,804	1,011	5.3	895	168
NJ	26.0	32,006	1,230	76.8	33,230	433
NM	11.5	11,478	994			
NY	37.5	44,064	1,176	107.8	16,687	155
NC	192.1	195,300	1,017	0.8	392	473
ND						
OH	100.6	114,519	1,138	48.8	19,526	400
OK	26.1	28,222	1,080	1.0	432	438
OR	20.6	25,368	1,229			
PA	840.8	1,026,338	1,221	153.1	62,753	410
SC						
SD						
TN	14.9	13,909	935	1.3	693	524
TX	6,238.2	7,482,947	1,200	10.4	5,802	556
UT	594.9	722,972	1,215			
VA	10.8	10,960	1,013			
WA	786.1	1,003,322	1,276			
WV	9.2	8,855	966			
WI	1,439.1	1,948,154	1,354	122.0	56,205	461
WY	7.6	9,082	1,189			
US[3]	32,387.7	40,688,693	1,256	734.4	259,371	353
PR	42.7			9.2		

[1] Includes slaughter in federally inspected and other slaughter plants; excludes animals slaughtered on farms. Average live weight is based on unrounded numbers. Totals may not add due to rounding. [2] CT, ME, MA, NH, RI, and VT. [3] States with no data printed are still included in the U.S. total. Data are not printed to avoid disclosing individual operations.

NASS, Iowa Agricultural Statistics Service, (515) 284–4340.

Table 7-15.—Cattle and calves: Number slaughtered under Federal inspection, and average live weight, 1996–2005

Year	Cattle		Calves	
	Number slaughtered	Average live weight	Number slaughtered	Average live weight
	Thousands	*Pounds*	*Thousands*	*Pounds*
1996	35,721	1,173	1,714	340
1997	35,567	1,177	1,534	335
1998	34,787	1,207	1,422	282
1999	35,486	1,212	1,251	288
2000	35,631	1,222	1,089	311
2001	34,771	1,224	981	318
2002	35,120	1,253	1,019	310
2003	34,907	1,234	976	316
2004	32,156	1,242	823	329
2005	31,832	1,259	718	352

NASS, Iowa Agricultural Statistics Service, (515) 284–4340.

Table 7-16.—Cattle and calves: Production, disposition, cash receipts, and gross income, by States, 2004 (preliminary)

State	Marketings[1] Cattle	Marketings[1] Calves	Cattle shipped in for feeding and breeding	Farm slaughter of cattle and calves[2]	Quantity produced (live weight)[3]	Value of production	Cash receipts from sales of cattle, calves, beef, and veal[4]	Value of cattle and calves slaughtered for home consumption	Gross income[5]
	1,000 head	*1,000 head*	*1,000 head*	*1,000 head*	*1,000 pounds*	*1,000 dollars*	*1,000 dollars*	*1,000 dollars*	*1,000 dollars*
AI	535.0	131.0	10.0	2.0	518,834	440,044	469,323	3,010	472,333
AK	1.4	0.3		0.5	2,914	2,641	1,413	405	1,818
AZ	552.0	122.0	494.0	1.0	574,578	556,547	770,066	2,963	773,029
AR	707.0	231.0	160.0	2.0	542,347	470,467	555,145	6,978	562,123
CA	1,886.0	472.0	750.0	12.0	1,982,875	1,267,226	1,633,740	10,110	1,643,850
CO	2,563.0	95.0	2,130.0	2.0	1,914,585	1,848,553	3,342,808	11,684	3,354,492
CT	10.6	9.0	2.0	1.0	14,064	9,484	8,587	1,278	9,865
DE	7.6	3.2	2.0	0.3	6,881	5,610	7,891	540	8,431
FL	263.0	661.0	75.0	2.0	457,908	415,081	443,145	1,617	444,762
GA	337.0	254.0	47.0	3.0	462,321	353,513	395,833	4,387	400,220
HI	24.0	34.0		1.0	36,770	22,754	22,125	806	22,931
ID	1,087.0	147.0	530.0	2.0	1,053,760	845,200	1,059,388	7,020	1,066,408
IL	430.0	86.0	130.0	7.0	560,618	490,145	486,612	17,022	503,634
IN	270.0	107.0	97.0	4.0	257,341	189,713	220,358	10,743	231,101
IA	1,924.0	102.0	1,250.0	4.0	1,611,439	1,329,170	2,124,600	8,732	2,133,332
KS	5,509.0	1.5	4,210.0	4.5	3,900,203	2,826,601	5,643,895	10,884	5,654,779
KY	581.0	493.0	75.0	6.0	598,892	551,559	620,650	12,549	633,199
LA	137.0	234.0	2.0	2.0	237,631	209,118	205,490	1,100	206,590
ME	18.4	16.0	3.0	1.0	21,587	16,853	16,667	1,501	18,168
MD	61.5	31.0	6.0	2.0	85,948	70,988	73,577	2,870	76,447
MA	8.0	10.0	2.0	1.0	10,252	7,624	7,571	1,377	8,948
MI	282.0	28.0	31.0	4.0	358,519	247,345	262,757	8,600	271,357
MN	892.0	127.0	325.0	6.0	1,066,671	809,517	989,285	18,001	1,007,286
MS	215.0	180.0	11.0	2.0	278,759	223,158	196,129	6,943	203,072
MO	865.0	946.0	40.0	4.0	1,194,794	1,212,128	1,131,621	35,259	1,166,880
MT	1,320.0	285.0	104.0	5.0	1,049,920	997,606	1,104,387	10,511	1,114,898
NE	5,453.0	85.0	4,000.0	2.0	4,392,322	3,605,457	6,196,896	11,392	6,208,288
NV	187.0	90.0	78.0	2.0	163,310	159,252	211,140	2,796	213,936
NH	7.0	8.7	1.0	0.5	11,239	8,691	7,001	1,369	8,370
NJ	7.1	8.9	0.5	0.5	10,855	7,334	7,656	802	8,458
NM	1,041.0	400.0	895.0	2.0	597,166	490,893	948,659	3,464	952,123
NY	121.5	415.5	6.0	2.0	183,959	121,275	127,331	2,606	129,937
NC	234.0	157.0	9.0	2.0	302,283	259,278	258,469	12,526	270,995
ND	870.0	161.0	85.0	2.0	763,410	702,022	738,975	7,408	746,383
OH	329.0	75.0	67.0	5.0	402,673	316,374	301,578	19,554	321,132
OK	2,210.0	295.0	1,120.0	10.0	1,977,824	1,951,003	2,362,342	22,219	2,384,561
OR	450.0	206.0	25.0	12.0	587,925	503,469	508,910	13,366	522,276
PA	442.0	235.0	110.0	10.0	534,035	410,288	459,569	18,948	478,517
RI	1.1	1.2	0.2	0.1	1,354	900	938	128	1,066
SC	153.0	25.0	8.0	2.0	179,667	148,295	145,504	3,500	149,004
SD	1,562.0	450.0	512.0	2.0	1,546,317	1,450,671	1,639,061	15,062	1,654,123
TN	617.0	390.0	45.0	3.0	557,705	475,024	514,388	7,009	521,397
TX	7,690.0	145.0	3,320.0	15.0	7,333,786	6,134,525	7,989,786	21,066	8,010,852
UT	369.0	95.0	120.0	4.0	384,190	358,715	431,201	8,424	439,625
VT	51.0	77.0	9.0	2.0	52,738	38,751	48,238	1,482	49,720
VA	360.0	192.0	12.0	5.0	465,832	385,641	317,677	13,901	331,578
WA	468.0	6.0	55.0	9.0	508,639	476,099	543,427	8,512	551,939
WV	105.0	74.0	38.0	4.0	120,659	83,990	87,386	5,752	93,138
WI	734.0	454.0	80.0	7.0	1,039,704	788,375	800,703	19,172	819,875
WY	860.0	213.0	315.0	1.0	583,300	592,854	855,676	10,517	866,193
US	44,808.2	9,065.3	21,396.7	185.4	41,501,303	34,887,821	47,295,574	427,865	47,723,439

[1] Includes custom slaughter for use on farms where produced and State outshipments, but excludes interfarm sales within the State. [2] Excludes custom slaughter for farmers at commercial establishments. [3] Adjustments made for changes in inventory and for inshipments. [4] Includes receipts from marketings and sales of farm-slaughter. [5] Includes cash receipts from sales of cattle, calves, beef, and veal plus value of cattle and calves slaughtered for home consumption.

NASS, Livestock Branch, (202) 720–3570.

Table 7-17.—Cattle: Number slaughtered under Federal inspection and percentage distribution, by classes, 1996–2005 [1]

Year	Steers	Heifers	Cows Dairy cows	Cows Other cows	Cows Total cows	Bulls and stags	Steers	Heifers	Cows Dairy cows	Cows Other cows	Cows Total cows	Bulls and stags
	Thousands	Thousands	Thousands	Thousands	Thousands	Thousands	Percent	Percent	Percent	Percent	Percent	Percent
1996 ...	17,400	10,502	3,037	4,068	7,105	715	48.7	29.4	8.5	11.4	19.9	2.0
1997 ...	17,172	11,287	2,926	3,498	6,424	683	48.3	31.7	8.2	9.8	18.1	1.9
1998 ...	17,101	11,228	2,620	3,245	5,865	593	49.2	32.3	7.5	9.3	16.9	1.7
1999 ...	17,608	11,648	2,573	3,030	5,603	627	49.6	32.8	7.3	8.5	15.8	1.8
2000 ...	17,758	11,835	2,632	2,796	5,427	612	49.8	33.2	7.4	7.8	15.2	1.7
2001 ...	17,097	11,379	2,582	3,092	5,674	621	49.2	32.7	7.4	8.9	16.3	1.8
2002 ...	17,523	11,342	2,607	3,051	5,658	598	49.9	32.3	7.4	8.7	16.1	1.7
2003 ...	17,177	11,078	2,860	3,163	6,023	629	49.2	31.7	8.2	9.1	17.3	1.8
2004 ...	16,192	10,345	2,363	2,706	5,069	550	50.4	32.2	7.3	8.4	15.8	1.7
2005 ...	16,797	9,761	2,252	2,523	4,775	498	52.8	30.7	7.1	7.9	15.0	1.6

[1] Totals and percentages based on unrounded data and may not equal sum of classes due to rounding.
NASS, Iowa Agricultural Statistics Service, (515) 284–4340.

Table 7-18.—Cattle and calves: Number of operations, 2004–2005, and inventory, Jan 1, 2005–2006, by States and United States [1]

State	Operations with cattle 2004	Operations with cattle 2005	January 1 cattle inventory 2005	January 1 cattle inventory 2006 [2]
	Number	Number	1,000 head	1,000 head
AL	25,000	25,000	1,320	1,280
AK	120	120	14.5	15.5
AZ	2,700	2,600	910	940
AR	30,000	30,000	1,860	1,750
CA	17,000	17,000	5,400	5,500
CO	13,000	12,900	2,500	2,650
CT	1,000	1,100	56	52
DE	420	420	23	23
FL	18,900	18,800	1,710	1,690
GA	22,000	21,000	1,210	1,180
HI	800	800	155	161
ID	10,600	10,400	2,060	2,120
IL	20,000	19,800	1,370	1,340
IN	19,000	19,000	850	900
IA	32,000	31,000	3,600	3,800
KS	33,000	32,000	6,600	6,650
KY	45,000	45,000	2,250	2,400
LA	14,500	14,500	860	820
ME	1,700	1,700	92	92
MD	4,000	4,000	235	230
MA	1,100	1,100	48	47
MI	14,500	14,400	1,000	1,040
MN	27,000	26,000	2,400	2,350
MS	20,000	21,000	1,070	1,000
MO	63,000	64,000	4,400	4,550
MT	12,700	12,600	2,350	2,400
NE	24,000	24,000	6,350	6,550
NV	1,600	1,600	500	500
NH	800	850	40	39
NJ	1,500	1,500	44	42
NM	7,700	7,600	1,500	1,550
NY	15,300	14,700	1,410	1,410
NC	22,000	21,000	870	860
ND	11,800	11,700	1,710	1,720
OH	27,000	27,000	1,300	1,280
OK	56,000	56,000	5,350	5,450
OR	15,500	15,300	1,430	1,440
PA	28,000	28,000	1,620	1,610
RI	220	220	5.5	5.0
SC	10,000	10,000	435	415
SD	17,500	17,000	3,700	3,750
TN	49,000	48,000	2,170	2,240
TX	150,000	150,000	13,700	14,100
UT	7,000	7,000	860	820
VT	2,500	2,500	275	280
VA	25,000	25,000	1,640	1,690
WA	12,700	13,000	1,080	1,120
WV	12,500	12,500	405	410
WI	37,000	36,000	3,350	3,400
WY	5,800	5,800	1,350	1,440
US	989,460	982,510	5,438.0	97,101.5
PR	4,100	4,300		

[1] An operation is any place having one or more head of cattle on hand at any time during the year.   [2] Preliminary.
NASS, Livestock Branch, (202) 720–3570.

Table 7-19.—Cattle and calves: Average dressed weight under Federal inspection, 1996–2005

Year	Cattle					Calves
	All cattle	Steers	Heifers	Cows	Bulls	
	Pounds	Pounds	Pounds	Pounds	Pounds	Pounds
1996	702	766	705	524	842	211
1997	706	764	703	539	851	208
1998	730	789	724	554	865	174
1999	736	793	727	560	881	176
2000	745	798	733	579	892	192
2001	744	798	734	584	893	196
2002	765	823	753	590	912	190
2003	746	803	732	590	904	194
2004	756	806	740	614	893	201
2005	769	817	750	621	905	216

NASS, Iowa Agricultural Statistics service (515) 284–4340.

Table 7-20.—Cattle and calves: Number of operations by size group, selected States, and United States, 2004–2005 [1]

State	Operations having									
	1–49 head		50–99 head		100–499 head		500–999 head		1000+ head	
	2004	2005	2004	2005	2004	2005	2004	2005	2004	2005
	Number									
AL	16,600	16,800	4,900	4,700	3,200	3,200	240	240	60	60
AZ	1,550	1,500	350	350	550	510	110	100	140	140
AR	19,500	19,200	5,700	5,900	4,500	4,550	230	270	70	80
CA	10,800	10,800	1,400	1,400	2,500	2,500	1,000	1,000	1,300	1,300
CO	7,800	7,500	1,700	1,900	2,600	2,600	550	550	350	350
FL	14,000	13,900	2,200	2,200	2,100	2,100	300	300	300	300
GA	14,300	14,000	4,200	3,600	3,200	3,090	240	240	60	70
ID	6,100	6,000	1,400	1,300	2,300	2,250	420	450	380	400
IL	12,900	12,400	3,600	3,700	3,280	3,480	170	170	50	50
IN	14,400	14,500	2,900	2,500	1,600	1,880	70	80	30	40
IA	14,800	14,300	6,500	6,200	9,600	9,400	780	780	320	320
KS	15,200	14,700	6,350	6,150	9,700	9,400	1,100	1,100	650	650
KY	30,000	30,000	8,500	8,600	6,100	6,070	340	255	60	75
LA	10,500	10,500	1,800	1,800	2,000	2,000	160	150	40	50
MD	2,700	2,700	550	560	710	700	30	30	10	10
MI	10,200	10,100	1,700	1,800	2,300	2,200	210	210	90	90
MN	14,300	13,500	5,600	5,400	6,600	6,600	360	360	140	140
MS	13,600	14,400	3,600	3,900	2,600	2,500	150	150	50	50
MO	39,000	39,000	12,200	13,000	11,000	11,130	650	720	150	150
MT	5,500	5,200	1,850	2,050	4,200	4,200	820	810	330	340
NE	9,000	8,500	4,700	4,700	8,000	8,400	1,400	1,400	900	1,000
NM	4,600	4,500	1,000	970	1,500	1,500	310	330	290	300
NY	8,100	7,600	2,900	3,000	3,900	3,700	260	250	140	150
NC	17,600	16,500	2,500	2,600	1,780	1,780	90	90	30	30
ND	3,400	3,300	2,550	2,500	5,200	5,200	550	590	100	110
OH	21,000	20,500	3,000	3,500	2,790	2,760	170	190	40	50
OK	32,000	32,000	10,500	10,500	11,800	11,800	1,100	1,100	600	600
OR	11,700	11,600	1,400	1,300	1,750	1,750	390	370	260	280
PA	17,200	17,700	6,300	5,900	4,300	4,200	155	155	45	45
SD	4,700	4,300	2,600	2,400	8,500	8,400	1,200	1,400	500	500
TN	36,000	35,000	7,700	8,000	5,100	4,800	165	160	35	40
TX	105,000	105,000	22,000	22,000	19,200	19,100	2,400	2,500	1,400	1,400
UT	3,900	4,000	1,100	1,100	1,600	1,500	270	280	130	120
VT	1,100	1,100	550	550	760	760	65	65	25	25
VA	16,100	16,100	4,500	4,500	4,100	4,080	240	255	60	65
WA	10,100	10,300	920	990	1,300	1,300	230	260	150	150
WI	15,800	15,400	9,000	8,700	11,400	11,100	620	600	180	200
WY	2,300	2,200	600	600	2,000	2,100	550	550	350	350
Oth Sts [2]	25,400	25,500	2,930	2,960	2,910	2,920	350	360	170	170
US	618,750	612,100	163,750	163,780	178,530	177,510	18,445	18,870	9,985	10,250

[1] An operation is any place having one or more cattle on hand during the year. [2] Individual State estimates not available for the 12 other States.

NASS, Livestock Branch, (202) 720–3570.

STATISTICS OF CATTLE, HOGS, AND SHEEP

Table 7-21.—Cattle and calves: Percent of inventory by size group, selected States, and United States, 2004–2005 [1]

State	Inventory on operations having									
	1–49 head		50–99 head		100–499 head		500–999 head		1,000+ head	
	2004	2005	2004	2005	2004	2005	2004	2005	2004	2005
	Percent									
AL	23.0	23.0	22.0	22.0	39.0	39.0	9.5	10.0	6.5	6.0
AZ	2.4	2.0	2.4	2.2	12.0	11.0	9.2	7.8	74.0	77.0
AR	21.0	20.0	21.0	21.0	43.0	43.0	8.5	9.0	6.5	7.0
CA	2.5	2.5	2.0	2.0	11.5	10.5	14.0	13.0	70.0	72.0
CO	5.0	4.2	5.0	5.3	23.0	21.5	15.0	15.0	52.0	54.0
FL	11.5	11.5	8.5	8.5	25.0	24.0	12.0	12.0	43.0	44.0
GA	15.5	16.0	21.0	19.0	45.0	45.0	11.0	11.0	7.5	9.0
ID	4.5	4.0	4.5	4.0	23.0	21.5	14.0	14.5	54.0	56.0
IL	20.0	19.0	19.0	19.0	46.0	46.0	8.5	8.2	6.5	7.8
IN	28.0	26.0	24.0	20.0	35.0	38.0	6.0	6.0	7.0	10.0
IA	8.0	9.0	13.0	12.0	50.0	49.0	14.0	14.0	15.0	16.0
KS	4.9	4.7	6.6	6.3	30.0	29.0	11.5	11.0	47.0	49.0
KY	23.0	23.0	22.0	23.0	44.0	43.0	8.0	7.0	3.0	4.0
LA	21.0	21.0	14.0	14.0	46.0	45.0	11.5	11.0	7.5	9.0
MD	16.0	15.0	15.0	16.0	52.0	53.0	8.0	8.0	9.0	8.0
MI	14.0	14.0	11.0	12.0	45.0	44.0	13.0	13.0	17.0	17.0
MN	13.0	12.5	17.0	16.0	51.0	52.0	10.0	10.0	9.0	9.5
MS	23.0	23.0	24.0	25.0	37.0	36.0	8.5	8.5	7.5	7.5
MO	18.0	17.0	20.0	20.0	46.0	46.0	10.0	11.0	6.0	6.0
MT	4.1	4.0	5.9	6.0	39.0	39.0	23.0	23.0	28.0	28.0
NE	3.2	2.8	5.3	5.2	27.0	27.0	15.5	15.0	49.0	50.0
NM	5.0	4.7	4.5	4.3	20.5	20.0	14.0	15.0	56.0	56.0
NY	10.0	9.0	15.0	15.0	45.0	45.0	12.0	12.0	18.0	19.0
NC	31.0	30.0	20.0	21.0	37.0	37.0	7.0	7.0	5.0	5.0
ND	4.5	4.6	9.5	9.4	57.0	55.0	19.0	20.0	10.0	11.0
OH	25.0	25.0	17.0	18.0	43.0	40.0	9.2	10.0	5.8	7.0
OK	12.0	12.0	13.0	13.0	40.0	40.0	12.0	12.0	23.0	23.0
OR	9.5	8.0	6.5	6.0	25.0	25.0	17.0	17.0	42.0	44.0
PA	16.0	17.0	26.0	25.0	47.0	47.0	6.4	6.5	4.6	4.5
SD	2.5	2.0	5.0	4.5	51.0	48.5	21.5	24.0	20.0	21.0
TN	31.0	30.0	23.0	25.0	39.0	38.0	4.8	4.7	2.2	2.3
TX	13.0	13.0	11.0	11.0	27.0	28.0	11.0	12.0	38.0	36.0
UT	7.0	7.0	9.0	9.0	39.0	36.0	20.0	23.0	25.0	25.0
VT	6.0	6.0	14.0	13.0	50.0	50.0	16.0	16.0	14.0	15.0
VA	19.0	19.0	18.0	19.0	46.0	44.0	10.5	11.0	6.5	7.0
WA	11.0	11.0	5.5	6.0	24.5	23.5	14.0	15.5	45.0	44.0
WI	8.5	8.5	17.0	17.0	54.0	53.0	12.0	12.0	8.5	9.5
WY	2.4	3.0	2.6	3.0	29.0	30.0	25.0	26.0	41.0	38.0
Oth Sts [2]	20.0	20.0	12.0	12.0	32.0	32.0	13.0	13.0	23.0	23.0
US	11.3	11.0	11.6	11.6	35.4	35.0	12.7	12.9	29.0	29.5

[1] Percents reflect average distributions of various probability surveys conducted during the year but are based primarily on beginning-of-year and mid-year surveys. [2] Individual State estimates not available for the 12 other States.

NASS, Livestock Branch, (202) 720–3570.

Table 7-22.—Beef cows: Number of operations, 2004–2005, and inventory, January 1, 2005–2006, by States and United States[1]

State	Operations with beef cows[2]		January 1 beef cow inventory	
	2004	2005[2]	2005	2006[3]
	Number	Number	1,000 head	1,000 head
AL	23,000	23,000	724	696
AK	90	90	5.4	6.1
AZ	1,900	1,900	175	190
AR	27,000	27,000	964	919
CA	11,500	11,500	720	700
CO	9,800	9,700	639	685
CT	700	770	7	5
DE	230	230	4	4
FL	15,600	15,400	932	926
GA	20,000	19,000	596	592
HI	650	650	81.3	87.4
ID	7,300	7,200	475	472
IL	15,000	14,800	460	446
IN	12,000	12,000	230	222
IA	25,000	25,000	1,013	1,053
KS	28,000	27,000	1,530	1,560
KY	38,000	38,000	1,100	1,128
LA	12,400	12,400	494	468
ME	1,000	1,000	12	12
MD	2,600	2,500	43	49
MA	700	750	7	8
MI	7,300	7,200	93	108
MN	15,200	15,000	395	390
MS	18,200	18,800	564	536
MO	54,000	54,000	2,121	2,236
MT	11,500	11,400	1,432	1,451
NE	20,000	20,000	1,909	1,930
NV	1,300	1,300	240	238
NH	500	530	4.0	4.0
NJ	700	700	10	9
NM	6,200	6,200	472	460
NY	6,600	6,200	80	78
NC	19,000	18,000	400	384
ND	10,500	10,600	947	937
OH	15,400	15,600	294	297
OK	48,000	48,000	2,015	2,075
OR	11,800	11,800	630	619
PA	11,900	12,000	154	152
RI	160	150	1.7	1.5
SC	9,000	9,000	222	213
SD	15,500	15,000	1,710	1,719
TN	43,000	42,000	1,078	1,110
TX	131,000	131,000	5,432	5,475
UT	5,200	5,200	347	335
VT	1,000	1,000	10	10
VA	22,000	22,000	705	747
WA	9,100	9,200	240	293
WV	10,900	10,900	197	204
WI	12,700	12,700	245	250
WY	4,800	4,800	756	763
US	774,930	770,170	32,915.4	33,253.0
PR	2,700	2,800		

[1] An operation is any place having one or more beef cows on hand at any time during the year. [2] Included in operations with cattle. [3] Preliminary.
NASS, Livestock Branch, (202) 720–3570.

STATISTICS OF CATTLE, HOGS, AND SHEEP

Table 7-23.—Beef cows: Number of operations by size group, selected States and United States, 2004–2005 [1] [2]

State	Operations having							
	1–49 head		50–99 head		100–499 head		500+ head	
	2004	2005	2004	2005	2004	2005	2004	2005
	Number							
AL	18,000	18,000	3,230	3,230	1,700	1,700	70	70
AZ	1,250	1,250	210	210	380	380	60	60
AR	21,500	21,500	3,630	3,600	1,820	1,840	50	60
CA	9,000	9,000	700	700	1,500	1,500	300	300
CO	6,600	6,300	1,400	1,400	1,600	1,780	200	220
FL	12,500	12,300	1,400	1,420	1,420	1,400	280	280
GA	16,100	15,400	2,400	2,200	1,460	1,350	40	50
ID	5,000	4,900	910	910	1,200	1,200	190	190
IL	12,700	12,500	1,620	1,630	670	660	10	10
IN	11,000	11,000	800	800	200	200		
IA	18,300	18,300	4,300	4,250	2,350	2,400	50	50
KS	18,100	17,400	5,400	5,200	4,340	4,230	160	170
KY	31,200	31,200	4,500	4,500	2,260	2,240	40	60
LA	9,900	10,000	1,400	1,300	1,050	1,050	50	50
MN	12,700	12,600	1,700	1,600	780	780	20	20
MS	14,700	15,300	2,560	2,500	900	960	40	40
MO	41,000	41,000	8,300	8,400	4,600	4,500	100	100
MT	5,200	5,000	2,000	2,050	3,800	3,900	500	450
NE	10,900	10,600	3,800	3,900	4,800	5,000	500	500
NM	4,200	4,100	830	820	1,000	1,100	170	180
NC	17,100	16,100	1,250	1,310	635	575	15	15
ND	4,300	4,400	2,810	2,620	3,300	3,500	90	80
OH	14,400	14,500	750	840	240	250	10	10
OK	36,000	36,000	7,200	7,200	4,600	4,600	200	200
OR	9,500	9,300	850	900	1,200	1,300	250	300
PA	11,300	11,400	470	470	130	130		
SD	6,000	5,700	3,560	3,350	5,500	5,500	440	450
TN	37,500	36,500	4,100	4,000	1,375	1,470	25	30
TX	103,000	103,000	15,800	15,800	11,400	11,400	800	800
UT	3,400	3,400	750	780	950	920	100	100
VA	18,000	18,000	2,700	2,700	1,260	1,260	40	40
WA	8,000	8,100	520	520	530	530	50	50
WY	2,300	2,100	700	800	1,500	1,600	300	300
Oth Sts [3]	51,300	50,800	3,100	3,130	1,570	1,580	160	160
US	601,950	596,950	95,650	95,040	72,020	72,785	5,310	5,395

[1] An operation is any place having one or more beef cows on hand at any time during the year. Missing data combined with other size groups to avoid disclosing individual operations. [2] Included in operations with cattle. [3] Individual State estimates are not available for 17 other States.

NASS, Livestock Branch, (202) 720-3570.

Table 7-24.—Beef cows: Percent of inventory by size group, selected States, and United States, 2004–2005[1]

State	Inventory on operations having							
	1–49 head		50–99 head		100–499 head		500+ head	
	2004	2005	2004	2005	2004	2005	2004	2005
	Percent							
AL	39.0	39.0	25.0	24.0	30.0	31.0	6.0	6.0
AZ	12.0	11.0	8.0	8.0	40.0	41.0	40.0	40.0
AR	40.0	39.5	25.0	25.0	31.0	31.0	4.0	4.5
CA	12.0	13.0	6.0	7.0	45.0	44.0	37.0	36.0
CO	15.0	15.0	16.0	15.0	47.0	48.0	22.0	22.0
FL	17.0	16.5	9.0	9.5	28.0	28.0	46.0	46.0
GA	36.0	36.0	23.0	22.0	37.0	36.8	4.0	5.2
ID	13.0	13.0	12.0	12.0	46.0	46.0	29.0	29.0
IL	51.0	52.0	24.0	23.0	23.0	22.5	2.0	2.5
IN	66.0	64.0	22.0	22.0	12.0	14.0	....	....
IA	32.0	31.0	30.0	30.0	35.0	36.0	3.0	3.0
KS	21.5	21.0	24.0	23.5	47.0	47.0	7.5	8.5
KY	42.0	43.0	25.7	24.5	30.0	29.0	2.3	3.5
LA	32.5	33.0	19.0	18.0	41.0	41.0	7.5	8.0
MN	43.5	44.0	27.0	26.0	26.0	26.0	3.5	4.0
MS	40.0	41.0	27.0	26.0	28.0	28.0	5.0	5.0
MO	37.0	37.0	26.0	26.0	33.0	33.0	4.0	4.0
MT	6.5	6.5	9.5	9.5	54.0	56.0	30.0	28.0
NE	12.0	12.0	14.0	14.0	50.0	50.0	24.0	24.0
NM	12.0	12.0	13.0	13.0	44.0	44.0	31.0	31.0
NC	55.5	55.0	20.0	21.0	22.0	21.0	2.5	3.0
ND	11.0	10.5	20.0	19.0	62.0	64.0	7.0	6.5
OH	64.0	65.0	18.0	18.0	15.5	14.6	2.5	2.4
OK	30.0	29.5	22.0	22.0	39.0	39.0	9.0	9.5
OR	16.0	13.0	9.0	9.0	39.0	37.0	36.0	41.0
PA	71.0	69.0	19.0	19.0	10.0	12.0	....	....
SD	8.0	7.0	14.0	13.5	60.0	61.0	18.0	18.5
TN	54.0	54.0	25.0	24.0	19.4	20.2	1.6	1.8
TX	30.0	30.0	20.0	20.0	36.0	36.0	14.0	14.0
UT	15.0	15.0	14.0	15.0	47.0	47.0	24.0	23.0
VA	40.0	42.0	25.0	24.5	30.0	29.0	5.0	4.5
WA	30.0	30.0	13.0	14.0	39.0	39.0	18.0	17.0
WY	4.5	4.5	6.5	7.5	47.0	50.0	42.0	38.0
Other States[2]	47.0	47.0	16.0	16.0	22.0	22.0	15.0	15.0
US	28.1	27.9	19.1	19.0	38.3	38.5	14.5	14.6

[1] Percents reflect average distributions of various probability surveys conducted during the year but are based primarily on beginning-of-year and mid-year surveys. Missing data combined with other size groups to avoid disclosing individual operations. [2] Individual State estimates are not available for 17 other States.

NASS, Livestock Branch, (202) 720–3570.

Table 7-25.—Hogs and pigs: Number and value, United States, Dec. 1, 1996–2005

Year	Number	Value	
		Per head	Total
	Thousands	Dollars	1,000 dollars
1996	56,124	94.00	5,280,742
1997	61,158	82.00	4,985,532
1998	62,204	44.00	2,765,745
1999	59,335	72.00	4,253,785
2000	59,110	77.00	4,540,410
2001	59,722	77.00	4,584,078
2002	59,554	71.00	4,230,728
2003	60,444	67.00	4,024,949
2004	60,975	102.00	6,211,851
2005¹	61,197	92.00	5,612,057

¹ Preliminary.
NASS, Livestock Branch, (202) 720–3570.

Table 7-26.—Hogs and pigs: Number and value, by States, Dec. 1, 2004–2005

State	Operations		Number		Value			
	2004	2005	2004	2005¹	Value per head		Total value	
					2004	2005¹	2004	2005¹
	Number	Number	Thou-sands	Thou-sands	Dollars	Dollars	1,000 dollars	1,000 dollars
AL	500	500	180	160	100	90	18,000	14,400
AK	50	50	1.7	1.6	190	180	323	288
AZ	180	180	150	142	110	99	16,500	14,058
AR	800	800	330	270	98	87	32,340	23,490
CA	870	840	140	145	130	120	18,200	17,400
CO	750	700	800	840	95	83	76,000	69,720
CT	180	200	4.2	3.5	130	120	546	420
DE	70	70	15.0	16.5	98	88	1,470	1,452
FL	1,400	1,000	20.0	20.0	110	99	2,200	1,980
GA	1,000	800	275	270	92	80	25,300	21,600
HI	250	230	22.0	19.0	160	150	3,520	2,850
ID	650	650	21.0	21.0	100	90	2,100	1,890
IL	3,400	3,100	4,100	4,000	100	90	410,000	360,000
IN	3,200	3,000	3,200	3,200	110	93	352,000	297,600
IA	9,200	8,900	16,300	16,400	110	100	1,793,000	1,640,000
KS	1,500	1,500	1,710	1,780	93	80	159,030	142,400
KY	1,000	800	350	370	79	69	27,650	25,530
LA	650	620	16.0	14.0	110	99	1,760	1,386
ME	350	370	5.0	5.0	110	99	550	495
MD	360	360	26.0	35.0	100	90	2,600	3,150
MA	250	270	12.0	13.0	110	99	1,320	1,287
MI	2,100	2,200	950	950	110	100	104,500	95,000
MN	5,000	4,900	6,500	6,600	120	110	780,000	726,000
MS	1,000	1,000	315	375	110	99	34,650	37,125
MO	2,200	2,200	2,900	2,750	85	78	246,500	214,500
MT	500	500	165	175	110	99	18,150	17,325
NE	2,600	2,600	2,850	2,850	110	97	313,500	276,450
NV	110	110	5.5	4.0	130	120	715	480
NH	250	270	3.6	3.2	120	110	432	352
NJ	300	300	11.0	9.0	120	110	1,320	990
NM	350	350	2.5	2.0	110	99	275	198
NY	1,100	1,300	84.0	83.0	100	90	8,400	7,470
NC	2,600	2,400	9,900	9,800	84	74	831,600	725,200
ND	420	450	169	157	110	99	18,590	15,543
OH	4,000	3,900	1,450	1,550	110	99	159,500	153,450
OK	2,500	2,500	2,390	2,370	81	79	193,590	187,230
OR	1,200	1,200	27.0	23.0	110	99	2,970	2,277
PA	3,400	3,200	1,080	1,090	99	84	106,920	91,560
RI	60	60	2.0	1.8	110	99	220	178
SC	900	1,000	300	315	90	81	27,000	25,515
SD	1,400	1,200	1,340	1,480	110	100	147,400	148,000
TN	1,400	1,300	215	190	87	78	18,705	14,820
TX	3,900	3,800	990	930	88	77	87,120	71,610
UT	500	450	690	690	110	99	75,900	68,310
VT	250	250	2.0	2.3	130	120	260	276
VA	700	850	375	490	87	77	32,625	37,730
WA	900	950	26.0	30.0	120	110	3,120	3,300
WV	800	800	10.0	8.0	110	99	1,100	792
WI	2,300	2,200	430	430	90	85	38,700	36,550
WY	150	150	114	113	120	110	13,680	12,430
US	69,500	67,330	60,975	61,197	102	92	6,211,851	5,612,057
PR	1,300	1,500	....	....	....	....	....	....

¹ Preliminary. Totals may not add due to rounding.
NASS, Livestock Branch, (202) 720–3570.

Table 7-27.—Sows farrowing and pig crop: Number, United States, 1996–2005

Year	Sows farrowing		Pig crop		
	Dec.-May	June-Nov.	Dec.-May	June-Nov.	Total
	Thousands	*Thousands*	*Thousands*	*Thousands*	*Thousands*
1996	5,665	5,449	47,888	46,571	94,459
1997	5,595	5,884	48,394	51,190	99,584
1998	6,014	6,046	52,469	52,535	105,004
1999	5,877	5,764	51,516	50,836	102,352
2000	5,683	5,726	50,086	50,656	100,742
2001	5,618	5,767	49,477	51,140	100,617
2002	5,776	5,716	50,858	50,820	101,678
2003	5,655	5,773	50,029	51,462	101,491
2004	5,706	5,793	50,738	52,043	102,067
2005 [1]	5,717	5,787	51,329	52,356	104,399

[1] Preliminary.
NASS, Livestock Branch, (202) 720–3570.

Table 7-28.—Hogs and pigs: Number for breeding and market, United States, 1996–2005

Year	All hogs and pigs	Kept for breeding	Market hogs by weight groups				
			Under 60 pounds	60 to 119 pounds	120 to 179 pounds	180 pounds and over	Total
			June 1				
	Thousands	*Thousands*	*Thousands*	*Thousands*	*Thousands*	*Thousands*	*Thousands*
1996	56,038	6,682	19,645	12,196	9,757	7,759	49,356
1997	57,366	6,789	19,988	12,574	10,002	8,013	50,577
1998	62,213	6,958	21,482	13,711	11,084	8,978	55,254
1999	60,894	6,515	20,532	13,500	11,075	9,271	54,379
2000	59,110	6,233	19,907	13,246	10,708	9,016	52,878
2001	58,525	6,178	19,900	12,945	10,531	8,971	52,347
2002	60,391	6,208	20,226	13,539	10,906	9,512	54,183
2003	59,602	6,026	20,433	12,952	10,828	9,363	53,576
2004	60,698	5,937	20,292	13,500	11,256	9,714	54,760
2005 [1]	60,732	5,977	20,423	13,376	11,143	9,813	54,754
			Dec. 1				
	Thousands	*Thousands*	*Thousands*	*Thousands*	*Thousands*	*Thousands*	*Thousands*
1996	56,124	6,578	18,503	12,193	10,209	8,641	49,546
1997	61,158	6,957	20,237	13,319	11,188	9,457	54,200
1998	62,204	6,682	20,140	13,630	11,584	10,167	55,522
1999	59,335	6,233	19,241	13,106	11,071	9,684	53,103
2000	59,110	6,267	19,413	12,926	10,841	9,663	52,843
2001	59,722	6,201	19,903	12,877	10,755	9,986	53,521
2002	59,554	6,058	19,485	13,033	10,875	10,103	53,496
2003	60,444	6,009	19,778	13,238	11,109	10,311	54,434
2004	60,975	5,969	19,980	13,439	11,186	10,401	55,005
2005 [1]	61,197	6,011	19,955	13,552	11,266	10,414	55,186

[1] Preliminary.
NASS, Livestock Branch, (202) 720–3570.

STATISTICS OF CATTLE, HOGS, AND SHEEP

Table 7-29.—Hogs: Number in specified countries, 2001–2004

Country	2001	2002	2003	2004 [1]
	Thousands	Thousands	Thousands	Thousands
China, Peoples Republic	446,815	457,430	462,915	466,017
European Union	152,825	152,473	154,311	152,793
United States	59,110	59,722	59,554	60,444
Brazil	32,440	32,710	32,655	32,081
Russian Federation	15,780	16,570	17,000	17,200
Canada	13,576	14,367	14,672	14,623
Philippines	11,715	11,816	12,218	12,518
Mexico	10,649	10,569	10,549	10,668
Japan	9,788	9,612	9,725	9,724
Ukraine	7,652	8,317	9,204	7,321
Korea, Republic of	7,350	7,856	8,110	8,367
Taiwan	7,495	7,165	6,794	6,779
Romania	4,797	4,477	5,058	5,145
Australia	2,748	2,563	2,940	2,658
Bulgaria	1,143	1,300	1,117	1,032
Total	783,883	796,947	806,822	807,370

[1] Preliminary.

FAS, Dairy, Livestock and Poultry Division, (202) 720–8031. Updated data available at http://www.fas.usda.gov/psd.

Table 7-30.—Hogs: Number slaughtered, United States, 1996–2005

Year	Commercial			Farm	Total
	Federally inspected	Other	Total [1]		
	Thousands	Thousands	Thousands	Thousands	Thousands
1996	90,534	1,860	92,394	175	92,569
1997	90,228	1,733	91,960	165	92,125
1998	99,285	1,745	101,029	165	101,194
1999	99,739	1,806	101,544	150	101,694
2000	96,436	1,540	97,976	130	98,106
2001	96,528	1,434	97,962	120	98,082
2002	98,915	1,348	100,263	115	100,378
2003	99,698	1,233	100,931	112	101,043
2004	102,361	1,103	103,463	110	103,573
2005	102,519	1,063	103,582	110	103,692

[1] Totals are based on unrounded numbers.

NASS, Iowa Agricultural Statistics Service, (515) 284–4340.

Table 7-31.—Sows farrowing and pig crop: Number by States, 2004 and 2005

State	Sows farrowing							
	Dec.–Feb.		Mar.–May		June–Aug.		Sept.–Nov.	
	2004	2005[1]	2004	2005[1]	2004	2005[1]	2004	2005[1]
	Thousands	Thousands	Thousands	Thousands	Thousands	Thousands	Thousands	Thousands
AR	38	43	41	42	43	43	42	41
CO	73	75	72	77	70	80	73	78
IL	205	210	215	210	210	205	210	200
IN	145	130	135	135	140	140	135	135
IA	410	440	430	440	440	450	460	455
KS	84	80	84	83	87	83	83	84
MI	45	44	44	45	48	46	46	47
MN	285	275	290	290	285	280	285	280
MO	170	170	170	170	170	170	170	175
NE	175	175	175	175	175	175	170	175
NC	550	540	550	550	560	560	550	550
OH	74	77	73	77	75	77	77	80
OK	185	190	190	190	195	200	190	185
PA	53	47	50	47	46	47	49	47
SD	70	72	70	75	68	68	71	80
TX	46	43	47	45	56	44	46	43
WI	24	24	26	25	24	25	25	27
Oth Sts[2] ...	204	200	208	206	213	206	206	206
US	2,836	2,835	2,870	2,882	2,905	2,899	2,888	2,888

State	Pig crop							
	Dec.–Feb.		Mar.–May		June–Aug.		Sept.–Nov.	
	2004	2005[1]	2004	2005[1]	2004	2005	2004	2005[1]
	Thousands	Thousands	Thousands	Thousands	Thousands	Thousands	Thousands	Thousands
AR	365	411	398	407	404	413	399	398
CO	606	641	583	678	588	704	617	686
IL	1,814	1,869	1,914	1,880	1,869	1,825	1,859	1,770
IN	1,283	1,144	1,202	1,195	1,260	1,246	1,195	1,215
IA	3,567	3,916	3,763	3,982	3,916	4,050	4,094	4,095
KS	731	696	739	730	766	730	730	731
MI	401	396	400	405	442	426	423	432
MN	2,579	2,503	2,596	2,654	2,579	2,604	2,594	2,590
MO	1,530	1,539	1,530	1,547	1,530	1,539	1,522	1,566
NE	1,558	1,575	1,558	1,584	1,558	1,584	1,530	1,584
NC	4,923	4,779	5,005	4,895	5,124	5,096	4,950	4,950
OH	651	693	650	693	668	701	693	728
OK	1,591	1,672	1,672	1,691	1,755	1,780	1,691	1,647
PA	451	447	438	428	423	442	466	461
SD	644	677	658	698	632	639	653	744
TX	386	396	404	423	507	400	412	383
WI	211	210	228	219	212	218	220	234
Oth Sts[2] ...	1,814	1,779	1,895	1,877	1,929	1,879	1,833	1,866
US	25,105	25,343	25,633	25,986	26,162	26,276	25,881	26,080

[1] Preliminary. Totals may not add due to rounding. [2] Individual State estimates not available for the 33 other States.
NASS, Livestock Branch, (202) 720–3570.

Table 7-32.—Hogs: Production, disposition, cash receipts, and gross income, United States, 1995–2004

Year	Mar-ketings [1]	Shipped in for feeding and breeding	Farm slaugh-ter [2]	Quantity produced (live weight) [3]	Value of production [4]	Cash re-ceipts from sales of hogs, pork, and lard [5]	Value of hogs slaugh-tered for home consump-tion	Gross income [6]	Average price per 100 pounds received by farmers
	1,000 head	*1,000 head*	*1,000 head*	*1,000 pounds*	*1,000 dollars*	*1,000 dollars*	*1,000 dollars*	*1,000 dollars*	*Dollars*
1995	103,007	7,557	188	24,426,543	9,829,498	10,254,866	41,849	10,296,715	40.50
1996	101,468	10,036	176	23,080,309	11,902,326	12,565,187	49,413	12,614,600	51.90
1997	104,301	14,935	161	23,979,220	12,551,845	13,053,680	48,320	13,102,000	52.90
1998	117,249	19,378	163	25,821,285	8,716,650	9,485,547	34,374	9,519,921	34.40
1999	121,138	22,634	141	25,856,590	7,770,907	8,624,295	28,381	8,652,676	30.30
2000	118,545	24,514	125	25,696,997	10,783,825	11,757,943	34,720	11,792,663	42.30
2001	119,272	26,745	119	25,866,250	11,416,397	12,394,560	35,462	12,430,022	44.40
2002	124,013	29,134	114	26,274,153	8,690,923	9,602,109	25,525	9,627,634	33.40
2003	124,383	31,543	116	26,260,140	9,663,024	10,618,027	27,774	10,645,801	37.20
2004 [7]	127,599	32,917	114	26,678,197	13,071,677	14,348,328	36,117	14,384,445	49.30

[1] Includes custom slaughter for use on farms where produced and State outshipments, but excludes interfarm sales within the State. [2] Excludes custom slaughtered for farmers at commercial establishments. [3] Adjustments made for changes in inventory and for inshipments. [4] Includes allowance for higher average price of State inshipments and outshipments of feeder pigs. [5] Receipts from marketings and sale of farm slaughter includes allowance for higher average price of State outshipments of feeder pigs. [6] Cash receipts from sale of hogs, pork, and lard plus value of hogs slaughtered for home consumption. [7] Preliminary.

NASS Livestock Branch, (202) 720–3570.

Table 7-33.—Hogs: Receipts at selected public stockyards and direct receipts at interior markets, 1996–2005 [1]

Year	Receipts at selected public stockyards				Direct receipts in interior Iowa and Southern Minnesota [4]
	South St. Joseph	South St. Paul	All others reporting	Total markets reporting [2][3]	
	Thousands	*Thousands*	*Thousands*	*Thousands*	*Thousands*
1996	259	330	937	1,972	27,199
1997	192	258	734	1,479	28,624
1998	158	265	565	988	34,082
1999	107	237	1,225	1,569	33,281
2000	59	203	998	1,260	36,504
2001	85	206	383	674	2,998
2002	16	154	376	546	4,486
2003	54	124	313	491	5,128
2004	44	113	197	354	19,760
2005	38	110	227	375	14,531

[1] Total rail and truck receipts. Saleable receipts 1978 on. [2] Rounded total of complete figures. [3] The number of stock-yards reporting varies from 25 to 55. [4] Covers receipts at 14 packing plants and 30 concentration yards. Prior to 1994 includes numbers from the following areas: Kansas City, National Stockyards and Fort Worth.

AMS, Livestock & Grain Market News, (202) 720–7316. Compiled from reports received from stockyard companies.

Table 7-34.—Hogs and corn: Hog-corn price ratio and average price received by farmers for corn, United States, 1995–2004

Year	Hog-corn price ratio [1]	Price of corn per bushel [2]
		Dollars
1995	16.4	2.56
1996	15.4	3.55
1997	20.1	2.60
1998	14.7	2.20
1999	17.3	1.89
2000	23.3	1.86
2001	23.4	1.89
2002	15.9	2.13
2003	16.6	2.27
2004	21.1	2.47

[1] Number of bushels of corn equal in value to buy 100 pounds of live hogs at local markets, based on average prices re-ceived by farmers for hogs and corn. Annual average is a simple average of monthly ratios for the calendar year. [2] Annual average is a simple average of entire month prices for the calendar year.

NASS, Environmental, Economics, and Demographics Branch, (202) 720–6146.

Table 7-35.—Hogs: Production, disposition, cash receipts, and gross income, by States, 2004 (preliminary)

State	Mar-ketings [1]	Shipped in for feeding and breeding	Farm slaughter [2]	Quantity produced (live weight) [3]	Value of produc-tion [4]	Cash receipts from sales of hogs, pork, and lard [5]	Value of hogs slaugh-tered for home con-sumption	Gross income [6]
	1,000 head	*1,000 head*	*1,000 head*	*1,000 pounds*	*1,000 dollars*	*1,000 dollars*	*1,000 dollars*	*1,000 dollars*
AL	329.0	106.0	1.0	26,956	13,797	16,408	242	16,650
AK	2.8	0.4	0.5	602	441	409	64	473
AZ	280.0	30.0	1.0	68,221	38,806	41,362	217	41,579
AR	1,624.0	100.0	1.0	125,107	89,727	99,741	503	100,244
CA	350.0	40.0	7.0	78,367	38,615	39,855	1,132	40,987
CO	2,428.0	195.0	1.0	374,785	198,364	206,057	267	206,324
CT	5.5	0.1	0.1	1,182	538	491	38	529
DE	54.0	2.5	0.1	5,383	2,714	2,954	41	2,995
FL	82.0	16.0	1.0	12,477	4,628	6,406	76	6,482
GA	893.0	99.0	2.0	148,837	74,147	84,790	499	85,289
HI	27.0	....	1.0	5,274	4,588	4,463	165	4,628
ID	38.0	1.0	1.0	9,153	4,649	4,825	215	5,040
IL	8,753.0	1,740.0	3.0	1,732,517	937,549	1,027,628	1,065	1,028,693
IN	6,149.0	1,530.0	1.0	1,437,014	684,757	738,470	752	739,222
IA	29,570.0	16,500.0	11.0	7,195,890	3,264,479	3,801,007	4,097	3,805,104
KS	3,326.0	626.0	1.0	754,774	355,598	379,048	486	379,534
KY	857.0	100.0	4.0	201,498	98,107	104,513	1,719	106,232
LA	33.9	2.0	1.0	6,473	2,852	3,065	87	3,152
ME	13.3	2.7	0.2	3,283	1,534	1,354	305	1,659
MD	106.5	40.0	0.2	18,387	7,975	9,577	165	9,742
MA	16.8	2.8	0.3	3,252	1,378	1,463	174	1,637
MI	1,931.0	345.0	4.0	480,741	217,539	234,992	465	235,457
MN	14,663.0	5,125.0	4.0	3,097,695	1,504,589	1,724,514	1,990	1,726,504
MS	467.0	41.0	2.0	112,968	52,278	54,331	423	54,754
MO	7,461.0	1,650.0	1.0	1,141,877	551,601	623,260	1,077	624,337
MT	317.5	18.0	2.0	69,654	33,607	38,789	502	39,291
NE	7,008.0	1,000.0	1.0	1,373,997	716,613	761,953	1,372	763,325
NV	10.3	5.0	0.3	2,105	958	1,046	33	1,079
NH	6.5	2.3	0.2	1,134	420	511	42	553
NJ	26.3	19.0	1.0	1,733	495	1,056	194	1,250
NM	4.6	3.0	1.0	1,363	608	522	291	813
NY	158.0	12.8	1.0	35,509	14,929	14,992	319	15,311
NC	18,633.0	280.0	14.0	3,849,081	2,065,482	2,078,800	2,222	2,081,022
ND	504.6	94.0	2.0	66,934	38,920	42,112	561	42,673
OH	3,186.0	570.0	8.0	768,946	372,859	402,719	3,263	405,982
OK	7,127.0	850.0	2.0	1,229,514	577,762	615,411	610	616,021
OR	45.0	....	1.0	10,880	5,614	5,325	356	5,681
PA	2,110.0	390.0	8.0	500,169	225,037	242,205	1,414	243,619
RI	3.6	0.1	0.1	823	374	344	19	363
SC	510.5	50.0	1.5	102,594	50,790	52,317	1,053	53,370
SD	3,429.5	1,030.0	2.5	629,293	317,710	371,009	1,445	372,454
TN	462.0	140.0	3.0	81,384	38,314	42,904	793	43,697
TX	1,427.3	19.3	6.0	202,064	90,748	88,675	1,793	90,468
UT	1,200.0	8.0	1.0	291,866	157,128	155,103	259	155,362
VT	3.8	0.5	0.2	799	342	287	88	375
VA	579.0	60.0	2.0	139,296	63,792	67,602	1,247	68,849
WA	43.0	3.0	1.5	10,099	4,982	4,926	364	5,290
WV	18.7	2.0	1.0	4,962	2,287	1,792	439	2,231
WI	949.0	64.0	2.0	207,084	111,991	118,615	618	119,233
WY	374.5	1.5	2.0	54,201	28,665	28,330	556	28,886
US	127,598.5	32,917.0	113.7	26,678,197	13,071,677	14,348,328	36,117	14,384,445

[1] Includes custom slaughter for use on farms where produced and State outshipments, but excludes interfarm sales within the State.　[2] Excludes custom slaughter for farmers at commercial establishments.　[3] Adjustments made for changes in inventory and for inshipments.　[4] Includes allowance for higher average price of State inshipments and outshipments of feeder pigs.　[5] Receipts from marketings and sale of farm-slaughter.　[6] Cash receipts from sales of hogs, pork, and lard plus value of hogs slaughtered for home consumption.

NASS, Livestock Branch, (202) 720–3570.

Table 7-36.—Hogs: Number slaughtered commercially, total and average live weight, by States, 2005 [1]

State	Number slaughtered	Total live weight	Average live weight
	Thousands	*1,000 pounds*	*Pounds*
AL	107.3	50,312	469
AK	1.1	283	252
AZ	4.0	710	178
AR	163.2	82,884	508
CA	2,580.2	619,686	240
CO	13.4	3,587	267
DE and MD	18.7	4,474	239
FL	95.5	12,657	133
GA	121.3	28,735	237
HI	23.5	5,045	215
ID	154.8	50,657	327
IL	9,433.6	2,611,010	277
IN	7,075.0	1,873,037	265
IA	29,832.3	8,078,870	271
KS			
KY	2,315.6	617,942	267
LA	21.0	3,987	190
MI	122.8	44,512	362
MN	9,245.8	2,442,526	264
MS			
MO			
MT	14.9	3,728	251
NE	7,185.8	1,933,729	269
NV			
N ENG [2]	20.6	4,315	209
NJ	114.1	11,334	99
NM	2.0	608	300
NY	32.6	7,218	221
NC	11,161.0	2,847,170	255
ND	163.9	39,052	238
OH	1,108.9	302,063	272
OK	4,973.4	1,400,142	282
OR	195.9	47,128	241
PA	2,841.1	695,632	245
SC			
SD	4,771.8	1,247,960	262
TN	644.4	311,039	483
TX	394.9	98,564	250
UT	46.0	10,916	237
VA	3,368.6	862,743	256
WA	23.5	5,973	254
WV	7.7	1,953	254
WI	513.2	229,324	447
WY	5.3	1,380	259
US [3]	103,581.5	27,827,948	269
PR	67.2		

[1] Includes slaughter in federally inspected and other slaughter plants; excludes animals slaughtered on farms. Average live weight is based on unrounded numbers. Totals may not add due to rounding.   [2] CT, ME, MA, NH, RI, and VT.   [3] States with no data printed are still included in US total. Data are not printed to avoid disclosing individual operations.
NASS, Iowa Agricultural Statistics Service, (515) 284-4340.

Table 7-37.—Hogs: Number slaughtered, average dressed and live weights, Federally inspected, 1996-2005 [1]

Year	Federally inspected											
	Barrows and gilts			Sows			Stags and boars			Total		
	Head	Percent of total	Avg. dressed weight	Head	Percent of total	Avg. dressed weight	Head	Percent of total	Avg. dressed weight	Head	Avg. dressed weight	Avg. live weight
	1,000		*Pounds*	*1,000*		*Pounds*	*1,000*		*Pounds*	*1,000*	*Pounds*	*Pounds*
1996	86,443	95.5	181	3,425	3.8	290	667	0.7	222	90,534	186	254
1997	86,587	96.0	185	3,064	3.4	291	577	0.6	220	90,228	189	257
1998	95,354	96.0	185	3,437	3.5	295	494	0.5	228	99,285	189	257
1999	96,000	96.3	187	3,336	3.3	296	404	0.4	232	99,739	191	259
2000	93,115	96.6	191	3,005	3.1	309	316	0.3	226	96,436	194	262
2001	93,201	96.6	193	3,009	3.1	316	318	0.3	226	96,528	197	265
2002	95,459	96.5	193	3,185	3.2	317	271	0.3	235	98,915	197	265
2003	96,242	96.5	195	3,215	3.2	315	241	0.2	241	99,698	199	267
2004	98,831	96.6	196	3,271	3.2	313	259	0.3	220	102,361	199	267
2005	99,123	96.7	197	3,116	3.0	310	280	0.3	213	102,519	201	269

[1] All weights calculated using unrounded totals. Totals and percentages based on unrounded data and may not equal sum of classes due to rounding.
NASS, Iowa Agricultural Statistics Service, (515) 284-4340.

Table 7-38.—Hogs and pigs: Number of operations and inventory by size groups, selected States, regions, and United States, 2004–2005 [1]

State	Operations having											
	1–99 head		100–499 head		500–999 head		1,000–1,999 head		2,000–4,999 head		5,000+ head	
	2004	2005	2004	2005	2004	2005	2004	2005	2004	2005	2004	2005
	Number											
AR	630	630	40	40	55	55	38	38	31	31	6	6
CO	685	635	28	26	10	10	4	6	11	11	12	12
IL	1,100	950	780	730	470	450	450	410	450	400	150	160
IN	1,450	1,400	680	550	400	350	250	270	310	310	110	120
IA	1,300	1,200	2,600	2,450	1,600	1,500	1,600	1,500	1,600	1,700	500	550
KS	940	940	250	260	100	100	90	85	75	70	45	45
MI	1,500	1,600	270	270	90	90	90	80	110	120	40	40
MN	1,300	1,300	1,300	1,300	780	700	600	600	730	710	290	290
MO	1,100	1,000	480	630	200	140	85	85	230	240	105	105
NE	800	800	850	850	380	380	290	290	190	190	90	90
NC	1,000	830	90	90	70	60	170	150	650	650	620	620
OH	2,700	2,600	600	610	260	210	240	250	170	200	30	30
OK	2,200	2,200	120	120	50	50	50	50	50	50	30	30
PA	2,500	2,300	370	360	170	170	130	130	195	210	35	30
SD	550	485	410	290	170	170	115	100	90	90	65	65
TX	3,720	3,634	140	130	10	8	7	5	5	5	18	18
WI	1,570	1,530	450	410	150	130	80	80	40	40	10	10
Oth Sts [2]	17,050	16,580	900	1,000	190	170	160	130	200	210	150	140
US	42,095	40,614	10,358	10,116	5,155	4,743	4,449	4,259	5,137	5,237	2,306	2,361

State	Inventory on operations having											
	1–99 head		100–499 head		500–999 head		1,000–1,999 head		2,000–4,999 head		5,000+ head	
	2004	2005	2004	2005	2004	2005	2004	2005	2004	2005	2004	2005
	Percent											
AR	2.8	3.0	2.7	2.8	13.0	12.7	15.5	15.5	26.5	26.5	39.5	39.5
CO	0.7	0.7	0.7	0.6	0.8	0.7	0.7	1.0	4.1	4.0	93.0	93.0
IL	1.0	1.0	6.0	5.0	8.0	8.0	16.0	14.0	34.0	31.0	35.0	41.0
IN	1.5	1.4	5.5	4.4	9.0	7.7	12.0	12.5	32.0	29.0	40.0	45.0
IA	0.3	0.3	4.5	4.2	7.0	6.5	13.7	13.2	31.5	32.3	43.0	43.5
KS	1.0	0.9	2.5	2.4	3.5	3.4	8.0	7.3	13.0	12.0	72.0	74.0
MI	2.0	2.0	5.0	6.0	7.0	7.0	13.0	11.0	37.0	38.0	36.0	36.0
MN	0.5	0.5	4.5	4.5	8.0	7.0	10.0	11.0	31.0	31.0	46.0	46.0
MO	1.1	1.0	3.9	4.5	5.0	3.5	4.0	4.0	23.5	24.0	62.5	63.0
NE	0.7	0.7	7.5	7.4	8.8	8.9	13.0	13.0	19.0	19.0	51.0	51.0
NC	0.1	0.1	0.2	0.2	0.6	0.4	2.6	2.3	21.5	21.0	75.0	76.0
OH	4.0	3.5	11.0	10.0	12.0	10.0	24.0	23.0	32.0	36.5	17.0	17.0
OK	1.0	1.0	1.0	1.0	1.5	1.5	2.5	2.5	6.0	6.0	88.0	88.0
PA	2.5	2.5	7.5	7.0	10.0	9.5	15.0	13.0	38.0	42.0	27.0	26.0
SD	1.0	1.0	7.5	5.0	9.0	8.0	10.0	8.0	21.0	20.0	51.5	58.0
TX	3.7	2.7	2.3	1.6	0.6	0.5	0.8	0.7	1.6	1.5	91.0	93.0
WI	7.0	7.0	19.0	17.0	19.0	18.0	19.0	21.0	21.0	20.0	15.0	17.0
Oth Sts [2]	5.0	5.0	5.0	5.0	3.5	3.0	5.5	5.0	17.0	18.0	64.0	64.0
US	1.0	1.0	4.0	4.0	6.0	6.0	10.0	10.0	26.0	26.0	53.0	53.0

[1] An operation is any place having one or more head of hogs and pigs on hand at any time during the year. Percents reflect average distributions based primarily on mid-year surveys.   [2] Individual State estimates not available for the 33 other States.
NASS, Livestock Branch, (202) 720–3570.

Table 7-39.—Lard, including rendered pork fat: Stocks Jan. 1, production, trade, and disappearance, United States, 1995–2004

Year	Supply						Disposition			
	Production				Stocks Jan. 1 [1]	Total [2]	Exports	Domestic disappearance		
	Feder-ally in-spected	Other commer-cial	Farm	Total				Total	Direct use	
									Total	Per civil-ian
	Million pounds	Million pounds	Mil-lion pounds	Mil-lion pounds	Million pounds	Million pounds	Mil-li on pounds	Mil-lion pounds	Mil-lion pounds	Pounds
1995	....	....	....	1,040	41	1,082	124	920	430	1.6
1996	....	....	....	998	38	1,038	101	918	468	1.7
1997	....	....	....	993	19	1,013	90	901	518	1.9
1998	....	....	....	1,091	22	1,116	131	956	541	2.0
1999	....	....	....	1,097	28	1,127	147	953	547	2.0
2000	....	....	....	1,058	27	1,087	174	897	561	2.0
2001	....	....	....	1,058	16	1,077	103	960	661	2.3
2002	....	....	....	1,083	14	1,105	84	1,010	709	2.5
2003	....	....	....	1,090	11	1,108	117	977	708	2.4
2004	....	....	....	1,117	13	1,136	288	834	566	1.9

[1] Factory and warehouse stocks as reported by the Bureau of the Census.   [2] Includes imports, which were less than 500,000 pounds.

ERS, Field Crops Branch, (202) 694–5300. Compiled from reports of the U.S. Department of Commerce and the U.S. Department of Agriculture. Totals and per capita estimates computed from unrounded numbers.

Table 7-40.—Lard: United States exports by country of destination, 2001–2004

Country	2001	2002	2003	2004 [1]
	Metric tons	Metric tons	Metric tons	Metric tons
Mexico	43,663	30,172	50,796	124,195
Canada	9,438	6,322	3,817	6,629
China, Peoples Republic	20	0	28	2,100
Taiwan	6,107	4,709	1,398	2,035
Honduras	611	50	139	1,830
El Salvador	117	100	94	1,364
Guatemala	384	442	201	1,246
New Zealand	63	63	0	718
Dominican Republic	0	0	0	692
Azerbaijan	0	0	0	393
Japan	2,174	311	245	287
Armenia	0	0	44	279
Cuba	0	0	0	221
United Arab Emirates	0	4	310	220
Russian Federation	909	219	176	151
Philippines	800	424	72	130
Australia	0	0	71	112
Belize	861	353	178	96
Bahamas, the	67	5	80	96
Hong Kong	192	203	108	63
Trinidad and Tobaggo	95	19	164	37
European Union	99	25	41	49
Saudi Arabia	40	103	141	21
Leeward-Windward	69	0	19	20
Bulgaria	0	0	0	71
South Africa, Republic	32	26	3,419	9
Haiti	62	187	183	7
Netherlands Antilles	30	25	10	5
Turkey	0	0	1,438	0
Korea, Republic of	324	950	325	0
India	0	0	50	0
Colombia	66	45	37	0
Costa Rica	62	612	19	0
Others	294	835	437	21
Grand total	66,579	46,203	64,040	143,097

[1] Preliminary.

FAS, Dairy, Livestock and Poultry Division, (202) 720–8031. Updated data available at http://www.fas.usda.gov/ustrade.

Table 7-41.—Sheep and lambs: Number and value, United States, Jan. 1, 1997–2006

Year	Number	Value	
		Per head	Total
	Thousands	Dollars	1,000 dollars
1997	8,024	96.00	761,650
1998	7,825	102.00	797,826
1999	7,247	88.00	640,819
2000	7,036	95.00	669,890
2001	6,908	100.00	690,489
2002	6,623	94.00	614,466
2003	6,321	104.00	656,638
2004	6,105	119.00	723,785
2005	6,135	130.00	799,288
2006 [1]	6,230	141.00	875,480

[1] Preliminary.
NASS, Livestock Branch, (202) 720–3570.

Table 7-42.—Sheep and lambs: Number by class, United States, Jan. 1, 1997–2006

Year	All sheep and lambs	Breeding sheep			
		Total [1]	Replacement lambs	1 year and over	
				Ewes	Rams
	Thousands	Thousands	Thousands	Thousands	Thousands
1997 [2]	8,024	5,919	787	4,912	220
1998 [2]	7,825	5,611	839	4,570	203
1999 [2]	7,247	5,306	768	4,336	203
2000 [2]	7,036	5,167	725	4,234	209
2001 [2]	6,908	4,952	679	4,071	202
2002 [2]	6,623	4,871	732	3,939	201
2003 [2]	6,321	4,670	703	3,773	194
2004 [2]	6,105	4,499	702	3,610	188
2005 [2]	6,135	4,533	771	3,573	190
2006 [2][3]	6,230	4,639	786	3,657	196

[1] Categories may not add to total due to rounding.   [2] Includes new crop lambs.   [3] Preliminary.
NASS, Livestock Branch, (202) 720–3570.

STATISTICS OF CATTLE, HOGS, AND SHEEP

Table 7-43.—Breeding sheep: Number by class, State and U.S., Jan. 1, 2005–2006

| State | Replacement lambs | | One year and over | | | |
| | Breeding | | Ewes | | Rams | |
	2005	2006[1]	2005	2006[1]	2005	2006[1]
	1,000 head	1,000 head	1,000 head	1,000 head	1,000 head	1,000 head
AZ	60.0	70.0	50.0	57.0	3.0	4.0
CA	325.0	325.0	275.0	270.0	10.0	10.0
CO	170.0	190.0	135.0	148.0	5.0	6.0
ID	225.0	220.0	182.0	178.0	5.0	5.0
IL	56.0	61.0	43.0	48.0	3.0	3.0
IN	44.0	46.0	33.0	33.0	2.5	2.0
IA	175.0	170.0	142.0	132.0	7.0	7.0
KS	65.0	60.0	54.0	50.0	3.0	2.5
KY	26.0	29.0	18.0	22.5	2.5	1.5
MD	17.5	16.0	14.0	12.5	1.5	1.0
MI	59.0	66.0	45.0	46.0	2.0	3.0
MN	100.0	110.0	80.0	85.0	5.0	5.0
MO	57.5	65.5	47.0	52.0	2.5	3.5
MT	280.0	270.0	215.0	208.0	7.0	7.0
NE	75.0	80.0	62.0	65.0	3.0	3.0
NV	66.0	64.0	54.0	51.0	2.0	2.0
N ENG[2]	39.0	40.5	29.0	30.5	2.5	2.5
NM	120.0	130.0	97.0	102.0	6.0	6.0
NY	58.0	54.0	43.0	41.0	3.0	3.0
NC	17.0	14.0	12.5	10.0	1.0	1.0
ND	81.0	76.0	63.0	59.0	2.0	2.0
OH	122.0	117.0	92.0	89.0	6.0	6.0
OK	55.0	62.0	41.0	46.0	3.0	4.0
OR	145.0	145.0	107.0	114.0	8.0	6.0
PA	83.0	94.0	63.0	73.0	5.0	6.0
SD	285.0	280.0	230.0	230.0	8.0	8.0
TN	20.0	22.0	14.0	15.0	1.5	2.0
TX	840.0	870.0	650.0	690.0	45.0	45.0
UT	245.0	260.0	200.0	210.0	8.0	11.0
VA	46.0	51.0	35.0	37.0	3.0	4.0
WA	41.0	41.0	32.0	33.0	2.0	2.0
WV	29.0	27.0	23.0	22.0	1.0	1.0
WI	69.0	72.0	52.0	54.0	3.0	3.0
WY	350.0	350.0	278.0	278.0	10.0	10.0
Other States[3]	87.0	91.0	62.0	65.5	8.0	8.0
US	4,533.0	4,639.0	3,572.5	3,657.0	190.0	196.0

[1] Preliminary. [2] N ENG includes CT, ME, MA, NH, RI, and VT. [3] Other States include AL, AK, AR, DE, FL, GA, HI, LA, MS, NJ, and SC.

NASS, Livestock Branch, (202) 720–3570.

Table 7-44.—Sheep and lambs: Average price per 100 pounds at San Angelo, 1996–2005[1]

| Year | Sheep | | | Slaughter lambs | | | |
| | | | | Shorn | | Spring | |
	Good	Utility	Cull	Prime	Choice	Prime	Choice
1996	34.50	35.38	23.96	85.68	85.68	82.00	82.00
1997	49.24	46.53	32.80	89.26	89.26	89.20	89.20
1998	40.11	39.01	26.53	71.79	71.79	74.37	74.37
1999	41.71	41.21	27.72	76.37	65.92	76.39	66.42
2000	45.37	42.53	29.84	80.36	80.36	80.10	80.10
2001	44.14	45.11	30.49	70.05	70.05	69.78	69.78
2002	38.04	39.26	24.51	71.69	71.69	72.09	72.09
2003	41.33	44.65	31.32	91.90	91.90	92.13	92.14
2004	46.67	47.54	34.51	....	96.25	....	96.31
2005	....	....	....	....	....	....	....

[1] Simple average of monthly bulk-of-sales prices from data of the livestock reporting service. 1995 to present price reflects wooled lamb as well as the weight range of 110-130.

AMS, Livestock & Grain Market News, (202) 720–7316.

Table 7-45.—Sheep and lambs: Number of breeding and market sheep, by State and U.S., Jan. 1, 2005–2006

State	Breeding sheep and lambs		Market sheep and lambs	
	2005	2006[1]	2005	2006[1]
	1,000 head	*1,000 head*	*1,000 head*	*1,000 head*
AZ	60.0	70.0	40.0	35.0
CA	325.0	325.0	345.0	325.0
CO	170.0	190.0	195.0	200.0
ID	225.0	220.0	45.0	40.0
IL	56.0	61.0	13.0	8.0
IN	44.0	46.0	6.0	4.0
IA	175.0	170.0	70.0	65.0
KS	65.0	60.0	41.0	35.0
KY	26.0	29.0	6.0	6.0
MD	17.5	16.0	5.5	6.0
MI	59.0	66.0	24.0	22.0
MN	100.0	110.0	45.0	45.0
MO	57.5	65.5	7.5	9.5
MT	280.0	270.0	25.0	25.0
NE	75.0	80.0	22.0	26.0
NV	66.0	64.0	4.0	10.0
N ENG[2]	39.0	40.5	6.5	6.5
NM	120.0	130.0	25.0	25.0
NY	58.0	54.0	17.0	16.0
NC	17.0	14.0	3.0	4.0
ND	81.0	76.0	24.0	28.0
OH	122.0	117.0	20.0	24.0
OK	55.0	62.0	15.0	18.0
OR	145.0	145.0	80.0	75.0
PA	83.0	94.0	17.0	16.0
SD	285.0	280.0	90.0	105.0
TN	20.0	22.0	3.0	5.0
TX	840.0	870.0	230.0	220.0
UT	245.0	260.0	25.0	20.0
VA	46.0	51.0	15.0	16.0
WA	41.0	41.0	5.0	9.0
WV	29.0	27.0	2.0	5.0
WI	69.0	72.0	16.0	17.0
WY	350.0	350.0	100.0	100.0
Other States[3]	87.0	91.0	14.5	20.0
US	4,533.0	4,639.0	1,602.0	1,591.0

[1] Preliminary. [2] N ENG includes CT, ME, MA, NH, RI, and VT. [3] Other States include AL, AK, AR, DE, FL, GA, HI, LA, MS, NJ, and SC.

NASS, Livestock Branch, (202) 720–3570.

STATISTICS OF CATTLE, HOGS, AND SHEEP

Table 7-46.—Lamb crop: Per 100 ewes 1+, number and percent of previous year, by State, and United States, 2004–2005

State	Breeding ewes 1 year & older, Jan. 1		Lambs per 100 ewes 1+, Jan. 1		Lamb crop[1]		
	2004	2005[2]	2004	2005[2]	2004	2005[2]	2005 as % of 2004
	1,000 head	*1,000 head*	*Percent*	*Percent*	*1,000 head*	*1,000 head*	*Percent*
AZ	48.0	50.0	83	84	40.0	42.0	105
CA	275.0	275.0	95	91	260.0	250.0	96
CO	139.0	135.0	119	130	165.0	175.0	106
ID	184.0	182.0	130	121	240.0	220.0	92
IL	40.0	43.0	145	144	58.0	62.0	107
IN	34.0	33.0	132	127	45.0	42.0	93
IA	147.0	142.0	146	144	215.0	205.0	95
KS	53.0	54.0	134	113	71.0	61.0	86
KY	16.0	18.0	138	128	22.0	23.0	105
MD	15.0	14.0	110	111	16.5	15.5	94
MI	43.0	45.0	128	131	55.0	59.0	107
MN	80.0	80.0	169	175	135.0	140.0	104
MO	45.0	47.0	133	143	60.0	67.0	112
MT	220.0	215.0	123	128	270.0	275.0	102
NE	67.0	62.0	125	142	84.0	88.0	105
NV	52.0	54.0	108	100	56.0	54.0	96
N ENG[3]	28.0	29.0	125	124	35.0	36.0	103
NM	108.0	97.0	79	93	85.0	90.0	106
NY	41.0	43.0	129	119	53.0	51.0	96
NC	12.5	12.5	124	100	15.5	12.5	81
ND	59.0	63.0	149	135	88.0	85.0	97
OH	94.0	92.0	128	124	120.0	114.0	95
OK	45.0	41.0	116	132	52.0	54.0	104
OR	112.0	107.0	129	140	145.0	150.0	103
PA	58.0	63.0	145	140	84.0	88.0	105
SD	240.0	230.0	121	126	290.0	290.0	100
TN	14.0	14.0	107	114	15.0	16.0	107
TX	675.0	650.0	77	86	520.0	560.0	108
UT	195.0	200.0	126	120	245.0	240.0	98
VA	34.0	35.0	138	140	47.0	49.0	104
WA	32.0	32.0	166	166	53.0	53.0	100
WV	25.0	23.0	116	122	29.0	28.0	97
WI	52.0	52.0	142	142	74.0	74.0	100
WY	270.0	278.0	111	108	300.0	300.0	100
Other States[4]	57.0	62.0	93	90	53.0	56.0	106
US	3,609.5	3,572.5	113	115	4,096.0	4,125.0	101

[1] Lamb crop defined as lambs born in the Eastern States and lambs docked or branded in the Western States. [2] Preliminary. [3] N ENG includes CT, ME, MA, NH, RI, and VT. [4] Other States include AL, AK, AR, DE, FL, GA, HI, LA, MS, NJ, and SC.

NASS, Livestock Branch, (202) 720–3570.

Table 7-47.—Sheep and lambs: Production, disposition, cash receipts, and gross income, United States, 1995–2004

Year	Lamb crop [1]	Marketings [2]		Inshipments	Farm slaughter	Production (live weight) [3]
		Sheep	Lambs			
	1,000 head	*1,000 head*	*1,000 head*	*1,000 head*	*1,000 head*	*1,000 pounds*
1995	5,643	1,052	6,286	2,141	76	602,011
1996	5,361	938	6,069	2,196	71	572,344
1997	5,356	1,015	5,676	2,055	71	602,535
1998	5,002	975	5,466	1,744	73	554,410
1999	4,754	799	5,258	1,891	67	537,119
2000	4,645	811	4,875	1,763	70	512,305
2001	4,520	740	4,838	1,543	65	501,483
2002	4,355	855	4,794	1,749	66	485,149
2003	4,140	828	4,387	1,558	67	470,108
2004 [4]	4,096	695	4,201	1,495	65	464,503

Year	Value of production	Cash receipts [5]	Value of home consumption	Gross income [6]	Average price per 100 pounds received by farmers	
					Sheep	Lambs
	1,000 dollars	*1,000 dollars*	*1,000 dollars*	*1,000 dollars*	*Dollars*	*Dollars*
1995	414,366	566,240	10,387	576,627	28.00	78.20
1996	440,686	612,345	10,690	623,035	29.90	88.20
1997	489,564	635,451	11,363	646,814	37.90	90.30
1998	355,150	481,592	8,550	490,142	30.60	72.30
1999	352,348	473,215	8,475	481,690	31.10	74.50
2000	365,183	476,131	9,532	485,663	34.30	79.80
2001	303,186	403,175	8,166	411,341	34.60	66.90
2002	313,946	429,125	8,560	437,685	28.20	74.10
2003	391,765	507,890	10,756	518,646	34.90	94.40
2004 [4]	412,333	520,998	11,464	532,462	38.80	101.00

[1] Lamb crop defined as lambs born in the native States. [2] Includes custom slaughter for use on farms where produced and state outshipments, but excludes interfarm sales within the State. [3] Adjustments made for changes in inventory and for inshipments. [4] Preliminary. [5] Receipts from marketings and sale of farm-slaughtered meat. [6] Cash receipts from sales of sheep, lambs, and mutton and lamb plus value of sheep and lambs slaughtered for home consumption.

NASS, Livestock Branch, (202) 720–3570.

Table 7-48.—Sheep and lambs: Receipts at selected public stockyards, 1996–2005 [1]

Year	Sioux Falls	South St. Joseph	South St. Paul	All others reporting	Total markets reporting [2][3]
	Thousands	*Thousands*	*Thousands*	*Thousands*	*Thousands*
1996	....	8	47	857	921
1997	100	8	47	998	1,060
1998	73	5	50	811	938
1999	77	5	54	962	1,097
2000	61	3	55	935	1,054
2001	48	4	56	913	1,021
2002	48	4	63	832	947
2003	45	3	44	658	750
2004	40	3	37	553	633
2005	41	2	36	543	622

[1] Total rail and truck receipts unloaded at public stockyards. Saleable receipts only, 1978. [2] Rounded totals of complete figures. [3] The number of stockyards reporting varies from 41 to 68.

AMS, Livestock & Grain Market News, (202) 720–7316. Compiled from reports received from stockyard companies.

Table 7-49.—Sheep and lambs: Production, disposition, cash receipts, and gross income, by States, 2004 (preliminary)

State	Marketings[1]		Inshipments	Farm slaughter[2]	Production (live weight)[3]	Value of production	Cash receipts[4]	Value of home consumption	Gross income[5]
	Sheep	Lambs							
	1,000 head	1,000 head	1,000 head	1,000 head	1,000 pounds	1,000 dollars	1,000 dollars	1,000 dollars	1,000 dollars
AZ	9.0	92.0	64.0	13.0	3,865	3,476	8,857	991	9,848
CA	39.0	445.0	245.0	5.0	45,285	38,504	57,612	968	58,580
CO	55.0	740.0	660.0	1.0	67,654	53,502	111,126	120	111,246
ID	33.0	185.0	11.0	1.0	24,401	21,147	20,949	232	21,181
IL	2.0	45.0	5.0	1.0	4,389	3,805	3,541	264	3,805
IN	11.7	27.3	5.0	1.5	3,611	2,685	2,769	372	3,141
IA	31.0	251.0	94.0	1.0	31,403	28,324	34,600	394	34,994
KS	4.5	62.5	15.0	1.0	7,219	6,459	6,677	239	6,916
KY	1.4	13.8	4.0	0.2	2,037	1,865	1,609	24	1,633
MD	4.9	11.5	2.9	2.5	1,326	1,225	1,343	210	1,553
MI	12.0	35.0	3.0	2.0	4,722	4,119	3,800	540	4,340
MN	13.0	113.0	22.0	1.0	14,215	12,512	13,316	223	13,539
MO	2.5	45.3	3.0	0.2	5,131	4,886	4,610	104	4,714
MT	48.0	194.0	5.0	1.0	26,181	24,113	23,288	278	23,566
NE	27.0	94.6	47.0	0.4	9,710	8,496	11,531	179	11,710
NV	5.5	51.5	15.5	1.5	3,058	2,822	4,210	257	4,467
N ENG[6]	6.9	22.9	1.8	0.9	3,057	3,193	2,672	376	3,048
NM	26.0	64.0	8.0	5.0	7,006	6,137	6,667	897	7,564
NY	4.5	34.5	4.0	2.0	3,083	3,035	2,746	236	2,982
NC	3.9	9.5	1.6	0.1	966	849	882	74	956
ND	12.0	68.7	10.0	0.3	7,705	6,617	7,249	80	7,329
OH	19.5	101.0	24.0	1.5	10,767	9,012	11,652	261	11,913
OK	17.0	37.0	9.0	1.0	4,350	3,431	3,974	221	4,195
OR	24.0	132.0	40.0	3.0	11,845	10,207	11,310	793	12,103
PA	1.6	58.7	1.0	1.7	6,959	7,205	6,399	240	6,639
SD	51.0	256.5	60.0	0.5	28,596	27,951	32,646	154	32,800
TN	4.0	7.7	1.0	0.3	1,362	1,063	1,007	36	1,043
TX	116.0	390.0	57.0	2.0	51,626	49,771	57,893	336	58,229
UT	28.0	193.0	15.0	5.0	20,985	18,947	18,782	768	19,550
VA	2.5	28.5	1.0	1.0	3,856	3,546	2,933	206	3,139
WA	3.3	45.1	1.4	1.0	4,591	4,180	3,930	370	4,300
WV	5.0	21.3	2.0	0.2	2,142	1,957	2,223	34	2,257
WI	10.8	48.6	1.4	1.0	8,117	6,597	6,323	253	6,576
WY	51.0	233.0	34.0	1.0	28,501	27,311	27,574	313	27,887
Oth Sts[7]	8.5	42.0	21.0	4.0	4,782	3,384	4,298	421	4,719
US	695.0	4,200.5	1,494.6	64.8	464,503	412,333	520,998	11,464	532,462

[1] Includes custom slaughter for use on farms where produced and State outshipments, but excludes interfarm sales within the State.   [2] Excludes custom slaughter for farmers at commercial establishments.   [3] Adjustments made for changes in inventory and for inshipments.   [4] Receipts from marketings and sale of farm-slaughter.   [5] Cash receipts from sales of sheep, lambs, and mutton and lamb plus value of sheep and lambs slaughtered for home consumption.   [6] N ENG includes CT, ME, MA, NH, RI, and VT.   [7] AL, AK, AR, DE, FL, GA, HI, LA, MS, NJ, and SC.

NASS, Livestock Branch, (202) 720–3570.

Table 7-50.—Sheep and lambs: Number slaughtered commercially, total and average
live weight, by States, 2005 [1]

State	Number slaughtered	Total live weight	Average live weight
	Thousands	1,000 pounds	Pounds
AL			
AK			
AZ	1.4	166	117
AR	0.6	67	109
CA			
CO	1,019.7	158,289	155
DE and MD	35.4	3,622	102
FL	2.4	161	66
GA			
HI			
ID	4.0	489	123
IL	199.5	27,320	137
IN	24.9	2,706	109
IA	413.8	61,680	149
KS	2.7	274	102
KY	13.4	1,790	133
LA	1.7	157	92
MI	174.3	23,008	132
MN	2.8	354	126
MS			
MO	6.0	522	87
MT	3.6	433	121
NE	2.1	282	133
NV	1.3	134	104
N ENG [2]	26.8	2,474	92
NJ	114.5	9,449	82
NM	14.8	1,769	120
NY	38.5	3,184	83
NC	5.7	479	85
ND			
OH	10.5	1,145	109
OK	0.9	101	109
OR	19.0	2,566	135
PA	58.6	5,618	96
SC			
SD	10.3	1,277	124
TN	10.6	845	80
TX	69.3	8,457	122
UT	30.2	4,168	138
VA	11.0	1,154	105
WA	2.5	333	133
WV			
WI	10.5	1,408	134
WY	1.7	229	135
US [3]	2,697.8	373,305	138
PR	1.0		

[1] Includes slaughter in federally inspected and in other slaughter plants; exludes animals slaughtered on farms. Average live weight is based on unrounded numbers. Totals may not add due to rounding.　[2] CT, ME, MA, NH, RI, and VT.　[3] States with no data printed are still included in US total. Data are not printed to avoid disclosing individual operations.
NASS, Iowa Agricultural Statistics Service, (515) 284–4340.

Table 7-51.—Sheep and lambs: Number slaughtered, United States, 1996–2005

Year	Commercial			Farm	Total
	Federally inspected	Other	Total [1]		
	Thousands	Thousands	Thousands	Thousands	Thousands
1996	4,032	152	4,184	65	4,249
1997	3,771	137	3,907	62	3,969
1998	3,670	133	3,804	57	3,861
1999	3,556	145	3,701	65	3,766
2000	3,308	152	3,460	67	3,527
2001	3,065	157	3,222	68	3,290
2002	3,092	194	3,286	65	3,351
2003	2,805	174	2,979	64	3,042
2004	2,676	163	2,839	67	2,906
2005	2,554	143	2,698	64	2,762

[1] Totals are based on unrounded numbers.
NASS, Iowa Agricultural Statistics Service, (515) 284–4340.

Table 7-52.—Sheep and lambs: Number slaughtered, average dressed and live weights, percentage distribution, by class, Federally inspected, 1996–2005 [1]

Year	Federally inspected								
	Lambs and yearlings			Mature sheep			Total		
	Head	Pct. of total	Avg. dressed weight	Head	Pct. of total	Avg. dressed weight	Head	Avg. dressed weight	Avg. live weight
	1,000		Pounds	1,000		Pounds	1,000	Pounds	
1996	3,763	93.3	65	270	6.7	58	4,032	64	129
1997	3,558	94.3	67	213	5.7	60	3,771	67	134
1998	3,460	94.3	66	211	5.7	61	3,670	66	133
1999	3,369	94.7	67	188	5.3	59	3,556	67	134
2000	3,141	95.0	68	167	5.0	63	3,308	68	137
2001	2,921	95.3	71	144	4.7	62	3,065	70	142
2002	2,944	95.2	68	148	4.8	63	3,092	68	135
2003	2,662	94.9	68	143	5.1	66	2,805	68	136
2004	2,529	94.5	69	147	5.5	66	2,676	69	138
2005	2,425	94.9	71	129	5.1	69	2,554	70	140

[1] All percents and weights calculated using unrounded totals.
NASS, Iowa Agricultural Statistics Service, (515) 284–4340.

Table 7-53.—Sheep and lambs: Operations with sheep, 2004–2005

State	Operations with sheep	
	2004	2005
	Number	Number
AZ	260	250
CA	2,600	2,500
CO	1,600	1,600
ID	1,300	1,300
IL	1,900	2,000
IN	2,000	2,000
IA	4,200	4,200
KS	1,400	1,500
KY	1,200	1,300
MD	750	750
MI	2,000	2,000
MN	2,600	2,600
MO	1,900	1,900
MT	1,800	1,700
NE	1,500	1,500
NV	300	300
N ENG [1]	1,900	2,000
NM	800	800
NY	1,900	2,000
NC	800	850
ND	770	830
OH	3,100	3,300
OK	1,900	1,900
OR	3,100	3,200
PA	3,400	3,600
SD	2,100	2,100
TN	1,200	1,100
TX	7,200	6,900
UT	1,400	1,400
VA	1,500	1,600
WA	1,500	1,500
WV	1,100	1,000
WI	2,400	2,400
WY	900	900
Other States [2]	3,300	3,500
US	67,580	68,280
PR	700	800

[1] N Eng includes CT, ME, MA, NH, RI, and VT.   [2] Other States include AL, AK, AR, DE, FL, GA, HI, LA, MS, NJ, and SC.
NASS, Livestock Branch, (202) 720–3570.

Table 7-54.—Breeding Sheep: Survey percent by size groups, United States, 2005–2006 [1]

Item	1–99 head		100–499 head		500–4,999 head		5,000+ head	
	2005	2006	2005	2006	2005	2006	2005	2006
	Percent	Percent	Percent	Percent	Percent	Percent	Percent	Percent
Operations	92.0	90.8	6.5	7.6	1.4	1.5	0.1	0.1
Inventory	30.3	28.7	22.0	24.0	33.5	33.8	14.2	13.5

[1] Percents reflect distributions from annual survey.
NASS, Livestock Branch, (202) 720–3570.

Table 7-55.—Wool: Number of sheep shorn, weight per fleece, production, average price per pound received by farmers, value of production, exports, imports, total new supply of apparel wool, and imports of carpet wool, United States, 1995–2004

Year	Sheep and lambs shorn [1]	Weight per fleece	Shorn wool production	Price per pound [2]	Value of production [3]
	Thousands	Pounds	1,000 pounds	Cents	1,000 dollars
1995	8,138	7.80	63,513	1.04	64,277
1996	7,215	7.78	56,669	0.70	39,270
1997	6,960	7.70	53,578	0.84	44,909
1998	6,428	7.66	49,255	0.60	29,415
1999	6,158	7.57	46,572	0.38	17,860
2000	6,135	7.56	46,446	0.33	15,377
2001	5,689	7.56	43,016	0.36	15,311
2002	5,462	7.52	41,078	0.53	21,689
2003	5,074	7.55	38,299	0.73	28,126
2004 [4]	5,073	7.42	37,622	0.80	29,931

Year	Shorn wool production	Raw wool supply (clean)				Total new supply [9]
		Domestic production [5]	Exports [6]	Imports for consumption		
				48's and Finer [7]	Not Finer than 46's [8]	
	1,000 pounds	1,000 pounds	1,000 pounds	1,000 pounds	1,000 pounds	1,000 pounds
1995	63,513	33,535	6,042	63,781	25,039	116,313
1996	56,159	29,921	5,715	54,073	21,296	99,575
1997	53,578	28,289	4,732	51,484	24,962	100,003
1998	49,255	26,007	1,721	45,805	24,702	94,793
1999	46,549	24,575	3,694	21,264	21,810	63,955
2000	46,446	24,413	6,629	23,902	21,099	62,785
2001	43,016	22,712	6,154	15,843	19,727	52,128
2002	41,078	21,689	8,461	10,526	14,159	37,913
2003	38,299	20,222	14,678	4,986	15,749	26,279
2004 [4]	37,622	19,864	14,023	6,204	16,455	28,500

[1] Includes sheep shorn at commercial feeding yards. [2] Price computed by weighting State average prices for all wool sold during the year by sales of shorn wool. [3] Production by States multiplied by annual average price. [4] Preliminary. [5] Conversion factor from grease basis to clean basis are as follows: Shorn wool production—52.8 percent (Stat. Bull. 616) from 1987-1997. [6] Includes carpet wool exports. [7] Prior to 1989, known as dutiable imports. [8] Prior to 1989, known as duty-free imports. In 1994 includes 24,645,306 pounds of imported raw wool not finer than 46's and 2,182,576 pounds of miscellaneous imported raw wool. [9] Production minus exports plus imports; stocks not taken into consideration.

ERS, Field Crops Branch, (202) 694-5300 and NASS. Imports and exports from reports of the U.S. Department of Commerce.

Table 7-56.—Wool: Price-support operations, United States, 1996–2005

Marketing year	Shorn wool price per pound		Payment rate		Marketings covered by payments [2]		Amount of payments [3]		
	Support	Season average received by producers	Average for shorn wool per pound	Unshorn lambs per cwt. [1]	Shorn wool	Unshorn lambs	Shorn wool	Unshorn lambs	Total
	Cents	Cents	Cents	Cents	Million pounds	Million pounds	Million dollars	Million dollars	Million dollars
1996	([4])	70.0							
1997	([4])	84.0							
1998	([4])	60.0							
1999	([4])	38.0	([5]) 20	([5]) 80	43.5		8.7		8.7
2000	([4])	33.0	([6]) 40	([6])	43.9		33.6		33.6
2001	([4])	35.0							
2002	100	53.0	([7]) 18	19.0	37.1	6.0	6.7	1.1	7.8
2003	100	73.0	20	21	30.9	4.3	6.1	0.9	7.0
2004	100	80.0	21	20	29.1	5.6	6.2	1.4	7.6
2005	100								

[1] For unshorn lambs sold. [2] Unadjusted for weight of unshorn lambs purchased. [3] Payments for wool marketed during the year shown are made after Mar. 31 of the following year, and include deductions for the American Sheep Industry Association. Figures for 1994 and 1995 reflect 20- and 50-percent reductions in payments, respectively, as required by Public Law 103-130 whcih also terminated price support for wool as of December 31, 1995. [4] Public Law 103-130 terminated price support for wool as of December 31, 1995. [5] Market Loss Assistance Payments for 1999-crop wool authorized by Public Law 106-224 on June 21, 2000. [6] Market Loss Assistance Payments authorized by Public Law 106-387 for marketing year 2000 production. Additional payments authorized by Public Law 107-25. [7] Nonrecourse Marketing Assistance Loan Program authorized by Public Law 107-171, enacted May 13, 2002.

FSA, Fibers Analysis, (202) 720-7954.

Table 7-57.—Wool: Mill consumption, by grades, on the woolen and worsted systems, scoured basis, United States, 1995–2004 [1][2][3]

Item	1995	1996	1997	1998	1999	2000	2001	2002	2003	2004
Apparel wool:										
Woolen system:	Mil. lb.	Mil. lb.	Mil. lb.	Mil. lb.	Mil. lb.	Mil. lb.	Mil. lb.	Mil. lb.	Mil. lb.	Mil. lb.
60's and finer	30.2	42.1	49.0	31.3	18.4	18.5	16.1	9.6	6.7	6.4
Coarser than 60's	27.1	27.6	21.3	15.1	10.8	13.4	9.8	8.5	5.3	8.1
Total	57.3	69.7	70.3	46.4	29.2	31.9	25.9	18.1	12.0	14.5
Worsted system:										
60's and finer	55.0	46.1	48.2	42.2	27.4	NA	NA	NA	NA	NA
Coarser than 60's	17.0	13.8	11.9	9.8	7.0	NA	NA	NA	NA	NA
Total	72.0	59.8	60.1	52.0	34.4	30.1	27.1	17.9	NA	NA
Total apparel:										
60's and finer	85.2	88.2	97.2	75.5	45.8	NA	NA	NA	NA	NA
Coarser than 60's	44.1	41.3	33.2	24.9	17.7	NA	NA	NA	NA	NA
Total	129.3	129.5	130.4	98.4	63.5	63.0	53.0	36.0	NA	NA
Carpet wool	12.7	12.3	13.6	16.3	13.9	15.2	13.3	6.9	6.0	6.9
Grand total mill	142.0	141.8	144.0	114.7	77.5	77.2	66.3	42.9	NA	NA

[1] Scoured wool, plus greasy wool converted to a scoured basis, using assumed average yields. Includes both pulled and shorn, foreign and domestic wool. Wool was considered as consumed (1) on the woolen system when laid in mixes and (2) on the worsted system as the sum of top and noil production. [2] Domestic, duty-paid, and duty-free foreign. [3] Excludes wool consumed on the cotton system and in the manufacture of felt, hat bodies, and other miscellaneous products.

ERS, Field Crops Branch, (202) 694–5300. Compiled from reports of the U.S. Department of Commerce.

Table 7-58.—Wool: United States imports (for consumption), clean content, by grades, 1995–2004 [1][2]

Grade	1995	1996	1997	1998	1999	2000	2001	2002	2003	2004
48's and finer:	Mil. lb.	Mil. lb.	Mil. lb.	Mil. lb.	Mil. lb.	Mil. lb.	Mil. lb.	Mil. lb.	Mil. lb.	Mil. lb.
Finer than 58's [3]	56.9	50.8	48.7	42.8	19.9	22.2	14.2	9.2	4.5	5.1
48's–58's [4]	6.9	3.3	2.8	3.0	1.4	1.7	1.6	1.3	0.5	1.1
Total	63.8	54.1	51.5	45.8	21.3	23.9	15.8	10.5	5.0	6.2
Not Finer than 46's:										
Wool for special use [5]	8.6	3.9	3.0	2.2	2.0	2.2	2.4	1.3	1.4	2.8
Not finer than 40's [6]	3.2	3.2	9.0	7.8	6.3	6.5	6.5	3.9	5.3	4.4
Finer than 40's– 44's [7]	9.4	10.1	9.3	10.0	8.1	5.7	6.7	7.1	6.3	5.8
46's [8]	3.9	4.1	3.7	4.7	5.4	5.6	4.1	1.9	2.7	3.4
Total	25.0	21.3	24.9	24.7	21.8	20.0	19.7	14.2	15.7	16.5
Miscellaneous [9]	38.8	0	0	0	0	0	0	0	0	0
Grand total	88.8	75.4	76.4	70.5	43.1	45.0	35.6	24.7	20.7	22.7

[1] Natural fiber grown by sheep or lambs. [2] Beginning 1989 the following Harmonized Tariff Schedule numbers are in the above 7 wool import groups: 5101.19.606060, 5101.19.6060, 5101.21.4000, 5101.21.4000, 5101.29.4060, 0.5(5101.30.4000). [4] 5101.11.6030, 5101.19.6030, 5101.21.4030, 5101.29.4030, 0.5(5101.30.4000). [5] 5101.11.1000, 5101.19.1000, 5101.21.1000, 5101.29.1000. [6] 5101.11.2000, 5101.19.2000, 5101.21.1500, 5101.29.1500, 5101.31.1000. [7] 5101.11.4000, 5101.19.4000, 5101.21.3000, 5101.29.3000, 5101.30.1500. [8] 5101.11.5000, 5101.19.5000, 5101.21.3500, 5101.29.3500, 5101.30.3000. [9] 5101.21.6000, 5101.29.6000, 5101.30.6000. They include wool not carded or combed but processed beyond the scoured or carbonized condition, e.g. dyed. This wool is not identified by use or grade. In 1989 this quantity was 48,074 pounds, 1990 was 32,979 pounds, 1991 was 47,245 pounds, and 1992 was 25,728 pounds.

ERS, Field Crops Branch, (202) 694–5300. Compiled from reports of the U.S. Department of Commerce.

Table 7-59.—Wool: United States imports (for consumption), clean content, by country of origin, 1995–2004[1]

Country of origin	1995	1996	1997	1998	1999	2000	2001	2002	2003	2004[2]
48's and finer:	Mil. lb.	Mil. lb.	Mil. lb.	Mil. lb.	Mil. lb.	Mil. lb.	Mil. lb.	Mil. lb.	Mil. lb.	Mil. lb.
Argentina	0.1	0.1	0.5	0.6	0.1	0.1	....	....	....	....
Australia	51.6	47.1	44.2	38.8	17.6	20.2	12.7	8.1	3.6	4.2
Canada	1.1	1.2	1.0	1.1	0.7	0.8	0.8	0.8	0.3	0.6
Chile	0.7	0.2	0.1	0.2	....	....	....	....	....	....
New Zealand	2.6	1.9	1.6	1.5	1.3	1.3	1.0	0.5	0.4	0.4
South Africa	0.9	0.7	1.2	1.9	1.1	0.8	0.6	0.5	0.4	0.5
United Kingdom	1.5	0.3	0.2	0.1	....	....	....	....	....	....
Uruguay	2.5	1.6	2.1	0.7	0.2	0.1	0.3	0.3	0.1	0.2
Other	2.8	2.0	1.1	0.9	0.3	0.6	0.2	0.3	0.2	0.3
Total	63.8	54.1	51.5	45.8	21.3	23.9	15.8	10.5	5.0	6.2
Not finer than 46's:										
Argentina	0.1	0.2	0.1	0.5	0.4	0.5	0.3	0.4	0.6	0.4
Australia	0.3	0.4	0.1	0.2	0.1	0.6	0.4	0.4	0.1	0.7
Canada	0.6	0.1	0.6	0.2	0.2	0.2	0.1	0.1	0.1	0.1
New Zealand	17.0	15.6	18.9	18.5	16.9	15.0	14.9	10.1	11.7	12.1
Uruguay	....	....	....	....	....	....	....	....	....	....
South Africa	0.0	0.0	0.0	....	....	0.3	0.2	0.1	0.1	0.2
United Kingdom	5.7	3.5	3.7	3.6	2.9	3.2	2.6	2.7	2.5	2.7
Other	....	1.2	0.0	0.0	0.1	1.2	0.4	0.4	0.6	0.3
Total	25.0	21.0	24.3	23.1	20.6	21.0	18.7	14.2	15.7	16.5
Grand total	88.8	75.4	75.8	68.9	41.8	44.9	35.6	24.7	20.7	22.7

[1] Wool not advanced in any manner or by any process of manufacture beyond washed, scoured, or carbonized condition. ERS, Field Crops Branch, (202) 694–5300. Compiled from reports of the U.S. Department of Commerce.

Table 7-60.—Wool: Average price per pound, clean basis, delivered to United States mills, 1995–2004[1]

Year	Territory[2]		Australian 64's good topmaking (in bond, American yield)
	64's (20.60–22.04 microns)	Avg. 58's–56's (24.95–27.84 microns)	
	Cents	Cents	Cents
1995	249	170	281
1996	193	137	234
1997	238	153	257
1998	162	113	184
1999	110	70	148
2000	108	61	150
2001	121	72	166
2002	190	130	268
2003	241	164	314
2004	235	162	275

[1] Beginning January 1976 the unit designation terminology for wool prices changed to microns. For example 64's (20.60–22.04 microns) formerly was fine good French combing and staple. Two designations 56's (26.40–27.84 microns) and 58's (24.95–26.39 microns) have been averaged in the price data shown here and together were formerly the category fleece 3/8 blood good French combing and staple.　[2] Wool grown in the range areas of California, Oregon, Washington, Texas, the intermountain States (including Arizona and New Mexico), and parts of the Dakotas, Kansas, Nebraska, and Oklahoma. These wools vary considerably in shrinkage and color.

ERS, Field Crops Branch, (202) 694–5300 and AMS.

Table 7-61.—Wool: Number of sheep shorn, weight per fleece, production, by State and U.S., 2004-2005

State	Sheep and lambs shorn		Weight per fleece		Shorn wool production	
	2004	2005[1]	2004	2005[1]	2004	2005[1]
	1,000 head	*1,000 head*	*Pounds*	*Pounds*	*1,000 pounds*	*1,000 pounds*
AZ	90.0	95.0	6.4	5.9	580	560
CA	480.0	500.0	7.1	7.0	3,400	3,500
CO	390.0	420.0	6.6	6.3	2,570	2,650
ID	225.0	210.0	9.4	9.0	2,125	1,890
IL	60.0	57.0	6.8	6.8	405	385
IN	43.0	42.0	6.4	6.4	275	270
IA	230.0	220.0	6.0	5.8	1,370	1,270
KS	66.0	63.0	7.3	6.8	485	430
KY	20.0	18.0	6.8	6.7	135	120
MD	15.0	17.0	6.9	7.0	103	119
MI	76.0	81.0	5.8	5.9	440	480
MN	140.0	150.0	6.5	6.5	910	970
MO	61.0	67.0	6.1	6.5	375	435
MT	267.0	260.0	9.3	9.6	2,472	2,490
NE	83.0	85.0	7.2	7.1	600	600
NV	55.0	54.0	9.3	9.3	510	500
N ENG[2]	40.0	41.0	7.1	7.1	284	293
NM	140.0	155.0	7.3	7.4	1,020	1,150
NY	53.0	49.0	6.7	6.7	356	330
NC	10.0	7.0	6.4	6.1	64	43
ND	82.0	78.0	9.1	8.8	745	690
OH	140.0	133.0	6.5	6.0	903	800
OK	50.0	45.0	6.2	6.0	310	270
OR	173.0	190.0	6.3	6.3	1,090	1,190
PA	68.0	71.0	6.5	6.5	440	460
SD	345.0	330.0	7.6	7.8	2,610	2,582
TN	16.0	17.0	6.2	6.2	99	105
TX	810.0	800.0	6.9	6.9	5,600	5,550
UT	245.0	235.0	9.2	9.3	2,250	2,180
VA	36.0	37.0	6.3	6.0	226	223
WA	40.0	40.0	8.2	8.1	326	324
WV	24.0	23.0	5.6	5.8	134	133
WI	70.0	67.0	7.1	7.3	500	490
WY	390.0	380.0	9.3	9.3	3,640	3,530
Other States[3]	40.0	35.0	6.8	6.3	270	220
US	5,073.0	5,072.0	7.4	7.3	37,622	37,232

[1] Preliminary.   [2] N ENG includes CT, ME, MA, NH, RI, and VT.   [3] Other States include AL, AK, AR, DE, FL, GA, HI, LA, MS, NJ, and SC.

NASS, Livestock Branch, (202) 720-3570.

Table 7-62.—Wool: Price and value, by State and U.S., 2004–2005

State	Price per pound		Value [1]	
	2004	2005 [2]	2004	2005 [2]
	Dollars	Dollars	Dollars	1,000 dollars
AZ	0.30	0.30	174	168
CA	0.82	0.70	2,788	2,450
CO	0.85	0.63	2,185	1,670
ID	0.88	0.75	1,870	1,418
IL	0.26	0.35	105	135
IN	0.21	0.17	58	46
IA	0.34	0.27	466	343
KS	0.57	0.40	276	172
KY	0.31	0.50	42	60
MD	0.41	0.56	42	67
MI	0.45	0.39	198	187
MN	0.37	0.38	337	369
MO	0.40	0.37	150	161
MT	1.17	0.98	2,892	2,440
NE	0.43	0.40	258	240
NV	0.94	0.87	479	435
N ENG [3]	0.45	0.45	128	132
NM	1.00	1.20	1,020	1,380
NY	0.21	0.19	75	63
NC	0.38	0.49	24	21
ND	0.75	0.60	559	414
OH	0.34	0.28	307	224
OK	0.50	0.40	155	108
OR	0.65	0.64	709	762
PA	0.27	0.27	119	124
SD	0.72	0.60	1,879	1,549
TN	0.53	0.47	52	49
TX	1.02	0.96	5,712	5,328
UT	0.83	0.71	1,868	1,548
VA	0.40	0.42	90	94
WA	0.80	0.68	261	220
WV	0.43	0.48	58	64
WI	0.30	0.30	150	147
WY	1.17	1.00	4,259	3,530
Other States [4]	0.65	0.70	176	154
US	0.80	0.71	29,921	26,272

[1] Production multiplied by marketing year average price. [2] Preliminary. [3] N ENG includes CT, ME, MA, NH, RI, and VT. [4] Other States include AL, AK, AR, DE, FL, GA, HI, LA, MS, NJ, and SC.
NASS, Livestock Branch, (202) 720–3570.

Table 7-63.—Mohair: Price-support operations, United States, 1996–2005

Marketing year begining January 1	Price per pound		Average payment rate per pound	Marketings covered by payments	Amount of payments[2]
	Loans[1]	Season average received by producers			
	Cents	Cents	Cents	Million pounds	Million dollars
1996	(3)	192.0			
1997	(3)	225.0			
1998	(3)	249.0			
1999	200.0	346.0	(4)40.0	(4) 3.9	(4) 1.6
2000	200.0	404.0	(5)40.0	(5) 2.7	(5) 1.1
2001	(3)	213.0			
2002	420.0	158.0	214	(6) 2.2	(6) 4.8
2003	420.0	166.0	208	2.0	4.2
2004	420.0	197.0	172	1.7	3.0
2005	420.0				

[1] The national average loan rate was also known as the price-support rate prior to enactment of the Farm Security and Riral Investment Act of 2002. [2] Payments for mohair marketed during the year shown are made after Mar. 31 of the following year and includes deductions for the Mohair Council of America. Figures for 1994 and 1995 relect 25- and 50-percent reductions in payments, respectively, as required by Public Law 103-130 which also terminated price support for mohair as of December 31, 1995. [3] No program. [4] Recourse loan program authorized by P.L. 105-277 for fiscal year 1999. No payments were involved. Market Loss Assistance Payments for 1999 crop mohair authorized by P.L. 106-224 on June 21, 2000. [5] Recourse loan program authorized by P.L. 106-78 for fiscal year 2000. No payments were involved. Market Loss Assistance Payments authorized by P.L. 106-387 for marketing year 2000 production. Additional payments authorized by P.L. 107-25. [6] Nonrecourse Marketing Assistance Loan Program authorized by Public Law 107-171, enacted May 12, 2002.

FSA, Fibers Analysis, (202) 720–7954.

Table 7-64.—Mohair: Goats clipped, production, price, and value by selected States, United States, 2004–2005

State	Goats clipped		Average clip per goat		Production		Price per pound		Value[1]	
	2004	2005	2004	2005	2004	2005	2004	2005	2004	2005
	Head	Head	Pounds	Pounds	1,000 pounds	1,000 pounds	Dollars	Dollars	1,000 dollars	1,000 dollars
AZ	25,000	25,000	5.0	4.8	125	120	1.10	1.10	138	132
CA	2,500	3,000	8.0	9.0	20	27	1.70	3.00	34	81
NM	10,000	10,000	6.4	7.0	64	70	1.10	3.25	70	228
TX	210,000	210,000	7.7	7.0	1,620	1,470	2.10	3.00	3,402	4,410
Oth Sts[2]	25,000	25,000	5.0	5.4	125	135	1.50	1.65	188	223
US	272,500	273,000	7.2	6.7	1,954	1,822	1.96	2.78	3,832	5,074

[1] Production multiplied by marketing year average price. U.S. value is summation of State values. [2] Other States include AL, AK, AR, CO, CT, DE, FL, GA, HI, ID, IL, IN, IA, KS, KY, LA, ME, MD, MA, MI, MN, MS, MO, MT, NE, NV, NH, NJ, NY, NC, ND, OH, OK, OR, PA, RI, SC, SD, TN, UT, VT, VA, WA, WV, WI, and WY.

NASS, Livestock Branch, (202) 720–3570.

Table 7-65.—Red meat: Production, by class of slaughter, United States, 1996–2005

Year	Commercial			Farm	Total	Commercial			Farm	Total
	Federally inspected	Other	Total[1]			Federally inspected	Other	Total[1]		
	Beef					Pork, excluding lard				
	Million pounds	Million pounds	Million pounds	Million pounds	Million pounds	Million pounds	Million pounds	Million pounds	Million pounds	Million pounds
1996	24,948	473	25,421	106	25,527	16,780	304	17,084	32	17,116
1997	24,964	420	25,384	106	25,490	16,962	283	17,245	30	17,275
1998	25,264	389	25,653	107	25,760	18,692	289	18,981	30	19,011
1999	25,998	387	26,385	107	26,492	18,977	301	19,278	28	19,306
2000	26,405	371	26,776	111	26,887	18,672	257	18,929	24	18,953
2001	25,743	365	26,108	105	26,213	18,899	240	19,139	22	19,161
2002	26,714	377	27,091	102	27,193	19,437	227	19,664	21	19,685
2003	25,880	358	26,238	101	26,340	19,739	207	19,946	21	19,967
2004	24,189	358	24,547	102	24,649	20,325	186	20,511	20	20,531
2005	24,328	354	24,682	101	24,784	20,506	179	20,685	21	20,706
	Veal					Lamb and Mutton				
	Million pounds	Million pounds	Million pounds	Million pounds	Million pounds	Million pounds	Million pounds	Million pounds	Million pounds	Million pounds
1996	355	13	368	11	379	258	7	265	4	269
1997	314	9	323	10	333	250	7	257	3	260
1998	243	8	251	10	261	242	7	249	3	252
1999	217	7	224	9	233	236	7	243	4	247
2000	205	10	215	10	225	224	8	232	4	236
2001	188	6	194	10	204	216	8	224	4	228
2002	190	6	196	9	205	209	9	218	4	222
2003	185	7	192	10	201	191	9	200	4	204
2004	162	5	167	9	176	185	9	194	5	199
2005	152	4	156	9	165	180	7	187	4	191
	All meat, excluding lard									
	Million pounds	Million pounds	Million pounds	Million pounds	Million pounds					
1996	42,340	798	43,138	153	43,291					
1997	42,491	718	43,209	149	43,358					
1998	44,441	692	45,133	150	45,283					
1999	45,428	702	46,130	148	46,278					
2000	45,506	645	46,151	149	46,300					
2001	45,045	619	45,664	141	45,805					
2002	46,549	620	47,169	137	47,305					
2003	45,995	581	46,576	136	46,712					
2004	44,861	557	45,418	136	45,554					
2005	45,166	545	45,711	135	45,845					

[1] Totals are based on unrounded data.
NASS, Iowa Agricultural Statistics Service, (515) 284–4340.

STATISTICS OF CATTLE, HOGS, AND SHEEP

Table 7-66.—Meat: Production by types in specified countries, 2003 and 2004 [1]

Country	Beef and veal		Pork [3]		Total production	
	2003	2004 [2]	2003	2004 [2]	2003	2004 [2]
	1,000 metric tons	1,000 metric tons	1,000 metric tons	1,000 metric tons	1,000 metric tons	1,000 metric tons
Argentina	2,800	3,130	NA	NA	2,800	3,130
Australia	2,073	2,114	419	394	2,492	2,508
Brazil	7,385	7,975	2,560	2,600	9,945	10,575
Bulgaria	NA	NA	142	91	142	91
Canada	1,190	1,496	1,882	1,936	3,072	3,432
China, Peoples Rep.	6,305	6,759	45,186	47,016	51,491	53,775
Egypt	440	455	NA	NA	440	455
European Union - 25	8,061	7,941	21,150	20,851	29,211	28,792
Hong Kong	13	14	145	161	158	175
India	1,960	2,130	NA	NA	1,960	2,130
Japan	496	513	1,260	1,271	1,756	1,784
Korea, Republic of	182	186	1,149	1,100	1,331	1,286
Mexico	1,950	2,099	1,100	1,150	3,050	3,249
New Zealand	693	720	NA	NA	693	720
Philippines	230	230	1,145	1,145	1,375	1,375
Romania	190	193	420	470	610	663
Russian Federation	1,670	1,590	1,710	1,725	3,380	3,315
Singapore	NA	NA	19	NA	19	0
South Africa, Rep.	613	655	NA	NA	613	655
Taiwan	6	5	893	898	899	903
Turkey	635	625	NA	NA	635	625
Ukraine	611	613	630	558	1,241	1,171
United States	12,039	11,261	9,056	9,312	21,095	20,573
Uruguay	450	544	NA	NA	450	544
Total meat	49,992	51,248	88,959	90,678	138,951	141,926

[1] Carcass weight equivalent: excludes offals, rabbit, and poultry meat.   [2] Preliminary.   [3] Includes edible pork fat, but excludes lard and inedible greases.   NA-not available.

FAS, Dairy, Livestock and Poultry Division, (202) 720–8031. Updated data available at http://www.fas.usda.gov/psd.

Table 7-67.—Meat: United States exports by type of product, 1995–2004 [1]

Year	Beef and veal			Lamb and mutton, fresh or frozen	Pork			Variety meats, fresh, chilled, or frozen	Other meats	Total
	Fresh and chilled	Frozen	Prepared and preserved		Fresh and chilled	Frozen	Prepared and preserved			
	Metric tons	Metric tons	Metric tons	Metric tons	Metric tons	Metric tons	Metric tons	Metric tons	Metric tons	Metric tons
1995	262,381	319,416	13,651	2,511	133,101	79,155	61,286	466,213	353,207	1,690,921
1996	273,276	324,329	14,577	2,478	125,220	72,650	58,336	495,343	434,759	1,800,967
1997	316,534	359,460	15,227	2,545	126,061	72,903	65,000	469,789	435,258	1,862,775
1998	346,403	352,050	17,966	2,528	146,965	76,230	70,227	495,643	423,980	1,931,993
1999	370,184	414,458	19,323	2,219	188,556	84,207	77,638	524,325	455,561	2,136,470
2000	395,588	417,538	21,791	2,184	229,395	92,672	91,446	601,738	503,942	2,356,294
2001	393,105	362,972	23,932	2,770	240,275	83,724	84,687	685,063	513,969	2,390,496
2002	407,599	393,836	27,232	3,042	276,639	91,379	90,423	592,185	619,491	2,501,826
2003	430,071	390,543	37,572	2,909	293,169	107,964	107,749	598,726	567,565	2,536,267
2004 [2]	114,966	20,458	9,068	3,671	266,583	96,840	109,915	456,480	189,127	1,267,108

[1] Product weight equivalent.   [2] Preliminary.
FAS, Dairy, Livestock and Poultry Division, (202) 720–8031. Updated data available at http://www.fas.usda.gov/ustrade.

Table 7-68.—Meat: United States exports and imports into the United States, carcass weight equivalent, 1997–2006 [1]

Year	Exports				Imports			
	Beef and veal	Lamb and mutton	Pork [2]	All meat	Beef & veal	Lamb and mutton	Pork [2]	All meat
	Million pounds	Million pounds	Million pounds	Million pounds	Million pounds	Million pounds	Million pounds	Million pounds
1997	2,136	5	1,044	3,185	2,343	83	634	3,061
1998	2,171	6	1,230	3,407	2,643	112	705	3,461
1999	2,412	5	1,283	3,700	2,873	112	827	3,813
2000	2,468	5	1,287	3,760	3,032	130	965	4,127
2001	2,269	7	1,559	3,835	3,163	146	951	4,260
2002	2,448	7	1,612	4,067	3,218	160	1,071	4,448
2003	2,518	7	1,717	4,242	3,006	168	1,185	4,359
2004	460	9	2,181	2,650	3,679	180	1,099	4,959
2005 [3]	644	9	2,683	3,336	3,587	179	1,002	4,768
2006 [4]	680	8	2,785	3,473	3,560	175	960	4,695

[1] Carcass weight equivalent of all meat, including the meat content of minor meats and of mixed products. Includes shipments to U.S. Territories are included in domestic consumption.   [2] The pork series has been revised to a dressed weight equivalent rather than "Pork, excluding lard."   [3] Preliminary.   [4] Forecast.

ERS, Market and Trade Economics Division, Animal Products Branch, (202) 694–5180. Data on imports and commercial exports are computed from records of the U.S. Department of Commerce, those on exports by the U.S. Department of Agriculture are separately estimated from deliveries and stocks.

Table 7-69.—Meat: United States imports, by country of origin, 2004[1][2]

Country of origin	Beef and veal			Lamb, mutton, and goat, except canned	Pork			Variety meats, fresh, chilled and frozen	Other livestock meats n.s.e.	Total
	Fresh	Frozen	Other prepared or preserved		Fresh and chilled	Frozen	Other prepared or preserved			
	Metric tons	Metric tons	Metric tons	Metric tons	Metric tons	Metric tons	Metric tons	Metric tons	Metric tons	Metric tons
Canada	330,345	25,085	1,040	153	266,195	54,129	66,791	21,532	4,336	779,993
Australia	25,030	347,549	575	55,727	0	9	0	7,467	233	436,013
New Zealand	5,642	206,798	2,233	26,940	0	0	1	1,834	1,095	245,539
European Union ...	0	0	3	0	24	55,416	28,504	507	4,451	89,653
Brazil	0	0	53,512	0	0	0	0	0	219	53,731
Argentina	0	25	28,138	0	0	0	0	0	22	28,185
Uruguay	14,337	113,868	4,487	0	0	0	1	1,836	17	132,966
Nicaragua	5,396	16,426	0	0	0	0	0	398	0	22,306
Costa Rica	2,520	5,364	0	0	0	0	0	65	4	8,146
Mexico	4,428	1,501	1,764	0	364	346	1,543	103	1,007	10,989
China	0	0	0	0	0	0	0	0	1,897	1,897
Honduras	58	1,223	0	0	0	0	1	2	0	1,284
Other	2	103	371	110	(0)	15	(0)	34	10,519	7,061
Total	387,758	717,942	92,123	82,930	266,583	109,915	96,840	33,776	23,797	1,811,664

[1] Preliminary. [2] Product weight equivalent.
FAS, Dairy, Livestock and Poultry Division, (202) 720–8031. Updated data available at http://www.fas.usda.gov/ustrade.

Table 7-70.—Meat: United States imports by type of product, 1995–2004[1]

Year	Beef and Veal			Lamb, mutton, and goat, except canned	Pork			Variety meats, fresh or frozen	Other livestock meats n.s.e.	Total
	Fresh	Frozen	Other prepared or preserved		Fresh	Frozen	Other prepared or preserved			
	Metric tons	Metric tons	Metric tons	Metric tons	Metric tons	Metric tons	Metric t ons	Metric tons	Metric tons	Metric tons
1995 ..	175,540	466,378	65,399	29,844	133,101	61,286	79,155	26,081	12,539	1,049,324
1996 ..	227,874	412,805	66,719	33,009	125,220	58,336	72,650	32,579	13,744	1,042,934
1997 ..	262,985	469,949	63,181	37,848	126,061	65,000	72,903	44,317	14,215	1,156,457
1998 ..	295,820	527,063	68,884	51,630	146,965	70,227	76,230	47,031	13,058	1,296,907
1999 ..	337,899	542,524	82,669	50,209	188,556	77,638	84,207	51,640	13,625	1,428,966
2000 ..	336,117	609,083	73,749	59,968	229,395	91,446	92,672	57,388	14,281	1,564,099
2001 ..	368,529	618,897	73,713	66,785	240,275	84,687	83,724	62,541	16,723	1,615,873
2002 ..	400,484	586,500	84,640	73,863	276,639	90,423	91,379	55,384	19,401	1,678,713
2003 ..	285,772	612,569	85,439	77,546	293,169	107,749	107,964	47,688	21,367	1,639,263
2004[2]	387,758	717,942	92,123	82,930	266,583	109,915	96,840	33,776	23,797	1,811,664

[1] Product weight equivalent. [2] Preliminary.
FAS, Dairy, Livestock and Poultry Division, (202) 720–8031. Updated data available at http://www.fas.usda.gov/ustrade.

Table 7-71.—Meat: International trade, selected countries, 2002–2004 [1]

Continent and country	2002		2003		2004 [2]	
	Exports	Imports	Exports	Imports	Exports	Imports
	1,000 metric tons	*1,000 metric tons*	*1,000 metric tons*	*1,000 metric tons*	*1,000 metric tons*	*1,000 metric tons*
Argentina	348	10	386	12	623	5
Australia	1,444	60	1,338	74	1453	87
Brazil	1,471	78	1,778	63	2249	53
Bulgaria	0	40	0	41	0	33
Canada	1,474	398	1,359	364	1531	216
China, Peoples Rep.	260	161	325	161	444	97
Egypt	0	162	0	93	0	114
El Salvador	0	16	0	0	0	0
European Union	1,744	518	1,762	539	1794	603
Hong Kong	0	346	0	383	0	81
India	417	0	439	0	0	499
Japan	0	1,840	0	1,943	0	1949
Korea, Republic	16	586	17	598	10	438
Mexico	71	814	60	741	70	745
New Zealand	505	21	578	13	606	12
Philippines	0	157	0	142	52	212
Romania	0	89	0	106	6	185
Russian Federation	6	1,460	6	1,270	29	1359
Singapore	2	40	0	40	0	0
South Africa, Rep.	11	17	7	14	12	22
Taiwan	0	117	0	145	0	141
Ukraine	147	5	180	12	116	57
United States	1,841	1,945	1,922	1,901	1198	2168
Uruguay	259	2	320	70	410	1
Total	10,016	8,882	10,477	8,725	10,603	9,077

[1] Carcass weight equivalent. Excludes fat, offals, and live animals. [2] Preliminary.
FAS, Dairy, Livestock and Poultry Division, (202) 720–8031. Updated data available at http://www.fas.usda.gov/psd.

Table 7-72.—Meats and lard: Production and consumption, United States, 1997–2006 [1]

Year	Beef			Veal			Lamb and mutton		
	Produc-tion	Consumption		Produc-tion	Consumption		Produc-tion	Consumption	
		Total	Per capita		Total	Per capita		Total	Per capita
	Million pounds	*Million pounds*	*Pounds*	*Million pounds*	*Million pounds*	*Pounds*	*Million pounds*	*Million pounds*	*Pounds*
1997	25,490	25,611	93.8	334	333	1.2	260	332	1.2
1998	25,760	26,305	95.2	262	265	1.0	251	360	1.3
1999	26,493	26,936	96.4	235	235	0.8	248	358	1.3
2000	26,888	27,338	96.8	225	225	0.8	234	354	1.3
2001	26,212	27,025	94.7	205	204	0.7	227	368	1.3
2002	27,192	27,877	96.7	205	204	0.7	223	381	1.3
2003	26,339	27,000	92.8	202	204	0.7	203	367	1.3
2004	24,650	27,750	94.4	176	177	0.6	200	372	1.3
2005 [2]	24,767	28,737	93.4	164	164	0.6	193	356	1.2
2006 [3]	25,952	28,867	96.3	184	180	0.6	209	374	1.2

Year	Pork			All meats			Lard		
	Produc-tion	Consumption		Produc-tion	Consumption		Produc-tion	Consumption	
		Total	Per capita		Total	Per capita		Total	Per capita
	Million pounds	*Million pounds*	*Pounds*	*Million pounds*	*Million pounds*	*Pounds*	*Million pounds*	*Million pounds*	*Pounds*
1997	17,274	16,823	61.6	43,358	43,099	157.9	NA	NA	NA
1998	19,010	18,308	66.3	45,283	45,238	163.8	NA	NA	NA
1999	19,308	18,948	67.8	46,284	46,477	166.4	NA	NA	NA
2000	18,952	18,642	66.0	46,299	46,559	164.9	NA	NA	NA
2001	19,160	18,493	64.8	45,804	46,090	161.5	NA	NA	NA
2002	19,685	19,146	66.4	47,305	47,608	165.2	NA	NA	NA
2003	19,966	19,436	66.8	46,710	47,006	161.5	NA	NA	NA
2004	20,529	19,537	66.1	45,555	47,735	162.4	NA	NA	NA
2005 [2]	20,727	19,054	64.2	45,851	47,311	159.4	NA	NA	NA
2006 [3]	21,145	19,310	64.4	47,486	48,731	162.6	NA	NA	NA

[1] Carcass weight equivalent or dressed weight. Beginning 1977, pork production was no longer reported as "pork, excluding lard." This series has been revised to reflect pork production in prior years on a dressed weight basis that is comparable with the method used to report beef, veal, and lamb and mutton. Edible offals are excluded. Shipments to the U.S. territories are included in domestic consumption.   [2] Preliminary.   [3] Forecast.   NA-not available.

ERS, Animal Products, (202) 694–5180.

Table 7-73.—Hides and skins: United States imports by country of origin, 2000–2004

Country of origin	2000	2001	2002	2003	2004[1]
	1,000 pieces	1,000 pieces	1,000 pieces	1,000 pieces	1,000 pieces
Cattle and buffalo hides:					
Canada	1,876	1,615	1,227	1,056	1,187
Mexico	48	71	56	85	94
EU	11	5	3	4	24
Dominican Republic	NA	NA	NA	1	8
New Zealand	0	7	1	0	2
Australia	0	7	0	1	1
Others	37	5	11	6	1
Total	1,972	1,710	1,298	1,153	1,317
Calf and kip:					
Canada	57	141	322	269	150
Mexico	0	12	15	21	11
Costa Rica	0	0	0	0	8
Brazil	0	2	3	20	3
Others	51	2	44	16	3
Total	108	157	384	326	175
Goat and kid:					
Canada	15	10	37	31	37
Cote d'Ivoire	0	1	2	1	4
New Zealand	2	0	0	1	1
Saudi Arabia	0	0	72	24	0
Mali	59	50	1	2	0
China, Peoples Rep.	10	50	0	0	0
Others	23	43	3	4	1
Total	109	155	115	63	43
Sheep and lambs:					
New Zealand	586	443	432	469	407
Canada	97	265	216	186	263
Saudi Arabia	48	93	43	96	216
Eritrea	12	12	21	0	24
EU	151	144	38	15	20
Argentina	0	0	0	16	9
Iceland	0	0	0	1	5
Australia	19	28	8	11	4
Pakistan	0	0	1	1	2
China, Peoples Rep.	0	1	43	88	0
Others	80	102	5	7	0
Total	993	1,088	807	890	950

[1] Preliminary.
FAS, Dairy, Livestock, and Poultry Division, (202) 720–8031. Updated data available at http://www.fas.usda.gov/ustrade.

Table 7-74.—Hides, packer: Average price per hundred pounds, Central U.S., 1996–2005

Year	Steers					Heifers		
	Heavy native	Light native	Heavy Texas	Butt branded	Colorado branded	Heavy native	Light native	Branded
	Dollars	Dollars	Dollars	Dollars	Dollars	Dollars	Dollars	Dollars
1996	87.62	....	63.76	79.75	73.72	92.15	....	82.27
1997	87.66	....	64.60	80.03	77.00	90.99	....	82.39
1998	76.39	....	49.65	62.14	56.54	75.45	....	63.12
1999	72.36	....	....	64.28	60.83	73.80	....	67.25
2000	80.17	....	....	73.67	71.24	83.41	....	77.54
2001	85.84	....	....	79.79	75.90	85.52	....	85.44
2002	82.25	....	....	75.97	71.07	85.73	....	78.75
2003	83.83	....	....	78.58	73.29	88.34	....	80.20
2004[1]	67.09	....	64.91	64.39	61.48	57.07	....	54.02
2005[1]	65.64	....	63.50	63.53	60.90	57.89	....	54.20

[1] Effective 2004, price is per piece not per hundred pounds.
AMS, Livestock & Grain Market News, (202) 720–7316.

Table 7-75.—Hides and skins: United States exports by country of destination, 2000–2004

Country of destination	2000	2001	2002	2003	2004 [1]
	1,000 pieces	*1,000 pieces*	*1,000 pieces*	*1,000 pieces*	*1,000 pieces*
Cattle and buffalo hides:					
China, Peoples Rep.	3,642	5,413	5,468	5,991	6,944
Korea, Republic of	8,470	7,981	5,817	4,908	4,265
Taiwan	2,754	2,756	2,228	2,011	1,842
Hong Kong	609	1,504	1,693	2,576	1,888
Mexico	2,496	2,329	1,529	1,439	1,672
Thailand	499	863	905	800	684
EU	2,094	2,247	1,704	1,075	649
Japan	1,790	1,573	604	742	635
Canada	977	794	839	566	357
South Africa	18	9	0	4	248
Others	371	741	271	518	523
Total	23,720	26,210	21,058	20,630	19,707
Sheep and lamb skins:					
Turkey	1,450	1,609	2,557	2,885	3,504
Russia	0	0	0	12	128
China, Peoples Rep.	66	18	37	104	58
Mexico	288	98	72	34	58
Canada	86	42	34	35	40
Bangladesh	0	0	0	0	16
EU	315	403	184	152	15
Dominican Republic	0	0	0	12	10
Hong Kong	112	67	14	5	4
Korea, Republic of	26	35	151	23	0
Others	62	134	7	12	5
Total	2,404	2,404	3,055	3,274	3,838
Calf and kip skins:					
China, Peoples Rep.	58	69	937	1,355	1,259
Korea, Republic of	976	424	1,138	1,257	1,021
Japan	172	136	760	798	871
EU	486	799	1,097	1,148	762
Taiwan	31	17	928	612	523
Hong Kong	45	99	149	443	530
Mexico	243	117	627	160	301
Thailand	24	1	127	198	200
Brazil	12	125	211	208	61
India	NA	NA	50	23	42
Others	144	204	169	203	200
Total	2,192	1,990	6,193	6,405	5,770

[1] Preliminary.

FAS, Dairy, Livestock, and Poultry Division, (202) 720–8031. Updated data available at http://www.fas.usda.gov/ustrade.

Table 7-76.—Hides and skins: United States imports and exports, 2000–2004

Year	Imports				Exports		
	Calf and kip	Cattle and buffalo	Goat and kid	Sheep and lamb	Calf and kip	Cattle and buffalo	Sheep and lamb
	1,000 pieces	*1,000 pieces*	*1,000 pieces*	*1,000 pieces*	*1,000 pieces*	*1,000 pieces*	*1,000 pieces*
2000	108	1,972	109	993	2,192	23,720	2,404
2001	157	1,710	155	1,088	1,990	26,210	2,404
2002	384	1,298	115	807	6,193	21,058	3,055
2003	326	1,153	63	890	6,405	20,630	3,274
2004 [1]	175	1317	43	950	5,770	19,707	3,838

[1] Preliminary.

FAS, Dairy, Livestock, and Poultry Division, (202) 720–8031. Updated data available at http://www.fas.usda.gov/ustrade.

Table 7-77.—Mink: Farms, pelts produced and value of mink pelts, United States, 1995–2004

Year	Mink farms	Pelts produced	Average marketing price	Value of mink pelts
	Number	Thousand	Dollars	Million dollars
1995	478	2,803	53.10	148.8
1996	449	2,783	35.30	98.2
1997	452	2,993	33.10	99.1
1998	438	2,938	24.80	72.9
1999	398	2,813	33.70	94.8
2000	350	2,666	34.00	90.6
2001	329	2,565	33.50	85.9
2002	324	2,607	30.60	79.8
2003	305	2,549	40.10	102.2
2004 [1]	296	2,563	48.40	124.0

[1] Preliminary.

NASS, Livestock Branch, (202) 720–3570.

Table 7-78.—Mink pelts: Number produced by color class, major States, and United States, 2004

State	Black	Demi wild	Pastel	Sapphire	Blue Iris	Mahogany	Pearl
	Number	Number	Number	Number	Number	Number	Number
ID	62,200	....	....	16,800	12,400	62,300	....
IL	46,900	....	....	....	....	....	....
IA	74,900	....	....	....	7,300	....	....
MI	....	....	....	....	....	....	....
MN	37,700	48,900	....	2,500	31,000	81,900	....
OH	40,900	....	....	6,300	....	18,000	....
OR	102,200	....	....	13,700	97,500	32,000	....
PA	18,200	....	....	11,700	13,000	....	....
SD	....	....	....	....	....	....	....
UT	245,000	37,000	....	....	7,500	210,000	....
WA	42,000	....	....	....	40,000	....	....
WI	450,000	....	7,000	55,300	81,800	94,000	....
Other States [1]	35,800	70,100	32,100	29,800	9,100	70,100	65,500
US	1,155,800	156,000	39,100	136,100	299,600	568,300	65,500

State	Lavender	Violet	White	Miscellaneous and unclassified	Total pelts
	Number	Number	Number	Number	Number
ID	....	....	....	....	174,000
IL	....	....	....	....	59,700
IA	....	....	....	....	132,700
MI	....	....	....	....	50,500
MN	....	....	700	13,900	220,600
OH	....	....	....	....	70,400
OR	....	....	....	....	247,100
PA	....	....	....	....	55,900
SD	....	....	....	....	67,200
UT	....	....	500	....	580,000
WA	....	....	....	....	97,500
WI	....	....	52,600	....	768,000
Other States [1]	4,600	21,600	38,200	10,600	39,500
US	4,600	22,300	105,200	10,600	2,563,100

[1] "Other States" include some pelts from the above listed States which were not published to avoid disclosing individual operations. Published color classes may not add to the State total.

NASS, Livestock Branch, (202) 720–3570.

Table 7-79.—Livestock: Number of animals slaughtered under Federal inspection and number of whole carcasses condemned, 1996–2005

Year	Cattle		Calves		Sheep and lambs	
	Total head	Condemned[1]	Total head	Condemned[1]	Total head	Condemned[1]
	1,000	*1,000*	*1,000*	*1,000*	*1,000*	*1,000*
1996	37,574	181.5	1,717	28.8	4,271	11.0
1997	35,859	176.2	1,583	27.1	3,747	8.8
1998	33,280	157.1	1,447	28.7	3,455	7.3
1999	33,680	155.3	1,368	26.6	3,563	6.5
2000	35,136	188.9	1,103	22.4	3,316	5.8
2001	37,641	198.2	1,333	25.2	3,463	5.6
2002	31,404	165.9	1,034	19.5	2,922	5.4
2003	NA	NA	NA	NA	NA	NA
2004	31,515	159.7	876	15.2	2,679	4.9
2005	31,847	145.8	757	12.1	2,582	5.4

Year	Goats		Hogs		Horses	
	Total head	Condemned[1]	Total head	Condemned[1]	Total head	Condemned[1]
	1,000	*1,000*	*1,000*	*1,000*	*1,000*	*1,000*
1996	417	2.3	93,182	320.7	112	0.6
1997	374	1.9	78,497	308.9	88	0.4
1998	396	1.5	93,259	395.9	71	0.5
1999	463	2.5	105,755	460.8	62	0.4
2000	530	1.2	93,385	410.8	50	0.3
2001	592	1.1	96,600	449.9	62	0.2
2002	553	1.0	89,855	379.0	43	0.2
2003	NA	NA	NA	NA	NA	NA
2004	582	1.2	98,416	391.2	58.7	0.1
2005	553	1.1	103,849	414.8	88	0.1

[1] Condemnations include ante-mortem and post-mortem inspection. Condemnations are for the fiscal year ending September 30. Data reported by Food Safety and Inspection Service, USDA.
NASS, Iowa Agricultural Statistics Service, (515) 284–4340.

Table 7-80.—Livestock: Number and value, United States, Jan. 1, 2004–2006

Class of livestock and poultry	Number			Value					
				Per head[2]			Total		
	2004	2005	2006[1]	2004	2005	2006[1]	2004	2005	2006[1]
	Thou-sands	*Thou-sands*	*Thou-sands*	*Dollars*	*Dollars*	*Dollars*	*1,000 dollars*	*1,000 dollars*	*1,000 dollars*
Cattle	94,888	95,438	97,102	818.00	916.00	1,008.00	77,594,700	87,385,945	97,872,945
Hogs[3]	60,444	60,975	61,197	67.00	102.00	92.00	4,024,949	6,211,851	5,612,057
Sheep and lambs	6,105	6,135	6,230	119.00	130.00	141.00	723,785	799,288	875,480
Angora goats[4]	260	246	246	63.30	67.20	76.10	16,460	16,540	18,673
Total[5]	....	....	....	....	....	....	82,359,894	88,833,624	104,360,482
Chickens[3]	449,764	453,599	452,816	2.48	2.48	2.50	1,116,273	1,122,923	1,133,558
Total[6]	....	....	....	....	....	....	83,476,167	89,956,547	105,512,713

[1] Preliminary. [2] Based on reporters' estimates of average price per head in their localities. [3] Dec. 1 of preceding year. [4] AZ, NM, and TX only for 2004. [5] Cattle, hogs, sheep, and goats. [6] Includes all cattle, hogs, sheep, goats, and chickens (excluding broilers).
NASS, Livestock Branch, (202) 720–3570.

Table 7-81.—Livestock: Average price per 100 pounds received by farmers, by States, 2003 and 2004

State	Cows[1]		Steers and heifers		Beef cattle[2]		Calves	
	2003	2004	2003	2004	2003	2004	2003	2004
	Dollars	*Dollars*	*Dollars*	*Dollars*	*Dollars*	*Dollars*	*Dollars*	*Dollars*
AL	36.80	46.60	82.70	96.40	68.60	80.70	96.10	125.00
AK	70.00	80.00	90.00	100.00	80.00	90.00	95.00	100.00
AZ	42.50	50.60	83.50	102.00	83.30	101.00	99.50	120.00
AR	39.00	46.10	82.80	97.90	70.10	82.60	94.20	117.00
CA	41.10	47.20	83.20	88.30	62.00	68.70	96.60	111.00
CO	45.60	51.30	88.40	105.00	88.10	104.00	104.00	126.00
CT	45.00	50.00	70.00	70.00	64.00	65.00	65.00	80.00
DE	47.80	52.00	81.80	82.80	78.60	80.00	89.00	101.00
FL	40.50	49.40	78.10	94.00	50.50	63.20	96.60	121.00
GA	41.20	51.20	77.10	92.70	53.60	66.10	95.80	118.00
HI	27.60	27.70	50.10	50.50	41.60	45.30	70.00	90.00
ID	41.60	47.70	82.30	88.20	72.40	78.00	100.00	117.00
IL	42.80	47.70	85.20	86.30	84.30	85.50	98.20	112.00
IN	42.70	49.80	84.50	83.00	70.40	72.20	86.10	103.00
IA	43.40	49.90	82.40	87.50	81.60	86.80	96.60	115.00
KS	42.80	49.80	83.90	85.70	82.90	84.80	107.00	130.00
KY	43.20	49.70	81.20	95.10	72.30	84.30	92.00	112.00
LA	40.50	48.20	78.20	95.60	53.20	63.70	93.00	113.00
ME	47.00	53.00	73.00	82.00	67.00	78.00	65.00	80.00
MD	47.80	52.00	81.80	82.80	78.60	80.00	89.00	101.00
MA	45.00	46.00	70.00	75.00	65.00	70.00	68.00	85.00
MI	41.60	50.40	72.00	76.60	63.00	68.70	92.50	109.00
MN	47.70	53.60	81.80	82.60	73.40	76.20	94.60	113.00
MS	38.90	46.40	78.10	95.30	63.70	70.60	93.40	116.00
MO	41.80	50.30	88.00	105.00	77.00	92.30	101.00	121.00
MT	45.60	51.40	90.40	106.00	82.20	91.00	106.00	125.00
NE	44.00	52.60	85.00	89.90	83.80	88.70	110.00	129.00
NV	41.60	42.30	94.00	107.00	80.90	89.40	112.00	126.00
NH	45.00	50.00	73.00	80.00	67.00	75.00	68.00	85.00
NJ	41.00	49.00	61.00	67.00	46.00	52.00	87.00	106.00
NM	43.30	51.50	88.20	104.00	69.50	82.00	101.00	119.00
NY	40.30	46.40	69.00	69.50	42.30	47.70	92.30	110.00
NC	42.50	50.00	79.80	96.10	65.70	79.80	86.70	108.00
ND	44.50	52.50	86.90	99.30	77.20	89.80	105.00	123.00
OH	38.80	44.90	77.40	81.40	73.80	77.70	84.00	103.00
OK	42.80	51.50	85.70	101.00	80.60	96.60	101.00	122.00
OR	44.70	48.60	82.00	96.40	70.70	82.30	94.30	107.00
PA	42.10	49.20	79.70	81.90	70.30	73.30	116.00	111.00
RI	45.00	46.00	70.00	70.00	64.00	65.00	65.00	75.00
SC	43.70	51.80	74.70	91.00	67.00	81.20	93.20	113.00
SD	44.80	51.20	85.30	97.50	78.70	89.30	107.00	125.00
TN	38.70	47.70	81.20	97.80	64.20	77.90	90.00	112.00
TX	40.30	49.30	83.10	89.90	79.50	86.50	102.00	123.00
UT	42.00	43.00	83.00	93.00	81.00	90.00	103.00	123.00
VT	45.00	50.00	73.00	75.00	67.00	70.00	68.00	80.00
VA	36.60	45.90	76.40	94.50	63.80	79.20	87.10	109.00
WA	43.90	50.40	89.00	102.00	83.80	94.00	97.30	114.00
WV	37.40	45.00	75.70	91.20	56.90	67.20	81.70	101.00
WI	44.50	51.40	76.60	83.70	58.00	65.00	122.00	140.00
WY	45.70	52.50	91.80	107.00	85.50	98.80	109.00	130.00
US	42.90	50.30	84.20	90.20	79.70	85.90	102.00	119.00

See footnotes at end of table.

Table 7-81.—Livestock: Average price per 100 pounds received by farmers, by States, 2003 and 2004—Continued

State	Hogs [3] 2003	Hogs [3] 2004	Lambs 2003	Lambs 2004	Sheep 2003	Sheep 2004
	Dollars	Dollars	Dollars	Dollars	Dollars	Dollars
AL	33.80	43.90				
AK	66.00	73.00				
AZ	44.70	59.30	89.00	95.00	38.00	40.00
AR	35.00	47.40				
CA	35.70	49.50	88.00	90.40	31.30	32.20
CO	40.30	52.70	96.10	101.00	33.00	38.50
CT	33.20	45.50				
DE	35.90	44.70				
FL	30.00	43.70				
GA	36.40	50.30				
HI	84.70	87.00				
ID	39.40	49.00	87.60	95.60	33.70	40.40
IL	38.70	50.80	92.90	100.00	35.90	41.90
IN	37.00	48.90	92.90	102.00	35.60	39.30
IA	36.40	49.90	89.50	94.90	34.10	41.60
KS	35.20	47.40	89.70	95.80	31.30	35.00
KY	36.40	48.10				
LA	31.40	44.10				
ME	33.20	45.50				
MD	35.90	44.60				
MA	33.20	45.50				
MI	35.00	45.90	86.00	94.00	35.00	40.00
MN	38.40	49.80	89.30	94.70	27.50	33.50
MS	35.20	46.70				
MO	34.10	46.10	91.00	101.00	35.00	40.00
MT	39.70	52.30	103.00	112.00	34.80	39.50
NE	39.30	50.80	90.60	98.30	33.20	37.60
NV	35.70	45.80	91.00	98.00	36.00	38.00
NH [4]	33.20	45.50	115.00	125.00	40.00	45.00
NJ	30.40	41.80				
NM	34.50	48.30	89.20	100.00	37.00	42.00
NY	33.20	43.80	102.00	114.00	40.50	44.50
NC	38.30	50.60				
ND	39.70	51.40	96.20	103.00	30.10	36.60
OH	38.10	49.30	91.40	98.50	33.50	39.10
OK	33.30	44.10	89.00	96.00	32.00	36.00
OR	41.00	51.60	88.20	94.40	31.90	36.10
PA	35.50	46.70	107.00	115.00	39.10	43.50
RI	33.20	45.50				
SC	36.50	49.00				
SD	38.60	50.40	106.00	115.00	37.20	40.90
TN	36.00	47.30				
TX	33.60	44.90	97.10	110.00	39.60	43.40
UT	45.40	53.90	92.00	101.00	29.90	33.80
VT	33.20	45.50				
VA	35.00	46.60	93.00	101.00	32.20	42.80
WA	39.30	48.90	90.40	96.00	34.00	38.00
WV	36.10	46.10	90.60	102.00	31.40	36.90
WI	36.90	46.30	87.10	92.50	31.20	37.00
WY	36.20	46.70	104.00	114.00	39.60	40.30
Other States [5]			88.00	96.00	35.00	38.00
US	37.20	49.30	94.40	101.00	34.90	38.80

[1] Includes cull dairy cows sold for slaughter, but not cows for dairy herd replacement. [2] Weighted average of prices for cows, and for steers and heifers. [3] December of preceding year through November. [4] For lambs and sheep, CT, ME, MA, NH, RI, and VT are included in NH. [5] AL, AK, AR, DE, FL, GA, HI, KY, LA, MD, MS, NJ, NC, SC, and TN.

NASS, Livestock Branch, (202) 720–3570.

Table 7-82.—Frozen meat: Cold storage holdings, end of month, United States, 2004 and 2005

Month	Boneless beef		Beef cuts		Total beef	
	2004	2005	2004	2005	2004	2005
	1,000 pounds	1,000 pounds	1,000 pounds	1,000 pounds	1,000 pounds	1,000 pounds
January	373,167	399,812	61,187	53,449	434,354	453,261
February	379,866	351,509	55,167	48,731	435,033	400,240
March	364,544	329,324	52,228	42,999	416,772	372,323
April	367,945	287,398	53,235	42,027	421,180	329,425
May	350,915	279,899	51,908	38,291	402,823	318,190
June	364,157	306,754	47,610	35,352	411,767	342,106
July	380,428	342,331	46,619	42,915	427,047	385,246
August	398,726	367,027	47,268	43,565	445,994	410,592
September ...	408,973	392,153	48,271	46,597	457,244	438,750
October	404,923	389,827	47,666	49,359	452,589	439,186
November	409,917	374,460	53,380	55,434	463,297	429,894
December	425,231	380,298	59,045	54,144	484,276	434,442

Month	Picnics		Bellies		Butts	
	2004	2005	2004	2005	2004	2005
	1,000 pounds	1,000 pounds	1,000 pounds	1,000 pounds	1,000 pounds	1,000 pounds
January	12,207	11,802	63,095	63,417	10,125	16,071
February	11,236	13,451	57,123	75,321	10,474	15,433
March	12,715	12,141	50,126	81,265	10,852	14,454
April	12,333	12,611	48,363	89,360	9,121	13,320
May	12,710	10,147	41,366	81,694	7,144	9,816
June	10,596	7,512	37,185	70,657	4,426	7,638
July	13,415	7,768	23,383	50,315	4,808	6,516
August	11,507	8,163	15,230	22,149	5,255	5,931
September ...	15,177	9,397	11,344	14,117	5,805	7,289
October	13,123	13,628	15,970	15,211	7,172	8,051
November	10,531	11,510	33,955	28,613	11,063	10,971
December	9,812	9,464	56,026	47,925	13,780	12,187

Month	Hams					
	Bone-in		Boneless		Total	
	2004	2005	2004	2005	2004	2005
	1,000 pounds	1,000 pounds	1,000 pounds	1,000 pounds	1,000 pounds	1,000 pounds
January	38,407	24,719	33,418	28,686	71,825	53,405
February	37,778	33,804	30,918	32,427	68,696	66,231
March	25,678	29,394	26,118	36,865	51,796	66,259
April	25,625	36,057	30,704	34,433	56,329	70,490
May	33,184	43,331	32,544	35,198	65,728	78,529
June	34,681	53,617	32,810	38,886	67,491	92,503
July	44,417	59,258	40,028	43,497	84,445	102,755
August	56,614	64,647	42,351	43,218	98,965	107,865
September ...	65,534	64,810	42,961	40,341	108,495	105,151
October	58,827	56,032	37,459	40,757	96,286	96,789
November	36,026	34,743	24,861	22,684	60,887	57,427
December	20,222	16,716	23,851	13,075	44,073	29,791

Month	Loins					
	Bone-in		Boneless		Total	
	2004	2005	2004	2005	2004	2005
	1,000 pounds	1,000 pounds	1,000 pounds	1,000 pounds	1,000 pounds	1,000 pounds
January	26,278	23,352	15,967	20,958	42,245	44,310
February	23,686	23,866	16,564	23,728	40,250	47,594
March	22,591	22,875	16,662	23,195	39,253	46,070
April	20,186	23,134	15,878	21,111	36,064	44,245
May	18,213	17,893	11,761	17,127	29,974	35,020
June	16,167	18,029	8,021	14,752	24,188	32,781
July	15,063	16,326	8,365	10,807	23,428	27,133
August	15,372	14,788	13,651	12,214	29,023	27,002
September ...	16,158	19,064	14,330	12,487	30,488	31,551
October	19,917	21,028	15,015	15,724	34,932	36,752
November	24,784	23,780	19,264	20,913	44,048	44,693
December	27,455	27,001	22,254	24,147	49,709	51,148

See end of table.

Table 7-82.—Frozen meat: Cold storage holdings, end of month, United States, 2004 and 2005—Continued

Month	Ribs		Trimmings		Other frozen pork	
	2004	2005	2004	2005	2004	2005
	1,000 pounds	*1,000 pounds*	*1,000 pounds*	*1,000 pounds*	*1,000 pounds*	*1,000 pounds*
January	79,130	99,071	48,623	36,106	80,278	86,119
February	75,261	101,372	44,637	41,492	72,677	87,273
March	74,032	93,471	40,814	45,845	71,947	90,448
April	68,126	95,604	45,719	51,457	73,011	88,299
May	54,162	76,710	41,268	48,429	66,941	78,344
June	39,128	60,180	28,769	44,964	67,688	82,065
July	38,489	49,888	23,526	37,727	68,271	76,027
August	40,680	48,096	19,434	32,719	70,244	74,698
September ...	46,804	55,473	23,533	33,232	74,119	84,229
October	57,674	60,973	26,515	31,159	78,734	87,122
November	83,989	68,894	26,457	35,049	78,727	90,185
December	97,444	63,982	34,543	39,881	85,310	90,175

Month	Variety meats		Unclassified pork		Total pork	
	2004	2005	2004	2005	2004	2004
	1,000 pounds	*1,000 pounds*	*1,000 pounds*	*1,000 pounds*	*1,000 pounds*	*1,000 pounds*
January	41,823	40,221	54,759	48,159	504,110	498,681
February	41,571	41,272	55,194	51,779	477,119	541,218
March	40,962	41,129	54,849	52,696	447,346	543,778
April	43,834	39,438	55,650	58,979	448,550	563,803
May	47,109	38,085	46,438	56,113	412,840	512,887
June	45,639	37,320	47,897	56,874	373,007	492,494
July	46,011	37,656	41,036	52,199	366,812	447,984
August	47,937	32,543	43,833	55,302	382,108	414,468
September ...	49,309	32,463	48,749	58,671	413,823	431,573
October	45,198	33,791	47,454	63,000	423,058	446,476
November	38,229	29,443	48,446	60,621	436,332	437,406
December	46,478	25,200	45,677	58,776	482,852	428,529

Month	Veal		Lamb & mutton		Canned hams	
	2004	2005	2004	2005	2004	2005
	1,000 pounds	*1,000 pounds*	*1,000 pounds*	*1,000 pounds*	*1,000 pounds*	*1,000 pounds*
January	5,284	2,849	3,671	7,549	5,872	2,942
February	5,197	2,729	3,355	7,585	5,689	2,300
March	5,889	2,613	3,164	7,650	5,035	2,393
April	5,843	2,951	3,251	8,739	4,387	2,345
May	5,306	1,635	3,504	9,719	4,432	2,129
June	5,618	3,045	3,872	9,362	3,457	2,191
July	5,675	3,303	3,376	11,756	3,832	4,750
August	4,257	3,561	3,878	11,790	4,070	4,224
September ...	4,106	3,688	4,179	10,942	5,092	4,058
October	3,768	4,144	4,166	10,137	4,350	3,560
November	3,616	4,544	3,715	9,332	4,360	3,093
December	3,542	5,064	3,497	9,967	3,184	2,703

Month	Other canned meat		Total red meat	
	2004	2005	2004	2005
	1,000 pounds	*1,000 pounds*	*1,000 pounds*	*1,000 pounds*
January	417	1,564	953,708	966,846
February	495	2,099	926,888	956,171
March	1,152	2,826	879,358	931,583
April	389	2,250	883,600	909,513
May	353	4,069	829,258	848,629
June	150	2,461	797,871	851,659
July	929	2,817	807,671	855,856
August	1,592	4,259	841,899	848,894
September ...	1,660	1,921	886,104	890,932
October	1,298	2,425	889,229	905,928
November	1,282	3,941	912,602	888,210
December	1,686	3,668	979,037	884,373

NASS, Livestock Branch, (202) 720–3570.

CHAPTER VIII
DAIRY AND POULTRY STATISTICS

Dairy statistics in this chapter include series relating to many phases of production, movement, prices, stocks, and consumption of milk and its products. Two series of number of milk cows on farms are included in this publication. One series is an inventory number of a specific classification estimated as one of the major groups making up the total cattle population on January 1. The other series identified as "milk cows" is an annual average number of milk cows during the year (excluding any not yet fresh) and is used in estimating milk production.

In comparing the several series of milk prices, it is important to note that prices received by farmers for all whole milk sold are for milk or milkfat content as actually sold, while certain prices paid by dealers for milk for fluid purposes or for specified manufacturing purposes may be quoted on a 3.5 percent butterfat basis, or for some types of manufacturing milk on the test of the milk used for that particular purpose.

Poultry and poultry products statistics include inventory numbers of chickens by classes; the production, disposition, cash receipts, and gross income from chickens and eggs; poultry and egg receipts at principal markets; commercial broiler production; turkey production, disposition, and gross income; poultry and eggs under Federal inspection; and the National Poultry Improvement Plan. Estimates relating to inventories, production, and income exclude poultry and eggs produced on places not classified as farms.

Table 8-1.—Milk cows and heifers: Number that have calved and heifers 500 pounds and over kept for milk cow replacements, United States, Jan. 1, 1997–2006

Year	Milk cows and heifers that have calved	Heifers 500 pounds and over kept for milk cow replacements
	Thousands	*Thousands*
1997	9,318	4,058
1998	9,199	3,986
1999	9,128	4,069
2000	9,183	4,000
2001	9,172	4,057
2002	9,106	4,055
2003	9,142	4,114
2004	8,990	4,020
2005	9,005	4,118
2006 [1]	9,058	4,278

[1] Preliminary.
NASS, Livestock Branch, (202) 720–3570.

Table 8-2.—Milk cows and heifers: Number that have calved and heifers 500 pounds
and over kept for milk cow replacements, by States, Jan. 1, 2005 and 2006

State	Milk cows and heifers that have calved		Heifers 500 pounds and over kept for milk cow replacements	
	2005	2006 [1]	2005	2006 [1]
	Thousands	Thousands	Thousands	Thousands
AL	16	14	7	6
AK	1.2	0.8	0.5	0.3
AZ	165	165	42	44
AR	26	21	10	8
CA	1,740	1,770	760	790
CO	101	105	50	55
CT	20	20	11.0	11.0
DE	8	7	2.5	3.0
FL	138	134	40	40
GA	84	78	24	23
HI	5.7	4.6	2	2
ID	435	473	230	250
IL	105	104	55	52
IN	155	158	56	65
IA	187	187	100	120
KS	110	110	45	55
KY	110	102	45	50
LA	36	32	9	9
ME	33	32	19.0	18.0
MD	73	70	32	30
MA	17	16	8.5	8.5
MI	307	312	120	137
MN	460	445	270	265
MS	26	24	15	11
MO	119	114	65	70
MT	18	19	9	8
NE	61	60	20	15
NV	25	27	10	12
NH	16.0	16.0	9.0	8.5
NJ	12	12	6.0	6.0
NM	318	340	100	105
NY	650	652	305	325
NC	55	52	25	22
ND	33	33	18	19
OH	266	273	125	120
OK	75	75	15	20
OR	120	121	60	75
PA	566	558	275	275
RI	1.1	1.0	0.8	0.7
SC	18	17	7	6
SD	80	81	40	45
TN	72	70	40	40
TX	318	325	130	130
UT	88	85	45	45
VT	143	143	58	61
VA	105	103	43	40
WA	235	237	102	98
WV	13	13	4	4
WI	1,235	1,240	650	670
WY	4	7	3	5
US	9,005.0	9,058.4	4,118.3	4,278.0
PR	92	90		

[1] Preliminary. NA=not available.
NASS, Livestock Branch, (202) 720–3570.

Table 8-3.—Milk-feed price ratios: All milk-price; dairy feed, 16%; Milk-feed price ratios
and value per 100 pounds of grain and concentrate rations fed to milk cows, United
States, annual 1995–2004

Year	All milk price cwt.	16% dairy feed price cwt [1]	Milk-feed price ratio [2]
	Dollars	Dollars	Pounds
1995	12.74	8.70	2.59
1996	14.75	11.25	2.44
1997	13.36	10.75	2.38
1998	15.46	9.70	3.34
1999	14.38	9.00	3.59
2000	12.40	8.75	3.05
2001	15.04	9.20	3.39
2002	12.20	9.50	2.60
2003	12.52	10.00	2.61
2004	16.04	10.90	3.09

[1] Commercially prepared 16%dairy ration: Annual average prior to 1995, April price 1995-current. [2] Annual ratios based on average of monthly ratios. Pounds of 16 % mixed dairy feed equal in value to one pound of whole milk. Effective January 1995, prices of commercial prepared feeds are based on current U.S. prices received for corn (51 lbs), soybeans (8 lbs), and alfalfa hay (41 lbs).
NASS, Environmental, Economics, and Demographics Branch, (202) 720–6146.

Table 8-4.—Milk cows: Number of operations, 2004–2005, and inventory, Jan. 1, 2005–2006, by selected States and United States[1]

State	Operations with milk cows		January 1 milk cow inventory	
	2004	2005	2005	2006[2]
	Number	*Number*	*1,000 head*	*1,000 head*
AL	190	190	16	14
AK	30	30	1.2	0.8
AZ	230	210	165	165
AR	380	320	26	21
CA	2,300	2,300	1,740	1,770
CO	670	660	101	105
CT	250	230	20	20
DE	90	85	8	7
FL	500	480	138	134
GA	630	610	84	78
HI	30	30	5.7	4.6
ID	900	850	435	473
IL	1,500	1,400	105	104
IN	2,300	2,200	155	158
IA	2,600	2,500	187	187
KS	950	900	110	110
KY	2,300	2,200	110	102
LA	420	390	36	32
ME	500	470	33	32
MD	850	850	73	70
MA	270	250	17	16
MI	2,900	2,800	307	312
MN	6,100	5,800	460	445
MS	390	350	26	24
MO	2,800	2,700	119	114
MT	600	650	18	19
NE	830	770	61	60
NV	120	110	25	27
NH	210	200	16.0	16.0
NJ	160	150	12	12
NM	450	450	318	340
NY	6,900	6,700	650	652
NC	800	680	55	52
ND	600	550	33	33
OH	4,500	4,400	266	273
OK	1,500	1,400	75	75
OR	780	790	120	121
PA	9,100	8,900	566	558
RI	30	30	1.1	1.0
SC	200	200	18	17
SD	1,000	800	80	81
TN	1,200	1,100	72	70
TX	1,700	1,500	318	325
UT	600	580	88	85
VT	1,300	1,300	143	143
VA	1,400	1,400	105	103
WA	820	810	235	237
WV	480	470	13	13
WI	15,900	15,300	1,235	1,240
WY	260	250	4	7
US	81,520	78,295	9,005.0	9,058.4
PR[3]	1,600	1,600	92	90

[1] An operation is any place having one or more milk cows on hand at any time during the year.   [2] Preliminary.   [3] Puerto Rico is not included in the U.S. total.

NASS, Livestock Branch, (202) 720–3570.

DAIRY AND POULTRY STATISTICS

Table 8-5.—Milk cows: Number of operations by size group, selected States, and United States, 2004–2005 [1]

State	1–29 Head		30–49 Head		50–99 Head		100–199 Head		200–499 Head		500+ Head	
	2004	2005	2004	2005	2004	2005	2004	2005	2004	2005	2004	2005
	Number	Number	Number	Number	Number	Number	Number	Number	Number	Number	Number	Number
AZ	120	100	....	....	....	....	10	10	10	10	90	90
CA	310	320	55	70	80	80	180	180	575	550	1,100	1,100
CO	500	500	10	5	25	25	30	25	45	45	60	60
FL	310	300	10	5	15	10	25	25	50	50	90	90
GA	290	280	20	20	80	80	130	130	80	70	30	30
ID	200	195	70	60	180	160	120	110	140	125	190	200
IL	400	350	265	250	500	480	250	240	70	65	15	15
IN	1,100	1,000	440	460	480	450	200	210	50	50	30	30
IA	550	530	600	550	905	890	410	400	110	100	25	30
KS	500	475	100	90	180	180	115	110	35	25	20	20
KY	1,000	980	420	410	570	520	250	240	55	45	5	5
MD	170	180	140	150	280	280	200	180	50	50	10	10
MI	950	870	440	420	660	660	540	510	225	245	85	95
MN	670	620	2,000	1,900	2,500	2,400	640	580	230	240	60	60
MO	1,300	1,250	400	350	700	700	330	340	65	55	5	5
NM	270	270	....	....	5	....	5	10	15	10	155	160
NY	1,400	1,400	1,300	1,300	2,600	2,500	1,000	890	430	440	170	170
NC	450	350	30	20	120	110	130	130	60	60	10	10
OH	2,000	1,900	800	850	1,000	950	525	500	140	150	35	50
OK	1,000	950	70	70	215	180	150	140	50	50	15	10
OR	380	390	30	30	80	75	145	140	100	110	45	45
PA	1,900	1,800	2,800	2,700	3,100	3,100	980	970	270	280	50	50
SD	350	170	170	150	270	270	145	140	45	45	20	25
TN	480	420	135	140	300	270	200	200	75	60	10	10
TX	880	760	70	60	150	140	210	180	210	180	180	180
UT	240	240	25	25	90	80	120	110	80	80	45	45
VT	160	190	230	220	510	510	240	220	125	120	35	40
VA	500	500	140	130	370	350	290	320	90	90	10	10
WA	230	250	30	20	90	85	155	150	180	170	135	135
WI	2,300	2,200	4,100	3,900	6,700	6,400	1,900	1,850	700	750	200	200
Oth Sts	2,900	2,750	600	530	1,300	1,200	820	815	340	340	80	90
US	23,810	22,490	15,500	14,885	24,055	23,135	10,445	10,055	4,700	4,660	3,010	3,070

[1] An operation is any place having one or more head of milk cows on hand at any time during the year.

NASS, Livestock Branch, (202) 720–3570.

Table 8-6.—Milk cows: Percent of inventory by size group, selected States, and United States, 2004–2005 [1]

State	1–29 head		30–49 head		50–99 head		100–199 head		200–499 head		500+ head	
	2004	2005	2004	2005	2004	2005	2004	2005	2004	2005	2004	2005
	Per-cent	Per-cent	Per-cent	Per-cent	Per-cent	Per-cent	Per-cent	Per-cent	Per-cent	Per-cent	Per-cent	Per-cent
AZ	0.2	0.2	....	....	....	....	0.7	0.7	2.1	2.1	97.0	97.0
CA	0.1	0.1	0.1	0.1	0.3	0.3	1.5	1.5	12.0	11.0	86.0	87.0
CO	1.1	0.9	0.4	0.2	2.0	1.7	4.0	3.2	15.5	15.0	77.0	79.0
FL	0.8	0.7	0.3	0.2	0.9	0.6	2.5	2.5	12.5	12.5	83.0	83.5
GA	1.1	1.1	0.9	0.9	7.0	7.0	23.0	23.0	26.0	23.0	42.0	45.0
ID	0.3	0.3	0.6	0.5	2.9	2.5	3.7	3.2	10.5	9.0	82.0	84.5
IL	3.0	2.5	8.5	8.5	30.0	30.5	30.0	30.5	17.5	17.0	11.0	11.0
IN	7.5	7.0	12.0	11.0	21.5	19.0	18.5	18.0	9.5	9.0	31.0	36.0
IA	3.0	3.0	11.0	10.0	30.0	30.0	27.0	26.0	15.0	15.0	14.0	16.0
KS	1.5	1.5	3.5	3.0	12.0	11.5	13.5	12.5	8.5	6.0	61.0	65.5
KY	8.0	8.0	14.0	14.0	35.0	34.0	28.0	30.0	12.5	11.0	2.5	3.0
MD	2.0	2.0	7.0	7.0	26.0	27.0	34.0	33.0	17.0	17.0	14.0	14.0
MI	3.0	2.5	5.5	5.0	15.0	15.0	24.5	23.0	21.0	22.5	31.0	32.0
MN	3.0	3.0	17.0	17.0	36.0	35.0	17.0	16.0	15.0	16.0	12.0	13.0
MO	5.0	5.0	12.0	11.0	34.0	35.0	29.0	31.0	13.5	12.0	6.5	6.0
NM	0.2	0.3	....	....	0.1	....	0.2	0.4	1.5	1.3	98.0	98.0
NY	2.0	2.0	7.5	7.5	27.0	25.5	20.5	18.5	19.0	20.5	24.0	26.0
NC	3.5	2.5	2.0	1.5	16.0	16.0	33.0	33.0	29.0	30.0	16.5	17.0
OH	8.0	8.0	12.0	12.0	26.0	24.0	26.0	24.0	16.0	17.0	12.0	15.0
OK	4.5	5.0	3.5	4.0	18.5	17.0	24.5	24.0	17.0	20.0	32.0	30.0
OR	1.0	1.0	1.0	1.0	5.0	5.0	17.0	17.0	25.0	27.0	51.0	49.0
PA	3.0	2.5	19.5	19.0	35.5	36.0	22.0	22.0	13.0	13.5	7.0	7.0
SD	3.0	2.0	8.0	7.0	22.0	22.0	22.5	22.0	16.5	17.0	28.0	30.0
TN	1.5	1.0	6.0	7.0	26.0	25.0	34.0	36.0	25.0	23.0	7.5	8.0
TX	1.1	0.9	0.9	0.8	3.5	3.3	9.5	8.5	20.0	17.5	65.0	69.0
UT	1.0	1.0	1.0	1.0	7.5	7.0	18.5	16.0	26.0	27.0	46.0	48.0
VT	2.0	2.0	6.0	6.0	24.0	24.0	22.0	21.0	25.0	23.0	21.0	24.0
VA	2.5	2.5	5.0	4.5	26.5	23.5	37.0	40.0	23.0	23.0	6.0	6.5
WA	0.2	0.3	0.5	0.3	2.3	2.4	10.0	9.0	24.0	23.0	63.0	65.0
WI	3.0	3.0	12.5	12.5	35.0	34.0	19.5	19.0	16.5	17.5	13.5	14.0
Oth Sts ...	3.0	3.5	5.5	5.0	22.0	21.0	26.0	27.0	22.5	23.5	21.0	20.0
US	2.1	2.0	6.6	6.4	17.8	17.1	15.1	14.6	15.5	15.4	42.9	44.5

[1] Percents reflect average distribution of various probability surveys conducted during the year but are based primarily on beginning-of-year and mid-year surveys.

NASS, Livestock Branch, (202) 720–3570.

Table 8-7.—Milk Production: Percent of production by size groups, selected States, and United States, 2004–2005 [1]

State	Production on operations having											
	1–29 Head		30–49 Head		50–99 Head		100–199 Head		200-499 Head		500+ Head	
	2004	2005	2004	2005	2004	2005	2004	2005	2004	2005	2004	2005
	Percent	Percent	Percent	Percent	Percent	Percent	Percent	Percent	Percent	Percent	Percent	Percent
AZ	0.1	0.1					0.4	0.4	2.0	2.0	97.5	97.5
CA	0.1	0.1	0.1	0.1	0.3	0.3	1.5	1.5	14.0	10.0	84.0	88.0
CO	0.5	0.4	0.5	0.1	1.5	1.0	3.5	2.5	14.0	15.0	80.0	81.0
FL	0.2	0.9	0.2	0.1	0.6	0.6	2.5	2.4	11.5	12.0	85.0	84.0
GA	0.5	0.5	0.5	0.5	6.0	6.0	22.0	22.0	26.0	22.0	45.0	49.0
ID	0.2	0.2	0.4	0.3	2.2	2.0	3.2	2.5	10.0	8.0	84.0	87.0
IL	2.5	2.0	7.5	7.0	28.0	28.0	30.0	31.0	19.0	18.0	13.0	14.0
IN	6.0	6.0	10.0	10.0	19.0	18.0	18.0	18.0	10.0	11.0	37.0	37.0
IA	2.0	2.0	10.0	8.0	28.0	27.0	27.0	26.0	16.0	16.0	17.0	21.0
KS	1.0	1.0	2.5	2.5	10.5	10.0	12.0	12.0	9.0	6.5	65.0	68.0
KY	6.0	7.0	14.0	14.0	34.0	33.0	29.0	32.0	14.0	11.0	3.0	3.0
MD	1.5	1.5	5.5	5.5	25.0	26.0	35.0	34.0	17.5	16.0	15.5	17.0
MI	2.0	1.5	4.5	3.5	13.0	14.0	23.0	20.0	22.5	24.0	35.0	37.0
MN	2.5	2.5	15.0	15.0	35.0	34.0	18.0	17.0	16.0	17.0	13.5	14.5
MO	3.0	3.0	11.0	10.0	33.0	34.0	30.0	32.0	15.0	13.0	8.0	8.0
NM	0.1	0.3			0.1		0.4	0.3	1.4	1.4	98.0	98.0
NY	1.5	1.0	5.5	5.5	24.0	22.0	20.0	18.5	20.0	22.0	29.0	31.0
NC	0.5	0.5	1.5	1.5	14.0	14.0	33.0	32.0	30.0	31.0	21.0	21.0
OH	5.0	6.0	10.0	10.0	24.0	22.0	27.0	24.0	18.0	19.0	16.0	19.0
OK	2.5	3.0	2.5	3.0	16.0	14.0	22.0	22.0	16.0	19.0	41.0	39.0
OR	0.5	1.0	0.5	1.0	4.0	4.0	16.0	15.0	25.0	26.0	54.0	53.0
PA	2.5	2.0	16.0	17.0	34.0	35.0	24.0	22.0	14.5	15.0	9.0	9.0
SD	3.0	2.0	7.0	6.0	20.0	19.0	22.0	22.0	18.0	19.0	30.0	32.0
TN	1.0	1.0	5.0	6.0	25.0	24.0	35.0	37.0	26.0	24.0	8.0	8.0
TX	0.4	0.4	0.6	0.6	3.0	3.0	8.0	7.0	19.0	19.0	69.0	70.0
UT	0.5	0.5	1.0	0.5	6.5	6.0	16.0	14.0	26.0	27.0	50.0	52.0
VT	1.0	1.0	5.0	4.0	21.0	21.0	22.0	22.0	27.0	26.0	24.0	26.0
VA	2.0	2.0	3.5	3.0	24.5	25.0	38.0	44.0	25.0	19.0	7.0	7.0
WA	0.1	0.2	0.4	0.2	2.0	2.0	8.5	7.6	24.0	22.0	65.0	68.0
WI	2.5	2.0	11.5	11.0	34.0	34.0	19.0	19.0	17.5	18.0	15.5	16.0
Oth Sts	2.0	2.0	5.0	4.0	19.0	19.0	24.0	26.0	26.0	25.0	24.0	24.0
US	1.4	1.3	5.4	5.1	15.7	15.2	14.2	13.5	16.3	15.4	47.0	49.5

[1] Percents reflect average distributions of various probability surveys conducted during the year but are based primarily on beginning-of-year and mid-year surveys.
NASS, Livestock Branch, (202) 720–3570.

Table 8-8.—Milk cows: Number of operations, percent of inventory and percent of milk production by size group, United States, 2004–2005 [1]

Head	Operations		Percent of inventory		Percent of production	
	2004	2005	2004	2005	2004	2005
	Number	Number	Percent	Percent	Percent	Percent
1-29	23,810	22,490	2.1	2.0	1.4	1.3
30-49	15,500	14,885	6.6	6.4	5.4	5.1
50-99	24,055	23,135	17.8	17.1	15.7	15.2
100-199	10,445	10,055	15.1	14.6	14.2	13.5
200-499	4,700	4,660	15.5	15.4	16.3	15.4
500-999	1,700	1,700	12.8	12.8	14.0	13.5
1,000-1,999	815	850	12.0	12.0	13.1	14.3
2,000+	495	520	18.1	19.7	19.9	21.7
Total	81,520	78,295	100.0	100.0	100.0	100.0

[1] An operation is any place having one or more head of milk cows on hand at any time during the year. Percents reflect average distributions of various probability surveys conducted during the year but are based primarily on beginning-of-year and mid-year surveys.
NASS, Livestock Branch, (202) 720–3570.

Table 8-9.—Official Dairy Herd Improvement test plans: Numbers of herds and cows and milk, fat, and protein production, United States, 1995–2004

Year	Herds	Cows	Cows per herd	Average production			Cows with protein informa-tion	Average protein production [1]	Average protein production [1]
				Milk	Fat	Fat			
	Number	Number	Number	Pounds	Percent	Pounds	Percent	Percent	Pounds
1995 ...	31,628	3,527,187	111.5	19,271	3.67	710	90	3.22	621
1996 ...	29,416	3,486,010	118.5	19,192	3.70	713	90	3.23	620
1997 ...	27,383	3,402,487	124.3	19,815	3.67	731	89	3.23	639
1998 ...	25,738	3,397,396	132.0	20,209	3.68	745	92	3.22	651
1999 ...	24,841	3,449,854	140.9	20,743	3.68	766	93	3.24	673
2000 ...	23,225	3,521,686	151.6	21,092	3.68	781	93	3.15	664
2001 ...	22,095	3,499,214	158.4	21,118	3.66	777	94	3.08	651
2002 ...	20,955	3,537,064	168.8	21,475	3.68	792	94	3.07	661
2003 ...	19,732	3,416,386	173.1	21,471	3.68	792	94	3.07	661
2004 ...	18,897	3,468,419	183.5	21,457	3.68	791	94	3.09	664

[1] The decline in protein production in 2000 reflects a measurement change by the dairy industry from crude to true protein beginning in May 2000. The percentage of milk that is true protein is lower than the percentage that is crude protein by an approximate difference of 0.19 percent.
ARS, Animal Improvement Programs Laboratory, (301) 504–8334, http://aipl.arsusda.gov.

Table 8-10.—Milk and milkfat production: Number of producing cows, production per cow, and total quantity produced, United States, 1995–2004

Year	Number of milk cows [1]	Production of milk and milkfat [2]				
		Per milk cow		Percentage of fat in all milk produced	Total	
		Milk	Milkfat		Milk	Milkfat
	Thousands	Pounds	Pounds	Percent	Million pounds	Million pounds
1995	9,466	16,405	600	3.66	155,292	5,681
1996	9,372	16,433	606	3.69	154,006	5,679
1997	9,252	16,871	617	3.66	156,091	5,706
1998	9,151	17,185	629	3.66	157,262	5,759
1999	9,153	17,763	652	3.67	162,589	5,970
2000	9,199	18,197	670	3.68	167,393	6,164
2001	9,103	18,162	667	3.67	165,332	6,073
2002	9,139	18,608	685	3.68	170,063	6,264
2003	9,083	18,760	688	3.67	170,394	6,249
2004 [3]	9,010	18,957	696	3.67	170,805	6,266

[1] Average number during year, excluding heifers not yet fresh.   [2] Excludes milk sucked by calves.   [3] Preliminary.
NASS, Livestock Branch, (202) 720–3570.

DAIRY AND POULTRY STATISTICS

Table 8-11.—Milk and milkfat production: Number of milk cows, production per cow, and total quantity produced, by States, 2003

State	Number of milk cows [1]	Per milk cow		Percent of fat			Total	
		Milk	Milkfat	Fluid grade	Manuf. grade	All milk	Milk	Milkfat
	Thousands	Pounds	Pounds	Percent	Percent	Percent	Million pounds	Million pounds
AL	18	14,000	510	3.64		3.64	252	9.2
AK	1.3	12,846	452	3.52		3.52	16.7	0.6
AZ	155	22,916	823	3.59		3.59	3,552	127.5
AR	29	12,207	442	3.62		3.62	354	12.8
CA	1,688	20,993	770	3.67	3.98	3.67	35,437	1,300.5
CO	100	21,530	756	3.51		3.51	2,153	75.6
CT	22	18,773	695	3.70		3.70	413	15.3
DE	8.3	15,904	596	3.75		3.75	132	5.0
FL	142	15,218	542	3.56		3.56	2,161	76.9
GA	85	16,988	615	3.62		3.62	1,444	52.3
HI	6.5	14,154	494	3.49		3.49	92	3.2
ID	404	21,718	780	3.59	3.79	3.59	8,774	315.0
IL	111	18,441	690	3.74	3.68	3.74	2,047	76.6
IN	149	19,725	724	3.67	3.72	3.67	2,939	107.9
IA	201	18,955	688	3.63	3.77	3.63	3,810	138.3
KS	111	19,189	702	3.66		3.66	2,130	78.0
KY	116	12,629	460	3.64		3.64	1,465	53.3
LA	43	12,070	420	3.48		3.48	519	18.1
ME	35	17,829	660	3.70		3.70	624	23.1
MD	78	15,577	584	3.75		3.75	1,215	45.6
MA	19	17,474	652	3.73		3.73	332	12.4
MI	302	21,109	764	3.62	3.64	3.62	6,375	230.8
MN	473	17,459	649	3.72	3.74	3.72	8,258	307.2
MS	31	13,645	487	3.57		3.57	423	15.1
MO	129	14,620	535	3.65	3.82	3.66	1,886	69.0
MT	18	19,167	690	3.60		3.60	345	12.4
NE	64	17,641	651	3.69		3.69	1,129	41.7
NV	25	19,400	681	3.51		3.51	485	17.0
NH	16	19,063	719	3.77		3.77	305	11.5
NJ	13	16,615	616	3.71		3.71	216	8.0
NM	317	21,028	751	3.57		3.57	6,666	238.0
NY	671	17,812	652	3.66		3.66	11,952	437.4
NC	61	17,115	625	3.65		3.65	1,044	38.1
ND	35	14,857	548	3.68	3.72	3.69	520	19.2
OH	260	17,269	651	3.76	3.81	3.77	4,490	169.3
OK	82	16,000	574	3.59		3.59	1,312	47.1
OR	119	18,294	670	3.66		3.66	2,177	79.7
PA	575	17,979	671	3.73	3.82	3.73	10,338	385.6
RI	1.3	17,000	636	3.74		3.74	22.1	0.8
SC	19	16,737	611	3.65		3.65	318	11.6
SD	82	16,220	607	3.74	3.76	3.74	1,330	49.7
TN	79	15,253	557	3.65		3.65	1,205	44.0
TX	319	17,649	646	3.66		3.66	5,630	206.1
UT	91	17,824	640	3.59	3.74	3.59	1,622	58.2
VT	149	17,698	662	3.74		3.74	2,637	98.6
VA	113	15,319	558	3.64		3.64	1,731	63.0
WA	245	22,780	834	3.66		3.66	5,581	204.3
WV	15	14,400	520	3.61		3.61	216	7.8
WI	1,256	17,728	659	3.71	3.79	3.72	22,266	828.3
WY	3.8	14,211	514	3.63	3.58	3.62	54	2.0
US [3]	9,083	18,760	688	3.66	3.80	3.67	170,394	6,248.7
PR	92	8,913	288	3.23	3.23	3.23	820	26.5

[1] Average number during year, excluding heifers not yet fresh. U.S. total may not add due to rounding. [2] Excludes milk sucked by calves. [3] Sum of parts may not equal due to rounding.
NASS, Livestock Branch, (202) 720–3570.

Table 8-12.—Milk and milkfat production: Number of milk cows, production per cow, and total quantity produced, by States, 2004 (preliminary)

State	Number of milk cows[1]	Per milk cow Milk	Per milk cow Milkfat	Percent of fat Fluid grade	Percent of fat Manuf. grade	All milk	Total Milk	Total Milkfat
	Thousands	Pounds	Pounds	Percent	Percent	Percent	Million pounds	Million pounds
AL	17	14,412	526	3.65	....	3.65	245	8.9
AK	1.2	12,167	411	3.38	....	3.38	14.6	0.5
AZ	160	22,788	823	3.61	....	3.61	3,646	131.6
AR	24	13,250	481	3.63	....	3.63	318	11.5
CA	1,725	21,139	776	3.67	3.87	3.67	36,465	1,338.3
CO	102	21,412	758	3.54	....	3.54	2,184	77.3
CT	20	19,600	717	3.66	....	3.66	392	14.3
DE	7.4	17,230	644	3.74	....	3.74	127.5	4.8
FL	138	16,326	591	3.62	....	3.62	2,253	81.6
GA	84	16,857	614	3.64	....	3.64	1,416	51.5
HI	6.1	13,197	465	3.52	....	3.52	80.5	2.8
ID	424	21,446	772	3.60	3.92	3.60	9,093	327.3
IL	107	18,486	690	3.74	3.67	3.73	1,978	73.8
IN	150	19,747	723	3.66	3.72	3.66	2,962	108.4
IA	193	19,912	723	3.63	3.80	3.63	3,843	139.5
KS	113	19,611	714	3.64	....	3.64	2,216	80.7
KY	110	12,936	470	3.63	....	3.63	1,423	51.7
LA	38	12,605	445	3.53	....	3.53	479	16.9
ME	34	18,000	666	3.70	....	3.70	612	22.6
MD	74	15,703	576	3.67	....	3.67	1,162	42.6
MA	17	17,412	648	3.72	....	3.72	296	11.0
MI	303	20,842	757	3.63	3.64	3.63	6,315	229.2
MN	463	17,499	658	3.76	3.75	3.76	8,102	304.6
MS	27	14,037	505	3.60	....	3.60	379	13.6
MO	122	15,139	554	3.65	3.91	3.66	1,847	67.6
MT	18	19,278	698	3.62	....	3.62	347	12.6
NE	61	17,230	636	3.69	....	3.69	1,051	38.8
NV	25	20,360	711	3.49	....	3.49	509	17.8
NH	16	18,875	706	3.74	....	3.74	302	11.3
NJ	12	16,667	610	3.66	....	3.66	200	7.3
NM	326	20,583	733	3.56	....	3.56	6,710	238.9
NY	655	17,786	649	3.65	....	3.65	11,650	425.2
NC	57	17,649	639	3.62	....	3.62	1,006	36.4
ND	34	15,471	577	3.72	3.75	3.73	526	19.6
OH	263	17,338	643	3.71	3.80	3.71	4,560	169.2
OK	78	16,192	586	3.62	....	3.62	1,263	45.7
OR	120	18,917	696	3.68	....	3.68	2,270	83.5
PA	562	17,904	655	3.66	3.76	3.66	10,062	368.3
RI	1.2	16,333	622	3.81	....	3.81	19.6	0.7
SC	17	16,882	618	3.66	....	3.66	287	10.5
SD	80	16,838	631	3.74	3.76	3.75	1,347	50.5
TN	75	15,400	557	3.62	....	3.62	1,155	41.8
TX	319	18,837	695	3.69	....	3.69	6,009	221.7
UT	88	18,284	660	3.61	3.74	3.61	1,609	58.1
VT	145	17,821	665	3.73	....	3.73	2,584	96.4
VA	105	16,486	590	3.58	....	3.58	1,731	62.0
WA	237	22,852	841	3.68	....	3.68	5,416	199.3
WV	13	14,923	545	3.65	....	3.65	194	7.1
WI	1,241	17,796	667	3.74	3.84	3.75	22,085	828.2
WY	4.3	14,744	543	3.68	3.70	3.68	63.4	2.3
US[3]	9,010	18,957	696	3.67	3.82	3.67	170,805	6,265.8
PR	94	8,277	270	3.26	3.26	3.26	778	25.4

[1] Average number during year, excluding heifers not yet fresh. U.S. total may not add due to rounding. [2] Excludes milk sucked by calves. [3] Sum of parts may not equal due to rounding.

NASS, Livestock Branch, (202) 720–3570.

DAIRY AND POULTRY STATISTICS

Table 8-13.—Milk: Quantities used and marketed by producers, by States, 2004 (preliminary)

State	Milk used where produced			Milk marketed by producers	
	Fed to calves [1]	Used for milk, cream, and butter	Total	Total quantity [2]	Fluid grade [3]
	Million pounds	Million pounds	Million pounds	Million pounds	Percent
AL	1	1	2	243	100
AK	0.6	0.3	0.9	13.7	100
AZ	12	1	13	3,633	100
AR	5	3	8	310	100
CA	31	5	36	36,429	98
CO	22	3	25	2,159	100
CT	3.5	0.5	4.0	388.0	100
DE	1.0	0.1	1.1	126.4	100
FL	4	1	5	2,248	100
GA	13	1	14	1,402	100
HI	1.4	0.6	2.0	78.5	100
ID	34	3	37	9,056	99
IL	9	2	11	1,967	98
IN	20	4	24	2,938	98
IA	29	10	39	3,804	98
KS	10	1	11	2,205	100
KY	29	2	31	1,392	100
LA	9	2	11	468	100
ME	4.5	0.5	5.0	607.0	100
MD	7	2	9	1,153	100
MA	2.5	0.5	3.0	293.0	100
MI	50	5	55	6,260	99
MN	95	5	100	8,002	97
MS	1	1	2	377	100
MO	20	5	25	1,822	96
MT	2	2	4	343	100
NE	10	1	11	1,040	99
NV	5	1	6	503	100
NH	2.5	0.5	3.0	299.0	100
NJ	2	1	3	197	100
NM	64	22	86	6,624	100
NY	40	2	42	11,608	100
NC	9	4	13	993	100
ND	10	1	11	515	77
OH	25	5	30	4,530	95
OK	13	1	14	1,249	100
OR	19	4	23	2,247	100
PA	10	1	11	10,051	99
RI	0.2		0.2	19.4	100
SC	2	1	3	284	100
SD	9	2	11	1,336	94
TN	3	1	4	1,151	100
TX	21	2	23	5,986	100
UT	12	2	14	1,595	99
VT	15	2	17	2,567	100
VA	6	2	8	1,723	100
WA	25	1	26	5,390	100
WV	2	1	3	191	100
WI	234	30	264	21,821	96
WY	1.0	0.2	1.2	62.2	83
US [4]	956	149	1,105	169,699	98
PR	10	3	13	765	99

[1] Excludes milk sucked by calves. [2] Milk sold to plants and dealers as whole milk and equivalent amounts of milk for cream. Includes milk produced by dealers' own herds and small amounts sold directly to consumers. Also includes milk produced by institutional herds. [3] Percentage of milk sold that is eligible for fluid use (grade A for fluid use in most States). Includes fluid-grade milk used in manufacturing dairy products. [4] May not add due to rounding.

NASS, Livestock Branch, (202) 720–3570.

Table 8-14.—Milk production: Marketings, income, and value, by States, 2004
(preliminary)

| State | Milk utilized | Average returns per cwt. [1] | | | Re-turns per lb milkfat | Cash receipts from mar-ketings | Used for milk, cream, and butter where produced | | Gross producer income [3] | Value of milk pro-duced [2] [4] |
		Fluid grade	Manuf. grade	All milk			Milk utilized	Value [2]		
	Million pounds	*Dollars*	*Dollars*	*Dollars*	*Dollars*	*1,000 dollars*	*Million pounds*	*1,000 dollars*	*1,000 dollars*	*1,000 dollars*
AL	243	17.90	....	17.90	4.90	43,497	1	179	43,676	43,855
AK	13.7	20.60	....	20.60	6.09	2,822	0.3	62	2,884	3,008
AZ	3,633	15.70	....	15.70	4.35	570,381	1	157	570,538	572,422
AR	310	16.80	....	16.80	4.63	52,080	3	504	52,584	53,424
CA	36,429	14.72	15.20	14.73	4.01	5,365,992	5	737	5,366,729	5,371,295
CO	2,159	15.90	....	15.90	4.49	343,281	3	477	343,758	347,256
CT	388	17.30	....	17.30	4.73	67,124	0.5	87	67,211	67,816
DE	126.4	17.00	....	17.00	4.55	21,488	0.1	17	21,505	21,675
FL	2,248	19.20	....	19.20	5.30	431,616	1	192	431,808	432,576
GA	1,402	16.80	....	16.80	4.62	235,536	1	168	235,704	237,888
HI	78.5	25.70	....	25.70	7.30	20,175	0.6	154	20,329	20,689
ID	9,056	15.00	16.00	15.00	4.17	1,358,400	3	450	1,358,850	1,363,950
IL	1,967	15.70	16.00	15.70	4.21	308,819	2	314	309,133	310,546
IN	2,938	16.80	14.60	16.70	4.56	490,646	4	668	491,314	494,654
IA	3,804	16.30	15.50	16.30	4.49	620,052	10	1,630	621,682	626,409
KS	2,205	15.40	....	15.40	4.23	339,570	1	154	339,724	341,264
KY	1,392	17.00	....	17.00	4.68	236,640	2	340	236,980	241,910
LA	468	16.50	....	16.50	4.67	77,220	2	330	77,550	79,035
ME	607	18.00	....	18.00	4.86	109,260	0.5	90	109,350	110,160
MD	1,153	17.00	....	17.00	4.63	196,010	2	340	196,350	197,540
MA	293	17.50	....	17.50	4.70	51,275	0.5	88	51,363	51,800
MI	6,260	16.30	14.70	16.30	4.49	1,020,380	5	815	1,021,195	1,029,345
MN	8,002	16.70	15.30	16.70	4.44	1,336,334	5	835	1,337,169	1,353,034
MS	377	16.80	....	16.80	4.67	63,336	1	168	63,504	63,672
MO	1,822	16.40	15.60	16.40	4.48	298,808	5	820	299,628	302,908
MT	343	15.50	....	15.50	4.28	53,165	2	310	53,475	53,785
NE	1,040	16.20	....	16.20	4.39	168,480	1	162	168,642	170,262
NV	503	14.90	....	14.90	4.27	74,947	1	149	75,096	75,841
NH	299	17.70	....	17.70	4.73	52,923	0.5	89	53,012	53,454
NJ	197	16.40	....	16.40	4.48	32,308	1	164	32,472	32,800
NM	6,624	15.10	....	15.10	4.24	1,000,224	22	3,322	1,003,546	1,013,210
NY	11,608	16.80	....	16.80	4.60	1,950,144	2	336	1,950,480	1,957,200
NC	993	17.20	....	17.20	4.75	170,796	4	688	171,484	173,032
ND	515	16.70	15.00	16.30	4.37	83,945	1	163	84,108	85,738
OH	4,530	16.70	15.10	16.60	4.47	751,980	5	830	752,810	756,960
OK	1,249	17.50	....	17.50	4.83	218,575	1	175	218,750	221,025
OR	2,247	16.00	....	16.00	4.35	359,520	4	640	360,160	363,200
PA	10,051	17.60	16.20	17.60	4.81	1,768,976	1	176	1,769,152	1,770,912
RI	19.4	17.70	....	17.70	4.65	3,434	....	....	3,434	3,469
SC	284	17.40	....	17.40	4.75	49,416	1	174	49,590	49,938
SD	1,336	16.60	15.10	16.50	4.40	220,440	2	330	220,770	222,255
TN	1,151	16.80	....	16.80	4.64	193,368	1	168	193,536	194,040
TX	5,986	16.30	....	16.30	4.42	975,718	2	326	976,044	979,467
UT	1,595	15.70	16.20	15.70	4.35	250,415	2	314	250,729	252,613
VT	2,567	16.90	....	16.90	4.53	433,823	2	338	434,161	436,696
VA	1,723	17.40	....	17.90	5.00	308,417	2	358	308,775	309,849
WA	5,390	15.90	....	15.90	4.32	857,010	1	159	857,169	861,144
WV	191	16.50	....	16.50	4.52	31,515	1	165	31,680	32,010
WI	21,821	16.90	15.80	16.90	4.51	3,687,749	30	5,070	3,692,819	3,732,365
WY	62.2	15.90	15.20	15.80	4.29	9,828	0.2	32	9,860	10,017
US	169,699	16.13	15.45	16.13	4.40	27,367,858	149	24,414	27,392,272	27,549,413
PR	765	24.00	15.60	23.90	7.33	182,835	3	717	183,552	185,942

[1] Cash receipts divided by milk or milkfat in combined marketings.   [2] Value at averaged returns per 100 pounds of milk in combined marketings of milk and cream.   [3] Cash receipts from marketings of milk and cream plus value of milk used for home consumption.   [4] Includes value of milk fed to calves.
NASS, Livestock Branch, (202) 720–3570.

Table 8-15.—Milk: Cows, yield per cow, and production in specified countries, 2002-2004

Country and continent	Milk cows			Per cow yield			Milk production		
	2002	2003	2004	2002	2003	2004	2002	2003	2004
	1,000 head	1,000 head	1,000 head	Kilo grams	Kilo grams	Kilo grams	1,000 metric tons	1,000 metric tons	1,000 metric tons
North America:.									
Canada	1,084	1,065	1,057	7,347	7,262	7,460	7,964	7,734	7,885
Mexico	6,800	6,800	6,800	1,406	1,439	1,452	9,560	9,784	9,874
United States	9,139	9,083	9,010	8,441	8,509	8,599	77,140	77,290	77,477
Total	17,023	16,948	16,867	....	....	....	94,664	94,808	95,236
South America:.									
Argentina	2,150	2,000	2,000	3,953	3,975	4,375	8,500	7,950	9,250
Brazil	15,600	15,300	15,200	1,451	1,494	1,520	22,635	22,860	23,317
Peru	620	630	650	1,926	1,946	1,969	1,194	1,226	1,280
Total	18,370	17,930	17,850	....	....	....	32,329	32,036	32,567
European Union-25:	25,140	24,456	23,963	5,212	5,391	5,459	131,040	131,847	130,825
Eastern Europe:.									
Romania	1,550	1,684	1,694	3,323	3,207	3,353	5,150	5,400	5,723
Total	1,550	1,684	1,694	3,323	3,207	3,353	5,150	5,400	5,723
Former Soviet Union:.									
Russia	12,200	11,700	11,200	2,746	2,821	2,857	33,500	33,000	32,000
Ukraine	4,918	4,715	4,330	2,818	2,842	3,067	13,860	13,400	13,787
Total	17,118	16,415	15,530	....	....	....	47,360	46,400	45,787
South Asia:.									
India	36,000	36,500	37,000	1,006	1,000	1,014	36,200	36,500	37,500
Total	36,000	36,500	37,000	1,006	1,000	1,014	36,200	36,500	37,500
Asia:.									
China	3,420	4,466	5,466	3,801	3,910	4,034	12,998	17,463	22,606
Japan	966	964	936	8,680	8,714	8,903	8,385	8,400	8,329
Total	4,386	5,430	6,402	....	....	....	21,383	25,863	30,935
Oceania:.									
Australia [1]	2,369	2,050	2,036	4,900	5,188	5,097	11,608	10,636	10,377
New Zealand [2]	3,749	3,842	3,920	3,714	3,734	3,827	13,925	14,346	15,000
Total	6,118	5,892	5,956	....	....	....	25,533	24,982	25,377
World total	125,705	125,255	125,262	....	....	....	393,659	397,836	403,950

[1] Year ending June 30 of the year shown.   [2] Year ending May 31 of the year shown.
FAS, Dairy, Livestock and Poultry Division, (202) 720-8870. Data from counselor/attaché reports and official statistics.

Table 8-16.—Milk: Quantities used and marketed by farmers, United States, 1995-2004

Year	Milk used on farms where produced			Milk marketed by producers	
	Fed to calves [1]	Consumed as fluid milk or cream	Total	Total [2]	Fluid grade [3]
	Million pounds	Million pounds	Million pounds	Million pounds	Percent
1995	1,216	340	1,556	153,737	96
1996	1,175	301	1,476	152,531	96
1997	1,138	256	1,394	154,697	97
1998	1,142	235	1,377	155,885	97
1999	1,107	219	1,326	161,263	98
2000	1,109	198	1,307	166,086	98
2001	1,036	173	1,209	164,123	98
2002	959	160	1,119	168,944	98
2003	964	155	1,119	169,276	98
2004 [4]	956	149	1,105	169,699	98

[1] Excludes milk sucked by calves.   [2] Milk sold to plants and dealers as whole milk and equivalent amounts of milk for cream. Includes milk produced by dealers' own herds and small amounts sold directly to consumers. Also includes milk produced by institutional herds.   [3] Percentage of milk sold that is eligible for fluid use (Grade A in most States). Includes fluid-grade milk used in manufacturing dairy products.   [4] Preliminary.
NASS, Livestock Branch, (202) 720-3570.

Table 8-17.—Federal milk order markets: Measures of growth, 1995–2004[1]

Year	Number of markets[2]	Population of Federal milk marketing areas	Number of Handlers[2]	Number of Producers[3]	Receipts of producer milk	Producer milk used in Class I	Percentage of producer milk used in Class I
	Number	Thousands	Number	Number	Million pounds	Million pounds	Percent
1995	33	207,548	571	88,717	108,548	45,004	41.5
1996	32	209,599	570	82,947	104,501	45,479	43.5
1997	31	208,379	570	78,422	105,224	44,917	42.7
1998	31	210,484	522	72,402	99,223	44,968	45.3
1999	31	212,118	487	69,008	104,479	45,216	43.3
2000	11	228,899	346	69,590	116,920	45,989	39.3
2001	11	231,487	350	66,423	120,223	45,887	38.2
2002	11	234,256	338	63,856	125,546	46,043	36.7
2003	11	236,180	331	58,110	110,581	45,843	41.5
2004	10	234,825	306	52,341	103,048	44,939	43.6

Year	Prices at 3.5 percent butterfat content per hundredweight[4]		Receipts as percentage of milk sold to plants and dealers		Daily deliveries of milk per producer	Gross value of receipts of producer milk[5]	
	Class I	Blend	Fluid grade	All milk		Per producer	All producers
	Dollars	Dollars	Percent	Percent	Pounds	Dollars	1,000 dollars
1995	14.19	12.79	75	71	3,350	157,754	13,995,454
1996	16.19	14.64	72	69	3,442	187,713	15,570,261
1997	14.36	13.10	71	69	3,676	178,424	13,992,366
1998	16.14	14.92	66	64	3,755	202,770	14,681,340
1999	16.24	14.09	67	65	4,148	216,794	14,960,544
2000	14.24	12.11	72	70	4,590	207,913	14,468,892
2001	16.96	14.90	75	73	4,959	275,642	18,308,968
2002	13.69	11.91	77	76	5,387	239,520	15,294,802
2003	14.10	12.12	67	65	5,178	242,066	14,066,672
2004	17.56	15.74	62	61	5,352	324,712	16,996,426

[1] Over this period, handlers elected periodically not to pool substantial volumes of milk that normally would have been pooled under Federal orders. This decision resulted from disadvantageous blend/class price relationships and qualification circumstances. This fact should be kept in mind if year-to-year comparisons are made using the various "producer deliveries" measures of growth. [2] End of year. [3] Average for year. [4] Prices are weighted averages. [5] Based on blend (uniform) price adjusted for butterfat content, and in later years, other milk components of producer milk.

AMS, Dairy Programs, (202) 720–7461.

Table 8-18.—Milk production: Marketings, income and value, United States, 1995–2004

Year	Combined marketings of milk and cream				Used for milk, cream, and butter on farms where produced		Gross farm income from dairy products[4]	Farm value of all milk produced[3][5]
	Milk utilized	Average returns[2]		Cash receipts from marketings	Milk utilized	Value[3]		
		Per 100 pounds milk	Per pound milkfat					
	Million pounds	Dollars	Dollars	1,000 dollars	Million pounds	1,000 dollars	1,000 dollars	1,000 dollars
1995	153,737	12.93	3.53	19,876,353	340	44,522	19,920,875	20,079,217
1996	152,531	14.94	4.05	22,781,435	301	45,304	22,826,739	23,002,715
1997	154,697	13.53	3.70	20,936,726	256	34,854	20,971,580	21,125,886
1998	155,885	15.46	4.22	24,105,134	235	36,487	24,141,621	24,318,718
1999	161,263	14.38	3.92	23,189,113	219	32,021	23,221,134	23,381,760
2000	166,086	12.40	3.37	20,586,629	198	24,777	20,611,406	20,749,871
2001	164,123	15.04	4.10	24,685,667	173	26,269	24,711,936	24,869,285
2002	168,944	12.20	3.31	20,582,238	160	19,816	20,602,054	20,720,482
2003	169,276	12.56	3.42	21,238,737	155	19,776	21,258,513	21,381,324
2004[1]	169,699	16.13	4.40	27,367,858	149	24,414	27,392,272	27,549,413

[1] Preliminary. [2] Cash receipts divided by milk or milkfat represented in combined marketings. [3] Valued at average returns per 100 pounds of milk in combined marketings of milk and cream. [4] Cash receipts from marketings of milk and cream plus value of milk used for home consumption. [5] Includes value of milk fed to calves.

NASS, Livestock Branch, (202) 720–3570.

Table 8-19.—Dairy products: Quantities manufactured, United States, 2000–2004

Product	2000	2001	2002	2003	2004[1]
	1,000 pounds	*1,000 pounds*	*1,000 pounds*	*1,000 pounds*	*1,000 pounds*
Butter	1,256,032	1,231,838	1,355,147	1,242,360	1,249,678
All American cheese	3,641,624	3,544,185	3,690,978	3,621,656	3,738,776
Cheddar cheese	2,819,023	2,746,691	2,822,099	2,701,064	3,004,427
Swiss cheese	229,322	245,504	254,096	264,707	281,176
Muenster cheese	85,475	82,222	81,088	79,360	72,215
Brick cheese	8,608	8,706	9,993	9,751	8,147
Limburger cheese	637	702	651	712	872
Cream and Neufchatel cheese	687,440	645,056	686,183	676,662	699,144
Hispanic cheese	96,303	108,810	124,481	133,676	142,400
Mozzarella	2,634,999	2,767,784	2,783,272	2,807,188	2,916,536
All Italian varieties of cheese	3,288,911	3,425,886	3,470,014	3,524,002	3,660,290
All other varieties of cheese	219,678	199,557	229,783	246,705	273,429
Total of all cheese	8,257,998	8,260,628	8,547,267	8,557,243	8,876,463
Cottage cheese:					
Curd[2]	460,974	453,195	436,618	447,981	458,144
Creamed[2]	371,460	371,623	374,162	385,156	377,229
Lowfat[2]	363,658	370,233	374,293	384,372	391,219
Sweetened condensed milk:					
Bulk goods:					
Skimmed	34,611	32,616	22,345	22,896	30,669
Unskimmed	70,803	70,212	76,892	76,091	76,132
Unsweetened condensed milk:					
Bulk goods:					
Skimmed	1,021,907	937,027	1,035,633	919,056	903,794
Unskimmed	74,841	70,132	56,028	128,258	116,856
Evaporated and condensed milk:					
Case goods:					
Skimmed	23,488	14,972	19,744	17,465	19,089
Unskimmed	441,986	452,846	573,231	577,840	529,909
Condensed or evaporated buttermilk	19,963	35,063	55,875	41,118	49,646
Dry buttermilk	56,245	51,712	54,886	52,220	54,979
Dry whole milk	111,377	41,201	47,411	38,620	41,587
Nonfat dry milk	1,451,751	1,413,777	1,595,939	1,589,041	1,406,390
Dry skim milk (animal feed)	5,567	5,507	7,565	5,601	5,243
Dry whey	1,187,903	1,045,655	1,115,321	1,085,165	1,034,898
Yogurt plain & fruit flavored	1,836,591	2,002,825	2,310,582	2,506,562	2,708,654
	1,000 gallons	*1,000 gallons*	*1,000 gallons*	*1,000 gallons*	*1,000 gallons*
Ice cream, regular[3]	979,645	970,121	1,004,992	992,876	943,659
Ice cream, lowfat[4]	373,383	380,165	338,538	398,265	415,474
Ice cream, nonfat	30,735	22,391	21,050	20,364	22,719
Sherbet (does not include water ices)	51,933	52,634	56,998	54,126	54,626
Frozen yogurt	94,478	71,153	70,771	70,394	67,672

[1] Preliminary.   [2] Cottage cheese curd includes pot and bakers' cheese. Creamed cottage cheese contains not less than 4 percent milkfat. Lowfat cottage cheese contains less than 4 percent milkfat.   [3] Contains minimum milkfat content of 10 percent and not less than 4.5 pounds per gallon.   [4] Includes freezer-made milkshake in most States. Contains less than 10 percent milkfat required for ice cream.

NASS, Livestock Branch, (202) 720–3570.

Table 8-20.—Dairy products: Average price per pound for specified products, 2000–2004

Item and market	2000	2001	2002	2003	2004
	Dollars	*Dollars*	*Dollars*	*Dollars*	*Dollars*
Butter, Chicago Mercantile Exchange:					
Grade AA:					
High[1]	1.8525	2.2250	1.3850	1.4850	2.3650
Low[1]	0.8750	1.1275	0.9250	1.0000	1.3900
Butter, National Agricultural Statistics Service, Grade AA:[2]	1.1408	1.6304	1.0931	1.1194	1.8239
Cheese, Cheddar, Chicago Mercantile Exchange, Barrels:					
High[1]	1.2900	1.6825	1.3500	1.5850	2.1700
Low[1]	0.9900	1.0650	1.0300	1.0250	1.2350
Cheese, Cheddar, Chicago Mercantile Exchange, 40-lb blocks:					
High[1]	1.3350	1.7800	1.3900	1.6000	2.2000
Low[1]	0.9800	1.0675	1.0175	0.9925	1.3000
Cheese, Cheddar, National Agricultural Statistics Service, Barrels:[2]	1.0985	1.4039	1.1575	1.2771	1.6216
Cheese, Cheddar, National Agricultural Statistics Service, 40-lb blocks:[2]	1.1332	1.4165	1.1808	1.2970	1.6325
Nonfat dry milk, National Agricultural Statistics Service:					
Low/medium heat	1.0115	0.9791	0.9043	0.8090	0.8405
Whey Powder, National Agricultural Statistics Service:					
Edible (nonhygroscopic)	0.1863	0.2700	0.1974	0.1667	0.2319

[1] Figures are the high and low prices for any trading day during the year.   [2] Prices used in Federal milk order price formulas. Averages were computed by Agricultural Marketing Service.

AMS, Dairy Programs, (202) 720–7461.

Table 8-21.—Dairy Products: Factory production of specified items, by States, 2003–2004

State	Butter		Total American cheese[2]		Total cheese[3]	
	2003	2004[1]	2003	2004[1]	2003	2004[1]
	1,000 pounds	*1,000 pounds*	*1,000 pounds*	*1,000 pounds*	*1,000 pounds*	*1,000 pounds*
CA	363,833	388,969	789,629	828,974	1,830,927	1,996,428
ID			481,045			718,245
IL					95,547	93,609
IA			128,445	122,695	162,712	154,757
MA					860	841
MN			592,385	594,146	616,853	623,968
MO					97,400	99,645
NY	24,773	20,216	88,266	83,113	706,684	699,560
OH			34,376	31,283	166,835	182,377
OR			100,619	104,879		
PA	61,928	60,327		98	368,945	376,730
SD			81,362		151,635	159,278
UT					74,055	67,294
WI	309,264	318,811	828,414	860,379	2,276,528	2,356,516
Other	482,562	461,355	497,115	1,113,209	2,008,262	1,347,215
US	1,242,360	1,249,678	3,621,656	3,738,776	8,557,243	8,876,463

State	Total ice cream, regular		Nonfat dry milk for human food	
	2003	2004[1]	2003	2004[1]
	Pounds	*Pounds*	*Pounds*	*Pounds*
CA	133,531	130,607	738,303	736,750
FL	28,005	27,557		
IL	48,719	45,772		
IN	92,925	98,032		
MD	19,229	20,552		
MI	17,322	18,897		
MN	41,718	42,078		
MO	22,934	24,652		
NY	33,436	31,414		
NC	17,723	15,658		
OH	32,157	24,880		
OR	14,014	14,837		
PA	52,574	44,633		79,879
TN	17,884			
TX	52,330	53,280		
UT	18,278	23,787	1,171	1,087
Other	369,326	327,023	849,567	588,674
US	992,876	943,659	1,589,041	1,406,390

[1] Preliminary. [2] Includes Colby, washed curd, high and low moisture Jack, and Monterey. [3] Includes full-skim American cheese; excludes cottage cheese.
NASS, Livestock Branch, (202) 720–3570.

Table 8-22.—Fluid milk and cream: Total and per capita consumption, United States, 1995–2004[1]

Year	Consumption	
	Total	Per capita
	Billion pounds	*Pounds*
1995	58.8	221
1996	59.2	220
1997	59.0	216
1998	58.9	213
1999	59.5	213
2000	59.3	210
2001	59.2	208
2002	59.5	207
2003	60.3	208
2004	59.8	204

[1] Sales of beverage, cream, and specialty fluid products plus farm household use.
ERS, Animal Products Branch, (202) 694–5180.

Table 8-23.—Milk cows, milk, and fat in cream: Average prices received by farmers, United States, 1995–2004

Year	Milk cows, per head [1]	Milk per 100 pounds [2]					
		Eligible for fluid market [3]		Of manufacturing grade		All milk wholesale	
		Price per 100 lb.	Fat test	Price per 100 lb.	Fat test	Price per 100 lb.	Fat test
	Dollars	Dollars	Percent	Dollars	Percent	Dollars	Percent
1995	1,130.00	12.80	3.65	11.79	3.75	12.78	3.66
1996	1,090.00	14.79	3.69	13.43	3.78	14.75	3.69
1997	1,100.00	13.40	3.65	12.17	3.77	13.36	3.66
1998	1,120.00	15.50	3.65	14.24	3.77	15.46	3.66
1999	1,280.00	14.42	3.67	12.84	3.79	14.38	3.67
2000	1,340.00	12.44	3.68	10.52	3.79	12.40	3.68
2001	1,500.00	15.08	3.67	13.44	3.78	15.04	3.67
2002	1,600.00	12.20	3.68	10.89	3.80	12.18	3.68
2003	1,340.00	12.56	3.66	11.71	3.80	12.55	3.67
2004	1,580.00	16.13	3.67	15.45	3.82	16.13	3.67

[1] Simple average of quarterly prices, by States, weighted by the number of milk cows on farms Jan. 1 of the current year.　[2] Average price at average fat test for all milk sold at wholesale to plants and dealers, based on reports from milk-market administrators, cooperative milk-market associations, whole-milk distributors, and milk-products manufacturing plants, f.o.b. plant or receiving station (whichever is the customary place for determining prices) before hauling costs are deducted and including all premiums.　[3] Includes fluid milk surplus diverted to manufacturing.

NASS, Livestock Branch, (202) 720–3570.

Table 8-24.—Dairy products: Manufacturers' average selling price [1] of specified products, United States, 1995–2004

Year	Dry skim milk for animal feed, per pound, f.o.b. factory	Dry whole milk, per pound, f.o.b. factory
	Cents	Cents
1995	50.13	112.70
1996	60.22	128.32
1997	59.49	117.60
1998	51.14	129.47
1999	51.92	125.59
2000	54.32	120.16
2001		134.48
2002		116.51
2003		108.45
2004		131.31

[1] Includes milk sold in bulk and in package.
NASS, Livestock Branch, (202) 720–3570.

Table 8-25.—Dairy products: Manufacturers' stocks, end of month, United States, 2003 and 2004

Month	Evaporated and sweetened condensed whole milk (case goods)		Dry whole milk		Nonfat dry milk (human food)	
	2003	2004	2003	2004	2003	2004
	1,000 pounds	1,000 pounds	1,000 pounds	1,000 pounds	1,000 pounds	1,000 pounds
January	52,615	43,577	2,614	3,305	102,449	104,197
February	53,244	45,832	2,737	3,755	114,406	96,026
March	53,681	49,792	3,353	2,736	123,418	80,330
April	57,233	50,755	4,453	4,469	129,585	104,241
May	63,464	55,640	3,088	4,093	127,940	127,307
June	76,267	68,280	2,959	3,588	120,189	146,588
July	86,926	82,202	2,878	2,601	86,090	161,842
August	89,024	87,157	1,440	2,132	72,268	150,681
September	77,020	77,637	2,009	1,278	62,261	127,591
October	58,877	54,331	899	1,365	70,072	115,844
November	38,595	37,541	1,726	1,198	87,145	94,802
December	37,257	35,966	1,981	1,556	110,822	98,195

NASS, Livestock Branch, (202) 720–3570.

Table 8-26.—Milk markets under Federal order program: Whole milk and fat-reduced milk products sold for fluid consumption within defined marketing areas, 2003 [1]

Federal milk order marketing area	Whole milk products [2]		Fat-reduced milk products [3]		Total fluid milk products	
	Quantity	Butterfat content	Quantity	Butterfat content	Quantity	Butterfat content
	Million pounds	Percent	Million pounds	Percent	Million pounds	Percent
Northeast	4,039	3.26	5,566	1.15	9,606	2.04
Appalachian	1,334	3.30	2,118	1.35	3,452	2.10
Southeast	2,065	3.28	2,744	1.36	4,809	2.18
Florida	1,316	3.31	1,572	1.21	2,888	2.17
Mideast	1,656	3.30	4,661	1.37	6,317	1.87
Upper Midwest	798	3.33	3,567	1.17	4,364	1.56
Central	1,241	3.28	3,419	1.30	4,660	1.83
Southwest	2,029	3.32	2,158	1.36	4,187	2.31
Arizona-Las Vegas [4]	469	3.29	834	1.38	1,303	2.07
Western	170	3.38	722	1.38	893	1.76
Pacific Northwest	429	3.43	1,731	1.36	2,159	1.77
Combined areas	5,547	3.30	29,091	1.28	44,638	1.98

[1] In-area sales include total sales in each of the areas by handlers regulated under the respective order, by handlers regulated under other orders, by partially regulated handlers, and by producer-handlers. Sales routes of handlers may extend outside defined marketing areas; therefore, some handlers' in-area sales are partially estimated. [2] Plain, flavored, and miscellaneous whole milk products, and eggnog. [3] Plain fortified and flavored reduced fat milk (2%), low fat milk (1%), and fat-free milk (skim), and miscellaneous fat-reduced milk products, and buttermilk. [4] The data for this order does not include all the sales in the marketing area due to the reporting exemption of the fluid milk processor located in Clark County, Nevada.

AMS, Dairy Programs, (202) 720–7461.

Table 8-27.—Milk markets under Federal order program: Whole milk and fat-reduced milk products sold for fluid consumption within defined marketing areas, 2004 [1]

Federal milk order marketing area	Whole milk products [2]		Fat-reduced milk products [3]		Total fluid milk products	
	Quantity	Butterfat content	Quantity	Butterfat content	Quantity	Butterfat content
	Million pounds	Percent	Million pounds	Percent	Million pounds	Percent
Northeast	3,889	3.26	5,672	1.16	9,562	2.01
Appalachian	1,291	3.30	2,191	1.34	3,482	2.07
Southeast	2,008	3.26	2,776	1.37	4,784	2.16
Florida	1,297	3.33	1,641	1.25	2,938	2.16
Mideast	1,597	3.30	4,693	1.37	6,290	1.86
Upper Midwest	779	3.31	3,563	1.16	4,342	1.54
Central	1,183	3.28	3,464	1.30	4,647	1.80
Southwest	1,958	3.31	2,262	1.37	4,220	2.27
Arizona-Las Vegas [4]	430	3.30	860	1.38	1,291	2.02
Western [5]	44	3.30	187	1.38	231	1.74
Pacific Northwest	433	3.44	1,723	1.36	2,156	1.78
Combined areas	14,910	3.29	29,032	1.29	43,942	1.97

[1] In-area sales include total sales in each of the areas by handlers regulated under the respective order, by handlers regulated under other orders, by partially regulated handlers, and by producer-handlers. Sales routes of handlers may extend outside defined marketing areas; therefore, some handlers' in-area sales are partially estimated. [2] Plain, flavored, and miscellaneous whole milk products and eggnog. [3] Plain, fortified and flavored reduced fat milk (2%), low fat milk (1%), fat-free milk (skim), miscellaneous fat-reduced milk products, and buttermilk. [4] The data for this order does not include all the sales in the marketing area due to the reporting exemption of the fluid milk processor located in Clark County, Nevada. [5] Effective 4/1/04, the order regulating this marketing area was terminated. Data are for January-March and are included in combined areas.

AMS, Dairy Programs, (202) 720–7461.

Table 8-28.—Supply and utilization, United States, 2003–2004

Product	Product pounds		Butterfat		Solids nonfat	
	2003	2004	2003	2004	2003	2004
	Million pounds					
Supply:						
Milk production	170,394	170,804	6,282	6,297	14,882	14,946
Net imports of ingredients	268	278	10	10	23	24
Net change in storage cream	0	0	0	0	0	0
Total supply	170,662	171,082	6,292	6,307	14,906	14,971
Utilization:						
Total butter[1]	1,242	1,250	1,008	1,013	12	12
Cheese:						
American	3,622	3,739	1,190	1,228	1,086	1,116
Other	4,936	5,138	1,159	1,267	1,207	1,321
Net cheese[2]	....	....	2,347	2,493	2,029	2,052
Total whey products[3]	2,296	2,274	17	17	2,084	2,085
Canned milk:						
Evaporated and condensed						
Whole and skim	595	549	50	46	374	344
Bulk milk:						
Condensed whole sweetened	76	76	7	7	49	49
Condensed whole unsweetened	128	117	10	9	23	21
Other condensed skim and condensed or evaporated buttermilk	1,111	1,101	13	12	324	325
Total evaporated and condensed	1,783	1,726	70	65	747	718
Dry whole milk	39	42	10	11	27	29
Nonfat dry milk	1,589	1,406	13	11	1,527	1,352
Dry buttermilk	52	55	3	3	48	50
Total dry products	1,680	1,503	26	26	1,602	1,431
Total yogurt[4]	2,507	2,709	61	66	279	302
Total sour cream[5]	935	980	196	205	36	37
Cottage cheese:						
Creamed	385	377	17	17	64	62
Low-fat	384	391	6	6	68	69
Total cottage cheese	....	....	23	23	132	131
Ice cream and other frozen dairy products.						
Ice cream:						
Regular, total	4,468	4,246	536	510	447	425
Lowfat, total	1,792	1,870	108	112	197	206
Nonfat, total	92	102	2	2	13	14
Sherbet, total	325	328	6	7	6	7
Frozen yogurt	422	406	7	7	38	37
Other frozen dairy products	33	36	2	2	3	3
Net frozen products[2]	....	....	574	550	387	430
Fluid milk[6]	54,981	54,524	1,114	1,083	4,933	4,831
Half and half	1,140	1,140	126	126	90	90
Light and heavy cream	720	720	235	235	40	40
Net fluid products[2]	....	....	1,475	1,444	5,276	4,953
Other unpublished dairy products[7]	90	111	44	39	31	37
Other food products[8]	1,549	1,534	57	56	134	133
Used where produced.						
Fed to calves	956	956	35	35	83	84
Consumed on farms	149	149	5	5	13	13
Total used by producers	1,105	1,105	41	41	97	97
Residual[9]	....	....	353	270	2,061	2,552
Residual as a percent of supply	....	....	5.6	4.3	13.8	17.0

[1] Including whey cream butter. [2] Adjustment made for duplication the use of dairy products in the manufacturing process of other dairy products. [3] Excluding whey cream butter. [4] Excludes frozen yogurt. [5] Sour cream data not available for 2002. [6] Total sales in U.S. (Source: USDA-AMS). [7] Includes anhydrous milkfat, butter oil, butterine, and other products. [8] Food products other than dairy (Source: USDA-ERS). [9] Residual, includes minor miscellaneous uses and any inaccuracies in production, utilization estimates, or milk equivalent conversions. Includes plant and shipping losses.

NASS, Livestock Branch, (202) 720–3570.

Table 8-29.—Milk markets under Federal order program: Uniform and Class I milk prices at 3.5 percent fat test, number of producers, producer milk receipts, producer milk used in Class I, Class I percentage, daily milk deliveries per producer, average fat test of producer milk receipts, by markets, 2003

Federal milk order marketing area	Class I price per cwt. [1]	Uniform price per cwt. [1][2]	Average number of producers	Receipts of producer milk	Producer milk used in Cl. I	Class I utilization	Daily milk delivery per producer	Average fat test
	Dollars	Dollars	Number	Million pounds	Million pounds	Percent	Pounds	Percent
Northeast [3]	14.69	12.93	16,114	24,038	10,701	44.5	4,087	3.70
Appalachian [4][5]	14.50	13.52	3,642	6,315	4,443	70.4	4,762	3.65
Southeast [5][6]	14.51	13.43	4,281	7,071	4,629	65.5	4,529	3.63
Florida [7]	15.34	14.69	298	2,833	2,412	85.2	26,285	3.57
Mideast [5][8]	13.43	11.83	10,376	15,750	6,546	41.6	4,165	3.69
Upper Midwest [5][9]	13.27	10.93	13,308	17,018	4,130	24.3	3,336	3.72
Central [5][10]	13.38	11.40	7,592	14,411	4,724	32.8	5,190	3.68
Southwest [5][11]	14.42	12.37	862	9,174	4,068	44.3	29,029	3.62
Arizona-Las Vegas [5][12]	13.75	11.78	106	3,061	976	31.9	79,400	3.59
Western [5][13]	13.32	10.94	705	4,573	1,109	24.3	16,975	3.60
Pacific Northwest [5][14]	13.32	11.16	827	6,336	2,105	33.2	20,792	3.66
All markets combined	14.10	12.12	58,110	110,581	45,843	41.5	5,178	3.67

[1] Prices are for milk of 3.5 percent fat content and for the principal pricing point of the market. See footnotes 3–14. [2] For those orders that use the component pricing system for paying producers (orders 1, 30, 32, 33, 124, 126, and 135), the figures are the statistical uniform price (the sum of the producer price differential and the Class III price). For those orders that use the skim milk/butterfat pricing system for paying producers (orders 5, 6, 7, and 131), the figures are the uniform price (the sum of the uniform butterfat price times 3.5 and the uniform skim milk price times 0.965). [3] Suffolk Co. (Boston), MA. [4] Mecklenburg Co. (Charlotte), NC. [5] Due to disadvantageous intraorder class and uniform price relationships in some months in these markets, handlers elected not to pool milk that normally woulld have been pooled under these orders. [6] Fulton Co. (Atlanta), GA. [7] Hillsborough Co. (Tampa), FL. [8] Cuyahoga Co. (Cleveland), OH. [9] Cook Co. (Chicago), IL. [10] Jackson Co. (Kansas City), MO. [11] Dallas Co. (Dallas), TX. [12] Maricopa Co. (Phoenix), AZ. [13] Salt Lake Co. (Salt Lake City), UT. [14] King Co. (Seattle), WA.

AMS, Dairy Programs, (202) 720-7461.

Table 8-30.—Milk markets under Federal order program: Uniform and Class I milk prices at 3.5 percent fat test, number of producers, producer milk receipts, producer milk used in Class I, Class I percentage, daily milk deliveries per producer, average fat test of producer milk receipts, by markets, 2004

Federal milk order marketing area	Class I price per cwt. [1]	Uniform price per cwt. [1][2]	Average number of producers	Receipts of producer milk	Producer milk used in Cl. I	Class I utilization	Daily milk delivery per producer	Average fat test
	Dollars	Dollars	Number	Million pounds	Million pounds	Percent	Pounds	Percent
Northeast [3][4]	18.15	16.46	15,039	22,670	10,692	47.2	4,122	3.66
Appalachian [4][5]	17.97	17.00	3,413	6,202	4,325	69.7	4,970	3.62
Southeast [4][6]	17.97	16.92	3,831	7,164	4,640	64.8	5,124	3.65
Florida [4][7]	18.88	18.39	300	2,873	2,440	84.9	26,456	3.61
Mideast [4][8]	16.85	15.33	9,940	15,940	6,493	40.7	4,364	3.67
Upper Midwest [4][9]	16.68	14.75	11,959	17,302	4,459	25.8	3,850	3.72
Central [4][10]	16.85	15.06	5,898	11,589	4,346	37.5	5,336	3.66
Southwest [4][11]	17.88	16.00	863	8,791	4,139	47.1	27,895	3.64
Arizona-Las Vegas [12]	17.16	15.51	96	2,901	967	33.3	82,723	3.61
Western [4][13][14]	13.70	12.68	689	1,096	286	26.1	17,069	3.70
Pacific Northwest [4][15]	16.80	14.75	832	6,518	2,153	33.0	21,330	3.68
All markets combined	17.56	15.74	52,341	103,048	44,939	43.6	5,352	3.67

[1] Prices are for milk of 3.5 percent butterfat content and for the principal pricing point of the market. See footnotes 3–14. [2] For those orders that use the component pricing system for paying producers (orders 1, 30, 32, 33, 124, 126, and 135), the figures are the statistical uniform price (the sum of the producer price differential and the Class III price). For those orders that use the skim milk/butterfat pricing system for paying producers (orders 5, 6, 7, and 131), the figures are the uniform price (the sum of the uniform butterfat price times 3.5 and the uniform skim milk price times 0.965). [3] Suffolk Co. (Boston), MA. [4] Due to disadvantageous intraorder class and uniform price relationships in some months in these markets, handlers elected not to pool milk that normally woulld have been pooled under these orders. [5] Mecklenburg Co. (Charlotte), NC. [6] Fulton Co. (Atlanta), GA. [7] Hillsborough Co. (Tampa), FL. [8] Cuyahoga Co. (Cleveland), OH. [9] Cook Co. (Chicago), IL. [10] Jackson Co. (Kansas City), MO. [11] Dallas Co. (Dallas), TX. [12] Maricopa Co. (Phoenix), AZ. [13] Salt Lake Co. (Salt Lake City), UT. [14] Effective 4/1/04, the order regulating this marketing area was terminated. Data are for January-March and are included in combined areas. [15] King Co. (Seattle), WA.

AMS, Dairy Programs, (202) 720-7461.

Table 8-31.—Dairy products: Total disappearance, and total and per capita consumption, United States, 1995–2004 [1]

Year	Butter			Cheese [2]			Condensed and evaporated milk [3]		
	Total dis-appear-ance	Consumption		Total dis-appear-ance	Consumption		Total dis-appear-ance	Consumption	
		Total	Per capita		Total	Per capita		Total	Per capita
	Million pounds	Million pounds	Pounds	Million pounds	Million pounds	Pounds	Million pounds	Million pounds	Pounds
1995	1,329	1,186	4.4	7,279	7,174	26.9	690	608	2.3
1996	1,190	1,148	4.3	7,478	7,365	27.3	696	611	2.3
1997	1,156	1,115	4.1	7,646	7,510	27.5	773	695	2.5
1998	1,229	1,220	4.4	7,799	7,664	27.8	638	553	2.0
1999	1,314	1,307	4.7	8,219	8,086	29.0	648	573	2.1
2000	1,289	1,277	4.5	8,580	8,406	29.8	596	560	2.0
2001	1,275	1,264	4.4	8,744	8,566	30.0	610	564	2.0
2002	1,288	1,281	4.4	8,949	8,779	30.5	706	661	2.3
2003	1,332	1,304	4.5	9,024	8,863	30.6	810	749	2.6
2004	1,356	1,354	4.6	9,369	9,171	31.2	745	649	2.2

Year	Ice cream (product weight)			Dry whole milk			Nonfat dry milk (human food)		
	Total dis-appear-ance	Consumption		Total dis-appear-ance	Consumption		Total dis-appear-ance	Consumption	
		Total	Per capita		Total	Per capita		Total	Per capita
	Million pounds	Million pounds	Pounds	Million pounds	Million pounds	Pounds	Million pounds	Million pounds	Pounds
1995	4,139	4,139	15.5	173	106	.40	1,280	910	3.4
1996	4,217	4,217	15.6	137	97	.36	1,081	1,005	3.7
1997	4,386	4,386	16.1	126	102	.37	1,171	908	3.3
1998	4,488	4,488	16.3	149	118	.43	1,120	884	3.2
1999	4,667	4,667	16.7	124	111	.40	1,275	787	2.8
2000	4,702	4,702	16.7	119	80	.28	1,073	741	2.6
2001	4,657	4,657	16.3	50	46	.16	1,156	927	3.2
2002	4,824	4,824	16.7	55	50	.17	1,362	886	3.1
2003	4,766	4,766	16.4	47	47	.16	1,758	983	3.4
2004	4,530	4,530	15.4	49	48	.16	1,877	1,251	4.3

[1] Total disappearance is based on production, imports, and change in stocks during the year. Production statistics for these commodities appear in other tables in this chapter. The total apparent consumption was obtained by subtracting ending stocks, shipments, and exports, from the total supply. The per capita consumption for each year was obtained by dividing the total apparent consumption by the number of persons.   [2] Includes all kinds of cheese except cottage and full-skim American.   [3] The evaporated milk is unskimmed, unsweetened, case goods. The condensed milk is unsweetened, unskimmed, bulk goods; and sweetened condensed milk, unskimmed, case and bulk goods.

ERS, Animal Products Branch, (202) 694–5180.

Table 8-32.—Dairy products: Dec. 31 stocks, United States, 1995–2004

Year	Butter [1][2]	Cheese [1][3]	Canned milk [1]	Dry whole milk	Nonfat dry milk for human consumption [1]
	1,000 pounds	1,000 pounds	1,000 pounds	1,000 pounds	1,000 pounds
1995	18,628	412,237	31,701	7,318	84,978
1996	13,707	487,174	19,937	6,422	71,414
1997	20,788	480,779	32,466	5,605	124,864
1998	25,910	517,647	36,495	5,161	152,172
1999	25,082	622,197	35,690	5,749	284,542
2000	24,115	708,597	41,228	4,390	662,182
2001	55,915	663,251	40,739	2,894	900,158
2002	157,820	732,551	54,428	3,244	1,145,689
2003	99,613	742,173	38,506	1,981	981,160
2004	44,988	709,715	36,363	1,556	511,549

[1] Includes Government holdings.   [2] Includes butter equivalent of butteroil held by CCC.   [3] Excludes cottage and full-skim American cheese. Includes process American cheese held by CCC.

ERS, Animal Products Branch, (202) 694–5180.

Table 8-33.—Butter: Production in specified countries, 2002–2004

Continent and country	2002	2003	2004
	1,000 metric tons	*1,000 metric tons*	*1,000 metric tons*
North America:			
Canada	77	84	86
Mexico	70	77	88
United States	615	563	567
Total	762	724	741
South America:			
Brazil	70	72	75
Total	70	72	75
Europe Union–25:	2,226	2,226	2,150
Eastern Europe:			
Romania	6	6	9
Total	6	6	9
Former USSR:			
Russia	280	280	270
Ukraine	131	148	138
Total	411	428	408
North Africa:			
Egypt	12	13	12
Total	12	13	12
Southeast Asia:			
India	2,400	2,450	2,600
Total	2,400	2,450	2,600
Asia:			
Japan	83	80	80
Total	83	80	80
Oceania:			
Australia [1]	164	163	132
New Zealand [2]	370	392	390
Total	534	555	522
Grand total	6,504	6,554	6,597

[1] Year ending June 30 of the year shown.   [2] Year ending May 31 of the year shown.

FAS, Dairy, Livestock, and Poultry Division, (202) 720–8870. Prepared or estimated on the basis of official statistics of foreign governments, other foreign source materials, reports of U.S. Agricultural Counselors, Attachés, and Foreign Service Officers, results of office research, and related information.

Table 8-34.—Cheese: Production in specified countries, 2002–2004

Continent and country	2002	2003	2004
	1,000 metric tons	*1,000 metric tons*	*1,000 metric tons*
North America:			
Canada	350	342	305
Mexico	145	126	134
United States	3,877	3,881	4,026
Total	4,372	4,349	4,465
South America:			
Argentina	370	325	370
Brazil	470	460	470
Total	840	785	840
European Union–25:	5,993	6,100	6,345
Total	5,993	6,100	6,345
Eastern Europe:			
Romania	88	23	26
Total	88	23	26
Former USSR:			
Russia	340	335	350
Ukraine	129	169	224
Total	469	504	574
North Africa:			
Egypt	410	450	455
Total	410	450	455
Asia:			
Japan	36	35	35
Korea	20	23	24
Total	56	58	59
Oceania:			
Australia [1]	413	368	389
New Zealand [2]	312	301	308
Total	725	669	697
World total	12,953	12,938	13,461

[1] Year ending June 30.   [2] Year ending May 31.

FAS, Dairy, Livestock and Poultry Division, (202) 720–8870. Prepared or estimated on the basis of official statistics of foreign governments, other foreign source materials, reports of U.S. Agricultural Counselors, Attachés, and Foreign Service Officers, results of office research, and related information.

Table 8-35.—Dairy products: United States imports by country of origin, 2002–2004

Commodity and country of origin	2002	2003	2004
	Metric tons	Metric tons	Metric tons
Cheese, all types:			
Canada	5,293	5,021	5,437
Argentina	7,555	8,037	8,931
Austria	805	573	933
Bel./Lux.	1,336	389	269
Denmark	14,073	15,013	14,380
Finland	6,838	7,996	8,618
France	18,132	18,614	21,717
Germany	10,564	7,341	8,084
Greece	2,338	2,235	2,241
Ireland	5,086	5,103	5,323
Italy	28,043	32,008	31,998
Netherlands	10,869	12,177	12,213
Portugal	545	596	540
Spain	1,580	1,898	2,217
Sweden	822	603	513
United Kingdom	7,677	6,153	6,650
Poland	2,917	4,898	2,952
Czech Republic	461	511	268
Hungary	901	511	510
Lithuania	13,080	9,511	3,874
Other EU-25	1,602	1,208	750
Total EU-2	127,669	127,338	124,050
Norway	6,159	7,032	7,237
Switzerland	6,090	6,904	7,108
Israel	492	505	513
Australia	8,210	10,721	10,074
New Zealand	45,517	37,877	35,873
Other countries	8,722	11,940	14,701
Total	215,707	215,375	213,924
Cheese, cheddar:[1]			
Canada	1,663	2,149	2,662
Germany	187	78	218
Ireland	371	374	534
United Kingdom	801	911	945
Australia	3,461	3,594	3,944
New Zealand	31,663	24,149	23,289
Other countries	493	492	557
Total	38,639	31,747	32,149
Cheese, Swiss:[2]			
Canada	90	135	479
Austria	528	285	447
Denmark	2,293	2,239	2,210
Finland	6,660	7,513	7,940
France	2,938	3,243	3,616
Germany	3,951	1,578	1,556
Ireland	1,222	1,207	1,580
Netherlands	502	690	499
Norway	5,722	6,623	6,863
Switzerland	3,319	3,753	3,812
Other countries	2,165	1,769	1,632
Total	29,390	29,035	30,634

[1] Includes American and Colby cheese. [2] Includes Emmenthaler with eye-formation.
FAS, Dairy, Livestock and Poultry Division, (202) 720–8870. Compiled from reports of the U.S. Department of Commerce.

Table 8-36.—Dairy products: United States imports by type of product, 1994–2003

Year	Dried milk[1]	Cheese				Butter[4]	Casein[5]
		Swiss[2]	Cheddar[3]	Other	Total		
	Metric tons	Metric tons	Metric tons	Metric tons	Metric tons	Metric tons	Metric tons
1995	1,128	28,047	9,472	116,877	154,396	697	93,433
1996	3,968	29,420	12,393	111,457	153,270	4,783	98,547
1997	6,080	25,094	11,566	104,825	141,485	10,956	102,404
1998	8,223	28,865	21,810	117,755	168,430	31,946	111,247
1999	10,557	34,023	30,748	132,826	197,597	18,056	108,382
2000	8,531	32,241	21,818	134,644	188,703	13,689	119,999
2001	8,077	31,403	32,215	138,153	201,771	34,614	106,827
2002	11,414	29,390	38,639	147,678	215,707	15,142	100,039
2003	8,583	29,035	31,747	154,593	215,375	14,187	116,829
2004[6]	8,760	30,634	32,149	151,141	213,924	23,726	111,299

[1] Includes whole and skimmed milk. [2] Includes Emmenthaler with eye-formation. [3] Includes American and Colby cheese. [4] Includes butter oil. [5] Includes caseinates. [6] Preliminary.
FAS, Dairy, Livestock, and Poultry Division, (202) 720–3761. Compiled from reports of the U.S. Department of Commerce.

Table 8-37.—Dairy products: Exports by principal exporting countries, 2002–2004

Commodity and country	2002	2003	2004
	Metric tons	Metric tons	Metric tons
Butter:			
United States	3	10	0
EU-25 [1]	222	307	352
Australia [2]	125	110	75
New Zealand [3]	343	386	374
Other	40	39	61
Total	733	850	868
Cheese: [4]			
United States	54	52	61
Canada	17	11	10
EU-25 [1]	516	514	515
Australia [2]	218	207	212
New Zealand [3]	277	290	289
Other	75	106	137
Total	1,157	1,180	1,239
Milk, dried whole:			
United States	0	0	0
EU-25 [1]	520	502	510
Australia [2]	213	142	173
New Zealand [3]	481	635	688
Other	270	198	248
Total	1,484	1,477	1,619
Milk, nonfat dry milk:			
Canada	49	36	16
United States	126	141	231
EU-25 [1]	267	339	282
Australia [2]	231	193	187
New Zealand [3]	248	314	305
Other	125	149	142
Total	1,046	1,172	1,164

[1] Within the European Union, exports to other members are not included. [2] Year ending June 30. [3] Year ending May 31. [4] Excludes fresh cheese.

FAS, Dairy, Livestock and Poultry Division, (202) 720–3761. Prepared on the basis of official statistics of foreign governments, other foreign source materials, reports of U.S. Agricultural Counselors, Attachés, and Foreign Service Officers, results of office research, and related information.

Table 8-38.—Dairy products: United States exports by type of product, 1995–2004

Year	Butter	Cheese	Milk and cream			
			Evaporated and condensed	WMP—Whole dried	Nonfat dry milk	Ice cream
	Metric tons	Metric tons	Metric tons	Metric tons	Metric tons	Metric tons
1995	37,689	29,519	41,378	64,297	59,311	37,827
1996	20,831	32,497	39,582	16,181	18,422	39,765
1997	14,989	37,559	9,347	48,609	62,134	36,767
1998	8,951	36,723	8,021	51,315	72,917	38,206
1999	3,208	38,341	4,821	17,656	141,315	39,701
2000	8,230	47,760	5,215	25,368	84,264	39,366
2001	3,816	52,366	10,672	46,070	96,081	40,003
2002	3,866	53,909	11,823	37,826	74,375	36,855
2003	11,626	52,101	16,707	25,352	113,333	29,201
2004	8,981	61,357	32,515	43,680	231,614	23,898

FAS, Dairy, Livestock and Poultry Division, (202) 720–3761. Compiled from reports of the U.S. Department of Commerce.

Table 8-39.—Dairy products: United States exports by country of destination, 2002–2004

Commodity and country of destination	2002	2003	2004
	Metric tons	Metric tons	Metric tons
Cheese, all types:			
Canada	6,684	5,947	5,968
Mexico	14,856	16,147	21,353
Brazil	147	40	14
Venzuela	1,042	399	238
United Kingdom	1,247	646	492
Saudi Arabia	601	621	856
Philippines	1,516	1,094	1,825
Korea	3,742	3,430	4,111
Hong Kong	495	544	837
Taiwan	639	770	904
Japan	10,145	7,950	9,432
Other countries	12,795	14,513	15,327
Total	53,909	52,101	61,357
Ice cream:			
Canada	4,961	4,175	4,229
Mexico	8,838	9,352	9,789
United Kingdom	8,151	5,142	1,170
Russia	243	120	92
Korea	1,283	530	642
Hong Kong	2,977	1,422	889
Japan	4,518	2,198	1,086
Others	5,884	6,262	6,000
Total	36,855	29,201	23,897
Milk, nonfat dry:			
Mexico	43,003	57,427	90,178
Dominican Rep.	763	965	2,762
Guatemala	305	693	6,217
El Salvador	238	233	6,132
Malaysia	3,665	640	11,431
Thailand	4,513	1,030	5,939
Vietnam	801	780	7,575
Indonesia	3,201	4,177	13,337
Philippines	3,725	11,147	22,788
Taiwan	113	255	2,600
Others	18,242	36,966	62,655
Total	74,375	113,333	231,614
Dry whey:			
Canada	30,238	23,054	22,847
Mexico	19,387	16,152	14,982
Thailand	10,388	11,041	9,546
Philippines	9,377	9,716	12,337
China	39,074	38,870	49,106
Korea	6,033	10,912	9,662
Taiwan	6,345	9,260	7,945
Japan	9,593	4,767	12,376
Other countries	20,241	15,212	22,664
Total	150,676	138,984	161,465

FAS, Dairy, Livestock and Poultry Division, (202) 720–3761. Compiled from reports of the U.S. Department of Commerce.

Table 8-40.—Dairy products: Price-support operations, United States, 1997–2006

Marketing year[1]	Manufacturing milk		Product purchase price per pound[2]		
	Support level at national average milkfat test, per cwt.	Average price received by farmers per cwt.	Butter[3]	Cheddar cheese[4]	Nonfat milk, spray process[5]
	Dollars	Dollars	Cents	Cents	Cents
1996–97	10.35	....	65.00	114.50	106.50
	([9])10.20	11.88	65.00	([9]) 113.00	([9]) 104.70
1997–98	10.20	....	65.00	113.00	104.70
	([10])10.05	13.28	65.00	([10]) 111.50	([10]) 102.80
1998–99	10.05	....	65.00	111.50	102.80
	([11]) 9.90	14.04	65.00	([11]) 110.00	([11]) 101.00
1999–2000	9.90	11.00	65.00	110.00	101.00
	....	....	([13]) 66.80	([13]) 112.20	101.00
2000–2001	9.90	12.85	([14]) 65.49	([14]) 113.14	([14]) 100.32
	....	....	([15]) 85.48	113.14	([15]) 90.00
2001–2002	9.90	11.46	85.48	113.14	90.00
2002–2003	9.90	11.10	85.48	113.14	90.00
	....	....	([16])105.00	....	([16]) 80.00
2003–2004	9.90	14.95	105.00	113.14	80.00
2004–2005	9.90	14.71	105.00	113.14	80.00
2005–2006	9.90	([12])12.55	105.00	113.14	80.00

[1] October 1–September 30. [2] Announced purchase prices for products in bulk containers. [3] U.S. Grade A or higher, salted, 25-kg blocks. [4] U.S. Grade A or higher, standard moisture basis 40-pound blocks. [5] U.S. Extra Grade, not more than 3.5 percent moisture content. Prices quoted are for product in 25-kg bags. [6] Effective July 7, 1993. [7] Effective January 1, 1996. [8] Basic Formula Price began May 1995 thru Sept. 1999. [9] Effective January 1, 1997. [10] Effective January 1, 1998. [11] Effective January 1, 1999. [12] Estimated value of milk used in manufactured products. [13] Effective July 31, 2000. [14] Effective January 31, 2001. [15] Effective June 13, 2001. [16] Effective December 1, 2002.

FSA, Dairy & Sweeteners Analysis, (202) 690–0050

Table 8-41.—Chickens: Inventory number and value, United States, Dec. 1, 1996–2005 [1]

Year	Layers 1 year old and older	Layers 20 weeks old but less than 1 year	Total layers	Pullets			Other chickens	All chickens	Value per head	Total value
				13 weeks to 20 weeks old	Under 13 weeks old	Total				
	Thou- sands	Thou- sands	Thou- sands	Thou- sands	Thou- sands	Thou- sands	Thou- sands	Thou- sands	Dollars	1,000 dol- lars
1996	138,048	165,874	303,922	33,518	48,054	....	7,243	392,737	2.65	1,039,071
1997	140,966	171,171	312,137	35,578	54,766	....	7,549	410,030	2.72	1,113,183
1998	151,298	170,350	321,828	39,864	55,981	....	7,682	425,355	2.69	1,143,041
1999	152,024	178,156	330,180	38,587	58,975	....	9,661	437,403	2.64	1,156,488
2000	153,439	180,154	333,593	38,395	56,764	....	8,088	436,840	2.44	1,064,171
2001	153,817	186,500	340,317	42,907	52,749	....	8,126	444,099	2.41	1,069,335
2002	153,884	186,325	340,209	39,865	55,424	....	8,353	443,851	2.38	1,055,316
2003	169,263	171,716	340,979	41,955	58,391	....	8,439	449,764	2.48	1,116,273
2004	([3])	([3])	343,922	([3])	([3])	101,429	8,248	453,599	2.48	1,122,923
2005 [2]	([3])	([3])	347,917	([3])	([3])	96,610	8,289	452,816	2.50	1,133,558

[1] Does not include commercial broilers. [2] Preliminary. [3] Not available due to program change.

NASS Livestock Branch, (202) 720-3570.

Table 8-42.—Chickens: Layer inventory, by State and United States, Dec. 1, 2004–2005 [1]

State	Total layers		Total pullets		Other Chickens	
	2004	2005	2004	2005	2004	2005
	Thousands	Thousands	Thousands	Thousands	Thousands	Thousands
AL	9,237	9,138	4,025	4,045	1,004	1,034
AR	15,126	14,757	7,303	7,861	1,583	1,604
CA	19,419	19,582	4,208	4,017	49	41
CO	3,960	3,932	967	656	64	66
CT	2,954	3,058	667	683	5	7
FL	10,826	11,344	2,216	2,200	45	51
GA	20,164	18,754	8,018	7,778	1,170	1,290
HI	507	486	91	61	0	0
ID	866	860	375	322	6	6
IL	4,358	4,289	569	430	21	18
IN	23,556	24,717	6,894	7,030	65	75
IA	46,592	49,951	10,877	8,434	65	70
KY	4,946	4,621	1,775	1,705	237	264
LA	1,883	1,864	585	568	106	103
ME	3,984	4,027	1,515	1,519	5	5
MD	3,2ᴗ5	2,957	1,381	1,210	30	20
MA	253	242	54	54	0	0
MI	7,720	8,357	1,615	1,752	1	1
MN	11,325	11,080	3,090	3,520	50	45
MS	6,754	6,579	3,659	3,313	735	757
MO	7,273	7,048	2,041	1,633	120	120
MT	350	340	129	138	1	2
NE	12,003	11,891	1,969	1,922	0	0
NH	162	149	72	65	2	2
NJ	1,986	1,976	98	64	0	0
NY	4,130	4,214	1,365	1,043	3	3
NC	10,901	10,941	5,619	5,940	1,200	1,080
OH	27,900	28,776	8,110	7,632	20	24
OK	3,321	3,173	1,174	1,183	252	232
OR	2,837	2,918	822	903	11	11
PA	23,290	24,305	4,532	4,511	110	110
SC	5,256	5,034	1,613	1,304	174	148
SD	3,181	3,301	542	413	0	0
TN	1,344	1,292	866	624	190	164
TX	18,539	18,688	5,563	5,593	458	466
UT	3,176	3,402	701	756	0	0
VT	198	212	25	25	2	1
VA	3,210	3,526	1,453	1,105	243	279
WA	4,892	4,873	1,040	1,192	1	1
WV	1,272	1,127	816	589	175	140
WI	5,130	4,798	1,320	1,257	30	38
WY	12	12	4	4	1	0
Other States [2]	5,899	5,326	1,671	1,556	14	11
US	343,922	347,917	101,429	96,610	8,248	8,289
PR	1,100	1,008	379	354	13	13

[1] Totals may not add due to rounding.　[2] AK, AZ, DE, KS, NV, NM, ND, and RI combined to avoid disclosing data for individual operations.

NASS, Livestock Branch, (202) 720–3570.

DAIRY AND POULTRY STATISTICS

Table 8-43.—Chicken inventory: Number, value per head, and total value, by State and United States, Dec. 1, 2004-2005 [1] [2]

State	Number		Value per bird		Total value	
	2004	2005	2004	2005	2004	2005
	1,000 head	1,000 head	Dollars	Dollars	1,000 dollars	1,000 dollars
AL	14,266	14,217	3.70	4.00	52,784	56,868
AR	24,012	24,222	4.40	4.70	105,653	113,843
CA	23,676	23,640	2.00	1.90	47,352	44,916
CO	4,991	4,654	2.00	2.00	9,982	9,308
CT	3,626	3,748	2.60	2.60	9,428	9,745
FL	13,087	13,595	2.30	2.30	30,100	31,269
GA	29,352	27,822	3.60	3.80	105,667	105,724
HI	598	547	1.90	2.00	1,136	1,094
ID	1,247	1,188	1.60	1.60	1,995	1,901
IL	4,948	4,737	0.91	1.10	4,503	5,211
IN	30,515	31,822	1.10	1.60	33,567	50,915
IA	57,534	58,455	1.90	1.40	109,315	81,837
KY	6,958	6,590	4.40	4.30	30,615	28,337
LA	2,574	2,535	3.90	2.80	10,039	7,098
ME	5,504	5,551	2.50	2.40	13,760	13,322
MD	4,641	4,187	3.00	3.00	13,923	12,561
MA	307	296	3.20	3.20	982	947
MI	9,336	10,110	1.50	1.80	14,004	18,198
MN	14,465	14,645	1.40	1.50	20,251	21,968
MS	11,148	10,649	4.70	4.50	52,396	47,921
MO	9,434	8,801	2.10	2.30	19,811	20,242
MT	480	480	3.30	2.70	1,584	1,296
NE	13,972	13,813	1.90	2.10	26,547	29,007
NH	236	216	4.40	5.40	1,038	1,166
NJ	2,084	2,040	1.00	1.00	2,084	2,040
NY	5,498	5,260	1.90	1.40	10,446	7,364
NC	17,720	17,961	4.40	5.40	77,968	96,989
OH	36,030	36,432	1.50	1.20	54,045	43,718
OK	4,747	4,588	4.10	4.80	19,463	22,022
OR	3,670	3,832	1.80	1.70	6,606	6,514
PA	27,932	28,926	1.90	1.90	53,071	54,959
SC	7,043	6,486	2.90	3.40	20,425	22,052
SD	3,723	3,714	2.10	2.00	7,818	7,428
TN	2,400	2,080	6.60	6.60	15,840	13,728
TX	24,560	24,747	2.40	2.70	58,944	66,817
UT	3,877	4,158	1.30	1.70	5,040	7,069
VT	225	238	1.90	1.90	428	452
VA	4,906	4,910	3.80	3.80	18,643	18,658
WA	5,933	6,066	2.40	2.00	14,239	12,132
WV	2,263	1,856	5.60	5.30	12,673	9,837
WI	6,480	6,093	1.80	1.90	11,664	11,577
WY	17	16	3.30	3.20	56	51
Oth Sts [3]	7,584	6,893	2.25	2.24	17,038	15,457
US	453,599	452,816	2.48	2.50	1,122,923	1,133,558
PR	1,492	1,375	4.30	4.60	6,416	6,325

[1] Excludes commercial broilers. [2] Totals may not add due to rounding. [3] AK, AZ, DE, KS, NV, NM, ND, and RI combined to avoid disclosing data for individual operations.
NASS, Livestock Branch, (202) 720-3570.

Table 8-44.—Broiler meat: Total imports by specified countries, 2001–2004

Continent and country	2001	2002	2003	2004 [1]
	1,000 tons	*1,000 tons*	*1,000 tons*	*1,000 tons*
Russian Federation	1,281	1,208	1,081	960
Japan	710	744	695	582
EU-25	190	197	407	441
Saudi Arabia	399	391	452	429
Mexico	245	267	338	326
Ukraine	64	61	88	277
Hong Kong	183	164	154	244
China, Peoples Republic of	448	436	453	174
United Arab Emirates	125	133	154	158
South Africa, Republic of	64	80	125	154
Kuwait	63	57	81	119
Romania	58	80	83	118
Canada	73	77	75	100
Taiwan	9	20	33	49
Korea, Republic of	83	94	89	32
Venezuela	0	0	6	25
Philippines	12	13	14	22
Malaysia	35	43	39	17
United States	221	199	219	201
Others	122	82	54	61
Grand total	4,102	4,091	4,381	4,244

[1] Preliminary.
FAS, Dairy, Livestock and Poultry Division, (202) 720–8031. Updated data available at http://www.fas.usda.gov/ustrade.

Table 8-45.—Broiler meat: Total exports by specified countries, 2001–2004

Continent and country	2001	2002	2003	2004 [1]
	1,000 tons	*1,000 tons*	*1,000 tons*	*1,000 tons*
Brazil	1,226	1,577	1,903	2,416
United States	2,520	2,180	2,232	2,170
EU-25	764	877	760	789
China, Peoples Republic	489	438	388	241
Thailand	392	427	485	200
Canada	69	84	76	74
Argentina	13	23	39	66
United Arab Emirates	20	37	40	15
Australia	19	15	15	13
Saudi Arabia	20	20	20	10
Others	33	29	36	25
Grand total	5,565	5,707	5,994	6,019

[1] Preliminary.
FAS, Dairy, Livestock and Poultry Division, (202) 720–8031. Updated data available at http://www.fas.usda.gov/ustrade.

Table 8-46.—Broiler meat: Production in specified countries, 2001–2004

Continent and country	2001	2002	2003	2004 [1]
	1,000 metric tons	*1,000 metric tons*	*1,000 metric tons*	*1,000 metric tons*
United States	14,033	14,467	14,696	15,286
China; Peoples Republic	9,278	9,558	9,898	9,998
Brazil	6,567	7,449	7,645	8,408
EU-25	7,883	7,788	7,439	7,656
Mexico	2,067	2,157	2,290	2,389
India	1,250	1,400	1,500	1,650
Japan	1,074	1,107	1,127	1,124
Canada	927	932	929	946
Argentina	870	640	750	910
Thailand	1,230	1,275	1,340	900
Malaysia	813	784	835	862
South Africa; Republic	730	760	808	808
Philippines	582	625	635	659
Australia	568	629	646	651
Russian Federation	430	500	560	650
Indonesia	522	632	735	627
Taiwan	622	612	597	600
Saudi Arabia	424	445	472	470
Korea; Republic of	413	437	429	432
Venezuela	360	320	300	315
Ukraine	49	102	130	208
Romania	194	150	185	205
United Arab Emirates	28	33	33	35
Others	1,245	1,200	88	57
Grand total	52,159	54,000	54,067	55,846

[1] Preliminary.
Prepared or estimated on the basis of official statistics of foreign governments, other foreign source materials, reports of U.S. Agricultural Counselors, Attachés, and Foreign Service Officers, inter-agency analysis, and related information.
FAS, Production Estimates and Crop Assessment Division, (202) 720–8031. Updated data available at http://www.fas.usda.gov/psd.

Table 8-47.—Mature chickens: Lost, sold for slaughter, and value of sales, 2004 (preliminary) [1]

State	Number lost [2]	Number sold [3]	Pounds sold [3]	Price per pound [3]	Value of sales [3]
	1,000 head	1,000 head	1,000 pounds	Dollars	1,000 dollars
AL	1,873	10,061	81,494	0.097	7,905
AR	3,861	15,880	127,040	0.096	12,196
CA	7,563	6,571	25,627	0.008	205
CO	1,056	2,560	10,496	0.034	357
CT	1,857	255	893	0.002	2
FL	3,417	5,885	24,128	0.029	700
GA	7,037	15,707	102,096	0.080	8,168
HI	102	147	529	0.199	105
ID	84	579	2,200	0.010	22
IL	1,161	1,548	6,192	0.010	62
IN	8,428	8,594	28,360	0.008	227
IA	22,292	11,524	38,029	0.002	76
KY	2,162	2,514	18,352	0.094	1,725
LA	619	1,143	8,458	0.090	761
ME	378	2,789	10,319	0.003	31
MD	535	1,130	3,955	0.010	40
MA	29	166	631	0.004	3
MI	1,270	4,546	15,456	0.001	15
MN	3,797	3,074	10,452	0.001	10
MS	1,223	7,839	56,441	0.087	4,910
MO	1,099	3,926	17,667	0.050	883
MT	132	133	479	0.006	3
NE	2,877	5,035	17,119	0.001	17
NH	22	180	810	0.020	16
NJ	429	744	2,455	0.001	2
NY	484	3,201	10,563	0.001	11
NC	1,601	11,663	93,304	0.090	8,397
OH	6,460	14,104	47,954	0.001	48
OK	1,129	3,081	22,799	0.090	2,052
OR	509	1,046	3,556	0.001	4
PA	2,367	15,105	52,868	0.009	476
SC	1,494	2,272	14,314	0.078	1,116
SD	1,210	459	1,607	0.001	2
TN	394	1,885	13,195	0.099	1,306
TX	6,259	13,759	70,171	0.049	3,438
UT	570	1,567	5,798	0.010	58
VT	22	177	690	0.011	8
VA	786	3,403	21,099	0.063	1,329
WA	706	2,577	8,762	0.001	9
WV	271	1,661	12,291	0.091	1,118
WI	1,280	1,671	6,851	0.025	171
WY	1	8	28	0.001	0
Other States [4]	2,233	1,897	6,537	0.004	26
Total US	101,079	192,066	1,002,065	0.058	58,010
PR	179	648	2,786	0.232	646

[1] Estimates cover the 12-month period, Dec. 1, previous year through Nov. 30 and excludes broilers. [2] Includes rendered, died, destroyed, composted, or disappeared for any reason (excluding sold for slaughter) during the 12-month period. [3] Sold for slaughter. [4] AK, AZ, DE, KS, ND, NM, NV, and RI combined to avoid disclosing data for individual operations.

NASS, Livestock Branch, (202) 720–3570.

Table 8-48.—Mature chickens: Lost, sold for slaughter, price, and value, United States, 1995–2004 [1]

Year	Number		Pounds (live weight) sold [3]	Price per pound live weight [3]	Value of sales [3]
	Lost [2]	Sold [3]			
	1,000 head	1,000 head	1,000 pounds	Dollars	1,000 dollars
1995	61,060	179,503	924,036	0.065	60,153
1996	60,435	174,299	900,652	0.066	59,187
1997	49,256	190,986	925,499	0.077	71,461
1998	53,428	200,286	977,060	0.081	79,987
1999	54,951	214,063	1,059,153	0.071	75,217
2000	50,907	218,411	1,112,604	0.057	63,988
2001	56,146	202,482	1,032,115	0.045	47,249
2002	55,330	199,931	1,039,118	0.048	49,931
2003	86,862	189,530	983,054	0.049	47,811
2004	101,079	192,066	1,002,065	0.058	58,010

[1] Estimates cover the 12-month period, Dec. 1, previous year through Nov. 30 and excludes broilers. [2] Includes rendered, died, destroyed, composted, or disappeared for any reason (excluding sold for slaughter) during the 12-month period. [3] Sold for slaughter.

NASS, Livestock Branch, (202) 720–3570.

Table 8-49.—Broilers: Production and value, United States, 1995–2004 [1]

Year	Production		Price per pound [2]	Value of production
	Number	Weight		
	Thousands	*1,000 pounds*	*Cents*	*1,000 dollars*
1995	7,325,670	34,222,000	34.4	11,762,222
1996	7,596,760	36,479,100	38.1	13,903,479
1997	7,764,200	37,540,750	37.7	14,158,926
1998	7,934,260	38,557,400	39.3	15,146,560
1999	8,146,410	40,829,600	37.1	15,128,509
2000	8,283,700	41,626,100	33.6	13,989,424
2001	8,389,770	42,452,400	39.3	16,696,089
2002	8,591,080	44,058,700	30.5	13,437,345
2003	8,492,850	43,958,200	0.346	15,214,947
2004 [3]	8,740,650	45,796,250	0.446	20,446,086

[1] Broilers are young chickens of the meat-type strains, raised for the purpose of meat production. These figures are not included in farm production of chickens. Estimates cover the 12-month period, Dec 1 previous year through Nov 30. Excludes States which produced less than 500,000 broilers.   [2] Live weight equivalent price.   [3] Preliminary.

NASS, Livestock Branch, (202) 720–3570.

Table 8-50.—Chickens: Supply, distribution, and per capita consumption, ready-to-cook basis, United States, 1997–2006

Year	Production			Commercial storage at beginning of year	Exports	Commercial storage at end of year	Consumption	
	Commercial broilers	Other chickens	Total [1]				Total [1][3]	Per capita
	Million pounds	*Million pounds*	*Million pounds*	*Million pounds*	*Million pounds*	*Million pounds*	*Million pounds*	*Pounds*
1997	27,041	510	27,570	647	4,787	614	22,802	84
1998	27,612	525	28,137	614	4,787	717	23,254	84
1999	29,468	554	30,022	717	4,978	804	24,965	90
2000	30,209	531	31,740	804	5,138	808	25,606	91
2001	30,938	515	31,453	808	5,737	720	25,819	90
2002	31,895	547	32,442	720	4,941	768	27,467	95
2003	32,399	502	32,901	768	5,015	611	28,058	96
2004	33,699	504	34,203	611	4,998	716	29,129	99
2005 [2]	34,816	515	35,331	716	5,463	728	29,891	101
2006 [4]	35,936	525	35,461	728	5,735	699	30,795	103

[1] Totals may not add due to rounding.   [2] Preliminary.   [3] Shipments to territories now included in total consumption.   [4] Forecast.

ERS, Animal Products Branch, (202) 694-5180.

Table 8-51.—Poultry: Feed-price ratios, United States, 1995–2004

Year	Ratios [1]		
	Egg-feed	Broiler-feed	Turkey-feed
	Pounds	*Pounds*	*Pounds*
1995	8.8	5.1	6.3
1996	8.6	4.4	5.3
1997	8.8	4.7	5.7
1998	9.7	6.3	6.7
1999	9.8	7.2	8.6
2000	10.5	6.6	8.7
2001	9.9	7.7	8.2
2002	8.6	5.3	6.8
2003	10.6	5.4	5.9
2004	8.3	5.9	6.2

[1] Number of pounds of poultry feed equivalent in value at local market prices to 1 dozen market eggs, or 1 pound of broiler or 1 pound of turkey live weight. Simple average of monthly feed-price ratios. Egg feed= corn (75 lbs) and soybeans (25 lbs); broiler feed= corn (58 lbs); soybeans (42 lbs); turkey feed= corn (51 lbs), soybeans (28 lbs), and wheat (21 lbs). Monthly equivalent prices of commercial prepared feeds are based on current U.S. prices received for corn, soybeans, and wheat.

NASS, Environmental, Economics, and Demographics Branch, (202) 720–6146.

Table 8-52.—Broilers: Production, price, and value, by States, 2003 and 2004 [1]

State	2003 Production Number	2003 Production Weight	Price per pound [2]	Value of production	2004 [3] Production Number	2004 [3] Production Weight	Price per pound [2]	Value of production
	Thousands	1,000 pounds	Dollars	1,000 dollars	Thousands	1,000 pounds	Dollars	1,000 dollars
AL	1,039,400	5,404,900	0.340	1,837,666	1,052,000	5,470,400	0.440	2,406,976
AR	1,192,400	5,842,800	0.340	1,986,552	1,241,500	6,207,500	0.440	2,731,300
DE	251,200	1,507,200	0.360	542,592	240,700	1,492,300	0.460	686,458
FL	91,300	511,300	0.350	178,955	78,500	463,200	0.450	208,440
GA	1,260,500	6,302,500	0.340	2,142,850	1,298,900	6,494,500	0.440	2,857,580
HI	750	2,950	0.570	1,682	....	....	....	....
KY	275,900	1,489,900	0.340	506,566	290,800	1,570,300	0.440	690,932
MD	292,400	1,374,300	0.360	494,748	284,600	1,366,100	0.460	628,406
MN	44,800	228,500	0.340	77,690	46,300	231,500	0.440	101,860
MS	790,300	4,188,600	0.340	1,424,124	827,800	4,387,300	0.440	1,930,412
NE	4,000	22,800	0.350	7,980	4,300	25,400	0.450	11,430
NY	2,600	14,600	0.350	5,110	2,600	14,600	0.450	6,570
NC	708,200	4,320,000	0.350	1,512,000	720,200	4,537,300	0.450	2,041,785
OH	41,000	225,500	0.350	78,925	41,600	224,600	0.450	101,070
OK	223,000	1,115,000	0.340	379,100	243,800	1,243,400	0.440	547,096
PA	129,600	686,900	0.360	247,284	133,500	707,600	0.460	325,496
SC	197,400	1,144,900	0.340	389,266	204,500	1,186,100	0.440	521,884
TN	182,300	948,000	0.340	322,320	195,900	999,100	0.440	439,604
TX	601,500	2,947,400	0.350	1,031,590	620,700	3,165,600	0.450	1,424,520
VA	265,100	1,299,000	0.340	441,660	263,000	1,341,300	0.440	590,172
WV	87,200	357,500	0.340	121,550	86,400	354,200	0.440	155,848
WI	34,400	154,800	0.350	54,180	33,800	152,100	0.450	68,445
Other States [4]	777,600	3,868,850	0.370	1,430,557	829,250	4,161,850	0.473	1,969,802
Total [5]	8,492,850	43,958,200	0.346	15,214,947	8,740,650	45,796,250	0.446	20,446,086

[1] Broilers are young chickens of the meat-type strains, raised for the purpose of meat production. Estimates cover the 12-month period, Dec. 1, previous year through Nov. 30. [2] Live weight equivalent price. [3] Preliminary. [4] CA, HI, IN, IA, LA, MI, MO, OR, and WA. [5] Excludes States producing less than 500,000 broilers.
NASS, Livestock Branch, (202) 720–3570.

Table 8-53.—Chicks hatched by commercial hatcheries: Number, average price, and value, United States, 1995–2004

Year	Chicks hatched Broiler-type	Chicks hatched Egg-type	Chicks hatched All	Average price of baby chicks per 100 Broiler-type	Average price of baby chicks per 100 Egg-type	Average price of baby chicks per 100 All	Value of chick production [1]
	Thousands	Thousands	Thousands	Dollars	Dollars	Dollars	1,000 dollars
1995	7,932,352	396,501	8,328,853	18.60	49.80	19.40	1,577,782
1996	8,078,159	401,640	8,479,799	18.60	53.80	19.50	1,611,380
1997	8,321,634	424,543	8,746,177	19.70	53.10	20.60	1,756,004
1998	8,491,938	438,273	8,930,211	19.50	53.70	20.30	1,771,713
1999	8,715,423	451,721	9,167,144	20.30	52.60	21.10	1,886,007
2000	8,846,185	430,412	9,276,597	20.50	48.00	21.10	1,913,453
2001	9,021,116	452,673	9,473,789	20.60	53.90	21.40	1,982,613
2002	9,079,092	421,549	9,500,641	21.10	52.00	21.80	2,025,371
2003	9,080,614	416,003	9,496,617	21.10	50.50	21.80	2,025,209
2004	9,333,251	437,304	9,770,555	20.50	53.60	21.30	2,035,027

[1] Excludes egg-type cockerels destroyed.
NASS, Environmental, Economics, and Demographics Branch, (202) 720–6146 and Livestock Branch, (202) 720-3570.

Table 8-54.—Poultry: Slaughtered under Federal inspection, by class, United States, 2003–2005

Class	Number inspected			Pounds inspected (live weight)		
	2003	2004	2005	2003	2004	2005
	Thousands	*Thousands*	*Thousands*	*Thousands*	*Thousands*	*Thousands*
Young chickens	8,536,865	8,752,436	8,853,809	44,317,531	46,109,201	47,578,696
Mature chickens	147,569	143,312	146,664	824,973	811,674	835,142
Total chickens	8,684,434	8,895,748	9,000,473	45,142,504	46,920,875	48,413,838
Young turkeys	264,753	251,563	245,642	7,093,431	6,822,172	6,883,300
Old turkeys	3,028	2,745	2,452	81,480	73,697	64,163
Total turkeys	267,781	254,308	248,094	7,174,911	6,895,869	6,947,463
Ducks	24,301	25,967	27,890	160,871	174,231	187,694
Other poultry	....	....	....	10,016	8,404	8,028
Total poultry	....	....	....	52,488,302	53,999,379	55,557,023

Class	Pounds certified (ready-to-cook)			Pounds condemned		
	2003	2004	2005	Ante-mortem (live weight)		
				2003	2004	2005
	Thousands	*Thousands*	*Thousands*	*Thousands*	*Thousands*	*Thousands*
Young chickens	32,748,996	34,063,339	35,364,834	170,725	188,010	169,073
Mature chickens	502,655	504,299	516,396	12,762	12,151	12,628
Total chickens	33,251,651	34,567,638	35,881,230	183,487	200,161	181,701
Young turkeys	5,589,037	5,399,031	5,455,683	25,910	23,976	19,201
Old turkeys	61,320	54,962	48,630	1,238	498	596
Total turkeys	5,650,357	5,453,993	5,504,313	27,148	24,474	19,797
Ducks	119,007	128,030	134,604	338	551	390
Other poultry	6,535	5,268	4,886	31	28	19
Total poultry	39,027,550	40,154,929	41,525,033	211,004	225,214	201,907

Class	Pounds condemned—Continued		
	Post-mortem (New York dressed weight)		
	2003	2004	2005
	Thousands	*Thousands*	*Thousands*
Young chickens	362,256	442,374	414,893
Mature chickens	41,907	36,662	37,425
Total chickens	404,163	479,036	452,318
Young turkeys	123,773	118,716	101,391
Old turkeys	4,735	4,210	3,558
Total turkeys	128,508	122,926	104,949
Ducks	3,503	4,249	4,505
Other poultry	136	96	82
Total poultry	536,310	606,307	561,854

NASS, Livestock Branch, (202) 720–3570.

Table 8-55.—Chickens and turkeys: Number classified as "U.S. Pullorum-Typhoid Clean," and number and percentage of reactors, United States, 1995–2004

Year be-ginning July	Chicken tests				Turkey tests			
	States reporting	Chickens in tested flocks (first test)	Reactors [1]		States reporting	Turkeys in tested flocks (first test)	Reactors	
	Number	Thousands	Number	Percent	Number	Number	Number	Percent
1995	48	58,019	0	0.0000	48	4,679,984	0	0.0000
1996	48	58,191	0	0.0000	48	5,905,799	0	0.0000
1997	48	62,402	0	0.0000	48	5,301,183	0	0.0000
1998	48	81,636	0	0.0000	48	5,548,802	0	0.0000
1999	48	79,037	0	0.0000	48	5,516,096	0	0.0000
2000	48	79,407	0	0.0000	48	4,956,140	0	0.0000
2001	48	79,397	0	0.0000	48	5,408,561	0	0.0000
2002	48	76,868	0	0.0000	48	5,733,250	0	0.0000
2003	48	77,952	0	0.0000	48	4,895,832	0	0.0000
2004	48	80,659	0	0.0000	48	4,009,155	0	0.0000

[1] Number of reacting birds and percent of birds tested. Testing year starting July 1, 1989.
APHIS, Veterinary Services, (770) 922–3496.

Table 8-56.—Turkeys: Supply, distribution, and per capita consumption, ready-to-cook basis, United States, 1997–2006

Year	Production	Commercial storage at beginning of year	Exports	Commercial storage at end of year	Consumption	
					Total [1][2]	Per capita
	Million pounds	Million pounds	Million pounds	Million pounds	Million pounds	Pounds
1997	5,412	328	606	415	4,720	17.3
1998	5,215	415	446	304	4,880	17.7
1999	5,230	304	378	254	4,902	17.5
2000	5,333	254	445	241	4,902	17.4
2001	5,489	241	487	241	5,004	17.5
2002	5,638	241	439	333	5,108	17.7
2003	5,576	333	484	354	5,074	17.4
2004	5,383	354	442	288	5,010	17.0
2005 [3]	5,426	288	580	250	4,893	16.5
2006 [4]	5,482	250	600	300	4,836	16.1

[1] Totals may not add due to rounding. [2] Shipments to territories now included in consumption. [3] Preliminary.
[4] Forecast.
ERS, Animal Products Branch, (202) 694–5180.

Table 8-57.—Turkey: Total imports by specified countries, 2001–2004

Continent and country	2001	2002	2003	2004 [1]
	1,000 tons	*1,000 tons*	*1,000 tons*	*1,000 tons*
Mexico	152	147	158	144
EU-25	64	64	85	84
Russian Federation	164	165	114	77
South Africa, Republic of	15	13	28	28
Taiwan	8	11	16	17
Canada	5	6	6	8
United States	0	0	1	2
Others	26	16	0	0
Grand total	434	422	408	360

[1] Preliminary.
FAS, Dairy, Livestock and Poultry Division, (202) 720–8031. Updated data available at http://www.fas.gov/ustrade.

Table 8-58.—Turkey: Total exports by specified countries, 2001–2004

Continent and country	2001	2002	2003	2004 [1]
	1,000 tons	*1,000 tons*	*1,000 tons*	*1,000 tons*
United States	221	199	219	201
EU-25	231	256	198	191
Brazil	69	90	112	136
Canada	14	16	16	16
Mexico	8	7	1	1
Others	1	0	0	0
Grand total	544	568	546	545

[1] Preliminary.
FAS, Dairy, Livestock and Poultry Division, (202) 720–8031. Updated data available at http://www.fas.gov/ustrade.

Table 8-59.—Turkey: Production in specified countries, 2001–2004

Continent and country	2001	2002	2003	2004 [1]
	1,000 tons	*1,000 tons*	*1,000 tons*	*1,000 tons*
United States	2,490	2,557	2,529	2,441
EU-25	2,098	2,102	2,025	2,038
Brazil	165	182	200	240
Canada	149	147	148	145
Russian Federation	7	9	12	15
Mexico	13	13	14	13
South Africa, Republic	4	3	4	5
Taiwan	5	5	4	4
Others	3	0	0	0
Grand total	4,934	5,018	4,936	4,901

[1] Preliminary.
FAS, Dairy, Livestock and Poultry Division, (202) 720–8031. Updated data available at http://www.fas.gov/psd.

Table 8-60.—Turkeys: Production, and value, United States, 1995-2004

Year	Number raised	Pounds (live weight) produced	Price per pound live weight	Value of production
	Thousands	1,000 pounds	Cents	1,000 dollars
1995	292,356	6,761,327	41.0	2,769,397
1996	302,713	7,222,834	43.3	3,124,496
1997	301,251	7,225,059	39.9	2,884,377
1998	285,603	7,061,925	38.0	2,683,473
1999	270,192	6,877,399	40.8	2,806,630
2000	270,466	6,959,833	40.6	2,828,489
2001	272,660	7,173,111	39.0	2,796,821
2002	275,477	7,494,861	36.5	2,732,481
2003	274,048	7,487,293	36.1	2,699,673
2004 [1]	264,207	7,304,813	42.0	3,065,417

[1] Preliminary.

NASS, Livestock Branch, (202) 720-3570.

Table 8-61.—Turkeys: Production and value, by State, 2004 [1]

State	Number raised [2]	Pounds produced	Price per pound [3]	Value of production
	1,000 head	1,000 pounds	Dollars	1,000 dollars
AR	28,500	527,250	0.43	226,718
CA	15,700	414,480	0.41	169,937
CT	5	127	1.35	171
IL	2,900	89,320	0.42	37,514
IN	13,300	409,640	0.42	172,049
IA	9,000	324,000	0.42	136,080
MD	750	13,275	0.43	5,708
MA	70	1,736	1.59	2,760
MI	5,000	188,000	0.37	69,560
MN	46,500	1,227,600	0.42	515,592
MO	21,500	666,500	0.42	279,930
NH	4	100	1.77	177
NJ	37	814	0.87	708
NY	580	13,746	0.43	5,911
NC	39,000	1,068,600	0.42	448,812
ND	1,000	26,400	0.40	10,560
OH	5,800	219,820	0.42	92,324
PA	12,000	234,000	0.48	112,320
SC	12,000	463,200	0.40	185,280
SD	4,500	150,750	0.38	57,285
VT	52	1,227	1.50	1,841
VA	19,700	435,370	0.42	182,855
WV	3,200	70,720	0.42	29,702
Other States [4]	23,109	758,138	0.42	321,623
US	264,207	7,304,813	0.42	3,065,417

[1] Preliminary. [2] Based on turkeys placed Sep. 1, 2003, through Aug. 31, 2004. Excludes young turkeys lost. [3] CA, CT, DE, MD, MA, MI, NE, NH, NJ, ND, OH, PA, SC, SD, and VT are actual live weight prices. All other states are equivalent live weight returns to producers. [4] CO, DE, KS, NE, OK, OR, TX, UT, and WI combined to avoid disclosing individual operations.

NASS, Livestock Branch, (202) 720-3570.

Table 8-62.—Turkeys: Poults placed by commercial hatcheries, United States, 1995-2004

Year	Total all breeds
	Thousands
1995	321,651
1996	327,213
1997	321,487
1998	297,798
1999	296,106
2000	297,299
2001	301,559
2002	297,051
2003	289,516
2004	277,238

NASS, Livestock Branch, (202) 720-3570.

Table 8-63.—Turkeys: Poults placed by commercial hatcheries, U.S. and regions, Monthly, 2003 and 2004 [1]

Month	United States			2004				
	2003	2004	2004 as percent of 2003	East North Central	West North Central	North and South Atlantic	South Central	West
	Thou-sands	*Thou-sands*	*Per-cent*	*Thou-sands*	*Thou-sands*	*Thou-sands*	*Thou-sands*	*Thou-sands*
All breeds:								
Jan	25,389	23,273	92	3,381	8,663	7,475	1,960	1,794
Feb	23,833	22,905	96	3,118	9,119	6,826	1,836	2,006
Mar	24,902	24,215	97	3,708	9,139	7,621	2,056	1,691
Apr	24,938	24,664	99	3,444	9,247	8,153	1,972	1,848
May	25,111	23,692	94	3,343	8,754	7,592	1,860	2,143
June	25,422	23,291	92	3,304	8,814	6,837	1,993	2,343
July	25,330	25,013	99	3,612	9,295	7,443	1,863	2,800
Aug	24,036	23,674	98	3,577	9,282	6,622	1,873	2,320
Sept	22,171	21,268	96	3,125	8,200	6,668	1,839	1,436
Oct	22,805	20,806	91	3,009	8,074	6,743	1,649	1,331
Nov	22,155	22,232	100	3,144	8,382	7,344	1,827	1,535
Dec	23,424	22,205	95	3,216	8,653	6,868	1,881	1,587
Total	289,516	277,238	96	39,981	105,622	86,192	22,609	22,834

[1] Regional placements refer to poults placed from hatcheries located in that region, not the actual location of the birds after placement. Excludes exported poults.

NASS, Livestock Branch, (202) 720–3570.

Table 8-64.—Eggs: Supply, distribution, and per capita consumption, United States, 1997–2006 [1]

Year	Total egg production	Storage at beginning of the year [1]	Imports [2]	Exports [2]	Eggs used for hatching	Consumption		
						Storage at end of the year [2]	Total [3]	Per capita
	Million dozen	*Million dozen*	*Million dozen*	*Million dozen*	*Million dozen*	*Million dozen*	*Million dozen*	*Number*
1997	6,473	9	7	228	895	7	5,359	236
1998	6,667	7	6	219	922	8	5,531	240
1999	6,933	8	7	162	942	8	5,838	251
2000	7,062	8	8	171	940	11	5,956	253
2001	7,187	11	9	190	964	10	6,043	254
2002	7,270	10	15	174	961	10	6,150	256
2003	7,297	10	13	146	959	14	6,201	256
2004	7,443	14	14	168	987	14	6,300	257
2005 [4]	7,500	14	9	200	998	14	6,311	255
2006 [5]	7,645	14	10	200	1,015	14	6,440	258

[1] Calendar years. [2] Shell eggs and the approximate shell-egg equivalent of egg product. [3] Shipments to territories now included in total consumption. [4] Preliminary. [5] Forecast.

ERS, Animal Products Branch, (202) 694–5180.

Table 8-65.—Eggs, shell: Average price per dozen on consumer Grade A cartoned white eggs to volume buyers, store-door delivery, New York metropolitan area, 1996–2005

Year	Large
	Cents
1996	73.00
1997	81.21
1998	75.80
1999	65.60
2000	68.90
2001	67.14
2002	67.06
2003	87.91
2004	82.18
2005	65.51

AMS, Poultry Division, Market News Branch, (202) 720–6911.

Table 8-66.—Eggs: Number of layers and pullets, rate of lay, and production, by State and United States, 2004 and 2005 [1] [2]

State	Average number of layers during year		Rate of lay per layer during year [3]		Eggs produced	
	2004	2005	2004	2005	2004	2005
	Thousands	Thousands	Number	Number	Millions	Millions
AL	9,345	9,141	225	227	2,099	2,071
AR	15,385	14,748	229	232	3,526	3,416
CA	20,222	19,336	265	263	5,352	5,082
CO	3,963	3,814	279	281	1,105	1,071
CT	2,853	3,026	287	280	818	846
FL	11,316	10,963	271	272	3,068	2,980
GA	20,323	19,489	248	249	5,038	4,850
HI	500	498	237	230	118.5	114.5
ID	853	862	279	280	238	241
IL	4,004	4,434	261	273	1,044	1,210
IN	23,532	23,596	266	265	6,256	6,254
IA	43,569	48,760	267	266	11,615	12,978
KY	4,982	4,781	247	257	1,232	1,228
LA	1,944	1,892	239	248	465	469
ME	4,147	4,138	279	248	1,156	1,025
MD	3,121	2,916	270	274	843	798
MA	264	254	284	280	75	71
MI	7,493	7,867	268	272	2,009	2,142
MN	10,859	11,038	270	270	2,927	2,985
MS	6,923	7,001	232	232	1,606	1,627
MO	7,043	7,204	265	265	1,865	1,910
MT	355	350	302	303	107	106
NE	11,766	11,987	270	268	3,174	3,217
NH	155	155	271	252	42	39
NJ	2,026	1,813	276	273	558	495
NY	4,021	4,167	289	286	1,163	1,190
NC	10,877	10,955	232	235	2,523	2,573
OH	27,938	28,026	263	268	7,355	7,506
OK	3,412	3,210	224	228	764	731
OR	2,964	2,916	276	282	818	823
PA	23,893	23,785	276	278	6,585	6,608
SC	5,243	5,042	258	256	1,351	1,289
SD	3,442	3,092	271	264	933	816
TN	1,401	1,299	228	243	319	316
TX	18,403	17,703	262	265	4,825	4,684
UT	3,182	3,285	261	267	831	878
VT	203	198	271	253	55	50
VA	3,241	3,326	235	248	761	823
WA	4,932	4,931	270	272	1,332	1,343
WV	1,259	1,165	217	224	273	261
WI	4,534	4,864	266	272	1,206	1,321
WY	12	12	300	300	3.6	3.6
Oth Sts [4]	6,061	5,466	273	278	1,657	1,519
US	341,956	343,501	261	262	89,091	89,960
PR	1,095	1,015	206	233	226	236

[1] Annual estimates cover the period December 1 previous year through November 30. [2] Totals may not add due to rounding. [3] Total egg production divided by average number of layers on hand. [4] AK, AZ, DE, KS, NV, NM, ND, and RI combined to avoid disclosing data for individual operations.

NASS, Livestock Branch, (202) 720–3570.

Table 8-67.—Eggs: Broken under Federal inspection, United States, 2004–2005

Item	Quantity	
	2004	2005
	1,000 dozen	1,000 dozen
Shell eggs broken	1,929,300	2,051,029
	1,000 pounds	1,000 pounds
Edible liquid from shell eggs broken:		
Whole	1,538,992	1,641,740
White	617,120	648,697
Yolk	329,006	356,327
Total	2,485,118	2,646,764
Inedible liquid from shell eggs broken	219,211	226,140

NASS, Livestock Branch, (202) 720–3570.

Table 8-68.—Eggs: Number, rate of lay, production, and value, United States, 1996–2005 [1]

Year	Layers average number during year	Rate of lay per layer during year[2]	Eggs, total produced	Price per dozen[3]	Value of production
	Thousands	Number	Millions	Dollars	1,000 dollars
1996	298,270	256	76,377	0.750	4,776,252
1997	303,604	255	77,532	0.703	4,539,929
1998	312,315	255	79,777	0.668	4,441,139
1999	323,251	257	82,946	0.621	4,292,371
2000	329,067	257	84,717	0.617	4,358,648
2001	336,330	256	86,093	0.622	4,460,701
2002	339,293	257	87,252	0.589	4,284,930
2003	338,393	259	87,473	0.732	5,333,014
2004	341,956	261	89,091	0.714	5,303,244
2005[4]	343,501	262	89,960	NA	NA

[1] Annual estimates cover the period December 1 previous year through November 30. [2] Total egg production divided by average number of layers on hand. [3] Average mid-month price of all eggs sold by producers including hatching eggs. [4] Preliminary. NA-not available.

NASS, Livestock Branch, (202) 720–3570.

Table 8-69.—Eggs: Production and value, by States, 2003–2004 [1] [2] [3]

State	Eggs produced		Price per dozen[4]		Value of production	
	2003	2004	2003	2004	2003	2004
	Millions	Millions	Dollars	Dollars	1,000 dollars	1,000 dollars
AL	2,190	2,099	1.620	1.650	295,101	287,956
AR	3,590	3,565	1.200	1.220	359,000	362,442
CA	5,439	5,380	0.623	0.643	282,458	288,412
CO	1,073	1,105	0.670	0.653	59,915	60,103
CT	795	818	0.667	0.674	44,189	45,944
FL	2,804	3,068	0.621	0.625	145,027	159,878
GA	5,047	5,038	0.941	0.939	395,769	394,223
HI	117.2	118.5	0.962	1.080	9,396	10,665
ID	243	238	0.717	0.730	14,525	14,479
IL	973	1,044	0.629	0.592	51,001	51,504
IN	6,035	6,256	0.612	0.560	307,785	291,947
IA	10,446	11,613	0.529	0.508	460,648	491,586
KY	1,122	1,231	0.889	0.858	83,153	88,067
LA	487	465	0.886	0.902	35,967	34,966
ME	1,121	957	0.755	0.770	70,530	61,408
MD	811	843	0.682	0.651	46,104	45,737
MA	77	74	0.802	0.810	5,149	4,995
MI	1,888	2,009	0.595	0.563	93,613	94,256
MN	3,028	2,930	0.592	0.583	149,381	142,349
MS	1,599	1,606	1.270	1.290	168,636	172,166
MO	1,861	1,865	0.645	0.652	99,989	101,395
MT	107	107	0.650	0.657	5,796	5,862
NE	3,126	3,174	0.535	0.525	139,368	138,863
NH	43	43	0.910	0.935	3,261	3,350
NJ	556	558	0.630	0.622	29,208	28,912
NY	1,048	1,163	0.645	0.617	56,330	59,798
NC	2,523	2,522	1.150	1.140	241,788	239,590
OH	7,642	7,355	0.588	0.545	374,458	334,040
OK	933	937	0.927	0.956	72,074	74,648
OR	783	818	0.667	0.714	43,549	48,693
PA	6,754	6,585	0.659	0.619	371,170	339,676
SC	1,373	1,351	0.762	0.735	87,186	82,749
SD	761	933	0.500	0.517	31,708	40,197
TN	290	319	1.320	1.340	31,922	35,511
TX	4,745	4,825	0.784	0.762	310,007	306,388
UT	866	831	0.520	0.520	37,556	36,012
VT	54	55	0.815	0.746	3,667	3,419
VA	744	761	1.180	1.100	73,160	69,758
WA	1,307	1,332	0.646	0.697	70,323	77,348
WV	271	273	1.510	1.420	34,128	32,325
WI	1,137	1,206	0.587	0.564	55,579	56,679
WY	3.6	3.6	0.620	0.607	186	182
Other States[5]	1,660	1,657	0.602	0.614	83,254	84,766
US	87,473	89,131	0.732	0.714	5,333,014	5,303,244
PR	227	230	0.943	0.838	17,811	16,067

[1] Revised data will be published in the "Poultry Production and Value" report. [2] Annual estimates cover the period December 1, previous year through November 30. [3] Totals may not add due to rounding. [4] Average mid-month price of all eggs sold by producers including hatching eggs. [5] AK, AZ, DE, KS, ND, NM, NV, and RI combined to avoid disclosing data for individual operations.

NASS, Livestock Branch, (202) 720–3570.

Table 8-70.—Poultry and poultry products: Cold storage holdings, end of month, United States, 2004 and 2005

Month	Frozen eggs							
	Whites		Yolks		Whole & mixed		Unclassified	
	2004	2005	2004	2005	2004	2005	2004	2005
	1,000 pounds	1,000 pounds	1,000 pounds	1,000 pounds	1,000 pounds	1,000 pounds	1,000 pounds	1,000 pounds
January	2,870	2,843	1,149	954	12,423	11,600	4,862	3,180
February	2,900	2,903	1,119	719	13,914	11,552	3,205	2,691
March	2,860	2,845	1,003	828	11,895	12,108	3,489	2,767
April	2,837	2,823	966	725	13,193	12,510	3,910	2,845
May	2,717	2,840	1,143	601	12,517	11,523	4,177	2,703
June	2,755	2,870	1,104	966	10,798	12,919	3,599	2,926
July	2,567	2,793	1,387	1,174	9,495	12,501	3,271	3,143
August	2,722	2,806	1,435	1,207	9,605	12,447	3,505	3,478
September	3,026	2,908	1,501	1,217	10,569	11,198	3,624	3,514
October	2,888	2,940	1,321	1,130	10,052	10,193	3,616	3,356
November	2,955	2,833	1,072	1,310	10,174	10,325	3,123	3,133
December	2,972	2,690	1,085	1,335	11,443	13,420	3,612	3,573

Month	Frozen eggs, total		Frozen chicken					
	2004	2005	Broilers (Whole)		Hens		Breast and breast meat	
			2004	2005	2004	2005	2004 [1]	2005
	1,000 pounds	1,000 pounds	1,000 pounds	1,000 pounds	1,000 pounds	1,000 pounds	1,000 pounds	1,000 pounds
January	21,304	18,577	20,767	27,880	4,182	2,711	86,265	149,741
February	21,138	17,865	23,560	22,942	4,109	2,112	86,391	151,453
March	19,247	18,548	24,837	27,202	4,681	2,876	88,847	150,851
April	20,906	18,903	23,295	29,259	4,387	2,073	91,045	149,783
May	20,554	17,667	23,601	27,960	3,540	1,968	88,457	145,135
June	18,256	19,681	22,830	31,751	4,603	3,307	97,841	140,939
July	16,720	19,611	21,723	29,885	3,276	2,835	104,526	143,016
August	17,267	19,938	21,945	20,533	2,861	1,665	106,699	134,933
September	18,720	18,837	24,132	21,989	3,799	1,885	119,834	136,933
October	17,877	17,619	25,783	23,205	3,862	2,183	135,275	149,593
November	17,324	17,601	20,727	24,435	3,696	1,720	149,607	150,187
December	19,112	21,018	23,308	23,388	2,924	2,150	151,876	169,318

Month	Frozen chicken							
	Drumsticks		Leg quarters		Legs		Thigh and thigh quarters	
	2004 [1]	2005	2004 [1]	2005	2004 [1]	2005	2004 [1]	2005
	1,000 pounds	1,000 pounds	1,000 pounds	1,000 pounds	1,000 pounds	1,000 pounds	1,000 pounds	1,000 pounds
January	7,840	13,235	72,439	51,846	13,344	8,388	5,751	8,489
February	12,651	11,807	77,065	50,183	10,487	9,287	5,952	9,561
March	17,171	10,473	106,200	59,798	13,997	10,958	7,028	13,226
April	16,508	10,680	126,925	53,833	11,631	11,563	8,156	9,282
May	8,927	9,838	115,094	55,923	9,057	12,248	6,781	12,937
June	12,219	8,264	153,725	69,046	11,470	10,470	8,097	9,288
July	16,808	10,411	100,972	73,390	9,700	12,363	7,037	10,315
August	15,024	10,750	99,106	89,182	7,481	9,522	11,641	8,177
September	15,710	11,779	111,986	109,658	7,725	12,021	13,048	9,321
October	14,860	13,016	116,900	132,852	9,227	13,473	15,425	9,957
November	21,585	17,911	104,300	156,766	9,186	20,199	11,808	12,614
December	16,497	19,734	76,091	176,718	9,965	24,127	11,733	12,273

Month	Frozen chicken							
	Thigh meat		Wings		Paws and feet		Other chicken	
	2004 [1]	2005	2004 [1]	2005	2004 [1]	2005	2004 [1]	2005
	1,000 pounds	1,000 pounds	1,000 pounds	1,000 pounds	1,000 pounds	1,000 pounds	1,000 pounds	1,000 pounds
January	17,712	13,234	18,800	33,809	10,554	14,457	277,247	323,464
February	15,131	11,987	20,001	33,854	15,886	12,373	271,557	319,219
March	15,516	10,656	24,281	42,510	18,306	12,667	283,874	332,253
April	15,018	11,901	26,500	44,369	19,453	11,303	302,903	358,107
May	15,967	12,818	29,711	42,035	23,493	10,182	345,218	333,962
June	19,419	9,961	36,147	36,916	21,087	11,607	387,115	359,291
July	21,419	11,627	38,823	43,947	20,973	9,023	385,033	375,380
August	17,498	11,300	34,628	42,078	18,733	8,209	390,170	369,179
September	14,716	9,368	34,580	42,746	15,150	10,146	417,194	389,036
October	15,141	10,429	39,695	46,419	16,608	10,690	407,518	392,375
November	14,742	14,050	38,867	49,012	17,853	15,849	382,340	413,830
December	14,382	12,933	32,429	41,102	11,464	13,167	365,511	430,853

See footnotes at end of table.

Table 8-70.—Poultry and poultry products: Cold storage holdings, end of month, United States, 2004 and 2005—Continued

Month	Frozen chicken, total		Frozen turkey					
	2004	2005	Toms		Hens		Total whole	
			2004	2005	2004	2005	2004	2005
	1,000 pounds	*1,000 pounds*	*1,000 pounds*	*1,000 pounds*	*1,000 pounds*	*1,000 pounds*	*1,000 pounds*	*1,000 pounds*
January	534,901	647,254	107,811	73,955	71,977	61,759	179,788	135,714
February	542,790	634,778	133,105	104,483	92,177	76,932	225,282	181,415
March	604,738	673,470	151,384	124,268	103,844	85,724	255,228	209,992
April	645,821	692,153	159,060	137,993	108,611	91,145	267,671	229,138
May	669,846	665,006	169,094	145,263	118,444	97,828	287,538	243,091
June	774,553	690,840	185,022	162,494	126,944	104,456	311,966	266,950
July	730,290	722,192	188,795	168,568	126,614	117,718	315,409	286,286
August	725,786	705,528	189,213	167,864	119,518	128,574	308,731	296,438
September ...	777,874	754,882	185,035	158,100	105,277	107,494	290,312	265,594
October	800,294	804,192	156,514	132,483	91,770	90,484	248,284	222,967
November	769,185	876,573	68,601	30,539	48,477	25,793	117,078	56,332
December	716,180	925,763	62,988	25,538	43,083	27,078	106,071	52,616

Month	Other		Total		Frozen ducks		Total frozen poultry	
	2004	2005	2004	2005	2004	2005	2004	2005
	1,000 pounds	*1,000 pounds*	*1,000 pounds*	*1,000 pounds*	*1,000 pounds*	*1,000 pounds*	*1,000 pounds*	*1,000 pounds*
January	240,678	197,180	420,466	332,894	1,484	401	956,851	980,549
February	246,400	197,993	471,682	379,408	1,328	3,236	1,015,800	1,017,422
March	249,333	204,167	504,561	414,159	1,246	1,250	1,110,545	1,088,879
April	281,102	210,984	548,773	440,122	1,020	1,199	1,195,614	1,133,474
May	283,586	222,829	571,124	465,920	1,043	2,277	1,242,013	1,133,203
June	285,610	239,352	597,576	506,302	936	2,475	1,373,065	1,199,617
July	284,180	232,593	599,589	518,879	972	2,523	1,330,851	1,243,594
August	291,454	226,682	600,185	523,120	969	2,622	1,326,940	1,231,270
September ...	237,076	212,207	527,388	477,801	918	2,452	1,306,180	1,235,135
October	224,059	194,638	472,343	417,605	823	2,714	1,273,460	1,224,511
November	177,790	138,383	294,868	194,715	582	2,880	1,064,635	1,074,168
December	182,286	153,550	288,357	206,166	663	2,615	1,005,200	1,134,544

NASS, Livestock Branch, (202) 720–3570.

Table 8-71.—Dairy products: Cold storage holdings, end of month, United States, 2004 and 2005

Month	Butter		American cheese	
	2004	2005	2004	2005
	1,000 pounds	*1,000 pounds*	*1,000 pounds*	*1,000 pounds*
January	152,448	77,219	518,113	484,227
February	159,066	110,876	532,616	504,979
March	158,118	132,436	520,803	527,275
April	155,718	164,501	526,389	553,814
May	178,744	178,045	558,823	582,686
June	189,183	179,648	590,544	590,113
July	193,520	176,666	615,904	603,471
August	161,025	148,878	568,783	582,015
September	133,008	124,061	553,884	554,760
October	107,152	98,112	528,234	541,625
November	57,177	60,430	481,244	516,701
December	44,988	58,649	481,077	536,905

Month	Swiss cheese		Other natural cheese	
	2004	2005	2004	2005
	1,000 pounds	*1,000 pounds*	*1,000 pounds*	*1,000 pounds*
January	25,244	26,010	213,575	203,562
February	25,356	27,462	208,118	191,240
March	23,277	22,512	215,466	199,441
April	26,910	21,677	214,269	205,318
May	26,079	23,540	219,557	209,409
June	23,958	22,737	227,545	210,594
July	28,012	22,483	226,086	211,227
August	26,759	21,847	215,979	209,065
September	26,992	24,263	209,841	189,978
October	26,155	25,853	201,729	188,452
November	25,934	23,699	197,073	180,372
December	25,956	26,039	198,757	195,216

Month	Total Natural cheese	
	2004	2005
	1,000 pounds	*1,000 pounds*
January	756,932	713,799
February	766,090	723,681
March	759,546	749,228
April	767,568	780,809
May	804,459	815,635
June	842,047	823,444
July	870,002	837,181
August	811,521	812,927
September	790,717	769,001
October	756,118	755,930
November	704,251	720,772
December	705,790	758,160

NASS, Livestock Branch, (202) 720–3570.

CHAPTER IX
FARM RESOURCES, INCOME, AND EXPENSES

The statistics in this chapter deal with farms, farm resources, farm income, and expenses. Many of the series are estimates developed in connection with economic research activities of the Department.

Table 9-1.—Economic trends: Data relating to agriculture, United States, 1996–2004

Year	Prices paid by farmers [1]			Farm income [2]		
	Total including interest, taxes, and wage rates	Production items	Prices received by farmers [1]	Gross farm income [6]	Production expenses	Net farm income
	Index numbers 1990–92=100	*Index numbers 1990–92=100*	*Index numbers 1990–92=100*	*Billion dollars*	*Billion dollars*	*Billion dollars*
1996	115	115	112	235.8	176.9	59.0
1997	118	119	107	238.0	186.7	51.3
1998	115	113	102	232.6	185.5	47.1
1999	115	111	96	235.0	187.2	47.7
2000	120	116	96	242.0	193.1	48.9
2001	123	120	102	248.7	197.1	51.5
2002	124	119	98	229.9	193.4	36.6
2003	128	124	107	259.8	200.3	59.5
2004	134	131	119	292.3	209.8	82.5

Year	National income [3][7]	Personal income [3][7]	Industrial production [4]	Consumer prices all items [5]	Producer prices consumer foods [5]
	Billion dollars	*Billion dollars*	*Index numbers 2002= 100*	*Index numbers 1982–84= 100*	*Index numbers 1982= 100*
1996	6,840.1	6,520.5	83.6	156.9	133.6
1997	7,292.2	6,915.1	89.7	160.5	134.5
1998	7,752.8	7,423.0	94.9	163.0	134.3
1999	8,236.6	7,802.4	99.3	166.6	135.1
2000	8,795.2	8,429.7	103.6	172.2	137.2
2001	8,979.8	8,724.1	99.9	177.1	141.3
2002	9,229.3	8,881.9	100.0	179.9	140.1
2003	9,660.9	9,169.1	100.6	184.0	146.0
2004	10,275.9	9,713.3	104.7	188.9	152.7
2005 [7]	10,812.5	10,230.2	109.4	195.3	155.6

[1] U.S. Department of Agriculture - NASS. [2] U.S. Department of Agriculture - ERS. [3] U.S. Department of Commerce, Bureau of Economic Analysis. [4] Federal Reserve Board. [5] U.S. Department of Labor, Bureau of Labor Statistics. [6] Includes cash receipts from farm marketings, government payments, nonmoney income (gross rental value of dwelling and value of home consumption), other income (machine hire custom work and recreational income), and value of change in farm inventories. [7] Forecast.

ERS, Farm and Rural Business Branch, (202) 694–5592. E mail contact is rogers@ers.usda.gov. For National Income, Personal Income, Industrial Production and Consumer Price Indexes, Contact David Torgerson at (202) 694-5334. E mail contact is dtorg@ers.usda.gov.

Table 9-2.—Farms: Number, land in farms, and average size of farm, U.S., 1996–2005 [1]

Year	Farms [2] [3]	Land in farms	Average size farm
	Number	1,000 acres	Acres
1996	2,190,500	958,675	438
1997	2,190,510	956,010	436
1998	2,192,330	952,080	434
1999	2,187,280	948,460	434
2000	2,166,780	945,080	436
2001	2,148,630	942,070	438
2002	2,135,360	940,300	440
2003	2,126,860	938,650	441
2004	2,112,970	936,295	443
2005 [4]	2,100,990	933,400	444

[1] The farm definition was changed in 1993 to include maple syrup, short rotation woody crops, and places with 5 or more horses. [2] A farm is any establishment from which $1,000 or more of agricultural products were sold or would normally be sold during the year. [3] Includes some accounting for individual farms on reservation land in AZ and NM from 1998 forward. [4] Preliminary.

NASS, Environmental, Economics, and Demographics Branch, (202) 720–6146.

Table 9-3.—Farms: Percent of farms, land in farms, and average size, by economic sales class, United States, 2004–2005

Economic sales class	Percent of total				Average size farm	
	Farms		Land		2004	2005 [1]
	2004	2005 [1]	2004	2005 [1]		
	Percent	Percent	Percent	Percent	Acres	Acres
$1,000–$2,499	26.7	26.5	4.1	4.0	68	67
$2,500–$4,999	15.2	15.2	3.9	3.8	114	111
$5,000–$9,999	14.0	13.9	4.9	4.9	155	156
$10,000–$24,999	11.5	11.5	7.5	7.2	289	279
$25,000–$49,999	8.6	8.7	9.5	9.2	490	471
$50,000–$99,999	8.3	8.2	11.5	11.5	614	624
$100,000–$249,999	7.9	7.9	20.7	20.6	1,170	1,158
$250,000–$499,999	4.2	4.3	16.2	16.3	1,723	1,683
$500,000–$999,999	2.1	2.1	10.8	10.8	2,297	2,283
$1,000,000+	1.5	1.7	10.9	11.7	3,246	3,056
Total	100.0	100.0	100.0	100.0	443	444

[1] Preliminary.
NASS, Environmental, Economics, and Demographics Branch, (202) 720–6146.

Table 9-4.—Number of farms: Economic sales class by region and United States, 2003–2005 [1]

Region and year	Economic Sales Class					Total
	$1,000–$9,999	$10,000–$99,999	$100,000–$249,999	$250,000–$499,999	$500,000 & over	
	Number	Number	Number	Number	Number	Number
NE: [2]						
2003	79,640	33,510	12,290	4,680	3,230	133,350
2004	78,340	33,760	11,880	4,840	3,430	132,250
2005	77,340	33,660	12,130	4,790	3,630	131,550
NC: [3]						
2003	353,100	266,500	96,300	46,700	28,200	790,800
2004	348,700	265,300	96,000	48,300	30,700	789,000
2005	344,600	263,700	95,600	49,000	32,400	785,300
South: [4]						
2003	604,650	214,720	34,900	23,260	25,370	902,900
2004	594,880	214,810	35,550	23,500	25,760	894,500
2005	588,600	214,410	35,800	23,700	25,990	888,500
West: [5]						
2003	161,880	85,810	23,730	11,910	16,480	299,810
2004	159,270	85,410	23,600	12,340	16,600	297,220
2005	157,680	84,670	23,350	12,550	17,390	295,640
US:						
2003	1,199,270	600,540	167,220	86,550	73,280	2,126,860
2004	1,181,190	599,280	167,030	88,980	76,490	2,112,970
2005	1,168,220	596,440	166,880	90,040	79,410	2,100,990
PR:						
2004	8,900	3,500	500	300	300	13,500
2005	8,800	3,600	600	300	300	13,600

[1] Number of farms estimated for 3 sales classes above $100,000 beginning in 2002 and set back to 1998 with the 5-year Census revision review. [2] CT, ME, MA, NH, NJ, NY, PA, RI, and VT. [3] IL, IN, IA, KS, MI, MN, MO, NE, ND, OH, SD, WI. [4] AL, AR, DE, FL, GA, KY, LA, MD, MS, NC, OK, SC, TN, TX, VA, WV. [5] AK, AZ, CA, CO, HI, ID, MT, NV, NM, OR, UT, WA, WY.

NASS, Environmental, Economics, and Demographics Branch, (202) 720–6146.

Table 9-5.—Land in farms: Economic sales class by region and United States, 2003–2005

Region and year	Economic Sales Class [1]					Total
	$1,000-$9,999	$10,000-$99,999	$100,000-$249,999	$250,000-$499,999	$500,000 & over	
	1,000 Acres	1,000 Acres	1,000 Acres	1,000 Acres	1,000 Acres	1,000 Acres
NE: [2]						
2003	5,860	5,555	3,775	2,135	2,855	20,180
2004	5,830	5,405	3,675	2,165	3,055	20,130
2005	5,710	5,345	3,675	2,115	3,205	20,050
NC: [3]						
2003	34,740	93,940	87,960	68,260	63,730	348,630
2004	33,300	91,300	86,450	68,500	68,550	348,100
2005	32,600	88,500	85,500	69,100	71,600	347,300
South: [4]						
2003	62,700	93,380	42,700	32,230	50,590	281,600
2004	61,070	92,985	43,100	31,920	51,450	280,525
2005	59,950	91,940	43,050	31,920	52,790	279,650
West: [5]						
2003	21,470	77,180	61,620	47,510	80,460	288,240
2004	20,690	76,910	60,440	48,790	80,710	287,540
2005	20,090	74,920	59,970	49,160	82,260	286,400
US:						
2003	124,770	270,055	196,055	150,135	197,635	938,650
2004	120,890	266,600	193,665	151,375	203,765	936,295
2005	118,350	260,705	192,195	152,295	209,855	933,400
PR:						
2004	140	230	80	60	90	600
2005	130	220	90	70	100	610

[1] Number of farms estimated for 3 sales classes above $100,000 beginning in 2002 and set back to 1998 with the 5-year Census revision review. [2] CT, ME, MA, NH, NJ, NY, PA, RI, and VT. [3] IL, IN, IA, KS, MI, MN, MO, NE, ND, OH, SD, WI. [4] AL, AR, DE, FL, GA, KY, LA, MD, MS, NC, OK, SC, TN, TX, VA, WV. [5] AK, AZ, CA, CO, HI, ID, MT, NV, NM, OR, UT, WA, WY.

NASS, Environmental, Economics, and Demographics Branch, (202) 720–6146.

Table 9-6.—Land in farms: Classification by tenure of operator, United States, 1910–2004

Year	Land in farms	Tenure of operator			
		Full owners	Part owners	Managers	All tenants
	Acres	Percent	Percent	Percent	Percent
1910	878,798,325	52.9	15.2	6.1	25.8
1920	958,676,612	48.3	18.4	5.7	27.7
1925	924,319,352	45.4	21.3	4.7	28.7
1930 [1]	990,111,984	37.6	24.9	6.4	31.0
1935	1,054,515,111	37.1	25.2	5.8	31.9
1940 [1]	1,065,113,774	35.9	28.2	6.5	29.4
1945	1,141,615,364	36.1	32.5	9.3	22.0
1950 [1]	1,161,419,720	36.1	36.4	9.2	18.3
1954	1,158,191,511	34.2	40.7	8.6	16.5
1959 [1]	1,123,507,574	31.0	44.0	9.8	14.8
1964 [1]	1,110,187,000	28.7	48.0	10.2	13.1
1969 [1]	1,062,892,501	35.3	51.8	....	13.0
1974 [1]	1,017,030,357	35.3	52.6	....	12.0
1978 [1]	1,014,777,234	32.7	55.3	....	12.0
1982 [1]	986,796,579	34.7	53.8	....	11.5
1987 [1]	964,470,625	32.9	53.9	....	13.2
1992 [1]	945,531,506	31.3	55.7	....	13.0
1997 [2]	932,475,414	26.7	62.2	....	11.2
1998 [2]	900,415,615	28.6	60.2	....	11.2
1999 [2]	870,720,495	25.6	61.6	....	12.8
2000 [2]	994,997,682	26.4	62.3	....	11.4
2001 [2]	959,163,331	24.7	61.2	....	14.2
2002 [2]	954,302,543	29.4	56.6	....	14.0
2003 [2]	926,985,610	28.9	59.7	....	11.4
2004 [2]	990,395,334	30.3	56.0	....	13.7

[1] Includes Alaska and Hawaii. [2] Excludes Alaska and Hawaii.

ERS, Resource and Rural Economics Division, (202) 694–5575. Data for 1910–1992 is from the Census of Agriculture, U.S. Department of Commerce. Data for 1997-2004 is from ERS Agricultural Resource Management Survey.

Table 9-7.—Farms: Classification by tenure of operator, United States, 1910–2004

Year	Farms	Tenure of operator			
		Full owners	Part owners	Managers	All tenants
	Number	Percent	Percent	Percent	Percent
1910	6,365,822	52.7	9.3	0.9	37.0
1920	6,453,991	52.2	8.7	1.1	38.1
1925	6,371,640	52.0	8.7	0.6	38.6
1930 [1]	6,295,103	46.3	10.4	0.9	42.4
1935	6,812,350	47.1	10.1	0.7	42.1
1940 [1]	6,102,417	50.6	10.1	0.6	38.8
1945	5,859,169	56.4	11.3	0.7	31.7
1950 [1]	5,388,437	57.4	15.3	0.4	26.9
1954	4,783,021	57.4	18.2	0.4	24.0
1959 [1]	3,710,503	57.1	21.9	0.6	20.5
1964 [1]	3,157,857	57.6	24.8	0.6	17.1
1969 [1]	2,730,250	62.5	24.6		12.9
1974 [1]	2,314,013	61.5	27.2		11.3
1978 [1]	2,257,775	57.5	30.2		12.3
1982 [1]	2,240,976	59.2	29.3		11.6
1987 [1]	2,087,759	59.3	29.2		11.5
1992 [2]	1,925,300	57.7	31.0		11.3
1997 [2]	2,049,384	55.3	35.4		9.3
1998 [2]	2,054,709	56.5	33.9		9.6
1999 [2]	2,186,950	58.3	33.9		7.8
2000 [2]	2,166,060	57.7	34.1		8.2
2001 [2]	2,149,683	57.2	34.9		8.0
2002 [2]	2,152,412	65.9	26.7		7.3
2003 [2]	2,121,107	62.1	31.7		6.1
2004 [2]	2,107,925	61.8	32.1		6.1

[1] Includes Alaska and Hawaii.   [2] Excludes Alaska and Hawaii.

ERS, Resource and Rural Economics Division, (202) 694–5575. Data for 1910-1992 is from the Census of Agriculture, U.S. Department of Commerce. Data for 1997-2004 is from ERS Agricultural Resource Management Survey.

Table 9-8.—Farmland Rented: Classification by Tenants and Part Owners, United States, 1900–2002

Year	Land in farms	Tenure of operator [1]			Percentage of land rented
		Tenants	Part-owners	Total	
	Million acres	Million acres	Million acres	Million acres	Percent
1900	841.8	195.1	[2] 71.1	266.2	31.6
1910	878.8	225.5	[3] 51.3	277.8	31.6
1920	958.7	[4] 265.0	[5] 54.7	319.7	33.3
1925	924.3	264.9	96.3	361.2	39.1
1930	990.1	307.3	125.2	432.5	43.7
1935	1,054.5	336.8	134.3	471.1	44.7
1940	1,065.1	313.2	155.9	469.1	44.0
1945	1,141.6	251.6	178.9	430.5	37.7
1950	1,161.4	212.2	196.2	408.4	35.2
1954	1,158.2	192.6	212.3	404.9	35.0
1959	1,123.0	166.8	234.1	400.9	35.7
1964	1,110.2	144.9	248.1	[6] 393.0	35.4
1969	1,063.3	137.6	241.8	379.4	35.7
1974	1,017.0	122.3	258.4	380.7	37.4
1978	1,029.7	124.1	282.2	406.2	39.4
1982	986.2	113.6	269.9	383.5	38.9
1987	964.5	126.9	275.4	402.3	41.7
1992	945.5	122.7	282.2	404.9	42.8
1997	931.8	108.1	270.0	378.1	40.6
2002 [7]	938.3	86.5	266.8	353.3	37.7

[1] Columns 3, 4, and 5 refer only to land rented from others and operated, so subleased land is not included. Acres of land rented are comparable in the same year, but definitions change over time. Basic sources are 1969 Census of Agriculture, table 5, p.14; 1974 Census of Agriculture, table 3, pp.1-6; 1978 Census of Agriculture, vol. 1, part 51, table 5, pp. 124-127; 1982 Census of Agriculture, vol. 1, part 51, table 48, p. 49; 1987 Census of Agriculture, vol. 1 part 51, table 48, p. 49; 1992 Census of Agriculture vol. 1, part 51, table 46, p. 53; 1997 Census of Agriculture, vol. 1, part 51, chapter 1, table 46, p. 57; 2002 Census of Agriculture, vol. 1, part 51, chapter 1, table 61, p. 214; and earlier census volumes as noted.   [2] Sum of part owners and owner/tenant, 1900 Census of Agriculture, table 20, pp.308.   [3] Assumes land leased by part-owners is the difference between the average size of full-owner and part-owner farms. Acreage leased by part-owners is this difference times the number of part-owners. 1910 Census of Agriculture, chapter 11, table 1 and 3, pp.97-99.   [4] 1920 Census of Agriculture, vol. VI, part 1, table 5, p. 19.   [5] Assumes same proportion of owner and part-owner as in 1910.   [6] 1964 Census of Agriculture, vol. II, chapter 8, p.757.   [7] The 2002 Census of Agriculture introduced new methodology to account for all farms in the United States. All 2002 published census items were reweighted for undercoverage. Strictly speaking, 2002 data are not fully comparable with data from earlier years.

ERS, Resource and Rural Economics Division, (202) 694–5572. Data from the Census of Agriculture, National Agricultural Statistics Service and Economic Research Service.

Table 9-9.—Farms: Number and land in farms, by States, 2004 and 2005

State	Farms [1]		Land in farms		Average per acre	
	2004	2005 [2]	2004	2005 [2]	2004	2005 [2]
	Number	Number	1,000 acres	1,000 acres	Acres	Acres
AL	44,000	43,500	8,750	8,600	199	198
AK	620	640	900	900	1,452	1,406
AZ	10,200	10,100	26,400	26,200	2,588	2,594
AR	47,500	47,000	14,400	14,400	303	306
CA	77,000	76,500	26,700	26,400	347	345
CO	30,900	30,500	30,900	30,700	1,000	1,007
CT	4,200	4,200	360	360	86	86
DE	2,300	2,300	525	520	228	226
FL	43,200	42,500	10,100	10,000	234	235
GA	49,000	49,000	10,700	10,500	218	214
HI	5,500	5,500	1,300	1,300	236	236
ID	25,000	25,000	11,800	11,800	472	472
IL	72,800	72,500	27,400	27,300	376	377
IN	59,300	59,000	15,000	15,000	253	254
IA	89,700	89,000	31,700	31,600	353	355
KS	64,500	64,500	47,200	47,200	732	732
KY	85,000	84,000	13,800	13,800	162	164
LA	27,000	26,800	7,850	7,800	291	291
ME	7,200	7,100	1,370	1,370	190	193
MD	12,100	12,100	2,050	2,040	169	169
MA	6,100	6,100	520	520	85	85
MI	53,200	53,000	10,100	10,100	190	191
MN	79,600	79,600	27,600	27,500	347	345
MS	42,200	42,200	11,050	11,050	262	262
MO	106,000	105,000	30,100	30,100	284	287
MT	28,000	28,000	60,100	60,100	2,146	2,146
NE	48,300	48,000	45,800	45,700	948	952
NV	3,000	3,000	6,300	6,300	2,100	2,100
NH	3,400	3,400	450	450	132	132
NJ	9,900	9,800	820	790	83	81
NM	17,500	17,500	44,700	44,500	2,554	2,543
NY	36,000	35,600	7,600	7,550	211	212
NC	52,000	50,000	9,000	8,900	173	178
ND	30,300	30,300	39,400	39,400	1,300	1,300
OH	77,200	76,500	14,500	14,300	188	187
OK	83,500	83,000	33,700	33,700	404	406
OR	40,000	40,000	17,200	17,100	430	428
PA	58,200	58,200	7,700	7,700	132	132
RI	850	850	60	60	71	71
SC	24,400	24,300	4,850	4,840	199	199
SD	31,600	31,400	43,800	43,700	1,386	1,392
TN	85,000	84,000	11,600	11,600	136	138
TX	229,000	230,000	130,000	129,800	568	564
UT	15,300	15,200	11,600	11,600	758	763
VT	6,400	6,300	1,250	1,250	195	198
VA	47,500	47,000	8,550	8,500	180	181
WA	35,000	34,500	15,200	15,100	434	438
WV	20,800	20,800	3,600	3,600	173	173
WI	76,500	76,500	15,500	15,400	203	201
WY	9,200	9,200	34,440	34,400	3,743	3,739
US	2,112,970	2,100,990	936,295	933,400	443	444
PR	13,500	13,600	600	610	44	45

[1] A farm is any establishment from which $1,000 or more of agricultural products were sold or would normally be sold during the year. [2] Preliminary.
NASS, Environmental, Economics, and Demographics Branch, (202) 720–6146.

FARM RESOURCES, INCOME, AND EXPENSES

Table 9-10.—Land: Utilization, by States, 2002

State	Cropland			Grassland pasture[2]	Forest land[3]	Special use areas[4]	Urban land	Other land[5]	Total land area[6]
	Used for crops[1]	Idle	Used only for pasture						
	1,000 acres	*1,000 acres*	*1,000 acres*	*1,000 acres*	*1,000 acres*	*1,000 acres*	*1,000 acres*	*1,000 acres*	*1,000 acres*
AL	2,097	459	1,181	1,905	22,922	1,522	1,139	1,252	32,476
AK	32	49	9	1,295	90,475	143,262	167	130,760	366,049
AZ	875	146	214	40,533	17,608	11,373	1,080	897	72,726
AR	7,570	349	1,727	2,312	18,373	1,534	582	876	33,324
CA	8,591	719	1,345	21,729	33,780	21,558	5,095	6,997	99,814
CO	8,054	2,155	1,835	28,158	18,925	6,022	814	417	66,380
CT	117	12	23	30	1,436	294	1,133	55	3,101
DE	436	11	7	10	376	112	195	104	1,251
DC	0	0	0	0	0	0	39	0	39
FL	2,351	258	1,107	4,701	14,636	4,509	3,960	2,992	34,513
GA	3,807	305	933	1,227	23,802	1,995	2,407	2,584	37,060
HI	88	62	37	1,002	1,552	769	227	374	4,111
ID	4,783	855	770	20,984	16,824	6,175	263	2,305	52,958
IL	22,940	1,055	536	1,950	4,087	1,972	2,302	732	35,574
IN	12,122	452	453	1,335	4,342	1,271	1,423	1,558	22,955
IA	24,248	1,943	1,365	2,134	1,944	1,857	521	1,744	35,756
KS	25,464	2,522	2,475	15,079	1,490	1,944	554	2,832	52,362
KY	5,040	712	2,601	2,304	11,947	1,607	7,893	431	25,426
LA	3,775	658	905	1,711	13,722	1,927	1,070	4,112	27,880
ME	344	86	48	42	16,951	590	223	1,468	19,752
MD	1,338	80	109	291	2,372	735	1,164	167	6,255
MA	152	14	30	31	2,330	555	1,807	99	5,018
MI	6,955	693	412	1,561	18,616	2,685	2,153	3,280	36,355
MN	20,362	2,157	745	1,830	14,722	4,679	966	5,489	50,950
MS	4,358	769	936	2,223	18,572	657	599	1,607	30,020
MO	13,297	1,594	4,194	6,419	13,364	1,938	1,170	2,111	44,087
MT	13,122	3,270	1,726	46,361	19,184	6,863	168	2,458	93,153
NE	19,758	1,099	1,908	21,941	897	1,652	293	1,651	49,198
NV	534	36	314	46,448	8,636	6,882	350	7,088	70,289
NH	77	11	20	47	4,503	336	357	388	5,740
NJ	465	30	57	40	1,476	811	1,794	73	4,747
NM	1,209	625	837	51,676	14,978	6,449	484	1,410	77,668
NY	3,763	220	511	1,359	15,389	3,971	2,535	2,468	30,217
NC	4,239	292	668	875	18,664	2,395	2,294	1,748	31,175
ND	23,373	2,969	1,343	11,529	441	1,687	94	2,708	44,145
OH	10,314	507	708	1,241	7,568	1,394	2,570	1,904	26,207
OK	9,060	961	5,209	17,488	6,234	1,635	736	2,625	43,947
OR	3,626	681	1,003	23,239	27,169	3,946	662	1,112	61,438
PA	4,227	297	600	962	15,853	2,593	2,745	1,404	28,683
RI	19	3	3	4	314	61	250	13	669
SC	1,585	321	420	448	12,301	1,074	1,201	1,920	19,270
SD	17,332	1,259	2,476	22,487	1,511	1,855	108	1,540	48,566
TN	4,415	545	2,124	1,194	13,656	2,370	1,566	509	26,379
TX	21,684	5,468	13,289	98,263	11,774	5,475	4,585	7,013	167,550
UT	1,194	248	602	24,339	14,905	4,958	444	5,882	52,572
VT	479	26	84	208	4,483	346	95	198	5,920
VA	2,712	189	1,267	1,373	15,372	1,590	1,526	1,312	25,340
WA	6,254	1,230	499	7,369	17,347	6,839	1,367	1,682	42,588
WV	680	37	491	507	11,900	929	364	502	15,410
WI	9,047	966	766	2,003	15,701	2,439	1,051	2,785	34,758
WY	1,583	364	913	44,323	5,739	6,416	109	2,697	62,144
US	339,949	39,768	61,834	586,521	651,163	296,807	59,587	228,334	2,263,962
PR	204	62	174	106	63	....	....	....	2,193

[1] Cropland harvested, crop failure, and cultivated summer fallow. [2] Grassland and other nonforest pasture and range. [3] Excludes reserved and other forest land duplicated in parks and other special uses of land. Includes forested grazing land. [4] Includes rural transportation areas, Federal and State areas used primarily for recreation and wildlife purposes, military areas, farmsteads, and farm roads and lanes. [5] Miscellaneous areas such as marshes, open swamps, bare rock areas, and deserts, including urban and other special uses not inventoried. [6] Approximate land area as established by the Bureau of the Census in conjunction with the 1990 Census of Population.

ERS, Resource Economics Division, (202) 694–5528. Estimates based on reports and records of the U.S. Departments of Agriculture and Commerce, and public land administering and conservation agencies. Estimates developed for years coinciding with a Census of Agriculture.

Table 9-11.—Land in farms:[1] Irrigated land, by States, 1964–2002

State	1964	1969	1974	1978[2]	1982	1987	1992	1997	2002
	1,000 acres	1,000 acres	1,000 acres	1,000 acres	1,000 acres	1,000 acres	1,000 acres	1,000 acres	1,000 acres
AL	12	11	14	59	66	84	82	80	109
AK	(3)	1	1	1	1	2	2	3	3
AZ	1,125	1,178	1,153	1,196	1,098	914	956	1,075	932
AR	974	1,010	949	1,683	2,022	2,406	2,702	3,785	4,150
CA	7,599	7,240	7,749	8,506	8,461	7,596	7,571	8,887	8,709
CO	2,690	2,895	2,874	3,431	3,201	3,014	3,170	3,374	2,591
CT	14	9	7	7	7	7	6	8	10
DE	18	20	20	34	44	61	62	75	97
FL	1,217	1,365	1,559	1,980	1,585	1,623	1,783	1,874	1,815
GA	64	79	112	463	575	640	725	773	871
HI	144	146	142	159	146	149	134	77	69
ID	2,802	2,761	2,859	3,475	3,450	3,219	3,260	3,544	3,289
IL	14	51	54	130	166	208	328	352	391
IN	17	34	33	75	132	170	241	256	313
IA	22	21	39	101	91	92	116	133	142
KS	1,004	1,522	2,010	2,686	2,675	2,463	2,680	2,696	2,678
KY	14	20	11	14	23	38	28	60	37
LA	581	702	702	681	694	647	898	961	939
ME	4	6	6	7	6	6	10	22	20
MD	16	22	23	28	39	51	57	69	81
MA	24	19	19	17	17	20	20	27	24
MI	49	77	97	226	286	315	366	407	456
MN	18	36	78	272	315	354	370	403	455
MS	123	150	162	309	431	637	883	1,110	1,176
MO	59	156	150	320	403	535	709	921	1,033
MT	1,893	1,841	1,759	2,070	2,023	1,997	1,978	2,102	1,976
NE	2,169	2,857	3,967	5,683	6,039	5,682	6,312	7,066	7,625
NV	825	753	778	881	830	779	556	764	747
NH	3	2	2	2	1	3	2	3	2
NJ	96	72	89	77	83	91	80	94	97
NM	813	823	867	891	807	718	738	852	845
NY	79	55	55	56	52	51	47	74	75
NC	97	59	51	90	81	138	113	156	264
ND	51	63	71	141	163	168	187	183	203
OH	17	22	22	25	28	32	29	35	41
OK	302	524	515	602	492	478	512	509	518
OR	1,608	1,519	1,561	1,881	1,808	1,648	1,622	1,963	1,908
PA	23	19	18	15	18	30	23	40	43
RI	1	2	2	3	2	4	3	3	4
SC	19	15	10	32	81	81	76	89	96
SD	130	148	152	335	376	362	371	367	401
TN	11	12	10	13	18	38	37	47	61
TX	6,385	6,888	6,594	6,947	5,576	4,271	4,912	5,764	5,075
UT	1,092	1,025	970	1,169	1,082	1,161	1,143	1,218	1,091
VT	2	4	(3)	1	1	1	2	2	3
VA	51	37	28	42	43	79	62	86	99
WA	1,150	1,224	1,309	1,639	1,638	1,519	1,641	1,787	1,823
WV	2	3	2	1	1	3	3	4	2
WI	62	106	128	235	259	285	331	358	386
WY	1,571	1,523	1,460	1,662	1,565	1,518	1,465	1,750	1,542
US	37,056	39,122	41,243	50,350	49,003	46,386	49,404	56,289	55,316
PR	89	91	70	54	42	36	46	35	46
VI	(3)	(3)	(3)	(4)	(4)	(4)	(4)	(4)	(3)
Total	37,145	39,213	41,313	50,350	49,002	46,386	49,404	55,058	55,363

[1] Data may not add because of rounding.   [2] Data for 1978 not directly comparable with earlier censuses as it includes estimates from the direct enumeration sample for farms not represented on the mail list.   [3] Less than 500 acres.   [4] Not available.   Note: Data from the Census of Agriculture, U.S. Department of Commerce. Beginning in 1997 Census of Agriculture, U.S. Department of Agriculture.

ERS, Resource Economics Division, (202) 694–5528.

FARM RESOURCES, INCOME, AND EXPENSES

Table 9-12.—Farm real estate: Value of farmland and buildings, by State, 2001–2005 [1]

State	Total value of land and buildings				
	2001	2002	2003	2004	2005
	Million dollars	Million dollars	Million dollars	Million dollars	Million dollars
AL	14,760	15,130	15,664	16,275	17,630
AZ [2]	10,054	10,386	10,556	10,846	11,156
AR	19,710	20,586	21,312	23,760	26,208
CA	89,600	94,520	97,560	101,460	109,824
CO	21,330	21,980	22,630	23,948	25,942
CT	2,772	3,060	3,420	3,672	3,888
DE	1,904	2,035	2,120	3,150	4,368
FL	27,040	28,016	29,580	31,310	37,000
GA	20,710	22,243	23,760	25,145	27,195
ID	14,280	14,632	15,104	16,048	17,464
IL	62,975	64,625	66,825	71,514	79,170
IN	35,720	37,146	38,653	41,550	45,750
IA	60,125	61,440	63,717	69,740	78,684
KS	30,638	31,455	32,332	33,748	37,760
KY	23,975	25,254	26,220	27,600	30,360
LA	11,081	11,390	11,775	12,403	13,104
ME	2,025	2,160	2,398	2,535	2,672
MD	8,094	8,400	8,549	11,685	16,116
MA	3,942	4,212	4,836	5,148	5,460
MI	23,142	24,996	27,041	29,492	31,815
MN	39,060	41,700	44,320	49,680	55,825
MS	14,173	14,803	15,554	16,354	17,459
MO	39,260	41,676	44,247	47,558	52,374
MT	20,755	22,052	23,439	24,641	26,745
NE	33,884	34,960	35,573	37,785	41,587
NV [2]	2,486	2,526	2,603	2,706	2,962
NH	1,122	1,232	1,395	1,463	1,553
NJ	6,723	7,138	7,462	7,995	8,137
NM [2]	11,210	11,428	11,774	11,959	12,829
NY	11,658	12,333	13,005	13,528	14,194
NC	24,683	26,448	28,210	29,700	31,773
ND	16,154	16,351	16,745	17,927	19,700
OH	36,482	38,168	40,004	42,485	45,474
OK	22,139	22,984	23,759	25,107	27,129
OR	19,030	19,780	20,640	21,500	23,085
PA	23,070	25,058	26,565	28,105	30,800
RI	462	498	558	612	672
SC	8,838	9,272	9,943	10,428	11,277
SD	17,820	18,877	20,148	21,900	24,909
TN	25,960	27,140	27,840	29,000	31,320
TX	95,557	101,293	105,705	111,150	120,065
UT [2]	7,861	8,332	8,767	9,131	9,712
VT	2,286	2,413	2,563	2,688	2,875
VA	20,730	21,960	23,220	27,360	33,150
WA	20,215	21,406	22,644	23,256	24,915
WV	4,572	4,788	5,040	5,400	5,760
WI	31,200	33,970	35,880	38,750	43,890
WY	9,315	9,833	10,332	10,849	12,040
48 States	1,050,582	1,102,083	1,151,985	1,230,043	1,353,775

[1] Total value of land and buildings is derived by multiplying average value per acre of farm real estate by the land in farms. [2] Value of all land and buildings adjusted to include American Indian reservation land value.

NASS, Environmental, Economics, and Demographics Branch, (202) 720–6146.

Table 9-13.—Land utilization, United States, selected years, 1945–2002

Major land uses	1945	1949	1959	1969	1978	1987	1992	1997	2002
	Million acres	Million acres	Million acres	Million acres	Million acres	Million acres	Million acres	Million acres	Million acres
Cropland used for crops[1] ...	363	383	359	333	369	331	338	349	340
Idle cropland ..	40	26	34	51	26	68	56	39	40
Cropland used for pasture ..	47	69	66	88	76	65	67	68	62
Grassland pasture[2]	659	632	633	604	587	591	591	580	587
Forest land[3] ..	602	760	745	723	703	648	648	641	651
Special uses[4]	85	87	115	143	158	279	281	286	297
Urban areas[5]	15	18	27	31	45	57	59	66	60
Other land[6]	93	298	293	291	301	227	224	236	228
Total land area[7]	1,905	2,273	2,271	2,264	2,264	2,265	2,263	2,263	2,264

[1] Cropland harvested, crop failure, and cultivated summer fallow. [2] Grassland and other nonforest pasture and range. [3] Excludes reserved and other forest land duplicated in parks and other special uses of land. Includes forested grazing land. [4] Includes rural transportation areas, Federal and State areas used primarily for recreation and wildlife purposes, military areas, farmsteads and farm roads and lanes. [5] The 2002 urban acreage estimate is not directly comparable to estimates in prior years due to a change in the definition of urban areas in the 2000 Census of Population and Housing. The apparent change in "urban" acreage between 1997 and 2002 reflects a definitional change, rather than a decline in acreage. [6] Miscellaneous areas such as marshes, open swamps, bare rock areas, deserts, and other uses not inventoried. [7] Remeasurement and changes in reservoirs account for changes in total land areas except for the major increase in 1949 when data for Alaska and Hawaii were added.
ERS, Resource Economics Division, (202) 694–5528. Estimates based on reports and records of the U.S. Department of Agriculture and Commerce, and public land administering and conservation agencies.

Table 9-14.—Farm real estate: Average value per acre of land and buildings, by State, Mar. 1, 1970, and Jan. 1, 2001–2005

State	Mar. 1, 1970	Jan. 1, 2001	Jan. 1, 2002	Jan. 1, 2003	Jan. 1, 2004	Jan. 1, 2005
	Dollars	Dollars	Dollars	Dollars	Dollars	Dollars
AL	200	1,640	1,700	1,760	1,860	2,050
AZ[1]	70	1,250	1,400	1,500	1,600	1,750
AR	260	1,350	1,410	1,480	1,650	1,820
CA	479	3,200	3,400	3,600	3,800	4,160
CO	95	675	700	730	775	845
CT	921	7,700	8,500	9,500	10,200	10,800
DE	499	3,400	3,700	4,000	6,000	8,400
FL	355	2,600	2,720	2,900	3,100	3,700
GA	234	1,900	2,050	2,200	2,350	2,590
ID	177	1,200	1,240	1,280	1,360	1,480
IL	490	2,290	2,350	2,430	2,610	2,900
IN	406	2,350	2,460	2,570	2,770	3,050
IA	392	1,850	1,920	2,010	2,200	2,490
KS	159	645	665	685	715	800
KY	253	1,750	1,830	1,900	2,000	2,200
LA	321	1,380	1,440	1,500	1,580	1,680
ME	161	1,500	1,600	1,750	1,850	1,950
MD	640	3,800	4,000	4,150	5,700	7,900
MA	565	7,300	8,100	9,300	9,900	10,500
MI	326	2,280	2,470	2,680	2,920	3,150
MN	226	1,400	1,500	1,600	1,800	2,030
MS	234	1,270	1,330	1,400	1,480	1,580
MO	224	1,300	1,380	1,470	1,580	1,740
MT	60	350	370	390	410	445
NE	154	735	760	775	825	910
NV[1]	53	450	465	480	500	550
NH	239	2,550	2,800	3,100	3,250	3,450
NJ	1,092	8,100	8,600	9,100	9,750	10,300
NM[1]	42	240	250	260	265	290
NY	273	1,520	1,610	1,700	1,780	1,880
NC	333	2,680	2,900	3,100	3,300	3,570
ND	94	410	415	425	455	500
OH	399	2,470	2,600	2,740	2,930	3,180
OK	173	655	680	705	745	805
OR	150	1,100	1,150	1,200	1,250	1,350
PA	373	3,000	3,250	3,450	3,650	4,000
RI	734	7,700	8,300	9,300	10,200	11,200
SC	261	1,800	1,900	2,050	2,150	2,330
SD	84	405	430	460	500	570
TN	268	2,200	2,300	2,400	2,500	2,700
TX	148	730	775	810	855	925
UT[1]	92	975	1,040	1,100	1,150	1,230
VT	224	1,800	1,900	2,050	2,150	2,300
VA	286	2,380	2,530	2,700	3,200	3,900
WA	224	1,300	1,390	1,480	1,530	1,650
WV	136	1,270	1,330	1,400	1,500	1,600
WI	232	1,950	2,150	2,300	2,500	2,850
WY	41	270	285	300	315	350
48 States[2]	196	1,150	1,210	1,270	1,360	1,510

[1] Excludes American Indian Reservation Land. [2] Excludes Alaska and Hawaii.
NASS, Environmental, Economics, and Demographics Branch, (202) 720–6146.

FARM RESOURCES, INCOME, AND EXPENSES

Table 9-15.—Land values, cropland and pasture: By State, 2004–2005

State	2004				2005			
	Cropland [1]	Irrigated cropland	Non-irrigated cropland	Pasture [2]	Cropland [1]	Irrigated cropland	Non-irrigated cropland	Pasture [2]
	Dollars	Dollars	Dollars	Dollars	Dollars	Dollars	Dollars	Dollars
AL	1,800			1,420	2,200			1,550
AZ	6,400	6,400		500	6,790	6,790		600
AR	1,290	1,450	1,150	1,300	1,420	1,600	1,270	1,570
CA	6,020	6,600	2,130	1,600	6,590	7,450	2,400	1,750
CO	1,060	2,100	580	470	1,110	2,350	625	555
CT								
DE	5,700				8,000			
FL	3,810	4,400	2,850	2,250	4,650	5,500	3,500	2,600
GA	2,260	2,100	2,300	2,950	2,730	2,650	2,750	3,150
ID	1,710	2,330	800	725	1,840	2,600	870	805
IL	2,700			1,110	3,030			1,240
IN	2,750			1,780	3,080			1,900
IA	2,320			880	2,650			1,000
KS	705	1,110	665	430	800	1,200	760	500
KY	2,230			1,530	2,400			1,700
LA	1,300	1,150	1,340	1,350	1,390	1,240	1,430	1,500
ME								
MD	5,600				5,500	7,600		7,300
MA								
MI	2,550			1,800	2,750			1,950
MN	1,690			700	1,850			810
MS	1,210	1,280	1,190	1,270	1,280	1,360	1,260	1,420
MO	1,690	2,250	1,650	1,130	1,890	2,470	1,850	1,260
MT	548	1,680	400	285	586	1,800	440	320
NE	1,290	1,750	1,050	275	1,430	1,890	1,200	310
NV	1,950	1,950		260	2,070	2,070		295
NH								
NJ	9,900			10,600	10,500			11,300
NM	1,450	3,000	270	170	1,450	3,370	300	195
NY	1,470			775	1,530			825
NC	3,150			3,200	3,400			3,430
ND	490			185	546			210
OH	2,940			2,100	3,230			2,240
OK	697	850	690	475	745	910	738	550
OR	1,690	2,350	1,250	470	1,800	2,580	1,360	510
PA	3,700			2,000	4,000			2,200
RI								
SC	1,850			2,000	2,050			2,150
SD	747	1,080	740	240	847	1,200	840	290
TN	2,420			2,450	2,600			2,620
TX	981	1,050	965	655	1,050	1,130	1,040	725
UT	2,900	3,800	850	520	2,900	4,200	950	630
VT								
VA	3,300			2,800	4,000			3,500
WA	1,510	3,300	990	540	1,610	3,550	1,060	585
WV	2,200			1,280	2,350			1,380
WI	2,350			1,200	2,600			1,400
WY	972	1,300	340	235	1,010	1,500	380	270
Other States ..	6,230			4,140	6,610			4,480
US	1,770			634	1,970			694

[1] Other cropland States include CT, ME, MA, NH, RI, and VT. [2] Other pasture States include CT, DE, ME, MA, NH, RI, and VT.

NASS, Environmental, Economics, and Demographics Branch, (202) 720–6146.

Table 9-16.—Cash rents, cropland and pasture: By State, 2004–2005

State	2004				2005			
	Cropland	Irrigated cropland	Non-irrigated cropland	Pasture	Cropland	Irrigated cropland	Non-irrigated cropland	Pasture
	Dollars	Dollars	Dollars	Dollars	Dollars	Dollars	Dollars	Dollars
AL	33.00			18.00	40.00			17.50
AZ		150.00				165.00		
AR	75.00	86.00	59.00		76.00	86.00	58.00	
CA		300.00		11.50		330.00		12.00
CO	58.00	91.00	22.00	3.70	61.00	100.00	23.00	4.30
CT								
DE	61.00				64.00			
FL			34.00	17.50			37.00	18.50
GA	58.00	110.00	42.00	24.00	58.00	115.00	41.00	22.00
ID	99.00	118.00	53.00		104.00	124.00	55.00	
IL	126.00			34.00	129.00			34.50
IN	107.00				109.00			
IA	126.00			32.50	131.00			36.00
KS	41.00	72.00	37.50	13.20	42.00	73.00	38.50	13.40
KY	72.00				73.00			
LA	66.00	76.00	62.00	15.50	66.00	70.00	62.00	16.50
ME								
MD	59.00				62.00			
MA								
MI	62.00				62.00			
MN	83.50			19.50	86.50			20.50
MS	66.00	85.00	58.00	16.50	69.00	96.00	60.00	16.50
MO			76.00	26.00			79.00	27.00
MT	24.50	49.00	18.90	5.00	25.00	53.00	19.50	5.90
NE	95.00	125.00	70.00	12.00	97.00	127.00	71.00	12.00
NV								
NH								
NJ	47.50				47.50			
NM				1.70				1.80
NY	40.00				41.00			
NC	53.00			23.00	55.00			25.00
ND	37.50			10.20	39.00			10.60
OH	80.00				82.00			
OK			30.00	9.00			29.00	9.00
OR	100.00	125.00	65.00		100.00	130.00	70.00	
PA	43.00			25.00	45.00			27.00
RI								
SC	28.50				29.00			
SD			47.50	11.60			50.40	12.30
TN	67.00			19.00	67.00			18.00
TX	29.80	56.00	23.70	7.80	29.70	57.50	23.00	8.30
UT		61.00		10.00		65.00		9.00
VT								
VA	39.00			17.50	40.00			20.00
WA		185.00				190.00		
WV	30.00				28.00			
WI	70.00			37.00	70.00			38.00
WY				4.00				4.00
48 Sts	76.50			9.60	78.00			10.30

NASS, Environmental, Economics, and Demographics Branch, (202) 720–6146.

FARM RESOURCES, INCOME, AND EXPENSES

Table 9-17.—Farm assets and claims: Comparative balance sheet of the farming sector, excluding operator households, United States, Dec. 31, 1994–2003 [1]

Item	1994	1995	1996	1997	1998
ASSETS	Billion	Billion	Billion	Billion	Billion
Physical assets:	dollars	dollars	dollars	dollars	dollars
Real estate	704.1	740.5	769.5	808.2	840.4
Non-real estate:					
Livestock [2]	67.9	57.8	60.3	67.1	63.4
Machinery and motor vehicles	86.8	87.6	88.0	88.7	89.8
Crops stored on and off farms [3]	23.3	27.4	31.7	32.7	29.7
Purchased inputs	5.0	3.4	4.4	4.9	5.0
Financial assets: [4]	47.6	49.1	49.0	49.7	54.8
Total [5]	934.7	965.7	1,002.9	1,051.3	1,083.1
CLAIMS					
Liabilities:					
Real estate debt	69.9	71.7	74.4	78.5	83.1
Non-real estate debt to—					
Reporting institutions [6]	54.3	55.6	57.2	60.4	62.8
Nonreporting creditors [7]	14.7	15.7	16.9	18.0	18.7
Total liabilities [5]	138.9	143.0	148.6	156.9	164.6
Proprietors' equity	795.8	822.8	854.3	894.4	918.5
Total [5]	934.7	965.8	1,002.9	1,051.3	1,083.1

Item	1999	2000	2001	2002	2003 [8]
ASSETS	Billion	Billion	Billion	Billion	Billion
Physical assets:	dollars	dollars	dollars	dollars	dollars
Real estate	887.0	946.4	996.2	1,045.7	1,378.8
Non-real estate:					
Livestock [2]	73.2	76.8	78.5	75.6	78.5
Machinery and motor vehicles	89.8	90.1	92.8	93.6	95.9
Crops stored on and off farms [3]	28.3	27.9	25.2	23.1	24.4
Purchased inputs	4.0	4.9	4.2	5.6	5.6
Financial assets: [4]	56.5	57.1	58.9	60.4	62.4
Total [5]	1,138.1	1,203.2	1,255.9	1,304.0	1,378.8
CLAIMS					
Liabilities:					
Real estate debt	87.2	91.1	96.1	103.4	108.0
Non-real estate debt to—					
Reporting institutions [6]	61.1	65.7	68.4	68.1	67.4
Nonreporting creditors [7]	19.4	20.8	21.3	21.9	22.6
Total liabilities [5]	167.7	177.6	185.7	193.3	198.0
Proprietors' equity	971.1	1,025.6	1,070.2	1,110.7	1,180.8
Total [5]	1,138.1	1,203.2	1,255.9	1,304.0	1,378.8

[1] Farms are defined as places with sales greater than $1,000 annually. [2] Horses and mules are excluded. [3] Excludes all crops held on farms including crops under loan to Commodity Credit Corporation, and crops held off farms as security for CCC loans. [4] Includes farm share of currency and demand deposits. [5] Total of rounded data. [6] Loans of all operating banks, the Farm Credit System, and direct loans of the Farm Service Agency. [7] Loans and credits extended by dealers, merchants, finance companies, individuals, and others. [8] Preliminary.

ERS, Farm Sector Performance Branch, (202) 694–5586.

Table 9-18.—Farm labor: Number of workers on farms and average wage rates,
United States, 1997-2006

Year	Total workers	Self-employed and unpaid workers[1]	Ag service workers[2]	Hired workers[2,3]	Hired workers[2,3]
	Number	Number	Number	Number	Wage rates
1997.					
Jan	(5)	(5)	131	624	7.20
Apr	(5)	(5)	207	808	7.03
July	(5)	(5)	340	1,069	6.88
Oct	(5)	(5)	283	1,004	7.31
Annual average		1,989.9	(4)	876.5	7.35
1998.					
Jan	(5)	(5)	141	661	7.61
Apr	(5)	(5)	202	803	7.49
July	(5)	(5)	379	1071	7.25
Oct	(5)	(5)	263	983	7.60
Annual average		1,946.6	(4)	879.5	7.47
1999.					
Jan	(5)	(5)	157	705	7.94
Apr	(5)	(5)	160	867	7.83
July	(5)	(5)	319	1,155	7.58
Oct	(5)	(5)	290	989	7.83
Annual average		2,048.4	(4)	929	7.77
2000.					
Jan		(5)	172	685	8.10
Apr		(5)	217	840	8.09
July		(5)	203	1,084	7.93
Oct		(5)	288	952	8.29
Annual average		2,062.3	(4)	890.3	8.10
2001.					
Jan		(5)	165	691	8.66
Apr		(5)	215	804	8.31
July		(5)	335	1,039	8.29
Oct		(5)	262	991	8.59
Annual average		2,049.8	(4)	873.3	8.45
2002.					
Jan		(5)	183	707	8.97
Apr		(5)	189	890	8.83
July		(5)	256	1,006	8.57
Oct		(5)	271	940	8.95
Annual average		(5)	(4)	884.5	8.80
2003.					
Jan		(5)	160	729	9.34
Apr		(5)	157	781	9.16
July		(5)	320	943	8.88
Oct		(5)	306	891	9.05
Annual average		(5)	(4)	836	9.08
2004.					
Jan		(5)	185	662	9.41
Apr		(5)	257	827	9.23
July		(5)	343	961	9.04
Oct		(5)	324	851	9.32
Annual average		(5)	(4)	825.2	9.22
2005.					
Jan		(5)	185	589	9.78
Apr		(5)	247	753	9.35
July		(5)	408	936	9.38
Oct		(5)	294	842	9.61
Annual average		(5)	(4)	779.5	9.50
2006.					
Jan		(5)	180	616	10.11

[1] Includes farm operators and partners doing 1 or more hours of farm work and other unpaid workers working 15 hours or more during the survey week without cash wages. [2] Includes all persons doing farm work for pay during the survey week. [3] Excludes agricultural service workers. [4] Annual average not computed. [5] Discontinued.

NASS, Economic, Environmental and Demographics Branch, (202) 720-6146.

FARM RESOURCES, INCOME, AND EXPENSES

Table 9-19.—Farm labor: Number of hired workers on farms and average wage rates, by States and regions, 2005 [1] [2]

State and region [3]	Workers on farms	Farm wage rates			
	Hired	Type of worker			
		Field	Livestock	Field and livestock	All hired workers [4]
Jan. 9–15, 2005	*Thousands*	*Dollars per hour*	*Dollars per hour*	*Dollars per hour*	*Dollars per hour*
Northeast I	23	9.47	9.17	9.32	10.37
Northeast II	18	8.47	8.76	8.62	9.66
Appalachian I	25	8.65	9.03	8.82	9.64
Appalachian II	28	8.46	8.04	8.25	9.02
Southeast	24	7.96	7.25	7.71	8.41
FL	48	8.50	8.60	8.51	9.52
Lake	43	9.65	9.67	9.66	10.61
Cornbelt I	28	9.40	8.95	9.18	10.06
Cornbelt II	21	9.16	10.28	10.07	10.63
Delta	18	9.52	7.63	8.97	9.29
N. Plains	27	10.26	8.60	9.20	9.82
S. Plains	50	8.01	9.35	8.75	9.56
Mountain I	12	9.42	8.82	8.95	9.76
Mountain II	17	7.37	9.65	8.83	9.93
Mountain III	19	7.70	8.41	8.02	8.61
Pacific	38	9.32	9.90	9.39	10.33
CA	143	8.56	9.93	8.86	9.82
HI	7	9.94	....	9.98	11.52
US (49 States)	**589**	**8.71**	**9.20**	**8.90**	**9.78**
Apr. 10–16, 2005					
Northeast I	34	9.01	8.51	8.83	9.47
Northeast II	26	9.24	8.62	9.05	9.65
Appalachian I	28	8.38	8.85	8.50	9.07
Appalachian II	32	8.38	7.69	8.08	8.59
Southeast	36	8.41	8.30	8.38	8.83
FL	49	8.20	9.90	8.37	9.31
Lake	55	8.99	10.05	9.45	9.95
Cornbelt I	41	8.84	9.17	8.91	9.51
Cornbelt II	27	8.85	9.27	9.06	9.38
Delta	28	7.37	7.18	7.34	7.64
N. Plains	27	9.33	9.69	9.46	9.70
S. Plains	55	8.13	9.15	8.53	9.28
Mountain I	24	7.89	8.49	8.23	8.43
Mountain II	20	7.70	8.41	8.02	8.50
Mountain III	18	7.95	9.40	8.51	9.18
Pacific	64	8.87	10.78	9.23	9.95
CA	*182	*8.62	*9.60	*8.76	9.48
HI	7	9.67	....	9.79	11.33
US (49 States)	***753**	***8.56**	***9.14**	***8.72**	***9.35**

See footnotes at end of table.

Table 9-19.—Farm labor: Number of hired workers on farms and average wage rates, by States and regions, 2005 [1] [2] —Continued

State and region [3]	Workers on farms	Farm wage rates			
		Type of worker			
	Hired	Field	Livestock	Field and livestock	All hired workers [4]
	Thousands	*Dollars per hour*	*Dollars per hour*	*Dollars per hour*	*Dollars per hour*
July 10–16, 2005					
Northeast I	46	8.88	9.55	9.11	9.70
Northeast II	50	8.71	9.37	8.90	9.79
Appalachian I	38	8.44	8.53	8.46	9.03
Appalachian II	24	8.46	8.19	8.30	8.68
Southeast	44	8.39	8.85	8.51	8.91
FL	41	8.75	9.15	8.81	9.70
Lake	75	8.66	9.52	8.97	9.66
Cornbelt I	54	9.20	8.77	9.10	9.56
Cornbelt II	31	8.86	9.14	9.05	9.56
Delta	24	7.59	7.80	7.65	7.85
No. Plains	45	8.15	9.49	8.60	9.05
So. Plains	63	8.07	9.06	8.50	9.27
Mountain I	29	8.39	8.51	8.44	8.79
Mountain II	26	8.62	8.49	8.58	9.20
Mountain III	24	7.90	8.11	7.98	8.53
Pacific	109	8.60	10.67	8.80	9.21
CA	*206	*8.76	*10.66	*9.00	*9.68
HI	7	10.00	....	10.05	11.76
US (49 States)	*936	*8.61	*9.26	8.78	*9.38
October 9–15, 2005					
Northeast I	38	9.42	9.42	9.42	10.19
Northeast II	39	9.21	8.62	9.09	10.00
Appalachian I	36	8.17	8.91	8.43	8.89
Appalachian II	24	8.48	8.26	8.40	9.03
Southeast	37	8.51	8.86	8.60	9.05
FL	42	8.60	8.45	8.58	9.33
Lake	72	9.96	9.49	9.80	10.35
Cornbelt I	50	9.88	8.68	9.57	10.10
Cornbelt II	29	9.21	10.89	10.02	11.16
Delta	34	7.04	7.95	7.26	7.70
No. Plains	35	10.14	9.37	9.86	10.12
So. Plains	64	7.60	7.84	7.68	8.38
Mountain I	29	8.26	9.27	8.55	8.91
Mountain II	22	7.94	8.39	8.14	8.75
Mountain III	25	7.27	8.87	7.67	8.28
Pacific	76	8.96	9.58	9.00	9.62
CA	*183	9.21	*10.45	*9.37	*10.13
HI	7	10.10	....	10.18	11.73
US (49 States)	*842	8.90	*9.15	8.96	9.61

[1] Excludes agricultural service workers. [2] Includes all persons doing work for pay during the survey week. [3] Regions consist of the following: Northeast I: CT, ME, MA, NH, NY, RI, VT; Northeast II: DE, MD, NJ, PA; Appalachian I: NC, VA; Appalachian II: KY, TN, WV; Southeast: AL, GA, SC; Lake: MI, MN, WI; Cornbelt I: IL, IN, OH; Cornbelt II: IA, MO; Delta: AR, LA, MS; No. Plains: KS, NE, ND, SD; So. Plains: OK, TX; Mountain I: ID, MT, WY; Mountain II: CO, NV, UT; Mountain III: AZ, NM; Pacific: OR, WA. [4] Includes field, livestock, supervisors, and other workers doing work for pay during the survey week. * Revised.

NASS, Economic, Environmental and Demographics Branch, (202) 720–6146.

Table 9-20.—Farm production and output: Index numbers of total output, and production of livestock, crops, and secondary output, by groups, United States, 1995–2004

[1996=100]

| Year | Farm output | Livestock and products | | | |
		All livestock and products [1]	Meat animals [2]	Dairy products [3]	Poultry and eggs [4]
1995	0.961	1.006	1.040	1.008	0.947
1996	1.000	1.000	1.000	1.000	1.000
1997	1.038	1.010	1.009	1.014	1.022
1998	1.052	1.038	1.038	1.022	1.041
1999	1.076	1.069	1.054	1.057	1.087
2000	1.083	1.081	1.061	1.090	1.106
2001	1.081	1.075	1.053	1.076	1.127
2002	1.072	1.105	1.057	1.106	1.172
2003	1.083	1.097	1.055	1.109	1.167
2004	1.123	1.095	1.040	1.111	1.194

| Year | Crops | | | | |
	All crops	Cereal crops	Forage crops	Industrial crops [5]	Vegetables and horticulture crops
1995	0.919	0.833	1.015	0.934	0.955
1996	1.000	1.000	1.000	1.000	1.000
1997	1.047	1.004	1.041	1.091	1.002
1998	1.036	1.042	1.030	1.043	0.989
1999	1.049	0.999	1.080	1.038	1.066
2000	1.068	1.018	1.046	1.057	1.076
2001	1.061	0.964	1.069	1.107	1.069
2002	1.024	0.880	1.027	1.018	1.103
2003	1.055	1.039	1.081	0.961	1.088
2004	1.139	1.148	1.084	1.176	1.105

| Year | Crops | |
	Fruits and nuts	Secondary output [6]
1995	0.993	1.082
1996	1.000	1.000
1997	1.183	1.109
1998	1.079	1.257
1999	1.103	1.333
2000	1.206	1.204
2001	1.143	1.262
2002	1.164	1.257
2003	1.160	1.216
2004	1.159	1.160

[1] Includes wool, mohair, horses, mules, honey, beeswax, bees, goats, rabbits, aquaculture, and fur animals. These items are not included in the separate groups of livestock and products shown. [2] Cattle and calves, sheep and lambs, and hogs. [3] Butter, butterfat, wholesale milk, retail milk, and milk consumed on farms. [4] Chicken eggs, commercial broilers, chickens, and turkeys. [5] Includes soybeans, peanuts harvested for nuts, sunflower seed, flaxseed, cottonseed, cotton lint, tobacco, sugar crops, forest products, legumes and grass seeds, hops, mint, broomcorn, popcorn, hemp fiber and seed, and flax fiber. [6] These activities are defined as activities closely linked to agriculture for which information on production and input use cannot be separately observed.

ERS, Resources, Technology and Production Branch (202) 694–5601.

Table 9-21.—Hired farmworkers: Number of Workers and Median Weekly Earnings, 2002–2004 [1]

Characteristics	Workers			Median Weekly Earnings [2]		
	2002	2003	2004 [3]	2002	2003	2004 [3]
	Thousands	Thousands	Thousands	Dollars	Dollars	Dollars
All workers	793	780	712	300	346	346
15–19 years old	104	82	99	156	120	120
20–24 years old	130	101	75	280	300	300
25–34 years old	161	201	166	320	350	375
35–44 years old	182	183	154	338	384	363
45–54 years old	122	120	113	350	420	400
55 years old and older	94	92	105	315	346	405
Male	624	639	595	320	350	350
Female	169	141	117	270	280	280
White [3]	413	417	387	315	385	385
Black and other races [3]	47	36	35	310	280	300
Hispanic	333	327	290	300	320	320
Schooling completed						
Less than 5th grade	88	70	66	315	327	300
5th-8th grade	158	169	148	290	300	300
9th-12th grade (no diploma) ..	168	167	151	280	290	270
High school diploma	219	228	195	338	392	420
Beyond high school	160	147	152	346	480	500
Full-time (35 or more hours per week) [4]	654	641	583	334	371	375
Part-time (less than 35 hours per week) [4]	139	136	127	120	160	120

[1] Represents average number of persons 15 years old and over in the civilian noninstitutional population who were employed per week as hired farmworkers. Based on the Current Population Survey microdata earnings file. [2] "Median weekly earnings" is the value that divides the earnings into two equal parts, one part having earnings above the median and the other part having earnings below the median. "Earnings" refers to the weekly earnings the farmworker usually earns at a farmwork job, before deductions, and includes any overtime pay or commissions. [3] Excludes persons of Hispanic origin. [4] The sum of full-time and part-time workers will not equal the total because usual hours worked varies for some individuals.

ERS, Farm and Rural Household Well-Being Branch, (202) 694–5423.

Table 9-22.—Crops: Area, United States, 1996–2005

Year	Principal crops			Area planted total [3]	Commercial vegetables, harvested area	Fruits and nuts, bearing area [6]
	Area harvested		Total [3]			
	Feed grains [1]	Food grains [2]				
	1,000 acres	1,000 acres	1,000 acres	1,000 acres	1,000 acres	1,000 acres
1996	93,817	65,968	313,202	333,682	3,371.8	3,920.5
1997	90,840	66,259	317,662	332,072	3,270.3	4,004.2
1998	88,918	62,677	311,475	329,970	3,284.2	4,029.4
1999	86,049	57,668	311,967	329,255	3,403.2	4,079.6
2000 [4]	87,691	56,398	307,955	328,685	3,488.8	4,114.9
2001 [4]	83,531	52,037	303,560	324,584	3,353.5	4,083.3
2002 [5]	82,636	49,248	299,146	327,283	3,270.2	4,071.4
2003 [5]	85,689	56,379	307,400	325,693	3,265.3	4,055.1
2004 [5]	85,956	53,624	304,581	322,378	3,236.9	4,015.1
2005 [5]	85,935	53,762	303,616	317,739	3,221.3	3,911.2

[1] Corn for grain, oats, barley, and sorghum for grain. [2] Wheat, rye, and rice. [3] Crops included in area planted and area harvested are corn, sorghum, oats, barley, winter wheat, rye, durum wheat, other spring wheat, rice, soybeans, peanuts, sunflower, cotton, dry edible beans, potatoes, canola, proso millet, and sugarbeets. Harvested acreage for all hay, tobacco, and sugarcane are used in computing total area planted. [4] For the 2000 crop year many changes occurred to the National Vegetable Estimation Program. Nine new commodities were added to the program. Additionally, States were added or dropped from the seasonal program. Some States were discontinued for the seasonal forecasts but remained in the program on an annual basis. When comparing 2001 and 2000 data to 1999 data, comparable States should be used. [5] For the 2002 crop year, many changes occured to the National Vegetable Estimation Program. Ten fresh market commodities and two processing commodities were removed from the program. States were removed from the program for certain commodities. When comparing 2000 and 2001 data to 2002 data, comparable States should be used. If you need assistance with these comparisons, please contact Debbie Flippin at (202) 720-2157. For details on the 2002 program changes see the following website: http://www.usda.gov/nass/events/programchg/vegprogchngs.htm. [6] Includes the following fruits and nuts: Citrus fruits—oranges, tangerines, Temples, grapefruit, lemons, limes, tangelos, and K-Early Citrus (area is for the year of harvest); limes and K-Early citrus were discontinued as of the 2002-03 crop; deciduous fruits—commercial apples, peaches, pears, grapes, cherries, plums, prunes, apricots, bananas, nectarines, figs, kiwifruit, olives, avocados, papayas, dates, berries, guavas, cranberries, pineapples and strawberries; nuts—almonds, hazelnuts, macadamias, pistachios, and walnuts.

NASS, Crops Branch, (202) 720–2127.

Table 9-23.—Crops: Area harvested and yield, United States, 2004–2005 [1]

Crop	Area harvested		Yield per harvested acre		
	2004	2005 [2]	Unit	2004	2005 [2]
	1,000 acres	*1,000 acres*			
Grains & Hay:					
Barley [3]	4,021.0	3,269.0	Bushel	69.6	64.8
Corn for Grain	73,631.0	75,107.0	Bushel	160.4	147.9
Corn for Silage	6,101.0	5,920.0	Ton	17.6	18.0
Hay, All	61,966.0	61,649.0	Ton	2.55	2.44
Alfalfa	21,707.0	22,389.0	Ton	3.48	3.38
All Other	40,259.0	39,260.0	Ton	2.06	1.91
Oats [3]	1,787.0	1,823.0	Bushel	64.7	63.0
Proso Millet	595.0	515.0	Bushel	25.3	26.3
Rice	3,325.0	3,364.0	Pound	6,988	6,636
Rye [3]	300.0	279.0	Bushel	27.5	27.0
Sorghum for Grain	6,517.0	5,736.0	Bushel	69.6	68.7
Sorghum for Silage	352.0	311.0	Ton	13.6	13.6
Wheat, All [3]	49,999.0	50,119.0	Bushel	43.2	42.0
Winter [3]	34,462.0	33,794.0	Bushel	43.5	44.4
Durum	2,363.0	2,716.0	Bushel	38.0	37.2
Other Spring	13,174.0	13,609.0	Bushel	43.2	37.1
Oilseeds:					
Canola	828.0	1,114.0	Pound	1,618	1,419
Cottonseed	....	....	Ton	....	....
Flaxseed	511.0	955.0	Bushel	20.3	20.6
Mustard Seed	68.7	44.6	Pound	819	787
Peanuts	1,394.0	1,629.0	Pound	3,076	2,960
Rapeseed	7.8	2.0	Pound	1,394	1,500
Safflower	159.0	160.0	Pound	1,204	1,203
Soybeans for Beans	73,958.0	71,361.0	Bushel	42.2	43.3
Sunflower	1,711.0	2,610.0	Pound	1,198	1,540
Cotton, Tobacco & Sugar Crops:					
Cotton, All	13,057.0	13,702.6	Pound	855	831
Upland	12,809.0	13,434.0	Pound	843	824
Amer-Pima	248.0	268.6	Pound	1,443	1,171
Sugarbeets	1,306.7	1,238.9	Ton	23.0	22.3
Sugarcane	938.2	922.9	Ton	30.9	30.2
Tobacco	408.1	298.0	Pound	2,161	2,147
Dry Beans, Peas & Lentils:					
Austrian Winter Peas	24.5	24.5	Pound	1,188	1,253
Dry Edible Beans	1,219.3	1,562.9	Pound	1,459	1,742
Dry Edible Peas	507.8	765.9	Pound	2,249	1,828
Lentils	329.0	439.0	Pound	1,271	1,176
Wrinkled Seed Peas	....	....	NA	....	....
Potatoes & Misc.:					
Coffee (HI)	5.8	6.1	Pound	965	1,050
Coffee (PR)	44.0	42.0	Pound	420	485
Ginger Root (HI)	0.2	0.1	Pound	40,000	42,500
Hops	27.7	29.5	Pound	1,990	1,791
Maple syrup	....	....	NA	....	....
Mushrooms	....	....	NA	....	....
Peppermint Oil	78.7	76.0	Pound	92	92
Potatoes, All	1,166.9	1,084.6	Cwt	391	388
Winter	18.5	19.8	Cwt	260	247
Spring	72.2	66.7	Cwt	314	281
Summer	53.9	48.6	Cwt	340	334
Fall	1,022.3	949.5	Cwt	401	401
Spearmint Oil	15.8	17.7	Pound	116	109
Sweet Potatoes	92.8	87.8	Cwt	174	179
Taro (HI) [4]	0.4	0.4	Pound	....	....

[1] Missing data are not available.   [2] Preliminary.   [3] Includes area seeded in preceding fall.   [4] Acreage is total acres in crop, not harvested acreage. Yield is not estimated.

NASS, Crops Branch, (202) 720–2127.

Table 9-24.—Crops: Production and value, United States, 2004–2005 [1]

Crop	Unit	Production 2004	Production 2005 [2]	Value of production 2004	Value of production 2005 [2]
		Thou-sands	Thou-sands	1,000 dollars	1,000 dollars
Grains & Hay:					
Barley [3]	Bushel	279,743	211,896	698,184	505,962
Corn for Grain	Bushel	11,807,086	11,112,072	24,381,294	21,040,707
Corn for Silage	Ton	107,293	106,311	....	....
Hay, All	Ton	158,247	150,590	12,211,868	12,491,263
Alfalfa	Ton	75,481	75,771	6,973,371	7,319,756
All Other	Ton	82,766	74,819	5,238,497	5,171,507
Oats [3]	Bushel	115,695	114,878	178,327	187,275
Proso Millet	Bushel	15,065	13,545	42,611	45,117
Rice	Cwt	232,362	223,235	1,701,822	1,789,225
Rye [3]	Bushel	8,255	7,537	26,551	25,053
Sorghum for Grain	Bushel	453,654	393,893	843,464	715,327
Sorghum for Silage	Ton	4,776	4,218	....	....
Wheat, All [3]	Bushel	2,158,245	2,104,690	7,283,324	7,140,357
Winter [3]	Bushel	1,499,434	1,499,129	4,948,510	4,924,953
Durum	Bushel	89,893	101,105	347,336	362,010
Other Spring	Bushel	568,918	504,456	1,987,478	1,853,394
Oilseeds:					
Canola	Pound	1,339,530	1,580,985	143,853	148,532
Cottonseed	Ton	8,242.1	8,501.0	877,372	808,598
Flaxseed	Bushel	10,368	19,695	83,767	116,305
Mustard Seed	Pound	56,290	35,114	8,550	4,737
Peanuts	Pound	4,288,200	4,821,250	813,551	845,873
Rapeseed	Pound	10,875	3,000	1,528	429
Safflower	Pound	191,365	192,545	23,092	24,268
Soybeans for Beans	Bushel	3,123,686	3,086,432	17,894,948	16,927,898
Sunflower	Pound	2,049,613	4,018,355	272,732	472,470
Cotton, Tobacco & Sugar Crops:					
Cotton, All	Bale	23,250.7	23,719.0	4,853,730	5,574,119
Upland	Bale	22,505.1	23,064.0	4,539,616	5,204,245
Amer-Pima	Bale	745.6	655.0	314,114	369,874
Sugarbeets	Ton	30,021	27,654	1,106,878	....
Sugarcane	Ton	29,013	27,897	821,118	....
Tobacco	Pound	881,973	639,709	1,752,335	1,053,430
Dry Beans, Peas & Lentils:					
Austrian Winter Peas	Cwt	291	307	2,764	2,400
Dry Edible Beans	Cwt	17,788	27,222	452,871	526,044
Chickpeas, all	Cwt	593	1,071	14,939	26,564
Large	Cwt	517	922	13,861	24,642
Small	Cwt	76	149	1,078	1,922
Dry Edible Peas	Cwt	11,419	14,003	66,476	63,167
Lentils	Cwt	4,182	5,163	60,893	58,940
Wrinkled Seed Peas	Cwt	899	755	12,719	12,593
Potatoes & Misc.:					
Coffee (HI)	Pound	5,600	6,400	19,880	24,320
Coffee (PR)	Pound	18,500	20,300	NA	NA
Ginger Root (HI)	Pound	6,000	5,100	5,400	4,080
Hops	Pound	55,203.9	52,914.5	103,969	103,294
Maple syrup	Gallon ...	1,507	1,242	42,795	....
Mushrooms	Pound	....	....	918,914	908,370
Peppermint Oil	Pound	7,236	6,980	86,421	83,791
Potatoes, All	Cwt	456,041	420,879	2,575,204	2,903,137
Spearmint Oil	Pound	1,839	1,933	17,700	19,966
Sweet Potatoes	Cwt	16,112	15,747	281,559	309,090
Taro (HI)	Pound	5,200	4,000	2,808	2,160

[1] Missing data are not available.　[2] Preliminary.　[3] Includes area seeded in preceding fall.
NASS, Crops Branch, (202) 720–2127.

FARM RESOURCES, INCOME, AND EXPENSES

Table 9-25.—Fruits and nuts: Bearing acreage and yield, United States, 2004–2005 [1]

Crop	Bearing acreage		Yield per bearing acre		
	2004	2005 [2]	Unit	2004	2005 [2]
	Acres	*Acres*			
Apples, commercial crop	385,560	381,160	Ton	13.55	12.95
Apricots	17,340	15,840	Ton	5.83	5.14
Avocados	68,670		Ton	2.61	
Bananas [3]	1,000		Ton	8.25	
Blackberries (OR) [3] [4]	6,300	6,400	Ton	3.72	3.42
Blueberries			Ton		
Cultivated [3]	44,430	48,310	Ton	2.56	2.41
Wild (ME) [5]			Ton		
Boysenberries [3]	1,050	910	Ton	2.95	2.80
Loganberries (OR) [3]	60	60	Ton	1.42	1.84
Raspberries [3]			Ton		
Black (OR)	1,100	1,300	Ton	1.08	1.80
Red	10,900	11,400	Ton	3.03	3.39
All (CA)	4,100	4,200	Ton	11.00	9.80
Cherries, sweet	78,275	79,010	Ton	3.62	3.18
Cherries, tart	36,950	37,100	Ton	2.88	3.65
Cranberries	39,200	39,100	Ton	7.88	7.96
Dates (CA)	4,700	4,600	Ton	3.64	3.61
Figs (CA)	12,800	12,300	Ton	3.99	4.14
Grapes	933,100	934,750	Ton	6.69	7.46
Guava (HI) [3]	500		Ton	8.10	
Kiwifruit (CA)	4,500	4,500	Ton	5.93	9.24
Nectarines	36,500	37,700	Ton	7.37	6.62
Olives (CA)	32,000	32,000	Ton	3.25	4.34
Papayas (HI) [3]	1,235	1,450	Ton	14.50	11.20
Peaches	146,170	140,360	Ton	8.94	8.43
Pears	64,450	63,350	Ton	13.60	12.80
Pineapples (HI) [6]	13,000	14,000	Ton		
Plums (CA)	36,000	36,000	Ton	4.33	4.75
Prunes, dried (CA)	70,000	67,000	Ton		
Prunes and plums, fresh basis (excluding CA)	3,960	3,860	Ton	6.31	2.25
Strawberries [3]	51,400	52,200	Ton	21.55	22.25
Oranges [7]	761,400	732,100	Ton	16.91	12.45
Grapefruit [7]	114,800	103,500	Ton	18.86	9.74
Lemons [7]	59,800	58,500	Ton	13.34	13.90
Tangerines [7]	36,200	35,600	Ton	11.52	9.30
Tangelos (FL) [7]	8,000	6,400	Ton	5.63	10.94
Temples (FL) [7]	3,400	2,900	Ton	18.53	10.00
Almonds (CA) [8]	570,000	580,000	Ton	1.52	1.34
Hazelnuts (OR)	28,400	28,300	Ton	1.32	0.99
Macadamia (HI)	17,800	18,000	Ton	1.59	1.67
Pecans [5]			Ton		
Pistachios (CA)	93,000	98,000	Ton	1.87	1.44
Walnuts (CA)	217,000	219,000	Ton	1.50	1.62

[1] Missing data are not available. [2] Preliminary. [3] Harvested acreage. Yield based on utilized production. [4] Cultivated. [5] Bearing acreage and yield not calculated. [6] Acreage is total acres in crop, not harvested acreage. Yield is not estimated. [7] Crop year begins with bloom in one year and ends with completion of harvest the following year. Citrus production is for the the year of harvest. [8] Yield based on in-shell basis. Shelling ratios are: 2004-0.580; 2005-0.580.

NASS, Crops Branch, (202) 720–2127.

Table 9-26.—Fruits and nuts: Production and value, United States, 2004–2005 [1]

Crop	Unit [2]	Total production 2004	Total production 2005 [3]	Value of production 2004	Value of production 2005 [3]
		Thou-sands	Thou-sands	1,000 dollars	1,000 dollars
Apples, commercial crop	Ton	5,225.3	4,934.8	1,647,983	1,786,674
Apricots	Ton	101.1	81.4	35,012	40,723
Avocados	Ton	179.4		292,754	
Bananas [4]	Ton	8.3		8,085	
Blackberries (OR) [5]	Ton	23.5	22.2	33,407	32,743
Blueberries	Ton				
Cultivated	Ton	114.4	116.5	275,963	323,788
Wild (ME)	Ton	23.0	29.3	20,970	35,370
Boysenberries	Ton	3.1	2.6	7,168	7,158
Loganberries (OR)	Ton	0.1	0.1	131	188
Raspberries	Ton				
Black (OR)	Ton	1.2	2.3	5,357	11,476
Red	Ton	33.1	38.7	51,723	45,052
All (CA)	Ton	45.0	41.3	188,100	164,175
Cherries, sweet	Ton	283.1	251.2	437,133	483,504
Cherries, tart	Ton	106.5	135.2	69,501	65,296
Cranberries	Ton	308.8	311.3	199,296	211,527
Dates (CA)	Ton	17.1	16.6	38,646	33,200
Figs (CA)	Ton	51.1	50.9	20,214	
Grapes	Ton	6,240.0	6,974.9	3,010,958	3,013,418
Guava (HI) [4]	Ton	4.1		1,166	
Kiwifruit (CA)	Ton	26.7	41.6	19,977	
Nectarines (CA)	Ton	269.0	249.4	86,184	129,969
Olives (CA)	Ton	104.0	139.0	59,379	76,126
Papayas (HI) [4]	Ton	17.9	16.3	12,361	10,971
Peaches	Ton	1,307.1	1,182.6	461,629	509,745
Pears	Ton	877.3	812.3	296,291	315,240
Pineapples (HI) [4]	Ton	220.0	212.0	83,104	79,288
Plums (CA)	Ton	156.0	171.0	74,347	94,163
Prunes, dried (CA)	Ton	143.9	274.2	72,000	130,500
Prunes and plums, fresh basis (excluding CA)	Ton	25.0	8.7	6,802	4,993
Strawberries [4]	Ton	1,106.9	1,161.1	1,460,077	1,383,064
Oranges [4] [6]	Ton	12,872	9,112	317,218	397,909
Grapefruit [4] [6]	Ton	2,165	1,008	269,753	351,897
Lemons [4] [6]	Ton	798	813	1,782,157	1,498,063
Tangelos (FL) [4] [6]	Ton	45	70	10,021	8,004
Tangerines [4] [6]	Ton	417	331	116,475	130,068
Temples (FL) [4] [6]	Ton	63	29	4,915	3,314
Almonds (CA) [4]	Ton	866.4	775.9	2,189,005	2,724,876
Hazelnuts (OR) [4]	Ton	37.5	28.0	54,000	57,120
Macadamia (HI) [4]	Ton	28.3	30.0	41,245	46,800
Pecans [4]	Ton	92.9	129.8	326,924	400,441
Pistachios (CA) [4]	Ton	173.5	141.5	464,980	574,490
Walnuts (CA) [4]	Ton	325.0	355.0	451,750	

[1] Missing data are not available.　[2] Ton refers to the 2,000 lb. short ton.　[3] Preliminary.　[4] Only utilized production estimated.　[5] Cultivated.　[6] Value of production is packinghouse-door equivalent.

NASS, Crops Branch, (202) 720–2127.

Table 9-27.—Vegetables: Area harvested and yield, United States, 2004–2005

Crop	Area harvested		Yield per harvested acre		
	2004	2005 [1]	Unit	2004	2005 [1]
	Acres	*Acres*			
Commercial Vegetables:					
Fresh Market					
Artichokes [2]	7,500	7,300	Cwt	110	115
Asparagus [2]	61,500	54,000	Cwt	34	33
Beans, snap	92,700	96,700	Cwt	62	56
Broccoli [2]	133,800	133,900	Cwt	148	148
Cabbage	75,550	73,700	Cwt	331	329
Cantaloups	86,950	89,160	Cwt	252	248
Carrots	82,600	83,700	Cwt	322	317
Cauliflower [2]	37,700	37,500	Cwt	170	174
Celery [2]	27,900	27,600	Cwt	698	695
Corn, sweet	242,700	238,900	Cwt	115	114
Cucumbers	57,170	57,170	Cwt	177	179
Garlic [2]	31,600	29,400	Cwt	165	158
Honeydew melons	21,900	21,200	Cwt	238	213
Lettuce, head	181,000	179,500	Cwt	366	354
Lettuce, leaf	61,500	62,600	Cwt	240	246
Lettuce, Romaine	75,200	82,400	Cwt	308	288
Onions [2]	168,950	161,520	Cwt	491	457
Peppers, bell [2]	52,900	57,000	Cwt	310	255
Peppers, Chile [2]	30,200	32,700	Cwt	172	159
Pumpkins [2]	45,500	44,700	Cwt	225	243
Spinach	39,600	44,200	Cwt	158	158
Squash [2]	52,600	54,600	Cwt	147	149
Tomatoes	131,100	129,800	Cwt	292	304
Watermelons	141,700	136,400	Cwt	260	278
Processing:					
Beans, lima	41,600	39,220	Ton	1.29	1.36
Beans, snap	200,990	210,620	Ton	4.16	3.90
Carrots	15,760	15,170	Ton	27.44	27.85
Corn, sweet	405,800	403,910	Ton	7.31	7.86
Cucumbers for pickles	113,000	113,700	Ton	5.23	5.02
Peas, green	206,900	211,500	Ton	1.92	1.79
Spinach	12,400	9,500	Ton	10.50	10.20
Tomatoes	300,620	282,040	Ton	40.80	36.17

[1] Preliminary.   [2] Includes processing total for dual usage crops.
NASS, Crops Branch, (202) 720–2127.

Table 9-28.—Vegetables: Production and value, United States, 2004–2005

Crop		Production		Value of production	
	Unit	2004	2005 [1]	2004	2005 [1]
		Thou-sands	Thou-sands	1,000 dollars	1,000 dollars
Commercial Vegetables:					
Fresh Market					
Artichokes [2]	Cwt	825	840	72,683	37,884
Asparagus [2]	Cwt	2,062	1,804	217,060	158,350
Beans, snap	Cwt	5,769	5,455	260,993	286,878
Broccoli [2]	Cwt	19,835	19,790	638,079	563,673
Cabbage	Cwt	24,973	24,246	344,719	325,462
Cantaloups	Cwt	21,876	22,120	322,188	300,388
Carrots	Cwt	26,630	26,559	538,337	556,318
Cauliflower [2]	Cwt	6,425	6,510	195,889	197,419
Celery [2]	Cwt	19,479	19,178	288,791	274,331
Corn, sweet	Cwt	27,885	27,266	580,320	601,519
Cucumbers	Cwt	10,101	10,232	223,602	234,516
Garlic [2]	Cwt	5,224	4,646	138,622	189,955
Honeydew melons	Cwt	5,221	4,505	92,133	69,010
Lettuce, head	Cwt	66,228	63,594	1,118,970	990,905
Lettuce, leaf	Cwt	14,790	15,405	454,677	533,324
Lettuce, Romaine	Cwt	23,155	23,725	442,863	458,068
Onions [2]	Cwt	83,007	73,769	777,339	922,369
Peppers, bell [2]	Cwt	16,400	14,509	558,863	482,960
Peppers, Chile [2]	Cwt	5,181	5,192	123,249	114,903
Pumpkins [2]	Cwt	10,219	10,856	103,742	105,705
Spinach	Cwt	6,266	7,001	233,037	157,473
Squash [2]	Cwt	7,756	8,145	222,718	210,155
Tomatoes	Cwt	38,346	39,462	1,439,197	1,637,394
Watermelons	Cwt	36,882	37,896	313,217	410,281
Processing:					
Beans, lima	Ton	53,550	53,510	22,772	21,940
Beans, snap	Ton	835,880	821,770	131,865	115,545
Carrots	Ton	432,400	422,530	34,698	30,616
Corn, sweet	Ton	2,968,180	3,174,120	213,993	217,096
Cucumbers for pickles	Ton	591,380	570,720	158,793	148,324
Peas, green	Ton	397,570	378,830	99,280	101,080
Spinach	Ton	130,220	96,870	15,088	10,521
Tomatoes	Ton	12,266,410	10,200,120	719,285	622,143

[1] Preliminary.　[2] Includes processing total for dual usage crops.
NASS, Crops Branch, (202) 720–2127.

Table 9-29.—Total farm input: Index numbers of farm input, by major subgroups, United States, 1995–2004

[1992=100]

Year	Total input	Farm labor	Capital	Land	Energy	Agricultural chemicals[1]	Feed, seed, and livestock[2]	Purchased services[3]
1995	1.047	1.074	1.031	0.996	1.035	0.936	1.107	1.040
1996	1.000	1.000	1.000	1.000	1.000	1.000	1.000	1.000
1997	1.027	0.990	0.994	1.002	1.040	1.032	1.075	1.060
1998	1.045	0.937	0.986	1.001	1.152	1.055	1.159	1.116
1999	1.055	0.933	0.981	0.998	1.039	1.037	1.224	1.148
2000	1.016	0.855	0.973	0.992	0.942	1.027	1.195	1.084
2001	1.011	0.873	0.970	0.985	0.991	1.001	1.164	1.108
2002	1.005	0.880	0.972	0.977	1.062	0.995	1.145	1.035
2003	0.975	0.826	0.971	0.968	0.846	0.931	1.164	0.997
2004	0.957	0.780	0.970	0.959	0.816	0.944	1.169	1.012

[1] Includes fertilizer, lime, and pesticide. [2] Includes broilers- and egg-type chicks and turkey poults and imports of livestock for purposes other than immediate slaughter. [3] Includes purchased services and miscellaneous inputs.

ERS, Resources, Technology and Productivity Branch (202) 694–5601.

Table 9-30.—Livestock and livestock products: Production and value, United States, 2002–2005

Product	Production[1]			Value of production		
	2002	2003	2004[2]	2002	2003	2004[2]
	1,000 pounds	*1,000 pounds*	*1,000 pounds*	*1,000 dollars*	*1,000 dollars*	*1,000 dollars*
Cattle and calves	42,384,722	42,242,705	41,501,303	27,083,342	32,112,931	34,887,821
Sheep and lambs	485,149	470,108	464,503	313,946	391,765	412,333
Hogs	26,274,153	26,260,140	26,678,197	8,690,923	9,663,024	13,071,677
Broilers[3]	44,058,700	43,958,200	45,796,250	13,437,345	15,214,947	20,446,086
Mature chickens	1,039,118	983,054	1,002,065	49,931	47,811	58,010
Turkeys	7,494,861	7,487,293	7,304,813	2,732,481	2,699,673	3,065,417
Milk	168,944,000	169,276,000	169,699,000	20,720,482	21,381,324	27,549,413
	Millions	*Millions*	*Millions*			
Eggs	87,252	87,473	89,131	4,281,416	5,333,014	5,303,244

Product	Production			Value of production		
	2003	2004	2005	2003	2004	2005
	1,000 pounds	*1,000 pounds*	*1,000 pounds*	*1,000 dollars*	*1,000 dollars*	*1,000 dollars*
Catfish[5]				425,024	480,175	482,125
Trout[6]				64,046	71,045	74,191
Honey	181,727	183,582	174,643	253,106	196,259	157,795
Wool (shorn)	38,299	37,622	37,232	28,126	29,921	26,272
Mohair[4]	1,880	1,954	1,822	3,127	3,832	5,074

[1] For cattle, sheep, and hogs, the quantity of net production is the live weight actually produced during the year, adjustments having been made for animals shipped in and changes in inventory. Estimates for broilers and eggs cover the 12-month period Dec. 1, previous year through Nov. 30. [2] Preliminary, except for wool shorn and mohair. [3] Young chickens of meat–type strains raised for meat production. [4] AZ, NM, and TX for 2003 only. [5] Value of fish sold. [6] Value of fish and eggs sold.

NASS, Livestock Branch, (202) 720–3570.

Table 9-31.—Agricultural productivity: Index numbers (1996=100) of farm output per unit of input, United States, 1995–2004

Year	Productivity[1]
1995	0.918
1996	1.000
1997	1.010
1998	1.007
1999	1.020
2000	1.065
2001	1.069
2002	1.066
2003	1.111
2004	1.174

[1] Productivity is the output-input ratio. The ratio is obtained by dividing the index of farm output in table 9–25 by the index of total input in table 9–26.

ERS, Resources, Technology and Productivity Branch (202) 694–5601.

Table 9-32.—U.S. farm foods: Marketing bill, farm value, and consumer expenditures, 1995–2004[1]

Year	Total marketing bill	Farm value	Expenditures for farm foods
	Billion dollars	Billion dollars	Billion dollars
1995	415.7	113.8	529.5
1996	424.5	122.2	546.7
1997	444.6	121.9	566.5
1998	465.4	119.6	585.0
1999	503.1	122.2	625.3
2000	537.8	123.3	661.1
2001	557.5	130.0	687.5
2002	576.9	132.5	709.4
2003	604.0	140.2	744.2
2004[2]	633.4	155.5	788.9

[1] The total marketing bill is the difference between total expenditures for domestic farm-originated food products and the farm value or payment farmers received for the equivalent farm products. It relates only to food purchased by consumers that is not imported or exported.   [2] Preliminary.

ERS, Food Markets Branch, (202) 694–5375.

Table 9-33.—Farm food products: Marketing costs, United States, 1995–2004

Year	Labor[1]	Packaging materials	Intercity transportation, rail and truck	Fuels and electricity	Corporate profits before taxes	Other[2]	Total marketing bill[3]
	Billion dollars	Billion dollars	Billion dollars	Billion dollars	Billion dollars	Billion dollars	Billion dollars
1995	196.6	48.2	22.3	18.6	19.5	110.5	415.7
1996	204.6	47.7	22.9	19.6	20.7	109.0	424.5
1997	216.9	48.7	23.6	20.2	22.3	112.9	444.6
1998	229.9	50.4	24.4	20.7	25.5	114.5	465.4
1999	241.5	50.9	25.2	22.0	29.2	134.3	503.1
2000	252.9	53.5	26.4	23.1	31.1	150.8	537.8
2001	263.8	55.0	27.5	24.1	32.0	155.1	557.5
2002	273.1	56.8	28.4	24.9	33.0	160.7	576.9
2003	285.9	59.5	29.7	26.1	34.6	168.2	604.0
2004[4]	303.7	63.1	31.6	27.6	35.5	171.9	633.4

[1] Includes employee wages or salaries, and their health and welfare benefits. Also includes imputed earnings of proprietors, partners, and family workers not receiving stated remuneration.   [2] Includes depreciation, rent, advertising and promotion, interest, taxes, licenses, insurance, professional services, local for-hire transportation, food service in schools, colleges, hospitals, and other institutions, and miscellaneous items.   [3] The marketing bill is the difference between the farm value or payments to farmers for foodstuffs and consumer expenditures for these foods both at foodstores and away from home eating places. Thus, it covers processing, wholesaling, transportation, and retailing costs and profits.   [4] Preliminary.

ERS, Food Markets Branch, (202) 694–5375.

Table 9-34.—Price components: Market basket of farm-originated food products by food group, United States, 1995–2004 [1]

Year	Market basket of food products				Bakery and cereal products			
	Retail cost [2]	Farm value [3]	Farm to retail spread [4]	Farm value share of retail cost	Retail cost	Farm value	Farm to retail spread	Farm value share of retail cost
	Index 1982–84=100	Index 1982–84=100	Index 1982–84=100	Percent	Index 1982–84=100	Index 1982–84=100	Index 1982–84=100	Percent
1995	149	103	175	24	168	110	176	8
1996	156	111	180	25	174	126	181	9
1997	160	106	189	23	178	108	187	7
1998	163	103	195	22	181	94	193	6
1999	167	98	205	21	185	83	199	6
2000	171	97	210	20	188	75	204	5
2001	177	106	215	21	194	79	210	5
2002	180	104	221	20	198	86	214	5
2003	185	110	226	21	203	94	218	6
2004 [5]	195	124	233	22	206	104	220	6

Year	Meat products				Fruits and vegetables, fresh			
	Index 1982–84=100	Index 1982–84=100	Index 1982–84=100	Percent	Index 1982–84=100	Index 1982–84=100	Index 1982–84=100	Percent
1995	136	94	178	35	210	133	248	21
1996	140	100	181	36	216	133	257	20
1997	144	101	189	36	220	128	265	20
1998	142	85	200	30	237	133	288	19
1999	142	82	205	29	252	136	308	18
2000	150	88	214	30	252	131	310	17
2001	159	97	223	31	261	138	321	17
2002	160	103	220	32	272	150	331	18
2003	169	108	231	33	280	157	339	19
2004 [5]	183	117	251	32	295	174	354	19

Year	Dairy products				Fats and oils			
	Index 1982–84=100	Index 1982–84=100	Index 1982–84=100	Percent	Index 1982–84=100	Index 1982–84=100	Index 1982–84=100	Percent
1995	133	92	170	33	137	121	143	24
1996	142	107	174	36	141	112	151	22
1997	146	98	189	32	142	109	154	21
1998	151	113	186	36	147	119	157	22
1999	160	108	207	32	148	89	170	16
2000	161	99	218	30	147	81	172	15
2001	167	119	212	34	156	77	185	13
2002	168	98	233	28	155	92	179	16
2003	168	99	231	28	157	113	174	19
2004 [5]	180	126	230	34	168	128	182	21

Year	Poultry				Fruits and vegetables, processed			
	Index 1982–84=100	Index 1982–84=100	Index 1982–84=100	Percent	Index 1982–84=100	Index 1982–84=100	Index 1982–84=100	Percent
1995	144	114	178	42	138	121	143	21
1996	152	126	183	44	144	122	152	20
1997	157	121	198	41	148	116	158	19
1998	157	126	193	43	151	115	162	18
1999	158	119	203	40	155	114	168	17
2000	160	117	209	39	154	106	168	17
2001	165	126	209	41	159	108	175	16
2002	167	102	242	33	166	111	184	16
2003	169	113	234	36	172	108	192	15
2004 [5]	182	143	226	42	183	125	201	16

[1] The market basket consists of foods that mainly originate on U.S. farms bought in foodstores in a base period, currently 1982–84.   [2] Indexes of retail cost are components of the Consumer Price Index published by the Bureau of Labor Statistics.   [3] Gross return or payment to farmers for the farm products equivalent to foods in the market basket.   [4] The spread between the retail cost and farm value is an estimate of the gross margin received by marketing firms for assembling, processing, transporting, and distributing the products.   [5] Preliminary.

ERS, Food Markets Branch (202) 694–5375.

AGRICULTURAL STATISTICS 2006   IX–27

Table 9-35.—Farm product prices: Marketing year average prices received by farmers;
Parity prices for January, United States, 2003 and 2004

Commodity and unit		Marketing year average price [1]		Parity price [3]	
		2003	2004 [2]	2003	2004
		Dollars	*Dollars*	*Dollars*	*Dollars*
Basic commodities:					
Cotton:					
American Upland	pound	0.517	0.543	1.67	1.95
Extra long staple	pound	1.210	0.878	2.34	2.49
Wheat	bushel	3.40	3.40	9.53	9.75
Rice	cwt	5.78	8.43	26.10	28.20
Corn	bushel	2.27	2.47	6.51	6.55
Peanuts	pound	0.193	0.189	0.650	0.640
Tobacco:					
Flue-cured, types 11–14	pound	1.851	1.845	4.29	4.48
Va., fire-cured, type 21	pound	1.641	1.798	4.34	4.44
Ky.-Tenn., fire-cured, types 22–23	pound	2.475	2.540	5.36	5.57
Burley, type 31	pound	1.977	1.994	4.60	4.79
Maryland, type 32 [4]	pound	1.463	1.308	3.65	3.72
Dark air-cured, types 35–36	pound	2.157	2.194	4.62	4.88
Sun-cured, type 37	pound	1.707	1.476	4.07	4.25
Pa., seedleaf, type 41	pound	1.400	1.450	3.33	3.56
Cigar binder type 51-52	pound	3.584	5.309	12.00	12.70
Puerto Rican filler, type 46	pound	....	....	2.73	2.80
Cigar filler types 54–55	pound	1.746	1.750	3.70	3.89
Designated nonbasic commodities:					
All milk, sold to plants	cwt	12.50	16.10	33.00	34.10
Fluid market	cwt	12.56	16.13	....	....
Manufacturing grade	cwt	11.71	15.45	....	....
Honey, all	pound	1.387	10.85	1.570	1.82
Wool and mohair:					
Wool [5]	pound	0.730	0.800	1.52	1.52
Mohair [6]	pound	1.66	1.96	5.44	5.77
Other nonbasic commodities:					
Field crops and miscellaneous:					
Barley	bushel	2.86	2.61	6.44	6.50
Beans, dry edible	cwt	17.70	21.20	49.70	50.50
Cottonseed	ton	117.00	107.00	261.00	273.00
Crude pine gum	barrel	....	....	233.00	240.00
Flaxseed	bushel	6.00	7.18	11.80	12.60
Hay, all, baled	ton	88.80	89.50	214.00	223.00
Hops	pound	1.86	1.88	4.27	4.43
Oats	bushel	1.71	1.52	4.07	3.85
Peas, dry edible	cwt	7.63	5.94	31.60	32.70
Peppermint oil	pounds	12.00	11.90	30.00	30.80
Popcorn, shelled basis	cwt	....	....	30.30	31.30
Potatoes	cwt	5.89	5.67	13.60	14.80
Rye	bushel	2.93	3.22	5.74	6.10
Sorghum grain	cwt	3.96	4.13	11.00	10.70
Soybeans	bushel	6.08	7.56	14.20	15.40
Spearmint oil	pound	9.29	9.48	26.60	26.50
Sweetpotatoes	cwt	19.20	17.50	36.70	38.90
Tobacco:					
Cigar wrapper, type 61	pound	26.00	....	42.00	43.40
Fruits:					
Citrus (equiv. on-tree): [7]					
Grapefruit	box	3.84	13.40	6.50	6.69
Lemons	box	8.57	12.00	18.40	19.00
Limes, Florida [21]	box	....	....	....	....
Oranges	box	3.75	4.57	9.54	10.70
Tangelos, Florida	box	2.60	7.48	....	....
Tangerines	box	8.90	13.70	24.40	24.60
Temples, Florida	box	1.07	2.48	7.75	7.80
Deciduous and other:					
Apples:					
For all sales	pound	....	....	....	....
For fresh consumption [8]	pound	0.185	0.200	0.480	0.512
For processing [9]	ton	131.00	107.00	301.00	309.00
Apricots:					
For all sales	ton	....	....	....	....
For fresh consumption [10]	ton	618.00	672.00	1,720.00	1,720.00
Dried, California (dried basis) [9]	ton	1,760.00	1,950.00	4,750.00	4,720.00
For processing (except dried) [9]	ton	262.00	279.00	673.00	692.00
Avocados [10]	ton	1,690.00	1,630.00	3,650.00	4,120.00

See footnotes at end of table.

FARM RESOURCES, INCOME, AND EXPENSES

Table 9-35.—Farm product prices: Marketing year average prices received by farmers; Parity prices for January, United States, 2003 and 2004—Continued

Commodity and unit		Marketing year average price [1]		Parity price [3]	
		2003	2004 [2]	2003	2004
		Dollars	*Dollars*	*Dollars*	*Dollars*
Deciduous and other—Con.					
Berries for processing:					
Blackberries (Oregon)	pound	0.695	0.712	1.040	1.07
Boysenberries (California & Oregon)	pound	0.866	1.160	1.470	1.52
Gooseberries	pound			0.624	0.645
Loganberries (Oregon)	pound	0.990	0.771	1.010	1.04
Raspberries, black (Oregon)	pound	1.360	2.250	1.77	1.83
Raspberries, red (Oregon & Washington)	pound	0.563	0.782	1.410	1.46
Cherries:					
Sweet	ton	1,400.00	1,570.00	3,030.00	3,180.00
Tart	pound	0.354	0.326	0.383	0.462
Cranberries [11]	barrel	33.90	32.30	104.00	102.00
Dates, California [10]	ton	2,300.00	2,260.00	2,700.00	2,940.00
Figs, California	ton	317.00	396.00		
Grapes:					
For all sales	ton	402.00	483.00		
Raisin varieties dried, California (dried basis) [9]	ton	563.00	1,210.00	2,180.00	2,090.00
Other dried grapes	ton	491.00	522.00	1,150.00	1,180.00
Kiwi	ton	853.00	809.00	1,240.00	1,400.00
Nectarines (California):					
For all sales	ton				
For fresh consumption [19]	ton	436.00	342.00	1,070.00	1,090.00
For processing [19]	ton			63.30	64.50
Olives (California): [12]					
For all sales	ton	409.00	571.00		
Crushed for oil	ton	238.00	361.00	28.30	334.00
For all sales (except crushed)	ton			1,410.00	1,430.00
For canning	ton	458.00	701.00	1,620.00	1,650.00
Papayas	pound	0.319	0.361	0.850	0.899
Peaches:					
For all sales	ton	377.00	375.00		
For fresh consumption [8]	ton	581.00	548.00	1,330.00	0.705
Dried, California (dried basis) [9]	ton	446.00	382.00	1,620.00	1,600.00
For processing California (except dried):					
Clingstone [12]	ton	215.00	263.00	552.00	567.00
Freestone [9]	ton	204.00	182.00	449.00	465.00
Pears:					
For all sales	ton	294.00	340.00		
For fresh consumption [8]	ton	358.00	437.00	865.00	892.00
Dried, California (dried basis) [9]	ton	1,350.00	1,030.00	2,700.00	2,870.00
For processing (except dried) [9]	ton	197.00	201.00	475.00	489.00
Plums (California):					
For all sales [10]	ton	418.00	516.00		
For fresh consumption [19]	ton			958.00	992.00
For processing [19]	ton			52.60	63.50
Prunes, dried (California) [9]	ton	772.00	1,500.00	2,210.00	2,230.00
Prunes and plums (excl. California):					
For fresh consumption [13]	ton	446.00	468.00	896.00	984.00
For processing (except dried) [9]	ton	254.00	228.00	417.00	455.00
Strawberries:					
For fresh consumption [14]	pound	0.749	0.781	1.590	1.690
For processing [9]	pound	0.281	0.263	0.686	0.707
Sugar crops:					
Maple syrup	gallon	28.30	28.40		
Sugarbeets	ton	41.40	36.90	94.60	96.70
Sugarcane for sugar	ton	29.50	28.30	67.80	70.00
Tree nuts: [15]					
Almonds	pound	1.57	2.21	3.53	3.53
Hazelnuts	ton	1,030.00	1,440.00	2,080.00	2,250.00
Pecans, all	pound	0.984	1.760	2.12	4,600.00
Improved	pound	1.100	1.920		
Seedling	pound	0.683	1.280		
Pistachios	pound	1.22	1.34	2.68	2.75
Walnuts	ton	1,160.00	1,390.00	3,030.00	3,040.00

See footnotes at end of table.

Table 9-35.—Farm product prices: Marketing year average prices received by farmers; Parity prices for January, United States, 2003 and 2004—Continued

Commodity and unit		Marketing year average price [1]		Parity price [3]	
		2003 [2]	2004 [2]	2003	2004
		Dollars	Dollars	Dollars	Dollars
Vegetables for fresh market: [14]					
Artichokes, California	cwt	75.10	88.10	76.20	78.70
Asparagus	cwt	105.00	122.00	275.00	287.00
Broccoli	cwt	31.60	32.20	68.10	71.20
Cabbage	cwt	13.20	14.00	23.10	23.90
Cantaloups	cwt	16.80	14.70	35.70	36.80
Carrots [16]	cwt	19.00	20.20	35.30	38.20
Cauliflower [16]	cwt	34.60	30.50	75.80	78.70
Celery [16]	cwt	13.40	14.80	33.50	34.20
Cucumbers	cwt	19.90	22.10	38.20	39.40
Eggplant [21]	cwt	....	....	40.30	41.70
Escarole/Endive [21]	cwt	....	....	51.90	53.60
Garlic	cwt	25.70	26.50	49.00	50.70
Green peppers [16]	cwt	30.70	34.10	60.30	62.20
Honeydew melons	cwt	18.80	17.60	46.50	48.20
Lettuce	cwt	18.10	16.90	41.30	43.20
Onions [16]	cwt	13.70	10.50	27.80	29.00
Snap beans	cwt	49.30	45.20	75.50	78.00
Spinach	cwt	37.20	37.20	66.10	68.30
Sweet corn	cwt	19.30	20.80	43.50	45.30
Tomatoes	cwt	37.40	37.50	72.10	75.90
Watermelons	cwt	8.98	8.49	14.20	14.60
Vegetables for processing: [9]					
Asparagus	ton	1,170.00	1,170.00	2,750.00	2,870.00
Beets [21]	ton	....	....	13.00	135.00
Cabbage [21]	ton	....	....	101.00	104.00
Cucumbers	ton	275.00	269.00	....	....
Green peas	ton	250.00	250.00	646.00	664.00
Lima beans	ton	442.00	425.00	1,120.00	1,160.00
Snap beans	ton	157.00	158.00	412.00	420.00
Spinach	ton	107.00	116.00	243.00	251.00
Sweet corn	ton	70.40	72.10	176.00	182.00
Tomatoes	ton	58.70	58.60	151.00	155.00
Livestock and livestock products:					
All beef cattle	cwt	80.00	85.60	158.00	165.00
Cows	cwt	42.90	50.30	....	....
Steers and heifers	cwt	84.20	90.20	....	....
Calves	cwt	103.60	121.60	213.00	221.00
Beeswax	pound			4.87	5.03
Chickens:					
Excluding broilers, live	pound	0.049	0.058	....	....
Broilers, live [20]	pound	0.346	0.446	....	....
All Eggs	dozen	0.746	0.698	1.56	1.64
Hogs	cwt	37.50	51.20	101.00	102.00
Lambs	cwt	94.80	101.50	182.00	195.00
Milk cows [17]	head	1.340	1.580	....	....
Sheep	cwt	35.90	38.80	77.80	81.30
Turkeys, live	pound	0.360	0.419	0.969	0.992

[1] Marketing year average prices for crops; weighted calendar year average for livestock and livestock products, except chickens, eggs, and hogs, which are on a Nov.-Dec. marketing year basis. Unless otherwise noted, these are averages for marketing season or calendar year computed by weighing State prices by quantities sold, or by production for those commodities for which virtually all the production is sold. [2] Preliminary. [3] Parity prices are for January of the year shown as published in the January issue of Agricultural Prices. [4] Previous year. [5] Average local market price for wool sold excluding incentive payment. [6] Average local market price for mohair sold excluding incentive payment. Texas only prior to 1988. [7] Crop year begins with bloom in one year and ends with completion of harvest the following year. Prices refer to the year harvest begins. Thus the prices shown for 1996 relate to the citrus crop designated as 1996–97 in the production reports. [8] Equivalent packinghouse-door returns for California, Oregon (pears only), Washington, and New York (apples only), and prices as sold for other States. [9] Equivalent returns at processing plant-door. [10] Equivalent returns at packing-house-door. [11] Weighted average of co-op and independent sales. Co-op prices represent pool proceeds excluding returns from non-cranberry products and before deductions for capital stock and other retains. [12] Equivalent per unit returns for bulk fruit at first delivery point. [13] Average price as sold. [14] FOB shipping point when available. Weighted average of prices at points of first sale when FOB shipping point price not available. [15] Prices are in-shell basis except almonds which are shelled basis. [16] Includes some processing. [17] Simple average of States weighted by estimated Jan. 1 head for U.S. average. [18] Sold by farmers directly to consumers. [19] Prices for fresh and processing breakdown no longer published to avoid disclosure of individual operations. [20] Live weight equivalent price. [21] Discontinued.

NASS, Environmental, Economics, and Demographics Branch (202) 720–6146.

Table 9-36.—Producer prices: Index numbers, by groups of commodities, United States, 1996–2005

[1982=100]

Year	Total finished goods	Consumer foods	Total consumer goods	Total intermediate materials	Total crude materials
1996	131.3	133.6	129.5	125.7	113.8
1997	131.8	134.5	130.2	125.6	111.1
1998	130.7	134.3	128.9	123.0	96.8
1999	133.0	135.1	132.0	123.2	98.2
2000	138.0	137.2	138.2	129.2	120.6
2001	140.7	141.3	141.5	129.7	121.0
2002	138.9	140.1	139.4	127.8	108.1
2003	143.3	145.9	145.3	133.7	135.3
2004	148.5	152.7	151.7	142.6	159.0
2005 [1]	155.7	155.6	160.5	153.9	182.1

[1] Final.

ERS, Food Marketing Branch, (202) 694–5349. Compiled from reports of the U.S. Department of Labor.

Table 9-37.—Prices received by farmers: Index numbers by groups of commodities and parity ratio, United States, 1996–2005 [1]

[1910–14=100]

Year	Food grains	Feed grains and hay	Cotton	Tobacco	Oil-bearing crops	Fruit & nuts [2]	Commer-cial vegeta-bles	Other crops
1996	497	521	626	1,592	700	824	740	532
1997	406	418	573	1,570	715	770	792	532
1998	328	356	546	1,572	588	781	818	532
1999	287	307	436	1,536	452	806	736	532
2000	272	308	421	1,614	467	681	808	541
2001	290	325	328	1,614	437	761	888	554
2002	331	356	284	1,641	480	734	914	561
2003	345	371	437	1,612	585	746	915	564
2004	380	392	459	1,421	731	865	878	569
2005 [4]	352	339	361	1,369	576	927	893	578

Year	Potatoes, and dry edi-ble beans	All crops	Meat animals	Dairy products	Poultry and eggs	Livestock and livestock products	All farm products	Parity ratio [3]
1996	576	624	882	914	337	761	712	47
1997	457	568	933	820	319	755	678	43
1998	500	526	804	953	329	741	645	42
1999	507	476	840	882	310	731	607	40
2000	472	473	955	757	299	744	611	38
2001	497	490	989	920	323	812	650	40
2002	652	517	884	744	265	692	620	38
2003	527	547	1,043	770	310	789	677	40
2004	515	576	1,179	986	372	933	757	43
2005 [4]	579	552	1,225	931	343	923	737	39

[1] These indexes are computed using the price estimates of averages for all classes and grades for individual commodities being sold in local farm markets. In computing the group indexes, prices of individual commodities have been compared with 1990–92 weighted average prices. The resulting ratios are seasonally weighted by average quantities sold for the most recent previous 5-year period. For example, 1994 indexes use quantities sold for the period 1988–92. Then, the 1990–92 indexes are adjusted to a 1910–14 reference.   [2] Fresh market for noncitrus, and fresh market and processing for citrus.   [3] Ratio of Index of Prices Received to the Index of Prices Paid by Farmers for Commodities and Services, Interest, Taxes, and Farm Wage Rates.   [4] Preliminary.

NASS, Environmental, Economics, and Demographics Branch, (202) 720–6146.

Table 9-38.—Prices received by farmers: Index numbers by groups of commodities and ratio, United States, 1996–2005 [1]

(1990–92=100)

Year	Food grains	Feed grains and hay	Cotton	Tobacco	Oil-bearing crops	Fruit & Nuts [2]	Commercial vegetables	Other Crops
1996	157	146	122	105	128	118	111	108
1997	128	117	112	104	131	110	118	108
1998	103	100	107	104	107	111	123	108
1999	91	86	85	102	83	115	110	108
2000	85	86	82	107	85	98	121	110
2001	91	91	64	107	80	109	133	112
2002	104	100	56	108	88	105	137	114
2003	109	104	85	107	107	107	137	114
2004	120	110	90	94	134	124	131	115
2005 [4]	111	95	70	91	105	133	134	117

Year	Potatoes and dry edible beans	All crops	Meat animals	Dairy products	Poultry and eggs	Livestock and live-stock products	All farm products	Ratio [3]
1996	114	127	87	114	120	99	112	98
1997	90	115	92	102	113	98	107	90
1998	99	107	79	119	117	97	102	89
1999	100	97	83	110	110	95	96	83
2000	93	96	94	94	106	97	96	80
2001	98	99	97	115	115	106	102	83
2002	129	105	87	93	94	90	98	79
2003	104	111	103	96	111	103	107	84
2004	102	117	116	123	132	122	119	89
2005 [4]	114	112	121	116	124	120	116	82

[1] These indexes are computed using the price estimates of averages for all classes and grades for individual commodities being sold in local farm markets. In computing the group indexes, prices of individual commodities have been compared with 1990–92 weighted average prices. The resulting ratios are seasonally weighted by average quantities sold for the most recent previous 5–year period. For example, 1994 indexes use quantities sold for the period 1988–92. [2] Fresh market for noncitrus, and fresh market and processing for citrus. [3] Ratio of Index of Prices Received (1990–92=100) to Index of Prices Paid by Farmers for Commodities & Services, Interest, Taxes, and Wage Rates (1990–92=100). [4] Preliminary.

NASS, Environmental, Economics, and Demographics Branch, (202) 720–6146.

Table 9-39.—Prices paid by farmers: Index numbers, by groups of commodities, United States, 1996–2005

(1990–92=100)

Year	Production indexes								
	Production (all commodities)	Feed	Livestock & Poultry	Seeds	Fertilizer	Agricultural chemicals	Fuels	Supplies and Repairs	Autos and trucks
1996	115	129	75	115	125	119	102	115	117
1997	119	125	94	119	121	121	106	118	119
1998	113	111	88	122	112	122	84	119	119
1999	111	100	95	122	105	121	93	121	119
2000	116	102	110	124	110	120	134	124	119
2001	120	109	111	132	123	121	119	128	118
2002	119	112	102	142	108	119	112	131	116
2003	124	114	109	154	124	121	140	134	115
2004	131	121	128	158	140	120	162	137	114
2005 [4]	139	117	140	168	162	120	224	144	114

Year	Production indexes - continued							Production, interest, taxes, and wage rates	Family living	Commodities, interest, taxes, and wage rates [2]
	Farm machinery	Building Materials	Farm services	Rent	Interest	Taxes	Wage rates [1]			
1996	125	115	116	128	106	112	117	115	116	115
1997	128	118	116	136	105	115	123	118	119	118
1998	132	118	115	120	104	119	129	114	121	115
1999	135	120	116	113	106	120	135	113	124	115
2000	139	121	119	110	113	123	140	118	128	120
2001	144	121	121	117	109	124	146	122	131	123
2002	148	122	120	119	104	126	153	121	133	124
2003	151	124	123	120	102	126	157	126	136	128
2004	162	134	124	120	104	125	160	132	140	134
2005 [3]	171	142	128	125	109	126	165	140	145	141

[1] Simple average of seasonally adjusted quarterly indexes. [2] Family Living component included. [3] Preliminary.

NASS, Environmental, Economics, and Demographics Branch, (202) 720–6146.

Table 9-40.—Prices paid by farmers: Index numbers, by groups of commodities, United States, 1996–2005 [1]

[1910–14=100]

| Year | Family living | Production indexes | | | | | | | |
		Production (all commodities)	Feed	Livestock and poultry	Seed	Fertilizer	Agricultural chemicals	Fuels	Supplies and repairs
1996	1,490	1,118	631	962	1,143	458	736	789	816
1997	1,525	1,151	612	1,200	1,180	443	745	816	835
1998	1,548	1,092	539	1,123	1,209	412	756	646	846
1999	1,582	1,078	486	1,217	1,201	385	746	720	862
2000	1,636	1,124	497	1,400	1,228	404	741	1,033	880
2001	1,682	1,161	530	1,419	1,306	451	745	915	906
2002	1,709	1,155	547	1,306	1,402	394	738	866	927
2003	1,747	1,203	554	1,394	1,521	454	747	1,083	949
2004	1,794	1,273	590	1,641	1,561	514	741	1,251	975
2005 [3]	1,855	1,351	569	1,787	1,662	595	742	1,733	1,019

| Year | Production indexes—Continued | | | | Interest | Taxes | Wage rates | Production, interest, taxes, and wage rates | Commodities, interest, taxes, and wage rates [2] |
	Autos and trucks	Farm machinery	Building materials	Farm services and rent					
1996	3,126	3,128	1,569	1,442	2,652	3,001	4,389	1,540	1,531
1997	3,161	3,216	1,602	1,477	2,621	3,093	4,591	1,585	1,574
1998	3,152	3,313	1,605	1,394	2,617	3,185	4,838	1,528	1,532
1999	3,166	3,394	1,628	1,364	2,663	3,214	5,037	1,520	1,531
2000	3,160	3,490	1,647	1,374	2,825	3,281	5,236	1,585	1,594
2001	3,141	3,601	1,646	1,422	2,738	3,330	5,468	1,633	1,642
2002	3,082	3,704	1,654	1,426	2,619	3,387	5,705	1,631	1,645
2003	3,044	3,789	1,679	1,447	2,566	3,368	5,885	1,690	1,700
2004	3,022	4,062	1,817	1,459	2,597	3,339	5,977	1,774	1,778
2005 [3]	3,032	4,291	1,930	1,510	2,731	3,383	6,158	1,874	1,871

[1] Based on Consumer Price Index-Urban of Bureau of Labor Statistics. [2] The index known as the Parity Index is the Index of Prices Paid by Farmers for Commodities and Services, Interest, Taxes, and Wage Rates expressed on the 1910–14=100 base. [3] Preliminary.

NASS, Environmental, Economics, and Demographics Branch, (202) 720–6146.

Table 9-41.—Prices paid by farmers: April prices, by commodities, United States, 2003–2005 [1]

Commodity	Unit	2003	2004	2005
		Dollars	Dollars	Dollars
Fuels and energy:				
Diesel fuel [2] [3]	Gal	1.238	1.310	1.968
Gasoline, service station, unleaded [4]	Gal	1.611	1.750	2.205
Gasoline, service station, bulk delivery [4]	Gal	1.601	1.760	2.225
L. P. gas, bulk delivery [2]	Gal	1.213	1.210	1.466
Feeds:				
Alfalfa Meal	Cwt	15.00	14.90	15.40
Alfalfa Pellets	Cwt	15.30	15.20	15.20
Bran	Cwt	13.70	14.80	15.00
Beef Cattle Concentrate.				
32-36% Protein	Ton	290	342	316
Corn Meal	Cwt	9.90	9.84	9.57
Cottonseed Meal, 41%	Cwt	16.60	18.40	17.20
Dairy Feed				
14% Protein	Ton	183	200	220
16% Protein	Ton	200	218	197
18% Protein	Ton	207	229	206
20% Protein	Ton	201	233	206
32% Protein Conc.	Ton	311	381	331
Hog Feed				
14-18% Protein	Ton	223	256	219
38-42% Protein Conc.	Ton	322	415	398
Molasses, Liquid	Cwt	13.30	14.20	14.80
Poultry Feed:.				
Broiler Grower	Ton	234	278	237
Chick Starter	Ton	241	299	268
Laying Feed	Ton	232	249	226
Turkey Grower	Ton	279	315	315
Soybean Meal, 44%	Cwt	14.50	19.60	16.10
Stock Salt	50 Lb	4.30	4.53	4.78
Trace Mineral Blocks	50 Lb	5.40	5.53	5.52

See footnotes at end of table.

Table 9-41.—Prices paid by farmers: April prices, by commodities, United States, 2003–2005 [1]—Continued

Commodity	Unit	2003	2004	2005
		Dollars	Dollars	Dollars
Fertilizer: [5]				
0-15-40	Ton	195	217	263
0-18-36	Ton	188	208	258
0-20-20	Ton	200	220	248
3-10-30	Ton	171	186	235
5-10-10	Ton	161	165	193
5-10-15	Ton	179	186	233
5-10-30	Ton	187	209	244
5-20-20	Ton	191	207	247
6- 6- 6	Ton	205	203	232
6- 6-18	Ton	212	223	247
6-12-12	Ton	169	209	223
6-24-24	Ton	227	248	287
8- 8- 8	Ton	179	194	204
8-20- 5	Ton	235	258	269
8-32-16	Ton	241	257	287
9-23-30	Ton	212	228	276
10- 3- 3	Ton	....	....	298
10- 6- 4	Ton	167	186	198
10-10-10	Ton	186	202	235
10-20-10	Ton	207	226	247
10-20-20	Ton	218	241	279
10-34- 0	Ton	255	261	271
11-52- 0	Ton	266	288	317
13-13-13	Ton	212	229	265
15-15-15	Ton	235	257	288
16- 0-13	Ton	185	195	229
16- 4- 8	Ton	239	249	289
16- 6-12	Ton	187	214	232
16-16-16	Ton	315	301	331
16-20- 0	Ton	253	263	293
17-17-17	Ton	229	251	293
18-46- 0 (DAP)	Ton	250	276	303
19-19-19	Ton	237	256	304
24- 8- 0	Ton	188	209	244
Ammonium Nitrate	Ton	243	263	292
Anhydrous Ammonia	Ton	373	379	416
Aqua Ammonia	Ton	130	132	139
Limestone, Spread on field	Ton	19.40	21.10	20.70
Muriate of Potash, 60–62% K2O	Ton	165	181	245
Nitrate of Soda	Ton	278	308	323
Nitrogen Solutions.				
28% N	Ton	166	179	218
30% N	Ton	161	178	215
32% N	Ton	184	197	243
Sulfate of Ammonia	Ton	195	205	244
Superphosphate, 44-46% P2O5	Ton	243	266	299
Urea, 44-46% Nitrogen	Ton	261	276	332
Farm Machinery:				
Baler, Pick-Up, Automatic Tie, P.T.O.				
Square Conventional, Under 200 Lb Bales	Each	17,300	17,400	18,200
Round, 1200-1500 Lb Bale	Each	18,300	19,500	20,300
Round, 1900-2200 Lb Bale	Each	25,600	27,000	28,200
Chisel Plow, Maxiumum 1 Foot Depth				
Tillage, Chisel or Sweep Type, Drawn.				
Mounted, 16-20 Foot	Each	13,100	15,300	15,600
Combine, Self Propelled with Grain head				
Extra-large capacity	Each	196,000	218,000	232,000
Large capacity	Each	159,000	180,000	192,000
Corn Head for combine				
6 Row	Each	25,900	27,400	28,900
8 Row	Each	33,900	35,900	37,700
Cotton Picker, Self Propelled, with sprindle,				
4-Row	Each	216,000	237,000	238,000
Cultivator, Row Crop				
6-Row	Each	6,330	6,920	7,490
12-Row, Flexible	Each	13,700	15,300	17,300
Disk Harrow, Tandem, Drawn [7]				
15-17 Foot	Each	15,200	14,300	15,700
18-20 foot	Each	19,300	19,400	21,600

See footnotes at end of table.

Table 9-41.—Prices paid by farmers: April prices, by commodities, United States, 2003-2005 [1]—Continued

Commodity	Unit	2003	2004	2005
		Dollars	*Dollars*	*Dollars*
Elevator, Portable, Without Power Unit,				
Auger Type, 8 Inch Diameter, 60 Foot	Each	4,180	4,130	4,680
Feed Grinder-Mixer, Trailer Mtd., P.T.O.	Each	15,600	16,800	18,900
Field Cultivator, Mounted or Drawn				
17-19 Foot	Each	11,600	12,400	14,400
20-25 Foot, Flexible	Each	15,900	17,500	19,600
Forage Harvester, P.T.O., Shear Bar,				
With Pick-Up Attachment	Each	31,600	32,700	33,400
With Row Crop Unit, 2-Row	Each	35,900	35,000	35,400
Forage Harvester, Self-propelled, Shear Bar				
With 4–6 row	Each	232,000	242,000	257,000
Front-End Loader, Hydraulic, Tractor Mounted				
1800-2500 Lb. Capacity, 60 Inch Bucket	Each	5,000	5,150	5,450
Grain Drill, Most Common Spacing				
Plain, 15-17 Openers	Each	14,000	14,500	16,800
Press, 23-25 Openers	Each	20,300	22,600	25,200
With Fertilizer Attachment, 20-24 Openers	Each	18,600	19,800	19,000
Min/No-Till W/Fert. Attach., 15 Foot	Each	27,600	29,400	30,500
Hayrake, Side-Delivery, or Wheel Rake,				
Traction Drive, 8-12 Foot Working Width	Each	5,200	5,380	5,940
Hay Tedder, 15-18 Foot	Each	4,900	5,130	5,380
Manure Spreader, Conveyor Type, P.T.O.,				
2-Wheel, with Tires.				
141-190 Bushel Capacity	Each	6,760	7,210	7,790
225-300 Bushel Capacity	Each	10,100	10,900	11,900
Mower-Conditioner, P.T.O., Pull Type, with				
8-10 Foot, Sickle (Cutter) Bar or Disc	Each	14,400	14,800	15,900
14-16 Foot, Sickle (Cutter) Bar or Disc	Each	22,700	23,000	24,600
Mower, Mounted or Drawn,				
7-8 ft Sickle (Cutter) Bar	Each	4,980	5,040	5,320
13-14 Foot, Sickle (Cutter) Bar or Disc	Each	14,000	15,400	16,300
Planter, Row Crop				
With Fertilizer Attachment, 4-Row	Each	15,200	16,100	16,900
With Fertilizer Attachment, 8-Row	Each	30,000	32,000	31,400
With Fertilizer Attachment, 24-Row	Each	95,700	102,000	108,000
12-Row Conservation (No-Till Cond), w/Fert	Each	52,400	53,100	57,900
Rotary Hoe, 20-25 Foot	Each	6,610	6,770	7,410
Rotary Cutter, 7-8 Foot	Each	3,130	3,480	3,470
Sprayer, Field Crop, Power, Boom Type				
(Excl. Self-Propelled and Orchard).				
Tractor Mounted, w/ 300 Gal. Spray Tank	Each	5,890	5,850	7,320
Trailer Type, w/ 500-700 Gal. Spray Tank	Each	13,100	13,300	15,100
Tractor, 2-Wheel Drive				
30-39 P.T.O. horsepower	Each	16,000	16,100	16,700
50-59 P.T.O. horsepower	Each	21,300	21,500	23,400
70-89 P.T.O. horsepower	Each	33,600	33,900	36,800
110 - 129 P.T.O. horsepower	Each	63,800	65,700	68,500
140 - 159 P.T.O. horsepower	Each	84,100	86,900	91,900
190 - 220 P.T.O. horsepower	Each	116,000	121,000	126,000
Tractor, 4-Wheel Drive				
200 - 280 P.T.O. horsepower	Each	133,000	141,000	142,000
Wagon, Gravity Unload, W/Box and Running				
Gear, and Tires,				
200-400 Bushel Capacity				
Without Side Extensions	Each	4,200	4,570	5,350
Wagon, Running Gear, W/O Box				
8-10 Ton Capacity	Each	1,720	1,810	2,060
Windrower, Self-Propelled,				
14-16 Foot	Each	64,200	67,300	72,100
Agricultural Chemicals: [8]				
Fungicides:				
Basic Copper Sulfate, 53% WP	Lb	1.20	1.30	1.51
Benomyl (Benlate), 50% WP	Lb	18.50	18.60	18.40
Calcium Polysulfide (Lime Sulfur) Liq.Conc	Gal	7.90	7.95	8.61
Captan 50% WP	Lb	3.50	3.52	3.65
Chlorothalonil (Bravo), 6#/Gal EC	Gal	47.20	47.40	45.20
Copper Hydroxide (Kocide 101), 77% WP	Lb	2.50	2.62	2.63
Dodine (Cyprex), 65% WP	Lb	11.60	11.70	9.35
Fenarimol (Rubigan), 1#/Gal EC	Gal	308	319	333
Ferbam (Carbamate), 76% WP	Lb	4.20	4.12	4.43
Fosethyl-AL (Aliette), 80% WP	Lb	12.60	12.10	12.80
Iprodione (Rovral), 50% WP	Lb	24.50	24.10	24.00
Mancozeb (Dithane 80% WP,Manzate 75% DF)	Lb	3.00	3.03	3.00
Maneb, 80% WP, 75% DF	Lb	2.70	2.76	2.77
Metalaxyl (Ridomil), 2#/Gal EC	Gal	191	223	281
Myclobutanil (Systhane, Nova, Rally), 40% WP	Lb	68.10	70.00	72.20
Oxytetraycline (Mycoshield), 17% WP	Lb	24.90	27.60	28.50
Sulfur, 95% WP	Lb	0.318	0.343	0.374
Triforine (Funginex), 1.6#/Gal EC	Gal	106	100	97.70
Triadimefon (Bayleton), 50% WP	Lb	70.70	70.70	73.00
Ziram, 76% WP	Lb	2.70	2.67	2.86

See footnotes at end of table.

Table 9-41.—Prices paid by farmers: April prices, by commodities, United States, 2003-2005 [1]—Continued

Commodity	Unit	2003	2004	2005
		Dollars	Dollars	Dollars
Fumigants:				
Methyl Bromide (Terr-o-gas 98)	Lb	7.30	6.67	5.10
Herbicides:				
2,4-D, 4#/Gal EC	Gal	15.20	15.20	15.90
Acetochlor (Harness, Surpass), ...				
6.4–7#/Gal EC	Gal	68.20	71.40	67.60
Alachlor (Lasso), 4#/Gal EC	Gal	24.50	24.50	25.70
Atrazine(AAtrex), 4#/Gal L	Gal	12.30	12.20	12.40
Bentazon (Basagran), 4#/Gal EC	Gal	83.70	84.20	85.20
Butylate (Sutan), 6.7#/Gal EC	Gal	23.30	26.80	28.70
Chlorimuron-ethyl (Classic), 25% DF	Oz	12.80	13.30	13.40
Chlorsulfuron (Glean), 75%	Oz	18.40	18.00	18.60
Cyanazine (Bladex), 4#/Gal EC	Gal	32.90	32.90	31.30
DCPA (Dacthal), 75% WP	Lb	13.80	15.10	15.90
Dicamba (Banvel), 4#/Gal EC	Gal	92.50	91.00	92.60
Diuron (Karmex, Diurex), 80% WP	Lb	4.90	4.93	4.78
EPTC (Eptan), 7E-(Eradicane),6.7#/Gal EC	Gal	35.60	37.90	37.90
Glyphosate (Roundup), 4#/Gal EC	Gal	43.30	39.70	33.80
Linuron (Lorox, Linex), 50% DF	Lb	12.50	14.30	14.50
MCPA, 4#/Gal, EC	Gal	17.70	17.60	18.00
Metolachlor (Dual), 8#/Gal EC	Gal	104	106	108
Metribuzin (Lexone or Sencor), 75% DF	Lb	20.80	21.70	22.80
MSMA (Super Arsonade), 4-6# Gal EC	Gal	21.20	19.10	18.70
Napropamide (Devrinol), 50% WP	Lb	9.10	9.49	9.26
Paraquat (Gramoxone Extra), 2.5#/Gal EC	Lb	40.70	42.40	43.80
Pendimethalin (Prowl),3.3#/Gal EC	Gal	22.70	23.10	23.50
Sethoxydim (Poast), 1.5#/Gal EC	Gal	73.90	72.80	72.10
Simazine (Princep), 4#/Gal EC	Gal	18.00	17.60	17.80
Terbacil (Sinbar), 80% WP	Lb	32.60	32.50	34.30
Trifluralin (Treflan), 4#/Gal EC	Gal	24.40	23.10	21.60
Insecticides:				
Acephate (Orthene), 75% SP	Lb	12.90	12.70	12.60
Aldicarb (Temik), 15% G	Lb	3.80	3.74	3.75
Azinphos-methyl (Guthion), 50% WP	Lb	10.60	10.70	10.80
Bt (Dipel 2X), WP	Lb	12.30	11.90	12.30
Carbaryl, (Sevin), 80% S, SP or WP	Lb	5.50	5.85	5.85
Carbofuran (Furadan), 4F	Gal	79.30	80.60	85.40
Chlorpyrifos (Lorsban), 4#/Gal EC	Gal	41.30	41.30	38.70
Cyfluthrin (Baythroid) 2#/Gal EC	Gal	388	362	379
Cypermethrin,(Ammo 2.5-Cymbush 3#G)EC	Gal	180	162	141
Diazinon, 4#/Gal EC	Gal	38.00	36.70	38.80
Dicofol (Kelthane), 35% WP	Lb	12.50	14.00	20.80
Dicrotophos (Bidrin), 8#/Gal EC	Gal	90.90	92.60	92.50
Dimethoate (Cygon), 2.67#/Gal EC	Gal	36.90	37.10	37.90
Disulfoton (Di-Syston), 8#/Gal EC	Gal	91.70	94.70	104
Endosulfon (Thiodan, Phaser), 3#/Gal EC	Gal	34.20	33.00	32.10
Esfenvalerate (Asana XL),0.66#/Gal EC	Gal	103	102	103
Ethion 4#/Gal EC	Gal	41.60	36.30	28.70
Fonofos (Dyfonate II), 20% G	Lb	3.20	2.03	(2)
Imidacloprid (Admire, Provado),.				
1.6–2#/Gal EC	Gal	573	578	577
Malathion, 5#/Gal EC	Gal	28.50	29.60	30.00
Methidathion (Supracide), 25% WP	Lb	7.50	7.03	7.56
Methomyl (Lannate) L), 1.81 #/Gal Liq.	Gal	55.60	52.60	52.70
Methyl Parathion, 4#/Gal EC	Gal	31.80	32.80	31.80
Oil, Superior Oil, Supreme, Volck	Gal	5.60	5.87	5.99
Oxamyl (Vydate-L), 2# L	Gal	69.80	68.90	82.20
Oxydemeton-Methyl (Metasystox-R).				
2#/Gal EC	Gal	76.10	84.30	86.00
Oxythioquinox (Morestan), 25% WP	Lb	21.90	20.10	(2)
Phorate (Thimet), 20% G	Lb	2.40	2.48	2.59
Phosmet (Imidan, Prolate), 50% WP	Lb	7.40	7.45	8.32
Propargite (Comite, Omite), 30% WP	Lb	6.60	6.43	6.99
Synthetic Pyrethroids,.				
(Pounce 2.0, Ambush 3.2 #/Gal) EC	Gal	133	130	124
Terbufos (Counter), 15% G	Lb	2.70	2.67	2.37
Zeta–Cyermethrin (Fury), 1.5#/Gal EC	Gal	202	204	215
Other:				
Gibberellic Acid,(Ry3Up,Pro-Gibb)4.0% L	Gal	173	174	174
Nad Napthalene Acetamide, 8.4 WP	Lb	65.80	65.40	72.30

[1] Prices paid by famers are collected, for the most part, from retail establishments located in smaller cities and towns in rural areas. Prior to 1995, recorded prices reflected a modified annual average based on frequency item was reported during the year. Recorded item values, 1995-99, are the U.S. April average price.   [2] Includes Federal, State, and local per gallon taxes where applicable.   [3] Excludes Federal excise tax.   [4] Includes Federal, State, and local per gallon taxes.   [5] Excludes cost of application, except for limestone.   [6] Discontinued in 2000.   [7] With hydraulic lift, transport wheels, and tires.   [8] Formulation abbreviations: EC–Emulsifiable Concentrate, DF–Dry Flowable, DG–Dry Granular, G–Granular, L– Liquid, S–Solution, SP–Soluble Powder, and WP–Wettable Powder.

NASS, Environmental, Economics, and Demographics Branch, (202) 720–6146.

Table 9-42.—Agricultural commodities: Support prices per unit, United States, 1996–2005 [1][2]

Commodity	Unit	1996	1997	1998	1999	2000
		Dollars	Dollars	Dollars	Dollars	Dollars
Basic commodities:						
Corn:						
Target price	Bushel	(9)	(9)	(9)	(9)	(9)
Loan rate	do	1.89	1.89	1.89	1.89	1.89
Cotton:						
American upland: [3]						
Target price	Pound	(9)	(9)	(9)	(9)	(9)
Loan rate	do	0.5192	0.5192	0.5192	0.5192	0.5192
Extra-long staple:						
Target price	do	(9)	(9)	(9)	(9)	(9)
Loan rate	do	0.7965	0.7965	0.7965	0.7965	0.7965
Peanuts: [4]						
Target price	do	....	....	....	....	....
Loan rate	do	0.3050	0.3050	0.3050	0.3050	0.3050
Rice:						
Target price	Cwt.	(9)	(9)	(9)	(9)	(9)
Loan rate	do	6.50	6.50	6.50	6.50	6.50
Wheat:						
Target price	Bushel	(9)	(9)	(9)	(9)	(9)
Loan rate	do	2.58	2.58	2.58	2.58	2.58
Tobacco:						
Flue-cured, types 11-14	Pound	1.601	1.621	1.628	1.632	1.640
Fire-cured, type 21	do	1.455	1.498	1.536	1.559	1.559
Fire-cured, types 22-23	do	1.557	1.623	1.681	1.716	1.716
Burley, type 31	do	1.737	1.760	1.778	1.789	1.805
Dark air-cured, types 35-36	do	1.339	1.398	1.450	1.481	1.481
Virginia sun-cured, type 37	do	1.288	1.326	1.360	1.380	1.380
Cigar filler, Puerto Rican, type 46	do	(8)	(8)	(8)	(8)	(8)
Ohio filler and Wisconsin binder, types 42-44 and 53-55	do	1.120	1.169	1.212	1.238	1.238
Barley: [6]						
Target price	Bushel	(9)	(9)	(9)	(9)	(9)
Loan rate	do	1.55	1.57	1.56	1.59	1.62
Sorghum grain: [6]						
Target price	Cwt.	(9)	(9)	(9)	(9)	(9)
Loan rate	do	3.23	3.14	3.11	3.11	3.05
Oats: [6]						
Target price	Bushel	(9)	(9)	(9)	(9)	(9)
Loan rate	do	1.03	1.11	1.11	1.13	1.16
Rye: [6]	do	(8)	(8)	(8)	(8)	(8)
Nonbasic commodities:						
Beans, dry edible	Cwt.	(8)	(8)	(8)	(8)	(8)
Cottonseed	Ton	(8)	(8)	(8)	(8)	(8)
Minor oilseeds: [7]						
Target price	Cwt.	....	....	....	....	....
Loan rate	do	8.91	9.30	9.30	9.30	9.30
Soybeans:						
Target price	Bushel	....	....	....	....	....
Loan rate	do	4.99	5.26	5.26	5.26	5.26
Dry Peas	Cwt.	....	....	....	....	....
Sugar, raw	Pound	0.1800	0.1800	0.1800	0.1800	0.1800
Milk for manufacturing	Cwt.	(10)10.35	(13)10.20	(14)10.05	(15) 9.90	9.90
Honey, extracted	Pound	(11)	(11)	(11)	0.59	0.59
Mohair	do	(12)	(12)	(12)	2.00	2.00
Wool	Pound	(12)	(12)	(12)	(12)	(12)

See footnotes at end of table.

Table 9-42.—Agricultural commodities: Support prices per unit, United States, 1996–2005 [1][2]—Continued

Commodity	Unit	2001	2002	2003	2004	2005
		Dollars	Dollars	Dollars	Dollars	Dollars
Basic commodities:						
Corn:						
Target price	Bushel	([9])	([16]) 2.60	2.60	2.63	2.63
Loan rate	do	1.89	1.98	1.98	1.95	1.95
Cotton:						
American upland: [3]						
Target price	Pound	([9])	([16]) 0.724	0.724	0.724	0.724
Loan rate	do	0.5192	0.5200	0.5200	0.5200	0.5200
Extra-long staple:						
Target price	do	([9])				
Loan rate	do	0.7965	0.7977	0.7977	0.7977	0.7977
Peanuts: [4]						
Target price	do		0.2475	0.2475	0.2475	0.2475
Loan rate	do	0.3050	0.1775	0.1775	0.1775	0.1775
Rice:						
Target price	Cwt.	([9])	([16])10.50	10.50	10.50	10.50
Loan rate	do	6.50	6.50	6.50	6.50	6.50
Wheat:						
Target price	Bushel	([9])	([16]) 3.86	([16]) 3.86	([16]) 3.92	3.92
Loan rate	do	2.58	2.80	2.80	2.75	2.75
Tobacco: [18]						
Flue-cured, types 11-14	Pound	1.660	1.656	1.663	1.690	
Fire-cured, type 21	do	1.572	1.603	1.636	1.636	
Fire-cured, types 22-23	do	1.736	1.767	1.817	1.863	
Burley, type 31	do	1.826	1.835	1.849	1.873	
Dark air-cured, types 35-36	do	1.499	1.526	1.571	1.612	
Virginia sun-cured, type 37	do	1.392	1.429	1.458	1.458	
Cigar filler, Puerto Rican, type 46	do	([8])	([8])			
Ohio filler and Wisconsin binder, types 42-44 and 53-55	do	1.252	1.286	1.323	1.357	
Barley: [6]						
Target price	Bushel	([9])	([16]) 2.21	([16]) 2.21	2.24	2.24
Loan rate	do	1.65	1.88	1.88	1.85	1.85
Sorghum grain: [6]						
Target price	Cwt.	([9])	([16]) 4.54	([16]) 4.54	4.59	4.59
Loan rate	do	3.05	3.54	3.54	3.48	3.48
Oats: [6]						
Target price	Bushel	([9])	([16]) 1.40	([16]) 1.40	([16]) 1.44	1.44
Loan rate	do	1.21	1.35	1.35	1.33	1.33
Rye: [6]	do	([8])	([8])	([8])	([8])	
Nonbasic commodities:						
Beans, dry edible	Cwt.	([8])	([8])	([8])	([8])	
Cottonseed	Ton	([8])	([8])	([8])	([8])	
Other oilseeds: [7]						
Target price	Cwt.		([16]) 9.80	([16]) 9.80	([16])10.10	10.10
Loan rate	do	9.30	9.60	9.60	9.30	9.30
Soybeans:						
Target price	Bushel		([16]) 5.80	([16]) 5.80	([16]) 5.80	5.80
Loan rate	do	5.26	5.00	5.00	5.00	5.00
Dry peas	Cwt.		([16]) 6.33	([16]) 6.33	([16]) 6.22	6.22
Small chick peas	do		([16]) 7.56	([16]) 7.56	([16]) 7.43	7.43
Lentils	do		([16])11.94	([16])11.94	([16])11.94	11.72
Sugar, raw cane	Pound	0.1800	0.1800	0.1800	0.1800	0.1800
Sugar, refined beet	do	0.229	0.229	0.229	0.229	
Sugar, in-process beet	do	0.1832	0.1832	0.1832	0.1832	
Milk for manufacturing	Cwt.	9.90	9.90	9.90	9.90	9.90
Honey, extracted	Pound	([18]) 0.65	0.60	0.60	0.60	0.60
Mohair	do		([17]) 4.20	([17]) 4.20	([17]) 4.20	([17]) 4.20
Wool	Pound	([12])	([17]) 0.40	([17]) 1.00	([17]) 1.00	([17]) 1.00

[1] National averages during the marketing years for the individual crops, beginning in the years shown. [2] The target price is known in the statute as the "established price". [3] 1 1/16 strict low middling, micronaire 3.5 through 4.9. [4] For quota portion of crop (1993 through 2001). Enactment of the Farm Security and Rural Investment Act of 2002 (2002 Act) repealed the peanut quota marketing program; and established payment rates for the 2002/2003 and subsequent crops according to the provisions of the Direct Payment Program. [5] Grade No. 2 or better except for oats which is Grade No. 3. [6] The rye price support program was terminated by the Federal Agriculture Improvement and Reform Act of 1996. Rye was not reestablished with the 2002 Act. [7] Includes flaxseed, sunflower seed (oil and other), safflower, rapeseed (industrial), canola, mustard seed and cambe and sesame. [8] No support program. [9] The Federal Agriculture Improvement and Reform Act of 1996 replaced the deficiency payment/production adjustment programs for the program crops with a Production Flexibility Contract program, making target prices no longer applicable beginning with the 1996/97 marketing year. [10] As of January 1, 1996. [11] The honey price support program was terminated by the Federal Agriculture Improvement and Reform Act of 1996. [12] The wool and mohair support programs terminated as of December 31, 1995, as required by Public Law 103-130. [13] As of January 1, 1997. [14] As of January 1, 1998. [15] As of January 1, 1999. [16] The Farm Security and Rural Investment Act of 2002 (2002 Act) reestablished target prices, now including soybeans and other oilseeds. The 2002 Act also established, for the first time, loan rates for dry peas, small chickpeas and lentils under the marketing loan program. [17] Wool and mohair programs were reestablished following enactment of the Farm Security and Rural Investment Act of 2002 (2002 Act). First wool number is for ungraded/second is graded. [18] Beginning with 2005 crop, all tobacco price support is discontinued.

FSA, Economic Policy and Analysis Staff, (202) 720–3451.

FARM RESOURCES, INCOME, AND EXPENSES

Table 9-43.—Farm income: Cash receipts by commodity groups and selected commodities, United States, 1997–2004 [1]

Commodity	1997	1998	1999	2000
	1,000 dollars	*1,000 dollars*	*1,000 dollars*	*1,000 dollars*
All commodities	207,789,893	196,457,059	187,849,232	192,113,281
Livestock and products	96,475,245	94,248,720	95,748,407	99,623,972
Cattle and calves	35,999,622	33,442,847	36,568,558	40,783,474
Hogs	13,053,680	9,485,547	8,624,295	11,757,943
Sheep and lambs	632,602	477,794	467,022	470,136
Dairy products	20,940,261	24,105,134	23,189,113	20,586,629
Broilers	14,158,926	15,146,560	15,127,787	13,989,424
Farm chickens	71,219	79,045	74,104	63,704
Chicken eggs	4,539,929	4,439,396	4,322,254	4,335,427
Turkeys	2,814,997	2,620,452	2,750,870	2,771,109
Miscellaneous livestock	3,589,390	3,769,522	3,947,333	4,182,455

Commodity	2001	2002	2003	2004
All commodities	200,058,312	194,984,340	216,592,032	241,241,402
Livestock and products	106,712,594	93,980,615	105,593,541	123,480,989
Cattle and calves	40,540,660	38,095,143	45,092,281	47,295,573
Hogs	12,394,562	9,602,110	10,618,028	14,348,331
Sheep and lambs	396,586	420,633	502,218	514,029
Dairy products	24,685,667	20,582,238	21,238,737	27,367,857
Broilers	16,694,515	13,437,700	15,214,945	20,446,085
Farm chickens	46,516	49,850	47,508	57,655
Chicken eggs	4,449,958	4,302,288	5,263,426	5,303,244
Turkeys	2,735,961	2,643,273	2,631,862	2,995,802
Miscellaneous livestock	4,058,513	4,121,338	4,259,494	4,425,373

Commodity	1997	1998	1999	2000
Crops	111,314,648	102,208,339	92,100,825	92,499,309
Food grains	10,410,552	8,808,374	6,931,289	6,507,596
Feed crops	27,086,775	22,578,380	19,527,517	20,535,169
Cotton	6,345,803	6,072,960	4,630,256	2,949,649
Tobacco	2,873,023	2,804,984	2,275,052	2,315,779
Oil crops	19,758,300	17,371,716	13,355,150	13,478,114
Vegetables	14,668,839	15,015,605	15,013,441	15,553,954
Fruits/nuts	12,957,987	11,982,796	12,016,211	12,458,118
All other crops	17,213,369	17,573,526	18,351,908	18,690,931

Commodity	2001	2002	2003	2004
Crops	93,345,718	101,003,725	110,989,491	117,760,413
Food grains	6,385,012	6,787,802	8,023,363	9,127,838
Feed crops	21,455,425	24,040,729	24,738,592	28,237,936
Cotton	3,639,446	3,418,096	6,527,296	5,405,215
Tobacco	1,894,764	1,743,429	1,552,586	1,519,104
Oil crops	13,337,865	15,049,124	18,671,097	19,787,369
Vegetables	15,450,237	17,177,230	17,401,367	17,256,235
Fruits/nuts	11,959,556	12,617,817	13,419,104	15,462,980
All other crops	19,223,412	20,169,498	20,665,086	20,963,737

[1] USDA estimates and publishes individual cash receipt values only for major commodities and major producing States. The U.S. receipts for individual commodities, computed as the sum of the reported States, may understate the value of sales for some commodities, with the balance included in the appropriate category labeled "other" or "miscellaneous." The degree of underestimation in some of the minor commodities can be substantial.

ERS, Farm and Rural Business Branch, (202) 694–5592. E-mail contact is rogers@ers.usda.gov.

Table 9-44.—Farm income: United States, 1997–2004 [1]

Item	1997	1998	1999	2000
	Billion dollars	Billion dollars	Billion dollars	Billion dollars
Total gross farm income	238.0	232.6	235.0	242.0
Value of Production [2]	230.5	220.2	213.5	219.1
Crops	112.5	102.1	92.8	94.9
Livestock and products	96.3	94.2	95.2	99.1
Services and forestry	21.7	23.9	25.4	25.0
Direct government payments	7.5	12.4	21.5	22.9
Total production expenses	186.7	185.5	187.2	193.1
Net farm income	51.3	47.1	47.7	48.9
Gross cash income	227.4	222.7	224.2	228.7
Cash expenses	166.4	165.0	166.3	171.8
Net cash income	60.9	57.7	58.0	57.0

Item	2001	2002	2003	2004
Total gross farm income	248.7	229.9	259.8	292.3
Value of production [2]	228.0	218.7	242.6	279.0
Crops	95.0	98.3	109.4	124.0
Livestock and product	106.4	93.5	104.9	124.6
Services and forestry	26.5	26.9	28.3	30.5
Direct government payments	20.7	11.2	17.2	13.3
Total production expenses	197.1	193.4	200.3	209.8
Net farm income	51.5	36.6	59.5	82.5
Gross cash income	235.6	221.0	249.5	271.7
Cash expenses	175.5	171.6	177.9	186.2
Net cash income	60.1	49.5	71.6	85.5

[1] Component values and additional details may be found in the value-added and cash income tables on the internet at http://www.ers.usda.gov/data/farmincome/finfidmu.htm. [2] Includes cash receipts, value of change in inventories, and home consumption. In the value-added table, value of production is synonymous with final output.

ERS, Farm and Rural Business Branch, (202) 694–5592. E-mail contact is rogers@ers.usda.gov

Table 9-45.—Expenses: Farm production expenses, United States, 1997–2004

Item	1997	1998	1999	2000
	Thousand dollars	Thousand dollars	Thousand dollars	Thousand dollars
Total production expenses	186,722,962	185,472,975	187,244,468	193,111,038
Feed purchased	26,334,286	25,031,151	24,501,398	24,486,129
Livestock and poultry purchased	13,820,160	12,588,563	13,763,754	15,852,020
Seed purchased	6,712,046	7,214,374	7,216,721	7,518,955
Fertilizer and lime	10,927,346	10,624,247	9,920,263	10,020,234
Pesticides	9,017,436	9,016,995	8,617,422	8,516,511
Fuel and oil	6,242,613	5,599,193	5,587,892	7,191,107
Electricity	3,043,921	2,908,113	2,986,512	2,999,261
Other [1]	41,825,203	42,883,744	44,065,002	43,004,997
Interest	13,247,317	13,452,555	13,780,635	14,670,920
Contract and hired labor expenses	18,410,252	19,122,503	19,811,764	20,642,044
Net rent to nonoperator landlords [2]	11,234,514	10,869,158	10,396,825	11,202,353
Capital consumption	19,250,026	19,560,651	19,789,551	20,099,777
Property taxes	6,657,842	6,601,728	6,806,729	6,906,730

Item	2001	2002	2003	2004
Total production expenses	197,132,791	193,358,084	200,333,485	209,778,055
Feed purchased	24,769,124	24,964,575	27,514,979	30,033,673
Livestock and poultry purchased	15,226,524	14,413,409	16,771,176	17,564,131
Seed purchased	8,221,541	8,924,511	9,425,847	9,525,841
Fertilizer and lime	10,322,222	9,619,305	10,022,069	11,428,147
Pesticides	8,616,218	8,316,338	8,416,767	8,516,120
Fuel and oil	6,890,100	6,603,708	6,801,504	8,197,282
Electricity	3,554,273	3,911,407	3,340,647	3,236,666
Other [1]	45,449,295	44,068,020	45,049,788	45,899,807
Interest	13,598,908	13,143,748	12,672,506	13,064,047
Contract and hired labor expenses	21,896,078	21,846,751	21,960,391	23,216,429
Net rent to nonoperator landlords [2]	11,127,448	9,756,157	10,284,449	9,749,856
Capital consumption	20,554,466	20,982,743	21,267,039	22,339,554
Property taxes	6,906,594	6,807,412	6,806,323	7,006,502

[1] Includes repair and maintenance, machine hire and customwork, marketing, storage and transportation, insurance premiums, and miscellaneous other expenses. [2] Includes landlord capital consumption.

ERS, Farm and Rural Business Branch, (202) 694–5592. E-mail contact is rogers@ers.usda.gov

FARM RESOURCES, INCOME, AND EXPENSES

Table 9-46.—Farm marketings, government payments, and principal commodities, 2004, by States

State	Cash receipts		Livestock and Products	Government payments	Rank
	Total	Crops			
	1,000 dollars	*1,000 dollars*	*1,000 dollars*	*1,000 dollars*	
AL	4,103,235	734,696	3,368,539	155,508	24-Broilers, cattle/calves, chicken eggs, greenhouse (84%).
AK	52,987	24,329	28,658	5,723	50-Greenhouse, hay, dairy, potatoes (45%).
AZ	3,065,604	1,628,576	1,437,028	99,959	29-Cattle/calves, lettuce, dairy, cotton (69%).
AR	6,604,401	2,431,732	4,172,669	599,457	12-Broilers, rice, soybeans, cattle/calves (74%).
CA	31,835,183	23,212,043	8,623,140	506,591	1-Dairy, greenhouse, grapes, almonds (43%).
CO	5,501,154	1,345,001	4,156,153	220,900	16-Cattle/calves, dairy, corn, greenhouse (77%).
CT	526,580	348,651	177,929	6,834	44-Greenhouse, dairy, chicken eggs, aquaculture (69%).
DE	933,842	191,185	742,657	16,135	39-Broilers, soybeans, corn, greenhouse (87%).
FL	6,843,731	5,359,595	1,484,136	214,409	10-Greenhouse, oranges, sugar cane, tomatoes (60%).
GA	6,107,025	2,036,173	4,070,852	344,290	12-Broilers, cotton, cattle/calves, chicken eggs (68%).
HI	549,830	457,079	92,751	2,392	43-Greenhouse, pineapples, sugar cane, macadamia nuts (51%).
ID	4,349,255	1,818,681	2,530,574	153,028	21-Cattle/Calves, dairy, potatoes, wheat (75%).
IL	9,708,305	7,769,390	1,938,915	1,163,440	6-Corn, soybeans, hogs, cattle/calves (88%).
IN	6,043,191	3,978,204	2,064,987	529,630	13-Corn, soybeans, hogs, dairy (77%).
IA	14,652,945	7,368,773	7,284,172	1,264,138	3-Corn, hogs, soybeans, cattle/calves (89%).
KS	9,502,727	3,082,658	6,420,069	645,081	7-Cattle/calves, wheat, corn, soybeans (86%).
KY	4,126,186	1,387,682	2,738,504	147,021	23-Horses/mules, broilers, cattle/calves, tobacco (65%).
LA	2,225,802	1,347,809	877,993	320,038	34-Sugar cane, rice, cattle/calves, soybeans (44%).
ME	553,830	223,221	330,609	10,713	42-Dairy, potatoes, chicken eggs, greenhouse (55%).
MD	1,743,357	732,691	1,010,666	52,286	36-Broilers, greenhouse, dairy, corn (75%).
MA	413,954	319,810	94,144	6,942	47-Greenhouse, cranberries, dairy, sweet corn (67%).
MI	4,312,320	2,566,437	1,745,883	215,232	22-Dairy, greenhouse, corn, soybeans (58%).
MN	9,794,911	4,860,595	4,934,316	704,496	5-Corn, hogs, soybeans, dairy (67%).
MS	4,089,158	1,377,005	2,712,153	401,413	20-Broilers, cotton, soybeans, aquaculture (77%).
MO	5,818,728	2,756,149	3,062,579	470,259	15-Soybeans, Cattle/calves, corn, hogs (64%).
MT	2,238,980	960,935	1,278,045	282,404	33-Cattle/calves, wheat, barley, hay (84%).
NE	11,779,728	4,441,545	7,338,183	728,310	4-Cattle/calves, corn, soybeans, hogs (92%).
NV	454,343	147,274	307,069	6,531	45-Cattle/calves, hay, dairy, onions (87%).
NH	168,871	95,222	73,649	4,590	48-Greenhouse, dairy, apples, cattle/calves (76%).
NJ	866,719	680,053	186,666	10,301	40-Greenhouse, horses/mules, blueberries, dairy (64%).
NM	2,564,862	565,345	1,999,517	80,649	31-Dairy, cattle/calves, hay, pecans (84%).
NY	3,653,430	1,351,115	2,302,315	82,064	28-Dairy, greenhouse, apples, cattle/calves (72%).
NC	8,210,496	2,859,152	5,351,344	212,611	8-Hogs, broilers, greenhouse, tobacco (69%).
ND	4,090,863	3,152,582	938,281	466,546	25-Wheat, cattle/calves, soybeans, corn (64%).
OH	5,459,380	3,387,276	2,072,104	332,574	17-Soybeans, corn, dairy, greenhouse (66%).
OK	5,054,570	1,172,866	3,881,704	220,307	18-Cattle/calves, hogs, broilers, wheat (80%).
OR	3,691,554	2,647,919	1,043,635	80,760	27-Greenhouse, cattle/calves, dairy, hay (56%).
PA	4,859,335	1,544,652	3,314,683	91,232	20-Dairy, cattle/calves, greenhouse, mushrooms agaricus (62%).
RI	63,826	54,014	9,812	1,499	49-Greenhouse, sweet corn, dairy, potatoes (78%).
SC	1,909,098	833,134	1,075,964	81,299	35-Broilers, greenhouse, turkeys, cattle/calves (60%).
SD	4,877,484	2,455,300	2,422,184	398,540	19-Cattle/calves, corn, soybeans, wheat (79%).
TN	2,561,984	1,263,003	1,298,981	153,214	32-Cattle/calves, broilers, soybeans, greenhouse (59%).
TX	16,498,398	5,391,411	11,106,987	1,152,040	2-Cattle/calves, cotton, broilers, greenhouse (75%).
UT	1,253,154	270,028	983,126	36,853	37-Cattle/calves, dairy, hogs, hay (76%).
VT	581,773	84,927	496,846	18,153	41-Dairy, cattle/calves, greenhouse, hay (90%).
VA	2,684,392	902,271	1,782,121	68,052	30-Broilers, cattle/calves, dairy, greenhouse (54%).
WA	5,868,195	4,132,390	1,735,805	197,011	14-Apples, dairy, cattle/calves, wheat (52%).
WV	422,872	74,359	348,513	7,615	46-Broilers, cattle/calves, chicken eggs, dairy (73%).
WI	6,864,150	1,781,723	5,082,427	298,182	9-Dairy, cattle/calves, corn, greenhouse (79%).
WY	1,104,702	153,746	950,956	36,343	38-Cattle/calves, hay, sugar beets, hogs (87%).
US	241,241,403	117,760,414	123,480,989	13,303,598	Cattle/calves, dairy, corn, broilers (47%).

ERS, Farm and Rural Business Branch, (202) 694 5592. Information contact: Larry Traub -- E-Mail: ltraub@ers.usda.gov or Roger Strickland -- E-Mail: rogers@ers.usda.gov.

Table 9-47.—Farm Operator Households: Average Income, United States, 2002–2005 [1]

Item	2002 [2]	2003	2004 [3]	2005F
	Dollars per farm			
Net cash farm business income [4]	11,331	14,979	20,638	19,640
Less depreciation [5]	8,189	7,334	8,085	NA
Less wages paid to operator [6]	758	695	747	NA
Less farmland rental income [7]	621	864	806	NA
Less adjusted farm business income due to other household(s) [8]	1,248	1,344	2,909	NA
	Dollars per farm operator household			
Equals adjusted farm business income	516	4,742	8,091	NA
Plus wages paid to operator	758	695	747	NA
Plus net income from farmland rental [9]	NA	NA	NA	NA
Equals farm self-employment income	1,273	5,437	8,838	NA
Plus other farm-related earnings [10]	2,199	2,447	5,363	NA
Equals earnings of the operator household from farming activities	3,473	7,884	14,201	13,258
Plus earnings of the operator household from off-farm sources [11]	62,285	60,713	67,279	70,401
Equals average farm operator household income comparable to U.S. average household income, as measured by the CPS	65,757	68,597	81,480	83,660
	Dollars per U.S. household			
U.S. average household income [12]	57,852	59,067	60,528	NA
	Percent			
Average farm operator household income as percent of U.S. average household income	113.7	116.1	134.6	NA
Average operator household earnings from farming activities as percent of average operator household income	5.3	11.2	16.9	15.3

F=Forecast. NA-not available. * The relative standard error exceeds 25 percent, but is no more than 50 percent. [1] This table derives farm operator household income estimates from the Agricultural Resource Management Study (ARMS) that are consistent with Current Population Survey (CPS) methodology. The CPS, conducted by the Census Bureau, is the source of official U.S. household income statistics. The CPS defines income to include any income received as cash. The CPS definition departs from a strictly cash concept by including depreciation as an expense that farm operators and other self-employed people subtract from gross receipts when reporting net cash income. [2] Prior to 2000, net cash income from operating another farm and net cash income from farm land rental were included in earnings from farming activities. However, because of a change in the ARMS survey design, net cash income from a farm other than the one being surveyed and net income from farm land rentals are not separable from total off-farm income. Although there is no effect upon estimates of farm operator household income in 2000, estimates of farm self-employment, other farm related earnings, earnings of the household from farming activities, and earnings of the farm from off-farm sources are not strictly comparable to those from previous years. [3] Starting in 2004, Farm operator household income specifically excludes net capital gains/losses. [4] A component of farm sector income. Excludes income of contractors and landlords as well as the income of farms organized as non-family corporations or cooperatives and farms run by a hired manager. Includes the income of farms organized as proprietorships, partnerships, and family corporations. [5] Consistent with the CPS definition of self-employment income, reported depreciation expenses are subtracted from net cash income. The ARMS collects farm business depreciation used for tax purposes. [6] Wages paid to the operator are subtracted here because they are not shared among other households that have claims on farm business income. These wages are added to the operator household's adjusted farm business income to obtain farm self-employment income. [7] Gross rental income is subtracted here because net rental income from the farm operation is added below to income received by the household. [8] More than one household may have a claim on the income of a farm business. On average,1.1 households share the income of a farm business. [9] Includes net rental income from the business. Also includes net income from farmland held by household members that is not part of the farm business. Beginning in 2000, net income from farmland rental is considered as part of off-farm income. (See footnote 2.) [10] Wages paid to other operator household members by the farm business and net cash income from a farm business other than the one being surveyed. In 2000 and 2001, however, net cash income from farm businesses other than the one being surveyed is included in off-farm earnings. In 2002, 2003 and 2004, also includes net cash income from farm land rental. (See footnote 2.) [11] Wages, salaries, net income from nonfarm businesses, interest, dividends, transfer payments, etc. In 2000 and 2001,also includes net cash income from other farm and net cash income from farm rental (See footnote 2.) [12] From the CPS.

Sources: U.S. Dept. of Agriculture, Economic Research Service, 2002, 2003, and 2004 Agricultural Resource Management Study (ARMS) for farm operator household data. U.S. Dept. of Commerce, Bureau of the Census, Current Population Survey (CPS), for U.S. average household income. For information on household income contact: Bob Green (202) 694-5568. Email rgreen@ers.usda.gov or Bob Hoppe (202) 694-5572. Email rhoppe@ers.usda.gov.

ERS, Farm Structure and Performance Branch, (202) 694-5568.

Table 9-48.—Grazing fees: Rates for cattle by selected States and regions, 2004–2005

State	Monthly lease rates for private non-irrigated grazing land [1]					
	Animal unit [2]		Cow-calf		Per head	
	2004	2005	2004	2005	2004	2005
	Dollars per month	Dollars per month	Dollars per month	Dollars per month	Dollars per month	Dollars per month
AZ	8.00	8.00	(7)	(7)	9.00	9.50
CA	14.50	15.40	19.50	20.50	15.50	17.00
CO	13.50	14.50	15.00	16.00	14.00	14.30
ID	12.20	12.50	14.20	14.60	12.60	13.00
KS	13.00	13.50	16.50	16.50	13.50	14.00
MT	15.90	16.20	17.40	18.70	16.20	17.30
NE	23.00	22.50	27.50	27.50	25.20	25.00
NV	10.60	12.20	12.00	12.50	12.00	12.50
NM	9.70	9.50	11.90	11.50	11.00	10.80
ND	13.00	13.70	14.20	16.00	13.50	14.50
OK	8.00	8.00	10.00	10.00	8.50	8.00
OR	13.00	13.00	15.10	15.70	12.50	12.80
SD	17.60	18.40	21.50	21.90	19.20	19.50
TX	10.00	9.40	10.80	9.00	9.80	9.90
UT	11.80	11.60	13.80	13.60	13.10	13.00
WA	10.80	9.70	12.50	12.50	10.80	12.20
WY	13.90	14.80	16.00	17.00	14.30	15.50
17-State [3]	13.10	13.20	15.30	15.20	13.70	14.00
16-State [4]	14.30	14.60	17.10	17.60	15.20	15.60
11-State [5]	13.30	13.70	15.50	16.20	13.80	14.60
9-State [6]	13.00	13.00	15.10	14.80	13.60	13.80

[1] The average rates are estimates (rates over $10.00 are rounded to the nearest dime) based on survey indications of monthly lease rates for private, non-irrigated grazing land from the January Cattle Survey. [2] Includes animal unit plus cow-calf rates. Cow-calf rate converted to animal unit (AUM) using (1 aum=cow-calf *0.833). [3] Seventeen Western States: All States listed. [4] Sixteen Western States: All States, except Texas. [5] Eleven Western States: AZ, CA, CO, ID, MT, NV, NM, OR, UT, WA, and WY. [6] Nine Great Plains States: CO, KS, NE, NM, ND, OK, SD, TX, and WY. [7] Insufficient data.

NASS, Environmental, Economics, and Demographics Branch, (202) 720–6146.

CHAPTER X
INSURANCE, CREDIT, AND COOPERATIVES

The statistics in this chapter deal with taxes, insurance, agricultural credit, and farm cooperatives. Some of the series were developed in connection with research activities of the Department, while others, such as data from agricultural credit agencies, are primarily records of operations.

Table 10-1.—Crop losses: Average percentage of indemnities attributed to specific hazards, by crops, 1948–2003

Crop	Year	Drought heat (excess)	Hail	Precip. (excess poor drainage)	Frost freeze, (other cold damage)	Flood	Cyclone, tornado, wind, hot wind	Insects	Disease	All others
		Percent	Percent	Percent	Percent	Percent	Percent	Percent	Percent	Percent
Adjusted gross rev-enue	2001-2004	10	9	20	28	0	1	0	0	32
Adjusted gross revenue-lite	1981-2004	0	0	3	3	0	0	0	0	94
Alfalfa seed	2002-2004	30	6	11	15	0	16	0	0	22
All other citrus trees	2000-2004	0	0	3	3	0	18	0	0	77
All other grapefruit	2001-2003	0	0	0	49	0	51	0	0	0
Almonds	1981-2004	2	5	57	31	0	6	0	0	0
Apples	1963-2004	7	26	4	53	0	3	0	0	6
Avocado trees	1996-2001	0	0	82	0	0	18	0	0	0
Avocados	1998-2005	4	0	7	21	0	47	10	0	11
Barley	1956-2004	39	18	26	4	1	2	2	4	4
Blueberries	1995-2004	16	10	14	54	0	1	0	0	4
Burley tobacco	1997-2004	20	8	34	4	8	5	0	20	1
Cabbage	1999-2004	13	1	11	9	1	13	50	2	1
Canola	1995-2004	14	15	47	17	0	2	3	2	0
Carambola trees	2001-2001	0	0	100	0	0	0	0	0	0
Cherries	1963-2004	4	9	45	35	0	5	0	0	0
Chile peppers	2000-2004	1	29	4	19	0	24	11	1	11
Cigar binder tobacco	1997-2004	3	6	19	2	0	0	0	70	1
Cigar filler tobacco	1998-2004	89	0	2	0	0	0	1	8	0
Cigar wrapper tobacco	1997-2003	0	0	47	5	0	0	0	48	0
Citrus	1989-1997	18	5	1	74	0	2	0	0	0
Citrus I	1998-2005	0	0	0	1	0	99	0	0	0
Citrus II	2000-2005	0	0	0	4	0	96	0	0	0
Citrus III	2001-2005	0	0	0	2	0	98	0	0	0
Citrus IV	1998-2005	0	6	0	15	0	79	0	0	0
Citrus trees	1990-1997	0	0	0	100	0	0	0	0	0
Citrus trees IV	2004-2004	0	0	100	0	0	0	0	0	0
Citrus V	1999-2005	0	1	0	7	0	92	0	0	0
Citrus VI	2005-2005	0	0	0	0	0	100	0	0	0
Citrus VII	1998-2005	0	6	0	4	0	90	0	0	0
Clams	2001-2004	0	0	0	6	0	6	0	0	87
Corn	1948-2004	23	11	10	12	9	14	3	4	14
Cotton	1948-2004	21	12	13	13	6	19	4	1	10
Cotton ex long staple	1984-2004	15	15	20	14	0	11	14	1	11
Crambe	1999-2003	22	12	28	9	0	23	0	5	0
Cranberries	1984-2004	14	13	11	50	1	1	5	1	4
Cultivated wild rice	1999-2004	9	25	4	6	0	19	2	1	34
Dark air tobacco	1997-2004	60	03	14	4	1	3	0	15	0
Dry beans	1948-2004	17	31	18	25	1	3	0	3	1
Dry peas	1963-2004	46	15	16	21	0	1	0	1	0
Early & midseason or-anges	1998-2005	0	18	1	54	0	26	0	0	1
Figs	1988-2003	9	0	50	30	0	3	0	0	8
Fire cured tobacco	1997-2004	28	10	18	16	1	4	0	9	15
Flax	1948-2004	32	10	48	6	0	1	1	1	1
Flue cured tobacco	1997-2004	25	14	7	3	1	20	0	30	0
Forage production	1979-2004	39	2	18	32	0	1	2	0	7
Forage seeding	1978-2004	42	0	25	32	0	0	0	0	0
Fresh apricots	1997-2004	3	50	24	20	0	2	0	0	0
Fresh freestone peaches	1997-2004	6	35	31	28	0	0	0	0	0
Fresh market beans	2000-2004	0	0	71	15	0	8	0	4	0
Fresh market sweet corn	1985-2004	22	2	28	38	0	9	0	0	0
Fresh market tomatoes	1984-2004	7	19	39	14	0	10	3	8	0
Fresh nectarines	1997-2004	10	49	27	13	0	0	0	0	0
Fresh plum	1990-1997	0	59	8	32	0	1	0	0	0
Grain sorghum	1959-2004	35	8	24	14	2	12	3	0	3
Grapefruit	1997-2005	28	5	4	43	0	20	0	0	0
Grapefruit trees	2000-2004	0	0	0	2	0	2	0	0	95
Grapes	1967-2004	23	4	19	53	0	1	0	0	1
Green peas	1962-2004	41	4	49	3	0	1	0	1	1
Hybrid corn seed	1983-2004	56	3	31	3	0	3	0	3	0
Hybrid sorghum seed	1988-2004	17	16	4	49	0	14	0	0	1
Income protection corn	1996-1996	3	0	93	3	0	0	0	0	0

See end of table.

Table 10-1.—Crop losses: Average percentage of indemnities attributed to specific hazards, by crops, 1948–2005—Continued

Crop	Year	Drought heat (excess)	Hail	Precip. (excess poor drainage)	Frost freeze, (other cold damage)	Flood	Cyclone, tornado, wind, hot wind	Insects	Disease	All others
		Percent	Percent	Percent	Percent	Percent	Percent	Percent	Percent	Percent
Income protection cotton	1996-1996	96	0	4	0	0	0	0	0	0
Income protection wheat	1996-1996	9	0	90	0	1	0	0	0	0
Late oranges	1998-2005	0	26	0	55	0	18	0	0	2
Lemons	1997-2005	2	0	0	97	0	0	0	0	0
Lime trees	1998-2003	0	0	0	1	0	0	0	0	99
Macadamia nuts	1996-2005	80	0	9	0	0	0	10	0	2
Macadamia trees	2000-2000	00	0	100	0	0	0	0	0	0
Mandarins	1997-2005	44	0	13	35	0	8	0	0	0
Mango trees	1997-1997	0	0	0	100	0	0	0	0	0
Maryland tobacco	1997-2004	60	8	7	4	0	6	0	14	0
Millet	1996-2004	83	14	1	0	0	1	0	0	0
Minneola tangelos	1998-2005	17	6	6	65	0	6	0	0	0
Mint	2000-2004	27	0	17	54	0	2	0	0	0
Mustard	1999-2004	32	47	6	12	0	2	0	0	0
Navel oranges	1998-2005	66	3	2	27	0	1	0	0	2
Nursery	1990-1999	24	0	7	11	7	42	3	5	0
Nursery (fg&c)	2001-2004	2	16	9	20	11	40	0	1	1
Oats	1956-2004	47	13	32	4	0	1	1	1	1
Onions	1988-2004	12	15	45	5	0	3	0	17	3
Orange trees	1996-2004	0	0	3	0	0	9	0	0	88
Oranges	1997-1997	30	0	0	37	0	21	0	0	11
Orlando tangelos	1998-2001	00	0	0	100	0	0	0	0	0
Peaches	1957-2004	4	20	2	71	0	0	0	0	3
Peanuts	1962-2004	42	0	21	6	1	5	0	22	3
Pears	1989-2004	1	23	4	69	0	0	0	0	2
Pecans	1998-2004	12	4	13	1	0	65	0	2	1
Peppers	1984-2004	0	6	59	29	0	4	0	1	0
Plums	1998-2004	13	38	33	17	0	0	0	0	0
Popcorn	1984-2004	59	7	24	4	0	2	1	3	0
Potatoes	1962-2004	23	6	22	25	0	1	0	21	1
Prevented planting endorse	1990-1994	31	0	11	0	53	0	0	0	5
Processing apricots	1997-2004	2	9	21	34	0	34	0	0	0
Processing beans	1988-2004	45	2	44	2	0	1	0	5	0
Processing cling peaches	1997-2004	25	4	34	36	0	1	0	0	0
Processing cucumbers	2000-2004	45	1	47	2	0	0	1	3	0
Processing freestone	1998-2004	7	11	8	74	0	0	0	0	0
Prunes	1986-2004	24	2	14	46	0	12	0	0	1
Raisins	1961-2003	0	0	100	0	0	0	0	0	0
Rangeland	1999-2004	0	0	0	0	0	0	0	0	100
Raspberry and blackberry	2002-2003	0	0	0	16	0	84	0	0	0
Revenue coverage corn ..	1996-1996	20	44	3	5	26	0	0	1	0
Revenue coverage soybeans	1996-1996	1	24	55	5	13	0	0	3	0
Rice	1960-2004	16	0	49	10	6	2	0	5	11
Rio red & star ruby	1998-2005	0	9	0	66	0	13	3	0	9
Ruby red grapefruit	1998-2005	0	16	6	72	0	7	0	0	0
Rye	1980-2004	43	13	30	11	0	2	0	0	0
Safflower	1964-2004	38	04	19	18	0	19	1	1	0
Soybeans	1955-2004	27	13	21	14	9	4	2	3	7
Special citrus	1992-1994	6	12	0	82	0	0	0	0	0
Stonefruit	1989-1996	1	28	44	19	0	2	0	0	6
Strawberries	2000-2004	9	0	68	6	0	0	0	17	0
Sugar beets	1965-2004	13	10	26	22	1	11	2	13	2
Sugarcane	1967-2004	19	0	13	18	0	3	2	16	28
Sunflowers	1976-2004	24	14	26	15	0	3	6	9	3
Sweet corn	1978-2004	44	0	36	16	0	2	0	0	1
Sweet oranges	1998-2005	22	0	10	55	0	12	1	0	0
Sweetpotatoes	1998-2004	39	0	43	0	0	13	4	0	1
Table grapes	1984-2004	37	07	33	21	0	0	0	0	1
Tangelos	1997-1997	3	0	0	97	0	0	0	0	0
Tobacco	1989-1996	17	20	20	1	2	18	0	20	2
Tomatoes	1963-2004	25	0	66	4	1	1	0	2	1
Valencia oranges	1998-2005	63	02	1	27	0	7	0	0	0
Walnuts	1984-2004	27	04	52	16	0	2	0	0	0
Watermelons	1999-1999	8	07	38	1	0	14	0	29	2
Wheat	1948-2004	20	10	11	26	4	10	3	8	7
Winter squash	1999-2004	6	14	79	0	0	0	0	0	0

GRP crops do not have any specific cause of loss.
RMA, Program Automation Branch, (816) 926–7910.

Table 10-2.—Crop insurance programs: Coverage, amount of premiums and indemnities, by crops, United States, 2003–2006 [1]

Commodity and year	Coverage					Indemnities		
	County programs	Insured units [2]	Area insured [3]	Maximum insured production	Amount of premium	Number	Area indemnified [3]	Amount
	Number	*Number*	*1,000 acres*	*1,000 dollars*	*1,000 dollars*		*1,000 acres*	*1,000 dollars*
Adjusted gross revenue:								
2002	214	749	0	244,746	8,962	145	....	10,879
2003	230	944	0	317,125	12,041	147	....	14,331
2004	230	864	0	306,084	11,373	155	....	21,759
Adjusted gross revenue-lite:								
2003	66	73	0	2,595	120	2		10
2004	286	88	0	3,222	148	11		147
Alfalfa seed:								
2002	10	179	11	5,353	402	54	3	552
2003	10	139	8	3,470	257	24	1	315
2004	10	175	14	4,372	320	46	4	760
All other citrus trees:								
2002	28	1,017	0	56,229	1,377	0	....	0
2003	28	983	0	52,009	1,254	2	3	92
2004	28	892	0	49,636	1,194	17	21	405
All other grape-fruit:								
2003	3	2	0	1	0	1	0	0
2004	3	2	0	2	0	0		0
2005	3	3	0	12	2	0	....	0
Almonds:								
2002	16	4,184	337	283,384	18,756	221	20	6,487
2003	16	4,339	395	384,939	23,802	364	24	6,169
2004	16	4,391	412	492,938	30,457	564	36	10,489
Apples:								
2002	332	5,090	254	321,021	22,437	1,354	44	33,187
2003	347	5,395	254	340,806	24,487	1,048	29	25,568
2004	349	5,568	255	370,784	27,394	642	17	18,417
Avocado trees:								
2002	1	210	0	5,523	161	0	....	0
2003	1	203	0	5,715	168	0	....	0
2004	1	195	0	6,131	180	0	....	0
Avocados:								
2004	1	135	2	2,453	177	8	0	24
2004	6	1,173	32	43,634	6,203	26	0	373
2005	6	1,162	32	41,114	5,921	4	0	24
Barley:								
2002	1,539	43,402	3,414	248,743	26,500	18,009	2,096	59,725
2003	1,539	49,405	3,832	356,106	41,938	13,128	1,828	49,662
2004	1,759	42,049	3,197	291,313	34,312	9,046	1,168	34,596
Blueberries:								
2002	32	525	32	22,182	1,819	50	2	872
2003	32	513	32	23,146	1,854	39	1	530
2004	47	600	34	26,288	2,095	54	2	620
Burley tobacco:								
2002	277	24,441	76	171,960	13,828	5,249	23	29,297
2003	277	22,122	71	176,177	15,014	5,654	23	32,262
2004	277	20,601	70	178,255	15,888	4,903	22	31,415
Cabbage:								
2002	27	446	15	12,756	831	76	2	970
2003	27	421	14	12,367	816	105	2	1,078
2004	27	441	16	13,602	867	73	1	904
Canola:								
2002	225	17,716	1,442	125,215	16,098	7,728	790	37,831
2003	225	13,631	1,105	107,202	16,539	3,673	315	13,191
2004	256	13,619	1,161	118,907	19,430	6,109	617	34,197
Carambola trees:								
2002	1	15	0	244	6	0	....	0
2003	1	14	0	202	5	0	....	0
2004	1	13	0	190	5	0	....	0
Cherries:								
2002	21	1,831	34	52,713	4,965	354	6	5,087
2003	21	1,960	37	57,100	5,303	291	4	3,470
2004	21	2,172	41	63,108	5,734	250	5	5,747
Chili peppers:								
2002	3	86	7	4,513	336	5	0	167
2003	3	108	9	5,321	390	4	0	78
2004	3	82	7	4,338	329	6	1	162
Cigar binder to-bacco:								
2002	16	764	4	16,440	1,139	92	1	2,024
2003	16	744	4	21,498	2,056	249	2	8,272
2004	16	729	4	19,314	1,791	206	1	3,722
Cigar filler to-bacco:								
2002	3	38	0	506	20	2	0	1
2003	3	40	0	766	36	1	0	9
2004	3	36	0	540	24	2	0	0
Cigar wrapper to-bacco:								
2002	5	39	1	21,457	1,335	19	1	3,207
2003	5	46	1	22,976	1,840	18	0	3,660
2004	5	51	1	21,695	1,773	0	0	....

See footnotes at end of table.

Table 10-2.—Crop insurance programs: Coverage, amount of premiums and indemnities, by crops, United States, 2003–2006 [1]—Continued

Commodity and year	Coverage				Amount of premium	Indemnities		
	County programs	Insured units[2]	Area insured[3]	Maximum insured production		Number	Area indemnified[3]	Amount
	Number	Number	1,000 acres	1,000 dollars	1,000 dollars		1,000 acres	1,000 dollars
Citrus I:								
2003	29	2,553	232	116,129	2,529	1	0	5
2004	29	2,467	231	115,750	2,535	0	....	0
2005	29	2,290	223	109,561	2,381	848	84	18,415
Citrus II:								
2003	29	1,955	235	143,515	3,647	0	....	0
2004	29	1,898	240	150,303	3,816	0	....	0
2005	29	1,843	239	142,058	3,620	746	94	25,133
Citrus III:								
2003	29	108	8	3,091	70	0	....	0
2004	29	94	4	1,778	39	0	....	0
2005	29	67	2	1,016	24	3	7 2	582
Citrus IV:								
2003	29	1,485	31	20,216	598	7	0	22
2004	29	1,373	27	17,641	521	5	0	61
2005	29	1,193	25	15,889	464	494	12	5,177
Citrus trees I:								
2002	3	584	6	17,752	764	0	....	0
2003	3	552	6	17,376	749	0	....	0
2004	3	520	6	16,475	708	0	....	0
Citrus trees II:								
2002	3	108	1	2,560	119	0	....	0
2003	3	112	1	2,680	125	0	....	0
2004	3	111	1	2,768	130	0	....	0
Citrus trees III:								
2002	3	7	0	109	6	0	....	0
2003	3	4	0	97	5	0	....	0
2004	3	5	0	153	8	0	....	0
Citrus trees IV:								
2002	3	851	14	37,412	2,060	0	....	0
2003	3	812	14	37,180	2,035	0	....	0
2004	3	772	14	36,267	1,980	1	0	10
Citrus trees V:								
2002	3	190	2	6,636	430	0	....	0
2003	3	163	2	5,753	379	0	....	0
2004	3	141	2	5,438	362	0	....	0
Citrus V:								
2003	29	465	13	18,907	647	7	0	58
2004	29	453	11	18,068	646	5	0	189
2005	29	412	11	16,279	571	217	5	5,646
Citrus VI:								
2003	5	8	1	915	19	0	....	0
2004	5	7	1	634	13	0	....	0
2005	5	6	1	798	16	1	0	6
Citrus VII:								
2003	29	1,468	123	65,196	1,940	2	0	1
2004	29	1,335	120	64,799	1,922	13	0	284
2005	29	1,111	110	60,101	1,767	646	77	31,777
Clams:								
2002	13	480	0	59,953	2,181	134	....	4,019
2003	13	431	0	51,177	1,860	97	....	2,775
2004	13	410	0	27,701	969	114	....	2,182
Corn:								
2002	8,524	903,996	58,693	11,422,896	909,613	278,163	20,297	1,259,870
2003	8,658	910,678	59,489	12,608,038	1,095,831	186,337	12,675	700,067
2004	9,085	925,932	62,084	15,543,673	1,406,636	156,051	13,859	814,352
Cotton ELS:								
2002	31	790	250	85,073	4,958	161	29	11,291
2003	31	792	200	63,492	4,381	298	37	12,522
2004	31	867	239	98,600	5,023	145	10	3,243
Cotton:								
2002	1,491	168,269	12,935	2,245,133	317,564	53,898	5,934	400,797
2003	1,646	165,689	12,631	2,325,606	346,954	41,314	4,621	411,791
2004	1,657	166,029	12,540	2,761,698	406,912	26,711	2,855	232,263
Crambe:								
2002	7	138	11	950	106	17	1	36
2003	7	7	0	27	3	0	....	0
2004	7	36	3	287	39	19	2	58
Cranberries:								
2002	30	593	30	36,860	1,279	80	3	1,119
2003	30	585	29	41,896	1,569	76	2	1,198
2004	30	615	30	58,366	2,276	48	2	1,208
Cultivated wild rice:								
2002	10	66	23	5,603	344	19	3	206
2003	10	71	21	5,562	336	28	5	495
2004	10	67	19	5,119	318	22	4	263

See footnotes at end of table.

Table 10-2.—Crop insurance programs: Coverage, amount of premiums and indemnities, by crops, United States, 2003–2005 [1]—Continued

Commodity and year	Coverage				Amount of premium	Indemnities		
	County programs	Insured units [2]	Area insured [3]	Maximum insured production		Number	Area indemnified [3]	Amount
	Number	Number	1,000 acres	1,000 dollars	1,000 dollars		1,000 acres	1,000 dollars
Dark air tobacco:								
2002	37	921	1	3,849	187	35	0	109
2003	37	808	1	3,863	183	34	0	105
2004	37	780	1	4,147	194	25	0	45
Dry beans:								
2002	286	27,601	1,667	279,842	38,383	7,235	548	41,692
2003	287	20,575	1,230	217,227	30,204	5,031	348	25,439
2004	301	19,855	1,208	206,215	28,540	7,394	604	44,690
Dry Peas:								
2002	97	4,718	362	27,315	2,983	1,454	141	4,840
2003	97	5,067	394	32,785	3,694	1,332	138	3,513
2004	102	8,144	688	52,640	6,852	1,593	167	6,271
Early and Midseason oranges:								
2003	3	323	6	2,763	150	23	0	49
2004	3	299	5	2,552	144	14	0	54
2005	3	270	5	2,354	133	14	0	26
Figs:								
2002	4	137	9	5,715	411	2	0	37
2003	4	130	9	6,794	508	53	4	725
2004	4	134	8	7,129	537	0		0
Fired cured tobacco:								
2002	43	2,106	6	20,880	1,002	136	0	792
2003	43	1,734	5	18,296	868	238	1	863
2004	43	1,776	6	20,704	949	113	0	493
Flax:								
2002	107	9,632	670	36,733	4,719	2,698	232	5,476
2003	108	7,897	517	27,818	3,589	1,913	141	2,710
2004	108	7,803	516	38,303	5,128	2,033	151	5,315
Flue cured tobacco:								
2002	173	28,339	223	652,098	25,531	5,324	57	69,577
2003	173	27,414	215	626,792	26,166	6,534	76	84,138
2004	173	26,744	211	625,084	27,819	2,920	36	45,341
Forage prod.:								
2002	570	29,408	2,848	236,704	17,891	14,119	1,571	49,515
2003	608	33,673	3,136	260,144	21,259	9,142	905	28,705
2004	641	48,373	4,030	300,857	27,876	21,956	2,123	61,165
Forage seeding:								
2002	263	3,786	149	15,459	1,952	1,198	62	4,243
2003	434	4,603	191	18,746	2,454	917	54	3,058
2004	448	4,217	170	14,826	2,004	1,144	63	3,298
Fresh apricots:								
2002	13	172	3	4,893	495	43	1	465
2003	29	220	4	5,018	574	42	1	362
2004	29	213	4	4,729	525	50	1	640
Fresh freestone peaches:								
2002	7	577	19	15,939	929	45	1	303
2003	23	624	22	18,942	1,274	36	1	298
2004	24	625	23	20,813	1,389	38	1	351
Fresh market beans:								
2002	5	676	29	25,980	3,322	292	14	10,710
2003	5	418	18	14,715	1,570	178	8	3,324
2004	5	433	18	14,707	1,588	141	6	2,380
Fresh market sweet corn:								
2002	227	1,443	69	26,578	2,526	297	5	1,460
2003	227	1,338	64	27,493	2,719	142	7	2,021
2004	227	1,320	63	27,957	2,982	90	3	1,022
Fresh market tomatoes:								
2002	49	807	59	119,793	10,625	180	5	8,319
2003	49	740	56	118,569	10,800	131	4	5,095
2004	51	722	60	128,112	11,069	188	6	9,929
Fresh nectarines:								
2002	7	677	23	22,926	1,562	45	1	207
2003	23	704	26	26,533	1,848	56	1	792
2004	23	661	25	26,244	1,826	57	1	620
Grain sorghum:								
2002	2,909	138,697	7,251	543,102	82,437	78,201	7,613	223,410
2003	2,936	141,730	7,055	595,024	87,624	73,483	6,086	167,808
2004	2,942	117,274	5,572	547,957	92,563	34,171	3,263	98,979
Grapefruit trees:								
2002	28	1,158	0	129,965	3,823	2	20	75
2003	28	1,057	0	116,923	3,376	7	13	683
2004	28	896	0	118,116	3,387	10	99	1,274

See footnotes at end of table.

Table 10-2.—Crop insurance programs: Coverage, amount of premiums and indemnities, by crops, United States, 2002–2005 [1]—Continued

Commodity and year	Coverage				Amount of premium	Indemnities		
	County programs	Insured units [2]	Area insured [3]	Maximum insured production		Number	Area indemnified [3]	Amount
	Number	*Number*	*1,000 acres*	*1,000 dollars*	*1,000 dollars*		*1,000 acres*	*1,000 dollars*
Grapefruit:								
2003	8	117	5	5,436	288	6	0	52
2004	8	103	4	4,267	222	1	0	61
2005	8	96	4	3,692	206	4	0	70
Grapes:								
2002	83	12,387	576	533,908	32,692	1,240	38	16,686
2003	91	11,910	552	522,103	31,868	1,307	36	16,776
2004	92	12,475	557	523,619	31,492	1,769	47	18,360
Green peas:								
2002	145	3,131	164	30,948	3,467	1,089	68	4,230
2003	150	3,344	181	35,812	4,066	660	48	3,155
2004	150	3,085	171	34,399	3,898	848	46	3,345
Hybrid corn seed:								
2002	384	5,917	346	107,125	11,096	ʋ16	48	3,693
2003	384	5,688	340	112,167	11,387	309	24	2,280
2004	384	4,597	274	105,563	10,461	184	13	909
Hybrid sorghum seed:								
2002	21	896	57	11,392	1,924	113	10	863
2003	21	772	52	12,052	2,402	144	16	2,309
2004	21	645	43	10,778	1,901	107	10	1,423
Late oranges:								
2003	3	84	1	487	65	1	0	0
2004	3	82	1	470	70	6	0	4
2005	3	84	1	520	76	9	0	11
Lemon trees:								
2002	4	2	0	1,957	39	0	....	0
2003	4	4	0	2,062	40	0	....	0
2004	4	2	0	1,957	39	0	....	0
Lemons:								
2003	15	692	38	69,405	3,875	13	1	413
2004	15	695	39	65,286	3,693	10	0	123
2005	15	674	40	66,401	3,773	6	1	223
Lime trees:								
2002	3	34	0	3,011	98	4	0	1,238
2003	3	25	0	1,341	41	4	0	256
2004	3	16	0	860	25	0	....	0
Macadamia nuts:								
2003	3	119	13	17,939	281	24	3	1,119
2004	3	126	13	17,723	302	13	1	520
2005	3	128	13	16,789	282	9	1	168
Macadamia trees:								
2002	3	124	14	64,287	764	0	....	0
2003	3	124	13	59,421	699	0	....	0
2004	3	126	13	54,416	633	0	....	0
Mandarins:								
2003	7	84	2	1,657	137	10	0	150
2004	7	97	2	2,295	187	12	0	160
2005	7	112	4	4,348	374	7	0	61
Mango trees:								
2002	1	29	0	429	11	0	....	0
2003	1	22	0	252	7	0	....	0
2004	1	20	0	273	8	0	....	0
Maryland tobacco:								
2002	6	20	0	192	6	4	0	40
2003	6	15	0	159	6	4	0	10
2004	6	16	0	161	6	2	0	10
Millet:								
2002	5	1,501	85	4,372	593	1,377	129	2,826
2003	55	7,092	540	27,842	3,739	3,628	436	8,354
2004	61	7,232	528	26,598	4,223	3,757	422	7,766
Minneola tangelos:								
2003	8	154	4	5,409	428	20	1	246
2004	8	162	5	5,847	497	13	1	274
2005	8	165	5	6,198	520	15	1	215
Mint:								
2002	9	240	14	7,054	361	71	2	208
2003	9	242	14	6,085	280	63	2	321
2004	9	217	13	6,737	318	51	2	528
Mustard:								
2002	19	857	91	6,724	1,011	420	63	2,185
2003	19	451	47	4,264	565	117	16	517
2004	19	292	41	3,455	484	86	20	889
Naval oranges:								
2003	16	2,818	109	139,699	7,516	341	9	4,609
2004	16	2,858	110	146,582	7,994	184	7	3,182
2005	16	2,838	111	151,807	8,290	199	8	3,821
Nursery:								
2002	3,087	4,203	0	3,006,447	53,897	99	....	9,175
2003	3,088	4,400	0	3,282,888	59,839	168	....	23,889
2004	3,088	4,692	0	3,600,218	64,054	463	....	81,000

See footnotes at end of table.

Table 10-2.—Crop insurance programs: Coverage, amount of premiums and indemnities, by crops, United States, 2002–2005 [1]—Continued

Commodity and year	Coverage					Indemnities		
	County programs	Insured units [2]	Area insured [3]	Maximum insured production	Amount of premium	Number	Area indemnified [3]	Amount
	Number	Number	1,000 acres	1,000 dollars	1,000 dollars		1,000 acres	1,000 dollars
Oats:								
2002	1,648	23,871	1,134	43,985	6,213	11,115	718	18,699
2003	1,648	21,976	1,008	43,113	6,623	4,016	233	4,924
2004	1,648	19,341	868	34,589	5,703	4,353	249	5,137
Onions:								
2002	95	1,776	77	99,105	10,876	564	24	19,011
2003	98	2,176	88	117,616	15,541	892	30	27,429
2004	98	2,337	98	136,156	19,702	986	32	30,402
Orange trees:								
2002	28	4,138	0	900,883	21,539	7	561	1,443
2003	28	4,231	0	877,692	20,567	20	842	2,087
2004	28	3,766	0	844,755	19,670	37	1,396	7,085
Orlando tangelos:								
2003	5	6	0	78	5	5	0	0
2004	5	9	0	79	5	0	0	0
2005	5	7	0	62	4	0	0	0
Peaches:								
2002	220	1,427	41	50,611	8,561	545	14	9,384
2003	258	1,498	42	48,639	8,537	471	12	10,534
2004	258	1,477	41	53,978	9,661	281	7	5,856
Peanuts:								
2002	326	28,438	1,243	337,268	29,938	10,245	459	56,580
2003	328	24,755	1,197	339,136	28,575	3,391	162	16,661
2004	345	26,427	1,271	375,572	31,815	4,364	203	24,985
Pears:								
2002	26	1,909	36	48,076	1,770	104	1	864
2003	26	1,994	37	49,570	1,848	42	0	288
2004	26	2,099	38	47,334	1,839	73	1	703
Pecans:								
2002	7	218	44	29,501	2,267	85	15	2,294
2003	86	625	81	48,369	4,265	172	18	3,449
2004	88	854	98	59,862	5,402	369	29	6,131
Peppers:								
2002	13	264	13	39,936	6,254	119	5	10,483
2003	13	267	14	44,566	6,740	144	6	12,395
2004	13	236	15	48,490	7,492	95	4	10,427
Plums:								
2002	7	1,395	21	21,782	1,783	139	2	802
2003	7	1,304	20	19,838	1,690	113	1	633
2004	7	1,221	19	17,657	1,657	285	4	1,605
Popcorn:								
2002	319	2,755	192	44,543	3,743	976	84	5,891
2003	320	3,137	224	62,984	5,295	580	46	3,062
2004	323	1,907	146	34,462	2,949	69	5	354
Potatoes:								
2002	325	8,468	988	821,069	67,789	1,619	141	55,938
2003	327	8,806	995	888,871	77,052	1,902	160	73,180
2004	327	8,496	958	885,201	76,915	1,793	127	59,867
Proc. apricots:								
2002	13	136	6	5,756	641	25	1	412
2003	13	111	6	5,119	549	27	1	343
2004	13	113	5	4,779	534	14	1	205
Processing beans:								
2002	124	1,265	89	19,875	2,025	271	14	1,721
2003	132	1,631	97	26,496	2,970	376	21	2,444
2004	132	1,743	107	27,018	2,881	324	17	1,960
Processing cling peaches:								
2002	10	1,301	20	24,613	1,389	54	1	263
2003	10	1,272	21	27,047	1,621	108	2	1,199
2004	10	1,381	22	30,406	1,838	76	1	635
Processing cucumbers:								
2002	11	378	19	8,251	669	264	1	5 2,425
2003	11	313	17	7,244	577	126	....	9 1,486
2004	11	308	16	7,481	706	133	1	0 2,198
Proc. freestone:								
2002	7	112	3	3,266	200	3	....	0 24
2003	7	110	3	3,333	209	7	....	0 53
2004	7	107	3	3,511	233	2	....	0 9
Prunes:								
2002	14	1,193	67	55,974	5,310	294	15	5,737
2003	14	1,071	63	52,857	5,483	95	4	1,257
2004	14	1,012	61	47,990	5,219	901	56	29,140
Raisins:								
2001	7	2,829	217	90,927	7,782	3	0	5
2002	7	2,466	264	115,794	10,171	2	0	31
2003	7	2,207	203	66,914	5,842	6	0	20
Rangeland:								
2002	12	902	8,085	46,498	2,211	722	6,740	16,786
2003	12	1,028	9,193	56,635	2,819	498	4,741	5,842
2004	12	1,062	10,098	62,247	3,023	804	8,124	21,399

See footnotes at end of table.

INSURANCE, CREDIT AND COOPERATIVES

Table 10-2.—Crop insurance programs: Coverage, amount of premiums and indemnities, by crops, United States, 2002–2005 [1]—Continued

Commodity and year	Coverage				Amount of premium	Indemnities		
	County programs	Insured units [2]	Area insured [3]	Maximum insured production		Number	Area indemnified [3]	Amount
	Number	Number	1,000 acres	1,000 dollars	1,000 dollars		1,000 acres	1,000 dollars
Raspberry and blackberry:								
2002	7	52	2	1,077	77	1	0	20
2003	7	84	3	1,638	113	1	0	4
2004	7	71	2	1,318	86	0		0
Rice:								
2002	276	20,423	2,437	391,161	19,271	1,652	250	14,927
2003	349	18,065	2,220	358,750	17,639	1,522	210	18,268
2004	349	19,097	2,369	429,525	21,816	1,041	129	9,163
Rio Red & Star Ruby:								
2003	3	476	12	9,243	1,368	15	0	105
2004	3	445	12	8,513	1,297	17	0	66
2005	3	402	12	8,124	1,237	12	0	46
Ruby red grapefruit:								
2003	3	110	2	1,152	146	5	0	10
2004	3	100	2	993	135	4	0	4
2005	3	90	2	894	121	4	0	4
Rye:								
2002	47	412	32	1,042	115	122	14	235
2003	48	434	33	1,391	171	67	8	146
2004	48	374	29	1,318	177	59	5	99
Safflower:								
2002	71	1,058	125	7,510	804	302	47	1,290
2003	71	1,109	124	7,481	860	244	42	948
2004	71	957	107	5,882	791	288	36	792
Soybeans:								
2002	6,422	881,729	56,009	6,917,660	495,032	205,981	14,352	488,693
2003	6,535	877,929	56,259	7,804,475	615,868	277,945	21,710	892,477
2004	6,849	898,054	58,650	10,033,607	943,445	202,215	18,222	739,800
Strawberries:								
2002	21	303	14	63,575	2,723	62	1	2,332
2003	21	302	13	63,253	2,810	128	2	7,146
2004	21	335	15	68,227	3,310	35	0	985
Sugarbeets:								
2002	176	17,454	1,191	544,404	31,018	5,811	471	51,872
2003	177	16,644	1,167	636,383	38,994	1,663	115	14,597
2004	179	16,780	1,162	647,313	40,478	3,348	214	30,150
Sugarcane:								
2002	31	6,006	858	208,675	6,836	248	14	2,007
2003	31	5,908	827	201,764	6,893	127	9	845
2004	31	5,481	727	186,557	6,604	288	17	1,901
Sunflowers:								
2002	329	30,065	2,362	200,685	26,378	15,291	1,583	68,876
2003	335	27,630	2,135	210,958	31,300	12,048	1,184	47,762
2004	512	23,160	1,842	201,552	32,695	12,526	1,281	66,799
Sweet corn:								
2002	170	3,721	265	58,781	3,775	360	21	2,190
2003	170	3,871	272	62,667	4,179	298	18	1,732
2004	170	3,683	271	59,586	3,860	329	31	1,471
Sweet oranges:								
2003	6	52	0	402	29	4	0	26
2004	6	50	0	419	31	5	0	19
2005	6	52	0	470	34	1	0	5
Sweet potatoes:								
2002	8	987	28	23,945	2,457	779	20	11,453
2003	8	857	24	19,929	2,029	462	10	5,423
2004	8	275	21	14,375	1,129	139	9	3,220
Table grapes:								
2002	12	1,189	88	155,597	7,027	77	2	2,791
2003	12	1,147	88	146,819	6,622	173	6	5,358
2004	12	1,144	84	145,975	6,740	102	3	2,821
Tomatoes:								
2002	86	2,973	280	266,207	12,347	158	13	3,410
2003	88	2,928	272	270,994	12,846	380	35	11,061
2004	88	3,014	286	291,649	14,269	166	11	4,255
Valencia oranges:								
2003	13	1,777	50	65,153	4,644	128	4	1,921
2004	13	1,633	47	60,492	4,386	211	9	4,636
2005	13	1,520	41	54,693	3,887	47	1	767
Walnuts:								
2002	26	1,371	84	55,408	2,309	90	4	1,152
2003	26	1,324	89	56,462	2,316	41	2	438
2004	26	1,301	89	56,383	2,312	40	2	363
Wheat:								
2002	6,702	596,749	45,485	3,431,811	423,097	234,839	29,997	862,668
2003	6,639	617,563	46,789	4,008,883	541,353	118,472	15,153	398,359
2004	6,886	610,198	46,022	3,904,981	560,071	155,835	17,576	509,261
Winter squash:								
2002	18	151	3	1,320	129	27	0	123
2003	18	128	3	1,035	107	34	0	168
2004	18	160	4	1,342	138	35	1	276

[1] Data for 2002 are preliminary. [2] Number of farms on which the insured crop was planted including duplication where both the landlord and tenant are insured. Insured farms on which no insured crop was planted are not included. [3] The insured's share of the planted area on the farm.

RMA, Program Automation Branch, (816) 926–7910.

Table 10-3.—Farm real estate debt: Amount outstanding by lender, United States, Dec. 31, 1994–2003 [1]

Year	Farm Credit System	Farm Service Agency [2]	Life insurance companies [3]	All operating banks [4]	Individuals and others [5]	CCC storage and drying facility	Total farm mortgage debt
	1,000 dol-lars	1,000 dol-lars	1,000 dol-lars	1,000 dol-lars	1,000 dol-lars	1,000 dol-lars	1,000 dol-lars
1994	26,300,421	5,852,920	9,562,841	22,555,042	18,700,000	0	82,971,224
1995	26,529,840	5,403,307	9,622,280	23,805,146	19,200,000	0	84,560,906
1996	27,462,253	5,025,262	10,021,976	24,870,048	19,700,000	0	87,079,339
1997	28,922,818	4,663,567	10,267,452	26,968,325	20,200,000	0	91,022,162
1998	30,824,704	4,352,326	11,353,694	29,029,001	20,000,000	0	95,559,725
1999	32,339,403	4,136,820	12,165,553	31,839,697	19,900,000	0	100,381,473
2000	33,907,358	3,907,122	12,514,220	34,016,207	19,543,500	58,000	103,946,407
2001	37,559,173	3,830,284	12,700,000	35,569,061	19,847,000	153,000	109,658,518
2002	43,416,173	3,655,172	13,000,000	37,995,001	20,243,000	257,000	118,566,346
2003 [6]	46,352,321	3,295,558	13,291,455	40,649,006	20,573,890	572,000	124,734,230

[1] Includes operator households. Includes regular mortgages, purchase-money mortgages, and sales contracts. [2] Includes farm ownership loans, soil and water loans to individuals, rural and labor housing loans, association loans for grazing, Indian tribe land acquisition loans, and one-half of economic emergency loans. [3] Compiled by American Council of Life Insurance. [4] Includes all operating commercial, savings, and private banks. [5] Estimated by ERS. [6] Preliminary.
ERS, Farm Sector Performance Branch, (202) 694–5586.

Table 10-4.—Nonreal estate farm debt: Amount outstanding, by lender, United States, Dec. 31, 1994–2003 [1]

Year	Debt owed to reporting institutions (excluding CCC)				Debts owed to individuals and others	Total excluding CCC loans	Price-sup-port loans made or guaranteed by CCC [2]	Total including CCC loans
	All operating banks	Farm Credit System [2]	Farm Service Agency	Total				
	Million dollars	Million dollars	Million dollars	Million dollars	Million dollars	Million dollars	Million dollars	Million dollars
1994	38,663	11,646	6,841	57,150	15,500	72,650	6,237	78,887
1995	39,735	12,992	5,786	58,513	16,500	75,013	2,979	77,992
1996	40,362	14,599	5,243	60,204	17,800	78,004	3,508	81,512
1997	43,908	15,878	4,899	64,685	19,200	83,885	1,982	85,867
1998	45,097	17,314	4,538	66,949	20,000	86,949	5,230	92,179
1999	44,203	16,579	4,557	65,340	20,700	86,040	5,681	91,721
2000	46,919	17,455	4,402	68,776	21,735	90,511	4,253	94,764
2001	46,741	19,962	4,309	71,012	22,100	93,112	5,464	98,576
2002	45,957	20,453	4,118	70,527	22,700	93,227	4,752	97,979
2003 [3]	45,533	21,018	3,982	70,532	23,608	94,140	6,342	100,482

[1] Includes operator households. [2] Although price-support loans of the Commodity Credit Corporation (CCC) are non-recourse loans, they are treated as income in the year received. They are not considered farm debt even though borrowers must either pay them or deliver the commodities on which they are based. [3] Preliminary.
ERS, Farm Sector Performance Branch, (202) 694–5586.

Table 10-5.—Farm Service Agency: Loans made to individuals and associations for farming purposes, and amount outstanding, United States and Territories, 1996–2005 [1]

Year	Loans to individuals						
	Farm ownership			Soil and water			Recreation
	New borrowers	Loans made	Outstanding Jan. 1	New borrowers	Loans made	Outstanding Jan. 1	Outstanding Jan. 1
	Number	*1,000 dollars*	*1,000 dollars*	*Number*	*1,000 dollars*	*1,000 dollars*	*1,000 dollars*
1996	3,630	624,316	6,816,032	0	0	118,484	
1997	3,482	613,877	7,008,911	0	0	98,774	3,874
1998	2,899	508,466	6,831,520	0	0	81,067	2,949
1999	4,308	944,694	6,675,272	0	0	76,042	2,694
2000	4,552	1,106,492	6,755,110	0	0	66,602	2,221
2001	3,704	1,015,634	7,287,728	0	0	52,883	1,784
2002	4,107	1,279,027	7,495,449	0	0	46,284	1,447
2003	4,174	1,399,740	7,749,043	0	0	38,484	1,263
2004	3,625	1,241,454	7,884,284	0	0	31,820	994
2005	4,199	1,298,943	8,190,313	0	0	27,341	875

Year	Loans to individuals					
	Operating			Emergency		
	New borrowers	Loans made	Outstanding Jan. 1	New borrowers	Loans made	Outstanding Jan. 1
	Number	*1,000 dollars*	*1,000 dollars*	*Number*	*1,000 dollars*	*1,000 dollars*
1996	10,377	1,882,431	5,945,331	2,163	176,500	3,046,279
1997	9,065	1,560,559	6,100,452	1,760	144,880	2,423,475
1998	9,433	1,568,071	6,040,488	1,045	97,569	2,039,657
1999	14,525	2,564,767	5,961,862	2,846	329,848	1,940,961
2000	12,979	2,464,802	6,570,523	1,557	150,852	1,915,780
2001	10,732	2,152,814	6,823,828	962	90,026	1,712,807
2002	10,476	2,217,735	6,639,837	501	57,608	1,523,438
2003	10,577	2,121,150	6,728,636	920	95,698	1,405,430
2004	9,157	1,832,093	6,405,468	430	29,789	1,437,464
2005	8,891	1,723,953	6,404,277	235	23,569	1,150,557

Year	Loans to associations			Grazing association	Irrigation, drainage, and soil conservation	Economic opportunity individual loans	Economic emergency loans
	Indian tribe land acquisition						
	New borrowers	Loans made	Outstanding Jan. 1	Outstanding Jan. 1	Outstanding Jan. 1	Outstanding Jan. 1	Outstanding Jan. 1
	Number	*1,000 dollars*	*1,000 dolllars*	*1,000 dollars*	*1,000 dollars*	*1,000 dollars*	*1,000 dollars*
1996	0	641	73,479	28,613	6,657	14	1,082,954
1997	0	224	62,603	23,878	6,229	14	874,601
1998	0	500	59,856	19,654	5,956	12	685,147
1999	0	0	58,461	17,855	5,666	11	653,953
2000	1	673	57,117	15,660	5,449	10	545,423
2001	1	590	62,738	12,785	5,177	8	427,176
2002	1	74	60,777	10,849	3,729	8	364,377
2003	1	110	55,421	8,947	3,330	7	315,601
2004	2	1,586	53,476	6,232	1,623	8	249,603
2005	0	0	55,205	4,883	1,471	8	249,039

[1] Includes loans made directly by FmHA and those guaranteed by the Agency. Amounts of loans made represent obligations and include loans to new borrowers and subsequent loans to borrowers who received an initial loan in a prior year. Amounts outstanding are loan advances less principal repayments for loans made directly by the Agency.

FSA, Loan Making Division, (202) 690–4006.

Table 10-6.—Farmers' marketing, farm supply, and related service cooperatives: Number, memberships, and business, United States, 1994–2003

Year [1]	Cooperatives [2]				Estimated memberships [4]				Estimated service receipts [5]
	Marketing	Farm supply	Related service [3]	Total	Marketing	Farm supply	Related service [3]	Total	
	Number	*Number*	*Number*	*Number*	*1,000 members*	*1,000 members*	*1,000 members*	*1,000 members*	*Million dollars*
1994	2,173	1,496	505	4,174	1,805	1,936	245	3,986	2,986
1995	2,074	1,458	474	4,006	1,712	1,846	210	3,767	3,284
1996	2,012	1,403	469	3,884	1,682	1,795	187	3,664	3,100
1997 [6]	1,941	1,386	464	3,791	1,498	1,706	183	3,387	3,647
1998	1,863	1,347	441	3,651	1,398	1,774	181	3,353	3,473
1999	1,749	1,313	404	3,466	1,283	1,731	159	3,173	3,905
2000	1,672	1,277	397	3,346	1,243	1,718	124	3,085	3,510
2001	1,606	1,234	389	3,229	1,160	1,746	128	3,034	3,471
2002	1,559	1,201	380	3,140	1,049	1,637	107	2,794	3,416
2003 [7]	1,551	1,156	379	3,086	1,054	1,590	113	2,758	4,118

Year [1]	Marketing volume		Farm supply volume		Total marketing and farm supply volume and service receipts	
	Estimated gross business [8]	Estimated net business [9]	Estimated gross business [8]	Estimated net business [9]	Estimated gross business [8]	Estimated net business [9]
	Million dollars	*Million dollars*	*Million dollars*	*Million dollars*	*Million dollars*	*Million dollars*
1994	72,148	65,545	30,405	20,779	105,539	89,309
1995	77,946	69,321	30,965	21,213	112,195	93,818
1996	90,270	79,429	34,728	23,653	128,098	106,182
1997 [6]	85,949	77,843	37,076	25,181	126,673	106,670
1998	84,524	76,642	32,964	24,551	120,961	104,667
1999	80,506	71,982	30,879	23,177	115,291	99,064
2000	80,400	72,065	36,809	24,085	120,719	99,659
2001	83,954	75,042	36,141	24,756	123,566	103,269
2002	76,618	69,656	31,519	23,679	111,553	96,750
2003 [7]	77,242	71,002	35,498	25,499	116,858	100,620

[1] Reports of cooperatives are included for the calendar year.　[2] Includes independent local cooperatives, centralized cooperatives, federations of cooperatives and cooperatives with mixed organizational structures. Cooperatives are classified according to their major activity. If, for example, more than 50 percent of a cooperative's business is derived from marketing activities, it is included as a marketing cooperative.　[3] Includes cooperatives whose major activity is providing services related to marketing and farm supply activities.　[4] Includes members (those entitled to vote for directors) but does not include nonvoting patrons. (Some duplication exists because some farmers belong to more than one cooperative.)　[5] Receipts for services related to marketing and purchasing activities, but not included in the volumes reported for these activities.　[6] Revised.　[7] Preliminary.　[8] Estimated gross business includes all business reported between cooperatives, such as the wholesale business of farm supply cooperatives with other cooperatives or terminal market sales for local cooperatives.　[9] Estimated net business represents the value at the first level at which cooperatives transact business for farmers. Figures are adjusted for duplication resulting from intercooperative business.

Rural Business-Cooperative Service (RBS), Statistics, (202) 690–1415. Based on records from cooperatives reporting to the Service.

Table 10-7.—Farmers' cooperatives: Business volume of marketing, farm supply, and related service cooperatives, United States, 2001 and 2002 (preliminary)

Item	Gross business		Net business [1]	
	2001	2002	2001	2002
	1,000 dollars	*1,000 dollars*	*1,000 dollars*	*1,000 dollars*
Products marketed:				
Beans and peas (dry edible)	101,448	132,155	98,788	129,032
Cotton and cotton products	2,526,938	2,616,669	2,461,250	2,535,634
Dairy products	25,891,132	25,984,368	23,037,708	23,453,909
Fruits and vegetables	8,435,417	7,452,538	7,337,916	6,900,007
Grain and oilseeds excluding				
cottonseeds	20,147,945	23,551,704	17,474,432	20,612,375
Livestock and livestock products	9,901,079	6,795,415	9,901,079	6,795,415
Nuts	947,172	851,233	935,526	851,233
Poultry products	2,467,207	2,666,692	2,402,767	2,652,546
Rice	750,892	968,947	748,361	967,595
Sugar products	2,440,433	3,512,517	2,440,433	3,512,517
Tobacco	226,733	85,451	226,733	85,451
Wool and mohair	7,760	10,504	7,760	10,504
Other [2]	2,773,701	2,613,816	2,583,066	2,495,894
Total farm products	76,617,857	77,242,009	69,655,819	71,002,111
Supplies purchased:				
Crop protectants	3,116,940	3,165,481	2,712,803	2,776,422
Feed	6,685,827	7,220,409	5,373,378	5,873,413
Fertilizer	5,150,680	6,481,527	4,314,685	4,640,048
Petroleum	11,383,867	12,826,671	7,157,086	7,511,780
Seed	1,592,441	1,827,974	1,085,500	1,234,102
Other supplies [3]	3,589,206	3,976,056	3,035,331	3,463,414
Total farm supplies	31,518,961	35,498,119	23,678,783	25,499,179
Receipts for services: [4]				
Trucking, cotton ginning, storage, grinding, locker plants, miscellaneous	3,415,786	4,118,188	3,415,786	4,118,188
Total business	111,552,604	116,858,317	96,750,388	100,619,478

[1] Represents value at the first level at which cooperatives transact business for farmers.   [2] Includes coffee, fish, forest products, hay, hops, seed marketed for growers, nursery stock, other farm products not separately classified, and sales of farm products not received directly from member-patrons. Also includes manufactured food products and resale items marketed by cooperatives.   [3] Includes automotive supplies, building materials, chicks, containers, farm machinery and equipment, hardware, meats and groceries, and other supplies not separately classified.   [4] Charges for services related to marketing or purchasing but not included in the volume reported for those activities, plus other income.

RBS, Statistics, (202) 690–1415. Based on records from cooperatives reporting to the Service.

Table 10-8.—Farmers' cooperatives: Types, numbers, and memberships, United States, 2002

Type	Year or date of data	Associations	Estimated memberships or participants
Marketing, farm supply, and related service: [1]			
Marketing	2002	1,559	1,049,091
Farm supply	2002	1,201	1,637,061
Related services [2]	2002	380	107,398
Service:			
Production credit associations [3]	Sept. 30, 2002	4	NA
Rural credit unions [4]	Dec. 31, 2002	512	3,655
Rural electric cooperatives [5]	Dec. 31, 2002	644	10,976
Production:			
Dairy herd improvement associations [6]	Jan. 01, 2002	NA	27,784

[1] Rural Business - Cooperative Service, U.S. Department of Agriculture.   [2] Includes trucking, storage, grinding, locker plant, and other services.   [3] Farm Credit Administration.   [4] Credit Union National Association, Inc.   [5] Rural Utilities Service, U.S. Department of Agriculture.   [6] Agriculture Research Service, U.S. Department of Agriculture.   NA-not available.

RBS, Statistics, (202) 690–1415.

Table 10-9.—Farmers' cooperatives: Number of cooperatives, memberships, and business volume of marketing, farm supply, and related service cooperatives, by States, 1999 to 2003 (preliminary)

State	Cooperatives headquartered in State		Memberships in State [1]		Net business [1]	
	2002	2003	1999	2001	1999	2001
	Number	Number	Number	Number	1,000 dollars	1,000 dollars
AL	61	61	53,886	42,222	1,169,594	949,788
AZ	9	9	3,376	2,710	927,983	607,896
AR	52	51	56,038	55,431	1,643,198	1,670,856
CA	164	162	53,604	49,553	7,823,548	7,572,686
CO	44	43	32,613	29,122	928,817	1,118,632
FL	39	39	26,050	29,753	2,419,178	2,413,535
GA	17	17	26,673	21,748	2,006,532	1,676,955
HI	20	20	2,768	904	103,807	28,829
ID	37	37	18,583	20,259	1,331,705	1,422,387
IL	169	160	195,258	175,871	5,316,036	5,470,384
IN	44	43	75,409	66,818	1,931,213	2,070,633
IA	157	148	180,168	162,249	7,932,170	9,232,231
KS	129	129	133,269	129,576	4,658,873	4,794,447
KY	42	42	231,862	242,873	773,265	707,833
LA	47	46	12,294	13,032	564,193	735,703
MD	17	16	71,953	74,404	352,277	411,482
MA	12	12	5,551	4,573	627,730	637,056
MI	62	60	29,788	28,660	1,925,074	2,198,121
MN	287	283	186,902	176,594	9,306,888	9,895,732
MS	70	70	105,239	111,043	949,425	1,055,738
MO	63	62	142,771	109,134	4,860,441	4,611,124
MT	65	63	28,720	33,066	685,725	719,154
NE	90	86	92,353	84,834	4,543,391	4,720,598
NJ	15	13	7,347	7,052	312,984	388,148
NM	9	9	2,974	2,705	554,906	507,304
NY	89	87	17,770	37,431	3,235,290	2,825,792
NC	20	19	98,404	96,351	825,510	975,355
ND	232	228	118,435	120,303	3,041,338	3,292,172
OH	70	69	55,970	53,407	2,277,303	2,792,014
OK	80	78	68,611	65,640	1,386,496	1,562,099
OR	32	32	24,638	27,043	1,716,532	1,703,893
PA	52	51	45,008	33,442	1,353,040	1,440,730
SD	122	118	107,046	98,395	2,451,052	2,699,501
TN	79	79	139,124	137,805	729,982	808,937
TX	230	231	118,064	109,926	3,400,868	3,876,461
UT	16	15	10,013	9,354	461,461	512,095
VA	57	58	174,308	193,470	996,687	846,781
WA	77	77	33,903	31,664	3,325,162	3,331,110
WV	26	26	75,083	83,051	88,703	99,281
WI	168	171	205,690	186,490	8,018,466	8,738,431
WY	12	12	5,145	5,485	258,813	241,042
Oth Sts	57	54	....	....	....	....
US	3,140	3,086	3,168,534	3,031,142	98,574,157	102,769,087
Foreign [3]	....	....	4,789	2,765	490,163	499,664
Total	3,140	3,086	3,173,323	3,033,907	99,064,320	103,268,751

[1] Represents value at the first level at which cooperatives transact business for farmers. These statistics for 2000 are presented on a national basis only. Totals may not add due to rounding. [2] Dollar volume or membership is not shown to avoid disclosing operations of individual cooperatives. [3] Sales outside the United States, sales to domestic military installations, and sales of certain products not received directly from member-patrons.

RBS, Statistics, (202) 690–1415.

Table 10-10.—Rural Utilities Service: Long-term electric financing approved by purpose, by States as of December 31, 2004

State	Borrowers	RUS loans [1]	Non-RUS financing — With RUS guarantee	Non-RUS financing — Without RUS guarantee [2]	Financing approved by purpose — Distribution	Financing approved by purpose — Generation and transmission [2]	Financing approved by purpose — Consumer facilities	Loan estimates — Miles of line	Loan estimates — Consumers
	Number	1,000 dollars	1,000 dollars	1,000 dollars	1,000 dollars	1,000 dollars	1,000 dollars	Number	Number
AL	27	848,572	1,058,168	280,177	928,114	1,257,445	0	65,916	0
AK	17	827,599	361,812	133,524	592,633	729,213	1,359	10,824	598,735
AZ	15	386,579	523,509	136,416	440,935	605,078	1,089	21,320	192,022
AR	20	1,043,691	1,231,594	527,914	1,268,822	1,529,905	490	75,229	200,900
CA	10	87,554	16,685	7,216	91,002	20,396	4,471	6,729	551,574
CO	25	1,252,554	2,692,757	683,600	1,206,661	3,422,176	56	72,995	78,790
CT	0	0	0	0	0	0	0	0	0
DE	1	68,630	8,000	24,066	99,832	861	74	6,246	448,568
FL	18	1,142,599	1,374,557	645,814	1,557,159	1,603,293	0	68,757	0
GA	51	2,310,661	6,289,858	1,842,266	3,675,274	6,764,553	3	167,196	76,143
HI	1	215,000	32,960	8,240	256,200	0	2,519	820	872,708
ID	10	196,308	20,134	36,612	224,792	27,066	2,959	12,859	1,822,038
IL	29	624,023	1,255,112	188,935	628,220	1,439,638	0	55,168	0
IN	46	474,044	1,690,518	534,176	607,385	2,090,636	0	57,523	30,500
IA	46	806,724	709,570	155,869	692,904	978,861	1,197	66,623	66,958
KS	30	670,121	705,085	107,019	627,773	854,037	212	72,172	252,924
KY	26	1,510,492	2,967,317	660,360	1,843,663	3,293,348	717	88,059	473,883
LA	20	641,971	2,825,637	445,986	892,274	3,021,142	397	52,858	211,006
ME	4	41,737	5,343	20,396	44,253	23,179	0	2,135	0
MD	2	270,343	18,355	140,356	320,841	108,213	415	13,960	218,485
MA	0	0	0	0	0	0	1,159	0	851,001
MI	10	614,733	827,416	92,692	582,622	951,746	177	37,965	494,540
MN	48	1,476,803	2,222,922	473,716	1,614,761	2,554,322	44	118,343	19,880
MS	29	871,773	1,076,296	283,796	1,020,802	1,210,369	0	86,074	165,040
MO	48	1,716,550	1,226,735	594,454	1,625,571	1,911,346	0	120,763	0
MT	27	405,784	39,326	53,566	420,007	78,439	0	45,987	0
NE	35	485,242	19,036	42,946	430,477	116,185	473	75,765	291,822
NV	8	75,008	1,241	10,441	65,315	21,127	4,357	6,265	653,736
NH	1	100,398	143,839	8,696	103,050	149,850	694	4,616	702,736
NJ	2	18,173	0	5,377	22,250	1,295	821	1,000	711,998
NM	17	519,943	90,899	77,382	625,304	59,910	0	44,388	0
NY	6	42,482	13,972	11,714	65,398	2,686	229	5,164	134,192
NC	34	1,640,859	1,588,006	485,907	1,942,518	1,767,619	562	96,670	177,162
ND	24	1,120,994	2,394,793	873,021	757,995	3,628,827	248	68,775	24,889
OH	27	572,496	514,249	472,666	707,502	851,691	32	47,651	64,601
OK	29	1,126,120	806,922	330,438	1,212,635	1,048,169	4	99,927	13,334
OR	18	296,703	69,128	85,897	328,132	123,360	0	24,535	0
PA	13	387,972	607,333	132,794	503,069	624,791	3,010	27,400	228,318
RI	1	0	3,420	0	340	3,080	85	4	27,653
SC	28	1,556,871	877,613	379,965	2,019,423	792,795	4,635	74,759	967,874
SD	31	734,553	89,388	108,800	738,765	193,095	1,986	66,244	143,577
TN	33	678,460	218,816	190,919	1,068,938	19,033	218	89,883	358,109
TX	100	2,317,867	2,470,628	978,918	2,693,887	3,071,003	0	266,709	0
UT	6	72,455	1,031,811	216,171	62,920	1,257,393	2,676	5,860	510,207
VT	3	75,437	50,726	7,310	57,949	74,620	237	2,956	141,936
VA	18	772,969	374,594	251,378	1,102,926	295,526	239	45,901	219,726
WA	23	242,695	7,234	40,181	270,851	19,007	0	20,945	160
WV	1	18,736	0	1,059	19,643	147	2,231	839	733,299
WI	26	530,587	716,237	158,684	439,280	964,430	0	45,961	0
WY	13	354,493	26,584	27,216	334,426	73,766	881	31,059	157,910
AS	1	0	3,000	0	0	3,000	223	0	1,067,727
MH	1	0	11,857	0		11,857	2,524	161	1,459,440
PA	1	0	35,000	0	3,518	31,482	124	7	25,033
PR	1	300,981	0	31,424	292,851	39,554	903	16,633	24,819
VI	1	430	0	0	234	197	0	85	0
US [3]	1,062	32,547,769	41,345,991	13,006,470	37,132,100	49,720,755	47,374	2,496,683	17,972,480

[1] Includes $628,296,535 discounted principal from 222 prepaid borrowers. [2] Includes loans obtained by RUS borrowers' affiliates specifically organized to facilitate non-RUS financing. [3] Includes figures not shown elsewhere in this table for two borrowers whose loans have been foreclosed. The total amount of these loans was $37,237.

RD, Planning and Policy Branch, (202)692-0341

Table 10-11.—Rural Utilities Service: Composite revenues and patronage capital, average number of consumers and megawatt-hour sales reported by RUS electric borrowers operating distribution systems—calendar years 2002–2004

Item	2002		2003 [1]		2004	
	Amount	Per-cent of total	Amount	Per-cent of total	Amount	Per-cent of total
Number of borrowers reporting	....	617	612	....	603	
Average number of consumers served:						
Residential service (farm & non-farm) ...	9,844,152	89.2	11,451,948	89.3	10,217,575	89.0
Commercial & industrial, small	1,026,816	9.3	1,201,124	9.4	1,095,209	9.5
Commercial & industrial, large	7,266	0.1	10,247	0.1	8,109	0.1
Irrigation	98,474	0.9	100,589	0.8	98,261	0.9
Other electric service	56,048	0.5	63,032	0.5	64,255	0.6
To others for resale	192	*	198	*	193	*
Total	11,032,948	100.0	12,827,138	100.0	11,483,602	100.0
Megawatt-hour sales:						
Residential service (farm & non-farm) ...	135,265,931	57.5	144,981,475	55.5	139,978,298	56.8
Commercial & industrial, small	41,490,388	17.6	49,030,015	18.8	44,035,996	17.9
Commercial & industrial, large	50,526,583	21.5	58,615,356	22.4	54,723,157	22.2
Irrigation	4,359,571	1.9	4,185,057	1.6	3,748,401	1.5
Other electric service	2,045,269	0.9	2,535,465	1.0	2,194,991	0.9
To others for resale	1,630,113	0.7	1,750,079	0.7	1,761,538	0.7
Total	235,317,829	100.0	261,097,471	100.0	246,442,386	100.0
	1,000 dollars		1,000 dollars		1,000 dollars	
Revenue and patronage capital:						
Residential service (farm & non-farm) ...	10,518,422	64.4	11,902,354	60.0	11,487,815	63.4
Commercial & industrial, small	2,989,355	18.3	4,088,462	20.6	3,357,964	18.5
Commercial & industrial, large	2,034,137	12.5	2,899,595	14.6	2,389,849	13.2
Irrigation	291,768	1.8	299,429	1.5	283,079	1.6
Other electric service	158,095	1.0	253,897	1.3	187,682	1.0
To others for resale	69,969	0.4	81,678	0.4	78,542	0.4
Total from sales of electric energy	16,061,746	98.3	19,525,416	98.4	17,784,931	98.1
Other operating revenue	270,571	1.7	324,770	1.6	339,916	1.9
Total operating revenue	16,332,317	100.0	19,850,186	100.0	18,124,847	100.0

[1] Includes data for the Puerto Rico Power Authority (PR-003) accounting for 1.4 milllion consumers served, 20 million mWh sold, and $2,6 billion in total operating revenue. *Less than 0.05 percent.
Rural Development, Planning and Policy Branch, (202) 692-0341

Table 10-12.—Rural Utilities Service: Annual revenues and expenses reported by electric borrowers, United States, 1995–2004

Year	Operating revenue	Operating expense	Interest expense	Depreciation and amortization expense	Net margins	Total utility plant
	1,000 dollars	1,000 dollars	1,000 dollars	1,000 dollars	1,000 dollars	1,000 dollars
1995	24,609,188	21,741,162	2,171,170	1,779,568	1,236,050	61,867,838
1996	24,438,558	19,777,504	2,054,058	1,788,164	1,328,237	61,443,968
1997	23,321,068	18,987,613	1,908,451	1,726,982	1,018,702	60,770,878
1998	23,987,773	19,491,238	1,890,050	1,732,056	1,274,324	61,720,967
1999	23,823,791	19,536,422	1,832,553	1,746,681	1,112,665	62,684,354
2000	25,628,917	21,161,991	1,905,043	1,819,616	1,164,076	66,353,227
2001	26,458,243	21,867,226	1,909,833	1,895,495	1,219,287	69,630,602
2002	27,458,144	22,568,763	1,867,431	1,992,415	1,382,964	72,481,696
2003 [1]	31,821,409	26,393,728	2,153,155	2,314,817	1,303,585	84,991,618
2004	30,649,839	25,646,411	1,919,835	2,181,541	1,340,627	79,508,979

[1] Revised.
RD, Planning and Policy Branch, (202) 692–0341.

Table 10-13.—Loans to farmers' cooperative organizations: Outstanding amounts held by the banks for cooperatives, and agricultural credit banks classified by type of loan, United States, Jan. 1, 1996–2005 [1]

Year	Operating capital loans	Facility loans
	1,000 dollars	1,000 dollars
1996	9,119,835	9,128,364
1997	7,332,313	10,657,055
1998	6,358,665	11,684,548
1999	6,227,194	10,911,590
2000	6,196,401	11,602,316
2001	7,293,142	11,348,179
2002	7,660,584	11,311,516
2003	8,907,313	12,317,966
2004	12,373,082	12,400,364
2005	11,549,929	12,406,423

[1] Includes Puerto Rico.
FCA, Office of Regulatory Policy, (703) 883-4073

CHAPTER XI

STABILIZATION AND PRICE-SUPPORT PROGRAMS

The statistics in this chapter relate to activities of the Commodity Credit Corporation (CCC), cropland diversion and production adjustment programs, and marketing agreement and order programs for fruits and vegetables. Statistics for Federal Milk Marketing Order programs are contained in chapter VIII.

Table 11-1.—Commodity Credit Corporation: Price-supported commodities owned as of Dec. 31, 1995–2004 [1] (Inventory quantity)

Year	Barley	Butter and butter oil	Cheese	Corn	Cotton upland	Sorghum grain	Nonfat dry milk	Oils and oilseeds
	Million bushels	*Million pounds*	*Million pounds*	*Million bushels*	*1,000 bales*	*Million bushels*	*Million pounds*	*Million cwt.*
1995	5	[3]	0	42	[7]	1	25	[2]
1996	[3]	0	0	30	[7]	0	[3]	0
1997	[3]	0	[3]	2	[2]	[2]	30	[2]
1998	1	0	0	15	[7]	1	111	0
1999	1	0	0	26	[7]	1	161	[3]
2000	[3]	0	0	36	[2]	[3]	602	[2]
2001	[3]	0	5	24	[2]	[2]	844	[2]
2002	[3]	0	4	18	[3]	1	1,201	0
2003	0	7	17	16	[2]	0	1,456	0
2004	0	[8]	7	12	[2]	0	605	[2]

Year	Oats	Rice [4]	Rye	Soybeans	Honey	Wheat	Value of all commodities owned [5]
	Million bushels	*Million cwt.*	*Million bushels*	*Million bushels*	*Million pounds*	*Million bushels*	*Million dollars*
1995	0	[2]	0	[2]	1	141	654
1996	0	0	0	0	0	96	435
1997	0	0	0	0	0	93	364
1998	[2]	0	0	3	0	107	363
1999	[3]	[3]	0	7	0	104	347
2000	[2]	[2]	0	10	0	109	790
2001	[2]	[3]	0	4	0	118	844
2002	0	[3]	0	3	0	93	656
2003	[2]	[3]	0	[3]	0	78	219
2004	[2]	[3]	0	[2]	0	81	116

[1] Commodities which were owned by CCC in some years but not shown in this table are as follows: blended foods, cottonseed and products, naval stores, wheat products, corn products, oat products, rice products, vegetable oil, mixed feed, linseed oil, evaporated milk, meat, tallow, egg mix-dry, foundation seeds, peanut products, and sugar.   [2] Less than 50,000 units.   [3] Less than 500,000 units.   [4] Rough basis; includes milled rice in rough equivalent and rice products.   [5] The total value of all commodities owned by CCC, including price-supported commodities not shown and commodities acquired under programs other than price-support programs, less reserve for losses on inventory.   [6] Includes extra long staple, cotton.   [7] Includes infant formula.   [8] Less than 500 units.

FSA, Financial Management Division, (703) 305–1277.

Table 11-2.—Commodity Credit Corporation: Loans made, United States and Territories, by crop years, 2000–2003 [1]

Commodity	Unit	2000		2001	
		Quantity pledged	Face amount	Quantity pledged	Face amount
			1,000 dollars		*1,000 dollars*
Barley	1,000 bushels	16,024	25,648	10,589	17,249
Corn	1,000 bushels	1,393,947	2,562,172	1,394,561	2,557,874
Cotton [3]	1,000 bales	8,959	2,216,075	14,111	3,597,980
Seed cotton	1,000 pounds ..	2,519	1,350	41,034	1,036
Sugar Cane and Beet	1,000 pounds ..	3,472,531	735,558	3,400,115	721,535
Flaxseed	1,000 Cwt	197	1,820	60	562
Honey	1,000 pounds ..	53,972	35,068	0	0
Oats	1,000 bushels ..	1,696	1,892	1,731	2,032
Peanuts	1,000 Pounds	477,285	47,173	961,568	226,012
Rice	1,000 cwt	97,430	625,715	128,019	835,285
Sorghum grain	1,000 bushels	15,303	25,781	17,172	28,891
Soybeans	1,000 bushels	312,916	1,627,897	311,706	1,625,176
Tobacco	1,000 pounds ..	51,801	109,246	41,071	80,676
Wheat	1,000 bushels	181,133	468,418	196,698	507,727
Sunflower Seed	1,000 cwt	2,385	21,836	1,539	14,078
Canola Seed	1,000 Cwt	748	7,144	511	4,859
Safflower Seed	1,000 Cwt	36	251	4	32
Mustard Seed	1,000 Cwt	23	208	14	128
Sunflower Seed (non-oil)	1,000 Cwt	294	2,736	285	2,653
Crambe Oilseed	1,000 Cwt	236	2,042	97	836
Mohair	1,000 Pounds	4,128	7,465	0	0

Commodity	Unit	2002		2003 [2]	
		Quantity pledged	Face amount	Quantity pledged	Face amount
			1,000 dollars		*1,000 dollars*
Barley	1,000 bushels	10,388	19,446	17,870	31,962
Corn	1,000 bushels	1,366,513	2,622,823	1,326,884	2,555,183
Cotton [3]	1,000 bales	13,101	3,301,244	10,607	2,798,423
Seed cotton	1,000 pounds ..	51,912	1,853	2,409	1,238
Sugar Cane and Beet	1,000 pounds ..	4,199,700	872,511	4,246,647	870,135
Flaxseed	1,000 Cwt	88	606	155,030	1,476
Honey	1,000 pounds ..	5,478	3,287	8,300	5,188
Oats	1,000 bushels ..	1,987	2,596	5,182	6,721
Peanuts	1,000 Pounds	1,336,908	247,794	3,313	594,244
Rice	1,000 cwt	132,817	870,748	91,150	595,452
Wool	1,000 cwt	35	19	24	10
Sorghum grain	1,000 bushels	6,670	13,281	6,191	12,223
Soybeans	1,000 bushels	384,326	1,901,506	156,561	779,259
Tobacco	1,000 pounds ..	97,064	178,672	41,350	235,876
Wheat	1,000 bushels	119,849	342,694	185,528	531,432
Sunflower Seed	1,000 cwt	1,701	15,408	1,430	13,656
Canola Seed	1,000 Cwt	1,306	12,544	333	3,253
Safflower Seed	1,000 Cwt	2	21	22	203
Mustard Seed	1,000 Cwt	10	98	28	268
Sunflower Seed (non-oil)	1,000 Cwt	550	6,740	287	2,765
Crambe Oilseed	1,000 Cwt	3	28	1	8
Mohair	1,000 Pounds	50	208	48	203
Chickpeas	1,000 Cwt	7	53	2	18
Dry Whole Peas	1,000 Cwt	174	1,100	9,485	608
Lentil Dry	1,000 Cwt	3,001	357	17,299	2,064

[1] Includes loans made directly by Commodity Credit Corporation.   [2] Loans through Sept. 30, 2003.   [3] Includes extra long staple cotton and upland cotton.

FSA, Financial Management Division, (703) 305–1277.

Table 11-3.—Commodity Credit Corporation: Loan transactions for fiscal year 2004, by commodities[1]

Commodity	Unit	Loans outstanding Oct. 1, 2003[2]	New loans made	Repayments	Collateral acquired in settlement	Loans written off[3]	Loans outstanding Sept. 30, 2004 Value[2]	Loans outstanding Sept. 30, 2004 Quantity collateral remaining pledged
		1,000 dollars	*1,000 dollars*	*1,000 dollars*	*1,000 dollars*	*1,000 dollars*	*1,000 dollars*	*1,000 units*
Basic commodities:								
Corn	Bushel	254,774	2,580,619	2,622,360	2,283	31,875	178,875	90,137
Cotton	Bale	65,565	2,755,954	2,675,608	3,475	27,879	114,557	439
Seed cotton	Pound	0	1,238	1,238	0	0	0	0
Peanuts	Pound	3,472	608,247	583,216	720	17,348	10,435	58,771
Rice	Cwt	102,201	724,853	462,180	53	76,422	288,399	43,462
Tobacco[4]	Pound	718,655	273,084	254,203	0	0	737,536	254,884
Wheat	Bushel	377,147	448,176	502,003	10,212	2,218	310,890	111,188
Total		1,521,814	7,392,171	7,100,808	16,743	155,742	1,640,692	XXXXX
Designated nonbasic commodities:								
Barley	Bushel	19,207	25,290	30,469	94	655	13,279	7,167
Sorghum	Bushel	2,687	12,551	11,247	234	215	3,542	1,737
Honey	Pound	1,380	7,124	4,389	0	47	4,068	6,434
Oats	Bushel	4,569	4,912	5,815	419	93	3,154	2,466
Sugar, beet	Pound	1,001	689,548	649,493	7,616	0	33,440	248,100
Sugar, cane	Pound	24,444	181,501	160,278	0	0	45,667	257,572
Sunflower seed.	Cwt	2,358	13,617	15,228	4	129	614	64
Flaxseed	Cwt	396	1,222	1,501	1	10	106	11
Canola seed	Cwt	3,683	1,319	4,137	9	145	711	76
Safflower seed	Cwt	7	196	188	0	0	15	2
Rapeseed	Cwt	0	0	0	0	0	0	0
Mustard seed	Cwt	35	274	268	0	0	41	4
Crambe Oilseed.	Cwt	14	0	8	0	6	0	0
Sunflower seed, non oil.	Cwt	1,150	2,765	3,809	33	0	73	7
Total		60,931	940,319	886,830	8,410	1,300	104,710	XXXXX
Other nonbasic commodities:								
Soybeans	Bushel	58,834	812,376	814,877	136	2,941	53,256	10,610
Mohair	Pound	193	299	157	0	123	212	51
Chickpeas	Pound	54	0	54	0	0	0	0
Lentils	Pound	1,307	3,300	2,078	7	0	2,522	21,732
Dry Whole Peas.	Pound	876	1,103	1,493	25	124	337	6
Wool	Pound	9	20	10	17	0	2	5
Total		61,273	817,098	818,669	185	3,188	56,329	XXXXX
Other loans:								
Farm Storage facility[5].		160,076	61,644	34,138	61	65	187,456	0
Bollweevil		9,975	0	25	0	0	9,950	0
Total		170,051	61,644	34,163	61	65	197,406	XXXXX
Grand total[5].		1,814,069	9,211,232	8,840,470	25,399	160,294	1,999,137	XXXXX

[1] Loans made directly by Commodity Credit Corporation. [2] Book value of outstanding loans; includes face amounts and any charges paid. [3] Includes transfers to accounts receivable. [4] Charge offs represents pre-No Net Cost Tobacco loans - 1981 and prior crop loans. [5] Table may not add due to rounding.

FSA, Financial Management Division, (703) 305–1277.

Table 11-4.—Commodity Credit Corporation: Selected inventory transactions, programs and commodity, fiscal year 2004

Program and commodity	Unit	Quantity				
		Inventory Oct. 1, 2003	Purchases	Collateral acquired from loans	Sales [1]	Inventory Sept. 30, 2004
		Thousands	*Thousands*	*Thousands*	*Thousands*	*Thousands*
Feed grains:.						
Barley	Bushel	0	377	52	412	17
Corn	Bushel	15,675	17,665	1,183	22,743	11,780
Corn products	Pound	7,564	387,551	0	395,116	0
Grain sorghum	Bushel	54	10,478	115	10,622	25
Sorghum grits	Pound	0	8,680	0	8,680	0
Oats	Bushel	0	0	334	294	40
Oats, rolled	Pound	0	0	0	0	0
Rye	Bushel	0	0	0	0	0
Total feed grains		xxx	xxx	xxx	xxx	xxx
Wheat (A)	Bushel	80,755	53,771	2,402	55,935	80,993
Wheat flour	Pound	10,990	323,136	0	334,126	0
Wheat products, other	Pound	4,502	390,198	0	391,116	3,585
Rice, milled	Cwt	251	3,076	0	3,076	251
Rice, rough	Cwt	18	48,199	8	48,202	23
Rice, brown	Pound	0	13,103	0	13,103	0
Rice, cereal	Pound	0	0	0	0	0
Cotton, extra long staple	Bale	1	0	2	1	2
Upland Cotton	Bale	97	2,209	11	2,317	0
Tobacco Products	Pound	96,620	0	0	94,655	1,965
Dairy products:						
Butter	Pound	10,482	(42)	0	10,302	138
Butter oil	Pound	0	0	0	0	0
Cheese	Pound	18,421	44,406	0	54,522	8,306
Milk, dried	Pound	0	0	0	0	0
Milk, UHT	Pound	0	0	0	0	0
Dry Whole Milk	Pound	0	0	0	0	0
Non fat dry milk	Pound	1,440,189	358,717	0	1,137,758	661,148
Total dairy products		xxx	xxx	xxx	xxx	xxx
Oils and oilseeds:						
Crambe oilseed	Cwt.	0	0	1	1	0
Canola seed	Cwt.	0	0	1	1	0
Mustardseed	Cwt.	0	0	0	0	0
Sunflower seed	Cwt.	0	10	0	10	0
Sunflower seed, non-oil	Cwt.	0	0	3	3	0
Sunflower seed oil, processed	Cwt.	0	22,465	0	22,465	0
Peanuts, farmers' stock	Pound	1,533	390	4,161	6,084	0
Peanut products	Pound	0	0	0	0	0
Peanut butter	Pound	0	0	0	0	0
Soybeans	Bushel	702	864	27	1,591	2
Soybean products	Pound	0	135,877	0	135,877	0
Flaxseed	Cwt.	0	0	0	0	0
Totals oils and oilseeds		xxx	xxx	xxx	xxx	xxx
Blended foods	Pound	63,059	453,898	0	482,149	34,807
Grains and seeds:						
Feed for Government facilities	Cwt	0	6	0	6	0
Foundation seeds	Pound	0	0	0	0	0
Total grains and seeds		xxx	xxx	xxx	xxx	xxx
Peas, dry whole	Pound	20,304	261,281	363	254,643	27,305
Honey	Pound	0	0	0	0	0
Denatured alcohol	Gallon	0	0	0	0	0
Sugar, cane and beet	Pound	0	0	32,000	0	32,000
Vegetable oil products	Pound	17,891	450,493	0	460,855	7,530
Potatoes	Pound	0	441	0	441	0
Veg Dehyd Vegetable Soup	Pound	0	96,143	0	96,143	0
Plants & Seeds	Pound	0	2,205	0	2,205	0
Tallow	Pound	0	56,214	0	56,214	0
Other (B)		254	1,653	0	1,679	228
Total inventory operations		xxx	xxx	xxx	xxx	xxx
Additional Adjustment for lag activity		0	0	0	0	0

See footnotes at end of table.

Table 11-4.—Commodity Credit Corporation: Selected inventory transactions, programs and commodity, fiscal year 2004—Continued

Program and commodity	Unit	Value				
		Inventory Oct. 1, 2003	Purchases	Collateral acquired from loans	Sales [1]	Inventory Sept. 30, 2004
		1,000 dollars	1,000 dollars	1,000 dollars	1,000 dollars	1,000 dollars
Feed grains:.						
Barley	Bushel	0	727	95	790	31
Corn	Bushel	28,654	49,769	2,353	59,116	21,660
Corn products	Pound	910	43,703	0	44,612	0
Grain sorghum	Bushel	109	33,796	244	34,100	50
Sorghum grits	Pound	0	1,205	0	1,205	0
Oats	Bushel	0	0	433	381	52
Oats, rolled	Pound	0	0	0	0	0
Rye	Bushel	0	0	0	0	0
Total feed grains		29,673	129,200	3,124	140,204	21,793
Wheat (A)	Bushel	290,022	229,939	10,231	239,111	291,081
Wheat flour	Pound	1,257	38,222	0	39,479	0
Wheat products, other	Pound	451	41,128	0	41,224	355
Rice, milled	Cwt	2,902	47,637	0	47,636	2,903
Rice, rough	Cwt	114	316,838	33	316,842	144
Rice, brown	Pound	0	2,272	0	2,272	0
Rice, cereal	Pound	0	0	0	0	0
Cotton, extra long staple	Bale	364	0	735	427	672
Upland Cotton	Bale	26,712	569,357	3,035	599,097	7
Tobacco products	Pound	278,051	0	0	275,556	2,495
Dairy products:						
Butter	Pound	11,038	(44)	0	10,830	164
Butter oil	Pound	0	0	0	0	0
Cheese	Pound	19,694	37,932	0	46,861	10,765
Milk, dried	Pound	0	0	0	0	0
Milk, UHT	Pound	0	0	0	0	0
Dry Whole Milk	Pound	0	0	0	0	0
Non fat dry milk	Pound	1,294,475	287,800	0	987,661	594,615
Total dairy products		1,325,207	325,688	0	1,045,352	605,543
Oils and oilseeds:						
Crambe Oilseed	Cwt.	0	0	7	7	0
Canola seed	Cwt.	0	0	10	10	0
Mustard seed	Cwt.	0	0	0	0	0
Sunflower seed	Cwt.	0	68	4	72	0
Sunflower seed, non-oil	Cwt.	0	0	33	33	0
Sunflower seed oil, processed	Cwt.	0	10,337	0	10,337	0
Peanuts, farmers' stock	Pound	280	71	720	1,071	0
Peanut products	Pound	0	0	0	0	0
Peanut butter	Pound	0	0	0	0	0
Soybeans	Bushel	3,606	6,102	133	9,829	11
Soybean meal	Pound	0	14,167	0	14,167	0
Flaxseed	Cwt.	0	0	1	1	0
Totals oils and oilseeds		3,886	30,744	907	35,526	11
Blended foods	Pound	8,360	69,965	0	73,877	4,448
Grains and seeds						
Feed for Government facilities	Cwt	0	83	0	83	0
Foundation seeds	Pound	0	0	0	0	0
Total grains and seeds		0	83	0	83	0
Peas, dry whole	Pound	3,333	47,300	21	46,631	4,023
Honey	Pound	0	0	0	0	0
Denatured alcohol	Gallon	0	0	0	0	0
Sugar, cane and beet	Pound	0	0	7,616	0	7,616
Vegetable oil products	Pound	7,718	190,884	0	195,290	3,312
Potatoes	Pound	0	165	0	165	0
Veg dehyd vegetable soup	Pound	0	138,425	0	138,425	0
Plants & Seeds	Pound	0	333	0	333	0
Tallow	Pound	0	12,784	0	12,784	0
Other (B)		6,105	35,356	0	35,662	5,799
Total inventory operations		1,984,156	2,226,320	25,701	3,285,973	950,204
Additional Adjustment for lag activity		1,984,156	2,226,320	25,701	3,285,973	950,204

[1] Includes sales, commodity donations, transfers to other government agencies and inventory adjustment.
(A) Excludes wheat set aside for Food Security Wheat Reserve (FSWR).
(B) Includes beans, dry edible, and fish, canned salmon.
Table may not add due to rounding.
FSA, Financial Management Division, (703) 305–1277.

Table 11-5.—Commodity Credit Corporation: Cost value of export and domestic commodity dispositions, by type of disposition, fiscal year 2004 [1]

(In Thousands)

Commodity		Domestic			
	Dollar sales	Transfers to other Government agencies	Donations [1]	Inventory adjustments and other recoveries (domestic)	Total domestic
	1,000 dollars	*1,000 dollars*	*1,000 dollars*	*1,000 dollars*	*1,000 dollars*
Feed grains:					
Barley	790,078	0	0	0	790,078
Corn	20,030,153	394,874	0	5,329	20,430,355
Corn products	0	0	0	103,097	103,097
Grain sorghum	1,375,704	0	0	1,366	1,377,070
Sorghum grits	0	0	0	5,536	5,536
Oats	380,974	0	0	0	380,974
Tobacco Products	3,290,527	0	0	272,265,130	275,555,657
Wheat	3,883,885	0	0	33,468	3,917,353
Wheat flour	588,266	0	0	152,020	740,287
Wheat product, Other	0	0	0	89,890	89,890
Rice, milled	0	0	0	0	0
Rice, rough	316,841,515	0	0	0	316,841,515
Rice,brown and Textured	0	0	0	0	0
Cotton, extra long staple	599,523,227	0	0	0	599,523,227
Veg dehyd vegetable soup	0	0	0	0	0
Dairy products:					
Butter oil	0	0	0	0	0
Butter	9,534,373	1,208,505	0	87,213	10,830,092
Cheese	0	3,756,008	39,236,749	4,890,239	47,882,995
Nonfat dry milk	343,509,161	0	253,053,136	32,064,194	628,626,492
Milk, dried	0	0	0	0	0
Oils and oilseeds:.					
Peanut products	0	0	0	0	0
Peanut butter	0	0	0	0	0
Peanuts, farmer's stock	1,070,884	0	0	0	1,070,884
Soya flour	0	0	0	0	0
Flaxseed	757	0	0	0	757
Sunflower Seed (oil & non-oil)	104,934	0	0	0	104,934
Soybeans	5,832,189	0	0	3,345	5,835,534
Fruit fresh apples	0	0	0	0	0
Blended foods	0	0	0	81,278	81,278
Potatoes	0	0	0	0	0
Grains and seeds:.					
Feed for Government facilities	0	82,588	0	0	82,588
Foundation seeds	0	0	0	0	0
Field Seeds	0	0	0	0	0
Vegetable Seeds	0	0	0	0	0
Canola seed	9,562	0	0	0	9,565
Crambe oilseed	6,668	0	0	0	6,668
Peas, dried whole	1,410,486	0	0	82,800	1,493,286
Dry edible beans	381,756	0	0	116,774	498,529
Honey	0	0	0	0	0
Sugar	0	0	0	0	0
Vegetable oil products	0	0	0	252,340	252,340
Meat (and products)	0	0	0	0	0
Pudding	0	0	138,425,265	0	138,425,265
Veg. canned tomato sauce	0	0	0	0	0
Raisins	0	0	0	0	0
Other	0	0	0	32,307	32,307
Total [2]	1,308,565,100	5,441,974	430,715,149	310,266,326	2,054,988,549

See footnotes at end of table.

Table 11-5.—Commodity Credit Corporation: Cost value of export and domestic commodity dispositions, by type of disposition, fiscal year 2004 [1]—Continued

(In Thousands)

Commodity	Export			Total export and domestic
	Public law 480 Title II/III	Donations [1]	Total export	
	1,000 dollars	*1,000 dollars*	*1,000 dollars*	*1,000 dollars*
Feed Grains:.				
Barley	0	0	0	790,078
Corn	37,864,662	789,707	38,654,368	59,084,723
Corn products	44,517,730	(8,494)	44,509,236	44,612,333
Grain sorghum	32,722,792	0	32,722,792	34,099,862
Sorghum grits	1,199,893	0	1,199,893	1,205,429
Oats	0	0	0	380,974
Oats, rolled	0	0	0	275,555,657
Wheat	216,515,008	8,401,841	224,916,849	228,834,202
Wheat flour	34,518,586	4,219,931	38,738,517	39,478,804
Wheat products, other	40,918,794	215,227	41,134,021	41,223,911
Rice, milled	0	0	0	0
Rice, rough	0	(15,793)	(15,793)	316,825,721
Rice, brown and textured soy	2,271,903	0	2,271,903	2,271,903
Cotton, extra long staple and upland	0	0	0	599,523,227
Veg dehyd vegetable soup	0	0	0	0
Dairy products:				
Butter oil	0	0	0	0
Butter	0	0	0	10,830,092
Cheese	0	0	0	47,882,995
Nonfat dry milk	148,923	135,034,332	135,183,255	763,809,747
Milk, dried	0	0	0	0
Oils and oilseeds:.				
Peanut products	0	0	0	0
Peanut butter	0	0	0	0
Peanuts, farmer's stock	0	0	0	1,070,884
Soya flour	0	0	0	0
Flaxseed	0	0	0	757
Sunflower Seed (oil & non-oil)	0	10,336,978	10,336,978	10,441,912
Soybeans	0	17,904,250	17,904,250	23,739,785
Fruit fresh apples	0	0	0	0
Blended foods	73,462,997	332,285	73,795,282	73,876,560
Potatoes	0	165,100	165,100	165,100
Grains and seeds:.				
Feed for Government facilities	0	0	0	98,817
Foundation seeds	0	0	0	0
Field Seeds	333,023	0	333,023	333,023
Vegetable Seeds	0	0	0	0
Canola seed	0	0	0	0
Crambe oilseed	0	0	0	0
Peas, dried whole	45,725,045	(587,695)	45,137,350	46,630,636
Dry edible beans	34,190,527	973,132	35,163,659	35,662,188
Honey	0	0	0	0
Sugar	0	0	0	0
Vegetable oil products	179,600,354	15,437,317	195,037,671	195,290,011
Meat (and products)	11,903,785	879,905	12,783,690	12,783,690
Pudding	0	0	0	138,425,265
Veg canned tomato sauce	0	0	0	0
Raisins	0	183,566	183,566	183,566
Other	37,665,204	9,938,268	47,603,472	47,635,778
Total [2]	793,559,227	204,199,856	997,759,083	3,052,747,632

[1] Includes donations under section 202,407,416, Section 210, P.L. 85-540 and miscellaneous donations under various other authorizations. [2] Totals may not add due to rounding.

FSA, Financial Management Division, (703) 305–1277.

Table 11-6.—Commodity Credit Corporation: Investment in price-support operations, by quarters, 1995–2004 [1]

Date	Inventory after revaluation	Loans after revaluation	Total investment
	Million dollars	Million dollars	Million dollars
1995:			
March	774.2	7,556.6	8,330.8
June	705.1	4,902.1	5,607.2
1996:			
March	649.6	3,710.1	4,359.7
June	574.6	2,172.4	2,747.0
1997:			
March	436.0	3,753.0	4,189.0
June	406.0	2,154.0	2,560.0
1998:			
March	386.1	5,058.9	5,445.0
June	436.1	3,376.7	3,812.8
1999:			
March	384.3	6,231.0	6,615.3
June	480.7	5,022.0	5,502.7
2000:			
March	500.8	5,160.5	5,661.3
June	650.7	4,545.3	5,196.0
2001:			
March	1,635.1	5,627.9	7,263.0
June	2,299.0	3,663.3	5,962.3
2002:			
March	875.6	5,323.8	6,199.4
June	920.1	2,723.7	3,643.8
2003:			
March	540.7	5,429.1	5,969.8
June	586.7	3,280.8	3,867.5
2004:			
March	134.6	4,971.6	5,106.2
June	91.5	2,839.8	2,931.3

[1] Reflects total CCC loans and inventories.
FSA, Financial Management Division, (703) 305–1277.

Table 11-7.—Farm Service Agency programs: Payments to producers, by program and commodity, United States, calendar years 2001–2005

Program and commodity	2001	2002	2003	2004	2005
	1,000 dollars	1,000 dollars	1,000 dollars	1,000 dollars	1,000 dollars
Production flexibility	4,040,639	3,499,648	(281,388)	(3,884)	(941)
Agricultural Management Assist	1,376	2,984	2,864	1,185	760
Quality Losses	52,478	96,956	81	76	1
Supl Oilseed Payment Program	422,392	209	(1)	(1)	(1)
WAMLAP III - Apportioned	16,442	343		5	
Peanut Marketing Asst Pgm III	53,911	14		0	0
Supplemental Tobacco Loss	128,259	71		0	0
Wamlap II - Apportioned	18,671	24	9	0	0
AMLAP - Apportioned	95,079			0	0
Bioenergy	5,294	33,104	150,861	146,519	60,245
Citrus Losses in California	2,154			0	0
Poultry Enteritis Sydnrome	1,788			0	0
Acreage Grazing Payments	6,245	481	4,756	353	155
Nursery Losses - Florida	6,581	760		0	0
AILFP -- Apportioned	6,020	480		0	(1)
Crop Disaster Program	1,841,870		2,332,277	236,337	2,440,097
Cottonseed Payment Program	81,385		49,835	15	0
Sugar PIK Diversion	44,193	44,288		0	0
Wool and mohair	2,838			0	0
Peanut Marketing Assistance	24,904	17		0	0
Oilseed Program	498,413	26	(13)	6	(2
Loan deficiency [1]	5,703,964	1,295,668	576,428	25,008	5,041,035
Emergency feed/livestock assistance	427,071			0	0
Lamb Meat Adjustment assistance	11,207	32,394	16,153	72	14,254
Tobacco loss/disaster assistance	1,018	4,920	(1)	0	0
Conservation reserve [2]	1,769,997			0	0
Pasture Recovery Program	26,365	786	(15)	(3)	0
Agricultural conservation	1,393	378	(22)	(4)	(2
Emergency conservation	31,183	38,869	32,067	24,015	65,478
Environ. quality incentives program	97,079	58,458	331	(324)	547
American Indian livestock assist. pro	885			0	0
Options pilot program	(172)			0	0
Potato diversion program	11,327			0	0
Market gains	707,909	458,230	197,943	130,398	365,586
Noninsured assistance program (NAP) [3]	55,571	224,623	205,897	142,261	85,803
Karnal bunt fungus	2,727	3,613	2,983	0	0
Marketing Loss Assistance	4,644,030			(712)	(331)
Dairy market loss assistance	123,668	152		0	6,824
Other [4]	24,687			0	0

See footnotes at end of table.

Table 11-7.—Farm Service Agency programs: Payments to producers, by program and commodity, United States, calendar years 2001–2005—Continued

Program and commodity	2001	2002	2003	2004	2005
	1,000 dollars	1,000 dollars	1,000 dollars	1,000 dollars	1,000 dollars
Direct and counter cyclical prog		570,512	9,002,616	6,502,679	9,268,407
Crp annual rental		1,552,851	1,587,169	1,632,047	1,652,588
Milk income loss contract		336,903	877,229	204,108	7,993
Livestock compensation program		836,063	272,521	(468)	(18)
Peanut quota buyout program		982,927	237,640	24,727	22,287
Crp incentives		116,055	99,642	77,803	78,716
Market access program		98,727	98,906	128,568	0
Auto crp - cost shares		46,048	98,690	119,693	92,187
NRCS environ qlty incentive		13,626	92,981	183,041	0
Apple market loss assistance		74,325	92,334	29	0
Auto environ qlty incentive pg		46,442	88,828	41,141	0
Livestock emergency assistance		(49)	72,419	(27)	(114)
Sugar cane payment program			51,721	0	(2)
Tobacco payment program			51,122	39	0
Sugar beet disaster program			48,302	832	16
ELS special provision program			42,421	64,547	177,526
Avian influenza indemnity prog		31,420	22,014	0	0
Milk inc loss contr transitional		523,698	21,657	4,992	1,846
Wetlands reserve		17,893	19,626	15,887	8,423
Hard white winter wheat			3,535	6,317	4,074
Soil/water conservation assist		4,721	2,315	1,511	1,090
NM tebuthiuron application			1,290	136	0
Interest payments		1,630	1,128	908	1,540
Emerging markets program		3,243	754	0	0
Crop disaster program		28,854	735		3,144
Grants for catfish producers			613	0	3
Dairy indemnity		90	494	517	377
Auto lta-conservation long term		329	226	115	101
Auto ag cons pg envirn long term		251	165	83	160
Dairy options pilot program		1,123	75		0
Dairy market loss assistance			32		0
2000 Florida nursery losses			29	17,325	
Wool & mohair market loss asst		(9)	18	0	0
Apple & potato quality loss		35,554	15	0	0
Cattle feed program		133,414	8	0	0
Grasslands reserve program			7	1,667	3,856
Nap-supplemental appropriations			6	0	0
Livestock indemnity program		302	2	(60)	(55)
Additional interest		38	1	3	0
Auto ana-conservation annual		1	1	0	0
Milk marketing fee			(2)	0	0
Finality rule		(2)	(3)	51	0
Ldp, non-contract pfc growers		17	(4)	(4)	0
Rice deficiency			(8)		
Tri valley growers program		(162)	(62)	(5)	(5)
Wheat deficiency		(117)	(89)		
National wool act			(100)		0
Crop loss disaster assistance			(120)	(86)	(95)
Cotton deficiency		(125)	(131)	(40)	0
Payment limitation refund		(334)	(174)	(70)	(101)
Feed grain deficiency		(229)	(179)	5,225	(2)
Disaster		(15)	(244)	(16)	(25)
Crp cost-shares		92,430	(1,067)	(7)	(836)
Marketing loss assistance		1,452	(1,111)		
Market loss onion producer program		10,000		0	0
Idaho oust program		4,889		0	0
Disaster reserve assistance		2,873		0	0
Crop loss disaster assistance		63			(89)
Pasture flood compensation		1		0	0
Klamath Basin water program		(4)		0	0
Small hog operation program		(5)		0	0
National wool act		(16)		0	
Rice/Wheat deficiency				2,835,237	0
Trade Adjustment Assistance Program				11,525	15,004
American Indian livestock-Feed					6,622
Crop Disaster - North Carolina					1,387
Crop Disaster - Virginia					7,142
Crop Disaster Program					(89)
Crop Hurricane Damage Program					5,542
Florida Hurricane Citrus Disaster					227,530
Florida Nursery Disaster					32,984
Florida Vegetable Disaster					10,786
Grassroots Source Water program					3,192
Livestock assistance program					260,079
Tree Assistance Program					4,706
Grand Total	20,990,842	11,365,194	16,177,044	12,581,287	19,977,473

[1] Includes Crop Special Grade Rice LDP, Rice Deficiency, Cotton Deficiency, Feed Grain Deficiency, Wheat Deficiency, Loan Deficiency, and LDP - Non-Contract. [2] Includes CRP Cost-Shares, CRP Incentives, and CRP Annual Rent. [3] Includes Nap-Supplemental and Noninsured Assistance Program. [4] Includes Wetlands Reserve Program, Soil and Water Conservation Program, Settlement Payments, Dairy Indemnity Program, Disaster and Disaster Reserve Refunds, Dairy Disaster Refunds, Small Hog Operation Program, Crop Loss Disaster Payments, Interest Payments, LIP Contract Growers, Flood Compensation, National Wool Act, Finanlity Rule, and Payment limitation refund.

FSA Budget/Corporate Programs Branch, (202) 720–5148.

Table 11-8.—Farm Service Agency programs: Payments received, by States, 2003–2005

State	Payments		
	2003	2004	2005
AL	219,214	128,668	201,718
AK	1,830	5,434	3,773
AZ	135,261	82,256	116,230
AR	819,994	404,890	441,349
CA	645,272	381,353	443,509
CO	316,893	216,185	354,637
CT	7,237	4,312	7,534
DE	17,096	13,067	21,837
FL	109,824	206,157	392,588
GA	549,155	278,131	475,947
HI	1,294	1,706	3,454
ID	151,620	150,504	176,191
IL	854,099	1,154,266	1,732,742
IN	438,053	521,365	860,064
IA	1,045,632	1,251,809	2,217,602
KS	807,415	640,189	1,049,611
KY	145,219	140,215	211,983
LA	422,076	230,532	294,438
ME	11,494	9,485	16,777
MD	66,299	48,307	71,779
MA	11,439	4,099	7,878
MI	251,608	208,631	363,951
MN	781,677	694,197	1,336,534
MS	470,694	297,698	383,202
MO	506,049	426,638	638,428
MT	353,350	276,013	353,820
NE	722,620	720,919	1,367,729
NV	11,953	6,379	8,425
NH	4,762	2,619	2,394
NJ	12,041	8,371	15,005
NM	92,390	76,908	99,156
NY	160,276	79,775	133,235
NC	357,543	176,422	323,217
ND	651,484	464,508	810,306
OH	395,322	326,313	570,082
OK	355,332	209,142	294,644
OR	106,595	73,414	86,460
PA	182,426	87,143	128,844
RI	611	877	548
SC	126,461	63,907	115,759
SD	547,920	395,774	780,050
TN	175,199	124,594	194,272
TX	1,661,141	998,199	1,607,027
UT	55,479	34,473	41,070
VT	28,479	14,991	17,054
VA	175,585	63,744	111,070
WA	263,950	192,665	219,860
WV	12,962	6,206	8,193
WI	475,696	291,465	549,187
WY	51,042	33,867	63,413
CM	(491)	(118)	(167)
KCCO	296,233	215,671	243,313
PR	12,777	7,453	5,191
VI	143	67	50
GU	268	771	705
MI	26	67	197
AS	....	26	414
Undistributed ..	101,027	128,568	3,193
Total [1]	16,177,044	12,581,287	19,977,473

[1] Total may not add due to rounding.
FSA, Budget, Corporate Programs Branch, (202) 720–5148.

AGRICULTURAL STATISTICS 2006   XI–11

Table 11-9.—Commodity Credit Corporation: Loans made in fiscal year 2004 for crop year 2003, by States and Territories [1]

State or Territory	Barley	Corn	Cotton	Flaxseed	Honey	Oats
	1,000 dollars	1,000 dollars	1,000 dollars	1,000 dollars	1,000 dollars	1,000 dollars
Alabama	0	1,071	129,497	0	173	7
Alaska	0	0	0	0	0	0
Arizona	0	2,331	307	0	51	0
Arkansas	0	6,416	47,142	0	10	0
California	126	3,455	475,852	0	848	23
Colorado	363	18,119	0	0	0	64
Connecticut	0	145	0	0	0	0
Delaware	152	2,171	0	0	0	0
Florida	0	0	607	0	528	0
Georgia	0	6,496	27,318	0	0	25
Hawaii	0	0	0	0	0	0
Idaho	6,253	104	0	0	490	85
Illinois	0	249,333	0	0	0	37
Indiana	3	193,428	0	0	0	14
Iowa	5	672,256	0	0	332	148
Kansas	32	28,054	0	0	194	35
Kentucky	53	25,802	0	0	0	0
Louisiana	0	3,975	54,423	0	147	0
Maine	164	0	0	0	0	316
Maryland	92	6,768	0	0	0	0
Massachusetts	0	146	0	0	0	0
Michigan	32	57,448	0	0	3	130
Minnesota	2,827	490,479	0	10	342	595
Mississippi	0	8,478	996,921	0	69	7
Missouri	5	51,679	36,930	0	0	0
Montana	2,569	0	0	84	1,279	30
Nebraska	27	324,866	0	0	571	191
Nevada	0	0	0	0	4	0
New Hampshire	0	0	0	0	0	0
New Jersey	0	1,214	0	0	0	14
New Mexico	0	1,912	1,617	0	0	0
New York	66	19,804	0	0	24	191
North Carolina	129	10,866	155,683	0	0	56
North Dakota	9,077	31,256	0	1,094	627	786
Ohio	0	80,061	0	0	0	32
Oklahoma	0	1,335	2,071	0	7	4
Oregon	330	0	0	0	177	38
Pennsylvania	27	6,918	0	0	22	45
Rhode Island	0	0	0	0	0	0
South Carolina	0	8,158	1,287	0	0	28
South Dakota	344	114,666	0	33	447	1,643
Tennessee	0	11,237	256,092	0	3	0
Texas	0	46,465	569,934	0	282	103
Utah	12	44	0	0	47	4
Vermont	0	0	0	0	0	0
Virginia	369	7,893	268	0	0	0
Washington	2,178	1,566	0	0	380	149
West Virginia	0	581	0	0	0	0
Wisconsin	13	82,864	0	0	65	103
Wyoming	42	758	0	0	0	8
Adjustments	0	0	0	0	0	0
Peanut Associations	0	0	0	0	0	0
Total [2]	25,290	2,580,618	2,755,949	1,221	7,122	4,911

See footnotes at end of table.

STABILIZATION AND PRICE-SUPPORT PROGRAMS

Table 11-9.—Commodity Credit Corporation: Loans made in fiscal year 2004 for crop year 2003, by States and Territories [1]—Continued

State or Territory	Oilseeds	Peanuts	Rice	Seed cottton	Sorghum	Soybeans
	1,000 dollars	*1,000 dollars*	*1,000 dollars*	*1,000 dollars*	*1,000 dollars*	*1,000 dollars*
Alabama	0	63,273	0	0	26	298
Alaska	0	0	0	0	0	0
Arizona	0	0	0	0	0	0
Arkansas	0	0	492,405	0	39	8,982
California	111	0	157,529	0	68	0
Colorado	649	0	0	0	227	52
Connecticut	0	0	0	0	0	0
Delaware	0	0	0	0	0	613
Florida	0	46,673	0	0	0	0
Georgia	0	291,632	0	0	107	475
Hawaii	0	0	0	0	0	0
Idaho	65	0	0	0	0	0
Illinois	0	0	0	0	296	94,898
Indiana	0	0	0	0	243	100,349
Iowa	12	0	0	0	0	231,335
Kansas	288	0	0	1,001	2,801	7,741
Kentucky	0	0	0	0	43	11,385
Louisiana	0	274	17,115	0	52	339
Maine	0	0	0	0	0	0
Maryland	0	0	0	0	0	2,482
Massachusetts	0	0	0	0	0	0
Michigan	47	0	0	0	0	17,993
Minnesota	1,674	0	0	0	0	109,069
Mississippi	0	4,092	30,158	0	10	3,467
Missouri	0	0	5,978	0	1,368	34,060
Montana	229	0	0	0	0	0
Nebraska	345	0	0	0	1,753	44,628
Nevada	0	0	0	0	0	0
New Hampshire	0	0	0	0	0	0
New Jersey	0	0	0	0	0	459
New Mexico	0	5,026	0	0	13	0
New York	0	0	0	0	0	4,770
North Carolina	0	37,836	0	0	5	4,979
North Dakota	8,731	0	0	0	0	9,775
Ohio	5	0	0	0	0	59,989
Oklahoma	0	6,892	0	33	303	576
Oregon	0	0	0	0	0	0
Pennsylvania	0	0	0	0	37	3,501
Rhode Island	0	0	0	0	0	0
South Carolina	0	8,369	0	0	8	2,385
South Dakota	5,918	0	0	0	665	38,316
Tennessee	0	0	229	0	0	4,031
Texas	46	119,026	21,437	116	4,485	688
Utah	4	0	0	0	0	0
Vermont	0	0	0	0	0	0
Virginia	0	25,155	0	88	1	2,314
Washington	17	0	0	0	0	0
West Virginia	0	0	0	0	0	383
Wisconsin	9	0	0	0	0	12,043
Wyoming	21	0	0	0	0	0
Adjustments	0	0	0	0	0	0
Peanut Associations	0	0	0	0	0	0
Total [2]	18,171	608,248	724,851	1,238	12,550	812,375

See footnotes at end of table.

Table 11-9.—Commodity Credit Corporation: Loans made in fiscal year 2004 for crop year 2003, by States and Territories[1]—Continued

State or Territory	Sugar	Tobacco	Wheat	Mohair	Dry edible peas	Wool	Total
	1,000 dollars	1,000 dollars	1,000 dollars	1,000 dollars	1,000 dollars	1,000 dollars	1,000 dollars
Alabama	0	0	85	0	0	0	194,430
Alaska	0	0	0	0	0	0	0
Arizona	0	0	955	0	0	0	3,644
Arkansas	0	0	190	0	0	0	555,184
California	0	0	10,470	0	0	0	648,482
Colorado	113,994	0	8,214	0	0	0	141,682
Connecticut	0	0	0	0	0	0	145
Delaware	0	0	15	0	0	0	2,951
Florida	128,766	0	86	0	0	0	176,660
Georgia	0	0	326	0	0	0	326,379
Hawaii	0	0	0	0	0	0	0
Idaho	0	0	23,886	0	305	0	31,188
Illinois	0	0	4,079	0	5	0	348,648
Indiana	0	0	1,156	0	0	0	295,193
Iowa	0	0	33	0	0	0	904,121
Kansas	0	0	51,151	0	0	0	91,297
Kentucky	0	78,429	1,877	0	0	0	117,589
Louisiana	52,735	0	0	0	0	0	129,060
Maine	0	0	0	0	0	0	480
Maryland	0	0	791	0	0	0	10,133
Massachusetts	0	0	0	0	0	0	146
Michigan	152,951	0	1,997	0	0	0	230,601
Minnesota	96,904	0	40,586	0	0	0	742,486
Mississippi	0	0	112	0	0	0	1,043,314
Missouri	0	0	3,433	0	0	0	133,453
Montana	0	0	52,561	0	1,760	1	58,513
Nebraska	0	0	10,588	0	38	0	383,007
Nevada	0	0	0	0	0	0	4
New Hampshire ...	0	0	0	0	0	0	0
New Jersey	0	0	17	0	0	0	1,704
New Mexico	0	0	864	0	0	0	9,432
New York	0	0	3,037	0	0	0	27,892
North Carolina	0	179,310	856	0	0	0	389,720
North Dakota	15,498	0	71,388	0	1,024	0	149,256
Ohio	0	0	1,707	0	0	0	141,794
Oklahoma	0	0	39,756	0	0	0	50,977
Oregon	0	0	12,244	0	181	19	12,989
Pennsylvania	0	0	436	0	0	0	10,986
Rhode Island	0	0	0	0	0	0	0
South Carolina	0	0	1,275	0	0	0	21,510
South Dakota	0	0	46,240	0	28	0	208,300
Tennessee	0	15,215	547	0	0	0	287,354
Texas	0	0	10,042	299	0	0	772,923
Utah	301,996	0	1,750	0	0	0	303,857
Vermont	0	0	0	0	0	0	0
Virginia	0	101	891	0	0	0	37,080
Washington	0	0	43,214	0	1,061	0	48,565
West Virginia	0	0	92	0	0	0	1,056
Wisconsin	0	29	736	0	0	0	95,862
Wyoming	8,206	0	491	0	0	0	9,526
Adjustments	0	0	0	0	0	0	0
Peanut Associa- tions	0	0	0	0	0	0	0
Total[2]	871,050	273,084	448,174	299	4,402	20	9,149,573

[1] Loans made directly by Commodity Credit Corporation. As far as possible, loans have been distributed according to the location of producers receiving the loans. Direct loans to cooperative associations for the benefit of members have been distributed according to the location of the association.　[2] Table may not add due to rounding.

FSA, Financial Management Division, (703) 305–1277.

Table 11-10.—Fruit, vegetable, and tree nut marketing agreement and order and peanut programs, 2004–2005

Program	Estimated number of commercial producers	Farm value
	Number	*1,000 dollars*
Citrus fruits (2004-05 season):		
Florida oranges, grapefruit, tangerines, and tangelos	8,500	300,033
Texas oranges and grapefruit	212	83,416
Deciduous fruits (2004 season):		
California fresh pears and peaches [1]	1,200	106,067
California nectarines	1,000	86,218
California olives	900	59,379
California desert grapes	50	132,650
California kiwifruit	275	19,977
Florida avocados	272	14,448
Washington apricots	272	6,260
Washington sweet cherries	1,953	219,780
Washington and Oregon winter pears	1,715	138,094
Tart cherries (7 States) [2]	900	69,591
Washington and Oregon Bartlett pears	1,850	41,371
Washington and Oregon fresh prunes [3]	215	3,696
Cranberries (10 States) [4]	1,200	214,096
Dried fruits (2004 season):		
California dates	125	41,000
California dried prunes	1,100	72,000
California raisins	4,500	327,303
Vegetables (2004-05 season):		
Florida tomatoes	75	622,636
Idaho and Eastern Oregon onions	233	91,048
South Texas melons	13	35,582
South Texas onions	110	87,575
Georgia onions (Vidalia)	103	50,030
Walla Walla onions	32	6,769
Potatoes (2004-05 season):		
Colorado	238	102,353
Idaho and eastern Oregon	987	216,856
Southeastern States (Virginia - North Carol	63	4,290
Washington	272	73,537
Nuts (2004 season):.		
California almonds	6,000	2,200,055
California Pistashios	740	444,160
California walnuts	5,500	438,750
Oregon and Washington Hazelnuts	703	52,992
Peanuts [5]	14,445	834,380
Spearmint oil (2004 season) [6]	150	14,793
(Total 34 programs) [7]		7,211,185

[1] Fresh value of non-Bartlett pears is not available.   [2] The tart cherry order covers the States of Michigan, New York, Pennsylvania, Oregon, Utah, Washington, and Wisconsin.   [3] Farm value is available only for fresh and processed combined.   [4] Massachusetts, Rhode Island, Connecticut, New Jersey, Wisconsin, Michigan, Minnesota, Oregon, Washington, and Long Island in New York. (Only top 5 are reported).   [5] The Farm Security and Rural Investment Act of 2002 terminated the Peanut Administrative committee (which locally administered marketing agreement No. 146). As a result, the agreement was terminated and new quality standards for all domestic and imported peanuts were established.   [6] The marketing order regulates the handling of spearmint oil produced in the States of Washington, Idaho, Montana, Nevada, Utah, Oregon, and California. The farm value is the sum of values for Idaho, Oregon, and Washington, the only significant producing States in the marketing order area.   [7] Total number of producers cannot be determined from totals for individual commodities; some producers produce more than one commodity.

AMS, Fruit and Vegetable Programs, (202) 720–2615.

CHAPTER XII
AGRICULTURAL CONSERVATION AND FORESTRY STATISTICS

Statistics in this chapter concern conservation of various natural resources, particularly soil, water, timber, wetlands, wildlife, and improvement of water quality. Forestry statistics include area of private and public-owned forest land, timber production, imports and exports, pulpwood consumption and paper and board production, area burned over by forest fires, livestock grazing, and recreational use of national forest lands.

Conservation Practices on Active CRP Contracts

Practice code	Practice	Acres
CP1	Introduced grasses and legumes	3,446,540
CP2	Native grasses	6,852,364
CP3	Tree planting	1,138,457
CP4	Wildlife habitat with woody vegetation	2,469,769
CP5	Field windbreaks	79,153
CP6	Diversions	834
CP7	Erosion control structures	540
CP8	Grass waterways	116,685
CP9	Shallow water areas for wildlife	51,059
CP10 [1]	Existing grasses and legumes	15,266,152
CP11	Existing trees	1,126,050
CP12	Wildlife food plots	81,610
CP13	Vegetative filter strips	12,991
CP15	Contour grass strips	80,328
CP16	Shelterbelts	31,233
CP17	Living snow fences	4,544
CP18	Salinity reducing vegetation	300,826
CP19	Alley cropping	52
CP20	Alternative perennials	13
CP21	Filter strips (grass)	1,001,262
CP22	Riparian buffers (trees)	752,121
CP23	Wetland restoration	1,748,508
CP24	Cross wind trap strips	724
CP25	Rare and declining habitat	983,188
CP26	Sediment retention	6
CP27	Farmable wetland (wetland)	40,717
CP28	Farmable wetland (upland)	98,815
CP29	Wildlife habitat buffer (marginal pasture)	21,677
CP30	Wetland buffer (marginal pasture)	14,765
CP31	Bottomland hardwood	21,442
CP32	Hardwood trees	7,287
CP33	Upland bird habitat buffers	73,396
....	Total	35,823,109

[1] Includes both introduced grasses and legumes and native grasses.
FSA, Conservation and Environmental Protection Division, (202) 720–0048.

CRP enrollment: By sign up and initial contract year [1], as of January 2006

Sign up	Before 1998	1998	1999	2000	2001	2002	2003	2004	2005	2006	Total
1-13	511,825	458,547										970,372
13		16,090,928	353,925								16,444,854
14 ..		1,760,379	4,050,553								5,810,932
15 ..		111,862	102,739								214,601
16 ..				4,706,678							4,706,678
17 ..			133,605	130,225							263,830
18 ..					2,225,649						2,225,649
19 ..				105,072	12,799						117,870
20 ..				33,186	169,483						202,669
21 ..					218,596	243,074					461,671
22 ..						288,981	152,677				441,658
23 ..							203,379	54,974			258,353
24 ..								1,652,305	164,412		1,816,718
25 ..							11,934	174,285			186,219
26 ..								155,458	101,513		256,971
27 ..										1,051,895	1,051,895
28 ..									203,202	160,520	363,721
30 ..										28,975	28,975
All	511,825	18,421,716	4,640,823	4,975,161	2,626,527	532,056	367,991	2,037,022	469,127	1,241,390	35,823,637

[1] For CRP, contract year is the same as fiscal year, which begins October 1.   Note: General Signup Numbers: 1-13, 15, 16, 18, 20, 26, 29. Continuous Signup Numbers: 14, 17, 19, 21-25, 27-28 and 30-31.
FSA, Conservation and Environmental Protection Division, (202) 720–0048.

Table 12-1.—Conservation Reserve Program (CRP): Enrollment by practice, under
contract, January 2006

(CP 1 and CP 2)

State	CP 1 Establishment of permanent introduced grasses and legumes			CP 2 Establishment of permanent native grasses		
	Total acres treated	Total cost share	Cost share per acre treated [1]	Total acres treated	Total cost share	Cost share per acre treated [1]
Alabama	4,752.9	203,672	60.57	3,824.8	250,884	82.99
Alaska	5,746.4	438,254	76.27	0.0	0	....
Arizona	0.0	0	....	0.0	0	....
Arkansas	3,189.4	183,159	61.95	3,445.5	310,721	96.55
California	5,391.3	328,384	70.37	1,645.9	330,182	222.96
Colorado	46,582.0	1,704,913	43.25	611,206.4	35,792,611	62.20
Connecticut	70.3	7,091	178.61	34.3	3,630	105.83
Delaware	53.1	4,419	83.22	23.3	1,967	84.41
Florida	147.4	19,864	144.57	150.5	335	67.00
Georgia	503.8	32,239	66.35	389.5	33,334	85.60
Hawaii	*	*	*	*	*	*
Idaho	95,252.4	2,736,963	34.01	23,178.1	1,439,072	73.03
Illinois	183,712.9	7,008,457	48.63	43,953.0	3,030,310	86.85
Indiana	39,125.7	2,045,873	64.69	29,963.1	1,980,886	77.41
Iowa	270,784.7	8,240,589	49.52	152,353.6	9,880,008	77.92
Kansas	17,274.1	344,657	42.09	835,741.4	28,315,618	45.79
Kentucky	88,644.8	6,228,773	74.26	43,647.7	4,117,216	100.18
Louisiana	51.8	1,899	47.00	3,242.9	275,170	86.36
Maine	1,685.5	204,963	134.92	112.4	13,681	123.92
Maryland	12,173.9	1,113,814	142.94	3,694.2	747,994	240.82
Massachusetts	0.0	0	....	0.0	0	....
Michigan	32,457.5	1,605,888	60.40	24,376.5	2,051,548	99.80
Minnesota	243,858.8	10,915,299	47.44	134,679.9	9,677,881	76.20
Mississippi	4,362.5	216,373	53.24	527.6	37,075	84.82
Missouri	360,947.7	14,299,466	48.09	174,602.1	11,132,378	67.73
Montana	716,153.0	15,525,650	22.70	852,583.9	29,408,855	36.20
Nebraska	38,943.2	845,421	26.38	397,101.5	16,514,705	49.87
Nevada	*	*	*	*	*	*
New Hampshire	10.0	1,200	120.00	0.0	0	....
New Jersey	1,132.9	186,097	164.27	362.6	69,681	192.17
New Mexico	2,333.9	71,084	45.15	180,641.4	7,598,205	43.60
New York	6,693.8	682,773	115.38	899.9	102,074	136.37
North Carolina	2,129.6	142,317	79.88	1,671.6	127,255	89.79
North Dakota	409,904.5	7,971,856	20.54	69,076.0	3,922,906	61.78
Ohio	25,630.2	1,131,532	53.32	38,778.3	3,063,179	88.61
Oklahoma	17,431.5	569,804	38.93	403,114.6	19,706,188	49.58
Oregon	114,456.1	3,566,922	35.74	75,427.4	4,417,300	63.21
Pennsylvania	96,052.1	13,477,772	143.11	31,470.3	4,999,045	163.76
Puerto Rico	108.0	17,550	162.50	0.0	0	....
Rhode Island	*	*	*	*	*	*
South Carolina	327.9	19,912	80.26	118.8	9,804	108.09
South Dakota	166,239.8	4,996,656	30.52	233,691.6	10,607,818	47.06
Tennessee	28,999.1	1,719,865	62.63	44,365.9	3,448,653	80.35
Texas	104,187.8	3,122,832	33.32	1,698,747.8	80,962,868	51.94
Utah	60,491.7	2,035,702	36.77	15,370.8	772,153	53.05
Vermont	0.0	0	....	0.0	0	....
Virginia	3,953.9	221,083	57.52	2,738.4	208,129	83.97
Washington	129,746.2	6,593,775	57.12	650,062.2	57,968,242	91.70
West Virginia	10.2	500	49.02	22.4	1,053	47.01
Wisconsin	45,121.3	2,373,320	60.93	57,456.3	4,621,354	85.34
Wyoming	59,697.7	1,955,206	33.94	7,869.7	257,282	34.08
United States, total	3,446,523	125,113,840	70.27	6,852,364	358,209,251	88.24

[1] Not including acres which receive no cost share.   * Data withheld to avoid disclosure of individual operations.   Note:
Total acres treated may not add due to rounding.

FSA, Conservation and Environmental Protection Division, (202) 720–0048.

Table 12-2.—Conservation Reserve Program (CRP): Enrollment by practice, under
contract, January 2006

(CP 3 and CP 4)

State	CP 3 Tree planting			CP 4 Permanent wildlife habitat		
	Total acres treated	Total cost share	Cost share per acre treated [1]	Total acres treated	Total cost share	Cost share per acre treated [1]
Alabama	141,933.3	46,497,856	362.10	10,802.4	51,440	41.13
Alaska	0.0	0	....	11.3	0	....
Arizona	0.0	0	....	0.0	0	....
Arkansas	41,705.0	39,352,207	1045.70	3,434.3	144,058	56.81
California	69.0	2,640	293.32	765.9	12,907	207.51
Colorado	134.6	121,998	911.11	319,826.6	32,527,673	110.42
Connecticut	0.0	0	....	0.0	0	....
Delaware	3,124.5	1,082,412	352.58	2,041.8	468,311	245.56
Florida	23,139.9	6,955,730	333.52	3,388.1	87,680	57.88
Georgia	160,387.9	71,425,326	452.57	6,562.7	477,503	159.92
Hawaii	*	*	*	*	*	*
Idaho	4,818.9	601,628	133.95	136,297.4	4,316,490	32.92
Illinois	53,301.4	32,809,812	702.65	129,423.4	18,301,300	216.09
Indiana	20,501.7	12,998,987	664.54	15,280.5	1,506,862	125.05
Iowa	16,553.1	3,379,927	218.76	321,024.0	14,028,227	78.08
Kansas	688.2	79,330	123.64	16,439.2	1,055,348	82.31
Kentucky	6,494.5	780,799	126.86	775.5	61,925	113.87
Louisiana	141,013.4	11,026,915	79.62	16,740.2	2,053,648	129.22
Maine	247.2	29,470	131.21	917.8	35,462	100.18
Maryland	1,256.0	257,758	257.55	2,146.5	388,855	246.41
Massachusetts	0.0	0	....	0.0	0	....
Michigan	9,328.6	2,536,804	290.13	26,311.1	2,627,264	138.01
Minnesota	36,120.0	4,620,869	131.45	346,905.8	26,165,467	79.71
Mississippi	279,948.4	46,134,440	174.33	8,285.2	260,262	85.55
Missouri	21,281.4	3,746,361	199.21	6,825.4	4,024,498	768.30
Montana	208.3	50,003	245.96	33,706.2	1,280,786	44.97
Nebraska	1,743.4	357,092	325.55	49,011.4	5,945,730	154.98
Nevada	*	*	*	*	*	*
New Hampshire	0.0	0	....	0.0	0	....
New Jersey	115.8	27,226	235.11	22.7	5,199	229.03
New Mexico	80.0	2,120	26.50	0.0	0	....
New York	1,571.6	311,067	213.78	582.6	60,604	131.75
North Carolina	19,429.4	1,689,691	113.43	3,026.4	1,450,161	574.14
North Dakota	416.2	72,233	184.17	569,736.2	14,608,754	30.06
Ohio	8,738.3	2,602,943	312.77	46,087.2	89,660,346	3074.49
Oklahoma	715.4	56,898	79.53	3,282.4	188,372	61.34
Oregon	2,141.0	184,110	142.52	12,650.0	723,960	66.97
Pennsylvania	1,443.8	1,170,435	839.80	3,390.1	1,396,808	443.09
Puerto Rico	51.0	7,701	151.00	0.0	0	....
Rhode Island	*	*	*	*	*	*
South Carolina	51,208.0	7,392,142	154.09	9,177.9	181,453	63.89
South Dakota	667.7	164,726	281.01	95,864.7	4,985,849	60.94
Tennessee	16,786.8	6,573,502	427.13	9,195.8	1,458,621	177.72
Texas	2,846.0	121,804	51.56	38,894.9	5,227,846	153.51
Utah	0.0	0	....	774.4	2,851	3.68
Vermont	0.0	0	....	0.0	0	....
Virginia	5,612.3	431,449	85.25	762.5	69,004	99.42
Washington	1,311.3	324,783	249.68	181,774.1	21,184,934	119.52
West Virginia	126.7	9,050	71.43	0.0	0	....
Wisconsin	61,185.5	64,264,187	1089.20	10,758.5	1,478,325	178.67
Wyoming	11.8	14,755	1261.11	26,714.8	493,733	39.24
United States, total	1,138,457	370,269,188	329.89	2,469,769	259,002,616	220.24

[1] Not including acres which receive no cost share.   * Data withheld to avoid disclosure of individual operations.   Note:
Total acres treated may not add due to rounding.

FSA, Conservation and Environmental Protection Division, (202) 720–0048.

Table 12-3.—Conservation Reserve Program (CRP): Enrollment by practice, under contract, January 2006
(CP 5, CP 6 and CP 7)

State	CP 5 Establishment of field windbreaks			CP 6 Diversions I			CP 7 Erosion control structures		
	Total acres reated	Total cost share	Cost share per acre treated[1]	Total acres treated	Total cost share	Cost share per acre treated[1]	Total acres treated	Total cost share	Cost share per acre treated[1]
AL	0.0	0		0.0	0		0	0	
AK	0.0	0		0.0	0		0	0	
AZ	0.0	0		0.0	0		0	0	
AR	0.0	0		0.0	0		2	729	364.50
CA	0.0	0		0.0	0		0	0	
CO	1,356.6	1,216,358	908.41	0.0	0		226.2	2,731	149.23
CT	0.0	0		0.0	0		0	0	
DE	0.0	0		0.0	0		0	0	
FL	0.0	0		0.0	0		0	0	
GA	0.0	0	*	0.0	0		7.7	0	
HI	*	*	*	*	*	*	*	*	*
ID	524.8	1,527,880	3,027.90	0.0	0		4	4,500	1,125.00
IL	2,535.1	608,643	247.84	16.0	6,950	1,878.38	14.1	34,991	3,464.46
IN	2,197.1	433,900	199.98	0.0	0		4.7	13,251	3,011.59
IA	6,146.5	1,622,948	279.79	10.0	1,500	150.00	12.7	15,270	2,279.10
KS	1,621.2	690,814	498.53	17.1	6,539	406.15	47.2	2,956	537.45
KY	7.7	2,071	268.96	0.0	0		4.7	8,472	1,802.55
LA	0.0	0		5.0	476	95.20	2	200	100.00
ME	0.0	0		0.0	0		0	0	
MD	0.4	1,375	3,437.50	7.0	276	2,760.00	0	0	
MA	0.0	0		0.0	0		0	0	
MI	2,151.8	609,555	300.96	3.0	6,250	2,083.33	9	23,500	2,611.11
MN	8,932.3	2,757,093	322.54	0.0	0		0.3	1,000	3,333.33
MS	0.0	0		0.8	1,750	2,187.50	1.1	2,175	21,750.00
MO	115.5	27,589	268.90	557.9	55,397	99.30	167	84,417	831.69
MT	389.1	156,147	433.26	0.0	0		0	0	
NE	27,881.6	13,652,493	661.56	0.0	0		9.9	0	
NV	*	*	*	*	*	*	*	*	*
NH	0.0	0		0.0	0		0.3	700	2,333.33
NJ	8.1	12,305	1,519.14	4.5	750	166.67	0	0	
NM	0.0	0		0.0	0		0	0	
NY	16.1	13,991	869.01	0.0	0		1	3,500	3,500.00
NC	21.7	2,353	108.43	0.0	0		0	0	
ND	4,428.0	2,055,961	475.74	0.6	143	238.33	0	0	
OH	2,298.5	771,466	341.95	0.0	0		0	0	
OK	43.1	12,713	391.17	59.3	13,928	240.97	20	1,741	87.05
OR	3.6	525	145.83	0.0	0		0	0	
PA	4.3	430	100.00	6.0	5,210	868.33	0.5	375	750.00
PR	0.0	0		0.0	0		0	0	
RI	*	*	*	*	*	*	*	*	*
SC	79.3	6,938	98.69	0.0	0		0	0	
SD	17,876.1	13,349,163	798.13	0.0	0		0	0	
TN	0.0	0		0.0	0		3	2,558	852.67
TX	43.1	47,898	1,111.32	0.0	0		0	0	
UT	4.4	9,311	2,116.14	0.0	0		0	0	
VT	5.0	1,010	202.00	0.0	0		0	0	
VA	3.0	117	39.00	0.0	0		0	0	
WA	12.7	8,403	840.30	0.0	0		0	0	
WV	0.0	0		0.0	0		0	0	
WI	217.2	52,189	258.49	0.5	600	1,200.00	2.1	14,400	6,857.14
WY	228.8	245,821	1,114.33	146.2	0		0	0	
US	79,153	39,897,460	712.86	834	99,769	951.86	540	217,466	2,933.70

[1] Not including acres which receive no cost share.　* Data withheld to avoid disclosure of individual operations.　Note: Total acres treated may not add due to rounding.

FSA, Conservation and Environmental Protection Division, (202) 720–0048.

Table 12-4.—Conservation Reserve Program (CRP): Enrollment by practice, under contract, January 2006
(CP 8, CP 9 and CP 10)

State	CP 8 Grass waterways			CP 9 Shallow water areas for wildlife			CP 10 Vegetative-cover-grass-already established		
	Total acres treated	Total cost share	Cost share per acre treated[1]	Total acres treated	Total cost share	Cost share per acre treated[1]	Total acres treated	Total cost share	Cost share per acre treated[1]
AL	47.5	11,553	394.30	161.9	123,822	789.18	114,853.7	0	
AK	0.0	0		4.6	56,864	12,361.74	23,454.1	0	
AZ	3.6	2,880	800.00	0.0	0		0.0	0	
AR	22.8	2,672	144.43	978.9	206,352	315.28	27,458.4	5,093	49.54
CA	0.0	0		163.8	101,386	618.96	126,510.4	0	
CO	916.3	277,111	314.11	48.8	11,870	243.24	1,377,788.5	18,278	69.84
CT	0.0	0		0.0	0		130.5	0	
DE	4.1	9,309	2,515.95	423.1	1,138,761	2,723.66	25.0	0	
FL	0.0	0		0.0	0		2,170.5	500	45.05
GA	90.2	57,096	698.85	27.8	29,577	1,063.92	7,164.0	0	
HI	*	*	*	*	*	*	*	*	*
ID	14.6	15,578	1,811.40	78.9	68,343	866.20	526,470.5	280,399	59.65
IL	30,096.5	41,929,305	1,463.72	5,694.2	2,756,492	514.80	253,628.4	1,091,165	300.26
IN	16,301.8	56,982,787	3,658.04	1,628.2	1,061,492	752.40	87,287.7	1,101	36.82
IA	31,977.4	35,129,664	1,214.98	17,196.5	3,855,069	261.07	583,428.3	8,616,219	572.61
KS	8,210.8	2,890,755	398.19	946.3	210,677	277.54	1,737,166.2	2,450,991	1,590.21
KY	3,800.3	4,839,517	1,307.91	3,020.8	1,497,973	525.92	138,738.4	136	30.22
LA	12.2	11,056	1,128.16	673.0	145,457	299.97	18,038.4	638	10.18
ME	26.2	203,352	8,069.52	0.0	0		19,603.3	0	
MD	248.2	778,812	4,062.66	1,377.6	1,738,346	1,435.58	4,027.2	4,940	200.00
MA	1.0	5	5.00	0.2	0		52.8	0	
MI	849.7	2,586,473	3,242.41	2,235.2	1,290,021	634.10	100,159.6	102,800	50.01
MN	4,627.5	5,059,826	1,171.93	969.0	139,787	171.64	295,586.0	257	5.02
MS	61.1	1,711	267.34	870.0	98,574	251.21	130,373.5	768	19.59
MO	1,882.2	1,742,144	1,009.06	2,709.8	712,152	296.62	834,009.5	23,246	58.39
MT	96.7	7,005	83.79	85.3	11,080	129.89	1,525,460.9	2,007	2.86
NE	1,921.3	696,327	391.39	255.5	70,807	291.63	588,232.1	2,545,051	2,564.54
NV	*	*	*	*	*	*	*	*	*
NH	0.0	0		0.0	0		0.0	0	
NJ	26.8	170,261	6,353.02	2.8	7,181	2,564.64	463.4	0	
NM	0.0	0		0.0	0		407,423.5	0	
NY	77.8	148,854	2,691.75	83.5	17,412	248.03	38,642.2	71,194	68.87
NC	181.8	350,272	2,446.03	3,282.1	2,152,385	774.91	18,006.3	7,081	89.52
ND	127.6	55,596	454.59	35.1	730	34.60	1,382,960.9	0	
OH	8,097.5	24,925,216	3,268.54	818.5	615,755	817.30	96,727.0	4,706,521	6,700.63
OK	322.3	84,251	272.04	103.4	16,442	381.48	596,312.5	267,341	119.45
OR	73.0	41,911	684.82	16.6	8,267	498.01	299,290.7	0	
PA	550.2	1,384,927	2,622.97	80.7	128,654	1,822.29	56,245.2	122,534	115.15
PR	0.0	0		0.0	0		316.0	0	
RI	*	*	*	*	*	*	*	*	*
SC	91.3	165,769	1,831.70	2,071.0	3,120,147	1,516.25	11,182.4	0	
SD	1,239.1	569,837	563.14	242.4	221,910	1,068.42	508,330.2	0	
TN	182.3	162,646	949.48	146.8	57,860	456.31	137,644.6	2,109	18.50
TX	2,215.3	1,521,988	745.16	157.9	84,329	551.53	2,139,485.8	2,413,514	710.82
UT	6.3	252	40.00	0.0	0		129,196.8	0	
VT	1.0	1,595	1,595.00	0.0	0		116.2	0	
VA	48.6	51,118	1,073.91	98.7	32,833	406.85	13,914.4	56,916	56.52
WA	457.2	159,720	384.50	54.7	48,473	902.66	391,975.2	89,926	20.64
WV	0.0	0		0.0	0		656.9	0	
WI	1,762.2	3,266,874	1,896.04	4,315.8	9,496,726	2,630.82	331,317.9	34,725	13.79
WY	12.6	5,611	445.32	0.0	0		184,126.5	0	
US	116,685	186,301,636	1,561.78	51,059	31,334,006	1,097.19	15,266,152	22,915,450	502.91

[1] Not including acres which receive no cost share.   * Data withheld to avoid disclosure of individual operations.   Note: Total acres treated may not add due to rounding.

FSA, Conservation and Environmental Protection Division, (202) 720–0048.

Table 12-5.—Conservation Reserve Program (CRP): Enrollment by practice, under contract, January 2006

(CP 11, CP 12 and CP 13)

State	CP 11 Vegetative-cover-trees-already established			CP 12 Wildlife food plots			CP 13 Filter strips		
	Total acres treated	Total cost share	Cost share per acre treated [1]	Total acres treated	Total cost share	Cost share per acre treated [1]	Total acres treated	Total cost share	Cost share per acre treated [1]
AL	179,804.6	346,099	42.17	1,682.6	0	....	108.3	8,154	91.52
AK	0.0	0	....	20.4	0	....	0.0	0	....
AZ	0.0	0	....	0.0	0	....	0.0	0	....
AR	58,293.5	172,878	43.24	555.5	0	....	80.2	3,847	48.51
CA	357.5	0	....	89.0	0	....	0.0	0	....
CO	232.5	0	....	1,038.8	0	....	76.0	10,733	141.22
CT	0.0	0	....	0.0	0	....	0.0	0	....
DE	38.4	0	....	35.6	0	....	0.0	0	....
FL	55,628.8	162,378	43.11	151.3	0	....	2.3	118	51.30
GA	125,700.0	308,881	42.86	1,904.5	0	....	542.5	21,700	40.00
HI	*	*	*	*	*	*	*	*	*
ID	2,890.1	32,457	43.23	1,135.2	0	....	5.9	2,427	411.36
IL	15,955.7	101,492	44.23	6,018.4	0	....	2,447.1	131,188	53.86
IN	8,790.9	76,939	43.11	1,221.5	0	....	779.5	69,712	92.36
IA	8,212.8	98,008	58.82	5,824.2	0	....	347.0	20,190	60.49
KS	1,351.9	1,718	34.63	5,660.4	0	....	739.8	30,800	41.63
KY	2,074.9	8,685	43.23	1,456.9	0	....	456.4	27,731	63.52
LA	43,205.9	136,374	34.74	1,756.6	0	....	11.3	475	42.04
ME	734.9	0	....	1.6	0	....	0.0	0	....
MD	620.9	3,394	43.23	138.0	0	....	341.2	121,828	357.06
MA	0.0	0	....	0.0	0	....	0.0	0	....
MI	6,858.2	40,427	43.22	1,907.9	0	....	274.0	11,076	44.54
MN	20,887.4	90,059	44.51	4,964.4	0	....	3,377.9	139,373	45.24
MS	356,205.7	849,699	45.58	4,884.7	0	....	396.2	16,928	43.38
MO	7,195.7	71,452	44.11	3,724.1	0	....	249.0	16,808	68.30
MT	947.3	575	43.23	3,314.9	0	....	0.0	0	....
NE	3,451.2	11,937	33.20	2,678.8	0	....	213.7	5,461	34.24
NV	*	*	*	*	*	*	*	*	*
NH	0.0	0	....	0.0	0	....	0.2	175	875.00
NJ	27.4	968	43.23	10.0	0	....	9.0	2,385	265.00
NM	79.7	0	....	38.0	0	....	0.0	0	....
NY	1,185.7	3,517	46.21	78.0	0	....	109.6	2,030	50.00
NC	41,464.3	129,177	44.98	55.4	0	....	30.4	5,793	190.56
ND	1,569.2	4,211	43.23	4,995.8	0	....	299.2	6,094	22.08
OH	5,952.0	31,389	43.23	1,006.3	0	....	405.2	26,670	66.86
OK	421.7	63	6.12	1,463.0	0	....	167.1	6,955	81.54
OR	1,474.3	695	24.30	197.9	0	....	0.0	0	....
PA	589.2	1,746	43.23	1,381.1	0	....	5.4	270	50.00
PR	121.0	0	....	0.0	0	....	0.0	0	....
RI	*	*	*	*	*	*	*	*	*
SC	101,503.7	68,386	43.23	943.2	0	....	488.6	20,746	42.46
SD	1,464.3	7,959	43.23	9,471.0	0	....	279.5	17,346	62.33
TN	17,892.5	42,231	43.23	412.6	0	....	107.4	8,847	84.90
TX	6,405.6	3,030	43.23	6,364.0	0	....	102.4	4,593	55.01
UT	0.0	0	....	41.7	0	....	0.0	0	....
VT	0.0	0	....	0.0	0	....	0.0	0	....
VA	14,236.8	32,926	48.79	101.1	0	....	5.0	250	50.00
WA	1,228.6	14,071	44.06	949.0	0	....	254.3	10,725	42.61
WV	9.0	0	....	0.3	0	....	0.0	0	....
WI	30,913.5	236,295	43.23	3,798.0	0	....	279.3	22,014	86.40
WY	72.6	0	....	137.9	0	....	0.0	0	....
US	1,126,050	3,090,115	41.64	81,610	0	....	12,991	773,442	113.80

[1] Not including acres which receive no cost share.   * Data withheld to avoid disclosure of individual operations.   Note: Total acres treated may not add due to rounding.

FSA, Conservation and Environmental Protection Division, (202) 720-0048.

Table 12-6.—Conservation Reserve Program (CRP): Enrollment by practice, under contract, January 2006
(CP 15, CP 16 and CP 17)

State	CP 15 Contour grass strips			CP 16 Shelter belts			CP 17 Living snow fences		
	Total acres treated	Total cost share	Cost share per acre treated[1]	Total acres treated	Total cost share	Cost share per acre treated[1]	Total acres treated	Total cost share	Cost share per acre treated[1]
AL	183.3	3,442	80.23	0.0	0	....	0.0	0	....
AK	0.0	0	....	0.0	0	....	0.0	0	....
AZ	0.0	0	....	4.3	1,262	293.49	0.0	0	....
AR	0.0	0	....	0.0	0	....	0.0	0	....
CA	0.0	0	....	0.0	0	....	0.0	0	....
CO	444.4	438	1.71	4,292.4	4,011,158	1,022.92	37.1	39,442	1,063.13
CT	0.0	0	....	0.0	0	....	0.0	0	....
DE	0.0	0	....	0.0	0	....	0.0	0	....
FL	0.0	0	....	0.0	0	....	0.0	0	....
GA	37.5	3,246	86.56	0.0	0	....	0.0	0	....
HI	*	*	*	*	*	*	*	*	*
ID	63.9	7,262	113.65	300.0	549,885	1,857.09	72.9	68,351	937.60
IL	2,015.0	106,649	62.23	140.4	30,493	245.32	38.2	14,658	383.72
IN	204.9	14,728	88.72	20.3	5,250	258.62	1.8	0	....
IA	30,695.0	964,458	48.02	1,989.0	1,629,716	885.04	360.2	74,537	236.40
KS	5,529.3	205,315	49.80	635.0	303,684	547.38	68.0	31,723	513.32
KY	72.3	6,223	98.47	0.0	0	....	0.0	0	....
LA	0.0	0	....	0.0	0	....	0.0	0	....
ME	0.0	0	....	0.0	0	....	0.0	0	....
MD	0.0	0	....	0.0	0	....	0.0	0	....
MA	0.0	0	....	0.0	0	....	0.0	0	....
MI	16.0	1,831	117.37	82.2	14,252	230.24	2.5	900	360.00
MN	1,290.5	88,088	72.98	3,636.1	1,315,399	382.69	3,061.7	603,044	209.24
MS	31.5	60	60.00	0.0	0	....	0.0	0	....
MO	2,200.2	70,522	52.47	35.9	2,767	96.41	0.0	0	....
MT	0.0	0	....	259.8	130,257	514.65	17.8	11,922	669.78
NE	620.6	25,394	54.51	2,298.7	1,134,844	513.76	144.0	43,744	481.76
NV	*	*	*	*	*	*	*	*	*
NH	0.0	0	....	0.0	0	....	0.0	0	....
NJ	4.4	1,045	237.50	0.3	175	583.33	0.0	0	....
NM	0.0	0	....	0.0	0	....	0.0	0	....
NY	4.0	495	123.75	0.2	422	2,110.00	0.0	0	....
NC	0.0	0	....	13.4	644	67.08	0.0	0	....
ND	0.0	0	....	4,060.4	2,544,145	640.65	333.7	168,204	530.44
OH	17.8	645	43.58	91.0	20,966	238.79	2.8	400	142.86
OK	1.8	0	....	37.1	7,118	191.86	3.7	0	....
OR	18.5	0	....	2.3	710	887.50	0.0	0	....
PA	144.1	21,820	177.69	0.0	0	....	0.0	0	....
PR	0.0	0	....	0.0	0	....	0.0	0	....
RI	*	*	*	*	*	*	*	*	*
SC	0.3	37	123.33	0.0	0	....	0.0	0	....
SD	131.5	8,446	97.42	13,231.9	9,842,331	774.88	356.2	226,686	674.26
TN	77.7	8,637	115.31	0.0	0	....	0.0	0	....
TX	272.7	15,073	56.69	33.9	15,445	475.23	0.0	0	....
UT	0.0	0	....	0.0	0	....	0.0	0	....
VT	0.0	0	....	0.0	0	....	0.0	0	....
VA	0.0	0	....	0.0	0	....	2.5	243	97.20
WA	35,021.8	2,460,278	81.67	9.2	28,638	4,773.00	0.0	0	....
WV	0.0	0	....	0.0	0	....	0.0	0	....
WI	1,227.9	104,464	107.20	26.4	9,295	352.08	36.7	9,258	252.26
WY	0.8	166	207.50	33.3	29,370	881.98	4.4	3,096	703.64
US	80,328	4,118,762	94.33	31,233	21,628,226	784.33	4,544	1,296,208	483.71

[1] Not including acres which receive no cost share.   * Data withheld to avoid disclosure of individual operations.   Note: Total acres treated may not add due to rounding.

FSA, Conservation and Environmental Protection Division, (202) 720–0048.

Table 12-7.—Conservation Reserve Program (CRP): Enrollment by practice, under contract, January 2006
(CP 18, CP 19 and CP 20)

State	CP 18 Salt tolerant grasses			CP 19 Alley cropping			CP 20 Alternative perennials		
	Total acres treated	Total cost share	Cost share per acre treated[1]	Total acres treated	Total cost share	Cost share per acre treated[1]	Total acres treated	Total cost share	Cost share per acre treated[1]
AL	0.0	0	....	0.0	0	....	0.0	0	....
AK	0.0	0	....	0.0	0	....	0.0	0	....
AZ	0.0	0	....	0.0	0	....	0.0	0	....
AR	0.0	0	....	0.0	0	....	0.0	0	....
CA	0.0	0	....	0.0	0	....	0.0	0	....
CO	140.2	11,411	81.39	0.0	0	....	0.0	0	....
CT	0.0	0	....	0.0	0	....	0.0	0	....
DE	0.0	0	....	0.0	0	....	0.0	0	....
FL	0.0	0	....	0.0	0	....	0.0	0	....
GA	0.0	0	....	0.0	0	....	0.0	0	....
HI	*	*	*	*	*	*	*	*	*
ID	0.0	0	....	0.0	0	....	0.0	0	....
IL	5.7	1,265	324.36	0.0	0	....	0.0	0	....
IN	0.5	35	170.00	0.0	0	....	0.0	0	....
IA	0.7	21	30.00	0.0	0	....	0.0	0	....
KS	2,272.3	50,174	25.13	0.0	0	....	13.2	0	....
KY	0.0	0	....	0.0	0	....	0.0	0	....
LA	0.0	0	....	0.0	0	....	0.0	0	....
ME	0.0	0	....	0.0	0	....	0.0	0	....
MD	0.0	0	....	0.0	0	....	0.0	0	....
MA	0.0	0	....	0.0	0	....	0.0	0	....
MI	0.0	0	....	0.0	0	....	0.0	0	....
MN	7,355.9	442,231	61.57	0.0	0	....	0.0	0	....
MS	0.0	0	....	0.0	0	....	0.0	0	....
MO	0.0	0	....	0.0	0	....	0.0	0	....
MT	150,224.0	1,610,332	12.79	0.0	0	....	0.0	0	....
NE	1,136.6	44,196	38.88	0.0	0	....	0.0	0	....
NV	*	*	*	*	*	*	*	*	*
NH	0.0	0	....	0.0	0	....	0.0	0	....
NJ	0.0	0	....	0.0	0	....	0.0	0	....
NM	0.0	0	....	0.0	0	....	0.0	0	....
NY	0.0	0	....	0.0	0	....	0.0	0	....
NC	0.0	0	....	0.0	0	....	0.0	0	....
ND	119,118.7	2,959,698	33.20	0.0	0	....	0.0	0	....
OH	0.0	0	....	0.0	0	....	0.0	0	....
OK	9,237.5	220,121	24.82	0.0	0	....	0.0	0	....
OR	0.0	0	....	0.0	0	....	0.0	0	....
PA	0.0	0	....	0.0	0	....	0.0	0	....
PR	0.0	0	....	0.0	0	....	0.0	0	....
RI	*	*	*	*	*	*	*	*	*
SC	0.0	0	....	0.0	0	....	0.0	0	....
SD	9,881.3	413,313	49.45	0.0	0	....	0.0	0	....
TN	0.0	0	....	0.0	0	....	0.0	0	....
TX	1,080.7	54,808	53.60	0.0	0	....	0.0	0	....
UT	0.0	0	....	0.0	0	....	0.0	0	....
VT	0.0	0	....	0.0	0	....	0.0	0	....
VA	0.0	0	....	0.0	0	....	0.0	0	....
WA	372.2	15,309	52.77	0.0	0	....	0.0	0	....
WV	0.0	0	....	0.0	0	....	0.0	0	....
WI	0.0	0	....	52.1	4,557	87.47	0.0	0	....
WY	0.0	0	....	0.0	0	....	0.0	0	....
US	300,826	5,822,964	73.69	52	4,557	87.47	13.2	0	....

[1] Not including acres which receive no cost share.   * Data withheld to avoid disclosure of individual operations.   Note: Total acres treated may not add due to rounding.
FSA, Conservation and Environmental Protection Division, (202) 720–0048.

Table 12-8.—Conservation Reserve Program (CRP): Enrollment by practice, under contract, January 2006
(CP 21, CP 22 and CP 23)

State	CP 21 Filter strips			CP 22 Riparian buffer			CP 23 Wetland restoration		
	Total acres treated	Total cost share	Cost share per acre treated [1]	Total acres treated	Total cost share	Cost share per acre treated [1]	Total acres treated	Total cost share	Cost share per acre treated [1]
AL	870.3	69,015	118.18	30,245.9	4,243,280	149.81	72.6	4,279	77.52
AK	7.8	780	100.00	197.9	36,696	185.43	0.0	0	
AZ	0.0	0		0.0	0		0.0	0	
AR	5,350.7	347,501	76.22	45,940.3	4,267,787	108.07	19,506.6	1,069,316	76.36
CA	0.0	0		5,345.3	1,159,295	262.28	5,108.9	125,262	24.57
CO	312.0	5,704	57.38	803.3	823,992	1,067.21	1,078.8	108,507	108.86
CT	33.9	3,336	98.41	63.1	30,292	480.06	0.0	0	
DE	1,420.0	369,516	273.09	157.6	57,891	367.33	319.7	243,991	835.01
FL	0.0	0		67.5	1,505	22.30	0.0	0	
GA	517.3	16,336	36.29	1,342.4	510,687	462.70	320.2	12,465	88.85
HI	*	*	*	*	*	*	*	*	*
ID	1,180.6	76,726	84.94	7,001.4	2,909,502	440.43	1,412.5	46,875	41.00
IL	144,762.1	7,954,105	58.49	105,215.3	18,353,702	186.69	46,680.5	5,531,098	148.19
IN	57,531.3	6,600,625	126.34	5,038.7	1,154,568	253.22	7,493.2	947,306	224.69
IA	240,734.0	13,311,675	63.86	63,207.4	18,100,757	303.72	42,592.9	4,784,731	151.79
KS	27,127.6	1,433,194	60.28	4,845.8	314,709	72.41	5,109.5	144,297	49.75
KY	33,922.3	3,780,983	129.33	14,927.7	7,576,424	529.69	133.3	18,635	168.49
LA	626.8	24,161	45.91	4,778.2	441,251	105.55	39,715.4	1,800,888	72.95
ME	126.2	29,873	461.72	199.1	470,923	2,365.26	0.0	0	
MD	39,704.5	5,888,679	169.00	16,621.0	5,596,729	408.33	2,213.1	1,551,506	801.98
MA	14.6	1,072	73.42	5.0	750	150.00	0.0	0	
MI	44,399.4	5,531,804	133.88	3,259.3	906,330	298.69	15,109.2	3,475,398	239.90
MN	151,994.2	10,139,983	71.57	44,799.2	8,076,329	191.93	326,317.8	20,712,306	72.93
MS	7,874.2	517,970	79.48	141,339.5	8,737,680	69.52	12,252.8	383,982	67.16
MO	42,539.6	2,493,291	69.75	26,535.2	5,107,742	234.30	7,170.3	454,428	126.97
MT	110.2	2,544	29.04	2,507.8	417,272	170.97	4,606.3	293,532	74.57
NE	20,964.0	1,169,785	60.50	3,161.0	736,235	243.23	15,100.0	296,587	36.26
NV	*	*	*	*	*	*	*	*	*
NH	162.9	2,554	16.53	19.1	24,075	1,594.37	0.0	0	
NJ	124.1	30,631	246.83	23.5	16,393	697.57	1.0	1,500	1,500.00
NM	0.0	0		7,711.7	2,080,975	274.98	0.0	0	
NY	446.4	134,616	309.68	11,121.4	7,725,983	765.84	50.5	10,925	225.26
NC	7,082.1	1,393,997	209.37	29,775.2	2,468,579	91.32	1,600.2	205,777	138.72
ND	8,591.0	323,479	45.45	586.0	156,750	296.93	770,780.2	11,672,860	24.03
OH	54,383.0	3,038,634	61.92	4,761.5	1,190,239	274.66	4,801.8	1,357,299	373.21
OK	884.2	44,547	61.65	1,631.2	305,993	206.95	1,414.6	32,230	46.95
OR	2,332.7	153,553	81.26	24,291.2	9,311,685	482.70	364.4	71,296	219.91
PA	1,924.5	408,599	230.89	14,869.0	19,457,310	1,333.96	711.0	981,041	1,389.58
PR	0.0	0		94.0	0		0.0	0	
RI	*	*	*	*	*	*	*	*	*
SC	4,989.0	102,919	48.65	27,680.6	1,427,335	70.21	283.6	4,391	18.86
SD	6,946.4	331,352	56.03	3,740.8	2,165,494	628.54	389,103.4	11,320,722	37.17
TN	9,515.0	954,807	120.98	5,899.7	917,959	166.01	856.5	12,485	44.64
TX	1,905.6	374,228	221.11	27,886.1	3,320,988	136.34	9,652.7	335,212	54.22
UT	38.6	4,465	115.67	204.8	83,915	409.74	0.0	0	
VT	149.2	31,615	211.90	1,394.8	1,167,109	886.32	0.0	0	
VA	4,425.5	433,258	103.47	18,544.8	19,160,681	1,109.23	287.4	192,576	825.44
WA	49,699.4	2,855,967	63.80	20,589.7	21,448,225	1,090.38	3,513.4	321,906	93.85
WV	48.5	23,519	484.93	2,286.0	1,262,818	619.30	0.0	0	
WI	25,480.9	2,358,936	113.57	16,422.7	4,950,383	308.77	12,773.6	1,307,340	167.31
WY	9.4	1,382	147.02	4,941.3	1,022,565	206.94	0.0	0	
US	1,001,262	72,771,716	126.23	752,121	189,764,456	498.84	1,748,508	69,832,949	247.06

[1] Not including acres which receive no cost share.   * Data withheld to avoid disclosure of individual operations.   Note: Total acres treated may not add due to rounding.

FSA, Conservation and Environmental Protection Division, (202) 720–0048.

Table 12-9.—Conservation Reserve Program (CRP): Enrollment by practice, under contract, January 2006

(CP 24, CP 25 and CP 26)

State	CP 24 Cross wind trap strips			CP 25 Rare and declining habitat			CP 26 Sediment retention		
	Total acres treated	Total cost share	Cost share per acre treated[1]	Total acres treated	Total cost share	Cost share per acre treated[1]	Total acres treated	Total cost share	Cost share per acre treated[1]
AL	0.0	0	....	509.8	1,754	55.51	0.0	0	....
AK	0.0	0	....	0.0	0	....	0.0	0	....
AZ	0.0	0	....	0.0	0	....	0.0	0	....
AR	0.0	0	....	0.0	0	....	0.0	0	....
CA	0.0	0	....	0.0	0	....	0.0	0	....
CO	29.6	23,065	779.22	381.1	40,774	108.93	0.0	0	....
CT	0.0	0	....	0.0	0	....	0.0	0	....
DE	0.0	0	....	0.0	0	....	0.0	0	....
FL	0.0	0	....	0.0	0	....	0.0	0	....
GA	0.0	0	....	0.0	0	....	0.0	0	....
HI	*	*	*	*	*	*	*	*	*
ID	0.0	0	....	41.8	0	....	0.0	0	....
IL	0.0	0	....	1,669.8	158,215	95.59	0.0	0	....
IN	0.0	0	....	1,342.3	199,180	148.88	0.0	0	....
IA	41.3	2,245	102.98	66,593.9	7,890,951	130.79	0.0	0	....
KS	208.9	7,472	41.91	385,831.5	26,149,243	73.60	0.0	0	....
KY	0.0	0	....	8,018.3	1,029,889	128.79	0.0	0	....
LA	0.0	0	....	0.0	0	....	0.0	0	....
ME	0.0	0	....	0.0	0	....	0.0	0	....
MD	0.0	0	....	0.0	0	....	0.0	0	....
MA	0.0	0	....	0.0	0	....	0.0	0	....
MI	0.0	0	....	212.5	12,787	72.86	6.3	10,967	1,740.79
MN	8.9	903	101.46	112,944.9	11,339,517	102.89	0.0	0	....
MS	0.0	0	....	0.0	0	....	0.0	0	....
MO	0.0	0	....	65,187.2	6,010,263	97.36	0.0	0	....
MT	22.9	110	20.00	188,264.2	9,746,502	53.10	0.0	0	....
NE	46.5	66	36.67	106,093.9	7,745,330	80.26	0.0	0	....
NV	*	*	*	*	*	*	*	*	*
NH	0.0	0	....	0.0	0	....	0.0	0	....
NJ	0.0	0	....	0.0	0	....	0.0	0	....
NM	0.0	0	....	0.0	0	....	0.0	0	....
NY	0.0	0	....	0.0	0	....	0.0	0	....
NC	0.0	0	....	0.0	0	....	0.0	0	....
ND	9.5	220	23.16	1,991.0	91,249	50.09	0.0	0	....
OH	3.5	1,656	473.14	2,142.9	268,173	147.50	0.0	0	....
OK	0.0	0	....	19,288.4	1,456,682	80.57	0.0	0	....
OR	0.0	0	....	13.3	10,241	770.00	0.0	0	....
PA	0.0	0	....	0.0	0	....	0.0	0	....
PR	0.0	0	....	0.0	0	....	0.0	0	....
RI	*	*	*	*	*	*	*	*	*
SC	0.0	0	....	0.0	0	....	0.0	0	....
SD	14.8	1,116	75.41	8,807.6	521,155	84.69	0.0	0	....
TN	0.0	0	....	0.0	0	....	0.0	0	....
TX	270.7	6,983	37.87	52.1	0	....	0.0	0	....
UT	0.0	0	....	0.0	0	....	0.0	0	....
VT	0.0	0	....	0.0	0	....	0.0	0	....
VA	37.7	16,362	434.01	0.0	0	....	0.0	0	....
WA	13.6	1,220	89.71	119.5	8,957	74.95	0.0	0	....
WV	0.0	0	....	0.0	0	....	0.0	0	....
WI	0.0	0	....	13,682.0	2,978,488	220.61	0.0	0	....
WY	16.5	1,764	106.91	0.0	0	....	0.0	0	....
US	724	63,182	178.65	983,188	75,659,353	135.63	6	10,967	1,740.79

[1] Not including acres which receive no cost share.　　* Data withheld to avoid disclosure of individual operations.　　Note: Total acres treated may not add due to rounding.

FSA, Conservation and Environmental Protection Division, (202) 720–0048.

Table 12-10.—Conservation Reserve Program (CRP): Enrollment by practice, under contract, January 2006
(CP 27, CP 28 and CP 29)

State	CP 27 Farmable wetland pilot (wetland)			CP 28 Farmable wetland pilot (buffer)			CP 29 Wildlife habitat buffer (marginal pastureland)		
	Total acres treated	Total cost share	Cost share per acre treated [1]	Total acres treated	Total cost share	Cost share per acre treated [1]	Total acres treated	Total cost share	Cost share per acre treated [1]
AL	0.0	0		0.0	0		0.0	0	
AK	0.0	0		0.0	0		0.0	0	
AZ	0.0	0		0.0	0		0.0	0	
AR	0.0	0		0.0	0		0.0	0	
CA	0.0	0		0.0	0		53.2	61,987	1,165.17
CO	0.0	0		0.0	0		187.9	44,083	462.09
CT	0.0	0		0.0	0		0.0	0	
DE	0.0	0		0.0	0		0.0	0	
FL	0.0	0		0.0	0		0.0	0	
GA	0.0	0		0.0	0		0.0	0	
HI	*	*	*	*	*	*	*	*	*
ID	0.0	0		0.0	0		92.9	34,917	385.40
IL	95.5	29,300	374.68	160.6	12,371	90.70	105.7	5,470	72.84
IN	226.1	439,403	2,107.45	464.3	57,939	129.65	57.0	23,169	468.06
IA	18,381.5	4,468,130	270.46	46,429.7	3,451,017	81.36	5,915.7	1,830,663	377.60
KS	82.0	7,699	538.39	181.4	2,727	34.65	19.4	4,056	209.07
KY	0.0	0		0.0	0		0.0	0	
LA	0.0	0		0.0	0		0.0	0	
ME	0.0	0		0.0	0		0.0	0	
MD	0.0	0		0.0	0		277.9	48,956	485.19
MA	0.0	0		0.0	0		0.0	0	
MI	3.2	0		2.6	184	92.00	0.0	0	
MN	9,378.0	1,303,904	164.00	22,352.5	1,990,119	101.46	832.8	49,144	83.68
MS	0.0	0		0.0	0		37.2	4,164	277.60
MO	0.0	0		0.0	0		307.1	128,823	476.24
MT	39.1	215	25.29	69.6	2,253	32.37	92.0	5,876	63.87
NE	1,307.1	66,452	172.07	2,210.2	137,421	68.96	760.0	182,430	258.33
NV	*	*	*	*	*	*	*	*	*
NH	0.0	0		0.0	0		0.0	0	
NJ	0.0	0		0.0	0		0.0	0	
NM	0.0	0		0.0	0		0.0	0	
NY	0.0	0		0.0	0		836.2	440,825	563.28
NC	0.0	0		0.0	0		0.0	0	
ND	3,697.3	192,851	62.63	9,987.8	537,755	55.57	0.0	0	
OH	9.1	16,437	3,222.94	21.3	697	67.02	191.5	38,101	234.04
OK	0.0	0		0.0	0		6.2	4,324	697.42
OR	0.0	0		0.0	0		6,916.7	572,541	111.81
PA	0.0	0		0.0	0		619.3	317,792	610.90
PR	0.0	0		0.0	0		341.9	0	
RI	*	*	*	*	*	*	*	*	
SC	0.0	0		0.0	0		45.7	89,075	1,988.28
SD	7,486.5	371,518	83.33	16,909.8	1,053,693	70.70	1,336.2	86,560	96.08
TN	0.0	0		0.0	0		8.0	400	50.00
TX	0.0	0		0.0	0		1,104.5	120,032	115.30
UT	0.0	0		0.0	0		26.0	4,463	375.04
VT	0.0	0		0.0	0		0.0	0	
VA	0.0	0		0.0	0		79.2	92,963	1,173.78
WA	0.0	0		0.0	0		198.2	159,095	802.70
WV	0.0	0		0.0	0		0.0	0	
WI	11.3	2,762	244.42	25.5	3,214	126.04	876.2	221,252	276.60
WY	0.0	0		0.0	0		352.6	56,307	174.49
US	40,717	6,898,671	660.52	98,815	7,249,390.0	79.21	21,677	4,627,468	446.48

[1] Not including acres which receive no cost share.   * Data withheld to avoid disclosure of individual operations.   Note: Total acres treated may not add due to rounding.

FSA, Conservation and Environmental Protection Division, (202) 720–0048.

Table 12-11.—Conservation Reserve Program (CRP): Enrollment by practice, under contract, January 2006
(CP 30, CP 31, CP 32 and CP 33)

State	CP 30 Wetland buffer (marginal pastureland)			CP 31 Bottomland hardwood			CP 32 Hardwood trees			CP 33 Upland bird habitat buffers		
	Total acres treated	Total cost share	Cost share per acre treated[1]	Total acres treated	Total cost share	Cost share per acre treated[1]	Total acres treated	Total cost share	Cost share per acre treated[1]	Total acres treated	Total cost share	Cost share per acre treated[1]
AL	0.0	0	....	168.9	18,288	113.80	0.0	0	....	628.1	15,809	70.54
AK	295.9	31,075	105.02	0.0	0	....	0.0	0	....	0.0	0	....
AZ	0.0	0	....	0.0	0	....	0.0	0	....	0.0	0	....
AR	0.0	0	....	3,395.4	353,745	110.79	382.7	0	....	1,307.3	130,878	108.32
CA	0.0	0	....	0.0	0	....	0.0	0	....	0.0	0	....
CO	5.8	2,364	407.59	0.0	0	....	0.0	0	....	0.0	0	....
CT	0.0	0	....	0.0	0	....	0.0	0	....	0.0	0	....
DE	0.0	0	....	0.0	0	....	0.0	0	....	0.0	0	....
FL	0.0	0	....	0.0	0	....	0.0	0	....	0.0	0	....
GA	0.0	0	....	19.3	2,000	103.63	0.0	0	....	897.0	45,470	64.05
HI	*	*	*	*	*	*	*	*	*	*	*	*
ID	89.0	3,000	33.71	0.0	0	....	0.0	0	....	0.0	0	....
IL	20.4	1,804	88.43	1,166.8	204,358	176.92	526.6	0	....	18,228.9	2,060,616	114.91
IN	8.0	1,500	187.50	890.1	186,115	230.88	506.2	9,252	52.57	5,354.7	622,709	122.36
IA	1,677.0	171,619	173.76	540.4	165,655	308.60	1,412.1	52,550	296.56	3,116.8	439,211	182.62
KS	0.0	0	....	26.4	495	27.35	0.0	0	....	15,111.8	589,296	43.32
KY	0.0	0	....	56.4	22,265	394.77	180.9	9	0.50	3,651.5	523,662	154.24
LA	0.0	0	....	9,552.3	812,733	85.08	681.9	0	....	93.1	12,611	135.46
ME	0.8	4,580	5,725.00	0.0	0	....	0.0	0	....	0.0	0	....
MD	11.9	0	....	0.0	0	....	0.0	0	....	274.4	31,780	115.82
MA	0.0	0	....	0.0	0	....	0.0	0	....	0.0	0	....
MI	154.1	78,563	547.48	10.8	7,300	675.93	0.0	0	....	163.3	14,789	129.05
MN	3,970.4	290,471	124.53	219.3	10,046	304.42	1,549.5	3,859	27.64	0.0	0	....
MS	9.5	746	78.53	2,589.2	185,047	76.97	558.1	0	....	1,170.9	100,929	89.53
MO	750.0	236,533	392.46	384.3	49,624	158.64	502.1	0	....	5,110.0	351,757	91.50
MT	0.0	0	....	0.0	0	....	0.0	0	....	0.0	0	....
NE	131.9	32,204	289.08	0.0	0	....	0.0	0	....	1,772.5	115,653	74.15
NV	*	*	*	*	*	*	*	*	*	*	*	*
NH	0.0	0	....	0.0	0	....	0.0	0	....	0.0	0	....
NJ	0.0	0	....	0.0	0	....	0.0	0	....	0.0	0	....
NM	0.0	0	....	0.0	0	....	0.0	0	....	0.0	0	....
NY	142.8	109,858	1,039.34	0.0	0	....	0.0	0	....	0.0	0	....
NC	0.0	0	....	1.7	260	152.94	0.0	0	....	3,147.1	212,921	82.37
ND	0.0	0	....	0.0	0	....	0.0	0	....	0.0	0	....
OH	10.0	4,000	869.57	53.4	17,589	329.38	29.6	0	....	3,063.5	292,094	98.96
OK	8.5	850	100.00	42.0	4,836	115.14	108.2	0	....	504.5	26,434	52.40
OR	0.0	0	....	0.0	0	....	0.0	0	....	0.0	0	....
PA	211.1	77,872	436.01	0.0	0	....	0.0	0	....	0.0	0	....
PR	0.0	0	....	0.0	0	....	0.0	0	....	0.0	0	....
RI	*	*	*	*	*	*	*	*	*	*	*	*
SC	33.8	54,478	1,611.78	0.0	0	....	0.0	0	....	4,167.7	276,033	71.68
SD	7,214.1	213,611	58.48	0.0	0	....	0.0	0	....	400.3	26,647	66.57
TN	0.0	0	....	1,902.8	195,311	103.70	0.7	0	....	2,676.1	179,476	75.11
TX	3.9	14,393	3,690.51	422.4	67,260	159.23	0.0	0	....	2,148.2	372,438	176.09
UT	0.0	0	....	0.0	0	....	0.0	0	....	0.0	0	....
VT	2.6	7,782	2,993.08	0.0	0	....	0.0	0	....	0.0	0	....
VA	0.0	0	....	0.0	0	....	0.0	0	....	408.2	30,389	76.03
WA	0.0	0	....	0.0	0	....	0.0	0	....	0.0	0	....
WV	0.0	0	....	0.0	0	....	0.0	0	....	0.0	0	....
WI	13.1	7,084	540.76	0.0	0	....	848.8	71	14.49	0.0	0	....
WY	0.0	0	....	0.0	0	....	0.0	0	....	0.0	0	....
US	14,765	1,344,387	928.22	21,442	2,302,927	201.57	7,287	65,741	78.35	73,395.9	6,471,602	99.78

[1] Not including acres which receive no cost share. * Data withheld to avoid disclosure of individual operations. Note: Total acres treated may not add due to rounding.
FSA, Conservation and Environmental Protection Division, (202) 720–0048.

Table 12-12.—Emergency Conservation Program: Assistance, by State and Caribbean area, fiscal years 1996–2005 [1]

Year	Emergency Conservation Program
	1,000 dollars
1996	26,867
1997	30,847
1998	20,533
1999	40,226
2000	97,970
2001	55,246
2002	32,601
2003	37,548
2004	22,480
2005	56,376

[1] Totals are from unrounded data.
FSA, Conservation and Environmental Protection Division, (202) 720–0048.

Table 12-13.—Conservation Reserve Program (CRP): Enrollment by State, January 2006

State [1]	Number of contracts	Number of farms	Acres	Annual rent ($1,000)	Payments [2]
AL	10,571	7,743	490,650.9	22,142	45.13
AK	67	46	29,738.4	999	33.58
AZ	2	1	7.9	0	36.71
AR	4,597	2,741	215,048.5	10,816	50.30
CA	569	444	145,500.2	4,653	31.98
CO	13,217	6,383	2,367,145.9	74,213	31.35
CT	27	25	332.1	23	68.78
DE	722	383	7,666.2	780	101.75
FL	1,934	1,578	84,846.3	3,191	37.61
GA	8,312	6,189	306,414.3	12,212	39.85
HI	*	*	*	*	93.39
ID	5,760	3,299	800,925.8	31,275	39.05
IL	72,082	41,483	1,047,622.9	106,933	102.07
IN	32,964	20,020	302,221.8	27,287	90.29
IA	99,033	51,060	1,937,578.2	202,912	104.72
KS	46,406	27,686	3,072,965.8	119,847	39.00
KY	15,433	9,167	350,085.3	26,160	74.72
LA	4,125	2,725	280,200.4	13,955	49.81
ME	869	579	23,655.0	1,176	49.71
MD	6,398	3,466	85,133.9	10,345	121.52
MA	14	11	73.6	7	91.59
MI	15,662	9,385	270,343.4	19,962	73.84
MN	58,360	32,044	1,790,618.8	106,171	59.29
MS	20,930	13,814	951,779.7	39,962	41.99
MO	35,016	21,432	1,564,989.2	103,863	66.37
MT	18,297	6,801	3,479,159.3	117,118	33.66
NE	27,423	15,765	1,267,185.8	71,139	56.14
NV	*	*	*	*	16.72
NH	16	14	192.5	10	51.88
NJ	140	101	2,339.3	118	50.30
NM	2,665	1,683	598,308.2	18,803	31.43
NY	2,809	2,091	63,284.0	3,235	51.11
NC	8,245	5,423	130,918.7	8,047	61.46
ND	36,403	17,716	3,362,705.0	111,382	33.12
OH	27,420	17,027	304,121.7	27,147	89.26
OK	9,230	6,285	1,056,624.2	34,430	32.58
OR	3,564	1,960	539,668.8	26,368	48.86
PA	9,925	6,319	209,713.3	18,050	86.07
RI	19	19	1,031.9	84	81.73
SC	*	*	*	*	81.98
SD	9,190	5,362	214,392.1	7,630	35.59
TN	27,302	13,644	1,500,927.2	62,114	41.38
TX	8,684	6,105	276,673.3	16,260	58.77
UT	25,176	18,195	4,044,050.0	142,708	35.29
VT	1,075	630	206,155.5	6,253	30.33
VA	175	138	1,668.8	130	78.09
WA	4,655	3,659	65,260.0	3,483	53.37
WV	11,629	4,722	1,467,362.5	77,751	52.99
WI	226	185	3,160.0	206	65.33
WY	31,784	19,994	618,605.3	42,961	69.45
PR	1,137	744	284,376.9	7,816	27.48
US	720,262	416,289	35,823,621	1,742,131	48.63

[1] State in which land is located.   [2] Payments scheduled to be made October 2006.   * Data withheld to avoid disclosure of individual operations.

FSA, Conservation and Environmental Protection Division, (202) 720–0048.

Table 12-14.—Forestry Incentives Program: Practices performed, by States and Caribbean area, 2004 and 2005 fiscal year and cumulative 1975–2005[1][2]

State	2004			2005			Cumulative 1975–2005		
	Planting trees	Improving a stand of forest trees	Site prep for natural regeneration	Planting trees	Improving a stand of forest trees	Site prep for natural regeneration	Planting trees	Improving a stand of forest trees	Site prep for natural regeneration
	Acres	Acres	Acres	Acres	Acres	Acres	Acres	Acres	Acres
AL	3,754	425	0	432	37	466	385,505	38,343	1,965
AK	174	41	14	0	0	0	509	85	8,721
AZ	0	0	0	0	0	0	13	1,807	0
AR	720	168	0	0	0	0	265,602	113,217	607
CA	0	0	0	0	0	0	10,731	12,377	326
CO	0	21	0	0	0	0	407	152	10
CT	0	0	0	0	0	0	12	7,864	0
DE	0	0	0	0	0	0	12,854	4,296	1,297
FL	496	0	0	0	0	0	359,739	360	233
GA	1,811	0	0	265	43	225	409,569	27,584	1,481
HI	11	0	68	0	0	0	285	0	68
ID	29	35	0	0	0	0	3,498	4,199	32
IL	0	195	0	0	0	0	133	42,267	397
IN	2	140	0	0	0	0	4,353	90,778	960
IA	0	34	0	0	0	0	3,461	10,880	31
KS	0	0	0	0	0	0	461	5,430	0
KY	0	0	0	0	0	0	5,567	59,455	1,986
LA	2,799	32	35	130	77	121	247,703	48,396	6,596
ME	0	55	0	0	0	0	6,878	21,797	10
MD	0	51	314	0	0	0	38,125	18,830	435
MA	0	123	9	0	0	0	703	38,897	253
MI	25	0	0	0	0	0	31,462	38,638	25
MN	45	65	0	0	0	0	18,872	14,106	1,881
MS	2,946	0	0	1,369	0	1,563	437,353	31,732	1,808
MO	0	0	0	0	0	0	12,275	86,576	398
MT	0	0	0	0	0	0	341	5,730	117
NE	0	0	0	0	0	0	576	242	0
NV	0	0	0	0	0	0	526	273	0
NH	0	0	0	0	0	0	315	32,351	1,557
NJ	0	0	0	0	0	0	15	12,348	20
NM	0	0	0	0	0	0	65	7,341	0
NY	0	60	0	0	0	0	112	68,857	142
NC	223	614	0	38	19	18	391,058	37,476	1,326
ND	0	0	0	0	0	0	207	141	0
OH	22	104	0	0	0	0	16,049	90,434	2,125
OK	419	262	0	0	40	40	27,654	37,514	658
OR	134	295	0	0	0	0	65,546	39,916	0
PA	0	94	0	0	0	0	6,197	41,552	351
PR	0	13	0	0	0	0	1,572	23	0
RI	0	0	0	0	0	0	1,153	2,653	13
SC	2,014	218	0	40	59	50	362,841	31,459	10,403
SD	0	32	0	0	0	0	29	5,980	28
TN	2,029	0	0	144	0	0	38,146	17,644	425
TX	0	0	0	0	0	0	256,683	59,360	1,243
UT	0	0	0	0	0	0	12	12	0
VT	0	7	0	0	0	0	491	23,396	280
VA	938	1,501	47	278	30	278	451,494	68,558	907
WA	131	362	276	30	178	0	51,897	25,033	296
WV	5	65	0	0	119	0	7,299	102,076	20
WI	137	0	95	0	0	0	39,057	36,240	5,228
WY	0	150	0	0	14	0	4	22,180	0
Total	18,864	5,162	858	2,726	616	2,761	3,975,397	1,486,855	54,659

[1] In 1974, the program was part of the Rural Environmental Conservation Program (now called the Agricultural Conservation Program). Data for the program year were published in the 1974 Rural Environmental Conservation Program Summary. In 1975 it became a separately funded program. [2] On May 13, 2002, the 2002 Farm Bill de-authorized the program, which was originally authorized in 1978. Funds remaining on May 13, 2002, will be exhausted through FIP closeout, primarily funding the existing contractual backlog.

NRCS, Conservation Operations Division (202) 720–1845.

Table 12-15.—Forestry Incentives Program: Participation and assistance, by States and Caribbean area, 2004 and 2005 fiscal years

State	2004			2005		
	Number of participants	Area served	Cost share paid	Number of participants	Area served	Cost share paid
	Number	Acres	Dollars	Number	Acres	Dollars
AL	196	0	18,584	53	0	111,244
AK	21	0	62,078	25	0	86,979
AZ	0	0	0	0	0	0
AR	21	0	49,740	17	0	36,478
CA	0	0	0	7	0	29,120
CO	2	0	4,763	0	0	0
CT	0	0	0	0	0	0
DE	0	0	0	1	0	2,250
FL	0	0	36,388	0	0	0
GA	75	0	157,054	19	0	40,283
HI	4	0	15,172	2	0	6,800
ID	3	0	1,918	0	0	0
IL	4	0	7,514	0	0	0
IN	6	0	4,365	3	0	1,217
IA	1	0	2,210	0	0	0
KS	0	0	0	0	0	0
KY	0	0	0	0	0	0
LA	46	0	180,634	50	0	166,684
ME	7	0	6,819	0	0	0
MD	8	0	13,063	0	0	0
MA	6	0	2,907	0	0	0
MI	2	0	3,906	3	0	2,474
MN	6	0	14,885	1	0	570
MS	110	0	169,694	224	0	459,242
MO	0	0	0	0	0	0
MT	0	0	0	0	0	0
NE	0	0	0	0	0	0
NV	0	0	0	0	0	0
NH	0	0	0	0	0	0
NJ	0	0	0	0	0	0
NM	0	0	0	0	0	0
NY	6	0	4,200	5	0	3,500
NC	23	0	50,059	11	0	16,153
ND	0	0	0	0	0	0
OH	4	0	5,469	3	0	2,800
OK	11	0	41,855	1	0	2,897
OR	15	0	62,309	0	0	0
PA	8	0	3,410	0	0	0
PR	2	0	5,010	1	0	8,503
RI	0	0	0	0	0	0
SC	74	0	150,261	10	0	13,480
SD	3	0	2,381	1	0	1,294
TN	29	0	120,852	16	0	48,026
TX	0	0	0	0	0	0
UT	0	0	0	0	0	0
VT	1	0	525	0	0	0
VA	91	0	78,991	26	0	32,771
WA	24	0	74,641	12	0	22,604
WV	26	0	15,474	10	0	4,603
WI	14	0	22,622	0	0	0
WY	4	0	8,000	5	0	7,845
Total	853	0	1,397,753	506	0	1,107,817

NRCS,Conservation Operations Division (202)720–1845.

Table 12-16.—Interim EQIP/GPCP [1]: Status of cost-share contracts, by States, year ending Sept. 30, 2005 [2]

State	Designated counties	Active land treatment contracts in operation	
		2004	2005
	Number	Number	1,000 acres
CO	38	40	172.0
KS	62	7	5.3
MT	46	9	99.8
NE	65	4	8.0
NM	27	12	102.8
ND	48	34	89.4
OK	44	13	8.4
SD	51	14	53.9
TX	156	44	93.7
WY	19	7	89.7
Total	556	184	723

[1] EQIP -- Environmental Quality Incentives Program; GPCP -- Great Plains Conservation Program.   [2] As authorized by the Act of Congress April 4, 1996, (Public Law 127, 104th Congress).
NRCS, Conservation Operations Division, (202) 720–1845.

Table 12-17.—Great Plains Conservation Program: Status of cost-share contracts, by States, year ending Sept. 30, 2005 [1]

State	Designated counties	Active land treatment contracts in operation	
		2004	2005
	Number	Number	1,000 acres
CO	38	22	130.0
KS	62	0	0.0
MT	46	0	0.0
NE	65	2	9.9
NM	27	0	0.0
ND	48	0	0.0
OK	44	5	4.0
SD	51	1	3.8
TX	156	1	1.2
WY	19	0	0.0
Total	556	31	148.9

[1] As authorized by the Act of Congress August 7, 1956 (Public Law 1021, 84th Congress).
NRCS, Conservation Operations Division, (202) 720–1845.

Table 12-18.—Small watershed protection and flood prevention projects: Accomplishments for years ending Sept. 30, 1994–98

Item	Unit of measure	1994	1995	1996	1997	1998
Small watershed protection: [1]						
Land treatment: [2]						
Forest land	Acres	38,322	16,806	1,905	2,193	8,402
Cropland	......do	501	626	0	1,160	741
Pastureland	......do	170	28	7,284	45	88
Total land treatment	......do	38,993	17,460	9,189	3,398	9,233
Land owners assisted	Number	3,534	1,483	1,465	1,348	1,186
Flood prevention: [3]						
Land treatment: [2]						
Forest land	Acres	2,196	6,335	63,028	8,682	6,541
Cropland	......do	....	....	575	1,668	20
Pastureland	......do	....	40	83	92	78
Total land treatment	......do	2,196	6,375	63,686	10,442	6,639
Land owners assisted	Number	1,452	1,528	2,461	2,265	1,183

[1] As authorized by the Watershed Protection and Flood Prevention Act of 1954 (Public Law 83–566), as amended. Accomplishments are limited to activities accomplished solely by small watershed protection program funds.   [2] Reported in land use categories consistent with those reported by the National Resources Conservation Service.   [3] As authorized by the Navigation and Flood Control Act of 1944 (Public Law 78–534), as amended. Accomplishments are limited to activities accomplished solely by small watershed protection program funds.
FS, Timber Demand and Technology Assessment, RWU-4851, (608) 231–9376.

Table 12-19.—Tree planting: Acres seeded and acres of tree planting, in States and Territories, fiscal year 2002

State or other area	Total	Federal lands			Non-federal public [1] lands	Private [2] lands
		Total	National Forest System	Other [3]		
	Acres	Acres	Acres	Acres	Acres	Acres
AL	69,725	1,691	986	705	30	68,004
AK	2,086	333	329	4	534	1,219
AZ	342	56	56	0	0	286
AR	25,768	1,919	1,919	0	5,696	18,153
CA	17,396	15,667	15,649	18	0	1,729
CO	4,493	774	773	1	0	3,719
CT	88	4	0	4	8	76
DE	1,772	0	0	0	45	1,727
FL	88,665	7,895	4,374	3,521	5,791	74,979
GA	193,905	2,371	266	2,105	446	191,088
HI	1,379	0	0	0	14	1,365
ID	18,224	11,464	11,464	0	3,496	3,264
IL	69,625	1,525	1,525	0	100	68,000
IN	8,096	97	42	55	143	7,856
IA	13,387	0	0	0	127	13,260
KS	1,863	7	0	7	0	1,856
KY	5,406	39	36	3	50	5,317
LA	117,608	953	908	45	11,224	105,431
ME	236	0	0	0	126	110
MD	20,849	3	0	3	136	20,710
MA	20	0	0	0	0	20
MI	6,772	3,844	3,812	32	2,499	429
MN	24,704	3,472	3,472	0	9,750	11,482
MS	222,401	3,179	3,036	143	1,896	217,326
MO	15,357	267	231	36	1,052	14,038
MT	9,386	8,651	8,651	0	735	0
NE	584	0	0	0	0	584
NV	346	40	40	0	104	202
NH	74	0	0	0	15	59
NJ	1,086	1	0	1	25	1,060
NM	1,262	135	135	0	0	1,127
NY	4,136	0	0	0	1,848	2,288
NC	85,049	1,293	467	826	440	83,316
ND	16,719	13	0	13	13	16,693
OH	1,962	138	138	0	73	1,751
OK	7,875	25	0	25	120	7,730
OR	38,638	13,914	13,914	0	4,694	20,030
PA	2,214	153	153	0	1,279	782
RI	0	0	0	0	0	0
SC	77,056	1,116	83	1,033	2,455	73,485
SD	10,301	0	0	0	68	10,233
TN	5,920	543	444	99	613	4,764
TX	40,474	493	183	310	840	39,141
UT	2,951	1,871	1,277	594	0	1,080
VT	263	43	40	3	0	220
VA	67,518	193	54	139	246	67,079
WA	45,771	7,974	7,637	337	13,227	24,570
WV	1,755	0	0	0	15	1,740
WI	17,529	1,435	1,410	25	1,208	14,886
WY	1,308	457	457	0	0	851
State totals	1,370,344	94,048	83,961	10,087	71,181	1,205,115
PR	5,283	0	0	0	0	5,283
Other [4]	158	0	0	0	53	105
Total	1,375,785	94,048	83,961	10,087	71,234	1,210,503

[1] State forest, other State, and other public agencies lands.   [2] Forest industry, other industry, and nonindustrial lands.   [3] U.S. Department of Interior and Indian Reservations, and other federal lands.   [4] Guam and the Trust Territories of the Pacific Islands.

FS, Timber Demand and Technology Assessment, RWU-4851, (608) 231-9376.

Table 12-20.—Forest land: Total forest land and area and ownership of timberland, by regions, Jan. 1, 2002 [1]

Region	Total forest land [2]	Timberland [3]							
		All ownerships	Federal			State, county, and municipal	Private		
			Total	National forest	Other		Total	Forest industry	Farmer and other private [4]
	1,000 acres	1,000 acres	1,000 acres	1,000 acres	1,000 acres	1,000 acres	1,000 acres	1,000 acres	1,000 acres
Northeast	85,031	85,834	10,085	2,164	7,921	7,464	68,285	10,855	57,430
North Central	84,653	94,164	22,462	7,676	14,786	13,821	57,881	3,793	54,088
North	169,684	179,998	32,547	9,840	22,707	21,285	126,166	14,648	111,518
Southeast	88,561	87,429	9,609	4,710	4,899	2,655	75,165	14,180	60,985
South Central	126,044	120,622	11,618	6,536	5,082	2,723	106,281	21,735	84,546
South	214,605	208,051	21,227	11,246	9,981	5,378	181,446	35,915	145,531
Great Plains	4,783	4,521	1,277	1,020	257	180	3,064	0	3,064
Intermountain	139,560	68,946	48,991	42,939	6,052	2,659	17,296	2,926	14,370
Rocky Mountains	144,343	73,467	50,268	43,959	6,309	2,839	20,360	2,926	17,434
Alaska	126,869	16,209	9,094	3,772	5,322	4,344	2,771	0	2,771
Pacific Northwest	51,441	44,386	23,505	17,911	5,594	3,207	17,674	9,174	8,500
Pacific Southwest [5]	41,981	18,987	10,637	9,916	721	506	7,844	2,932	4,912
Pacific Coast	220,291	79,582	43,236	31,599	11,637	8,057	28,289	12,106	16,183
All regions	748,923	541,098	147,278	96,644	50,634	37,559	356,261	65,595	290,666

[1] Data may not add to totals because of rounding. [2] Forest land is land at least 10 percent stocked by forest trees of any size, including land that formerly had such tree cover and that will be naturally or artificially regenerated. Forest land includes transition zones, such as areas between heavily forested and nonforested lands that are at least 10 percent stocked with forest trees, and forest areas adjacent to urban and built-up lands. Also included are pinyon-juniper and chaparral areas in the West and afforested areas. The minimum area for classification of forest land is 1 acre. Roadside, streamside, and shelterbelt strips of timber must have a crown width at least 120 feet wide to qualify as forest land. Unimproved roads and trails, streams, and clearings in forest areas are classified as forest if less than 120 feet in width. [3] Timberland is forest land that is producing or is capable of producing crops of industrial wood and that is not withdrawn from timber utilization by statute or administrative regulation. Areas qualifying as timberland have the capability of producing more than 20 cubic feet per acre per year of industrial wood in natural stands. Currently inaccessible and inoperable areas are included. [4] Includes Indian lands. [5] Includes Hawaii.

FS, Timber Demand and Technology Assessment, RWU-4851, (608) 231–9376.

Table 12-21.—Timber volume: Net volume of growing stock and sawtimber on timberland, by softwoods and hardwoods, and regions, Jan. 1, 2002 [1]

Region	Growing stock [2]			Sawtimber [3]		
	All species	Softwoods	Hardwoods	All species	Softwoods	Hardwoods
	Million cubic feet	Million cubic feet	Million cubic feet	Million board feet	Million board feet	Million board feet
Northeast	123,667	31,476	92,191	333,627	94,203	239,424
North Central	93,957	18,402	75,555	264,218	55,054	209,164
North	217,624	49,878	167,746	597,845	149,257	448,588
Southeast	124,002	52,758	71,244	396,131	177,171	218,960
South Central	143,963	55,260	88,703	498,656	222,763	275,893
South	267,965	108,018	159,947	894,787	399,934	494,853
Great Plains	4,260	1,880	2,380	15,210	6,925	8,285
Intermountain	127,399	118,957	8,442	501,147	484,967	16,180
Rocky Mountains	131,659	120,837	10,822	516,357	491,892	24,465
Alaska	31,997	29,124	2,873	146,117	141,506	4,611
Pacific Northwest	148,635	135,591	13,044	842,513	795,370	47,143
Pacific Southwest [4]	58,181	48,355	9,826	319,246	291,670	27,576
Pacific Coast	238,813	213,070	25,743	1,307,876	1,228,546	79,330
All regions	856,061	491,803	364,258	3,316,865	2,269,629	1,047,236

[1] Data may not add to totals because of rounding. [2] Live trees of commercial species meeting specified standards of quality or vigor. Cull trees are excluded. Includes only trees 5.0-inches diameter or larger at 4½ feet above ground. [3] Live trees of commercial species containing at least one 12-foot sawlog or two noncontiguous 8-foot logs, and meeting regional specifications for freedom from defect. Softwood trees must be at least 9.0-inches diameter and hardwood trees must be at least 11.0-inches diameter at 4½ feet above ground. [4] Includes Hawaii.

FS, Timber Demand and Technology Assessment, RWU-4851, (608) 231–9376.

Table 12-22.—Timber removals: Roundwood product output, logging residues and other removals from growing stock and other sources, by softwoods and hardwoods, 2002[1]

Roundwood products, logging residues, and other removals	All sources			Growing stock[2]			Other sources[3]		
	All species	Soft-woods	Hard-woods	All species	Soft-woods	Hard-woods	All species	Soft-woods	Hard-woods
	Million cubic feet	Million cubic feet	Million cubic feet	Million cubic feet	Million cubic feet	Million cubic feet	Million cubic feet	Million cubic feet	Million cubic feet
Roundwood products:									
Sawlogs	7,237	5,218	2,019	6,793	4,962	1,831	444	256	187
Pulpwood	4,977	2,865	2,112	4,352	2,528	1,824	625	337	288
Veneer logs	1,353	1,183	170	1,285	1,124	160	68	58	9
Other products[4]	814	444	370	728	396	331	86	48	38
Fuelwood[5]	1,621	397	1,224	592	156	436	1,029	241	788
Total	16,001	10,107	5,894	13,750	9,167	4,583	2,251	941	1,310
Logging residues[6]	3,354	1,316	2,038	1,362	605	757	1,992	711	1,281
Other removals[7]	1,333	380	953	899	291	608	434	89	345
Total	4,687	1,696	2,991	2,261	897	1,365	2,425	799	1,626

[1] Data may not add to totals because of rounding.   [2] Includes live trees of commercial species meeting specified standards of quality or vigor. Cull trees are excluded. Includes only trees 5.0-inches diameter or larger at 4½ feet above ground.   [3] Includes salvable dead trees, rough and rotten trees, trees of noncommercial species, trees less than 5.0-inches diameter at 4½ feet above ground, tops, and roundwood harvested from nonforest land (for example, fence rows).   [4] Includes such items as cooperage, pilings, poles, posts, shakes, shingles, board mills, charcoal and export logs.   [5] Downed and dead wood volume left on the ground after trees have been cut on timberland.   [6] Net of wet rot or advanced dry rot, and excludes old punky logs; consists of material sound enough to chip; excludes stumps and limbs.   [7] Unutilized wood volume from cut or otherwise killed growing stock, from nongrowing stock sources on timberland (for example, precommercial thinnings), or from timberland clearing. Does not include volume removed from inventory through reclassification of timberland to reserved timberland.

FS, Timber Demand and Technology Assessment, RWU-4851, (608) 231–9376.

Table 12-23.—Timber growth, removals and mortality: Net annual growth, removals, and mortality of growing stock on timberland by softwoods and hardwoods and regions, 2002[1]

Region	Growth[2]			Removals[3]			Mortality[4]		
	All species	Soft-woods	Hard-woods	All species	Soft-woods	Hard-woods	All species	Soft-woods	Hard-woods
	Million cubic feet	Million cubic feet	Million cubic feet	Million cubic feet	Million cubic feet	Million cubic feet	Million cubic feet	Million cubic feet	Million cubic feet
Northeast	2,833	658	2,175	1,275	414	861	810	275	536
North Central	2,585	525	2,061	1,590	266	1,324	873	183	690
North	5,418	1,167	4,184	2,865	680	2,185	1,683	457	1,226
Southeast	5,157	3,097	2,059	4,363	2,881	1,482	987	451	536
South Central	6,365	3,370	2,995	5,763	3,625	2,138	1,090	467	622
South	11,522	6,467	5,055	10,126	6,506	3,620	2,077	919	1,158
Great Plains	87	42	45	37	21	16	45	7	38
Intermountain	1,975	1,816	159	495	481	14	1,112	1,012	100
Rocky Mountains	2,062	1,858	204	532	502	30	1,157	1,019	138
Alaska	207	122	85	140	137	3	164	155	9
Pacific Northwest	3,154	2,841	313	1,721	1,621	99	904	784	120
Pacific Southwest[5]	1,326	1,196	131	628	618	10	320	262	57
Pacific Coast	4,687	4,159	528	2,489	2,376	113	1,388	1,201	186
All regions	23,689	13,651	9,971	16,012	10,064	5,948	6,304	3,596	2,708

[1] Data may not add to totals because of rounding.   [2] The net increase in the volume of trees during a specified year. Components include the increment in net volume of trees at the beginning of the specific year surviving to its end, plus the net volume of trees reaching the minimum size class during the year, minus the volume of trees that died during the year, and minus the net volume of trees that became cull trees during the year.   [3] The net volume of trees removed from the inventory during a specified year by harvesting, cultural operations such as timber stand improvement, or land clearing.   [4] The volume of sound wood in trees that died from natural causes during a specified year.   [5] Includes Hawaii.

FS, Timber Demand and Technology Assessment, RWU-4851, (608) 231–9376.

Table 12-24.—Timber volume: Net volume of sawtimber on timberland in the West, by regions and species, Jan. 1, 2002 [1]

*Species	Total West	Inter-mountain	Alaska	Pacific Northwest	Pacific South-west [2]	Great Plains
	Million board feet	*Million board feet*	*Million board feet*	*Million board feet*	*Million board feet*	*Million board feet*
Softwoods:						
Douglas-fir	652,505	134,711	0	429,296	88,498	0
Ponderosa and Jeffrey pines	200,835	78,084	0	63,597	54,492	4,662
True fir	242,254	73,952	25	93,016	75,261	0
Western hemlock	174,719	4,835	55,613	114,111	160	0
Sugar pine	22,567	3	0	4,605	17,959	0
Western white pine	6,955	2,802	0	2,185	1,968	0
Redwood	27,576	0	0	193	27,383	0
Sitka spruce	49,283	0	46,949	2,334	0	0
Engelmann and other spruces	109,925	76,554	16,790	15,967	227	387
Western larch	29,382	17,775	0	11,607	0	0
Incense cedar	19,723	21	0	3,966	15,736	0
Lodgepole pine	88,657	68,613	317	12,611	7,116	0
Western Red Cedar	44,700	10,804	5,257	28,636	3	0
Other	51,359	16,813	16,555	13,246	2,868	1,877
Total	1,720,440	484,967	141,506	795,370	291,671	6,926
Hardwoods:						
Cottonwood and aspen	21,889	15,977	1,890	3,744	271	7
Red alder	29,147	0	196	28,336	615	0
Oak	13,002	2	0	795	12,205	0
Other	39,603	201	2,524	14,269	14,331	8,278
Total	103,641	16,180	4,610	47,144	27,422	8,285
All species	1,824,081	501,147	146,116	842,514	319,093	15,211

[1] International ¼-inch rule. Data may not add to totals because of rounding.　[2] Includes Hawaii.
FS, Timber Demand and Technology Assessment, RWU-4851, (608) 231–9376.

Table 12-25.—Timber volume: Net volume of sawtimber on timberland in the East, by regions and species, Jan. 1, 2002 [1]

Species	Total East	North			South		
		Total	Northeast	North Central	Total	Southeast	South Central
	Million board feet	*Million board feet*	*Million board feet*	*Million board feet*	*Million board feet*	*Million board feet*	*Million board feet*
Softwoods:							
Longleaf and slash pines	53,910	0	0	0	53,910	34,304	19,606
Loblolly and shortleaf pines	281,019	4,666	2,025	2,641	276,353	98,398	177,955
Other yellow pines	32,854	5,027	4,014	1,013	27,827	18,919	8,908
White and red pines	69,932	59,935	38,545	21,390	9,997	7,930	2,067
Jack pine	4,186	4,186	35	4,151	0	0	0
Spruce and balsam fir	30,060	29,980	20,439	9,541	80	80	0
Eastern hemlock	30,627	27,733	22,634	5,099	2,894	1,723	1,171
Cypress	25,766	127	24	103	25,639	14,580	11,059
Other	20,836	17,602	6,487	11,115	3,234	1,237	1,997
Total	549,190	149,256	94,203	55,053	399,934	177,171	222,763
Hardwoods:							
Select white oaks	94,894	39,578	13,508	26,070	55,316	23,569	31,747
Select red oaks	81,671	50,118	28,357	21,761	31,553	12,007	19,546
Other white oaks	56,821	14,567	10,486	4,081	42,254	18,457	23,797
Other red oaks	138,097	36,029	16,051	19,978	102,068	40,858	61,210
Hickory	51,125	17,470	7,051	10,419	33,655	10,502	23,153
Yellow birch	9,608	9,341	7,085	2,256	267	235	32
Hard maple	56,561	51,428	30,864	20,564	5,133	1,304	3,829
Soft maple	72,612	54,788	36,800	17,988	17,824	12,318	5,506
Beech	28,119	18,724	14,488	4,236	9,395	3,125	6,270
Sweetgum	51,709	2,496	1,925	571	49,213	22,040	27,173
Tupelo and black gum	31,757	1,742	1,264	478	30,015	18,587	11,428
Ash	37,132	23,640	11,994	11,646	13,492	4,938	8,554
Basswood	15,725	13,550	4,761	8,789	2,175	1,096	1,079
Yellow-poplar	87,928	25,638	18,770	6,868	62,290	37,512	24,778
Cottonwood and aspen	41,506	37,969	7,655	30,314	3,537	321	3,216
Black walnut	4,828	3,386	709	2,677	1,442	490	952
Black cherry	19,868	18,239	14,227	4,012	1,629	526	1,103
Other	63,476	29,884	13,430	16,454	33,592	11,072	22,520
Total	943,437	448,587	239,425	209,162	494,850	218,957	275,893
All species	1,492,627	597,843	333,628	264,215	894,784	396,128	498,656

[1] International ¼-inch rule. Data may not add to totals because of rounding.
FS, Timber Demand and Technology Assessment, RWU-4851, (608) 231–9376.

Table 12-26.—National Forest System: National Forest System lands and other lands in States and Territories, Sept. 30, 2004

State or other area	Gross acreage	National Forest System acreage [1]	Other acreage [2]
	1,000 acres	*1,000 acres*	*1,000 acres*
AL	1,276	667	609
AK	24,359	21,974	2,386
AZ	11,891	11,263	628
AR	3,522	2,593	929
CA	24,430	20,770	3,660
CO	16,019	14,499	1,520
CT	24	24	0
FL	1,434	1,157	277
GA	1,858	865	992
HI	0	1	1
ID	21,652	20,716	1,189
IL	857	293	564
IN	644	200	445
KS	116	108	8
KY	2,212	811	1,397
LA	1,025	604	420
ME	93	53	40
MI	4,894	2,859	2,021
MN	5,467	2,840	2,626
MS	2,320	1,171	1,146
MO	3,060	1,487	1,573
MT	19,110	16,924	2,193
NE	442	352	90
NV	6,275	5,836	439
NH	828	731	97
NM	10,455	9,417	1,038
NY	16	16	0
NC	3,168	1,255	1,911
ND	1,106	1,106	0
OH	834	234	600
OK	772	398	375
OR	17,502	15,667	1,835
PA	743	513	230
SC	1,379	620	759
SD	2,369	2,013	355
TN	1,276	700	576
TX	1,995	755	1,239
UT	9,209	8,190	1,019
VT	817	386	431
VA	3,224	1,661	1,563
WA	10,111	9,261	849
WV	1,869	1,034	835
WI	2,023	1,525	498
WY	9,703	9,238	466
PR	56	28	28
VI	0	0	0
Total	232,436	192,817	39,857

[1] *National Forest System acreage.*—A nationally significant system of Federally owned units of forest, range, and related land consisting of national forests, purchase units, national grasslands, land utilization project areas, experimental forest areas, experimental range areas, designated experimental areas, other land areas; water areas, and interests in lands that are administered by USDA Forest Service or designated for administration through the Forest Service.

National forests.—Units formally established and permanently set aside and reserved for national forest purposes.

Purchase units.—Units designated by the Secretary of Agriculture or previously approved by the National Forest Reservation Commission for purposes of Weeks Law Acquisition.

National grasslands.—Units designated by the Secretary of Agriculture and permanently held by the Department of Agriculture under Title III of the Bankhead-Jones Farm Tenant Act.

Land utilization projects.—Units designated by the Secretary of Agriculture for conservation and utilization under Title III of the Bankhead-Jones Farm Tenant Act.

Research and experimental areas.—Units reserved and dedicated by the Secretary of Agriculture for forest or range research and experimentation.

Other areas.—Units administered by the Forest Service that are not included in the above groups. [2] *Other acreage.*— Lands within the unit boundaries in private, State, county, and municipal ownership and Federal lands over which the Forest Service has no jurisdiction. Areas of such lands which have been offered to the United States and have been approved for acquisition and subsequent Forest Service administration, but to which title had not yet been accepted by the United States.

FS, Timber, Demand and Technology Assessment, RWU-4851, (608) 231–9376.

Table 12-27.—Forest products cut on National Forest System lands: Volume and value of timber cut and value of all products, United States, fiscal years 1995–2004

Year [1]	Timber cut [2]		Value of miscellaneous forest products [4]	Total value including free-use timber [5]
	Volume	Value [3]		
	Million bd. ft.	*1,000 dollars*	*1,000 dollars*	*1,000 dollars*
1995	3,866	616,117	2,935	619,732
1996	3,725	544,349	3,262	547,428
1997	3,285	497,957	3,262	500,896
1998	3,298	445,774	3,262	448,752
1999	2,939	339,471	3,262	342,599
2000	2,542	302,934	3,262	305,921
2001	1,938	177,634	3,262	180,708
2002	1,728	164,051	3,262	167,313
2003	1,818	157,323	3,262	160,585
2004	2,032	217,534	3,262	220,796

[1] Fiscal years Oct. 1–Sept. 30.　[2] Commercial and cost sales and land exchanges.　[3] Includes collections for forest restoration or improvement under the Knutson-Vandenberg Act, 1930.　[4] Includes materials not measurable in board feet, such as Christmas trees, tanbark, turpentine, seedlings, Spanish moss, etc.　[5] Total value including free-use timber from 1996-2002 has been estimated.

FS, Timber Demand and Technology Assessment, RWU-4851, (608) 231–9376.

Table 12-28.—National Forest System lands: Receipts, United States and Puerto Rico, fiscal years 1994–2003

Year [1]	From the use of timber [2]	From the use of grazing	From special land uses, water power, etc.	Total [2]
	1,000 dollars	*1,000 dollars*	*1,000 dollars*	*1,000 dollars*
1994	431,615	11,056	72,196	514,867
1995	303,046	8,756	74,943	386,745
1996	195,000	7,352	71,183	273,535
1997	197,194	6,972	80,588	284,754
1998	207,938	6,992	78,869	293,799
1999	NA	NA	NA	NA
2000	NA	NA	NA	NA
2001	NA	NA	NA	NA
2002	NA	NA	NA	NA
2003	NA	NA	NA	NA

[1] Fiscal years Oct. 1–Sept. 30.　[2] Includes receipts from Oregon and California Railroad Grant Lands.

FS, Timber Demand and Technology Assessment, RUW-4851, (608) 231–9376.

Table 12-29.—National forests: Payments to States and Puerto Rico from receipts from timber sales, grazing fees, and miscellaneous uses, fiscal years 2000–02 [1] [2]

State or other areas	2000	2001	2002
	1,000 dollars	1,000 dollars	1,000 dollars
AL	617	2,032	2,015
AK	2,304	8,796	8,875
AZ	1,781	7,002	7,057
AR	6,707	6,410	5,988
CA	26,418	61,909	60,937
CO	4,530	5,595	5,434
FL	945	2,381	2,366
GA	53	1,221	1,231
ID	7,584	20,202	20,022
IL	167	285	287
IN	5	122	123
KY	72	418	391
LA	1,839	3,644	3,518
ME	27	39	39
MI	3,856	3,036	2,456
MN	4,072	3,908	3,852
MS	6,504	7,619	7,311
MO	1,168	2,387	2,499
MT	7,051	13,446	12,464
NE	34	40	40
NV	295	422	428
NH	397	445	220
NM	681	1,894	2,022
NY	8	8	8
NC	455	956	964
ND	3	3	3
OH	([3])	40	61
OK	1,250	1,303	1,214
OR	76,323	141,075	140,987
PA	2,982	4,831	3,665
SC	577	3,080	3,104
SD	3,070	3,669	3,699
TN	374	525	529
TX	666	4,447	4,435
UT	1,900	1,865	1,913
VT	328	336	283
VA	487	790	718
WA	24,658	41,229	40,191
WV	1,285	1,861	1,869
WI	1,788	2,230	1,596
WY	1,592	2,184	2,193
PR	21	21	8
Total	194,869	363,702	357,009

[1] Fiscal years Oct. 1–Sept. 30.   [2] Payments under the acts of May 23, 1908 (as amended), July 24, 1956, and Oct. 22, 1976, are 25 percent of total receipts remaining after deducting (a) payments to Arizona and New Mexico on account school section lands administered by Forest Service, (b) appropriations of receipts under laws authorizing such appropriations for acquisition of lands in specified national forests or portions thereof, and (c) receipts from an area of the Superior National Forest, Minnesota, on account of which the State (for the counties) is paid 0.75 percent of the appraised valuation in lieu of 25 percent of the receipts. Payments made in the following year.   [3] Less than $500.

FS, Timber Demand and Technology Assessment, RWU-4851, (608) 231–9376.

Table 12-30.—Livestock on National Forest System lands: Number grazed and grazing receipts, United States, 1993–2002

Year	Number grazed [1]		Receipts from grazing [2]
	Cattle, horses, and burros	Sheep and goats	
	Thousands	Thousands	1,000 dollars
1993	1,318	1,111	10,518
1994	1,229	941	11,056
1995	1,227	940	8,756
1996	1,174	868	7,352
1997	1,225	932	6,972
1998	1,208	909	6,992
1999	NA	NA	NA
2000	1,246	954	NA
2001	1,233	960	NA
2002	1,079	916	NA

[1] Calendar year data for number actually grazed.   [2] Fiscal years Oct. 1–Sept. 30.
FS, Timber Demand and Technology Assessment, RWU-4851, (608) 231–9376.

Table 12-31.—Livestock on National Forest System lands: Number grazed and grazing receipts for fiscal year 2002, by State

State or other area	Head months [1]		Receipts from grazing, 1992 [2]
	Cattle, horses, and burros	Sheep and goats	
	Number	*Number*	*Dollars*
AL	480	0	619
AZ	558,191	89,180	1,647,142
AR	9,899	0	39,045
CA	284,407	88,905	592,633
CO	614,397	376,106	1,310,957
FL	600	0	18,885
GA	4,392	0	8,080
ID	396,690	520,200	977,514
IL	2,066	0	167
KS	NA	0	0
KY	NA	0	25,701
LA	5,998	0	25,719
MI	8	0	1,557
MN	119	0	136
MS	42	0	3,426
MO	14,014	0	27,154
MT	394,778	40,248	820,929
NE	90,208	0	158,367
NV	171,652	194,060	402,127
NM	601,754	35,782	1,235,864
NY	7,706	0	57
ND	8,348	0	443,840
OH	NA	0	961
OK	24,934	0	28,620
OR [3]	313,932	62,307	618,786
SD	408,008	10,996	547,595
TX	43,046	0	97,788
UT	313,166	503,723	836,937
VT	NA	0	0
VA	8,682	0	10,694
WA	61,092	20,623	169,044
WV	4,499	284	18,585
WY	295,168	274,204	709,486
National forests	4,638,276	2,216,618	9,464,412
National grasslands and land utilization project land			1,314,323
Total	4,638,276	2,216,618	10,778,735

[1] A head month is the billing unit for permitted grazing and is equal to 1 month's occupancy. [2] 1992 is the most recent year for which grazing receipts by state are available. [3] Figure does not include $2,285.44 receipts from Oregon & CA Railroad Grant.

FS, Timber Demand and Technology Assessment, RWU-4851, (608) 231–9376.

Table 12-32.—Timber prices: Average stumpage prices for sawtimber sold from national forests, by selected species, 1995–2004

Year	Douglas-fir [1]	Southern pine [2]	Ponderosa pine [3]	Western hemlock [4]	All eastern hard-woods [5]	Oak, white, red, and black [5]	Maple, sugar [6]
	Dollars per 1,000 bd. ft.	*Dollars per 1,000 bd. ft.*	*Dollars per 1,000 bd. ft.*	*Dollars per 1,000 bd. ft.*	*Dollars per 1,000 bd. ft.*	*Dollars per 1,000 bd. ft.*	*Dollars per 1,000 bd. ft.*
1995	453.54	248.49	149.94	297.09	313.92	296.59	285.57
1996	453.04	251.05	269.97	289.30	312.57	264.44	213.20
1997	331.40	307.30	270.20	211.30	286.88	264.50	357.12
1998	254.20	287.80	204.90	161.40	240.90	270.20	394.80
1999	314.70	268.50	181.00	95.70	195.10	317.40	448.10
2000	433.40	258.10	154.60	46.12	368.61	265.63	445.80
2001	255.38	153.49	115.47	33.98	530.45	326.38	587.22
2002	184.83	166.4	117.75	73.19	382.04	273.73	484.97
2003	279.00	148.00	32.00	95.00	279.00	236.00	586.00
2004	114.00	84.00	60.00	32.00	351.00	291.00	618.00

[1] Western Washington and western Oregon. [2] Southern region. [3] Pacific Southwest region. Includes Jeffrey pine. [4] Pacific Northwest region. [5] Eastern and Southern regions. [6] Eastern region.

Forest Service National Forest prices in this table are for timber sold on a Scribner Decimal C log rule basis, except in the Northeastern States where International 1/4-inch log rule is used. Prices include KV payments; exclude timber sold by land exchanges and from land utilization project lands. Data for 1983 are statistical high bid prices; beginning in 1984, data are high bid prices which include specified road costs.

FS, Timber Demand and Technology Assessment, RWU-4851, (608) 231–9376.

Table 12-33.—National Forest System lands: Number of visitor estimates, by region, national forest visit, site visit[1] and viewing corridors, 2000–02[2][3]

Region	Year		
	2000	2001	2002
	Millions	*Millions*	*Millions*
Region 1:.			
National forest visits	12.4	12.2	11.6
Site visits	14.5	13.6	13.2
Viewing corridors	NA	1.4	NA
Region 2:.			
National forest visits	38.6	32.6	34.9
Site visits	48.6	38.3	41.6
Viewing corridors	NA	55.6	NA
Region 3:.			
National forest visits	17.3	18.6	22.6
Site visits	20.9	22.6	26.7
Viewing corridors	NA	46.9	NA
Region 4:.			
National forest visits	20.5	22.0	19.9
Site visits	22.7	24.0	22.4
Viewing corridors	NA	9.7	NA
Region 5:.			
National forest visits	20.2	28.7	30.5
Site visits	24.5	34.7	39.2
Viewing corridors	NA	16.3	NA
Region 6:.			
National forest visits	34.0	29.4	29.1
Site visits	40.1	35.7	37.3
Viewing corridors	NA	37.2	NA
Region 8:.			
National forest visits	24.9	32.3	31.8
Site visits	31.6	42.2	39.7
Viewing corridors	NA	33.0	NA
Region 9:.			
National forest visits	34.2	29.0	26.1
Site visits	46.1	35.5	33.0
Viewing corridors	NA	1.5	NA
Region 10:.			
National forest visits	7.0	9.4	4.4
Site visits	7.8	10.4	5.0
Viewing corridors	NA	13.7	NA
National total:.			
National forest visits [4]	209.0	214.1	210.9
Site visits [5]	256.9	256.2	258.1
Viewing corridors [6]	258.0	215.4	NA

[1] Includes wilderness visits.   [2] National forest visits are composed of multiple site visits, the average person goes to 1.2 sites while on their NF visit.   [3] National Visitor Use Monitoring (NVUM) changed data compilation techniques.   [4] The entry of one person onto national forest lands regardless of how long they stay.   [5] The entry of one person onto a National Forest site or area regardless of how long they stay.   [6] People who view National Forest scenery from non-Forest Service managed roads and waterways.

FS, Timber Demand and Technology Assessment, RWU -4851 (608) 231–9376.

Table 12-34.—Timber products: Production, imports, exports, and consumption, United States, 1993–2002 [1]

Year	Lumber				Plywood and veneer				Pulp products			
	Production	Imports	Exports	Consumption	Production	Imports	Exports	Consumption	Production	Imports[2]	Exports[2]	Consumption
	Million cu. ft.³	Million cu. ft.³	Million cu. ft.³	Million cu. ft.³	Million cu. ft.³	Million cu. ft.³	Million cu. ft.³	Million cu. ft.³	Million cu. ft.³	Million cu. ft.³	Million cu. ft.³	Million cu. ft.³
1995	6,857	2,545	462	8,939	1,303	107	89	1,321	6,079	1,248	905	6,422
1996	6,975	2,664	454	9,185	1,281	97	87	1,291	5,908	1,144	891	6,161
1997	7,210	2,675	457	9,428	1,213	114	103	1,224	6,101	1,250	930	6,422
1998	7,222	2,791	354	9,658	1,201	131	55	1,277	6,230	1,293	835	6,688
1999	7,533	2,888	410	10,011	1,208	160	45	1,323	5,984	1,394	794	6,584
2000	7,345	2,924	434	9,835	1,187	155	42	1,300	6,021	1,493	865	6,649
2001	7,110	3,071	362	9,819	1,067	173	32	1,208	5,853	1,499	827	6,524
2002	7,293	3,170	359	10,103	1,074	206	31	1,249	5,708	1,472	785	6,395
2003	7,079	3,193	360	9,912	1,044	218	35	1,227	5,883	1,563	657	6,789
2004	7,435	3,704	652	10,488	1,072	295	42	1,325	6,040	1,614	694	6,960

Year	Other industrial products,[4] production and consumption	Logs		Pulpwood chip imports	Pulpwood chip exports	Total				Fuelwood production and consumption	Production, all products	Consumption, all products
		Imports	Exports			Production	Imports	Exports	Consumption			
	Million cu. ft.³	Million cu. ft.³	Million cu. ft.³	Million cu. ft.³	Million cu. ft.³	Million cu. ft.³	Million cu. ft.³	Million cu. ft.³	Million cu. ft.³	Million cu. ft.³	Million cu. ft.³	Million cu. ft.³
1995 ..	387	*	451	19	377	15,454	3,917	2,285	17,101	2,150	17,604	19,251
1996 ..	342	*	422	12	416	15,344	3,899	2,269	17,010	1,924	17,268	18,934
1997 ..	330	*	384	4	424	15,662	3,864	2,298	17,428	1,700	17,362	19,128
1998 ..	305	*	316	7	414	15,687	3,979	1,974	17,963	1,632	17,319	19,595
1999 ..	298	*	326	2	409	15,758	4,231	1,982	18,265	1,625	17,383	19,890
2000 ..	300	*	422	2	354	15,630	4,310	2,117	18,158	1,622	17,252	19,780
2001 ..	270	*	403	1	264	14,966	....	1,888	17,896	1,640	16,606	19,536
2002 ..	263	*	388	2	189	14,915	....	1,753	18,099	1,618	16,533	19,717
2003 ..	318	80	356	4	155	14,835	5,058	1,563	18,330	1,515	16,350	19,845
2004 ..	318	73	366	5	168	15,398	5,691	1,921	19,168	1,540	16,938	20,708

[1] Data may not add to totals because of rounding. [2] Includes both pulpwood and the pulpwood equivalent of woodpulp, paper, and board. [3] Roundwood equivalent. [4] Includes cooperage logs, poles and piling, fence posts, hewn ties, round mine timbers, box bolts, excelsior bolts, chemical wood, shingle bolts, and miscellaneous items.

FS, Timber Demand and Technology Assessment, RWU-4851, (608) 231–9376.

Table 12-35.—Timber products: Pulpwood consumption, woodpulp production, and paper and board production and consumption, United States, 1993–2002 [1]

Year	Pulpwood consumption [2]	Woodpulp production [3]	Paper and board [4]		
			Production	Consumption or new supply [5]	Per capita consumption
	1,000 cords [6]	1,000 tons	1,000 tons	1,000 tons	Pounds
1995	97,052	67,103	89,509	96,126	731
1996	90,190	65,503	90,381	94,287	710
1997	95,247	66,650	95,029	99,175	740
1998	96,305	65,163	94,510	100,978	747
1999	94,265	62,914	97,020	104,873	768
2000	95,904	62,758	94,491	103,147	731
2001	92,181	58,198	88,913	97,303	683
2002	90,500	58,069	89,636	97,227	676
2003	85,436	53,197	80,712	94,422	629
2004	87,110	54,301	83,612	95,068	627

[1] Revised to match data from American Forest and Paper Association and American Pulpwood Association. [2] Includes changes in stocks. [3] Excludes defibrated and exploded woodpulp used for hard pressed board. [4] Excludes hardboard. [5] Production plus imports and minus exports (excludes products); changes in inventories not taken into account. [6] One cord equals 128 cubic feet.

FS, Timber Demand and Technology Assessment, RWU-4851, (608) 231–9376. Compiled from U.S. Department of Commerce and American Forest and Paper Association.

Table 12-36.—Timber products: Producer price indexes, selected products, United States, 1993–2002

[1982=100]

Year	Lumber	Softwood plywood	Woodpulp	Paper	Paperboard
1993	94.2	96.8	81.2	86.1	89.8
1994	97.0	100.9	90.7	87.5	96.8
1995	89.1	107.3	142.6	110.6	126.5
1996	92.4	99.2	104.2	104.0	107.8
1997	100.0	100.0	100.0	100.0	100.0
1998	90.9	99.9	95.5	101.6	104.9
1999	95.4	118.1	93.2	98.6	105.8
2000	90.6	98.9	113.1	104.2	122.1
2001	87.0	95.7	98.0	104.8	118.9
2002	86.5	93.6	90.6	100.8	113.7

FS, Timber Demand and Technology Assessment, RWU-4851, (608) 231–9376. Compiled from reports of the U.S. Department of Labor, Bureau of Labor Statistics.

Table 12-37.—Timber products: Structual panels, LVL, and lumber production, United States, 1995–2004

Year	Laminated veneer lumber [1]	Oriented strand board	Plywood	Medium-density fiberboard	Lumber	
					Hardwood	Softwood [2]
	Million cubic meters	Million cubic meters	Million cubic meters	Million cubic meters	Million cubic meters	Million cubic meters
1995	0.79	6.99	17.14	1.96	29.80	54.80
1996	0.91	8.24	16.98	2.21	29.50	56.60
1997	1.08	9.32	15.90	2.45	29.90	58.90
1998	1.16	9.94	15.73	2.48	29.97	59.00
1999	1.36	10.28	15.77	2.50	30.44	62.39
2000	1.35	10.54	15.47	2.63	29.74	61.20
2001	1.51	11.09	13.38	2.45	27.93	58.78
2002	1.58	11.88	13.45	2.87	27.73	60.86
2003	1.63	12.05	13.01	2.88	25.02	61.71
2004	1.69	12.63	12.98	2.91	25.72	65.28

[1] Prior to 1994, data are estimates from various articles and reports.   [2] Revised due to softwood conversion factor of 1.7 (2.36 was previously used).
FS, Timber Demand and Technology Assessment, RWU-4851, (608) 231–9376.

Table 12-38.—Lumber: Production, United States, 1994–2003

Year	Total	Softwoods	Hardwoods
	Million bd. ft.	Million bd. ft.	Million bd. ft.
1995	44,877	32,233	12,644
1996	45,754	33,266	12,488
1997	47,340	34,667	12,673
1998	47,407	34,677	12,730
1999	49,532	36,605	12,927
2000	48,565	35,967	12,598
2001	46,411	34,577	11,834
2002	47,580	35,830	11,750
2003	46,784	36,290	10,494
2004	49,314	38,360	10,954

FS, Timber Demand and Technology Assessment, RWU-4851, (608) 231–9376. From data published by the American Forest and Paper Association.

CHAPTER XIII

CONSUMPTION AND FAMILY LIVING

The statistics in this chapter deal with the consumption of food by both rural and urban people, retail price levels, and other aspects of family living of farm people. Data presented here on quantities of food available for consumption are based on material presented in the earlier commodity chapters, but they are shown here at the retail level, a form that is more useful for an analysis of the demand situation faced by the producer. Data on quantities of farm-produced food consumed directly by farm households are presented in the commodity chapters. Its value and the rental value of the farm home are given in the section on farm income.

Table 13-1.—Population: Number of people eating from civilian food supplies, United States, Jan. 1 and July 1, 1996–2005 [1]

Year	Jan. 1	July 1
	Millions	*Millions*
1996	266.6	268.1
1997	269.8	271.4
1998	273.1	274.6
1999	276.3	277.8
2000	279.5	281.0
2001	282.5	283.9
2002	285.4	286.7
2003	288.2	289.5
2004	291.0	292.4
2005 [1]	293.9	295.3

ERS, Farm and Rural Household Well-Being Branch (202) 694–5436. Compiled from reports of the U.S. Department of Commerce, Census Bureau. [1] 2005 estimates are short term projections.

Table 13-2.—Macronutrients: Quantities available for consumption per capita per day, United States, 1970–2004 [1]

| Year | Food energy | Protein | Fat | | | | Cholesterol | Carbohydrate | Dietary fiber |
			Total fat	Monounsaturated	Saturated	Polyunsaturated			
	Kilocalories	*Grams*	*Grams*	*Grams*	*Grams*	*Grams*	*Milligrams*	*Grams*	*Grams*
1970	3,200	98	145	58	51	25	460	394	19
1971	3,300	99	149	59	53	26	480	395	19
1972	3,200	99	146	58	51	26	460	393	19
1973	3,200	97	142	56	49	27	430	399	20
1974	3,200	97	144	57	49	27	440	389	19
1975	3,100	95	140	55	47	27	420	389	20
1976	3,300	98	145	58	49	29	430	402	20
1977	3,200	98	143	57	48	28	420	403	20
1978	3,200	97	144	58	48	29	420	395	20
1979	3,200	97	145	58	48	29	430	401	20
1980	3,200	97	146	59	49	29	420	401	20
1981	3,200	97	146	59	49	30	420	399	20
1982	3,200	96	146	59	48	30	410	400	20
1983	3,300	98	150	60	50	31	420	404	21
1984	3,300	99	153	62	51	31	420	409	21
1985	3,500	102	158	64	52	32	420	424	22
1986	3,500	104	157	64	52	31	420	429	22
1987	3,500	104	154	63	51	31	420	441	22
1988	3,500	106	155	63	51	32	410	448	23
1989	3,500	105	150	61	49	31	410	444	23
1990	3,500	106	151	62	49	31	400	457	24
1991	3,500	107	149	63	48	31	400	459	24
1992	3,600	109	152	65	49	32	400	468	24
1993	3,700	109	155	67	49	32	400	477	24
1994	3,700	110	151	65	48	31	400	483	24
1995	3,600	109	148	64	47	31	400	481	24
1996	3,700	110	149	64	47	31	400	492	25
1997	3,700	109	147	63	46	31	400	495	25
1998	3,700	110	149	64	48	31	410	495	25
1999	3,800	112	155	66	50	32	420	498	25
2000	3,900	113	173	77	54	36	420	497	25
2001	4,000	115	174	77	54	36	420	508	27
2002	3,900	112	180	79	57	37	420	484	24
2003	3,900	112	178	78	56	37	420	482	25
2004	3,900	113	179	79	56	37	430	481	25

Center for Nutrition Policy and Promotion (CNPP), (703) 305–2563.

Table 13-3.—Vitamins: Quantities available for consumption per capita per day, United States, 1970–2000 [1]

Year	Vita-min A	Caro-tenes	Vita-min E	Vita-min C	Thia-min	Ribo-flavin	Niacin	Vita-min B_6	Total Folate	Folate DFE	Vita-min B_{12}
	Micro-grams retinol activity equiv-alent	Micro-grams retinol equiv-alent	Milli-grams alpha-to-copherol	Milli-grams	Milli-grams	Milli-grams	Milli-grams	Milli-grams	Micro-grams	Micro-grams	Micro-grams
1970	1,240	510	13.3	106	2.0	2.3	22	2.0	298	300	9.5
1971	1,280	520	13.1	108	2.1	2.4	22	2.0	301	303	9.5
1972	1,240	560	13.5	108	2.0	2.3	23	2.0	300	302	9.4
1973	1,220	590	14.0	107	2.0	2.3	22	1.9	306	308	8.9
1974	1,280	610	13.9	112	2.4	2.7	26	2.1	332	358	9.1
1975	1,270	630	14.1	117	2.4	2.7	26	2.0	343	370	8.6
1976	1,300	630	14.5	118	2.5	2.7	27	2.1	348	376	8.9
1977	1,260	590	14.1	117	2.5	2.7	27	2.1	349	377	8.8
1978	1,240	580	14.4	113	2.4	2.7	27	2.1	337	365	8.5
1979	1,250	620	14.5	114	2.5	2.7	28	2.1	349	377	8.2
1980	1,240	600	14.4	117	2.5	2.7	27	2.1	344	373	8.2
1981	1,240	610	14.6	115	2.5	2.7	28	2.1	342	371	8.2
1982	1,220	630	14.8	116	2.5	2.6	27	2.1	348	377	7.9
1983	1,220	600	15.2	121	2.5	2.7	28	2.2	352	382	8.1
1984	1,240	640	15.6	118	2.5	2.7	28	2.2	347	376	8.2
1985	1,230	630	16.1	119	2.6	2.8	29	2.2	362	393	8.3
1986	1,230	610	16.0	123	2.7	2.8	29	2.3	367	398	8.2
1987	1,240	640	16.0	120	2.7	2.9	30	2.3	357	390	8.2
1988	1,200	610	16.6	121	2.8	2.9	30	2.3	372	406	8.0
1989	1,230	650	16.2	122	2.8	2.9	30	2.3	366	400	8.0
1990	1,240	670	16.4	118	2.9	3.0	31	2.4	374	410	8.0
1991	1,220	640	16.9	122	2.9	2.9	31	2.4	385	421	7.9
1992	1,250	680	17.1	125	3.0	3.0	32	2.5	396	432	7.9
1993	1,280	750	17.6	129	3.0	2.9	32	2.5	393	422	7.7
1994	1,320	830	16.8	129	3.0	3.0	32	2.5	392	421	7.9
1995	1,280	770	16.4	125	2.9	2.9	31	2.4	383	411	8.0
1996	1,300	820	16.7	131	3.0	2.9	32	2.4	384	414	8.0
1997	1,310	870	16.6	130	3.0	2.9	32	2.4	383	412	7.8
1998	1,250	730	16.5	131	3.0	2.9	32	2.4	695	911	8.0
1999	1,250	710	17.4	130	3.0	2.9	33	2.5	704	920	8.0
2000	1,250	720	20.0	130	3.0	2.9	33	2.5	706	925	8.2
2001	1,080	680	20.4	119	3.1	2.9	34	2.5	703	918	8.2
2002	1,070	650	21.0	114	2.9	2.8	32	2.4	679	889	8.2
2003	1,080	690	20.9	118	2.9	2.8	33	2.4	687	899	8.2
2004	1,080	680	21.0	119	2.9	2.9	33	2.4	687	898	8.2

[1] Computed by Center for Nutrition Policy and Promotion (CNPP), USDA. Based on Economic Research Service estimates of per capita quantities of food available for consumption (retail weight) and on CNPP estimates of quantities of produce from home gardens and certain other foods. No deduction is made in food supply estimates for loss of food or nutrients in further processing, in marketing, or in the home. Data include iron, thiamin, riboflavin, niacin, vitamin A, vitamin B_6, vitamin B_{12}, ascorbic acid, and zinc added by enrichment and fortification.　[2] Sodium levels do not reflect sodium from most processed foods and therefore underestimate total sodium available in the U.S. food supply.

Center for Nutrition Policy and Promotion (CNPP), (703) 305–2563.

Table 13-4.—Minerals: Quantities available for consumption per capita per day, United States, 1970–2004[1]

Year	Calcium	Phos- phorus	Magne- sium	Iron	Zinc	Copper	Potas- sium	So- dium[2]	Sele- nium
	Milli- grams	*Milli- grams*	*Milli- grams*	*Milli- grams*	*Milli- grams*	*Milli- grams*	*Micro- grams*	*Milli- grams*	*Milli- grams*
1970	960	1,550	340	15.9	12.7	1.7	3,670	1,260	124.4
1971	970	1,560	340	16.1	12.8	1.7	3,670	1,280	125.4
1972	960	1,560	350	16.2	12.7	1.7	3,660	1,280	126.3
1973	970	1,540	350	16.4	12.4	1.7	3,650	1,260	122.8
1974	940	1,540	340	16.7	13.8	1.7	3,590	1,260	117.4
1975	920	1,490	340	16.9	13.6	1.7	3,580	1,240	136.2
1976	930	1,540	350	17.4	14.0	1.8	3,650	1,290	139.5
1977	930	1,530	350	17.3	14.0	1.8	3,590	1,280	133.5
1978	920	1,510	340	16.8	13.7	1.7	3,510	1,270	135
1979	920	1,530	350	17.3	13.8	1.8	3,590	1,270	134
1980	910	1,510	340	17.2	13.7	1.7	3,550	1,240	131.9
1981	900	1,510	340	17.3	13.8	1.8	3,510	1,220	132
1982	910	1,510	350	17.5	13.8	1.8	3,520	1,230	134.5
1983	920	1,530	350	19.9	14.0	1.8	3,590	1,240	137.1
1984	930	1,560	360	20.0	14.2	1.8	3,610	1,270	137.3
1985	960	1,600	370	20.9	14.5	1.9	3,700	1,290	140.7
1986	970	1,620	380	21.1	14.8	1.9	3,760	1,300	143
1987	960	1,630	380	21.4	14.6	1.9	3,700	1,290	143.6
1988	960	1,650	380	21.9	14.9	1.9	3,740	1,260	145
1989	950	1,640	380	22.0	14.9	1.9	3,730	1,270	146
1990	980	1,670	390	22.7	15.3	2.0	3,760	1,300	147.9
1991	970	1,670	400	23.0	15.4	2.0	3,810	1,300	156.9
1992	990	1,700	400	23.4	15.8	2.0	3,860	1,320	160.7
1993	970	1,690	400	23.3	15.5	2.0	3,850	1,310	161.1
1994	1,000	1,700	400	23.2	15.4	2.0	3,890	1,310	161.6
1995	970	1,680	390	22.8	15.2	2.0	3,800	1,290	158.5
1996	980	1,690	390	23.2	15.1	2.0	3,870	1,280	162.9
1997	980	1,680	390	23.0	14.8	2.0	3,850	1,280	162.9
1998	980	1,690	390	23.1	15.1	2.0	3,860	1,270	176.2
1999	980	1,710	400	23.6	15.4	2.1	3,910	1,270	177.2
2000	980	1,720	400	23.7	15.4	2.1	3,920	1,280	178.9
2001	970	1,770	430	24.3	15.9	2.1	3,900	1,240	197
2002	950	1,680	390	23.1	15.2	2.0	3,750	1,250	182.5
2003	950	1,690	400	23.3	15.3	2.0	3,810	1,240	186
2004	970	1,710	400	23.4	15.4	2.1	3,820	1,240	189.7

[1] Computed by Center for Nutrition Policy and Promotion (CNPP), USDA. Based on Economic Research Service estimates of per capita quantities of food available for consumption (retail weight) and on CNPP estimates of quantities of produce from home gardens and certain other foods. No deduction is made in food supply estimates for loss of food or nutrients in further processing, in marketing, or in the home. Data include iron, thiamin, riboflavin, niacin, vitamin A, vitamin B_6, vitamin B_{12}, ascorbic acid, and zinc added by enrichment and fortification.　[2] Sodium levels do not reflect sodium from most processed foods and therefore underestimate total sodium available in the U.S. food supply.

Center for Nutrition Policy and Promotion (CNPP), (703) 305–2563.

Table 13-5.—Food nutrients: Percentage of total contributed by major food groups, 1970 [1]

Nutrient	Meat, poultry, fish	Dairy[2] products	Eggs	Fats,[3] oils	Fruits Citrus	Fruits Non-citrus	Fruits Total[5]
	Percent	Percent	Percent	Percent	Percent	Percent	Percent
Food energy	19.6	11.0	2.0	17.9	0.9	2.0	2.9
Carbohydrate	0.1	6.7	0.1	0	1.8	4.2	6.0
Protein	39.9	22.0	5.6	0.2	0.5	0.7	1.2
Total fat	34.8	12.7	2.9	43.3	0.1	0.3	0.4
Saturated fat	37.9	22.6	2.5	33.3	0	0.2	0.2
Monounsaturated fat	38.5	9.1	2.7	44.3	0	0.4	0.4
Polyunsaturated fat	19.1	2.5	2.2	64.5	0.1	0.4	0.5
Cholesterol	39.2	15.5	39.5	5.7	0	0	0
Dietary fiber	0	0.4	0	0	3.0	10.0	13.0
Vitamin A (retinol activity equivalents)	36.0	21.8	6.7	11.6	0.3	1.6	1.9
Carotene (retinol equivalents)	0	3.3	0	4.8	1.6	8.0	9.6
Vitamin E	5.4	3.9	3.4	65.7	0.9	3.2	4.1
Vitamin C	2.4	4.2	0	0	25.5	14.5	40.0
Thiamin	25.1	8.8	1.3	0.1	2.3	1.9	4.2
Riboflavin	21.6	38.4	9.5	0.3	0.6	1.7	2.3
Niacin	43.9	2.2	0.1	0	0.7	2.0	2.7
Vitamin B[6]	38.3	12.1	2.9	0.1	1.6	7.2	8.7
Folate	9.8	9.5	7.0	0.1	6.6	2.7	9.3
Folate DFE	9.8	9.1	7.0	0.1	6.6	2.7	9.3
Vitamin B[12]	73.4	20.3	4.6	0.2	0	0	0
Calcium	2.8	75.6	2.3	0.6	1.2	1.1	2.3
Phosphorus	25.6	36.6	5.1	0.3	0.6	1.0	1.6
Magnesium	12.9	20.8	1.3	0.1	2.0	3.9	5.9
Iron	22.7	2.4	4.0	0.1	0.6	2.5	3.1
Zinc	46.9	19.4	3.9	0.1	0.3	0.9	1.3
Copper	19.8	3.5	0.3	0	1.7	5.0	6.7
Selenium	18.5	16.1	10.5	0.1	0.2	0.4	0.6
Potassium	16.6	23.7	1.5	0.2	3.3	6.0	9.4
Sodium	24.7	26.0	4.0	13.2	0	1.4	1.4

Nutrient	Vegetables White potatoes	Vegetables Dark green, deep yellow	Vegetables Other	Vegetables Total[5]	Legumes, nuts, soy	Grain products	Sugars, sweeteners	Miscellaneous[4]	Total[5]
	Percent	Percent	Percent	Percent	Percent	Percent	Percent	Percent	Percent
Food energy	2.8	0.4	1.5	5.2	3.0	19.5	18.2	0.8	100
Carbohydrate	5.3	0.7	2.7	9.8	2.2	34.3	39.6	1.2	99.9
Protein	2.4	0.4	2.2	5.7	5.4	18.6	0	1.5	100
Total fat	0.1	0	0.2	0.4	3.5	1.4	0	0.7	100.2
Saturated fat	0.1	0	0.1	0.2	1.9	0.6	0	0.7	99.9
Monounsaturated fat	0	0	0.1	0.1	3.8	0.5	0	0.6	100
Polyunsaturated fat	0.2	0.1	0.6	1.1	6.2	3.2	0	0.6	99.9
Cholesterol	0	0	0	0	0	0	0	0	100
Dietary fiber	11.4	3.5	13.8	32.9	14.3	30.5	0	8.9	100
Vitamin A (retinol activity equivalents)	0	12.8	2.3	16.6	0	0.2	0	5.2	100
Carotene (retinol equivalents)	0	62.5	11.7	78.4	0.1	0.5	0	3.3	100
Vitamin E	0.3	1.1	2.7	8.1	6.3	2.7	0	0.4	100
Vitamin C	18.5	6.4	14.7	48.9	0	0	0	4.5	100
Thiamin	5.5	0.8	3.9	11.7	5.3	42.7	0.1	0.6	100
Riboflavin	1.2	0.9	2.7	5.9	1.5	18.5	0.7	1.2	99.8
Niacin	6.2	0.8	3.1	12.3	4.9	29.2	0.2	4.6	99.9
Vitamin B[6]	14.0	2.2	4.8	24.2	3.4	9.1	0.3	1.1	100.1
Folate	5.5	2.8	15.9	27.3	19.5	15.2	0	2.3	100
Folate DFE	5.5	2.9	15.9	27.4	19.5	15.5	0	2.3	100
Vitamin B[12]	0	0	0	0	0	1.6	0	0	99.9
Calcium	1.2	0.9	3.5	6.5	3.6	3.5	0.6	2.2	100
Phosphorus	3.6	0.6	3.0	8.2	5.1	14.2	0.3	2.9	100.1
Magnesium	6.6	1.3	6.4	16.6	12.1	16.7	0.7	13	99.9
Iron	4.9	1.2	5.5	13.7	9.3	36.4	1.1	7.1	100.1
Zinc	2.6	0.5	3.0	6.8	5.9	12.0	0.5	3.2	100
Copper	6.3	1.5	5.3	18.0	17.1	17.7	4.2	13.1	99.9
Selenium	1.7	0.2	0.7	2.8	9.6	39.5	0.8	1.5	100.1
Potassium	12.3	1.8	6.8	25.3	7.8	6.4	0.5	8.6	99.9
Sodium	2.8	1.0	15.4	28.0	0.2	0.5	2.1	0.3	100

[1] Percentages of food groups are based on aggregate data.   [2] Excludes butter.   [3] Includes butter.   [4] Coffee, tea, spices, chocolate liquor equivalent of cocoa beans, and fortification not assigned to a specific group.   [5] Components may not add to total due to rounding.

Center for Nutrition Policy and Promotion, (703) 305-2563.

Table 13-6.—Food nutrients: Percentage of total contributed by major food groups, 2000[1]

Nutrient	Meat, poultry, fish	Dairy products[2]	Eggs	Fats, oils[3]	Fruits Citrus	Fruits Non-citrus	Fruits Total[5]
	Percent	Percent	Percent	Percent	Percent	Percent	Percent
Food energy	14.0	9.1	1.4	21.8	0.9	2.2	3.1
Carbohydrate	0.1	4.5	0.1	0	1.8	4.2	5.9
Protein	39.7	19.4	4.0	0.1	0.5	0.7	1.3
Total fat	22.9	11.8	2.1	55.7	0	0.4	0.5
Saturated fat	25.6	23.4	2.0	43.7	0	0.3	0.3
Monounsaturated fat	25.2	7.9	1.9	59.0	0	0.4	0.5
Polyunsaturated fat	13.6	1.8	1.3	72.4	0	0.4	0.5
Cholesterol	43.5	16.0	35.2	5.3	0	0	0
Dietary fiber	0	0.4	0	0	2.3	9.1	11.4
Vitamin A (retinol activity equivalents)	27.0	22.1	5.3	9.1	0.3	1.8	2.1
Carotene (retinol equivalents)	0	2.1	0	2.6	1.1	6.4	7.5
Vitamin E	4.0	2.4	1.9	71.6	0.8	2.4	3.2
Vitamin C	2.0	2.5	0	0	25.8	16.1	41.9
Thiamin	17.5	4.7	0.7	0	1.8	1.7	3.5
Riboflavin	16.7	26.3	6.1	0.2	0.4	1.8	2.3
Niacin	35.7	1.2	0.1	0	0.5	1.5	2.1
Vitamin B[6]	34.7	8.7	1.9	0	1.5	8.3	9.8
Folate	3.6	3.4	2.4	0	4.2	1.7	5.8
Folate DFE	2.7	2.6	1.8	0	3.2	1.3	4.5
Vitamin B[12]	75.2	20.3	4.3	0.2	0	0	0
Calcium	3.2	72.2	1.8	0.4	1.2	1.3	2.5
Phosphorus	24.8	32.7	3.8	0.2	0.7	1.1	1.9
Magnesium	12.6	15.8	0.9	0.1	2.1	4.3	6.4
Iron	15.6	1.9	2.2	0.1	0.4	2.0	2.4
Zinc	37.8	16.8	2.6	0.1	0.3	0.9	1.2
Copper	14.4	2.7	0.2	0	1.7	4.8	6.4
Selenium	28.1	10.9	6.1	0	0.1	0.4	0.5
Potassium	16.8	18.1	1.1	0.1	3.8	7.5	11.3
Sodium	19.4	32.8	3.3	11.1	0.1	1.4	1.4

Nutrient	Vegetables White potatoes	Vegetables Dark-green, deep-yellow	Vegetables Other	Vegetables Total[5]	Legumes, nuts, soy	Grain products	Sugars, sweeteners	Miscellaneous[4]	Total[5]
	Percent	Percent	Percent	Percent	Percent	Percent	Percent	Percent	Percent
Food energy	2.4	0.4	1.2	4.5	3.0	23.6	18.7	0.8	100.1
Carbohydrate	4.3	0.8	2.0	8.1	2.1	38.8	38.9	1.2	100.1
Protein	2.2	0.6	1.8	5.2	6.3	22.2	0	1.8	100
Total fat	0.1	0.1	0.2	0.4	3.5	2.2	0	0.9	100
Saturated fat	0.1	0	0.1	0.2	2.2	1.5	0	1.1	100
Monounsaturated fat	0	0	0.1	0.1	3.7	1.1	0	0.7	100.1
Polyunsaturated fat	0.1	0.1	0.4	0.8	5.0	3.9	0	0.7	100
Cholesterol	0	0	0	0	0	0	0	0	100
Dietary fiber	8.6	4.4	9.9	26.8	15.0	35.1	0	11.3	100
Vitamin A (retinol activity equivalents)	0	20.9	1.7	24.2	0	4.3	0	5.9	100
Carotene (retinol equivalents)	0	72.9	5.9	82.2	0.1	0.5	0	5.0	100
Vitamin E	0.2	1.8	1.5	6.7	5.3	4.3	0	0.5	99.9
Vitamin C	15	12.7	9.8	45.1	0.1	4.3	0	4.2	100.1
Thiamin	4.4	1.0	2.4	8.8	4.6	59.4	0.1	0.6	99.9
Riboflavin	1.0	1.2	2.2	5.4	1.6	39.2	0.8	1.4	100
Niacin	4.6	0.9	2.1	9.3	3.8	44.8	0	3.0	100
Vitamin B[6]	11.3	2.9	4.7	21.5	3.8	17.8	0.2	1.6	100
Folate	2.3	2.4	5.9	11.7	9.9	61.8	0	1.3	99.9
Folate DFE	1.7	1.8	4.5	8.9	7.6	70.9	0	1.0	100
Vitamin B[12]	0	0	0	0	0	0.1	0	0	100.1
Calcium	1.1	1.5	3.4	6.9	4.5	4.8	0.6	3.0	100
Phosphorus	3.0	1.0	2.7	7.7	6.2	18.6	0.3	3.8	100
Magnesium	5.4	1.9	4.6	14.0	13.5	22.3	0.7	13.6	100
Iron	3.6	1.3	3.4	9.8	7.9	52.1	0.9	7.1	100
Zinc	2.2	0.8	2.4	5.9	5.7	25.6	0.5	3.8	100
Copper	5.0	1.7	4.3	14.9	20.8	22.5	3.6	14.5	99.9
Selenium	1.3	0.2	0.7	2.4	6.1	43.8	0.9	1.2	100
Potassium	11.3	3.2	6.2	25.2	9.6	9.1	0.5	8.2	100
Sodium	3.0	1.0	11.8	27.0	0.3	0.9	3.3	0.4	100

[1] Percentages of food groups are based on aggregate nutrient data [2] Excludes butter. [3] Includes butter. [4] Coffee, tea, spices, chocolate liquor equivalent of cocoa beans, and fortification not assigned to a specific food group. [5] Components may not add to total due to rounding.

Center for Nutrition Policy and Promotion, (703) 305-2563.

Table 13-7.—Consumption: Per capita consumption of major food commodities, United States, 1997–2004[1]

Commodity	1997	1998	1999	2000	2001	2002	2003	2004[2]
	Pounds	*Pounds*	*Pounds*	*Pounds*	*Pounds*	*Pounds*	*Pounds*	*Pounds*
Red meats[3][4]	109.0	113.2	115.1	113.7	111.4	114.0	111.6	112.0
Beef	62.6	63.6	64.3	64.5	63.1	64.5	61.9	62.9
Veal	0.8	0.7	0.6	0.5	0.5	0.5	0.5	0.4
Lamb and mutton	0.8	0.9	0.8	0.8	0.8	0.9	0.8	0.8
Pork	44.7	48.2	49.3	47.8	46.9	48.2	48.4	47.8
Fish[3]	14.3	14.5	14.8	15.2	14.7	15.6	16.3	16.5
Canned	4.3	4.3	4.5	4.7	4.2	4.3	4.7	4.5
Fresh and frozen	9.7	9.9	10.1	10.2	10.2	11.0	11.3	11.7
Cured	0.3	0.3	0.3	0.3	0.3	0.3	0.3	0.3
Poultry[3][4]	63.6	64.3	67.4	67.9	67.8	70.7	71.2	72.7
Chicken	50.0	50.4	53.6	54.2	54.0	56.8	57.5	59.2
Turkey	13.6	13.9	13.8	13.7	13.8	14.0	13.7	13.4
Eggs	30.2	30.8	32.2	32.4	32.5	32.8	32.8	33.0
Dairy products[5]								
Total dairy products	567.2	572.4	584.1	592.2	586.5	585.4	588.8	591.8
Fluid milk and cream	216.4	213.3	213.1	210.1	207.6	206.7	206.0	204.9
Plain and flavored whole milk	71.0	69.5	70.1	69.2	67.2	66.5	65.5	62.7
Plain reduced fat and light milk(2%, 1%, and 0.5%)	87.0	85.0	84.5	83.8	82.9	82.0	81.0	80.4
Plain fat free milk (skim)	33.5	33.4	32.2	29.9	28.9	27.9	26.8	26.5
Flavored lower fat free milk	7.9	8.2	8.4	8.7	9.0	10.5	10.8	11.7
Buttermilk	2.5	2.5	2.4	2.2	2.1	2.0	1.9	1.8
Eggnog	0.4	0.4	0.4	0.3	0.4	0.4	0.5	0.4
Yogurt (excl. frozen)	5.8	5.9	6.2	6.5	7.0	7.4	8.2	9.2
Heavy cream, light cream and half and half	5.5	5.6	6.0	6.2	6.8	6.5	7.4	7.9
Sour cream and dip	2.9	3.0	3.0	3.2	3.5	3.6	4.0	4.2
Cheese (excluding cottage)[6]	27.5	27.8	29.0	29.8	30.0	30.5	30.5	31.3
American	11.8	11.9	12.6	12.7	12.8	12.8	12.5	12.9
Cheddar	9.4	9.4	9.8	9.7	9.9	9.6	9.2	10.3
Italian	10.8	11.1	11.6	12.1	12.4	12.5	12.6	12.9
Mozzarella	8.2	8.6	9.0	9.3	9.7	9.7	9.6	9.9
Cottage cheese	2.6	2.7	2.6	2.6	2.6	2.6	2.7	2.6
Condensed and evaporated milk	6.5	6.1	6.5	5.8	5.4	6.0	5.9	5.5
Ice cream	16.1	16.3	16.7	16.7	16.3	16.7	16.4	15.4
Fats and oils[7]	63.7	64.2	66.7	82.3	84.0	88.6	88.0	87.5
Butter	4.1	4.4	4.7	4.5	4.4	4.4	4.5	4.6
Margarine	8.4	8.2	7.9	7.5	7.0	6.5	5.3	5.3
Shortening	20.5	20.5	21.1	31.6	32.6	33.3	32.8	32.6
Lard (direct use)	0.8	0.7	0.7	0.8	1.1	1.3	1.3	0.7
Edible tallow (direct use)	2.1	3.1	3.6	4.0	3.1	3.4	3.8	4.0
Salad and cooking oils	29.2	28.4	29.8	34.8	36.5	40.3	40.8	40.8
Fruits and vegetables[4][8]	710.6	697.0	706.2	711.7	685.3	685.4	704.0	694.3
Fruits	294.3	285.0	291.1	288.7	273.0	273.7	282.2	271.4
Fresh	130.1	129.3	130.4	128.7	126.1	127.0	128.3	127.1
Citrus	26.5	26.6	20.4	23.5	23.9	23.4	23.9	22.7
Noncitrus	103.6	102.7	110.1	105.2	102.1	103.7	104.4	104.4
Processing	164.2	155.7	160.6	160.0	146.9	146.6	153.9	144.3
Citrus	97.8	90.2	91.7	95.2	79.9	82.2	85.3	77.2
Noncitrus	66.4	65.5	68.9	64.8	67.1	64.5	68.6	67.1
Vegetables	416.3	412.1	415.2	423.0	412.3	411.8	421.8	422.8
Fresh	190.4	185.7	192.3	198.7	195.7	194.7	199.8	204.6
Processing	225.8	226.3	222.9	224.3	216.6	217.0	222.0	218.2
Flour and cereal products[4]	197.4	194.0	196.1	199.2	195.0	191.7	193.1	191.5
Wheat flour[9]	146.8	143.0	144.0	146.3	141.0	136.8	136.7	134.3
Rice (milled basis)	18.2	18.0	18.6	18.9	19.3	19.5	20.3	20.4
Corn products	26.5	27.2	27.8	28.4	29.0	29.7	30.3	30.9
Oat products	4.7	4.5	4.4	4.4	4.5	4.5	4.7	4.7

See footnotes at end of table.

Table 13-7.—Consumption: Per capita consumption of major food commodities, United States, 1997–2004 [1]—Continued

Commodity	1997	1998	1999	2000	2001	2002	2003	2004 [2]
	Pounds	Pounds	Pounds	Pounds	Pounds	Pounds	Pounds	Pounds
Barley and rye products	1.3	1.3	1.2	1.2	1.2	1.2	1.2	1.2
Caloric sweeteners (dry weight basis) [4]	147.7	148.9	151.3	148.8	147.0	146.1	141.4	141.0
Sugar (refined)	64.9	64.9	66.3	65.5	64.5	63.2	60.9	61.5
Corn sweeteners [10]	81.5	82.7	83.5	81.8	81.3	81.5	79.1	78.1
Honey and edible syrups	1.3	1.3	1.4	1.5	1.3	1.4	1.4	1.4
Other.								
Coffee (green bean equivalent)	9.1	9.3	9.8	10.3	9.5	9.2	9.5	9.6
Cocoa(chocolate liquor equivalent) [11]	4.0	4.3	4.5	4.7	4.5	3.9	4.2	4.8
Tea (dry leaf equivalent)	0.8	0.9	0.9	0.8	0.9	0.8	0.8	0.8
Peanuts (shelled)	5.9	5.9	6.1	5.9	5.9	5.9	6.4	6.7
Tree nuts (shelled)	2.2	2.3	2.8	2.6	2.9	3.2	3.5	3.6

[1] Quantity in pounds, retail weight unless otherwise shown. [2] Preliminary. [3] Boneless, trimmed weight equivalent. [4] Total may not add due to rounding. [5] Total dairy products reported on a milk-equivalent, milkfat basis. All other dairy categories reported on a product weight basis. [6] Natural equivalent of cheese and cheese products. [7] Total fats and oils reported on a fat content basis. All other fats and oils categories reported on a product weight basis. [8] Farm weight. [9] White, whole wheat, semolina, and durum flour. [10] High fructose, glucose, and dextrose. [11] Chocolate liquor is what remains after cocoa beans have been roasted and hulled; it is sometimes called ground or bitter chocolate. NA=Not available.

ERS, Food Economics Division, (202) 694-5400. Historical consumption and supply-utilization data for food may be found at,www.ers.USDA.gov/data/food consumption/, ERS, USDA, 2006.

Table 13-8.—Food plans: Food cost at home, at four cost levels, for families and individuals in the United States, for week and month, June 2005 [1]

Age-gender groups	Weekly cost [2]				Monthy cost [2]			
	Thrifty plan	Low-cost plan	Mod-erate-cost plan	Liberal plan	Thrifty plan	Low-cost plan	Mod-erate-cost plan	Liberal plan
	Dollars	Dollars	Dollars	Dollars	Dollars	Dollars	Dollars	Dollars
Individuals: [3].								
Child:.								
1 year	17.70	22.30	26.10	31.10	76.80	96.60	113.10	134.90
2 year	17.70	22.00	26.30	31.70	76.90	95.50	113.80	137.50
3-5 years	19.60	24.40	30.10	36.00	85.10	105.50	130.40	156.10
6-8 years	24.60	32.70	40.40	47.10	106.80	141.90	174.90	204.20
9-11 years	28.90	36.90	47.20	54.80	125.20	159.80	204.60	237.40
Male:.								
12-14 years	30.20	41.70	51.80	60.70	130.70	180.70	224.30	263.10
15-19 years	31.10	43.00	53.90	62.50	134.90	186.40	233.50	270.60
20-50 years	33.30	43.00	53.70	65.50	144.10	186.20	232.50	283.60
51 years and over	30.30	40.90	50.50	60.60	131.10	177.20	218.80	262.50
Female:.								
12-19 years	30.10	36.00	43.90	52.80	130.60	156.10	190.00	228.60
20-50 years	30.20	37.40	45.80	59.00	130.70	162.10	198.60	255.60
51 years and over	29.60	36.40	45.40	54.40	128.30	157.60	196.60	235.60
Families:.								
Family of 2: [4].								
20-50 years	69.80	88.40	109.40	136.90	302.30	383.10	474.20	593.10
51 years and over	65.80	85.00	105.50	126.50	285.30	368.20	456.90	547.90
Family of 4:.								
Couple, 20-50 years and children.								
2 and 3-5 years	100.80	126.80	155.80	192.20	436.80	549.30	675.30	832.80
6-8 and 9-11 years	117.00	150.00	187.10	226.40	506.80	649.90	810.60	980.80

[1] Basis is that all meals and snacks are purchased at stores and prepared at home. For specific foods and quantities of foods in the Thrifty Food Plan, see Family Economics and Nutrition Review, Vol. 13, No. 1 (2001), pp. 50-64; for specific foods and quantities of foods in the Low-Cost, Moderate-Cost, and Liberal Plans, see The Low-Cost, Moderate-Cost, and Liberal Food Plans, 2003 Administrative Report (2003). All four Food Plans are based on 1989-91 data and are are updated to current dollars using the Consumer Price Index for specific food items. [2] All costs are rounded to nearest 10 cents. [3] The costs given are for individuals in 4–person families. For individuals in other size families, the following adjustments are suggested: 1 person-add 20 percent; 2 person-add 10 percent; 3 person-add 5 percent; 4 person-no adjustment; 5 or 6 person-subtract 5 percent; 7 (or more) person-subtract 10 percent. To calculate overall household food costs, (1) adjust food costs for each person in household and them (2) sum these adjusted food costs. [4] Ten added for family size adjustment.

Center for Nutrition Policy and Promotion, (703) 305–7600.

Table 13-9.—Food Stamp Program: Participation and Federal costs, fiscal years 1995–2004

Fiscal year[1]	Average monthly participation[2]		Recipient benefits	Total cost[3]	Average monthly benefit	
	Persons	Housholds			Per person	Per household
	1,000	1,000	1,000 dollars	1,000 dollars	Dollars	Dollars
1995	26,619	10,879	22,764,067	24,619,544	71.27	174.37
1996	25,543	10,549	22,440,108	24,330,990	73.21	177.27
1997	22,858	9,455	19,548,863	21,485,345	71.27	172.30
1998	19,791	8,250	16,890,487	18,888,051	71.12	170.62
1999	18,183	7,668	15,769,397	17,710,400	72.27	171.37
2000	17,194	7,351	14,983,319	17,054,017	72.62	169.85
2001	17,318	7,449	15,547,390	17,789,482	74.81	173.93
2002	19,096	8,195	18,256,204	20,642,707	79.67	185.65
2003	21,259	9,154	21,404,276	23,862,125	83.90	194.86
2004[4]	23,858	10,279	24,629,786	27,148,535	86.03	199.69

[1] October 1 to September 30. [2] Participation data are 12-month averages. [3] Total cost includes matching funds for state administrative expenses (e.g., certification of households, quality control, anti-fraud activities; employment and training); and for other Federal costs (e.g., benefit redemption processing; computer support; electronic benefit transfer systems; retailer redemption and monitoring; certification of SSI recipients; nutrition education and program information). [4] Preliminary.

FNS, Budget Division/Program Reports, Analysis and Monitoring Branch, (703) 305–2163.

Table 13-10.—Food and Nutrition Service Programs: Federal costs of the National School Lunch, School Breakfast, Child Care Food, Summer Food Service, WIC, Special Milk, and Food Distribution Programs, fiscal years 1995–2004 [1]

Fiscal year[2]	Child Nutrition				Cost of food distributed[5]	WIC[6]	Special Milk	Food Distribution Programs[7]
	Cash payments[3]			Summer Food				
	School Lunch	School Breakfast	Child & Adult Care[4]					
	1,000 dollars	1,000 dollars	1,000 dollars	1,000 dollars	1,000 dollars	1,000 dollars	1,000 dollars	1,000 dollars
1995	4,466,186	1,048,244	1,411,144	235,477	732,967	3,436,184	16,982	513,694
1996	4,661,542	1,118,738	1,478,988	248,499	733,709	3,695,382	16,755	410,472
1997	4,934,059	1,214,279	1,514,226	242,594	661,280	3,843,802	17,432	518,174
1998	5,101,576	1,272,226	1,489,438	261,045	774,268	3,890,360	16,837	557,099
1999	5,314,723	1,345,546	1,555,814	266,654	753,623	3,940,327	16,493	601,371
2000	5,493,528	1,393,366	1,618,758	265,597	704,159	3,981,717	15,440	538,210
2001	5,612,297	1,450,126	1,666,147	268,402	917,016	4,149,736	15,593	716,498
2002	6,049,711	1,566,645	1,776,612	260,545	862,271	4,340,165	16,116	802,945
2003	6,340,645	1,651,653	1,845,885	255,171	908,807	4,526,834	14,470	663,138
2004[8]	6,662,878	1,774,306	1,932,659	260,397	1,030,334	4,892,592	14,411	676,891

[1] See table 13-7 for Food Stamp Program costs. [2] October 1–September 30. [3] Includes sponsor administrative costs for the Child and Adult Care Food Program (CACFP) and the Summer Food Service Programs (SFS), and State administrative and health clinic expenses for SFS. Excludes CACFP audit and startup costs and School Breakfast startup costs. [4] The Adult Care component was initiated in fiscal year 1989. [5] Includes entitlement commodities, bonus commodities, and cash-in-lieu for the National School Lunch, School Breakfast, Child and Adult Care Food, and Summer Food Service Programs. [6] Includes food costs, administrative costs, program evaluation funds, special grants, and Farmers Market projects for the Special Supplemental Food Program for Women, Infants and Children. [7] Includes entitlement and bonus commodities, cash-in-lieu of commodities, and administrative costs of the following programs: Indian Reservations (Needy Family), Nutrition for the Elderly, Commodity Supplemental Food, Charitable Institutions, Summer Camps, Emergency Food Assistance Program (TEFAP), Soup Kitchens/Food Banks, Disaster Feeding, Bureau of Federal Prisons, Veteran Affairs Administration, and the Food Stamp Program Elderly Pilot Project. [8] Preliminary. Note: Prior years incorporate revisions and corrections due to a major database upgrade.

FNS, Budget Division/Program Reports, Analysis and Monitoring Branch, (703) 305–2163.

Table 13-11.—Food and Nutrition Service program benefits: Cash payments made under the National School Lunch, School Breakfast, Child and Adult Care, Summer Food and Special Milk Programs and the value of food benefits provided under the Food Stamp, WIC, Commodity Distribution and the Emergency Feeding Food Programs, fiscal year 2004 [1]

State/Territory	Child Nutrition Program (cash payments only) [2]					Special Supplemental Food (WIC) [3]	Commodity distribution [4]	Food Stamp Program [5]	Emergency food assistance (TEFAP)	Total [5]
	Child and Adult Care Food	Summer Food	Special Milk	National School Lunch	Breakfast					
	1,000 dollars	1,000 dollars	1,000 dollars	1,000 dollars	1,000 dollars	1,000 dollars	1,000 dollars	1,000 dollars	1,000 dollars	1,000 dollars
Alabama	31,023	3,921	58	127,504	34,366	58,932	17,026	512,604	6,387	791,821
Alaska	6,017	306	9	19,175	3,750	14,207	2,106	64,405	1,408	111,383
Am. Samoa [5] ...	0	0	0	0	0	4,677	0	0	0	4,677
Arizona	38,898	1,101	120	143,902	36,147	84,408	27,977	577,868	7,232	917,653
Arkansas	22,902	1,707	24	76,440	23,705	39,324	9,766	346,881	3,038	523,788
California	216,032	12,817	674	893,349	230,214	628,080	122,704	1,989,813	49,945	4,143,629
Colorado	17,771	699	138	67,410	13,821	35,944	13,734	252,942	4,766	407,224
Connecticut	8,852	781	355	54,519	11,606	26,030	13,001	197,530	3,405	316,080
Delaware	7,715	1,138	34	14,141	3,853	7,233	3,197	56,542	771	94,624
District of Col.	2,966	2,306	6	14,435	3,832	8,414	3,709	97,508	1,039	134,215
Florida	96,511	14,504	77	382,577	104,568	174,720	51,310	1,268,549	16,490	2,109,307
Georgia	66,545	9,775	33	263,801	84,643	110,618	49,351	923,905	9,872	1,518,453
Guam	39	0	0	4,339	1,353	4,054	8	48,115	227	58,134
Hawaii	4,359	721	8	28,788	6,500	20,662	2,854	151,809	1,800	217,502
Idaho	4,014	2,004	194	30,519	7,068	13,464	4,752	90,972	1,678	154,665
Illinois	85,105	8,315	2,676	265,981	44,003	135,797	44,013	1,211,362	14,976	1,812,226
Indiana	26,513	3,458	301	121,284	26,869	52,201	28,739	549,501	6,750	815,615
Iowa	15,960	881	100	55,892	11,440	26,533	13,908	176,334	2,051	303,099
Kansas	25,087	1,244	135	57,151	13,645	26,679	11,327	158,017	3,580	296,865
Kentucky	22,640	6,494	103	111,761	37,627	52,167	22,199	542,744	6,305	802,039
Louisiana	44,689	6,298	44	155,412	49,379	67,505	43,353	753,905	8,379	1,128,965
Maine	8,550	712	79	19,953	5,126	7,400	3,477	139,619	2,244	187,158
Maryland	29,156	3,058	412	84,028	21,106	42,920	13,978	286,695	3,714	485,068
Massachusetts	37,921	4,158	444	91,848	22,929	47,678	21,073	304,421	5,093	535,565
Michigan	44,473	3,657	760	170,245	43,427	87,093	48,622	896,140	9,277	1,303,693
Minnesota	48,052	2,544	942	83,661	18,216	48,373	26,265	248,990	3,600	480,643
Mississippi	24,805	3,856	6	113,170	38,963	44,996	15,918	360,952	4,141	606,807
Missouri	33,696	5,803	429	121,829	34,298	54,902	26,067	663,426	8,145	948,594
Montana	7,710	669	36	15,900	3,817	9,340	5,797	79,197	1,025	123,492
Nebraska	20,183	698	83	35,949	7,075	16,735	13,423	108,691	1,454	204,291
Nevada	3,539	639	93	38,117	9,460	18,615	7,572	119,520	1,394	198,949
New Hampshire	2,388	554	197	13,588	2,697	6,156	6,450	43,549	1,284	76,863
New Jersey	40,536	6,770	796	130,756	22,382	66,940	24,749	377,526	7,316	677,770
New Mexico	29,067	4,685	14	60,256	19,746	28,094	14,299	217,424	3,494	377,079
New York	127,914	33,763	940	430,384	101,806	254,770	73,176	1,876,078	24,398	2,923,229
North Carolina	67,988	4,291	171	207,164	63,476	97,637	36,742	753,200	11,111	1,241,781
North Dakota ...	7,782	444	89	11,346	2,361	6,919	5,927	40,286	691	75,846
Ohio	53,842	4,872	717	192,526	46,417	108,401	43,441	1,009,262	15,185	1,474,663
Oklahoma	44,060	2,009	53	94,014	31,466	45,063	30,713	397,777	5,639	650,793
Oregon	19,078	1,212	150	64,328	22,917	45,029	9,872	415,267	6,706	584,559
Pennsylvania ...	47,903	10,302	702	199,191	44,497	96,053	42,630	933,274	9,750	1,384,302
Puerto Rico [5] ...	18,968	7,584	0	110,529	28,036	144,452	12,781	0	5,883	328,233
Rhode Island ...	6,053	911	94	19,432	4,562	9,726	3,215	73,551	1,837	119,382
South Carolina	22,450	7,041	10	118,264	38,441	45,956	18,547	501,205	3,968	755,882
South Dakota ..	5,546	586	37	17,801	4,140	9,073	11,142	53,934	1,613	103,872

See footnotes at end of table.

Table 13-11.—Food and Nutrition Service program benefits: Cash payments made under the National School Lunch, School Breakfast, Child and Adult Care, Summer Food and Special Milk Programs and the value of food benefits provided under the Food Stamp, WIC, Commodity Distribution and the Emergency Feeding Food Programs, fiscal year 2004 [1]—Continued

State/Territory	Child Nutrition Program (cash payments only) [2]					Special Supplemental Food (WIC) [3]	Commodity distribution [4]	Food Stamp Program [5]	Emergency Food Assistance (TEFAP)	Total [5]
	Child and Adult Care Food	Summer Food	Special Milk	National School Lunch	Break-fast					
	1,000 dollars	1,000 dollars	1,000 dollars	1,000 dollars	1,000 dollars	1,000 dollars	1,000 dollars	1,000 dollars	1,000 dollars	1,000 dollars
Tennessee	37,211	5,884	40	142,303	40,063	73,963	23,104	811,798	7,384	1,141,749
Texas	152,214	22,109	75	744,566	249,357	328,712	90,234	2,306,786	31,620	3,925,674
Utah	17,917	1,535	71	52,248	8,797	23,323	11,685	123,127	2,320	241,023
Vermont	3,317	306	111	8,845	2,736	8,151	2,779	40,076	966	67,286
Virginia	569	577	2	4,116	698	3,384	614	19,215	47	29,222
Virgin Islands	25,254	4,845	252	124,167	30,825	54,893	23,102	476,166	10,220	749,723
Washington	33,323	2,192	264	108,883	26,688	79,367	19,090	455,273	7,962	733,043
West Virginia	12,827	1,452	34	42,653	15,058	21,976	6,200	231,721	4,983	336,903
Wisconsin	30,699	2,747	1,190	88,526	12,388	46,830	23,589	269,439	5,577	480,985
Wyoming	4,139	206	23	8,588	1,933	4,401	2,151	24,981	460	46,883
Dpt. of Defense	0	0	0	5,324	0	0	1,117	0	0	6,453
Outlying Areas [6]	0	0	0	0	0	0	63	0	0	63
Total	1,812,766	231,137	14,411	6,662,878	1,774,306	3,578,967	1,204,638	24,629,786	360,597	40,269,486

[1] Preliminary. Excludes all administrative and program evaluation costs. [2] Excludes $1.1 million for Food Safety Education,$10.3 million for Team Nutrition. [3] Includes $17.8 million for WIC Farmers Market Nutrition Program benefits. [4] Includes distribution of bonus and entitlement commodities to the National School Lunch, Child and Adult Care, Summer Food Service, Charitable Institutions, Summer Camps, Food Distribution on Indian Reservations, Nutrition Services Incentive Program (NSIP, formerly Nutrition Program for the Elderly), Commodity Supplemental Food, Food Stamp Elderly Pilot Project and Disaster Feeding programs. Also includes cash-in-lieu of commodities for the National School Lunch and the Child and Adult Care Food programs. Effective FY 2003, NSIP cash-in-lieu was transferred to the Agency on Aging (DHHS). [5] Excludes Nutrition Assistance grants of $1,413 million for Puerto Rico, $7.7 million for the Northern Marianas, $5.5 million for American Samoa, and $0.6 million for nuclear affected areas of the Marshall Islands. [6] Dept. of Defense represents food service to children of armed forces personnel in overseas schools. [7] Outlying Areas include the Northern Marianas and the Marshall Islands.

FNS, Budget Division/Program Reports, Analysis and Monitoring Branch (703) 305–2163.

Table 13-12.—Food and Nutrition Service Programs: Persons participating, fiscal years 1995–2004

Fiscal year	National School Lunch Program [1]	School Breakfast Program [1]	Child and Adult Care Program [2]	Summer Food Service [3]	WIC Program [4]
	Thousands	Thousands	Thousands	Thousands	Thousands
1995	25,684	6,318	2,338	2,107	6,894
1996	25,942	6,583	2,404	2,213	7,186
1997	26,340	6,922	2,489	2,176	7,407
1998	26,598	7,141	2,599	2,308	7,367
1999	26,957	7,371	2,681	2,172	7,311
2000	27,307	7,554	2,707	2,103	7,192
2001	27,513	7,794	2,726	2,090	7,306
2002	27,998	8,144	2,850	1,923	7,491
2003	28,387	8,428	2,917	2,070	7,631
2004 [5]	28,962	8,903	3,006	1,994	7,904

[1] Average monthly participation (excluding summer months). [2] Average daily attendance (data reported quarterly). [3] Average daily attendance for peak month (July). [4] Average monthly participation. WIC is an abbreviation for the Special Supplemental Food Program for Women, Infants and Children. [5] Preliminary.
FNS, Budget Division/Program Reports, Analysis and Monitoring Branch (703) 305–2163.

Table 13-13.—Consumers' prices: Index number of prices paid for goods and services, United States, 1996–2005 [1]

[1982–84=100]

Year	Food	Nonfood items						All items
		Apparel and upkeep	Housing		Transportation	Medical care		
			Total	Rent				
1996	153.3	131.7	152.8	178.0	143.0	228.2		156.9
1997	157.3	132.9	156.8	183.4	144.3	234.6		160.5
1998	160.7	133.0	160.4	189.6	141.6	242.1		163.0
1999	164.1	131.3	163.9	195.0	144.4	250.6		166.6
2000	167.8	129.6	169.6	201.3	153.3	260.8		172.2
2001	173.1	127.3	176.4	208.9	154.3	272.8		177.1
2002	176.2	124.0	180.3	216.7	152.9	285.6		179.9
2003	180.0	120.9	184.8	221.9	157.6	297.1		184.0
2004	186.2	120.4	189.5	227.9	163.1	310.1		188.9
2005 [1]	190.7	119.5	195.7	233.7	173.9	323.2		195.3

[1] Reflects retail prices of goods and services usually bought by average families in urban areas of the United States. This index is the official index released monthly by the U.S. Department of Labor. Beginning 1978 data are for all urban consumers; earlier data are for urban wage earners and clerical workers.
ERS, Food Markets Branch, (202) 694–5349. Compiled from data of the U.S. Department of Labor.

CHAPTER XIV
STATISTICS OF FERTILIZERS AND PESTICIDES

This chapter contains statistics on percentages of crop acres treated by various types of fertilizers and pesticides. Nitrogen, phosphate, and potash are the most common fertilizers; herbicides, insecticides, fungicides, and other chemicals are the main categories of pesticides. Other chemicals include soil fumigants, vine killers, and dessicants. The tables show data for field crops for 1999–2002, fruits for 2001, and vegetables for 2002. NASS collects data for field crops on an annual basis and data for fruits and vegetables on a bi-yearly alternating basis. The surveyed States are generally the major producing States for each crop shown in the tables and represent 65–95 percent of the U.S. planted acres, depending on the selected crop. Quantities and rates of active chemical ingredients applied to each crop at State levels are available in the NASS series of "Agricultural Chemical Usage" reports.

Table 14-1.—Field crops: Fertilizer, and percent of area receiving applications, all States surveyed, 2001–2004 [1]

Crop	Nitrogen	Phosphate	Potash
	Percent	*Percent*	*Percent*
2001:			
Corn	96	79	65
Cotton, Upland	76	48	41
Potatoes, Fall	98	95	86
Soybeans	11	17	20
2002:			
Corn	96	79	68
Soybeans	20	26	29
Wheat, Durum	88	58	5
Wheat, Other Spring	86	74	27
Wheat, Winter	86	55	15
2003:			
Barley	93	79	29
Corn	96	79	64
Fall Potatoes	100	94	88
Sorghum	82	49	9
Upland Cotton	82	62	52
2004:			
Peanuts	60	66	63
Soybeans	21	26	23
Wheat, Durum	95	73	7
Wheat, Other Spring	93	79	25
Wheat, Winter	84	55	16

[1] Refers to acres receiving one or more applications of a specific fertilizer ingredient. See tables 14-2 through 14-17 for surveyed States.   Note: Acreage estimates are on page I–24 for corn, page II–1 for cotton, page III–13 for soybeans, and page I–1 for wheat.   [2] Data not available for all states for all years.

NASS, Environmental, Economics, and Demographics Branch, (202) 720–6146.

STATISTICS OF FERTILIZERS AND PESTICIDES

Table 14-2.—Barley: Pesticide usage, 2003 [1]

| State and Year | Percent treated and amount applied | | | | | | | |
| | Herbicide | | Insecticide | | Fungicide | | Other Chemicals | |
	Area applied	Pounds applied	Area applied	Pounds applied	Area applied	Pounds applied	Area applied	Pounds applied
	Percent	*Thousands*	*Percent*	*Thousands*	*Percent*	*Thousands*	*Percent*	*Thousands*
CA: 2003	67	32	*	*	*	*	*	*
ID: 2003	94	573	3	16	*	*	5	9
MN: 2003	89	88	8	3	39	9	....	....
MT: 2003	93	1,005	2	5	*	*	*	*
ND: 2003	98	1,067	4	12	11	20	....	....
PA: 2003	32	8	*	*	*	*	....	....
SD: 2003	86	34	....	....	*	*	....	....
UT: 2003	75	17	*	*	....	....	....	....
WA: 2003	94	358	....	....	....	....	· *	*
WI: 2003	21	5	....	....	....	....	....	....
WY: 2003	83	57	10	(2)	....	....	....	....

[1] Data not available for all States for all years. [2] Amount applied is less than 500 lbs. Note: Planted acres are on page I-36. * Insufficient number of reports to publish data.
NASS, Environmental, Economics, and Demographics Branch, (202) 720–6146.

Table 14-3.—Barley: Fertilizer usage, 2003 [1]

| State and Year | Percent treated and amount applied | | | | | |
| | Nitrogen | | Phosphate | | Potash | |
	Area applied	Pounds applied	Area applied	Pounds applied	Area applied	Pounds applied
	Percent	*Millions*	*Percent*	*Millions*	*Percent*	*Millions*
CA: 2003	72	5.2	32	0.6	2	0
ID: 2003	91	56.2	58	15.4	25	5.7
MN: 2003	91	11.4	87	5.6	66	4
MT: 2003	92	44.2	88	30.2	52	9.7
ND: 2003	98	116.5	91	50.7	20	4.2
PA: 2003	69	2.2	39	1.1	40	1.2
SD: 2003	82	2.6	78	1.9	13	0.2
UT: 2003	58	2.1	14	0.3	0	0
WA: 2003	99	22.5	58	2.5	8	0.5
WI: 2003	37	0.5	36	0.7	44	1.8
WY: 2003	78	7.3	60	2.4	22	0.7

[1] Data not available for all States for all years. Note: Planted acres are on page I-36.
NASS, Environmental, Economics, and Demographics Branch, (202) 720–6146.

Table 14-4.—Corn: Pesticide usage, 2000–2003 [1]

State and Year	Percent treated and amount applied			
	Herbicide [2]		Insecticide [3]	
	Area applied	Pounds applied	Area applied	Pounds applied
	Percent	Thousands	Percent	Thousands
CO:				
2000	97	1,501	59	505
2001	92	1,506	51	431
2003	77	1,099	39	278
GA:				
2000	94	31,723	31	1,996
2001	95	398	34	57
IL:				
2000	100	28,190	43	3,131
2001	100	31,868	42	1,787
2002	90	25,157	36	1,088
2003	98	28,926	58	1,640
IN:				
2000	99	15,460	30	797
2001	99	16,007	47	1,103
2002	90	11,535	39	729
2003	93	13,064	52	1,323
IA:				
2000	100	24,158	16	635
2001	99	20,627	7	864
2002	91	22,485	12	432
2003	96	25,328	14	623
KS:				
2000	93	7,765	31	287
2001	95	9,958	24	657
2003	97	6,041	29	337
KY:				
2000	95	2,600	26	65
2001	97	2,834	18	43
2003	97	2,716	16	52
MI:				
2000	99	5,658	10	131
2001	88	4,944	22	288
2003	98	4,934	14	206
MN:				
2000	99	10,597	8	369
2001	99	13,446	*	*
2002	96	10,002	6	212
2003	95	10,927	13	454
MO:				
2000	87	5,988	20	114
2001	97	7,232	37	167
2003	98	7,733	33	139
NE:				
2000	97	16,862	55	1,470
2001	99	15,159	48	1,104
2002	83	12,869	38	986
2003	93	15,209	36	742
NY:				
2000	92	2,312	31	204
2001	96	2,610	19	69
2003	96	2,107	28	141
NC:				
2000	93	1,732	46	363
2001	96	1,558	37	181
2003	97	1,854	28	213
ND:				
2000	71	1,284	*	*
2001	90	745	*	*
2003	96	1,564	*	*
OH:				
2000	99	10,339	24	603
2001	99	9,986	26	647
2002	91	8,424	14	125
2003	96	9,198	11	110
PA:				
2000	100	4,419	57	302
2001	99	4,484	60	550
2003	92	3,620	31	179
SD:				
2000	100	5,790	15	44
2001	96	5,622	87	87
2003	96	6,003	*	*
TX:				
2000	81	2,039	55	426
2001	90	1,990	76	664
2003	87	2,273	53	594
WI:				
2000	95	6,410	20	365
2001	98	6,265	16	155
2002	81	5,304	20	356
2003	98	6,533	22	273

[1] Data not available for all States for all years. [2] Insufficient number of reports to publish data for fungicides and other chemicals. [3] Amount applied excludes Bt (bacillus thuringiensis). Note: Planted acres are on page I-24. * Insufficient number of reports to publish data.
NASS, Environmental, Economics, and Demographics Branch, (202) 720–6146.

Table 14-5.—Corn: Fertilizer usage, 2000–2003 [1]

State and Year	Nitrogen		Phosphate		Potash	
	Area applied	Pounds applied	Area applied	Pounds applied	Area applied	Pounds applied
	Percent	*Millions*	*Percent*	*Millions*	*Percent*	*Millions*
CO:						
2000	95	182.0	78	42.2	17	7.4
2001	93	141.5	65	32.1	24	10.8
2003	89	138.2	59	30	31	8.3
GA:						
2001	97	28.6	91	12.6	87	20.8
IL:						
2000	99	1,797.7	83	739.3	82	1,028.5
2001	99	1,682.8	81	720.6	85	1,092.2
2002	94	1,698.3	77	754.1	77	1,028.7
2003	98	1,758.5	83	751.4	78	963.9
IN:						
2000	99	864.8	90	366.1	85	625.9
2001	98	837.4	85	331.7	86	660.0
2002	99	786.7	92	350.4	84	567.1
2003	99	854.4	85	376.4	83	640.0
IA:						
2000	95	1,533.0	74	503.2	74	630.9
2001	87	1,272.8	62	415.8	60	482.4
2002	94	1,408.0	72	515.8	69	607.4
2003	93	1,544.3	59	468.6	65	670.6
KS:						
2000	100	506.0	78	97.3	39	37.1
2001	97	444.4	71	93.5	19	24.8
2003	99	453.9	81	92.7	30	33.5
KY:						
2000	99	198.7	81	88.3	80	92.0
2001	91	173.4	87	92.5	82	99.9
2003	98	189.0	83	81	78	76.1
MI:						
2000	99	240.1	96	96.9	83	154.3
2001	91	251.3	78	85.9	78	175.2
2003	99	281.8	86	95.3	88	201.6
MN:						
2000	97	786.4	91	404.2	76	377.9
2001	97	750.2	90	283.4	81	340.5
2002	95	839.9	86	330.1	78	344.8
2003	95	835.9	89	309.2	73	349.2
MO:						
2000	100	422.7	82	136.3	82	169.1
2001	99	411.6	82	129.6	83	161.2
2003	99	482.2	91	162	88	210.7
NE:						
2000	99	1,260.7	82	243.2	22	21.5
2001	100	1,067.0	77	219.4	25	42.8
2002	97	1,195.5	70	220.3	21	32.3
2003	95	1,005.1	76	232.1	25	39.3
NY:						
2000	99	71.2	89	45.6	78	41.8
2001	100	76.8	98	49.4	90	45.6
2003	98	81.7	81	43.3	75	50.9
NC:						
2000	96	86.0	88	37.5	86	52.7
2001	98	81.8	85	41.6	84	56.6
2003	99	95.9	89	37.9	86	61.8
ND:						
2000	98	103.0	80	38.8	29	8.7
2001	94	89.9	83	33.8	38	10.1
2003	98	157.2	87	62.8	37	20.0
OH:						
2000	100	572.8	92	224.2	83	287.0
2001	100	572.1	92	210.8	89	338.9
2002	99	500.1	85	183.2	78	283.1
2003	100	538.6	91	225.7	85	284.6
PA:						
2000	95	103.8	87	59.9	67	35.9
2001	98	130.2	79	55.8	76	43.4
2003	91	98.6	72	52.2	66	33.5
SD:						
2000	99	418.9	92	153.6	39	36.1
2001	95	393.8	69	119.4	32	38.9
2003	92	396.5	78	159.8	25	27.9
TX:						
2000	98	304.0	85	80.3	27	15.9
2001	100	245.6	83	66.3	40	18.4
2003	98	261.4	85	70.9	37	17.1
WI:						
2000	97	300.7	89	120.6	90	161.0
2001	98	355.3	95	120.9	89	169.5
2002	98	325.0	87	102.2	88	202.2
2003	99	380.1	90	138.6	89	233.6

[1] Data not available for all States for all years.   Note: Planted acres are on page I-24.
NASS, Environmental, Economics, and Demographics Branch, (202) 720–6146.

Table 14-6.—Upland Cotton: Pesticide usage, 2000–2003 [1]

State and Year	Percent treated and amount applied							
	Herbicide		Insecticide [2]		Fungicide		Other Chemicals	
	Area applied	Pounds applied	Area applied	Pounds applied	Area applied	Pounds applied	Area applied	Pounds applied
	Percent	Thousands	Percent	Thousands	Percent	Thousands	Percent	Thousands
AL:								
2000	97	1,435	67	270	16	84	58	398
2003	99	1,336	84	260	15	44	93	930
AZ:								
2000	94	497	66	455	10	31	79	670
2003	94	382	74	374	*	*	80	323
AR:								
2000	95	1,993	82	1,610	17	57	89	1,459
2001	96	2,312	53	2,038	8	9	78	1,395
2003	96	2,703	89	3,575	17	64	92	1,947
CA:								
2000	99	1,475	90	1,051	1	9	99	2,714
2001	*	*	*	*	*	*	*	*
2003	97	1,005	95	899	7	13	96	2,091
GA:								
2000	98	3,526	81	725	**	**	78	3,258
2001	93	2,958	59	366	*	*	65	1,902
2003	96	2,994	73	746	4	43	91	2,709
LA:								
2000	96	1,825	98	4,795	23	229	88	749
2001	95	2,552	93	2,217	16	70	88	931
2003	100	1,448	97	2,007	17	11	99	690
MS:								
2000	98	3,557	99	6,112	15	131	99	1,986
2001	99	3,913	92	3,306	5	22	95	2,461
2003	100	3,475	94	1,534	17	63	99	1,590
MO:								
2000	94	677	90	360	*	*	97	695
2003	96	636	74	146	*	*	95	822
NC:								
2000	99	2,375	94	510	4	19	91	1,921
2001	*	*	*	*	*	*	*	*
2003	97	2,118	88	420	7	41	90	2,041
SC:								
2003	92	470	97	141	3	4	79	307
TN:								
2000	99	1,347	100	4,333	20	77	93	691
2003	98	1,270	88	422	20	33	90	863
TX:								
2000	92	7,847	69	20,639	*	*	29	1,593
2001	90	21,098	68	23,810	4	212	55	13,435
2003	99	7,701	36	3,102	2	22	31	1,400

[1] Data not available for all States for all years. [2] Amount applied excludes Bt (bacillus thuringiensis). * Insufficient number of reports to publish data. ** No reports received for this pesticide class. Note: Planted acres are on page II-1.

NASS, Environmental, Economics, and Demographics Branch, (202) 720–6146.

Table 14-7.—Upland Cotton: Fertilizer usage, 2000–2003 [1]

State and Year	Percent treated and amount applied					
	Nitrogen		Phosphate		Potash	
	Area applied[2]	Pounds applied	Area applied[2]	Pounds applied	Area applied[2]	Pounds applied
	Percent	*Millions*	*Percent*	*Millions*	*Percent*	*Millions*
AL:						
2000	100	60.5	95	35.2	91	46.7
2003	97	51.9	84	31.2	83	33.4
AZ:						
2000	98	35.6	30	4.7	8	0.9
2003	93	35.3	35	4.6	11	0.8
AR:						
2000	100	84.2	78	30.5	84	66.1
2001	93	80.3	63	24.6	68	54.0
2003	97	89.7	84	33.5	90	79.9
CA:						
2000	98	105.4	29	12.6	12	5.3
2001	*	*	*	*	*	*
2003	94	72.9	47	14.3	25	11.6
GA:						
2000	96	124.9	94	77.6	93	117.7
2001	99	116.2	92	71.9	93	119.3
2003	100	124.5	90	65.8	91	105.8
LA:						
2000	100	60.7	64	20.1	66	33.0
2001	95	70.8	50	18.4	52	35.1
2003	99	45.1	45	8.8	59	16.1
MS:						
2000	100	147.7	44	29.5	68	86.1
2001	99	179.9	31	25.8	46	72.5
2003	99	119.8	45	23.0	70	82.2
MO:						
2000	100	40.4	86	11.7	95	33.5
2003	100	35.5	73	11.6	81	26.2
NC:						
2000	96	76.0	80	34.9	91	98.5
2001	*	*	*	*	*	*
2003	97	59.9	74	24.4	93	79.7
SC:						
2003	95	16.0	78	7.9	90	21.6
TN:						
2000	99	47.5	93	29.8	98	50.4
2003	97	50.0	92	27.3	96	46.4
TX:						
1999	71	281.8	45	112.8	23	26.6
2000	63	263.4	54	136.9	26	31.1
2001	52	195.9	37	85.2	14	16.4
2003	61	258.0	50	141.7	20	28.6

[1] Data not available for all States for all years.   [2] Planted acres are on page II-1.
NASS, Environmental, Economics, and Demographics Branch, (202) 720–6146.

Table 14-8.—Peanuts: Pesticide usage, 2004 [1]

State and Year	Percent treated and amount applied							
	Herbicide		Insecticide		Fungicide		Other Chemicals	
	Area applied	Pounds applied	Area applied	Pounds applied	Area applied	Pounds applied	Area applied	Pounds applied
	Percent	Thousands	Percent	Thousands	Percent	Thousands	Percent	Thousands
AL: 2004	100	277	81	200	100	896	....	....
FL: 2004	100	298	88	199	100	835	....	....
GA: 2004	99	878	77	569	99	2,275	....	....
NC: 2004	100	221	92	161	96	164	43	1,404
TX: 2004	94	258	3	2	67	154	....	....

[1] Planted acres are on page III-8.
NASS, Environmental, Economics, and Demographics Branch, (202) 720–6146.

Table 14-9.—Peanuts: Fertilizer usage, 2004 [1]

State and Year	Percent treated and amount applied					
	Nitrogen		Phosphate		Potash	
	Area applied	Pounds applied	Area applied	Pounds applied	Area applied	Pounds applied
	Percent	Millions	Percent	Millions	Percent	Millions
AL: 2004	70	4.3	79	8.6	75	12.4
FL: 2004	71	3.3	80	5.4	94	12.7
GA: 2004	48	5.3	59	17.5	51	23.7
NC: 2004	37	1	35	1.2	64	6.7
TX: 2004	86	14.4	77	10.6	62	9.3

[1] Planted acres are on page III-8.
NASS, Environmental, Economics, and Demographics Branch, (202) 720–6146.

Table 14-10.—Fall potatoes: Pesticide usage, 2000–2003 [1]

State and Year	Herbicide		Insecticide [2]		Fungicide		Other Chemicals	
	Area applied [3]	Pounds applied	Area applied [3]	Pounds applied	Area applied [3]	Pounds applied	Area applied [3]	Pounds applied
	Percent	Thousands	Percent	Thousands	Percent	Thousands	Percent	Thousands
CO:								
2003	84	168	71	40	90	122	57	14,815
ID:								
2001	75	714	93	853	70	691	59	46,698
2003	89	693	78	458	78	606	57	31,892
ME:								
2001	92	28	88	13	98	530	97	405
2003	100	34	88	18	100	576	21	52
MI:								
2003	94	68	99	19	96	382	48	696
MN:								
2001	78	53	95	18	97	431	56	456
2003	94	42	69	6	98	461	4	1,294
ND:								
2001	*	*	*	*	*	*	*	*
2003	82	57	80	29	99	1,350	3	311
OR:								
2001	*	*	*	*	*	*	*	*
2003	95	71	83	140	94	169	70	3,626
PA:								
2003	91	28	99	23	96	126	6	3
WA:								
2001	92	290	95	647	91	1,108	78	14,470
2003	94	339	97	701	99	1,704	77	20,847
WI:								
2001	88	73	100	110	97	1,193	86	2,644
2003	94	72	99	133	99	1,038	38	1,846

[1] Data not available for all States for all years. [2] Amount applied excludes Bt (bacillus thuringiensis). [3] Planted acres are on page IV-19.
NASS, Environmental, Economics, and Demographics Branch, (202) 720–6146.

Table 14-11.—Fall potatoes: Fertilizer usage, 2000–2003 [1]

State and Year	Nitrogen		Phosphate		Potash	
	Area applied	Pounds applied	Area applied	Pounds applied	Area applied	Pounds applied
	Percent	Millions	Percent	Millions	Percent	Millions
CO:						
2003	98	15.9	96	9.7	90	7.0
ID:						
2001	99	79.6	97	63.2	77	35.1
2003	100	81.4	95	63.2	86	37.3
ME:						
2001	98	11.0	98	11.4	98	11.8
2003	100	12.0	100	12.3	100	13.8
MI:						
2003	100	8.5	98	4.0	98	9.1
MN:						
2001	93	6.4	89	4.5	89	7.6
2003	100	8.6	94	4.9	92	8.5
ND:						
2001	*	*	*	*	*	*
2003	97	16.5	92	10.0	84	13.7
OR:						
2001	*	*	*	*	*	*
2003	100	10.7	96	7.4	84	8.8
PA:						
2003	100	1.9	99	1.3	99	1.4
WA:						
2001	97	37.6	92	33.0	92	37.4
2003	100	43.1	85	33.2	82	30.7
WI:						
2001	100	22.0	98	13.7	100	24.3
2003	100	19.9	99	12.2	100	25.5

[1] Data not available for all states for all years.   Note: Planted acres are on page IV-19.
NASS, Environmental, Economics, and Demographics Branch, (202) 720–6146.

Table 14-12.—Sorghum: Pesticide usage, 2003 [1]

| State and Year | Percent treated and amount applied | | | |
| | Herbicide | | Insecticide [2] | |
	Area applied	Pounds applied	Area applied	Pounds applied
	Percent	Thousands	Percent	Thousands
CO: 2003	52	132	*	*
KS: 2003	90	9,014		
MO: 2003	98	571	6	4
NE: 2003	98	2,030	4	29
OK: 2003	84	329	*	*
SD: 2003	87	430	*	*
TX: 2003	78	2,881	20	208

[1] Data not available for all States for all years. [2] Insufficient number of reports to publish data for other chemicals. [3] Amount applied excludes Bt (bacillus thuringiensis). Note: Planted acres are on page I-41. * Insufficient number of reports to publish data.
NASS, Environmental, Economics, and Demographics Branch, (202) 720–6146.

Table 14-13.—Sorghum: Fertilizer usage, 2003 [1]

| State and Year | Percent treated and amount applied | | | | | |
| | Nitrogen | | Phosphate | | Potash | |
	Area applied	Pounds applied	Area applied	Pounds applied	Area applied	Pounds applied
	Percent	Millions	Percent	Millions	Percent	Millions
CO: 2003	61	7.8	39	5.5	0	0
KS: 2003	97	261.8	55	57.5	4	4.7
MO: 2003	100	25.0	75	9.1	72	10.8
NE: 2003	99	56.7	40	6.1	1	0.1
OK: 2003	69	15.5	36	3.6	11	0.8
SD: 2003	84	13.0	54	4.4	3	0.1
TX: 2003	63	182.8	43	45.5	14	5.5

[1] Data not available for all States for all years. Note: Planted acres are on page I-41.
NASS, Environmental, Economics, and Demographics Branch, (202) 720–6146.

Table 14-14.—Soybeans: Pesticide usage, 2000–2003 [1]

| State and Year | Percent treated and amount applied [2] | | | |
| | Herbicide | | Insecticide [3] | |
	Area applied	Pounds applied	Area applied	Pounds applied
	Percent	Thousands	Percent	Thousands
AR: 2000	86	2,918	3	4
2001	80	2,440	*	*
2002	90	2,945	14	112
2004	92	3,642	7	57
IL: 2000	98	10,582	1	3
2001	96	10,102	*	*
2002	100	12,939	*	*
2004	98	10,832	1	15
IN: 2000	99	5,414	*	*
2001	98	5,612	*	*
2002	100	7,853	*	*
2004	99	7,037		
IA: 2000	98	13,053	*	*
2001	95	11,704	*	*
2002	99	13,143	9	58
2004	98	11,964	1	5
KS: 2000	94	2,953	*	*
2002	98	2,931	*	*
2004	97	3,225		

See footnotes at end of table.

Table 14-14.—Soybeans: Pesticide usage, 2000–2004 [1]—Continued

State and Year	Percent treated and amount applied [2]			
	Herbicide		Insecticide [3]	
	Area applied	Pounds applied	Area applied	Pounds applied
	Percent	Thousands	Percent	Thousands
KY:				
2000	88	1,151	1	6
2002	100	1,479	*	*
LA:				
2000	96	1,091	56	173
2002	98	1,257	72	470
MD:				
2002	98	753	3	*
MI:				
2000	98	2,094	*	*
2002	98	2,496	*	*
MN:				
2000	95	7,151	*	*
2001	99	6,363	*	*
2002	99	7,073	*	*
2004	98	8,289		
MS:				
2000	99	2,096	5	23
2002	98	2,392	24	24
MO:				
2000	98	5867	**	**
2001	95	4,691	*	*
2002	99	5,924	*	*
2004	98	5,394		
NE:				
2000	98	5,795	*	*
2001	96	5,336	*	*
2002	100	6,014	4	36
2004	94	5,625	15	274
NC:				
2000	92	1,016	7	20
2002	95	1,361	25	89
ND:				
2000	99	2,046	**	**
2002	100	3,350	*	*
2004	99	4,460		
OH:				
2000	98	4,586	1	*
2001	96	4,216	*	3
2002	100	6,365	*	*
2004	98	5,597	3	6
SD:				
2000	98	4,863	**	*
2002	100	5,117	19	97
2004	96	4,763	19	70
TN:				
2000	95	1,319	1	*
2002	100	1,496	10	1
VA:				
2002	94	591	46	25
WI:				
2000	85	1,169	**	**
2002	86	1,253	*	*

[1] Data not available for all States for all years. [2] Insufficient number of reports to publish data for fungicides and other chemicals. [3] Amount applied excludes Bt (bacillus thuringiensis). * Insufficient number of reports to publish data. ** No reports received for this pesticide class. Note: Planted acres are on page III-13.
NASS, Environmental, Economics, and Demographics Branch, (202) 720–6146.

Table 14-15.—Soybeans: Fertilizer usage, 2000–2004 [1]

State and Year	Percent treated and amount applied					
	Nitrogen		Phosphate		Potash	
	Area applied	Pounds applied	Area applied	Pounds applied	Area applied	Pounds applied
	Percent	Millions	Percent	Millions	Percent	Millions
AR:						
2000	10	21.0	30	43.4	31	73.0
2001	3	3.4	30	42.8	24	54.9
2002	7	5.2	36	57.8	35	66.1
2004	10	9.3	38	67.2	38	98.4
IL:						
2000	11	16.8	16	77.5	29	286.0
2001	10	42.8	12	95.8	22	250.5
2002	18	37.5	25	143.1	38	422.6
2004	14	49.5	18	185.1	32	525.2
IN:						
2000	7	11.0	15	53.9	33	207.8
2001	12	11.4	20	58.1	36	222.4
2002	18	17.4	24	67.9	46	276.0
2004	15	30.7	25	121.4	40	331.5
IA:						
2000	15	81.0	22	110.1	22	138.0
2001	5	9.9	9	47.9	10	71.3
2002	3	9.3	7	48.3	12	163.7
2004	10	38.4	11	99.8	15	157.2
KS:						
2000	18	10.3	16	16.9	*	*
2002	24	12.2	25	28.7	8	5.9
2004	22	22.0	25	34.2	5	7.1
KY:						
2000	13	7.7	40	31.7	39	37.7
2002	21	9.6	37	30.3	38	46.6
LA:						
2000	6	1.5	20	7.3	26	15.6
2002	2	0.1	18	5.5	18	7.5
MD:						
2002	23	2.7	17	2.9	26	7.0
MI:						
2000	37	11.1	40	44.8	72	131.2
2002	44	24.4	34	32.0	67	119.1
MN:						
2000	8	10.2	9	24.1	24	118.6
2001	13	15.3	13	32.3	12	41.5
2002	11	16.1	12	34.2	10	39.1
2004	19	41.3	18	81.2	16	85.6
MS:						
2000	9	3.4	19	14.3	20	23.5
2002	12	3.7	20	15.8	20	25.7
MO:						
2000	20	27.5	28	98.1	27	94.2
2001	6	5.4	24	52.2	22	61.7
2002	13	11.8	29	62.9	36	158.1
2004	20	23.4	35	128.1	38	206.3

See footnotes at end of table.

Table 14-15.—Soybeans: Fertilizer usage, 2000–2004 [1]—Continued

State and Year	Percent treated and amount applied					
	Nitrogen		Phosphate		Potash	
	Area applied	Pounds applied	Area applied	Pounds applied	Area applied	Pounds applied
	Percent	*Millions*	*Percent*	*Millions*	*Percent*	*Millions*
NE:						
2000	30	19.8	20	36.7	15	6.2
2001	22	23.4	21	38.3	10	13.2
2002	31	23.1	36	79.9	11	14.6
2004	25	24.6	28	76.8	7	12.4
NC:						
2000	38	12.6	62	64.7	47	47.7
2002	36	14.4	36	25.0	41	51.3
ND:						
2000	46	27.8	41	25.3	*	*
2002	64	44.1	59	50.5	11	3.3
2004	64	61.3	63	113.1	11	15.7
OH:						
2000	25	21.7	32	70.2	47	192.8
2001	17	19.1	30	63.9	41	164.7
2002	20	14.1	27	62.6	56	276.4
2004	20	19.0	24	73.0	43	282.0
SD:						
2000	38	24.3	43	66.0	12	12.2
2002	37	32.5	41	102.0	15	24.4
2004	42	38.6	45	116.0	8	12.5
TN:						
2000	18	3.0	29	14.3	31	22.2
2002	42	14.5	47	31.1	57	48.6
VA:						
2002	25	3.6	33	7.3	46	18.4
WI:						
2000	24	6.5	30	16.6	40	46.2
2002	40	9.2	35	18.9	48	54.7

[1] Data not available for all States for all years.   Note: Planted acres are on page III-13.   * Insufficient number of reports to publish data.
NASS, Environmental, Economics, and Demographics Branch, (202) 720–6146.

Table 14-16.—Wheat: Pesticide usage, 2000–2004 [1]

State and Year	Percent treated and amount applied [2]			
	Herbicide		Insecticide [3]	
	Area applied	Pounds applied	Area applied	Pounds applied
	Percent	*Thousands*	*Percent*	*Thousands*
Winter				
AR:				
2000 ..	41	239	**	**
CO:				
2000 ..	23	281	*	*
2002 ..	12	68	*	*
2004 ..	54	908	....	....
ID:				
2000 ..	89	411	4	15
2004 ..	94	380	1	2
IL:				
2000 ..	44	21	**	**
2002 ..	39	10	*	*
2004 ..	35	41	....	....
IN:				
1999 ..	39	28	*	*
KS:				
2000 ..	31	478	8	395
2002 ..	32	347	7	30
2004 ..	38	1,138	....	....
KY:				
2000 ..	51	57	8	15
MI:				
2004 ..	50	94	11	3
MO:				
2000 ..	51	47	*	*
2002 ..	12	12	*	*
2004 ..	35	109	8	9
MT:				
2000 ..	91	745	*	*
2002 ..	80	433	*	*
2004 ..	95	2,533	....	....
NE:				
2000 ..	26	248	**	**
2002 ..	49	225	*	*
2004 ..	51	537	....	....
NC:				
2000 ..	65	206	19	3
OH:				
2000 ..	18	53	**	**
2002 ..	31	72	*	*
2004 ..	29	96	....	....
OK:				
2000 ..	25	94	*	*
2002 ..	36	155	32	285
2004 ..	34	267	24	511
OR:				
2000 ..	99	550	**	**
2004 ..	98	694	3	7
SD:				
2004 ..	66	646	....	....
TX:				
2000 ..	12	441	1	26
2002 ..	34	274	21	291
2004 ..	19	810	7	189
WA:				
2000 ..	95	847	**	**
2002 ..	87	856	*	*
2004 ..	88	1,007	....	....

See footnotes at end of table.

Table 14-16.—Wheat: Pesticide usage, 2000–2004 [1]—Continued

State and Year	Percent treated and amount applied [2]			
	Herbicide		Insecticide [3]	
	Area applied	Pounds applied	Area applied	Pounds applied
	Percent	*Thousands*	*Percent*	*Thousands*
Durum				
MT:				
2004 ..	99	508		
ND:				
2000 ..	97	2,807	*	*
2002 ..	100	1,238	*	*
2004 ..	99	1,216		
Other Spring				
ID:				
2004 ..	92	288	4	6
MN:				
2000 ..	92	1,845	*	*
2002 ..	84	858	*	*
2004 ..	99	1,054	10	28
MT:				
2000 ..	92	2,955	**	**
2002 ..	89	2,171	*	*
2004 ..	95	1,652		
ND:				
2000 ..	97	4,205	*	*
2002 ..	95	3,749	*	*
2004 ..	97	3,452		
OR:				
2004 ..	95	133	4	1
SD:				
2000 ..	93	619	**	**
2004 ..	89	702		
WA:				
2004 ..	99	364	4	8

[1] Data not available for all States for all years.   [2] Insufficient number of reports to publish data for fungicides and other chemicals.   [3] Amount applied excludes Bt (bacillus thuringiensis).   * Insufficient number of reports to publish data.   ** No reports received for this pesticide class.   Note: Planted acres are on page I-2.

NASS, Environmental, Economics, and Demographics Branch, (202) 720–6146.

Table 14-17.—Wheat: Fertilizer usage, 2001-2004 [1]

State and Year	Percent treated and amount applied					
	Nitrogen		Phosphate		Potash	
	Area applied	Pounds applied	Area applied	Pounds applied	Area applied	Pounds applied
	Percent	*Millions*	*Percent*	*Millions*	*Percent*	*Millions*
Winter						
AR:						
2000	92	110.1	28	12.3	28	16.0
CO:						
2000	87	85.2	14	5.6	*	*
2002	64	55.1	31	18.2	*	0.0
2004	59	51.2	31	15.8	5	2.7
ID:						
2000	90	75.5	54	12.1	13	2.7
2004	89	89.2	62	18.5	31	6.1
IL:						
2000	98	80.1	82	55.5	78	65.7
2002	96	59.4	76	37.0	74	46.8
2004	98	103.2	85	74.2	77	92.3
IN:						
1999	97	46.3	91	31.6	90	39.0
KS:						
2000	94	522.9	65	178.7	6	11.2
2002	91	487.4	64	162.2	8	24.5
2004	90	788.6	62	281.8	6	23.4
KY:						
2000	80	52.0	62	25.9	60	29.2
MI:						
2004	97	73.5	71	27.5	77	38.4
MO:						
2000	96	86.8	76	39.9	84	59.1
2002	97	65.9	75	31.8	74	40.8
2004	97	125.9	84	52.9	86	70.0
MT:						
2000	82	74.2	77	34.0	43	8.2
2002	88	38.4	81	18.5	46	4.8
2004	92	83.0	83	47.3	21	3.9
NE:						
2000	90	76.5	68	31.5	*	*
2002	79	57.6	45	22.6	4	2.1
2004	73	76.4	42	24.3	3	1.2
NC:						
2000	88	78.3	48	15.8	56	30.9
OH:						
2000	94	107.0	81	64.1	82	74.0
2002	98	66.4	89	46.8	88	51.4
2004	100	91.6	95	65.8	90	69.5
OK:						
2000	97	393.3	62	148.4	5	8.3
2002	92	203.6	59	65.9	4	6.4
2004	92	571.0	62	147.8	13	22.0
OR:						
2000	99	46.1	11	1.8	7	1.4
2004	96	64.7	11	5.3	6	2.5
SD:						
2000	91	60.8	61	26.6	12	1.3
2004	77	105.8	58	44.6	7	5.1
TX:						
2000	55	280.2	35	79.7	14	32.0
2002	62	124.0	28	30.3	7	5.4
2004	64	347.7	35	116.6	9	9.6
WA:						
2000	100	111.7	30	10.2	6	1.3
2002	99	126.5	39	12.3	11	3.5
2004	97	161.2	24	11.6	3	1.4

See footnotes at end of table.

Table 14-17.—Wheat: Fertilizer usage, 2001–2004 [1]—Continued

State and Year	Percent treated and amount applied					
	Nitrogen		Phosphate		Potash	
	Area applied	Pounds applied	Area applied	Pounds applied	Area applied	Pounds applied
	Percent	*Millions*	*Percent*	*Millions*	*Percent*	*Millions*
Durum MT:						
2004	96	32.5	84	11.8	10	0.6
ND:						
2000	86	173.8	66	47.6	5	2.1
2002	88	116.1	58	31.6	5	1.2
2004	95	115.3	73	46.9	7	1.7
Other Spring: ID:						
2004	93	56.1	63	12.7	23	4.4
MN:						
2000	94	169.8	85	51.8	73	29.3
2002	89	129.0	83	60.8	68	44.7
2004	98	180.1	91	75.5	54	34.8
MT:						
2000	90	167.6	84	75.5	36	15.6
2002	66	97.8	54	47.0	21	14.9
2004	79	134.6	69	72.6	13	9
ND:						
2000	97	501.8	83	170.1	12	13.3
2002	97	499.8	83	197.7	19	30.6
2004	98	691.9	86	269	27	39.9
OR:						
2004	91	9.7	28	1.7	9	0.5
SD:						
2000	86	173.8	66	47.6	5	2.1
2004	92	132.5	68	53.2	19	8.5
WA:						
2004	100	45.4	67	7.4	9	2.1

[1] Data not available for all States for all years.   * Insufficient number of reports to publish data.   Note: Planted acres are on page I-2.

NASS, Environmental, Economics, and Demographics Branch, (202) 720–6146.

Table 14-18.—Fruits: Percent of acres receiving applications, for surveyed States, 2003 [1]

Crop	Herbicide	Insecticide	Fungicide	Other
		Percent		
Apples	42	94	90	20
Apricots	46	78	78	4
Avocados	22	49	*	5
Blackberries	83	84	89	*
Blueberries	60	89	86	4
Cherries, Sweet	35	83	84	18
Cherries, Tart	40	90	93	41
Dates	*	12	*	*
Figs	8	....	*	*
Grapefruit	64	83	76	7
Grapes, All	47	42	68	8
Grapes, Raisin	42	30	61	19
Grapes, Table	46	56	89	24
Grapes, Wine	52	45	74	8
Kiwifruit	8	15	*	*
Lemons	45	61	27	16
Nectarines	53	78	77	17
Olives	31	41	24	7
Oranges	59	84	61	8
Peaches	51	84	80	9
Pears	32	89	85	20
Plums	56	71	54	10
Prunes	41	69	50	11
Raspberries	87	87	93	*
Tangelos	62	84	73	*
Tangerines	46	68	60	4
Temples	58	96	95	*

[1] Refers to acres receiving one or more applications of a specific agricultural chemical.　*Insufficient number of reports to publish data.

NASS, Environmental, Economics, and Demographics Branch, (202) 720–6146.

Table 14-19.—Vegetables: Percent of acres receiving applications, for surveyed States, 2004[1]

Crop	Herbicide	Insecticide	Fungicide	Other
		Percent		
Asparagus	69	69	37	66.4
Beans, Lima, Processing	91	88	94	....
Beans, Snap, Fresh	60	76	79	2
Beans, Snap, Proc	91	87	65	....
Broccoli	34	74	12	....
Cabbage, Fresh	57	85	58	4
Cantatoupes	37	54	51	22
Carrots, Fresh	46	15	53	23
Carrots, Proc	81	50	63	36
Cauliflower	26	81	8	....
Celery	39	57	38	....
Sweet Corn, Fresh	79	88	36	2
Sweet Corn, Proc	92	71	17	....
Cucumbers, Fresh	49	77	88	17
Cucumbers, Pickles	84	32	37	2
Garlic	75	57	63	....
Honeydews	17	84	29	11
Head Lettuce	38	89	63	1
Other Lettuce	43	85	66	4
Onions, Bulb	78	77	76	18
Green Peas, Proc	88	21	2	....
Peppers, Bell	27	89	80	50
Pumpkins	74	68	76	1
Spinach	24	66	50	....
Squash	39	71	74	9
Strawberries	16	72	77	44
Tomatoes, Fresh	64	90	89	51
Tomatoes, Proc	70	53	63	22
Watermelons	46	51	85	15

[1] Refers to acres receiving one or more applications of a specific agricultural chemical.   * Insufficient number of reports to publish data.

NASS, Environmental, Economics, and Demographics Branch, (202) 720–6146.

CHAPTER XV

MISCELLANEOUS AGRICULTURAL STATISTICS

This chapter contains miscellaneous data which do not fit into the preceding chapters. Included here are summary tables on foreign trade in agricultural products; statistics on fishery products; tables on refrigerated warehouses; and statistics on crops in Alaska.

Foreign Agricultural Trade Statistics

Agricultural products, sometimes referred to as food and fiber products, cover a broad range of goods from unprocessed bulk commodities like soybeans, feed corn and wheat to highly-processed, high-value foods and beverages like sausages, bakery goods, ice cream, or beer sold in retail stores and restaurants. All of the products found in Chapters 1-24 (except for fishery products in Chapter 3) of the U.S. Harmonized Tariff Schedule are considered agricultural products. These products generally fall into the following categories: grains, animal feeds, and grain products (like bread and pasta); oilseeds and oilseed products (like canola oil); livestock, poultry and dairy products including live animals, meats, eggs, and feathers; horticultural products including all fresh and processed fruits, vegetables, tree nuts, as well as nursery products and beer and wine; unmanufactured tobacco; and tropical products like sugar, cocoa, and coffee. Certain other products are considered "agricultural," the most significant of which are essential oils (Chapter 33), raw rubber (Chapter 40), raw animal hides and skins (Chapter 41), and wool and cotton (Chapters 51-52). Manufactured products derived from plants or animals, but which are not considered "agricultural" are cotton yarn, textiles and clothing; leather and leather articles of apparel; and cigarettes and spirits.

U.S. foreign agricultural trade statistics are based on documents filed by exporters and importers and compiled by the Bureau of the Census. Puerto Rico is a Customs district within the U.S. Customs territory, and its trade with foreign countries is included in U.S. export and import statistics. U.S. export and import statistics include merchandise trade between the U.S. Virgin Islands and foreign countries even though the Virgin Islands of the United States are not officially a part of the U.S. Customs territory.

Data on trade of other U.S. outlying possessions with foreign countries is not compiled by the United States. Export statistics are fully compiled on shipments to all countries, except Canada, where the value of commodities classified under each individual Schedule B number is over $2,500. Value data for such commodities valued under $2,501 are estimated for individual countries using factors based on the ratios of low-valued shipments to individual country totals for past periods. The estimates for low-valued shipments are shown under a single Schedule B number and are omitted from the statistics for the detailed commodity classifications. Shipments valued under $2,501 to all counties, except Canada, represent slightly less that 2.5 percent of the monthly value of U.S. exports to those countries. As a result of the data exchange between the United States and Canada, the United States has adopted the Canadian import exemption level for its export statistics on shipments to Canada. The Canadian import exemption level is based on total value per shipment rather than value per commodity classification line item.

The export value, the value at the port of exportation, is based on the selling price and includes inland freight, insurance, and other charges to the port. The country of destination is the country of ultimate destination or where the commodities are consumed or further processed. When the shipper does not know the ultimate destination, the shipments are credited to the last country, as known at the time of shipment from the United States.

Agricultural products, like manufactured goods, are often transhipped from the one country to another. Shippers are asked to identify the ultimate destination of a shipment. However, transhipment points are often recorded as the ultimate destination even though the actual point of consumption may be in a neighboring state. Thus, exports to countries which act as transhipment points are generally overstated, while exports to neighboring countries are often understated. Major world transhipment points include the Netherlands, Hong Kong, and Singapore. In such cases, exports are over reported for the Netherlands, but under reported for Germany, Belgium and the United Kingdom. They are overstated to Hong Kong, but under reported to China, and they overstated to Singapore, but understated to Malaysia and Indonesia. After the collapse of communism in Eastern Europe and Russia, Germany and the Baltic countries became important transhipment points to those countries further east.

Imports for consumption are a combination of entries for immediate consumption and withdrawals from warehouses for consumption. The import value, defined generally as the market value in the foreign country, excludes import duties, ocean freight, and marine insurance. The country of origin is defined as the country where the commodities were grown or processed. Where the country of origin is not known, the imports are credited to the country of shipment.

Import statistics are fully compiled on shipments valued over $1,250. Value data for shipments valued under $1,251 are not required to be reported on formal entries. They are estimated for individual countries using factors based on the ratios of low-valued shipments to individual country totals for past periods. The estimates for low-valued shipments are shown under a single HTS number. The total value excluded represents slightly less than 1 percent of the monthly import value.

Table 15-1.—Foreign trade: Value of total agricultural exports and imports, United States, fiscal years 1996–2005

Fiscal year ending Sep. 30[1]	U.S. total domestic exports			U.S. total imports for consumption, customs value			Surplus agricultural exports over agricultural imports
	Total merchandise exports	Agricultural exports[2]	Agricultural exports share of total exports	Total merchandise imports	Agricultural imports	Agricultural imports share of total imports	
	Million dollars	Million dollars	Percent	Million dollars	Million dollars	Percent	Million dollars
1996	574,646	59,752	10	795,289	32,444	4	27,308
1997	629,317	57,269	9	865,346	35,654	4	21,615
1998	639,556	53,653	8	895,900	36,837	4	16,816
1999	635,754	49,043	8	976,258	37,293	4	11,750
2000	701,651	50,744	7	1,167,768	38,857	3	11,887
2001	690,634	52,698	8	1,152,642	39,027	3	13,671
2002	628,241	53,302	8	1,120,317	40,956	4	12,346
2003	637,152	56,183	9	1,222,573	45,686	4	10,497
2004	712,287	62,369	9	1,397,117	52,656	4	9,713
2005[3]	783,221	62,385	8	1,617,569	57,716	4	4,669

[1] Fiscal years Oct. 1–Sept. 30 revised.   [2] Includes food exported for relief or charity by individuals and private agencies.   [3] Fiscal 2005 is nonrevised data.
ERS, Market and Trade Economics Division, (202) 694–5211.

Table 15-2.—Foreign Trade: Value and quantity of bulk commodity exports, United States, fiscal years, 2001–2005[1]

Fiscal year	Wheat, unmilled	Rice, milled	Feed grains[2]	Oilseeds[3]	Tobacco unmanufactured	Cotton and linters	Bulk commodities
	Value						
	Million dollars	Million dollars	Million dollars	Million dollars	Million dollars	Million dollars	Million dollars
2001	3,248	754	5,239	6,097	1,181	2,093	18,611
2002	3,498	734	5,292	6,711	1,148	2,052	19,434
2003	3,909	925	5,147	7,270	1,001	2,854	21,107
2004	5,095	1,198	6,611	8,375	1,050	4,534	26,863
2005	4,236	1,252	5,299	8,021	983	3,872	23,663
	Quantity						
	1,000 metric tons	1,000 metric tons	1,000 metric tons	1,000 metric tons	1,000 metric tons	1,000 metric tons	1,000 metric tons
2001	25,275	3,058	55,164	27,748	177	1,686	113,108
2002	25,411	3,536	53,625	30,303	163	2,206	115,243
2003	24,295	4,469	46,055	29,815	150	2,514	107,297
2004	31,179	3,690	53,770	26,130	163	3,021	117,954
2005	26,406	4,290	50,382	31,603	151	3,368	116,199

[1] Fiscal years, Oct. 1–Sept. 30.   [2] Corn, barley, sorghum, rye, and oats.   [3] Soybeans, peanuts, rapeseed, cottonseed, sunflowerseed, safflowerseed, and others.
ERS, Market and Trade Economics Division, (202) 694–5211.

Table 15-3.—Agricultural exports: Value to top 50 countries of destination, United States, fiscal years 2003–2005 [1]

Country	2003	2004	2005
	Million dollars	Million dollars	Million dollars
Canada	9,132.8	9,606.9	10,349.8
Mexico	7,609.7	8,407.6	9,196.9
Japan	8,811.0	8,523.8	7,832.2
European Union-25	6,310.1	6,799.0	6,930.5
China (Mainland)	3,483.7	6,095.1	5,289.6
China (Taiwan)	1,946.0	2,141.6	2,197.0
South Korea	2,760.7	2,776.7	2,179.3
Turkey	876.8	916.2	1,022.5
Indonesia	917.6	977.7	982.0
Russia	500.8	735.6	900.9
Hong Kong	1,059.8	991.4	882.2
Philippines	652.8	684.9	835.7
Egypt	832.4	976.6	808.4
Thailand	627.3	678.6	758.7
Colombia	522.9	599.5	598.1
Nigeria	319.7	422.0	515.4
Dominican Rep.	457.2	456.2	502.0
Australia	388.2	398.4	466.9
Guatamala	362.6	355.7	451.8
Malaysia	379.0	376.4	381.5
United Arab Emirates	244.6	273.0	378.9
Venezuela	388.3	390.5	350.7
Saudi Arabia	320.4	350.1	345.1
Israel	363.5	521.1	343.4
Cuba	186.6	400.7	329.6
Singapore	261.6	253.3	291.3
India	317.3	243.4	287.8
Costa Rica	247.0	280.2	279.4
Pakistan	198.9	287.3	264.2
Peru	225.8	289.4	237.0
El Salvador	235.5	236.9	234.4
Honduras	195.0	212.2	223.6
Brazil	360.6	324.8	219.8
Algeria	180.8	291.4	217.3
Iraq	75.0	62.4	200.2
Haiti	178.8	211.5	196.0
Panama	183.0	172.9	191.3
Vietnam	93.2	141.0	191.0
Jamaica	188.5	197.3	190.9
Romania	42.6	153.4	172.9
Syria	77.1	122.4	165.4
Bahamas	127.7	143.4	160.3
Switzerland	271.3	166.8	153.5
Morocco	116.3	200.0	150.9
Trinidad and Tobago	118.6	142.3	149.1
Republic of South Africa	116.4	186.2	142.6
New Zealand	131.9	114.8	134.0
Ukraine	12.0	79.8	133.3
Ecuador	112.6	129.2	127.6
Chile	138.0	120.3	119.3
Other	2,326.5	2,750.2	2,723.1
Total World agricultural exports [2]	55,986.6	62,368.1	62,385.3

[1] Fiscal years Oct. 1–Sept. 30.   [2] Totals may not add due to rounding.
ERS, Market and Trade Economics Divison, (202) 694–5211.

Table 15-4.—Foreign trade in agricultural products: Value of exports by principal commodity groups, United States, fiscal years 2002–2005[1]

Commodity	2001/2002	2002/2003	2003/2004	2004/2005
	1,000 dollars	1,000 dollars	1,000 dollars	1,000 dollars
Total Merchandise Exports	628,241,414	637,152,152	712,286,605	783,221,305
Nonagricultural U. S. Exports	574,939,734	582,345,814	649,917,564	720,836,031
Total Agricultural exports	53,291,233	56,182,727	62,368,055	62,385,274
Animals and animal products[3]	11,615,328	11,934,151	10,599,065	11,902,072
Animals, live excluding poultry	696,552	617,586	373,753	454,645
Cattle and calves-live	226,501	82,256	14,510	6,728
Horses,mules,burrors-live	413,091	491,474	311,191	412,567
Swine-live	34,522	28,298	40,306	28,094
Sheep-live	20,439	13,534	6,492	6,799
Other live Animals	1,999	2,024	1,255	457
Red meat and products	5,114,232	5,546,425	3,699,674	4,112,401
Beef and Veal	2,578,160	3,027,147	1,128,777	832,122
Beef or veal-fr or frozen	2,484,897	2,922,845	1,063,002	760,810
Beef prep or pres	93,263	104,303	65,775	71,312
Horsemeat fr chill. Froz	28,670	26,138	34,214	56,388
Lamb, mut or goat-fr. ch, frz	6,148	7,221	6,842	10,103
Pork	1,384,244	1,356,625	1,699,665	2,235,693
Pork-fr or froz	1,237,974	1,168,115	1,478,040	1,991,100
Pork prep or pres	146,270	185,510	221,625	244,593
Variety meats, ed. offals	831,512	830,000	626,875	732,345
Beef variety meats	659,036	680,803	350,955	405,949
Pork variety meats	127,906	109,034	227,708	281,107
Other variety meats	37,517	40,163	48,213	45,289
Other meats-all prep	285,499	299,294	203,301	245,750
Poultry and poultry products	2,280,123	2,103,673	2,519,170	3,011,363
Poultry - live	107,830	103,142	77,880	99,105
Baby chicks	97,398	91,528	66,883	88,672
Other live poultry	10,432	11,614	10,997	10,432
Poultry meats	1,880,037	1,678,116	2,111,201	2,504,764
Chickens - fresh or frozen	1,431,663	1,241,224	1,623,381	1,904,824
Turkeys - fresh or frozen	178,156	182,106	247,986	320,629
Other poultry - fresh or frozen	7,470	8,152	8,824	16,480
Poultry meat-prep or pres	262,747	246,633	231,011	262,831
Poultry, misc	119,098	145,598	132,420	159,288
Eggs	173,159	176,818	197,669	248,206
Dairy products	1,031,403	1,030,156	1,325,017	1,743,474
Evaporated and condensed milk	11,383	18,669	34,605	17,295
Nonfat dry milk	167,171	211,103	362,700	638,415
Butter and anhydrous milkfat	4,632	6,749	14,363	11,044
Cheese	164,868	143,909	186,015	200,678
Whey,fluid or dried	145,053	129,991	141,336	207,911
Other dairy products	538,295	519,735	585,998	668,130
Fats, oils and greases	428,032	539,374	574,065	477,303
Lard	24,792	28,835	58,353	48,663
Tallow, inedible	233,956	303,055	318,442	268,677
Other animal fats and oils	169,285	207,484	197,269	159,962
Hides and skins, including furskins	1,777,298	1,785,354	1,763,275	1,745,969
Bovine hides, whole	1,120,393	1,070,435	1,071,820	1,101,114
Other cattle hides-pieces	42,752	67,018	43,301	28,634
Calf skins, whole	221,439	320,836	307,899	194,164
Horse hides, whole	122,299	73,212	48,851	55,235
Sheep and lamb skins	23,953	19,646	20,125	21,726
Other hides and Skin, Ex furs	84,672	91,476	105,614	177,080
Furskins	161,790	142,731	165,666	168,015
Mink pelts	121,433	102,580	125,069	134,621
Other furskins	40,357	40,152	40,596	33,394
Wool and mohair	22,394	26,359	27,576	32,820
Sausage casings	57,876	79,307	88,975	100,770
Bull semen	51,117	44,768	47,612	55,236
Misc animal prods - Other	156,301	161,149	179,950	167,188
Grains and feeds	14,094,521	14,740,845	17,997,271	16,112,917
Wheat,unmilled	3,498,078	3,909,267	5,094,895	4,236,091
Wheat flour	115,754	94,129	71,906	55,143
Bulgur wheat	15,120	23,113	6,720	3,789
Other wheat products	88,796	95,580	106,167	104,399
Rice-paddy, milled, parb	733,836	925,428	1,197,924	1,252,188
Feed grains and products	5,676,420	5,603,032	7,081,456	5,821,196
Feed grain	5,291,631	5,147,400	6,611,157	5,299,329
Barley	77,454	83,514	56,720	90,643
Corn	4,599,161	4,534,375	5,984,189	4,729,961
Grain sorghum	609,874	524,247	565,003	473,792
Oats	4,785	5,055	5,025	4,698
Rye	357	210	220	234

See footnotes at end of table.

Table 15-4.—Foreign trade in agricultural products: Value of exports by principal groups, United States, fiscal years 2002–2005 [1]—Continued

Commodity	2001/2002	2002/2003	2003/2004	2004/2005
	1,000 dollars	*1,000 dollars*	*1,000 dollars*	*1,000 dollars*
Feed grains and products--Continued				
Feed grain products	384,790	455,631	470,300	521,867
Popcorn	60,292	54,815	75,953	98,916
Blended food prods	87,170	64,140	137,169	109,168
Other grain prods	1,263,924	1,349,863	1,521,116	1,695,328
Feed and fodders, ex oilcake	2,555,132	2,621,479	2,703,965	2,736,699
Corn by-products	571,243	576,211	636,585	498,419
Alfalfa meal and cubes	22,632	37,560	29,483	29,935
Beef pulp	74,372	79,812	83,892	81,124
Citrus pulp pellets	31,063	22,000	43,041	17,738
Other feeds and fodders	1,855,822	1,905,895	1,910,964	2,109,484
Fruit and prep, ex juice	2,739,400	2,891,903	3,128,807	3,315,668
Fruits-fresh	2,097,519	2,231,682	2,364,001	2,555,622
Fruits-fresh-citrus	585,502	631,441	704,659	626,719
Grapefruit-fresh	201,265	187,238	242,282	151,104
Lemons and limes-fresh	71,130	83,535	71,744	75,323
Oranges and tanger-fresh	311,837	359,013	388,694	398,190
Other citrus-fresh	1,269	1,655	1,940	2,102
Fruit fresh-noncitrus	1,512,017	1,600,241	1,659,342	1,928,903
Apples-fresh	361,323	342,155	329,354	454,210
Berries-fresh	184,994	241,147	260,285	312,634
Cherries-fresh	149,962	180,281	187,308	205,493
Grapes-fresh	385,771	401,678	435,271	467,675
Melons-fresh	99,015	101,558	98,361	119,620
Peaches-fresh	112,690	109,472	106,463	113,435
Pears-fresh	99,257	99,460	106,590	105,866
Plums-fresh	57,084	56,408	47,837	54,684
Other noncitrus-fresh	61,920	68,083	87,873	95,284
Fruits dried	344,031	348,736	401,692	378,329
Raisin dried	152,867	154,219	190,792	208,957
Prunes-dried	133,529	130,091	137,188	106,691
Other dried fruits	57,635	64,426	73,713	62,682
Fruits-canned ex juice	139,513	155,617	198,280	206,475
Fruits-frozen ex juice	56,871	52,490	56,629	52,727
Other fruits-prep or pres	101,465	103,397	108,203	122,516
Fruits juices incl frozen	694,106	657,583	704,662	765,087
Apple juice	22,575	17,040	15,199	18,027
Grape juice	57,819	54,625	55,452	54,927
Grapefruit juice	71,939	71,610	67,430	52,314
Orange juice	290,416	247,510	260,992	265,294
Other fruit juices	251,356	266,798	305,590	374,525
Wine	501,326	584,390	674,346	675,179
Nuts and prep	1,467,262	1,627,838	2,077,960	2,606,815
Almonds (shelled basis)	745,593	982,188	1,298,341	1,625,842
Filberts	35,309	17,642	38,602	47,416
Peanuts, shelled of prep	234,471	137,912	190,709	189,252
Pistachios	126,050	142,767	144,048	277,157
Walnuts Shelled/unshelled	188,987	189,476	233,135	257,009
Pecans shelled or unshelled	58,372	76,501	104,126	104,934
Other nuts shelled or prepared	78,480	81,351	68,999	105,205
Vegetables and preparations	4,545,255	4,667,701	5,213,290	5,606,111
Vegetables fresh	1,226,369	1,248,277	1,289,195	1,529,588
Aspargus-fresh	37,632	37,908	35,959	33,995
Broccoli-fresh	105,482	103,463	106,239	114,330
Carrots-fresh	84,505	89,680	93,968	103,031
Cabbage-fresh	18,934	21,371	19,466	27,244
Celery-fresh	44,000	41,631	53,974	57,669
Cauliflower-fresh	58,577	60,363	60,426	66,497
Corn sweet-fresh	25,086	27,623	27,840	34,675
Cucumbers-fresh	15,496	13,136	12,313	14,128
Garlic-fresh	10,643	9,799	6,831	7,126
Lettuce-fresh	219, 927	233,498	258,421	342,184
Mushrooms-fresh	16,140	11,203	17,371	17,326
Onions and shallots-fresh	89,656	100,240	98,824	106,588
Peppers-fresh	71,127	76,884	82,748	95,112
Potatoes-fresh	122,247	90,703	71,386	98,889
Tomatoes-fresh	135,643	148,065	147,553	187,236
Other fresh vegetables	171,273	182,712	195,878	223,558
Vegetables-frozen	513,464	499,880	527,453	587,648
Corn, sweet, frozen	52,315	64,938	59,027	59,304
Potatoes frozen	362,191	332,857	383,167	431,531
Other frozen vegetables	98,958	102,085	85,259	96,813
Vegetables-canned	328,439	317,251	319,725	312,259
Pulses	238,389	242,991	229,540	257,043
Dried Beans	163,597	163,746	138,085	137,127
Dried Peas	35,906	40,942	48,738	77,414
Dried Lentils	38,886	38,302	42,716	42,502
Hops, including hop extract	97,410	90,481	123,125	106,009
Other vegetables-prep or pres	2,141,184	2,268,820	2,724,252	2,813,565

See footnotes at end of table.

Table 15-4.—Foreign trade in agricultural products: Value of exports by principal groups, United States, fiscal years 2002–2005 [1]—Continued

Commodity	2001/2002	2002/2003	2003/2004	2004/2005
	1,000 dollars	*1,000 dollars*	*1,000 dollars*	*1,000 dollars*
Oilseeds and products	9,681,957	10,138,789	11,120,382	10,960,603
Oilcake and meal	1,335,836	1,165,255	1,178,861	1,256,749
Bran and residues, legum. Veg.	17,945	10,446	18,052	12,430
Corn oilcake and meal	1,054	272	234	640
Soybean meal	1,276,895	1,116,002	1,096,245	1,165,445
Other oilcake and meal	39,942	38,535	64,331	78,234
Oilseeds	6,710,850	7,270,411	8,375,443	8,020,606
Rapeseed	59,142	77,999	80,595	38,140
Safflower seeds	4,149	4,668	1,784	1,161
Soybeans	5,474,499	6,514,217	7,462,896	7,019,238
Sunflowerseeds	103,594	80,249	94,105	88,849
Peanuts including oilstock	19,765	17,175	23,579	23,666
Other oilseeds	178,167	177,453	291,060	404,915
Protein substances	871,535	398,650	421,424	444,638
Vegetable oils	1,635,271	1,703,123	1,566,078	1,683,248
Soybean oil	454,248	557,939	287,953	351,251
Cottonseed oil	31,732	30,924	33,738	17,142
Sunflower oil	111,689	36,598	74,497	53,503
Corn oil	276,731	275,495	264,688	271,646
Peanut oil	3,489	31,656	4,200	4,135
Rapeseed oil	52,287	42,822	95,243	73,546
Safflower oil	17,955	14,705	16,820	19,875
Other vegetables and waxes	687,139	712,984	788,939	892,151
Tobacco-unmfg	1,148,440	1,001,237	1,049,598	983,162
Tobacco-light air cured	340,901	354,508	333,522	427,829
Tobacco-flue-cured	608,238	483,452	535,783	388,816
Other-tobacco-unmfg	199,302	163,277	180,293	166,517
Cotton, ex linters	2,036,122	2,840,521	4,508,192	3,861,200
Cotton linters	15,444	13,217	26,007	10,912
Essential oils	763,703	957,088	938,528	970,383
Seeds, field and garden	833,272	802,902	865,421	915,409
Sugar and tropical products	1,626,871	1,758,812	1,895,572	2,103,093
Sugar and related products	614,015	621,177	679,109	759,780
Sugar-cane or beet	40,532	25,034	44,746	67,526
Related sugar products	573,483	596,142	634,363	692,254
Coffee	244,518	285,255	276,241	345,988
Cocoa	116,091	129,117	143,051	134,912
Chocolate and prep	437,517	474,549	528,093	565,512
Tea and mate	112,135	112,511	151,900	177,671
Spices	65,287	72,400	76,840	79,900
Rubber-crude natural	33,125	60,557	37,547	36,969
Fibers ex cotton	4,183	3,245	2,791	2,361
Other misc Veg prods	501,303	500,426	541,086	559,733
Nursery and Greenhouse Prods	253,511	258,811	286,509	315,899
Beverages ex juices	773,412	806,514	741,361	721,030

[1] Fiscal years, Oct. 1–Sept. 30. Totals may not add due to rounding.

ERS, Market and Trade Economics Division, (202) 694–5211. Compiled from reports of the U.S. Department of Commerce.

Table 15-5.—Foreign trade in agricultural products: Value of imports by principal groups, United States, fiscal years 2002–2005 [1]

Product	2001-2002	2002-2003	2003-2004	2004-2005
	1,000 dollars	1,000 dollars	1,000 dollars	1,000 dollars
Total merchandise imports	1,120,317,246	1,222,572,821	1,397,116,634	1,617,569,100
Non-agricultural U.S. imports	1,079,361,593	1,176,887,255	1,344,460,620	1,548,457,293
Total agricultural imports	40,953,715	45,685,566	52,656,013	57,716,135
Animals & prods.	9,066,314	8,593,655	10,351,005	11,117,279
Animals - live ex. poultry	1,994,877	1,673,027	1,318,962	1,577,031
Cattle and calves	1,376,114	1,085,764	582,150	713,032
Horses, mules, burros	286,395	230,186	239,674	271,748
Swine	320,092	342,711	495,892	590,236
Sheep, live	8,941	11,016	16	156
Other live animals	3,334	3,349	1,230	1,859
Red meat & products	4,186,597	4,020,151	5,526,532	5,716,909
Beef & veal	2,748,789	2,391,880	3,505,597	3,772,305
Beef & veal - fr. or froz.	2,527,257	2,140,465	3,198,834	3,431,969
Beef & veal - prep. or pres.	221,532	251,415	306,762	340,336
Pork	991,647	1,149,562	1,329,881	1,297,755
Pork - fr. or froz.	711,168	802,905	942,941	951,578
Pork - prep. or pres.	280,480	346,656	386,940	346,177
Mutton, goat & lamb	274,491	316,579	435,509	470,727
Horsemeat - fr. or froz.	97	11	2	0
Variety meats - fr. or froz.	100,111	81,950	80,211	85,582
Other meats - fr. or froz.	22,079	25,191	25,529	31,198
Other meats & prods.	49,382	54,979	149,803	59,342
Poultry and prods.	316,503	293,588	401,149	375,387
Poultry - live	32,282	30,534	37,013	34,814
Poultry meat	81,612	94,790	126,583	135,426
Eggs	27,284	22,129	32,327	23,666
Poultry, misc.	175,325	146,135	205,226	181,480
Dairy products.	1,840,645	1,865,604	2,332,577	2,609,323
Milk & cream, fr. or dried	43,824	39,399	49,672	64,931
Butter & butterfat mixtures	53,853	39,046	85,694	79,449
Cheese	808,045	821,100	970,664	1,014,782
Casein & mixtures	461,615	435,322	534,092	613,343
Other dairy prods.	473,308	530,737	692,454	836,817
Fats, oils, & greases	63,034	67,497	72,632	87,599
Hides & skins	135,977	132,619	142,632	146,959
Sheep & lamb skins	3,797	3,904	4,439	4,149
Other hides & skins	76,470	68,838	72,189	75,281
Furskins	55,710	59,878	66,004	67,529
Wool - unmfg.	30,688	39,715	37,895	37,680
Apparel grade wool	16,226	19,048	14,879	16,422
Carpet grade wool	14,461	20,667	23,016	21,258
Sausage casings	70,305	91,012	86,241	80,578
Bull semen	15,604	15,461	18,650	27,553
Misc. animal prods	426,546	415,648	436,751	479,519
Silk, raw	164	531	176	50
Grains & feeds	3,599,087	3,892,032	4,191,919	4,428,697
Wheat, ex. seed	312,943	140,433	149,869	164,313
Corn, unmilled	21,561	35,821	39,399	27,969
Oats, unmilled	186,955	231,118	151,425	192,296
Barley, unmilled	70,947	40,586	72,298	18,387
Rice	161,624	200,566	239,032	216,759
Biscuits & wafers	1,154,521	1,326,812	1,500,247	1,667,318
Pasta & noodles	262,694	287,202	292,599	335,526
Other grains & preps.	924,805	1,091,798	1,203,321	1,259,568
Feeds & fodders, ex. oilcake	503,036	537,696	543,729	546,562
Fruits & preps.	4,277,837	4,602,313	5,006,652	5,652,844
Fruits - fr. or froz.	3,462,006	3,661,809	3,963,794	4,486,461
Apples, fresh	104,506	135,597	181,918	100,621
Avocados	120,553	148,467	157,452	304,902
Berries, excl. strawberries	158,017	202,963	275,244	354,239
Bananas & plantains - fr. or froz.	1,161,811	1,138,871	1,088,264	1,144,529
Citrus, fresh	205,618	283,420	315,047	335,409
Grapes, fresh	669,167	659,619	730,338	947,926
Kiwifruit, fresh	33,354	34,063	33,173	35,954
Mangoes	152,833	168,205	172,256	180,421
Melons	267,156	234,428	286,152	304,573
Peaches	52,573	55,265	56,462	61,385
Pears	71,252	76,536	65,627	87,271
Pineapples - fr. or froz.	182,073	220,006	243,243	229,416
Plums	32,037	28,459	32,093	38,668
Strawberries - fr. or froz.	107,513	110,881	126,514	152,179
Other fruits - fr. or froz.	143,544	165,032	200,013	208,967
Fruits - prep. or pres.	815,831	940,504	1,042,858	1,166,383
Bananas & plantains - prep. or pres.	25,905	25,222	31,443	31,099
Pineapples - canned or prep.	189,753	206,108	210,277	236,789
Other fruits - prep. or pres.	600,173	709,174	801,138	898,495
Fruit juices	652,671	776,289	782,740	1,003,197
Apple juice	225,453	256,597	305,051	331,514
Grape juice	51,864	41,068	57,323	93,723
Grapefruit juice	667	711	820	18,562
Lemon juice	13,405	16,030	13,168	15,234
Lime juice	4,923	7,105	7,930	10,297
Orange juice	158,819	239,846	147,327	257,050
Pineapple juice	79,536	87,072	90,150	81,504
Other fruit juice	118,004	127,861	160,970	195,313

See footnotes at end of table.

Table 15-5.—Foreign trade in agricultural products: Value of imports by principal groups, United States, fiscal years 2002-2005 [1]—Continued

Product	2001-2002	2002-2003	2003-2004	2004-2005
	1,000 dollars	*1,000 dollars*	*1,000 dollars*	*1,000 dollars*
Nuts & preps	648,127	723,801	951,540	1,194,417
Brazil nuts	19,665	19,677	42,684	58,199
Cashew nuts	368,187	402,904	519,760	635,646
Chestnuts	13,077	11,436	11,580	9,072
Coconut meat	56,371	61,677	54,094	63,664
Filberts	21,159	19,316	25,247	31,511
Macadamia nuts	33,437	43,211	78,321	104,940
Pecans	42,799	60,515	106,168	165,487
Pistachio nuts	1,764	3,130	4,652	3,944
Other nuts	91,668	101,935	109,035	121,955
Vegetables & preps.	5,442,109	6,200,768	6,845,772	7,543,717
Vegetables - fr. or froz.	3,120,036	3,700,844	4,087,003	4,499,283
Tomatoes	739,657	1,044,210	952,285	1,113,442
Asparagus - fr. or froz.	125,010	156,291	176,946	209,275
Beans - fr. or froz.	49,120	56,030	66,909	78,682
Cabbage	11,552	12,077	9,817	17,056
Carrots - fr. or froz.	27,314	25,780	32,189	33,435
Cauliflower & broccoli - fr. or froz.	186,861	193,141	221,146	236,989
Celery, fresh	14,687	8,380	8,409	10,397
Cucumbers	198,815	216,850	355,198	314,668
Eggplant	28,119	28,793	44,430	50,904
Endive, fresh	4,447	4,126	5,063	4,980
Garlic	54,428	39,304	50,504	75,497
Lettuce	37,157	26,079	33,999	47,592
Okra - fr. or froz.	12,771	16,627	21,209	24,754
Onions	150,639	154,468	184,888	236,274
Peas, incl. chickpeas	39,603	51,229	53,507	59,197
Peppers	440,258	541,341	597,425	687,070
Potatoes - fr. or froz.	495,544	543,065	623,197	601,776
Radishes, fresh	11,057	14,248	17,874	16,307
Squash	130,893	181,712	189,638	160,268
Other vegs. - fr. or froz.	362,104	387,092	442,369	520,718
Vegetables - prep. or pres.	2,323,766	2,499,924	2,758,769	3,044,434
Bamboo shoots, preserved	15,514	13,945	12,685	14,078
Bean cake, Miso	16,450	17,632	19,489	21,668
Cucumbers, preserved	25,733	38,099	41,786	42,109
Garlic, dried	13,629	14,487	16,798	20,259
Olives - prep. or pres.	209,645	230,977	285,867	314,467
Mushrooms, canned	100,286	112,126	121,320	102,819
Mushrooms, dried	15,463	21,728	18,872	20,249
Onions, preserved	7,144	9,079	10,531	17,444
Hops, incl. extract	23,365	32,428	19,067	36,352
Artichokes - prep.	69,853	91,567	104,098	105,082
Asparagus- prep.	2,265	6,362	9,777	13,991
Tomatoes, incl. paste & sauce	125,841	111,496	113,313	132,003
Waterchestnuts	19,777	17,247	21,520	22,147
Beans & peas, dried	86,391	65,742	75,293	102,799
Mustard	20,143	18,864	20,006	20,013
Peppers & pimentos, prep.	33,260	33,006	38,459	50,314
Soy sauce	43,643	47,427	47,666	48,651
Starches, excl. wheat & corn	79,067	88,084	97,595	115,666
Soups & sauces	385,942	365,036	387,047	438,586
Vinegar	64,226	65,601	70,332	74,093
Yeasts	90,464	98,395	101,597	114,654
Other vegetables - prep. or pres.	875,664	1,000,867	1,125,650	1,216,986
Sugar & related prods.	1,709,922	2,111,138	2,123,290	2,305,764
Sugar - cane & beet	518,303	584,045	568,049	713,157
Molasses	98,759	93,672	92,094	112,679
Confectionery prods.	844,618	1,046,281	1,148,152	1,158,280
Other sugar & related prods.	248,243	387,242	314,994	321,648
Cocoa & products	1,713,833	2,272,743	2,578,819	2,632,757
Coffee & products	1,609,509	1,949,455	2,162,630	2,828,348
Tea	279,329	299,935	328,016	368,383
Spices & herbs	569,959	664,607	722,269	528,891
Pepper	217,491	224,270	234,154	240,163
Other spices & herbs	352,468	440,337	487,115	288,728
Drugs, crude & natural	505,933	541,723	552,848	572,116
Essential oils	339,274	906,304	1,825,017	2,334,622
Fibers, excl. cotton	23,616	21,458	21,726	24,567
Rubber & gums	654,816	1,032,406	1,330,968	1,504,545
Tobacco - unmfg.	736,363	670,236	760,852	641,132
Tobacco - filler	677,600	622,049	710,800	583,824
Tobacco - scrap	22,845	20,362	18,421	17,969
Other tobacco	35,918	27,826	31,632	39,339
Beverages, ex. fruit juice	5,572,105	6,406,286	7,006,282	7,757,137
Wine	2,521,873	3,185,774	3,315,844	3690,939
Malt beverages	2,526,172	2,591,364	2,804,661	2,993,578
Other beverages	524,060	629,148	885,777	1,072,621

See footnotes at end of table.

Table 15-5.—Foreign trade in agricultural products: Value of imports by principal groups, United States, fiscal years 2002–2005 [1]—Continued

Product	2001-2002	2002-2003	2003-2004	2004-2005
	1,000 dollars	*1,000 dollars*	*1,000 dollars*	*1,000 dollars*
Oilseeds & prods.	1,674,740	2,009,330	2,924,421	2,946,558
Oilseeds & oilnuts	216,795	236,598	334,177	362,386
Flaxseed	16,407	32,462	42,289	58,664
Mustardseed	16,473	17,437	21,926	21,556
Rapeseed	17,878	44,363	93,953	103,180
Sesame seed	41,778	36,615	54,463	54,236
Soybeans	15,006	30,067	53,758	50,899
Sunflower seeds	20,977	31,537	31,533	22,443
Other oilseeds & oilnuts	88,276	44,116	36,255	51,408
Oils & waxes - vegetables	1,309,848	1,594,776	2,244,982	2,367,931
Castor oil	21,394	20,933	33,148	37,129
Coconut oil	165,702	173,771	205,925	271,513
Cottonseed oil	25	6,685	46	567
Olive oil	420,647	508,088	713,267	815,618
Palm oil	62,842	68,437	133,216	146,683
Palm kernel oil	54,882	113,725	147,085	161,393
Peanut oil	19,958	11,119	66,920	26,509
Rapeseed oil	226,225	275,312	397,070	330,720
Soybean oil	9,378	12,090	80,346	7,502
Sesame oil	26,923	26,277	30,431	35,090
Other vegetable oils	301,871	378,339	437,529	535,208
Oilcake & meal	148,098	177,955	345,261	216,240
Cotton, excl. linters	15,480	27,226	19,607	14,332
Cotton, linters	7,928	1,603	1,681	4,891
Seeds - field & garden	417,213	442,697	445,525	516,707
Cut flowers	543,850	584,759	701,767	702,087
Nursery stock, bulbs, etc.	590,900	631,556	661,017	678,166
Other vegetable prods.	286,482	301,947	336,458	393,674

[1] Fiscal years, Oct. 1–Sept. 30.
ERS, Market and Trade Economics Division, (202) 694–5211. Compiled from reports of the U.S. Depart. of Commerce.

Table 15-6.—Agricultural exports: Value of U.S. exports to the top market, Canada, by commodity, fiscal years 2002/2003–2004/2005 [1]

Commodity	Value		
	2002/2003	2003/2004	2004/2005
	1,000 dollars	*1,000 dollars*	*1,000 dollars*
Total agricultural exports	9,132,841	9,606,867	10,349,838
Animals and animal products	1,413,290	1,341,293	1,455,776
Animals Live-Ex Poultry	67,115	29,643	36,955
Cattle and calves-live	36,625	5,396	4,915
Horses, Mules, Burros-live	28,871	22,770	31,278
Swine-Live	1,256	1,087	523
Sheep-Live	31	24	37
Other live animals	332	366	202
Red meat and Products	598,549	463,210	609,419
Beef and Veal	333,835	118,394	168,723
Beef and Veal-fresh or frozen	261,937	72,328	105,629
Beef-prep or pres	71,898	46,067	63,094
Lamb-mutton or goat-fr-ch-froz	529	709	1,293
Pork	165,105	247,664	344,728
Pork-fresh or frozen	104,557	163,121	251,098
Pork-prep or pres	60,548	84,543	93,630
Variety meats, Ed Offals	29,919	40,205	37,672
Beef variety meats	11,136	5,132	6,712
Pork variety meats	15,080	29,152	22,735
Other variety meats	3,703	5,921	8,315
Other meats-fr or prep	69,161	56,237	56,914
Poultry and poultry products	346,352	421,416	392,919
Poultry-Live	25,878	27,144	25,426
Baby chicks	18,489	19,048	18,069
Other live poultry	7,389	8,097	7,357
Poultry meats	271,889	323,348	295,645
Chickens-fresh or frozen	147,337	194,492	136,565
Turkeys-fresh or frozen	10,361	10,626	21,897
Other poultry-fresh or frozen	2,678	4,141	5,629
Poultry meats-prep or pres.	111,514	114,089	131,554
Poultry misc.	2,818	3,092	5,313
Eggs	45,767	67,832	66,535
Dairy prods	241,530	268,386	270,814
Evap and condensed milk	610	715	396
Nonfat dry milk	1,889	4,485	7,664
Butter and Anhydrous Milkfat	392	9,661	3,435
Cheese	25,185	24,279	29,716
Whey,fluid or dried	25,113	31,063	31,894
Other dairy products	188,342	198,183	197,709
Fats, oils and greases	26,064	31,456	30,089
Lard	1,408	2,536	3,042
Tallow-inedible	3,425	5,266	4,648
Other animal fats and oils	21,230	23,654	22,400
Hides and skins include furs	77,666	66,974	55,957
Bovine hides, whole	26,235	15,692	6,027
Other cattle hides-pieces	998	524	113
Calf skins, whole	420	241	227
Horse hides whole	1,089	257	225
Sheep and lamb skins	326	384	594
Other hides and skins, ex.furs	1,304	500	232
Furskins	47,295	49,375	48,539
Mink pelts	30,522	28,392	31,282
Other furskins	16,773	20,983	17,257
Wool and Mohair	552	238	150
Sausage casings	9,612	11,294	9,998
Bull semen	3,829	3,315	3,885
Misc animal products-other	42,021	45,361	45,589
Grains and feeds	1,846,472	1,800,565	1,950,450
Wheat, unmilled	3,093	1,762	2,175
Wheat flour	6,936	12,478	14,966
Other wheat products	53,677	56,258	57,901
Rice-paddy,milled parb	67,790	95,333	91,922
Feed grains and products	465,272	271,537	270,733
Feed grains	412,831	220,546	215,439
Barley	19,898	3,795	10,951
Corn	390,009	212,775	201,279
Grain sorghums	711	1,126	751
Oats	2,144	2,815	2,444
Rye	69	35	14
Feed grain products	52,440	50,991	55,294
Popcorn	5,403	26,861	47,094
Other grain prods	797,240	882,614	974,125
Feeds and fodders, ex.oilcakes	447,060	453,723	491,534
Corn by-products	23,665	27,333	24,707
Alfalfa meal and cubes	142	37	123
Beet pulp	3,633	3,241	3,730
Other feeds and fodders	419,621	423,112	462,974

See footnotes at end of table.

Table 15-6.—Agricultural exports: Value of U.S. exports to the top market, Canada, by commodity, fiscal years 2002/2003–2004/2005 [1]—Continued

Commodity	Value		
	2002/2003	2003/2004	2004/2005
	1,000 dollars	*1,000 dollars*	*1,000 dollars*
Fruits and prep. ex.juice	959,242	1,080,460	1,200,606
Fruits-fresh	800,532	877,528	987,220
Fruits-fresh-citrus	144,688	164,194	171,594
Grapefruit-fresh	24,564	26,215	26,227
Lemons and limes-fresh	18,370	21,080	25,752
Oranges and tangerines fresh	101,074	116,309	119,203
Other citrus-fresh	680	591	413
Fruits-fresh-noncitrus	655,844	713,334	815,626
Apple-fresh	90,478	90,442	92,483
Berries-fresh	172,865	190,602	229,321
Cherries-fresh	42,239	44,616	57,839
Grapes, fresh	117,183	139,814	152,082
Melon-fresh	88,838	86,187	104,552
Peaches-fresh	55,266	57,767	66,295
Pears-fresh	34,346	39,026	40,066
Plums-fresh	23,192	24,278	27,022
Other noncitrus-fresh	31,437	40,601	45,965
Fruits, dried	52,539	54,127	55,920
Raisins, dried	22,643	24,398	26,785
Prunes,dried	10,147	9,224	8,120
Other dried-fruits	19,749	20,506	21,015
Fruits-canned excl. juice	51,923	79,997	75,200
Fruits-froz. excl. juice	18,584	21,192	26,502
Other fruits-prep. or pres	35,664	47,615	55,765
Fruit juices incl. frozen	266,283	302,859	343,604
Apple juice	6,308	6,182	8,714
Grapejuice	27,578	29,753	30,949
Grapefruit juice	13,364	12,373	10,531
Orange juice	150,245	160,832	177,669
Other fruit juices	68,788	93,719	115,742
Wine	95,423	109,050	118,726
Nuts and prep	176,669	234,167	287,993
Almonds(shelled basis)	51,228	76,310	102,775
Filbert	1,672	2,796	4,431
Peanuts,shelled or prep	49,034	70,093	72,204
Pistachios	12,760	14,566	21,486
Walnuts, shelled or unshelled	17,357	19,841	22,978
Pecan, shelled or unshelled	20,875	33,058	34,928
Other nuts, shelled or prep	23,742	17,503	29,191
Vegetables and prep	1,822,092	1,917,686	2,241,709
Vegetables-fresh	956,226	995,477	1,197,407
Asparagus-fresh	16,501	17,224	18,881
Broccoli-fresh	49,296	57,078	62,373
Carrots-fresh	81,832	85,734	92,400
Cabbage-fresh	19,869	18,042	25,402
Celery-fresh	30,754	42,160	45,260
Cauliflower-fresh	35,383	42,867	51,382
Corn, sweet-fresh	17,987	19,780	22,327
Cucumber-fresh	13,033	12,085	14,055
Garlic-fresh	3,050	2,543	2,879
Lettuce-fresh	188,933	217,515	292,093
Mushroom-fresh	7,397	7,735	8,971
Onion and Shallots-fresh	65,819	61,225	62,322
Peppers-fresh	71,885	74,685	87,046
Potatoes-fresh	76,271	51,620	64,149
Tomatoes-fresh	129,446	123,987	162,101
Other fresh vegetables	148,768	161,198	185,766
Vegetables-frozen	83,295	80,972	90,948
Corn, sweet-frozen	10,075	4,195	2,701
Potatoes-frozen	27,382	35,198	45,394
Other frozen vegetables	45,838	41,579	42,854
Vegetables-canned	95,206	103,369	105,086
Pulses	23,474	12,827	23,041
Dried beans	11,893	6,627	8,688
Dried peas	10,515	5,297	12,370
Dried lentils	1,066	903	1,984
Hops,incl hop ext	6,186	6,250	6,583
Other veg-prep or pres	657,706	718,791	818,643

See footnotes at end of table.

Table 15-6.—Agricultural exports: Value of U.S. exports to the top market, Canada, by commodity, fiscal years 2002/2003–2004/2005 [1]—Continued

Commodity	Value		
	2002/2003	2003/2004	2004/2005
	1,000 dollars	*1,000 dollars*	*1,000 dollars*
Oilseeds and prods	846,679	1,007,033	829,448
Oilcake and meal	234,956	309,741	254,078
Bran and residues, legum.veg.	800	1,260	1,205
Corn oilcake and meal	173	29	11
Soybean meal	224,248	304,805	236,681
Other oilcake and meal	9,735	3,647	16,180
Oilseeds	292,162	318,634	194,374
Rapeseed	67,389	67,608	28,467
Safflowers seeds	520	554	560
Soybeans	150,273	175,058	83,896
Sunflowerseeds	8,328	6,936	16,238
Peanuts, including oilstock	2,615	4,188	5,111
Other oilseeds	15,497	12,578	14,241
Protein substances	47,541	51,713	45,861
Vegetable oils	319,561	378,658	380,996
Soybean oil	69,734	68,718	45,727
Cottonseed oil	23,549	20,463	9,280
Sunflower oil	12,532	15,495	33,983
Corn oil	10,419	18,198	19,868
Peanut oil	2,196	2,709	2,503
Rapeseed oil	16,312	33,678	32,056
Safflower oil	853	1,031	1,336
Other Vegetable oils & Waxes	183,921	218,367	236,242
Tobacco-unmfg	1,820	1,021	1,810
Tobacco-light air cured	177	19	0
Tobacco-flue cured	219	419	602
Other tobacco-unmfg	1,423	583	1,207
Cotton, ex. linters	102,867	106,267	75,334
Cotton linters	846	594	380
Essential oils	248,727	276,168	242,338
Seeds-field and garden	110,804	125,083	134,642
Sugar and tropical prods	852,284	909,739	1,015,296
Sugar and related products	274,508	277,251	305,950
Sugar cane or beet	2,265	3,243	11,024
Related sugar product	272,243	274,008	294,926
Coffee	212,015	194,376	252,978
Cocoa	82,641	90,562	95,221
Chocolate and prep	215,163	274,283	279,801
Tea and Mate	44,719	48,561	52,025
Spices	19,826	20,151	24,698
Ruber-crude-natural	2,728	3,727	3,831
Fibers Ex Cotton	684	828	791
Other misc veg prods	68,314	73,303	81,696
Nursery and greenhouse prods	142,276	152,711	168,647
Beverages ex juices	178,753	168,869	201,384

[1] Fiscal years Oct. 1–Sept. 30.
ERS, Market and Trade Economics Division, (202) 694–5211.

Table 15-7.—Agricultural imports for consumption: Value of Top 50 countries of origin,
United States, fiscal years 2003–2005 [1]

Country	2003	2004	2005
	Million dollars	Million dollars	Million dollars
European Union-25	10,301.3	12,061.6	13,235.9
Canada	10,251.8	11,274.9	11,816.8
Mexico	5,994.8	7,023.1	8,096.8
Australia	1,975.7	2,387.0	2,487.5
Brazil	1,465.1	1,636.5	1,837.7
China (mainland)	1,184.4	1,570.7	1,789.7
New Zealand	1,287.0	1,572.5	1,617.9
Indonesia	1,157.5	1,445.2	1,604.2
Chile	1,200.3	1,313.6	1,531.1
Colombia	1,030.9	1,132.7	1,378.6
Thailand	889.5	1,025.2	1,085.9
India	692.2	810.0	901.7
Guatemala	772.2	779.3	894.5
Costa Rica	844.9	899.2	880.2
Argentina	584.1	568.5	766.0
Malaysia	422.0	559.4	628.5
Ecuador	544.6	575.0	584.0
Philippines	491.7	524.1	575.5
Ivory Coast	440.4	497.2	565.4
Uruguay	67.6	294.0	462.0
Peru	274.9	307.8	426.2
Japan	369.9	436.7	417.1
Vietnam	229.2	340.8	413.6
Turkey	312.8	351.9	360.3
Honduras	224.5	256.8	289.6
Dominican Republic	275.1	261.0	251.4
Switzerland	184.7	220.5	237.2
Israel	165.2	189.5	209.4
South Korea	152.6	224.5	203.3
China(Taiwan)	165.6	175.7	186.5
Nicaragua	107.0	147.0	169.3
Rep.S. Africa	148.3	153.1	166.1
El Salvador	92.3	101.6	139.4
Liberia	50.9	73.4	91.3
Finland	104.7	84.2	84.5
Jamaica	63.0	66.6	73.7
Bulgaria	42.4	51.3	68.1
Singapore	83.3	78.9	67.3
Hong Kong	78.5	64.4	64.5
Morocco	71.8	72.5	55.2
Nigeria	27.1	30.9	54.1
Kenya	37.9	45.1	53.8
Venezuela	37.6	51.7	52.7
Norway	46.7	46.6	52.3
Malawi	50.3	63.2	52.3
Ethiopia	27.9	30.7	51.1
Panama	38.0	00.0	47.3
Former Soviet Union-12	42.7	59.6	45.9
Belize	27.3	27.8	44.5
Egypt	49.7	44.5	42.5
Other	505.6	647.8	505.5
Total U. S. Agricultural Imports [2]	45,685.6	52,656.0	57,716.1

[1] Fiscal years Oct. 1–Sept. 30. [2] Totals may not add due to rounding.

ERS, Market and Trade Economics Division, (202) 694–5211. Compiled from reports of the U.S. Department of Commerce.

Table 15-8.—European Union: Value of agricultural imports by origin, 1994–2003 [1]

Year[2]	United States	EU countries	Other countries	Total
	Million dollars	Million dollars	Million dollars	Million dollars
1994	8,405	109,913	51,108	169,426
1995	8,567	122,003	54,104	184,674
1996	9,026	127,148	54,939	191,112
1997	9,105	129,520	52,439	191,064
1998	7,961	133,739	52,482	194,182
1999	6,603	132,666	49,032	188,301
2000	6,312	117,228	48,673	172,213
2001	6,429	117,910	48,004	172,343
2002	6,290	133,948	59,540	191,778
2003	6,450	167,970	61,746	236,168

[1] EU-15. Based on bilateral import data from the United Nations.   [2] Data on calendar year basis.
ERS, Market and Trade Economics Division, (202) 694–5273.

Table 15-9.—Fisheries: Landings and value of principal species: 1997–2004 [1]

[Preliminary]

Species	Landings							
	1997	1998	1999	2000	2001	2002	2003	2004
	Mil. lbs.	Mil. lbs.	Mil. lbs.	Mil. lbs.	Mil. lbs.	Mil. lbs.	Mil. lbs.	Mil. lbs.
Fish:								
Cod, Atlantic	29	25	21	25	33	29	24	16
Flounder	566	391	331	413	352	373	365	362
Haddock	3	6	7	9	13	17	15	18
Halibut	70	73	80	75	78	82	80	79
Herring, sea	348	272	267	235	300	214	287	256
Jack mackerel	3	3	2	3	8	2	1	3
Menhaden	2,028	1,706	1,989	1,760	1,741	1,751	1,599	1,498
Ocean perch, Atlantic	1	1	1	1	1	1	1	1
Pollock	2,522	2,729	2,336	2,616	3,188	3,349	3,372	3,362
Salmon, Pacific	568	644	815	629	723	567	674	738
Tuna	83	85	58	51	52	49	62	57
Whiting	34	33	31	27	28	18	19	19
Shellfish:								
Clams (meats)	114	108	112	118	123	130	128	119
Crabs	430	553	458	299	272	308	332	314
Lobsters, American	84	80	87	83	74	82	74	75
Oysters (meats)	40	34	27	41	33	34	37	39
Scallops (meats)	15	13	27	33	47	53	56	65
Shrimp	317	290	278	304	324	317	315	308
	Value							
	Mil. dol.	Mil. dol.	Mil. dol.	Mil. dol.	Mil. dol.	Mil. dol.	Mil. dol.	Mil. dol.
Fish:								
Cod, Atlantic	24	25	24	26	32	31	28	22
Flounder	131	97	90	110	105	102	94	124
Haddock	4	8	9	12	15	19	17	18
Halibut	117	104	125	144	115	136	172	177
Herring, sea	23	22	26	22	26	21	26	29
Jack mackerel	(2)	(2)	(2)	(2)	(2)	(2)	(2)	(2)
Menhaden	112	104	113	112	103	105	96	72
Ocean perch, Atlantic	(2)	(2)	(2)	3	3	2	(2)	(2)
Pollock	248	198	171	168	237	210	208	277
Salmon, Pacific	270	257	360	270	209	155	201	273
Tuna	110	94	86	95	93	84	87	91
Whiting	15	13	14	11	13	7	9	10
Shellfish:								
Clams (meats)	130	135	135	154	162	167	162	159
Crabs	430	473	521	405	382	398	481	448
Lobsters, American	267	254	323	301	254	293	292	315
Oysters (meats)	117	89	73	91	81	89	104	111
Scallops (meats)	94	80	129	165	175	204	229	322
Shrimp	544	516	561	690	569	461	421	426

[1] Data exclude landings by U.S. flag vessels at Puerto Rico and other ports outside the 50 States, and production of artificially cultivated fish and shellfish.   [2] Less than $500.000.
U.S. Department of Commerce, NOAA, NMFS, Fisheries Statistics Division. (301) 713–2328.

Table 15-10.—Fresh and frozen fishery products: Production and value, 1997–2004 [1]

[2002 is preliminary]

Product	Production							
	1997	1998	1999	2000	2001	2002	2003 [4]	2004
	Mil. lb.	Mil. lb.	Mil. lb.	Mil. lb.	Mil. lb.	Mil. lb.	Mil. lb.	Mil. lb.
Fish fillets and steaks [2]	410	422	362	368	480	517	612	590
Cod	79	67	61	56	40	50	56	37
Flounder	27	24	23	27	30	25	21	20
Haddock	7	6	5	6	6	8	8	10
Ocean perch, Atlantic	1	1	1	1	(3)	(3)	1	1
Rockfish	17	16	11	11	7	7	5	4
Pollock, Atlantic	1	4	2	2	2	4	7	4
Pollock, Alaska	112	161	144	160	271	308	367	384
Other	166	143	115	105	124	115	147	130
	Value							
	Mil. dol.	Mil. dol.	Mil. dol.	Mil. dol.	Mil. dol.	Mil. dol	Mil. dol.	Mil. dol.
Fish fillets and steaks [2]	961	961	807	823	914	981	1,133	969
Cod	179	161	108	167	123	155	171	81
Flounder	79	70	67	71	74	73	62	66
Haddock	24	22	20	24	27	32	35	44
Ocean perch, Atlantic	2	2	2	1	1	1	3	3
Rockfish	33	33	23	25	17	15	12	9
Pollock, Atlantic	2	7	4	4	8	11	10	6
Pollock, Alaska	129	190	169	178	296	330	395	366
Other	513	476	414	353	368	364	445	394

[1] Excludes Alaska and Hawaii, except frozen products includes Alaska and Hawaii. [2] Fresh and frozen. [3] Less than 500,000 lb. [4] Revised.

U.S. Department of Commerce, NOAA, NMFS, Fisheries Statistics Division (301) 713–2328.

Table 15-11.—Canned fishery products: Production and value, 1997–2004 [1]

[2002 is preliminary]

Product	Production							
	1997	1998	1999	2000	2001	2002	2003 [6]	2004
	Mil. lb.	Mil. lb.	Mil. lb.	Mil. lb.	Mil. lb.	Mil. lb.	Mil. lb.	Mil. lb.
Total [2]	1,565	1,533	1,897	1,747	1,664	1,317	1,295	1,106
Tuna	627	681	694	671	507	547	529	434
Salmon	162	159	234	171	185	224	188	199
Clam products	127	113	123	127	126	140	123	109
Sardines, Maine	16	12	12	(3)	(3)	(3)	(3)	(3)
Shrimp	1	2	2	2	2	2	1	1
Crabs	(5)	(5)	(5)	(5)	(5)	(5)	(5)	(5)
Oysters [4]	(5)	(5)	(5)	(5)	1	(5)	(5)	(5)
	Value							
	Mil. dol.	Mil. dol.	Mil. dol.	Mil. dol.	Mil. dol.	Mil. dol.	Mil. dol.	Mil. dol.
Total [2]	1,593	1,765	1,861	1,626	1,400	1,290	1,239	1,099
Tuna	919	983	946	856	658	675	669	569
Salmon	253	274	393	288	259	296	242	251
Clam products	115	105	110	120	125	118	132	113
Sardines, Maine	29	19	20	(3)	(3)	(3)	(3)	(3)
Shrimp	5	11	10	11	10	9	5	5
Crabs	(5)	(5)	(5)	(5)	(5)	(5)	(5)	(5)
Oysters [4]	(5)	(5)	(5)	1	1	(5)	(5)	1

[1] Natural pack only. [2] Includes other products not shown separately. [3] Confidential data. [4] Includes oyster specialties. [5] Less than 500,000 pounds or $500,000. [6] Revised.

U.S. Dept. of Commerce, NOAA, NMFS, Fisheries Statistics Division (301) 713–2328.

Table 15-12.—Fisheries: Fishermen and craft, 1977, and catch, 1999–2004 by area

[1998–2003 are preliminary]

Area	1977 [1]			1999		2000	
	Fishermen	Fishing vessels	Fishing boats [2]	Total catch	Value	Total catch	Value
	1,000	*Number*	*1,000*	*Mil. lb.*	*Mil. dol.*	*Mil. lb.*	*Mil. dol.*
United States	182.1	17,545	89.2	9,339	3,467	9,069	3,549
New England States	31.7	929	15.4	584	655	571	681
Middle Atlantic States	17.3	573	11.3	225	181	220	173
Chesapeake Bay States	27.9	2,086	19.0	527	172	492	172
South Atlantic States	11.6	1,463	6.7	230	198	221	204
Gulf States	29.3	5,328	11.0	1,945	758	1,760	911
Pacific Coast States	54.0	7,643	15.4	5,766	1,472	5,750	1,321
Great Lakes States	1.2	217	0.5	24	16	22	19
Hawaii	2.7	101	1.3	37	65	33	68

	2001		2002		2003		2004	
	Total catch	Value	Total catch	Value	Total catch	Value	Total catch	Value
	Mil. lb.	*Mil. dol.*	*Mil. lb.*	*Mil. dol.*	*Mil. lb.*	*Mil. dol.*	*Mil. lb.*	*Mil. dol.*
United States	9,492	3,228	9,397	3,092	9,507	3,347	9,643	3,652
New England States	635	646	584	685	661	691	686	758
Middle Atlantic States	217	173	207	170	215	177	224	191
Chesapeake Bay States	617	175	496	172	496	180	531	210
South Atlantic States	200	176	215	173	197	153	197	152
Gulf States	1,606	798	1,716	693	1,600	683	1,474	667
Pacific Coast States	6,174	1,187	6,138	1,131	6,291	1,382	6,483	1,587
Great Lakes States	19	18	18	16	17	13	17	12
Hawaii	24	55	24	52	24	52	24	57
Utah	....	....	....	....	6	16	7	18

[1] Exclusive of duplication among regions. Computation of area amounts will not equal U.S. total. Mississippi River data included with total.   [2] Refers to craft having capacity of less than 5 net tons.   Note: Table may not add due to rounding.

U.S. Department of Commerce, NOAA, NMFS, Fisheries Statistics Division (301) 713–2328.

Table 15-13.—Fisheries: Quantity and value of domestic catch, 1995–2004

[1994–2003 are preliminary]

Year	Quantity [1]			Ex-vessel value	Average price per lb.
	Total	For human food	For industrial products [2]		
	Mil. lb.	*Mil. lb.*	*Mil. lb.*	*Mil. dol.*	*Cents*
1995	9,788	7,667	2,121	3,770	38.5
1996	9,565	7,474	2,091	3,487	36.5
1997	9,842	7,244	2,598	3,448	35.0
1998	9,194	7,173	2,021	3,128	34.0
1999	9,339	6,832	2,507	3,467	37.1
2000	9,069	6,912	2,157	3,550	39.1
2001	9,495	7,314	2,178	3,228	34.0
2002	9,397	7,205	2,192	3,092	32.9
2003	9,507	7,521	1,986	3,347	35.2
2004	9,643	7,768	1,875	3,652	37.9

[1] Live weight.   [2] Meals, oil, fish solubles, homogenized condensed fish, shell products, bait, and animal food.

U.S. Department of Commerce, NOAA, NMFS Fisheries Statistics Division (301) 723–2328.

Table 15-14.—Fishery products: Supply, 1995–2004 [1]
[1994–2003 are preliminary]

Item	1995	1996	1997	1998	1999
	Mil. lbs	Mil. lbs	Mil. lbs	Mil. lbs	Mil. lbs
Total	16,484	16,474	17,133	16,898	17,378
For human food	13,584	13,626	13,740	14,175	14,462
Finfish	10,692	10,699	10,580	10,837	10,831
Shellfish [2]	2,891	2,927	3,160	3,338	3,630
For industrial use	2,900	2,848	3,393	2,723	2,916
Domestic catch	9,788	9,565	9,843	9,194	9,339
Percent of total	59.4	58.1	57.4	54.4	53.7
For human food	7,667	7,476	7,245	7,174	6,832
Finfish	6,414	6,205	5,969	5,935	5,490
Shellfish [2]	1,252	1,271	1,277	1,238	1,341
For industrial use	2,121	2,089	2,598	2,021	2,507
Imports [3]	6,696	6,909	7,290	7,704	8,039
Percent of total	40.6	41.9	42.5	45.6	46.3
For human food	5,917	6,150	6,495	7,001	7,630
Finfish	4,278	4,494	4,612	4,901	5,341
Shellfish [2]	1,639	1,656	1,883	2,100	2,289
For industrial use [4]	779	759	795	702	409

Item	2000	2001	2002	2003	2004
	Mil. lbs	Mil. lbs	Mil. lbs	Mil. lbs	Mil. lbs
Total	17,339	18,119	19,028	19,850	20,373
For human food	14,740	15,306	16,007	17,187	17,622
Finfish	11,006	11,330	11,770	12,617	12,954
Shellfish [2]	3,734	3,977	4,237	4,570	4,668
For industrial use	2,599	2,812	3,022	2,663	2,751
Domestic catch	9,068	9,492	9,397	9,507	9,643
Percent of total	52.3	52.4	49.4	47.9	47.3
For human food	6,912	7,314	7,205	7,521	7,768
Finfish	5,637	6,162	6,013	6,388	6,636
Shellfish [2]	1,275	1,152	1,192	1,133	1,132
For industrial use	2,157	2,178	2,193	1,986	1,875
Imports [3]	8,271	8,627	9,631	10,343	10,730
Percent of total	47.7	47.6	50.6	52.1	52.7
For human food	7,828	7,992	8,802	9,666	9,854
Finfish	5,369	5,168	5,757	6,229	6,318
Shellfish [2]	2,459	2,825	3,045	3,437	3,536
For industrial use [4]	442	634	829	677	876

[1] Live weight, except percent. May not add due to rounding.   [2] For univalve and bivalves mollusks (conchs, clams, oysters, scallops, etc.), the weight of meats, excluding the shell is reported.   [3] Excluding imports of edible fishery products consumed in Puerto Rico; includes landings of tuna caught by foreign vessels in American Samoa.   [4] Fish meal and sea herring.

U.S. Department of Commerce, NOAA, NMFS Fisheries Statistics Division (301) 713–2328.

Table 15-15.—Fisheries: Disposition of domestic catch, 1995–2004 [1]
[1994–2003 are preliminary]

Disposition	1995	1996	1997	1998	1999	2000	2001	2002	2003	2004
	Mil. lbs.	Mil. lbs.	Mil. lbs.	Mil. lbs.	Mil. lbs.	Mil. lbs.	Mil. lbs.	Mil. lbs.	Mil. lbs.	Mil. lbs.
Fresh and frozen	7,099	7,054	6,873	6,870	6,416	6,657	7,085	6,826	7,266	7,448
Canned	769	678	648	516	712	530	536	652	498	552
Cured	90	93	108	129	133	119	123	117	119	137
Reduced to meal, oil, etc	1,830	1,740	2,213	1,679	2,078	1,763	1,748	1,802	1,624	1,506
Total	9,788	9,565	9,842	9,194	9,339	9,069	9,492	9,397	9,507	9,643

[1] Live weight catch. In addition to whole fish, a large portion of waste (400–500 mil. lb.) derived from canning, filleting, and dressing fish and shellfish is utilized in production of fish meal and oil in each year shown.

U.S. Department of Commerce, NOAA, NMFS Fisheries Statistics Division (301) 713–2328.

Table 15-16.—Processed fishery products: Production and value, 1997–2004 [1]

Item	Production							
	1997	1998	1999	2000	2001	2002	2003	2004 [3]
	Mil. lb.	*Mil. lb.*	*Mil. lb.*	*Mil. lb.*	*Mil. lb.*	*Mil. lb.*	*Mil. lb.*	*Mil. lb.*
Fresh and frozen:.								
Fillets	355	391	337	336	450	495	588	575
Steaks	55	31	25	32	30	22	25	16
Fish sticks	69	69	65	40	43	48	31	60
Fish portions	196	185	203	183	189	187	162	135
Breaded shrimp	117	109	119	121	152	147	152	110
Canned products [2]	1,565	1,533	1,897	1,747	1,664	1,317	1,295	1,106
Fish and shellfish	953	989	1,100	1,008	885	953	858	763
Animal feed	612	544	797	739	779	365	437	343
Industrial products	NA	NA	NA	NA	NA	NA	NA	NA
Meal and scrap	725	613	672	627	644	638	603	575
Oil (body and liver)	283	223	286	192	279	211	196	179
Other	NA	NA	NA	NA	NA	NA	NA	NA

Item	Value							
	1997	1998	1999	2000	2001	2002	2003	2004 [3]
	Mil. dol.	*Mil. dol.*	*Mil. dol.*	*Mil. dol.*	*Mil. dol.*	*Mil. dol.*	*Mil. dol.*	*Mil. dol.*
Fresh and frozen:.								
Fillets	845	887	739	741	845	920	1,064	918
Steaks	116	74	68	82	70	62	69	51
Fish sticks	64	63	63	43	42	51	35	71
Fish portions	285	211	269	233	235	237	227	210
Breaded shrimp	335	333	352	375	540	464	465	306
Canned products [2]	1,593	1,775	1,861	1,626	1,400	1,290	1,239	1,099
Fish and shellfish	1,361	1,425	1,522	1,334	1,110	1,150	1,076	966
Animal feed	232	350	340	292	290	140	163	132
Industrial products	347	233	268	219	237	233	222	203
Meal and scrap	174	117	147	115	126	140	134	153
Oil (body and liver)	55	56	42	21	48	41	34	35
Other	118	60	79	83	83	52	54	15

[1] Includes cured fish.   [2] Includes salmon eggs for baits.   [3] Preliminary.   NA=not available.
U.S. Department of Commerce, NOAA, NMFS, Fisheries Statistics Division (301) 713–2328.

Table 15-17.—Selected fishery products: Imports and exports, 1997-2004 [1]

Product	Quantity							
	1997	1998	1999	2000	2001	2002	2003	2004
Imports	Mil. lb.	Mil. lb.	Mil. lb.	Mil. lb.	Mil. lb.	Mil. lb.	Mil. lb.	Mil. lb.
Edible	3,339	3,647	3,888	3,978	4,102	4,427	4,907	4,951
Fresh or frozen	2,861	3,119	3,227	3,310	3,449	3,670	4,032	4,075
Salmon [2]	163	152	156	151	159	182	163	153
Tuna	438	571	491	445	405	358	462	407
Groundfish fillets, blocks [3]	384	376	410	393	310	347	332	361
Other fillets and steaks ...	339	392	429	510	601	691	760	813
Scallops (meats)	60	52	44	54	40	48	52	45
Lobster, American and spiny	65	64	81	95	92	100	99	97
Shrimp and prawn	645	692	728	757	878	942	1,108	1,138
Canned	387	428	546	556	539	632	748	745
Sardines, in oil	13	15	16	26	19	15	16	18
Sardines and herring, not in oil	31	33	38	46	42	42	45	43
Tuna	212	240	335	313	292	378	459	443
Oysters	10	14	13	43	12	13	15	15
Pickled or salted	37	42	38	42	43	46	49	49
Cod, haddock, hake, pollock, cusk	5	7	7	9	8	8	8	8
Nonedible scrap and metal	142	125	73	79	113	148	121	156
Exports								
Canned salmon	82	77	114	81	110	99	96	118
Fish oil, nonedible	215	197	233	142	249	213	147	110

Product	Value							
	1997	1998	1999	2000	2001	2002	2003	2004
Imports	Mil. dol.	Mil. dol.	Mil. dol.	Mil. dol.	Mil. dol.	Mil. dol.	Mil. dol.	Mil. dol.
Edible	7,754	8,173	9,014	10,054	9,864	10,121	11,095	11,331
Fresh or frozen	7,022	7,356	8,043	9,120	8,832	8,948	9,815	9,916
Salmon [2]	344	319	345	333	323	344	324	307
Tuna	494	556	550	520	515	417	543	551
Groundfish fillets, blocks [3]	534	579	674	589	479	544	505	537
Other fillets and steaks ...	727	837	982	1,233	1,263	1,383	1,580	1,726
Scallops (meats)	237	218	193	212	128	144	157	146
Lobster, American and spiny	481	476	628	712	727	825	883	876
Shrimp and prawn	2,943	3,102	3,131	3,749	3,617	3,414	3,753	3,675
Canned	525	588	682	670	774	907	1,010	1,123
Sardines, in oil	25	28	28	39	30	23	28	30
Sardines and herring, not in oil	26	29	36	44	39	38	41	40
Tuna	250	289	336	258	314	399	455	483
Oysters	25	26	27	26	24	24	28	32
Pickled or salted	47	57	59	60	61	68	72	72
Cod, haddock, hake, pollock, cusk	8	13	15	19	16	18	16	16
Nonedible scrap and metal	36	34	17	18	27	39	32	43
Exports								
Canned salmon	135	143	198	146	168	141	148	177
Fish oil, nonedible	54	60	36	24	42	49	38	32

[1] Includes Puerto Rico.　[2] Excludes fillets.　[3] Includes cod, cusk, haddock, hake, pollock, ocean perch, and whiting.
U.S. Dept. of Commerce, NOAA, NMFS, Fisheries Statistics Division (301) 713-2328.

Table 15-18.—Fishery products: Imports and exports, 1995–2004 [1]

Year	Imports [2]				Exports			
	Total value	Edible products		Non-edi-ble, value	Total value	Edible products		Non-edi-ble, value
		Quantity	Value			Quantity	Value	
	Mil. lb.	Mil. lb.	Mil. lb.	Mil. lb.	Mil. lb.	Mil. lb.	Mil. lb.	Mil. lb.
1995	12,452	3,066	6,792	5,660	8,268	2,047	3,262	5,006
1996	13,060	3,170	6,730	6,331	8,653	2,112	3,032	5,621
1997	14,528	3,339	7,754	6,774	9,354	2,019	2,713	6,640
1998	15,633	3,647	8,173	7,459	8,697	1,664	2,260	6,437
1999	17,040	3,888	9,014	8,026	10,007	1,961	2,849	7,158
2000	19,013	3,978	10,054	8,959	10,782	2,165	2,952	7,830
2001	18,547	4,102	9,864	8,683	11,834	2,565	3,195	8,639
2002	19,691	4,427	10,121	9,570	11,713	2,398	3,120	8,593
2003	21,283	4,907	11,095	10,187	11,999	2,396	3,268	8,731
2004	22,949	4,951	11,331	11,618	13,592	2,888	3,708	9,884

[1] Includes Puerto Rico.   [2] Includes landings of tuna by foreign vessels in American Samoa.
U.S. Department of Commerce, NMFS, Fisheries Statistics Division (301) 713–2328.

Table 15-19.—Fish trips: Estimated number of fishing trips taken by marine recreational fishermen by subregion and year, Atlantic and Gulf and Pacific Coasts, 2001–2004

Subregion	2001	2002	2003	2004
	Thousands	Thousands	Thousands	Thousands
Atlantic and Gulf: [1]				
North Atlantic	9,035	8,591	8,578	8,713
Mid-Atlantic	21,206	16,645	19,852	18,712
South Atlantic [2]	21,596	17,763	21,246	20,778
Gulf [2]	22,890	19,666	22,957	24,583
Total	74,727	62,665	72,633	72,786

Subregion	2001	2002	2003	2004
	Thousands	Thousands	Thousands	Thousands
Pacific: [3]				
Southern California	4,052	4,313	3,826	3,083
Northern California	2,208	2,290	2,723	1,366
Oregon	1,170	993	502	223
Washington	2,191	1,786	614	198
Total	9,621	9,382	7,665	4,870

[1] Data do not include recreational trips in Texas.   [2] Does not include trips from headboats (party boats) in the South Atlantic and Gulf of Mexico.   [3] Data do not include recreational trips in Hawaii or Alaska. Pacific state estimates do not include salmon data collected by recreational surveys.
U.S. Department of Commerce, NOAA, NMFS, Fisheries Statistics Division (301) 713–2328.

Table 15-20.—Fish harvested: Estimated number of fish harvested by marine recreational anglers by subregion and year, Atlantic, Gulf Coasts, and Pacific Coasts, 2001-2004

Subregion	2001	2002	2003	2004
Atlantic and Gulf: [1]	Thousands	Thousands	Thousands	Thousands
North Atlantic	12,153	11,132	11,559	10,031
Mid-Atlantic	34,704	30,802	39,591	31,285
South Atlantic [2]	43,824	42,928	50,787	48,344
Gulf [2]	76,571	79,015	74,814	91,009
Total	167,252	163,877	176,751	180,669

Subregion	2001	2002	2003	2004
Pacific: [3]	Thousands	Thousands	Thousands	Thousands
Southern California	7,726	8,950	8,418	7,503
Northern California	4,799	6,884	6,228	2,524
Oregon	2,123	3,392	1,033	605
Washington	4,798	4,841	1,198	685
Total	19,446	24,067	16,877	11,318

[1] Data do not include recreational catch in Texas. [2] Does not include catch for headboats (party boats) in the South Atlantic and Gulf of Mexico. [3] Data do not include recreational catch in Hawaii or Alaska. Pacific estimates do not include salmon data collected by State recreational surveys. Note: "Harvested" includes dead discards and fish used for bait but does not include fish released alive.

U.S. Department of Commerce, NOAA, NMFS, Fisheries Statistics Division (301) 713-2328.

Table 15-21.—Fish harvested: Estimated number of fish harvested by marine recreational anglers by mode and year, Atlantic, Gulf Coasts, and Pacific Coasts, 2001-2004

Mode	2001	2002	2003	2004
Atlantic and Gulf: [1]	Thousands	Thousands	Thousands	Thousands
Shore	53,092	44,702	52,551	47,519
Party/charter [2]	11,628	10,575	11,596	11,735
Private/rental	102,532	108,600	112,604	121,415
Total	167,252	163,877	176,751	180,669

Mode	2001	2002	2003	2004
Pacific: [3]	Thousands	Thousands	Thousands	Thousands
Shore	8,040	7,807	7,277	5,652
Party/charter	5,176	8,973	3,193	3,533
Private/rental	6,230	7,287	6,407	2,133
Total	19,446	24,067	16,877	11,318

[1] Data do not include recreational catch in Texas. [2] Does not include catch for headboats (party boats) in the South Atlantic or Gulf of Mexico. [3] Data do not include recreational catch in Hawaii or Alaska. Pacific estimates do not include salmon data collected by State recreational surveys. Note: "Harvested" includes dead discards and fish used for bait but does not include fish released alive.

U.S. Department of Commerce, NOAA, NMFS, Fisheries Statistics Division (301) 713-2328.

Table 15-22.—Fish harvested: Estimated number of fish harvested by marine recreational anglers by species group and year, Atlantic and Gulf coasts, 2001–2004 [1]

Species group	2001	2002	2003	2004
	Thousands	*Thousands*	*Thousands*	*Thousands*
Barracudas	163	122	158	92
Bluefish	7,016	5,495	6,243	7,249
Dogfish sharks	49	109	66	69
Other sharks	343	208	214	195
Skates/rays	57	67	73	59
Freshwater catfishes	118	160	830	383
Saltwater catfishes	629	533	592	474
Atlantic cod	1,118	644	707	650
Other cods/hakes	168	121	181	364
Pollock	356	239	158	227
Red hake	58	25	48	30
Dolphins	2,088	1,727	1,822	1,327
Other croaker	0	0	0	0
Atlantic croaker	14,681	12,389	11,509	11,812
Black drum	990	941	1,161	930
Kingfishes	7,456	4,121	5,655	6,366
Other drum	278	661	307	416
Red drum	3,475	2,827	3,151	3,334
Sand seatrout	3,308	3,074	3,062	2,312
Silver perch	404	216	314	344
Spot	7,308	5,336	9,274	8,552
Spotted seatrout	10,200	8,143	10,496	11,810
Weakfish	1,527	1,172	498	770
Eels	54	10	53	37
Gulf flounder	212	173	200	249
Other flounders	88	53	45	51
Southern flounder	1,128	903	1,202	1,387
Summer flounder	5,307	3,281	4,578	4,565
Winter flounder	964	469	624	421
Other grunts	463	541	686	586
Pigfish	1,552	1,323	1,193	682
White grunt	2,772	2,560	2,239	2,195
Herrings	33,473	47,399	46,563	53,795
Blue runner	3,160	2,339	2,586	2,356
Crevalle Jack	812	684	525	530
Florida pompano	614	528	892	827
Greater amberjack	135	157	180	125
Other jacks	2,965	3,166	1,909	2,235
Mullets	7,435	9,764	9,680	10,303
Other fishes	4,769	4,088	4,457	3,729
Other porgies	228	201	204	326
Pinfishes	9,469	8,868	6,772	8,898
Red porgy	75	72	97	143
Scup	5,099	3,647	9,452	4,918
Sheepshead	2,267	1,972	3,095	2,979
Puffers	346	354	255	140
Sculpins	(2)	9	0	(2)
Black sea bass	3,932	4,223	4,023	2,737
Epinephelus groupers	248	292	248	547
Mycteroperca groupers	540	577	578	737
Other sea basses	337	352	420	538
Searobins	143	200	195	215
Gray snapper	1,199	1,156	1,545	1,373
Lane snapper	392	204	318	311
Other snappers	79	131	136	148
Red snapper	900	1,159	1,029	1,104
Vermilion snapper	613	443	483	704
Yellowtail snapper	189	271	332	494
Other temperate basses	1	1	0	0
Striped bass	2,039	1,841	2,515	2,456
White perch	664	1,382	2,700	1,743
Toadfishes	7	19	18	14
Triggerfishes/filefishes	308	443	475	619
Atlantic mackerel	4,127	3,663	2,460	1,565
King mackerel	691	690	810	662
Little tunny/Atlantic bonito	260	268	197	299
Other tunas/mackerels	737	699	664	603
Spanish mackerel	3,747	3,334	2,695	3,188
Cunner	56	64	33	161
Other wrasses	79	73	141	96
Tautog	792	1,501	731	1,111
Total [3]	167,252	163,877	176,751	180,668

[1] Data does not include recreational catch in Texas or headboats (party boats) in the South Atlantic and the Gulf of Mexico. [2] Less than one thousand. [3] Totals may not add due to rounding. Note: "Harvested" includes dead discards and fish used for bait but does not include fish released alive.

U.S. Department of Commerce, NOAA, NMFS, Fisheries Statistics Division (301) 713–2328.

Table 15-23.—Fish harvested: Estimated number of fish harvested by marine recreational anglers by species group and year, Pacific coast[1], 2001–2004

Species group	2001	2002	2003	2004
	Thousands	*Thousands*	*Thousands*	*Thousands*
Northern anchovy	579	176	137	430
Other anchovies	0	5	0	1
California scorpionfish	293	251	171	88
Dogfish sharks	20	12	13	1
Other sharks	36	26	33	17
Skates/rays	29	20	16	13
Other cods/hakes	0	0	2	2
Pacific cod	1	1	3	6
Pacific hake	0	2	0	1
Pacific tomcod	2	4	2	4
California corbina	14	20	2	13
Other croakers	172	152	99	97
Queenfish	76	579	314	344
White croaker	389	388	425	216
Dolphins	0	0	0	0
Other drum	4	22	8	0
California halibut	202	251	199	39
Other flounders	80	253	49	39
Rock sole	12	41	158	(2)
Sanddabs	451	3,316	493	369
Starry flounder	14	14	12	4
Kelp greenling	153	182	132	31
Lingcod	113	270	369	80
Other greenlings	28	22	31	3
Herrings	799	2,216	1,737	1,452
Other jacks	21	58	51	40
Yellowtail	87	54	82	82
Mullets	5	1	13	0
Other fishes	2,305	1,523	1,179	1,023
Pacific barracuda	311	440	193	246
Black rockfish	1,119	1,117	1,198	644
Blue rockfish	464	772	479	363
Bocaccio	199	121	8	53
Brown rockfish	185	151	208	41
Canary rockfish	78	47	32	16
Chilipepper rockfish	77	45	0	15
Copper rockfish	78	75	56	34
Greenspotted rockfish	83	35	1	35
Olive rockfish	159	151	73	79
Other rockfishes	994	1,157	975	596
Quillback rockfish	26	26	19	10
Gopher rockfish	272	352	225	85
Widow rockfish	19	21	1	28
Yellowtail rockfish	162	201	61	64
Sablefishes	1	14	2	(1)
Cabezon	69	62	70	30
Sculpins	49	50	28	12
Barred sand bass	1,119	1,776	1,019	778
Kelp bass	633	569	514	499
Other sea basses	24	7	15	1
Spotted sand bass	361	52	66	10
Halfmoon	132	165	40	33
Opaleye	59	48	25	40
Jacksmelt	614	333	585	354
Other silversides	46	113	634	352
Other smelts	0	0	2	6
Surf smelt	3,661	4,174	1,595	2
Sturgeons	17	15	42	2
Barred surfperch	147	166	366	256
Black perch	54	50	70	74
Other surfperches	108	122	103	77
Pile perch	32	43	33	7
Redtail surfperch	123	53	120	28
Shiner perch	183	226	80	191
Silver surfperch	16	21	33	23
Striped seaperch	96	101	88	24
Walleye surfperch	163	93	151	103
White seaperch	32	26	18	12
Striped bass	44	61	64	25
Other tunas/mackerels	1,755	1,064	1,729	1,078
Pacific bonito	31	6	70	569
California sheephead	75	74	48	21
Other wrasses	5	13	8	5
Total[3]	19,446	24,067	16,877	11,318

[1] Data do not include recreational catch in Hawaii or Alaska. Pacific estimates do not include salmon data collected by State recreational surveys. [2] Less than one thousand. [3] Totals may not add due to rounding. Note: "Harvested" includes dead discards and fish used for bait but does not include fish released alive.

U.S. Department of Commerce, NOAA, NMFS, Fisheries Statistics Division. (301) 713–2328.

Table 15-24.—Fish harvested: Estimated number of fish harvested by marine recreational anglers, by area of fishing and year, Atlantic and Gulf and Pacific Coast, 2001–2004

Area	2001	2002	2003	2004
	Thousands	Thousands	Thousands	Thousands
Atlantic and Gulf: [1]				
Inland	96,875	97,493	104,423	106,086
State Territorial Sea [2]	49,450	45,894	50,378	51,322
Federal Exclusive Ecomomic Zone [3]	20,928	20,490	21,950	23,260
Total	167,253	163,877	176,751	180,668

Area	2001	2002	2003	2004
	Thousands	Thousands	Thousands	Thousands
Pacific: [4]				
Inland	7,528	7,998	3,557	1,051
State Territorial Sea [2]	9,182	11,551	11,249	9,287
Federal Exclusive Ecomomic Zone [3]	2,736	4,518	2,071	980
Total	19,446	24,067	16,877	11,318

[1] Data do not include recreational catch in Texas or headboats (party boats) in the South Atlantic and the Gulf of Mexico. [2] Open Ocean extending 0 to 3 miles from shore, except West Florida (10 miles). [3] Open ocean extending to 200 miles offshore from the outer edge of the State Territorial Sea. [4] Data do not include recreational catch in Hawaii or Alaska. Pacific state estimates do not include salmon data collected by recreational surveys. Note: "Harvested" includes dead discards and fish used for bait but does not include fish released alive.

U.S. Department of Commerce, NOAA, NMFS, Fisheries Statistics Division. (301) 713–2328.

Table 15-25.—Farm-raised catfish: Processed, sales, inventory, and imports, 1996–2005

Year	Round [1] weight processed	Prices paid to producer	Fresh sales	Frozen sales	Total sales	Inventory end of year	Imports [2]
	(000) pounds	Cents per pounds	(000) pounds	(000) pounds	(000) pounds	(000) pounds	(000) pounds
1996	472,123	77.3	96,722	140,458	237,180	11,894	2,482
1997	524,949	71.2	106,512	155,248	261,760	11,911	942
1998	564,355	74.3	113,092	168,306	281,398	10,807	1,386
1999	596,628	73.7	116,697	175,968	292,665	12,551	3,451
2000	593,603	75.1	116,734	180,422	297,156	13,598	8,236
2001	597,108	64.7	120,775	175,592	296,367	14,997	18,079
2002	603,601	56.8	123,451	194,198	317,649	12,283	10,201
2003	661,504	58.1	126,841	192,486	319,327	13,592	5,430
2004	630,601	69.7	117,599	189,180	306,779	15,172	9,224
2005	600,670	72.5	107,984	191,984	299,968	13,707	NA

[1] Price for fish delivered to processing plant door. [2] Data furnished by U.S. Bureau of Census. NA=not available.

NASS, Livestock Branch, (202) 720–3570.

Table 15-26.—Farm-raised catfish: Prices received by processors, 1996–2005

Year	Fresh			Frozen		
	Whole fish [1]	Fillets [2]	Other [3]	Whole fish [1]	Fillets [2]	Other [3]
	Dollars per/lb	Dollars per/lb	Dollars per/lb	Dollars per/lb	Dollars per/lb	Dollars per/lb
1996	1.68	2.87	1.80	1.99	2.78	1.88
1997	1.55	2.75	1.67	1.93	2.63	1.76
1998	1.59	2.80	1.72	1.94	2.69	1.73
1999	1.59	2.81	1.64	1.99	2.76	1.69
2000	1.66	2.86	1.68	2.03	2.83	1.65
2001	1.57	2.74	1.60	1.98	2.61	1.63
2002	1.32	2.52	1.51	1.84	2.39	1.54
2003	1.35	2.48	1.52	1.84	2.41	1.44
2004	1.56	2.71	1.71	1.95	2.62	1.46
2005	1.59	2.83	1.69	2.00	2.67	1.50

[1] Dressed weight, (head, visera, and skin removed). [2] Includes regular, shank, and strip fillets; excludes any breaded product. [3] Includes nuggets, steaks, and all other products not already reported, includes weight of breading and added ingredients.

NASS, Livestock Branch, (202) 720–3570.

Table 15-27.—Catfish: Number of operations and water surface acres used for production, 2005–06, and total sales, 2004–05, by State and United States

State	Number of operations on Jan. 1		Water surface accres used for production during Jan 1 - Jun 30		Total sales	
	2005 [1]	2006	2005 [1]	2006	2004 [1]	2005
	Number	*Number*	*Acres*	*Acres*	*1,000 dollars*	*1,000 dollars*
AL	230	194	25,100	23,500	101,198	97,602
AR	153	132	31,500	32,600	66,618	77,556
CA	31	36	1,700	1,500	7,482	7,308
FL	46	31	650	520	1,139	1,120
GA	55	60	1,090	1,300	1,475	2,066
KY	60	34	600	390	1,151	887
LA	38	28	7,600	6,600	14,316	14,936
MS	410	390	101,000	99,000	274,971	268,303
MO	24	24	1,320	1,360	1,358	1,723
NC	49	49	2,000	2,100	7,021	6,077
TX	62	57	1,030	1,500	3,446	4,547
US	1,158	1,035	173,590	170,370	480,175	482,125

[1] Revised.

NASS, Livestock Branch, (202) 720-0585.

Table 15-28.—Catfish production: Water surface acre usage by State and United States, 2005–06

State	Acres intended for utilization during Jan 1-Jun 30 for:					Acres taken out of production during Jul 1-Dec 31 prev. year
	Foodsize	Fingerlings	Broodfish	Currently under or scheduled for:		
				Renovation	New construction	
2005 [1]						
AL	23,300	1,300	430	450	250	290
AR	27,400	3,300	550	980	*	4,800
CA	1,310	250	75	165	....	*
FL	530	*	25	70	*	60
GA	655	290	75	25	*	*
KY	500	65	15	*	*	20
LA	6,500	1,000	100	400	*	730
MS	81,500	14,500	2,500	3,300	400	3,500
MO	790	*	25	15	*	*
NC	1,700	200	60	*	*	*
TX	870	70	70	60	50	30
Oth Sts	....	520	....	35	96	115
US	145,055	21,495	3,925	5,500	796	9,545
2006						
AL	21,600	920	320	860	160	1,140
AR	27,500	4,300	570	760	100	970
CA	1,240	150	50	60	60	*
FL	480	*	10	*	....	*
GA	840	325	120	70	*	*
KY	340	30	15	*	....	200
LA	5,500	970	*	550	....	360
MS	77,300	15,500	3,500	4,000	400	3,700
MO	970	*	25	65	*	*
NC	1,700	100	*	200	300	40
TX	1,350	90	40	20	250	30
Oth Sts	....	365	130	30	60	240
US	138,820	22,750	4,780	6,615	1,330	6,680

[1] Revised.　　* Included in other States to avoid disclosure of individual operations.

NASS, Livestock Branch, (202) 720–0585.

Table 15-29.—Catfish: Sales by size category, by State and United States, 2004–05

Size category and State	Number of fish		Live weight		Sales			
					Total		Average per pound	
	2004[1]	2005	2004[1]	2005	2004[1]	2005	2004[1]	2005
	1,000	*1,000*	*1,000 pounds*	*1,000 pounds*	*1,000 dollars*	*1,000 dollars*	*Dollars*	*Dollars*
Foodsize:								
AL	82,000	78,000	145,000	142,000	95,700	92,300	0.66	0.65
AR	58,000	64,000	103,000	104,000	63,860	73,840	0.62	0.71
CA	1,620	1,720	3,400	3,180	7,242	7,187	2.13	2.26
FL	1,100	1,400	1,900	1,640	1,102	1,066	0.58	0.65
GA	1,140	990	1,200	1,580	900	1,248	0.75	0.79
KY	530	820	1,020	1,225	775	833	0.76	0.68
LA	15,800	14,800	23,000	21,100	14,260	14,770	0.62	0.70
MS	220,000	235,000	388,000	348,000	256,080	247,080	0.66	0.71
MO	650	510	1,130	1,500	904	1,230	0.80	0.82
NC	5,300	4,600	10,000	8,100	6,900	5,994	0.69	0.74
TX	3,180	3,570	4,500	6,100	3,150	4,331	0.70	0.71
US	389,320	405,410	682,150	638,425	450,873	449,879	0.66	0.70
Broodfish:								
AL	*	300	*	2,400	*	1,440	*	0.60
AR	*	70	*	331	*	265	*	0.80
CA	*	*	*	*	*	*	*	*
FL	....	....	....	....	....	....	....	....
GA	*	*	*	*	*	*	*	*
KY	*	*	*	*	*	*	*	*
LA	*	*	*	*	*	*	*	*
MS	300	230	1,000	530	620	345	0.62	0.65
MO	*	*	*	*	*	*	*	*
NC	*	*	*	*	*	*	*	*
TX	*	*	*	*	*	*	*	*
Oth Sts	57	47	205	154	247	95	1.20	0.62
US	357	647	1,205	3,415	867	2,145	0.72	0.63

See footnotes at end of table.

Table 15-29.—Catfish: Sales by size category, by State and United States, 2004–05—Continued

Size category and State	Number of fish		Live weight		Sales			
					Total		Average per pound	
	2004 [1]	2005	2004 [1]	2005	2004 [1]	2005	2004 [1]	2005
	1,000	1,000	1,000 pounds	1,000 pounds	1,000 dollars	1,000 dollars	Dollars	Dollars
Stockers:								
AL	*	13,600	*	3,600	*	2,700	*	0.75
AR	*	8,400	*	1,230	*	1,095	*	0.89
CA	*	*	*	*	*	*	*	*
FL	*	*	*	*	*	*	*	*
GA	*	*	*	*	*	*	*	*
KY	....	....	....	....	....	....	....	....
LA	....	*	....	*	....	*	....	*
MS	33,000	10,000	3,100	1,300	3,379	1,430	1.09	1.10
MO	*	*	*	*	*	*	*	*
NC	*	*	*	*	*	*	*	*
TX	*	*	*	*	*	*	*	*
Oth Sts	25,435	2,947	3,092	806	2,881	769	0.93	0.95
US	58,435	34,947	6,192	6,936	6,260	5,994	1.01	0.86
Fingerlings and fry:								
AL	87,000	34,600	2,500	830	3,800	1,162	1.52	1.40
AR	60,700	121,000	1,000	1,260	1,840	2,356	1.84	1.87
CA	1,520	620	38	27	131	80	3.45	2.96
FL	*	*	*	*	*	*	*	*
GA	7,840	7,800	235	340	541	694	2.30	2.04
KY	*	*	*	*	*	*	*	*
LA	*	*	*	*	*	*	*	*
MS	400,000	511,000	10,200	13,600	14,892	19,448	1.46	1.43
MO	*	*	*	*	*	*	*	*
NC	1,100	1,200	60	60	96	64	1.60	1.07
TX	1,170	770	60	35	211	125	3.52	3.57
Oth Sts	8,330	2,650	264	72	664	178	2.52	2.47
US	567,660	679,640	14,357	16,224	22,175	24,107	1.54	1.49

[1] Revised.   * Included in other States to avoid disclosure of individual operations.
NASS, Livestock Branch, (202) 720–0585.

Table 15-30.—Trout: Number of operations by State and United States, 2004–05

State	Total		Selling trout		Distributing trout[1]	
	2004[2]	2005	2004[2]	2005	2004[2]	2005
	Number	Number	Number	Number	Number	Number
AR	5	4			5	4
CA	26	33	14	15	16	21
CO	30	27	9	8	21	19
CT	6	6	3	3	3	3
GA	9	14	6	10	3	4
ID	46	42	29	26	17	17
ME	15	15	7	8	8	8
MA	13	15	9	9	5	7
MI	28	20	26	17	7	5
MO	11	15	8	8	3	7
NY	37	36	28	22	14	14
NC	51	49	48	43	3	6
OR	49	48	18	18	32	31
PA	58	57	44	42	19	19
TN	14	14	7	7	8	7
UT	27	21	16	9	12	13
VA	16	20	11	16	6	7
WA	59	69	16	17	44	54
WV	31	26	21	17	16	14
WI	61	70	45	51	20	23
US	592	601	365	346	262	283

[1] Trout distributed for restoration, conservation or recreational purposes.   [2] Revised.
NASS, Livestock Branch, (202) 720-0585.

Table 15-31.—Trout: Value of fish sold and distributed, by State (excluding eggs), and United States (including and excluding eggs), 2004–05

State	Total value of fish sold		Total value of distributed fish	
	2004[1]	2005	2004[1]	2005
	1,000 dollars	1,000 dollars	1,000 dollars	1,000 dollars
AR			*	*
CA	5,130	6,077	8,585	8,127
CO	991	1,480	5,393	5,215
CT	360	411	*	*
GA	935	844	*	*
ID	34,564	35,387	2,579	2,993
ME	212	178	*	*
MA	363	396	*	*
MI	790	793	*	*
MO	2,637	2,649	1,044	2,236
NY	478	507	*	*
NC	5,909	6,590	614	729
OR	807	803	3,529	6,898
PA	4,223	4,807	9,128	10,813
TN	181	291	*	*
UT	760	540	*	*
VA	924	1,256	818	2,071
WA	4,792	4,124	4,989	4,630
WV	694	348	2,155	1,818
WI	1,465	1,573	1,774	2,120
Oth Sts			19,964	24,057
US[2]	66,215	69,054	60,572	71,707
US[3]	71,045	74,191	62,516	74,258

[1] Revised.   [2] Excludes value of eggs.   [3] Includes value of eggs.   *Included in other States to avoid disclosure of individual operations.
NASS, Livestock Branch, (202) 720-0585.

Table 15-32.—Trout: Sales by size category, by State and United States, 2004–05

Size category and State	Number of fish		Live weight		Sales			
					Total		Average per pound	
	2004 [1]	2005	2004 [1]	2005	2004 [1]	2005	2004 [1]	2005
	1,000 dollars	1,000 dollars	1,000 dollars	1,000 dollars	1,000 dollars	1,000 dollars	Dollars	Dollars
12 inch or longer:								
CA	1,840	2,020	2,200	2,450	4,312	5,317	1.96	2.17
CO	275	410	235	346	576	952	2.45	2.75
CT	*	*	*	*	*	*	*	*
GA	300	450	420	500	827	830	1.97	1.66
ID	39,000	44,000	42,900	43,600	34,320	35,316	0.80	0.81
ME	*	*	*	*	*	*	*	*
MA	37	33	33	29	158	157	4.80	5.40
MI	285	255	305	295	601	634	1.97	2.15
MO	*	*	*	*	*	*	*	*
NY	90	80	87	83	262	251	3.01	3.03
NC	3,510	3,530	3,940	4,130	5,437	5,699	1.38	1.38
OR	220	110	200	145	486	405	2.43	2.79
PA	1,160	1,290	1,150	1,320	3,335	3,960	2.90	3.00
TN	46	84	54	90	134	247	2.48	2.74
UT	180	166	165	157	421	466	2.55	2.97
VA	400	680	400	670	808	992	2.02	1.48
WA	740	910	4,050	4,150	3,969	3,403	0.98	0.82
WV	363	161	378	172	658	330	1.74	1.92
WI	400	530	387	484	1,072	1,297	2.77	2.68
Oth Sts	745	792	732	1,051	2,021	2,298	2.76	2.19
US	49,591	55,501	57,636	59,672	59,397	62,554	1.03	1.05
6 inch-12 inch:								
CA	550	480	250	180	680	565	2.72	3.14
CO	*	*	*	*	*	*	*	*
CT	*	*	*	*	*	*	*	*
GA	*	15	*	8	*	14	*	1.80
ID	*	*	*	*	*	*	*	*
ME	*	*	*	*	*	*	*	*
MA	*	*	*	*	*	*	*	*
MI	165	*	65	*	167	*	2.57	*
MO	*	*	*	*	*	*	*	*
NY	105	115	38	46	179	207	4.70	4.49
NC	300	610	140	270	217	437	1.55	1.62
OR	310	*	110	*	310	*	2.82	*
PA	450	490	216	190	821	760	3.80	4.00
TN	*	*	*	*	*	*	*	*
UT	*	61	*	25	*	68	*	2.71
VA	*	160	*	80	*	248	*	3.10
WA	880	560	277	171	693	402	2.50	2.35
WV	*	*	*	*	*	*	*	*
WI	400	260	123	88	332	244	2.70	2.77
Oth Sts	2,358	2,034	960	778	2,453	2,235	2.56	2.87
US	5,518	4,785	2,179	1,836	5,852	5,180	2.69	2.82

See footnotes at end of table.

Table 15-32.—Trout: Sales by size category, by State and United States, 2004–05— Continued

Size category and State	Number of fish/eggs		Live weight		Sales			
					Total		Average per 1,000 fish eggs	
	2004[1]	2005	2004[1]	2005	2004[1]	2005	2004[1]	2005
	1,000 dollars	1,000 dollars	1,000 dollars	1,000 dollars	1,000 dollars	1,000 dollars	Dollars	Dollars
1 inch-6 inch:								
CA	330	510	5	10	138	195	418.00	382.00
CO	*	*	*	*	*	*	*	*
CT	*	*	*	*	*	*	*	*
GA	*	....	*	....	*	....	*	....
ID	*	*	*	*	*	*	*	*
ME	*	*	*	*	*	*	*	*
MA	*	*	*	*	*	*	*	*
MI	55	*	3	*	22	*	408.00	*
MO	*	*	*	*	*	*	*	*
NY	110	110	3	3	37	49	332.00	444.00
NC	2,830	3,440	45	60	255	454	90.00	132.00
OR	40	*	1	*	11	*	265.00	*
PA	190	290	5	5	67	87	355.00	300.00
TN	*	*	*	*	*	*	*	*
UT	*	22	*	2	*	6	*	259.00
VA	*	90	*	1	*	16	*	183.00
WA	410	1,330	13	40	130	319	317.00	240.00
WV	*	*	*	*	*	*	*	*
WI	290	200	4	4	61	32	209.00	159.00
Oth Sts	1,295	1,067	34	43	245	162	189.00	152.00
US	5,550	7,059	113	168	966	1,320	174.00	187.00
Trout eggs	....	....	....	....	....	....	....	....
Region[2][3]	....	....	....	....	....	....	....	....
North East	940	712	....	....	27.70	19.70	26	14
South and Central	1,080	1,550	....	....	22.20	21.30	24	33
West	287,600	305,210	....	....	16.60	16.70	4,781	5,089
WA	277,000	295,000	....	....	16.50	16.90	4,571	4,986
US	289,620	307,472	....	....	16.70	16.70	4,831	5,136

[1] Revised.　[2] Data published at the regional level to avoid disclosure of individual operations.　[3] Regions are defined as follows - North East: CT, MA, ME, NY, PA, and WV; South: AR, GA, NC, TN, and VA; Central: MI, MO, and WI; West: CA, CO, ID, OR, UT, and WA.　*Included in other States to avoid disclosure of individual operations.

NASS, Livestock Branch, (202) 720-3570.

Table 15-33.—Refrigerated warehouses: Gross refrigerated space by type of plant, United States, biennially, October 1987–2005 [1] [2]

Type	1987	1989	1991	1993	1995
	1,000 Cubic Feet				
General:					
Public	1,285,860	1,391,901	1,572,879	1,678,461	1,741,585
Private and Semiprivate	676,369	603,402	624,005	658,893	674,649
Total	1,962,229	1,995,303	2,196,884	2,337,354	2,416,234
Apple:					
Public	19,750	21,945	27,227	21,645	23,419
Private and Semiprivate	494,404	554,150	584,296	613,093	647,993
Total	514,154	576,095	611,523	634,737	671,412
Total, all	2,476,384	2,571,397	2,808,407	2,972,092	3,087,646

Type	1997	1999	2001	2003	2005
	1,000 Cubic Feet				
General:					
Public	2,043,908	2,146,643	2,251,943	2,357,080	2,435,773
Private and Semiprivate	683,372	756,505	788,853	802,454	771,725
Total	2,727,280	2,903,152	3,040,796	3,159,535	3,207,497
Apple:					
Public	23,907	21,690	14,183	12,517	9,270
Private and Semiprivate	675,838	680,736	712,412	723,499	711,951
Total	699,745	702,426	726,595	736,016	721,221
Total, all	3,427,025	3,605,578	3,767,394	3,895,551	3,928,718

[1] Warehouse space is defined as all space artificially cooled to temperatures of 50 degrees F. or less, in which food commodities are normally held for 30 days or longer. [2] Totals may not add due to rounding.
NASS, Livestock Branch, (202) 720–8784.

Table 15-34.—Apple and pear storages: Number of refrigerated warehouses, gross and usable refrigerated space, regular and CA capacity, by State and United States, October 1, 2005 [1] [2]

State	Number of warehouses	Refrigerated space		Apple & pear storage capacity		
		Gross	Usable	Regular	Controlled atmosphere	Total
		1,000 Cubic feet	1,000 Cubic feet	1,000 Bushels	1,000 Bushels	1,000 Bushels
AZ	1					
CA	35	32,239	24,310	6,458	2,154	8,612
CT	28	1,240	1,042	245	115	360
DE	2					
ID	7	4,777	4,092	1,059	821	1,880
IL	11	1,098	946	262	8	270
IN	34	2,169	1,702	358	193	551
IA	1					
KS	1					
KY	3	117	97	33		33
ME	22	3,111	2,760	522	728	1,250
MD	6	1,390	1,115	173	354	527
MA	54	4,275	3,601	781	451	1,232
MI	155	33,717	29,278	4,853	7,175	12,028
MN	11	763	573	224	33	257
MO	4					
NE	1					
NH	22	1,767	1,547	306	397	703
NJ	22	2,275	1,930	536	130	666
NM	1					
NY	136	32,839	29,032	4,432	7,717	12,149
NC	14	3,769	3,409	946	390	1,336
OH	61	3,570	2,918	824	333	1,157
OR	67	55,672	44,922	8,654	3,889	12,543
PA	150	19,619	15,746	4,008	2,066	6,074
RI	7	96	87	24	4	28
SC	2					
UT	17	3,310	2,639	430	414	844
VT	12	2,192	1,899	301	512	813
VA	30	12,194	10,508	1,832	1,951	3,783
WA	241	488,733	406,858	55,215	142,612	197,827
WV	13	6,744	5,559	1,689	338	2,027
WI	16	1,041	870	246	112	358
Oth Sts		2,505	1,932	457	7	464
US	1,187	721,221	599,371	94,865	172,904	267,769

[1] Totals may not add due to rounding. [2] Firms in this table store only apples or pears.
NASS, Livestock Branch, (202) 720–8784.

Table 15-35.—General storages: Gross and usable cooler and freezer space, by State and United States, October 1, 2005[1][2]

State	Cooler		Freezer		Total	
	Gross	Usable	Gross	Usable	Gross	Usable
	1,000 Cubic Feet					
AL	4,689	4,236	23,259	19,461	27,948	23,697
AK	163	495	1,883	1,382	2,046	1,877
AZ	*	*	*	*	14,050	11,729
AR	*	*	*	*	79,233	66,449
CA	137,696	110,792	261,799	211,897	399,495	322,690
CO	5,294	3,838	12,005	9,604	17,298	13,442
CT	*	*	*	*	5,348	3,855
DE	*	*	*	*	29,031	20,711
FL	93,047	79,291	159,798	128,419	252,845	207,710
GA	37,506	30,482	87,442	69,074	124,948	99,557
HI	*	*	*	*	*	*
ID	4,885	3,585	56,887	48,334	61,771	51,919
IL	26,370	19,824	125,119	95,064	151,489	114,888
IN	13,920	12,600	68,811	59,468	82,731	72,068
IA	18,129	13,458	74,522	61,845	92,651	75,304
KS	8,362	5,408	36,128	26,802	44,490	32,211
KY	6,384	4,883	16,637	15,007	23,021	19,890
LA	1,629	1,450	17,064	15,905	18,693	17,355
ME	261	226	12,094	6,784	12,355	7,011
MD	2,295	1,842	25,263	19,232	27,558	21,074
MA	15,297	12,327	71,512	61,636	86,808	73,963
MI	13,804	11,304	78,447	62,733	92,252	74,037
MN	15,395	11,319	59,313	45,882	74,708	57,201
MS	*	*	*	*	21,814	17,632
MO	31,562	26,105	82,075	67,512	113,637	93,617
MT	*	*	*	*	987	764
NE	8,833	6,190	40,307	31,635	49,141	37,825
NV	*	*	*	*	*	*
NH	*	*	*	*	*	*
NJ	27,300	21,523	54,909	44,308	82,209	65,831
NM	*	*	*	*	2,400	1,976
NY	27,518	22,220	67,821	56,924	95,339	79,144
NC	5,184	4,005	56,083	46,365	61,267	50,370
ND	*	*	*	*	9,246	7,045
OH	8,689	7,093	54,323	44,215	63,012	51,308
OK	4,327	3,555	10,861	8,428	15,188	11,983
OR	5,290	3,911	103,417	84,743	108,707	88,655
PA	25,964	21,049	123,126	104,987	149,090	126,036
RI	*	*	*	*	*	*
SC	769	581	21,573	17,551	22,342	18,133
SD	*	*	*	*	11,460	5,782
TN	429	322	33,476	25,582	33,905	25,904
TX	31,016	23,876	121,816	91,231	152,832	115,107
UT	2,881	2,224	29,654	24,791	32,535	27,015
VT	*	*	*	*	*	*
VA	20,731	16,989	39,508	34,617	60,239	51,606
WA	10,435	7,642	184,989	145,937	195,424	153,579
WV	*	*	*	*	2,420	1,096
WI	74,712	55,079	104,040	87,777	178,753	142,856
WY	*	*	*	*	*	*
Oth Sts	25,850	17,699	174,922	138,481	24,783	19,139
US	716,615	567,423	2,490,883	2,013,613	3,207,497	2,581,036

[1] Totals may not add due to rounding.   [2] Excludes storages used exclusively for storing apples and pears. Includes frozen juice tank storage capacity.   * Not published to avoid disclosure of individual operations. Included in "Other States" and U.S. totals.

NASS, Livestock Branch, (202) 720–8784.

Table 15-36.—Alaska crops: Acreage harvested, volume harvested, and value of production, 1996–2005

Year	Oats for grain	Barley for grain	All hay	Potatoes	Other vegetables [1]
			Acreage harvested		
	Acres	*Acres*	*Acres*	*Acres*	*Acres*
1996	700	6,900	20,200	630	343
1997	1,500	7,000	22,500	820	337
1998	1,500	6,500	22,000	820	340
1999	1,500	4,600	20,300	850	357
2000	300	3,300	18,000	840	370
2001	1,200	5,100	23,000	910	361
2002	1,200	3,800	23,000	850	368
2003	1,200	3,500	22,000	800	359
2004	1,300	4,200	21,000	810	328
2005 [2]	900	4,300	21,000	780	NA

Year	Oats for grain	Barley for grain	All hay	Potatoes	Other vegetables [1]
			Volume harvested		
	Bushels	*Bushels*	*Tons*	*Cwt.*	*Cwt.*
1996	31,500	283,000	14,400	126,000	43,232
1997	65,300	164,500	26,000	168,000	46,723
1998	45,000	122,900	23,700	150,000	41,846
1999	62,100	154,800	23,200	185,000	53,745
2000	7,000	102,500	17,000	129,000	58,042
2001	61,000	208,000	30,000	230,000	49,989
2002	48,000	149,000	26,000	154,000	51,762
2003	34,000	135,000	29,000	168,000	52,690
2004	41,000	145,000	28,000	177,000	47,762
2005 [2]	58,000	208,000	30,000	166,000	NA

Year	Oats for grain	Barley for grain	All hay	Potatoes	Other vegetables [1]
			Value of production		
	Dollars	*Dollars*	*Dollars*	*Dollars*	*Dollars*
1996	79,000	891,000	2,736,000	2,494,000	1,443,000
1997	163,000	526,000	4,940,000	3,360,000	1,620,000
1998	117,000	442,000	4,740,000	3,105,000	1,397,000
1999	152,000	581,000	4,524,000	3,830,000	1,897,000
2000	22,000	369,000	3,740,000	2,670,000	2,080,000
2001	153,000	707,000	6,300,000	4,669,000	2,169,000
2002	125,000	529,000	5,590,000	3,080,000	2,318,000
2003	87,000	479,000	6,525,000	3,310,000	2,619,000
2004	100,000	500,000	6,440,000	3,469,000	2,439,000
2005 [2]	148,000	759,000	7,200,000	3,403,000	NA

[1] Excludes greenhouse-grown vegetables.   [2] Preliminary.   NA-not available.
NASS, Crops Branch, (202) 720–2127.

Table 15-37.—Crop ranking: Major field crops, rank by production, major States, 2005

Rank	State	Corn, grain	State	Soybeans	State	All wheat
		1,000 Bushels		*1,000 Bushels*		*1,000 Bushels*
1	IA	2,162,500	IA	532,650	KS	380,000
2	IL	1,708,850	IL	444,150	ND	303,765
3	NE	1,270,500	MN	306,000	MT	192,480
4	MN	1,191,900	IN	263,620	WA	139,300
5	IN	888,580	NE	235,330	SD	133,420
6	SD	470,050	OH	201,600	OK	128,000
7	KS	465,750	MO	183,520	ID	100,590
8	OH	464,750	SD	138,600	TX	96,000
9	WI	429,200	ND	107,300	MN	71,470
10	MO	329,670	KS	105,450	NE	68,640
	US	11,112,072	US	3,086,432	US	2,104,690

Rank	State	Winter wheat	State	Durum wheat	State	Other spring wheat
		1,000 Bushels		*1,000 Bushels*		*1,000 Bushels*
1	KS	380,000	ND	68,250	ND	224,400
2	OK	128,000	MT	16,380	MT	81,600
3	WA	120,600	AZ	7,900	MN	70,930
4	TX	96,000	CA	6,555	SD	67,600
5	MT	94,500	ID	1,760	ID	32,400
6	NE	68,640	SD	260	WA	18,700
7	ID	66,430		....	OR	5,980
8	SD	65,560		....	CO	1,235
9	OH	58,930		....	UT	754
10	CO	52,800		....	WY	315
	US	1,499,129	US	101,105	US	504,456

Rank	State	Sorghum, grain	State	Barley	State	Oats
		1,000 bushels		*1,000 bushels*		*1,000 bushels*
1	KS	195,000	ND	57,240	ND	14,160
2	TX	111,000	ID	52,200	WI	13,760
3	NE	21,750	MT	39,200	SD	12,960
4	OK	12,480	WA	12,505	MN	12,710
5	MO	9,880	CO	7,670	IA	9,875
6	LA	8,712	WY	5,580	PA	6,050
7	IL	7,636	VA	3,915	TX	4,730
8	AR	4,960	MN	3,870	MI	4,575
9	SD	4,420	CA	3,780	NE	4,380
10	NM	4,365	MD	3,526	NY	4,050
	US	393,893	US	211,896	US	114,878

Rank	State	All cotton	State	Peanuts	State	Rice
		1,000 bales		*1,000 pounds*		*1,000 cwt.*
1	TX	8,245.0	GA	2,152,500	AR	108,792
2	AR	2,190.0	TX	910,000	CA	38,836
3	MS	2,160.0	AL	613,250	LA	30,983
4	GA	2,150.0	FL	410,400	MS	16,832
5	CA	1,630.0	NC	288,000	MO	14,124
6	NC	1,430.0	SC	168,000	TX	13,668
7	LA	1,120.0	OK	105,600		....
8	TN	1,120.0	VA	66,000		....
9	MO	885.0	NM	62,700		....
10	AL	850.0	MS	44,800		....
	US	23,719.0	US	4,821,250	US	223,235

Rank	State	All hay, baled	State	Alfalfa hay, baled	State	Other hay, baled
		1,000 tons		*1,000 tons*		*1,000 tons*
1	TX	9,140	CA	6,900	TX	8,330
2	CA	8,935	SD	5,160	MO	5,503
3	SD	7,560	IA	5,125	KY	4,945
4	NE	6,945	ID	4,788	TN	4,255
5	MO	6,718	MN	4,725	OK	3,900
6	KS	6,680	NE	4,625	KS	3,280
7	MN	6,055	MT	3,850	VA	3,146
8	IA	5,860	WI	3,720	SD	2,400
9	MT	5,850	KS	3,400	ND	2,346
10	KY	5,777	ND	3,300	NE	2,320
	US	150,590	US	75,771	US	74,819

Rank	State	All tobacco	State	Dry edible beans	State	Potatoes
		1,000 pounds		*1,000 cwt.*		*1,000 cwt.*
1	NC	278,900	ND	8,588	ID	116,975
2	KY	167,260	MI	3,910	WA	95,480
3	TN	51,670	NE	3,870	WI	27,880
4	SC	42,000	MN	2,430	CO	24,044
5	VA	39,840	CO	1,898	OR	22,023
6	GA	27,760	ID	1,862	ND	20,500
7	PA	10,700	CA	1,385	MN	17,630
8	OH	6,732	WA	792	ME	15,736
9	FL	5,500	WY	776	CA	14,964
10	CT	4,067	SD	301	MI	13,920
	US	639,709	US	27,222	US	420,879

NASS, Crops Branch, (202) 720–2127.

Table 15-38.—U.S. crop progress: 2005 crop and 5-year average

[In percent]

Week-end-ing date	Winter wheat							
	Planted		Emerged		Headed		Harvested	
	2004	Avg	2004	Avg	2004	Avg	2004	Avg
2004: [1]								
Sep 5	7	5						
Sep 12	16	12						
Sep 19	29	23	7	7				
Sep 26	42	38	17	15				
Oct 3	58	54	29	27				
Oct 10	70	68	44	40				
Oct 17	78	78	57	53				
Oct 24	85	85	68	64				
Oct 31	89	89	77	74				
Nov 7	91	92	83	81				
Nov 14	93	94	87	85				
Nov 21	95	96	90	89				
Nov 28			93	91				
Dec 5			93	91				
2005:								
Apr 17					8	9		
Apr 24					18	17		
May 1					30	31		
May 8					44	47		
May 15 ...					59	62		
May 22 ...					71	73		
May 29 ...					81	81		
Jun 5					88	87		
Jun 12					93	92	12	16
Jun 19					97	96	22	29
Jun 26							48	46
Jul 3							62	61
Jul 10							72	71
Jul 17							79	77
Jul 24							85	83
Jul 31							90	89
Aug 7							94	93
Aug 14							96	96

Week-end-ing date	Spring wheat							
	Planted		Emerged		Headed		Harvested	
	2004	Avg	2004	Avg	2004	Avg	2004	Avg
2005:								
Apr 10	12	9						
Apr 17	23	17						
Apr 24	40	28						
May 1	61	47	20	18				
May 8	80	62	38	32				
May 15 ...	89	74	55	46				
May 22 ...	94	87	75	63				
May 29 ...	97	94	88	78				
Jun 5			96	90				
Jun 12					4	4		
Jun 19					9	14		
Jun 26					30	29		
Jul 3					57	50		
Jul 10					80	71		
Jul 17					91	88		
Jul 24					98	95		
Jul 31							7	7
Aug 7							23	20
Aug 14							42	38
Aug 21							59	56
Aug 28							76	71
Sep 4							90	81
Sep 11							96	89

See footnote at end of table.

Table 15-38.—U.S. crop progress: 2005 crop and 5-year average—Continued
[In percent]

Week-ending date	Rice Planted 2005	Avg	Rice Emerged 2005	Avg	Rice Headed 2005	Avg	Rice Harvested 2005	Avg	Sorghum Planted 2005	Avg	Sorghum Headed 2005	Avg	Sorghum Coloring 2005	Avg	Sorghum Mature 2005	Avg	Sorghum Harvested 2005	Avg
2005:																		
Apr 3 ...	5	12							10	11								
Apr 10	14	21							13	12								
Apr 17	22	34	11	15					15	14								
Apr 24	46	50	19	26					17	15								
May 1 ..	65	65	36	41					18	19								
May 8 ..	79	78	50	56					21	24								
May 15	87	86	63	70					26	31								
May 22	94	92	78	79					37	41								
May 29	97	97	86	87					51	56								
Jun 5 ...			93	93					63	69								
Jun 12			96	97					72	79	11	11						
Jun 19					1	5			82	87	13	13						
Jun 26					4	9			92	93	14	15						
Jul 3					7	13			97	96	15	17	12	12				
Jul 10 .					13	18					19	21	13	13				
Jul 17 ..					19	27					24	28	15	15				
Jul 24 ..					29	38					31	38	16	17				
Jul 31 .					45	53					52	51	19	21				
Aug 7 ...					67	69	3	8			69	63	21	26				
Aug 14					82	82	7	12			80	75	30	33				
Aug 21					90	91	13	16			86	82	39	43	18	21		
Aug 28					97	96	16	20			92	89	48	55	20	26		
Sep 4 ...							24	27			96	93	61	67	23	35		
Sep 11							33	38					74	76	32	44	22	29
Sep 18							43	52					82	84	42	53	26	34
Sep 25							60	66					89	90	53	64	30	40
Oct 2 ..							72	77					94	94	67	73	36	47
Oct 9 ...							86	87					97	96	75	81	43	54
Oct 16							93	92							82	86	50	61
Oct 23							97	95							89	91	61	68
Oct 30															95	94	71	74
Nov 6 ...																	79	81
Nov 13																	88	86
Nov 20																	92	90
Nov 27																	96	93

Week-ending date	Corn Planted 2005	Avg	Emerged 2005	Avg	Silked 2005	Avg	Dough 2005	Avg	Dent 2005	Avg	Mature 2005	Avg	Harvested 2005	Avg
2005:														
Apr 17	14	10												
Apr 24	30	22												
May 1	52	45	13	12										
May 8	79	67	23	26										
May 15 ...	89	79	41	48										
May 22 ...	95	88	66	66										
May 29 ...			85	80										
Jun 5			95	90										
Jun 12														
Jun 19														
Jun 26					4	5								
Jul 3					11	11								
Jul 10					25	22	3	3						
Jul 17					49	41	6	6						
Jul 24					79	65	14	13						
Jul 31					92	84	27	23	4	5				
Aug 7					97	93	44	38	10	11				
Aug 14							65	56	23	21				
Aug 21							80	72	40	35	6	6		
Aug 28							91	84	61	52	11	12		
Sep 4							96	93	79	69	20	22		
Sep 11									89	82	36	36	6	7
Sep 18									96	91	57	53	11	11
Sep 25											76	71	18	18
Oct 2											90	84	26	26
Oct 9											96	92	36	36
Oct 16													49	48
Oct 23													65	62
Oct 30													80	74
Nov 6													90	84
Nov 13													95	91

See footnote at end of table.

Table 15-38.—U.S. crop progress: 2005 crop and 5-year average—Continued

[In percent]

Week-end-ing date	Soybeans											
	Planted		Emerged		Blooming		Pods set		Leaf drop		Harvested	
	2005	Avg	2005	Avg	2005	Avg	2005	Avg	2005	Avg	2005	Avg
2005:												
May 1	8	9										
May 8	26	23										
May 15 ...	46	39	11	14								
May 22 ...	65	56	27	28								
May 29 ...	81	71	50	45								
Jun 5	90	82	70	63								
Jun 12	94	90	85	78								
Jun 19	96	94	92	88								
Jun 26			96	93	6	5						
Jul 3					21	15						
Jul 10					43	30	6	4				
Jul 17					63	50	16	13				
Jul 24					81	68	36	26				
Jul 31					91	82	55	44				
Aug 7					95	90	76	63				
Aug 14							89	78				
Aug 21							94	89				
Aug 28							97	95	6	7		
Sep 4									15	15		
Sep 11									37	31		
Sep 18									64	53	8	6
Sep 25									83	72	19	14
Oct 2									93	85	36	30
Oct 9									97	93	60	51
Oct 16											76	67
Oct 23											87	79
Oct 30											92	86
Nov 6											96	91

Week-end-ing date	Cotton									
	Planted		Squaring		Bolls set		Bolls open		Harvested	
	2005	Avg	2005	Avg	2005	Avg	2005	Avg	2005	Avg
2005:										
Apr 10	7	8								
Apr 17	11	12								
Apr 24	18	18								
May 1	27	28								
May 8	39	43								
May 15 ...	55	57								
May 22 ...	68	70								
May 29 ...	83	81								
Jun 5	90	88	9	13						
Jun 12	94	94	16	22						
Jun 19	97	97	28	36	4	7				
Jun 26			41	52	6	11				
Jul 3			55	65	13	19				
Jul 10			67	77	22	30				
Jul 17			82	86	35	45				
Jul 24			89	91	49	62				
Jul 31			94	95	69	75				
Aug 7			98	98	78	84	7	8		
Aug 14					86	91	9	13		
Aug 21					91	95	14	20		
Aug 28					97	98	19	29		
Sep 4							30	40		
Sep 11							43	52	9	9
Sep 18							56	64	11	12
Sep 25							68	75	15	16
Oct 2							77	82	20	23
Oct 9							83	87	28	30
Oct 16							90	91	36	38
Oct 23							94	94	44	47
Oct 30							96	97	53	55
Nov 6									61	63
Nov 13									73	69
Nov 20									77	76
Nov 27									84	81

See footnote at end of table.

Table 15-38.—U.S. crop progress: 2005 crop and 5-year average—Continued

[In percent]

Week-ending date	Oats Planted 2005	Avg	Emerged 2005	Avg	Headed 2005	Avg	Harvested 2005	Avg	Barley Planted 2005	Avg	Emerged 2005	Avg	Headed 2005	Avg	Harvested 2005	Avg
2005:																
Apr 10	43	37	....	......	....	......	....	......	11	11	....	......	....	......	....	......
Apr 17	55	46	....	......	....	......	....	......	19	18	....	......	....	......	....	......
Apr 24	67	57	40	36	....	......	....	......	34	29	....	......	....	......	....	......
May 1	79	69	51	46	....	......	....	......	52	44	14	18	....	......	....	......
May 8	91	81	66	59	....	......	....	......	74	60	28	31	....	......	....	......
May 15	96	88	78	72	....	......	....	......	83	73	48	45	....	......	....	......
May 22	....	......	88	83	....	......	....	......	90	87	68	61	....	......	....	......
May 29	....	......	95	91	....	......	....	......	96	95	85	78	....	......	....	......
Jun 5	....	......	....	......	28	29	....	......	....	......	94	90	....	......	....	......
Jun 12	....	......	....	......	37	36	....	......	....	......	97	96	3	6	....	......
Jun 19	....	......	....	......	49	49	....	......	....	......	....	......	10	15	....	......
Jun 26	....	......	....	......	65	64	....	......	....	......	....	......	28	28	....	......
Jul 3	....	......	....	......	84	78	....	......	....	......	....	......	48	47	....	......
Jul 10	....	......	....	......	92	89	....	......	....	......	....	......	74	69	....	......
Jul 17	....	......	....	......	98	95	22	19	....	......	....	......	89	86	....	......
Jul 24	....	......	....	......	....	......	34	29	....	......	....	......	96	96	....	......
Jul 31	....	......	....	......	....	......	51	43	....	......	....	......	....	......	7	6
Aug 7	....	......	....	......	....	......	71	59	....	......	....	......	....	......	23	18
Aug 14	....	......	....	......	....	......	84	74	....	......	....	......	....	......	49	35
Aug 21	....	......	....	......	....	......	93	86	....	......	....	......	....	......	65	57
Aug 28	....	......	....	......	....	......	98	93	....	......	....	......	....	......	78	74
Sep 4	....	......	....	......	....	......	....	......	....	......	....	......	....	......	90	85
Sep 11	....	......	....	......	....	......	....	......	....	......	....	......	....	......	95	92

Week-end-ing date	Peanuts Planted 2005	Avg	Pegging 2005	Avg	Harvested 2005	Avg	Sunflower Planted 2005	Avg	Harvested 2005	Avg	Sugarbeets Planted 2005	Avg	Harvested 2005	Avg
2005:														
Apr 10	....	....	....	....	....	....	....	....	....	....	11	12	....	....
Apr 17	....	....	....	....	....	....	....	....	....	....	28	25	....	....
Apr 24	....	....	....	....	....	....	....	....	....	....	49	40	....	....
May 1	5	9	....	....	....	....	....	....	....	....	80	65	....	....
May 8	11	24	....	....	....	....	....	....	....	....	98	79	....	....
May 15	32	47	....	....	....	....	....	....	....	....	....	....	....	....
May 22	62	70	....	....	....	....	18	17	....	....	....	....	....	....
May 29	83	86	....	....	....	....	40	40	....	....	....	....	....	....
Jun 5	94	94	....	....	....	....	59	62	....	....	....	....	....	....
Jun 12	96	98	2	7	....	....	72	81	....	....	....	....	....	....
Jun 19	....	....	7	14	....	....	84	91	....	....	....	....	....	....
Jun 26	....	....	18	25	....	....	93	96	....	....	....	....	....	....
Jul 3	....	....	32	41	....	....	97	99	....	....	....	....	....	....
Jul 10	....	....	50	58	....	....	....	....	....	....	....	....	....	....
Jul 17	....	....	67	72	....	....	....	....	....	....	....	....	....	....
Jul 24	....	....	78	83	....	....	....	....	....	....	....	....	....	....
Jul 31	....	....	88	91	....	....	....	....	....	....	....	....	....	....
Aug 7	....	....	92	95	....	....	....	....	....	....	....	....	....	....
Aug 14	....	....	97	98	....	....	....	....	....	....	....	....	....	....
Aug 21	....	....	....	....	....	....	....	....	....	....	....	....	....	....
Aug 28	....	....	....	....	....	....	....	....	....	....	....	....	....	....
Sep 4	....	....	....	....	....	....	....	....	....	....	....	....	....	....
Sep 11	....	....	....	....	1	4	....	....	....	....	....	....	....	....
Sep 18	....	....	....	....	5	10	....	....	....	....	....	....	3	3
Sep 25	....	....	....	....	14	19	....	....	....	....	....	....	6	6
Oct 2	....	....	....	....	23	31	....	....	6	11	....	....	10	20
Oct 9	....	....	....	....	37	45	....	....	13	21	....	....	29	44
Oct 16	....	....	....	....	48	60	....	....	28	36	....	....	57	67
Oct 23	....	....	....	....	65	71	....	....	47	53	....	....	79	80
Oct 30	....	....	....	....	78	81	....	....	69	67	....	....	88	89
Nov 6	....	....	....	....	86	87	....	....	85	77	....	....	96	95
Nov 13	....	....	....	....	94	91	....	....	92	87	....	....	....	....
Nov 20	....	....	....	....	98	95	....	....	97	93	....	....	....	....

[1] Planted the preceding fall.   NA-not available.

NASS, Crops Branch, (202) 720–2127.

Appendix I

Telephone Contact List

Appreciation is expressed to the following agencies for their help in this publication. The information offices are listed to provide help to those users who require additional information about specific tables in this publication.

Agricultural Marketing Service:
USDA/AMS
Room 3510 South Bldg.
Washington, DC 20250
202–720–8998

Agricultural Research Service:
USDA/ARS
5601 Sunnyside Ave
Bldg 1, Rm 2250
Beltsville, MD 20705–5128
301–504–1638

Animal and Plant Health Inspection Service:
USDA/APHIS
4700 River Rd
Riverdale, MD 20737
301–734–7280

Center for Nutrition Policy and Promotion:
USDA/CNPP
3101 Park Center Drive
Alexandria, VA 22302
703–605–4266

Economic Research Service:
USDA/ERS
1800 M St, NW
Washington, DC
202–694–5050

Farm Credit Administration:
FCA
1501 Farm Credit Dr.
McLean, VA 22102
703–883–4000

Farm Service Agency:
USDA/FSA
Room 3624 South Bldg.
Washington, DC 20250
202–720–7809

Food and Nutrition Service:
USDA/FNS
3101 Park Center Drive, Room 914
Alexandria, VA 22302
703–305–2286

Foreign Agricultural Service:
USDA/FAS
Room 5074 South Bldg.
Washington, DC 20250
202–720–7115

Forest Service:
USDA/FS
2nd Floor Central Wing, Aud. Bldg.
Washington, DC 20250
202–205–1273

National Agricultural Statistics Service:
USDA/NASS
Room 5829 South Bldg.
Washington, DC 20250
202–720–3878

National Marine Fisheries Service:
USDC/NOAA/NMFS
1315 East/West Highway,
SSMC III - Room 12340
Silver Spring, MD 20910–3282
301–713–2328

Natural Resources Conservation Service:
USDA/NRCS
Room 6121 South Bldg.
Washington, DC 20250
202–720–3210

Rural Business-Cooperatives Service:
USDA/RECD/RBS
Room 5801 South Bldg.
Washington, DC 20250
202–720–4323

Rural Utilities Service:
USDA/RD/RUS
Room 5144 South Bldg.
Washington, DC 20250
202–720–1255

INDEX

INDEX

INDEX

INDEX

INDEX